# Texts in Applied Mathematics

Volume 79

The mathematization of all sciences, the fading of traditional scientific boundaries, the impact of computer technology, the growing importance of computer modelling and the necessity of scientific planning all create the need both in education and research for books that are introductory to and abreast of these developments. The aim of this series is to provide such textbooks in applied mathematics for the student scientist. Books should be well illustrated and have clear exposition and sound pedagogy. Large number of examples and exercises at varying levels are recommended. TAM publishes textbooks suitable for advanced undergraduate and beginning graduate courses, and complements the Applied Mathematical Sciences (AMS) series, which focuses on advanced textbooks and research-level monographs.

Rolf Brigola

# Fourier Analysis and Distributions

## A First Course with Applications

 Springer

Rolf Brigola
Applied Mathematics, Physics and
Humanities
TH Nürnberg Georg Simon Ohm
Nürnberg, Germany

ISSN 0939-2475 ISSN 2196-9949 (electronic)
Texts in Applied Mathematics
ISBN 978-3-031-81313-9 ISBN 978-3-031-81311-5 (eBook)
https://doi.org/10.1007/978-3-031-81311-5

Mathematics Subject Classification: 35E05, 42A16, 46F12, 65R10, 42A15, 42A38



This Springer imprint is published by the registered company Springer Nature Switzerland AG
The registered company address is: Gewerbestrasse 11, 6330 Cham, Switzerland

If disposing of this product, please recycle the paper.

# Preface

Fourier analysis and distribution theory are fundamental mathematical tools for describing and solving a wide array of technical and scientific problems. These include areas of physics, mechanical engineering, electrical engineering, and signal and control theory. This text is intended for future scientists seeking to understand the theoretical foundation of mathematical modeling, Fourier analysis methods, and their practical applications. It builds on the author's earlier work, published by Vieweg in 1996, and incorporates insights from many years of lectures delivered to students of applied mathematics, physics, electrical engineering, and communications engineering at the Technische Hochschule Nürnberg Georg Simon Ohm, starting in their fourth semester.

The book is tailored for undergraduate and early master's students in mathematics, physics, and engineering. A basic understanding of differential and integral calculus is a prerequisite. The text is divided into chapters covering the mathematical foundations of Fourier series, distributions, and Fourier transforms, each complemented by examples of practical applications. The fundamentals of distribution theory, widely used in engineering disciplines, enable and simplify numerous calculations for physical and technical problems.

The book is structured such that theoretical and application-oriented chapters each account for half of the content. The theoretical chapters introduce Fourier series and integrals, distributions, and the z-Transform.

The application chapters are designed to be read independently, depending on the reader's interest. These chapters provide an introduction to the fundamental types of linear partial differential equations and essential principles of linear systems theory. Topics include methods for linear filter design and sampling, offering an accessible introduction to modern signal processing. The unified representation of analog and discrete linear systems within the framework of distributions is also explored.

The sections on discrete Fourier and wavelet transforms, along with their applications in signal processing, and the introductory discussion on the finite element method, provide a glimpse into the numerical aspects of practical applications. Specific examples utilize physical SI units to ground the concepts in real-world scenarios.

Through this book, I aim to offer mathematics and physics students a clear introduction to widely used techniques in technology and engineering. Simultaneously, I hope to provide technically oriented students with a comprehensible mathematical presentation that supports their work. To facilitate learning, the text includes 175 illustrations. I trust that the presentation will encourage readers to apply their newly acquired knowledge in practice and continue their learning journey with other available literature if necessary. The appendices include essential theorems from function theory and Lebesgue integration, which are used throughout the text.

I would like to express my gratitude to my colleagues Herbert Leinfelder, Rudolf Rupp, and Jörg Steinbach in Nürnberg, and Peter Wagner in Innsbruck, for their valuable discussions during the preparation of this book. Special thanks go to Donna Chernyk and Kirithiga Nandini Gnanasekaran at Springer Nature for their professional assistance in publishing this text. I am also indebted to Charlott Caroline, Krefeld, and Friederike Laus, Niederzissen, for translating and typesetting several chapters from German into English, and to my students, whose motivated collaboration and constructive feedback in lectures and seminars contributed significantly to the development of this book.

An important part of this book are the exercises, which I encourage serious readers to complete independently. Most exercises are designed to reinforce the content of the book and develop strong calculation techniques. Exercises marked with an asterisk are primarily intended for mathematicians and sometimes explore topics not discussed in detail within the text. For reference, solutions to all exercises are provided in Appendix C.

Nürnberg, Germany                                                            Rolf Brigola
January 2025

# Contents

# List of Symbols and Physical Quantities

| | |
|---|---|
| $\mathbb{N}, \mathbb{Z}$ | The sets of natural and integer numbers, 5 |
| $\mathbb{R}, \mathbb{C}$ | The sets of real and complex numbers, 9 |
| $\mathbb{N}_0, \mathbb{R}_0^+$ | $\mathbb{N} \cup \{0\}$ and the set of nonnegative real numbers, 2 |
| $j, j^2 = -1$ | Complex unit, 9 |
| $\Re(z), \Im(z)$ | Real and imaginary parts of $z$, 9 |
| $\overline{z}$ | Complex conjugate of $z$, 10 |
| $f(t+) = \lim\limits_{x \to t+} f(x)$ | Right-hand limit of $f$ as $x \to t, x > t$, 28 |
| $f(t-) = \lim\limits_{x \to t-} f(x)$ | Left-hand limit of $f$ as $x \to t, x < t$, 28 |
| $\mathbf{x} \cdot \mathbf{y}$ | Inner product of two vectors $\mathbf{x}$ and $\mathbf{y}$, 60 |
| $\langle f|g \rangle$ | Inner product of $f$ and $g$, 9, 108 |
| $\langle T, \varphi \rangle$ | Value of a distribution $T$ for a test function $\varphi$, 164 |
| $\|f\|_2$ | $L^2$-norm of $f$, 54 |
| $(f * g)_T$ | $T$-periodic convolution, 63 |
| $f * g$ | Convolution of $f$ and $g$, 193 |
| $T * G$ | Convolution of two distributions $T$ and $G$, 194 |
| $s(t)$ | Unit step function, 157 |
| $1_A$ | Indicator function of a set A, 198, 394 |
| $\delta, \delta(t)$ | Dirac distribution, Dirac impulse, 159 |
| $\delta_n, \delta(t - na)$ | Dirac impulse at $na$, 209, 323 |
| $\mathrm{pv}(t^{-1}), \mathrm{vp}(t^{-1})$ | Cauchy principal value of $t^{-1}$, 167 |
| $\mathrm{pf}(t^{-m})$ | Pseudofunction of $t^{-m}$, 168 |
| $\dot{T}, T^{(k)}$ | Generalized derivatives of $T$, 171 |
| $\mathrm{supp}(T)$ | Support of a distribution $T$, 193 |
| $\partial_i, \partial^k$ | Partial derivatives, 187 |
| $P(D)$ | Ordinary linear differential operator with constant coefficients, 220 |
| $P(\partial)$ | Partial linear differential operator with constant coefficients, 438 |
| $\Delta u$ | Laplace operator applied to $u$, 233 |
| $\mathcal{F}$ | Fourier transform, 270 |
| $\widehat{f}, \mathcal{F}f$ | Fourier transform of $f$, 270 |

| | |
|---|---|
| $\mathcal{G}_w f = \widetilde{f}$ | Windowed Fourier transform of $f$ for a given window function $w$, 410 |
| $C^k(\Omega)$ | Space of k-times continuously differentiable functions, 477 |
| $L^2([0, T])$ | Space of square-integrable functions on $[0, T]$, 62 |
| $L^2(\mathbb{R})$ | Space of square-integrable functions on $\mathbb{R}$, 409 |
| $L^p(\Omega)$ | Space of $p$-integrable functions on $\Omega$, 500 |
| $L^\infty(\Omega)$ | Space of essentially bounded functions on $\Omega$, 500 |
| $l_d^p$ | Space of discrete signals with $p$-summable coefficients, 323 |
| $l_d^\infty$ | Space of discrete signals with bounded coefficients, 323 |
| $\mathcal{D}$ | Space of test functions on $\mathbb{R}$, 155 |
| $\mathcal{D}'$ | Space of distributions on $\mathbb{R}$, 164 |
| $\mathcal{D}(\Omega)$ | Space of test functions on a domain $\Omega$, 187 |
| $\mathcal{D}'(\Omega)$ | Space of distributions on a domain $\Omega$, 188 |
| $\mathcal{D}'_R$ | Space of causal distributions on $\mathbb{R}$, 322 |
| $\mathcal{D}'_+$ | Space of distributions with support in $[0, \infty[$, 322 |
| $\mathcal{S}$ | Space of rapidly decreasing functions on $\mathbb{R}$, 286 |
| $\mathcal{O}_M$ | Space of slowly increasing functions on $\mathbb{R}$, 290 |
| $\mathcal{S}'$ | Space of tempered distributions on $\mathbb{R}$, 289 |
| $\mathcal{S}'_R$ | Space of causal tempered distributions on $\mathbb{R}$, 322 |
| $\mathcal{S}'_+$ | Space of tempered distributions on $\mathbb{R}$ with support in $[0, \infty[$, 322 |
| $\mathcal{E}'$ | Space of distributions on $\mathbb{R}$ with compact support, 202 |
| $\mathcal{O}'_C$ | Space of rapidly decreasing distributions, 301 |
| $\mathcal{X}$ | Space of discrete signals, 323 |

## Physical Quantities

| | |
|---|---|
| N | Force, Newton N, $\mathrm{kg\,m/s^2}$ |
| $\gamma$ | Gravitational constant, Gamma, $\mathrm{m^3/(kg\,s^2)}$ |
| $U$ | Voltage, Volt, V |
| $u$ | Electric potential, V<br>or gravitational potential, $\mathrm{m^2/s^2}$ |
| $I$ | Electric current, Ampere, A |
| $R$ | Resistance, Ohm, $\Omega$, V/A |
| $C$ | Capacitance, Farad, F, As/V |
| $L$ | Inductance, Henry, H, Vs/A |
| $q$ | Electric charge, Coulomb, C, As |
| $\varrho$ | Electric charge density per $\mathrm{m^3}$, $\mathrm{C/m^3}$<br>per $\mathrm{m^2}$ or m, $\mathrm{C/m^2}$ or C/m<br>or mass density, analogous with kg instead of C |
| $\varepsilon_0$ | Permittivity, F/m, As/(Vm) |
| $k$ | Thermal diffusivity, $\mathrm{m^2/s}$ |
| $\hbar$ | The reduced Planck constant $h/(2\pi)$, Js |

# Chapter 1
# Introduction

**Abstract** As an introduction, the initial value problem for a vibrating string is treated as an application of Fourier expansions to a differential equation. The solutions are first prototypes of signals that will be studied in later chapters. Theoretical questions are discussed for approximate solutions by trigonometric polynomials and series solutions, which are subsequently answered.

## 1.1 Preliminary Remarks on History

Historically, trigonometric series such as

$$\frac{a_0}{2} + \sum_{n=1}^{\infty} \left( a_n \cos(n\omega t) + b_n \sin(n\omega t) \right)$$

were initially used to describe periodical events in astronomy and to work on motion equations for vibrating strings. These types of series were later—under suitable conditions for the series coefficients—named *Fourier series*. As early as 1753, D. Bernoulli (1700–1782) was convinced that "almost every" vibrational shape of a string could be expressed as a superposition of a fundamental vibration with an angular frequency $\omega$ and harmonics with angular frequency multiples $n\omega$, $n = 2, 3, 4, \ldots$ In 1807, French mathematician Jean-Baptiste Joseph Fourier (1768–1830) used such trigonometric series to express solutions for the heat equation (Fourier (2009)). For a thin bar of length $l$ with a thermal diffusivity $k$, where the bar ends are kept at temperature zero, the temperature $u(x, t)$ in $x \in [0, l]$ at time $t \geqslant 0$ is the solution of the following homogeneous partial differential equation:

$$\frac{\partial u}{\partial t}(x, t) = k \frac{\partial^2 u}{\partial x^2}(x, t) \quad \text{(no external energy input),}$$

$$u(x, 0) = f(x) \qquad \text{(initial temperature distribution } f\text{),}$$

$$u(0, t) = u(l, t) = 0 \quad \text{(the bar ends are chilled with ice).}$$


R. Brigola, *Fourier Analysis and Distributions*, Texts in Applied Mathematics 79,
https://doi.org/10.1007/978-3-031-81311-5_1

The Fourier series solution was expressed as

$$u(x,t) = \sum_{n=1}^{\infty} b_n \, e^{-k(n\pi/l)^2 t} \sin\left(\frac{n\pi}{l}x\right),$$

$$b_n = \frac{2}{l} \int_0^l f(y) \sin\left(\frac{n\pi}{l}y\right) dy.$$

Since Fourier argued in part intuitively, his theory of heat conduction was met with concerns and reservations that were only addressed after decades of ultimately very fruitful discussions. The exact clarification of fundamental mathematical concepts for Fourier's arguments essentially goes back to the work of mathematician P. L. Dirichlet (1805–1859). After almost a century, it became clear that Fourier's work provided important impact for many mathematical subdisciplines. Questions derived from Fourier analysis, i.e., the representation of functions through trigonometric functions, led Dirichlet to the modern concept of functions, stood at the origin of G. Cantor's (1845–1918) set theory, and were starting points for B. Riemann (1826–1866) and H. Lebesgue's (1875–1941) measure and integration theory. Fourier series theory, with its abstract terms and the resulting new methods for solving specific application problems, provides strong impulses to functional analysis and modern numerical mathematics still today. Even during their initial discussion period, Fourier's ideas rapidly entered natural and engineering sciences and are considered among the most effective mathematical tools in these fields today.

To explain Bernoulli and Fourier's basic ideas, let us first look at the problem of a vibrating string. One way a string's vibration can be understood is as an elementary example of an acoustic signal. From this, we can already develop essential terms for many Fourier analysis applications.

## 1.2   The Problem of the Force-Free Vibrating String

Let us look at the force-free motion of a thin homogeneous string of length $l$, fixed in place at both ends. How will the string move if it is displaced from its at-rest state and then released? To deal with this question mathematically, we introduce a coordinate system and designate the transversal displacement of the string at position $x$ at time $t$ as the function $u(x,t)$.

We are searching for a function $u(x,t)$ on $[0,l] \times \mathbb{R}_0^+$ that is twice continuously differentiable and which fulfills *boundary value conditions* (see Fig. 1.1)

$$u(0,t) = u(l,t) = 0 \quad \text{for } t \geqslant 0,$$

**Fig. 1.1** Initial displacement of a thin string

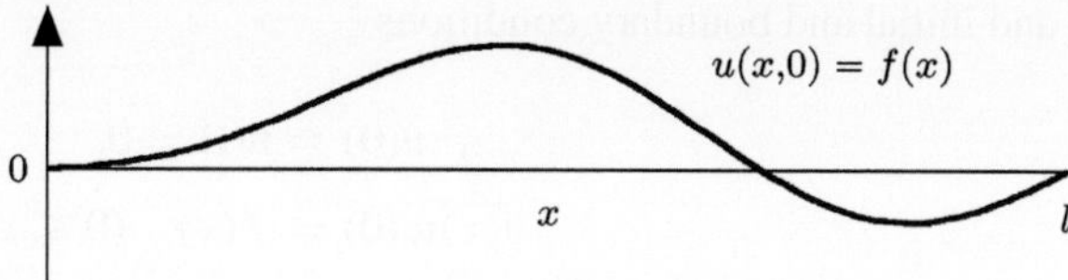



as well as *initial value conditions*

$$u(x, 0) = f(x) \quad \text{for } 0 \leqslant x \leqslant l, \ f(0) = f(l) = 0,$$

$$\lim_{t \to 0+} \frac{\partial u}{\partial t}(x, t) = g(x) \quad \text{for } 0 \leqslant x \leqslant l, \ g(0) = g(l) = 0.$$

In our example, "to let go off the string" means that $g = 0$ on $[0, l]$.

To determine $u(x, t)$ for times $t > 0$, physics tells us that during forceless motion, and with the previously mentioned boundary and initial conditions in place, for small transversal displacements the function $u(x, t)$ approximately satisfies the one-dimensional wave equation:

$$\frac{\partial^2 u}{\partial t^2} = c^2 \frac{\partial^2 u}{\partial x^2} \qquad (0 < x < l, \ t > 0).$$

In this equation, the constant $c^2 = P/\varrho$ is the quotient of the string's tension $P$ and mass density $\varrho$. $P$ is the quotient of tension force $F$ and the string's cross-sectional area $A$. In this context, the constant $c$ has the physical dimension of a velocity.

If we additionally assume that $u(x, t)$ is of the form

$$u(x, t) = v(x) \cdot w(t) \qquad \text{(separation of variables)},$$

substitution in the wave equation is as follows:

$$v\ddot{w} = c^2 v'' w.$$

For this, we use the notation $\dot{w} = \dfrac{dw}{dt}$, $v' = \dfrac{dv}{dx}$, $\ddot{w} = \dfrac{d^2 w}{dt^2}$, and $v'' = \dfrac{d^2 v}{dx^2}$.

Division through $c^2 vw$ (under the condition $c^2 vw \neq 0$) results in

$$\frac{\ddot{w}}{c^2 w} = \frac{v''}{v}.$$

Because the left side is only a function of $t$ and does not depend on $x$, the right side can also not depend on $x$, and it has to remain constant. If we name this constant $\lambda$, we end up with two ordinary linear differential equations:

$$v'' - \lambda v = 0,$$

$$\ddot{w} - \lambda c^2 w = 0$$

and initial and boundary conditions

$$v(0) = v(l) = 0, \tag{1.1}$$

$$v(x)w(0) = f(x) \quad (0 \leqslant x \leqslant l), \tag{1.2}$$

$$v(x) \lim_{t \to 0+} \dot{w}(t) = g(x) \quad (0 \leqslant x \leqslant l). \tag{1.3}$$

As a calculus reminder of ordinary linear differential equations, we will determine the solutions of $v'' - \lambda v = 0$:

The $v(x) = e^{sx}$ approach leads to the equation $e^{sx}(s^2 - \lambda) = 0$. Since $e^{sx} \neq 0$ is always true, we find solutions by determining the zeros of the characteristic polynomial $P(s) = s^2 - \lambda$. The zeros of the characteristic polynomial $P(s) = s^2 - \lambda$ are

$$
\begin{array}{ll}
\pm\sqrt{\lambda} & \lambda > 0, \\
0 & \text{if } \lambda = 0, \\
\pm j\sqrt{-\lambda} & \lambda < 0,
\end{array}
$$

where $j$ designates the imaginary unit with $j^2 = -1$. For readers accustomed to the expression $i$, $i^2 = -1$, which is more common in mathematics than $j$, it should be noted that $j^2 = -1$ is the notation widely used in electrical engineering and signal processing, because in these fields the letter $i$ is firmly used to designate the electric current. Mathematician readers should easily be able to deal with this notational variant in the text.

1. **Case:** $\lambda > 0$: Let us assume that one of the solutions $v(x) = c_1 e^{\sqrt{\lambda}x} + c_2 e^{-\sqrt{\lambda}x}$ fulfills the boundary conditions. It follows that for the corresponding $c_1$ and $c_2$

$$c_1 e^{\sqrt{\lambda} \cdot 0} + c_2 e^{-\sqrt{\lambda} \cdot 0} = c_1 + c_2 = 0,$$

$$c_1 e^{\sqrt{\lambda} \cdot l} + c_2 e^{-\sqrt{\lambda} \cdot l} = 0$$

is valid. Because the determinant $\det \begin{pmatrix} 1 & 1 \\ e^{\sqrt{\lambda} \cdot l} & e^{-\sqrt{\lambda} \cdot l} \end{pmatrix} \neq 0$, the result is $c_1 = c_2 = 0$, i.e., only the zero solution is obtained. However, the zero solution does not fulfill the initial condition (1.2) for $f \neq 0$ and therefore cannot be an option.

2. **Case:** $\lambda = 0$: $c_1 + c_2 x = 0$ for $x = 0$ and $x = l$ also result in $c_1 = c_2 = 0$ here, i.e., this case constitutes another solution to the homogeneous differential equation $v'' - \lambda v = 0$ that does not meet our initial condition for $f \neq 0$.

3. **Case:** $\lambda < 0$: It follows that the general solution of $v'' - \lambda v = 0$ is given through

$$v(x) = c_1 \cos(\sqrt{-\lambda} \cdot x) + c_2 \sin(\sqrt{-\lambda} \cdot x).$$

To fulfill the boundary value conditions, $c_1, c_2$ must be chosen so that

$$c_1 \cos(\sqrt{-\lambda} \cdot 0) + c_2 \sin(\sqrt{-\lambda} \cdot 0) = 0 = c_1 \cos(\sqrt{-\lambda} \cdot l) + c_2 \sin(\sqrt{-\lambda} \cdot l),$$

meaning $c_1 = 0$ and $c_2 \sin(\sqrt{-\lambda} \cdot l) = 0$. This is possible for any given $c_2 \in \mathbb{R}$ and $\sqrt{-\lambda} \cdot l = n \cdot \pi$, $n \in \mathbb{Z}$, therefore for $\lambda$ of the form $\lambda_n = -(n\pi/l)^2$, $n \in \mathbb{Z}$.

To summarize, we can say that for every $n \in \mathbb{N}$ the functions

$$v_n(x) = c_n \sin\left(\frac{n\pi}{l} x\right), \quad c_n \in \mathbb{R} \text{ arbitrary,}$$

are solutions of $v'' - \lambda_n v = 0$ that fulfill the boundary value conditions (1.1).

We subsequently determine the general solution of $\ddot{w} - \lambda_n c^2 w = 0$ for every value $\lambda_n = -(n\pi/l)^2$ analogously resulting in the so-called $n$th *string eigensolution*

$$u_n(x,t) = \sin\left(\frac{n\pi}{l} x\right)\left(a_n \cos\left(\frac{cn\pi}{l} t\right) + b_n \sin\left(\frac{cn\pi}{l} t\right)\right)$$

($n \in \mathbb{N}$; factors $c_n$ of $v_n$ are included into $a_n$, $b_n$).

The $n$th eigensolution has angular frequency $\omega_n = cn\pi/l$. By inserting initial conditions, we observe that an eigensolution $u_n(x,t)$ is a solution for the problem, if the following is true:

$$f(x) = a_n \sin\left(\frac{n\pi}{l} x\right) \quad \text{und} \quad g(x) = \frac{cn\pi}{l} b_n \sin\left(\frac{n\pi}{l} x\right).$$

In the mathematical model, trigonometric polynomials, i.e., linear combinations of the form

$$f(x) = \sum_{n=1}^{N} a_n \sin\left(\frac{n\pi}{l} x\right) \quad \text{und} \quad g(x) = \sum_{n=1}^{N} \frac{cn\pi}{l} b_n \sin\left(\frac{n\pi}{l} x\right),$$

are approximations for the exact initial conditions of a string's vibration. The resulting linear combination of eigensolutions with coefficients $a_n$ and $b_n$ of the initial conditions

$$u(x,t) = \sum_{n=1}^{N} \sin\left(\frac{n\pi}{l} x\right)\left(a_n \cos\left(\frac{cn\pi}{l} t\right) + b_n \sin\left(\frac{cn\pi}{l} t\right)\right)$$

is then an approximate solution for the exact string displacement. To end up with good approximations for different physical conditions, we want to deal with initial conditions $f$ and $g$ that are as general as possible. The more trigonometric functions we use to represent $f$ and $g$, the better approximations we can expect. We therefore set up $f$ and $g$ as *infinite trigonometric series*

$$f(x) = \sum_{n=1}^{\infty} a_n \sin\left(\frac{n\pi}{l} x\right) \quad \text{and} \quad g(x) = \sum_{n=1}^{\infty} \frac{cn\pi}{l} b_n \sin\left(\frac{n\pi}{l} x\right)$$

and try to find a solution using the *superposition of infinitely many eigensolutions* in the form of

$$u(x, t) = \sum_{n=1}^{\infty} \sin\left(\frac{n\pi}{l} x\right) \left(a_n \cos\left(\frac{cn\pi}{l} t\right) + b_n \sin\left(\frac{cn\pi}{l} t\right)\right).$$

When the series converges to a sufficiently smooth function, it expresses a possible vibration satisfying the boundary value conditions $u(0, t) = u(l, t) = 0$. The values $a_n$ and $b_n$ are determined by the fact that the initial conditions should be fulfilled. Their physical unit is the same as of $u(x, t)$. Inserting the series into the initial conditions using term-by-term differentiation and interchanging the limiting operation $t \to 0+$ with the series infinite summation operation result in

$$u(x, 0) = \sum_{n=1}^{\infty} a_n \sin\left(\frac{n\pi}{l} x\right) = f(x),$$

$$\lim_{t \to 0+} \frac{\partial u}{\partial t}(x, t) = \sum_{n=1}^{\infty} \frac{cn\pi}{l} b_n \sin\left(\frac{n\pi}{l} x\right) = g(x).$$

In order to definitively solve the problem, some inevitable questions arise at this point:

**Question 1:**   Which functions $f$ and $g$ on $[0, l]$ can be expressed as trigonometric series at all?

Bernoulli and Fourier's fundamental thought was that through suitable selection of the infinitely many coefficients $a_n$ and $b_n$, nearly every practically relevant function could be expressed as a superposition of harmonic oscillations. This would make the string problem solvable using the series method, for nearly "any" set of initial conditions.

**Question 2:**   If we can assume that the given functions $f$ and $g$ can be represented as such trigonometric series, how can we then calculate the required coefficients $a_n$ and $b_n$?

We could only explicitly solve the vibration problem by determining these coefficients. We will answer Question 2 in the next two chapters. This will also require an answer to the following question: Is the calculated series representation for the wanted function $u(x, t)$ actually a unique, twice differentiable solution to the initial value problem? The solution above was calculated assuming a very special solution form $u(x, t) = v(x)w(t)$ with separate variables. We also once learned that we cannot differentiate function series simply term by term or interchange limiting operations without care but did just that in calculating our solutions. This problem therefore leads to the following question:

**Question 3:**   Dependent on initial conditions $f$ and $g$—in what sense do trigonometric series converge to the expression of $u(x, t)$ at all? Is the series actually twice differentiable? Is the solution found for the initial boundary value problem for the wave equation unique?

These questions immediately show the concerns raised in the early nineteenth century against Fourier's approach to a solution. A. L. Cauchy (1789–1857) only developed a convergence theory for infinite series during Fourier's time around 1821.

Satisfying arguments regarding the solvability of linear partial differential equations only appeared around the middle of the twentieth century with the treatment of such problems within the theory of generalized functions, or as we also say, of distributions.

To answer the questions raised in a step-by-step manner, we will begin the next sections with some fundamentals on trigonometric polynomials. In this regard, we find that frequently expressions using complex numbers are very useful. Recommended preparatory readings for readers, which have so far only been accustomed to real analysis, are respective sections in E. Kreyszig (2011) or G. Strang (2017). A collection of formulas such as that by L. Råde, B. Westergren (2004) can be equally helpful. We will particularly use complex exponential functions and their close link to trigonometric functions frequently.

**Six Important Protagonists in 250 Years History of Fourier Analysis**

Daniel Bernoulli, Jean Baptiste Fourier, Peter G. Lejeune Dirichlet, Bernhard Riemann, David Hilbert, and Laurent Schwartz

Ⓢ All pictures from Wikimedia Commons, in the public domain everywhere.

# Chapter 2
# Trigonometric Polynomials and Fourier Coefficients

**Abstract** Representations of trigonometric polynomials are given as a preparation for the following chapters, in terms of their Fourier coefficients and as a convolution with a Dirichlet kernel. The computation of the complex coefficients is shown, and the number of zeros of a trigonometric polynomial is calculated. The orthogonality relation is deduced for sine and cosine functions with period T, but different frequencies n/T and m/T. Furthermore, properties of the Dirichlet kernels are discussed, which provide an initial insight into periodic pulse sequences, which play an essential role in discrete signal processing.

## 2.1 Representation of Trigonometric Polynomials

A trigonometric polynomial with period $T$ is a function $f$ with values in $\mathbb{R}$ or $\mathbb{C}$ of the form

$$f(t) = \frac{a_0}{2} + \sum_{n=1}^{N} (a_n \cos(n\omega_0 t) + b_n \sin(n\omega_0 t)),$$

with $N \in \mathbb{N}$, $t \in \mathbb{R}$, $\omega_0 = 2\pi/T$. The maximum of $n$ with $|a_n| + |b_n| \neq 0$ is called the *degree of the trigonometric polynomial $f$*.

For calculation purposes most often a complex representation of trigonometric polynomials is useful. With the complex unit $j$, $j^2 = -1$ and the formulas for the real and imaginary parts of $e^{jn\omega_0 t}$, we have

$$\cos(n\omega_0 t) = \frac{1}{2}\left(e^{jn\omega_0 t} + e^{-jn\omega_0 t}\right) = \Re\left(e^{jn\omega_0 t}\right),$$

$$\sin(n\omega_0 t) = \frac{1}{2j}\left(e^{jn\omega_0 t} - e^{-jn\omega_0 t}\right) = -\frac{j}{2}\left(e^{jn\omega_0 t} - e^{-jn\omega_0 t}\right) = \Im\left(e^{jn\omega_0 t}\right).$$

© The Author(s), under exclusive license to Springer Nature Switzerland AG 2025

R. Brigola, *Fourier Analysis and Distributions*, Texts in Applied Mathematics 79,

https://doi.org/10.1007/978-3-031-81311-5_2

With $b_0 = 0$, it follows by insertion above

$$f(t) = \frac{a_0}{2} + \sum_{n=1}^{N} \left( \frac{1}{2} a_n e^{jn\omega_0 t} + \frac{1}{2} a_n e^{-jn\omega_0 t} - \frac{j}{2} b_n e^{jn\omega_0 t} + \frac{j}{2} b_n e^{-jn\omega_0 t} \right)$$

$$= \sum_{n=0}^{N} \underbrace{\left( \frac{a_n - jb_n}{2} \right)}_{=c_n} e^{jn\omega_0 t} + \sum_{n=1}^{N} \underbrace{\left( \frac{a_n + jb_n}{2} \right)}_{=c_{-n}} e^{-jn\omega_0 t}$$

$$= \sum_{n=0}^{N} c_n e^{jn\omega_0 t} + \sum_{n=-N}^{-1} c_n e^{jn\omega_0 t} = \sum_{n=-N}^{N} c_n e^{jn\omega_0 t}.$$

The constants $c_n$, or alternatively $a_n$ and $b_n$, are called the *Fourier coefficients* of $f$. For

$$f(t) = \frac{a_0}{2} + \sum_{n=1}^{N} (a_n \cos(n\omega_0 t) + b_n \sin(n\omega_0 t)) = \sum_{n=-N}^{N} c_n e^{jn\omega_0 t},$$

we find the following conversion formulas between the Fourier coefficients:

$$c_n = \frac{a_n - jb_n}{2}, \quad c_{-n} = \frac{a_n + jb_n}{2},$$

$$b_0 = 0, \qquad\qquad a_0 = 2c_0, \qquad\qquad a_n = c_n + c_{-n}, \quad b_n = j(c_n - c_{-n}).$$

## 2.2  Fourier Coefficients of Trigonometric Polynomials

### *Computation of Fourier Coefficients*

The answer to the issue of computing Fourier coefficients results from the following so-called *orthonormality relations for trigonometric functions*:

For all $n, k \in \mathbb{Z}$, the complex conjugate function $\overline{e^{jk\omega_0 t}} = e^{-jk\omega_0 t}$ of $e^{jk\omega_0 t}$ ($j^2 = -1$) gives us

$$\frac{1}{T} \int_0^T e^{jn\omega_0 t} \overline{e^{jk\omega_0 t}} \, dt = \begin{cases} 1 & \text{for } n = k \\ 0 & \text{for } n \neq k, \end{cases}$$

because

$$\int_0^T e^{jn\omega_0 t} e^{-jk\omega_0 t}\, dt = \int_0^T 1\, dt = T \quad \text{for } k = n;$$

$$\int_0^T e^{j(n-k)\omega_0 t}\, dt = \frac{1}{j(n-k)\omega_0}\left[\underbrace{e^{j(n-k)\cdot 2\pi}}_{1} - 1\right] = 0 \quad \text{for } k \neq n.$$

When $f(t)$ has the form $f(t) = \sum_{k=-N}^{N} c_k e^{jk\omega_0 t}$, one computes $c_k$ by

$$c_k = \frac{1}{T}\int_0^T f(t)e^{-jk\omega_0 t}\, dt,$$

since

$$\frac{1}{T}\int_0^T f(t)e^{-jk\omega_0 t}\, dt = \frac{1}{T}\sum_{n=-N}^{N} c_n \underbrace{\int_0^T e^{jn\omega_0 t} e^{-jk\omega_0 t}\, dt}_{\substack{T \text{ for } n = k \\ 0 \text{ otherwise}}} = c_k.$$

Furthermore, for the Fourier coefficients $a_n, b_n, n = 1, \ldots, N$, we obtain

$$\frac{a_0}{2} = c_0 = \frac{1}{T}\int_0^T f(t)\, dt,$$

$$a_n = c_n + c_{-n} = \frac{1}{T}\int_0^T f(t)\underbrace{\left[e^{-jn\omega_0 t} + e^{jn\omega_0 t}\right]}_{2\cos(n\omega_0 t)}\, dt = \frac{2}{T}\int_0^T f(t)\cos(n\omega_0 t)\, dt,$$

$$b_n = j(c_n - c_{-n}) = \frac{j}{T}\int_0^T f(t)\underbrace{\left[e^{-jn\omega_0 t} - e^{jn\omega_0 t}\right]}_{-2j\sin(n\omega_0 t)}\, dt = \frac{2}{T}\int_0^T f(t)\sin(n\omega_0 t)\, dt.$$

## *Equality of Trigonometric Polynomials*

For each pair of continuous $T$-periodic functions $f : \mathbb{R} \to \mathbb{C}$ and $g : \mathbb{R} \to \mathbb{C}$, we set

$$\langle f | g \rangle = \langle f(t) | g(t) \rangle = \frac{1}{T} \int_0^T f(t)\overline{g(t)}\,\mathrm{d}t.$$

This defines an *inner product* in the vector space $V$ of continuous $T$-periodic functions. It has the same properties as the inner product for vectors in $\mathbb{R}^n$ order $\mathbb{C}^n$ and allows to transfer geometric terms like orthogonality of vectors to functions. For continuous $T$-periodic functions $f$, $g$, and $h$, we have

$$\langle f + g | h \rangle = \langle f | h \rangle + \langle g | h \rangle$$

$$\langle f | g + h \rangle = \langle f | g \rangle + \langle f | h \rangle$$

$$\langle \alpha f | g \rangle = \alpha \langle f | g \rangle \qquad (\alpha \in \mathbb{C})$$

$$\langle f | \beta g \rangle = \overline{\beta} \langle f | g \rangle \qquad (\beta \in \mathbb{C})$$

$$\langle f | g \rangle = \overline{\langle g | f \rangle}$$

$$\langle f | f \rangle \geqslant 0$$

$$\langle f | f \rangle = 0 \iff f(t) = 0 \text{ for all } t \in [0, T].$$

Therefore, the orthonormality relations $\langle e^{jn\omega_0 t} | e^{jk\omega_0 t} \rangle = \begin{cases} 1 & \text{for } n = k \\ 0 & \text{for } n \neq k \end{cases}$ show that the functions $\left( e^{jn\omega_0 t} \right)_{n \in \mathbb{Z}}$ build a *linearly independent system* in the vector space $V$. The subspace of all $T$-periodic trigonometric polynomials of maximum degree $N$ has dimension $2N + 1$ and is spanned by the functions $e^{jn\omega_0 t}$, $-N \leqslant n \leqslant N$, $\omega_0 = 2\pi/T$. They form an orthonormal basis of that subspace with respect to the inner product introduced above. With that notation, the $k$th Fourier coefficient $c_k$ of a $T$-periodic trigonometric polynomial $f$ is given by[1]

$$c_k = \langle f(t) | e^{jk\omega_0 t} \rangle \qquad (\omega_0 = 2\pi/T).$$

---

[1] The notation $f(t)$ will be used hereafter not only for the value of $f$ at $t$ but also for a function (and later a distribution) $f$, to show its parameter $t$. In spite of this ambivalent notation—in place of $f$ or perhaps $f(.)$—the meaning will be readily apparent from the respective context.

Thus, the Fourier coefficients of a $T$-periodic trigonometric polynomial are just its coordinates with respect to that orthonormal basis. They are uniquely defined, in other words

$$f(t) = \sum_{k=-N}^{N} c_k e^{jk\omega_0 t} = \frac{a_0}{2} + \sum_{n=1}^{N} (a_n \cos(n\omega_0 t) + b_n \sin(n\omega_0 t)) = 0$$

*for all $t$ if and only if all $c_k = 0$ and correspondingly all $a_n = b_n = 0$.*
Two $T$-periodic trigonometric polynomials are equal if and only if all their Fourier coefficients corresponding to the same basis functions are equal.

Additionally, the formula for the Fourier coefficients shows that every $T$-periodic trigonometric polynomial $f$ of maximum degree $N$ has the following *integral representation*:

$$f(t) = \frac{1}{T} \int_0^T f(s) D_N(t-s)\, ds \quad \text{mit} \quad D_N(t-s) = \sum_{k=-N}^{N} e^{jk\omega_0(t-s)}.$$

### *Real-Valued Trigonometric Polynomials and Complex Amplitudes*

For real-valued $T$-periodic trigonometric polynomials, we have $f(t) = \overline{f(t)}$, and thus

$$f(t) = \sum_{k=-N}^{N} c_k e^{jk\omega_0 t} = \sum_{k=-N}^{N} c_{-k} e^{-jk\omega_0 t} = \sum_{k=-N}^{N} \overline{c_k} e^{-jk\omega_0 t} = \overline{f(t)},$$

with $\omega_0 = 2\pi/T$. Equating the coefficients shows

$$f \text{ is real-valued if and only if } c_k = \overline{c_{-k}} \quad (-N \leqslant k \leqslant N).$$

Since $c_k = |c_k| e^{j\,\arg(c_k)}$ and $\arg(c_{-k}) = -\arg(c_k)$ hold for $k \neq 0$, we obtain

$$f(t) = c_0 + \sum_{k=1}^{N} |c_k| e^{j(k\omega_0 t + \arg(c_k))} + \sum_{k=1}^{N} |c_{-k}| e^{-j(k\omega_0 t - \arg(c_{-k}))}$$

$$= c_0 + \sum_{k=1}^{N} |c_k| \cdot 2\,\Re\left(e^{j(k\omega_0 t + \arg(c_k))}\right).$$

Therefore, in that case we get

$$f(t) = c_0 + 2 \cdot \sum_{k=1}^{N} |c_k| \cos(k\omega_0 t + \arg(c_k)),$$

as a *representation in polar form*.

The complex Fourier coefficients include the information on the amplitudes and phases of the oscillations that build up $f$ in linear combination. These values $c_k$ are called *complex amplitudes*. $\mathbb{C}$-valued trigonometric polynomials can be visualized as circular waves (see p. 17) or as curves in $\mathbb{C}$ (Nyquist plots), either through separate views of their real and imaginary parts or through visualization of amplitude and phase progressions versus (time) parameter $t$.

## *Number of Zeros of Trigonometric Polynomials*

We can generate a $T$-periodic trigonometric polynomial of degree $N > 0$

$$f(t) = \sum_{k=-N}^{N} c_k e^{jk\omega_0 t} \quad \left(\omega_0 = \frac{2\pi}{T}\right)$$

with $|c_N| + |c_{-N}| \neq 0$ by substitution of $z = e^{j\omega_0 t}$ into the rational function

$$F(z) = \sum_{k=-N}^{N} c_k z^k = \frac{c_{-N} + c_{-N+1}z + \ldots + c_N z^{2N}}{z^N}.$$

*Since $|z| = 1$, the function $f$ cannot have more than $2N$ zeros per period $T$.* Specifically, it follows that two trigonometric polynomials $P$ and $Q$ of degree $N$ are identical when they have the same values at $2N + 1$ points in $[0, T[$. In that case, it is clear that $P - Q$ has a maximum degree $N$, but more than $2N$ zeros in $[0, T[$, i.e., $P - Q$ is identically zero (cf. Appendix, Fundamental Theorem of Algebra).

## 2.3  Dirichlet Kernels

Let us consider the trigonometric polynomial $D_N(t) = \sum_{k=-N}^{N} e^{jkt}$. This function is called *Dirichlet kernel of degree $N$*, and it plays  an important role in answering question 1 from Sect. 1.2 regarding the possibility of representing periodic functions as trigonometric series. In polar form we have for $t \in \mathbb{R}$

$$D_N(t) = \sum_{k=-N}^{N} e^{jkt} = 1 + 2\cos(t) + 2\cos(2t) + \ldots + 2\cos(Nt).$$

Observe that $D_N$ is an even function. We can now substitute $z = e^{jt}$ and use the common geometric sum formula for $z \neq 1$

$$\sum_{k=0}^{2N} z^k = \frac{z^{2N+1} - 1}{z - 1}.$$

For $t \neq 2\pi n, n \in \mathbb{Z}$, we obtain

$$D_N(t) = \sum_{k=-N}^{N} z^k = \frac{1 + z + \ldots + z^{2N}}{z^N} = \frac{z^{2N+1} - 1}{(z-1)z^N}$$

$$= \frac{z^{N+1} - z^{-N}}{z - 1} = \frac{z^{N+1/2} - z^{-(N+1/2)}}{z^{1/2} - z^{-1/2}}$$

$$= \frac{e^{j(N+1/2)t} - e^{-j(N+1/2)t}}{e^{jt/2} - e^{-jt/2}} = \frac{\sin\left(\left(N + \frac{1}{2}\right)t\right)}{\sin\left(\frac{t}{2}\right)}.$$

We can therefore write the function $D_N$ equivalently in the following form:

$$D_N(t) = \sum_{k=-N}^{N} e^{jkt} = \begin{cases} 2N + 1 & \text{for } t = 2\pi n, \ n \in \mathbb{Z} \\[2ex] \dfrac{\sin\left((N + \frac{1}{2})t\right)}{\sin(\frac{t}{2})} & \text{for } t \neq 2\pi n, \ n \in \mathbb{Z}. \end{cases}$$

**Theorem 2.1** *The following assertion holds true for the Dirichlet kernels $D_N(t)$:*

$$\lim_{N \to \infty} \sum_{k=-N}^{N} e^{jkt} = \lim_{N \to \infty} \left(1 + \sum_{k=1}^{N} 2\cos(kt)\right) = \begin{cases} +\infty \, \text{for } t = 2\pi n \,, \ n \in \mathbb{Z} \\[1.5ex] \text{otherwise indefinitely divergent.} \end{cases}$$

*There is no single point $t \in \mathbb{R}$ for which we have a limit of the trigonometric series*

$$\sum_{k=-\infty}^{+\infty} e^{jkt} = 1 + \sum_{k=1}^{\infty} 2\cos(kt).$$

If the series were convergent at a point $t_0$, this would give us $\lim_{k\to\infty} \cos(kt_0) = 0$, and for $k \to \infty$ it would lead to the following contradiction:

$$\sin^2(kt_0) = 1 - \cos^2(kt_0) \longrightarrow 1,$$

and also

$$\sin^2(kt_0) = \frac{1}{2}(1 - \cos(2kt_0)) \longrightarrow \frac{1}{2}.$$

However, if we interpret $D_N(t)$ as a signal, say the voltage change output of an electric transmission system, it appears intuitive that increasing $N$ would result in $D_N(t)$ as a model for an impulse sequence. The impulses will appear at the "times" $2\pi n$, $n \in \mathbb{Z}$, but the signal will never disappear between these impulses; it will increase its oscillations more and more (see Fig. 2.1). Signal processing in causal linear systems mathematically leads to integral transforms of a signal (convolution with the system's impulse response, cf. p. 66 and p. 219 later on). For increasing $N$, the oscillations between the points $2\pi n$ lead to annihilation in integrals over intervals $[2\pi n+\varepsilon, 2\pi(n+1)-\varepsilon]$, $0 < \varepsilon < \pi$, because, with increasing frequencies, an increasing part of the area between the graph of $D_N$ and the $t$-axis will alternate above and below the $t$-axis, thus adding to zero in the integral. We will work out and confirm this intuition more precisely in computations and a proof on p. 23, p. 48, and p. 131 later on and in Sect. 9.1 as well.

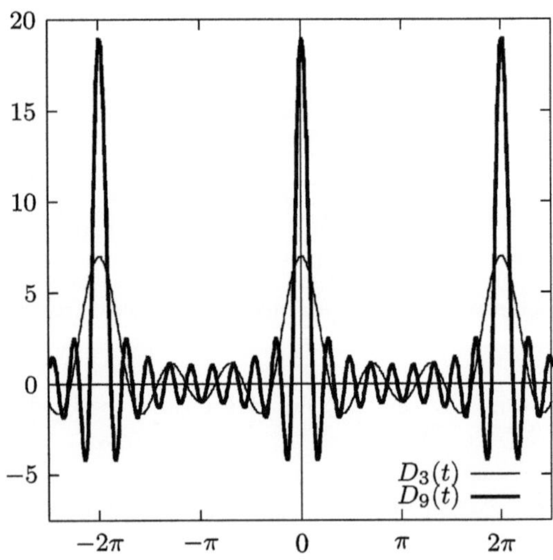


**Fig. 2.1** Dirichlet kernels $D_3$ and $D_9$

## 2.4  Summary on Trigonometric Polynomials

For trigonometric polynomials $f$ of degree $N > 0$ with a minimum period $T$, the following relations hold:

| | |
|---|---|
| *Basic angular frequency* | $\omega_0 = \dfrac{2\pi}{T}.$ |
| *Complex form* | $f(t) = \displaystyle\sum_{k=-N}^{N} c_k e^{jk\omega_0 t} = \frac{1}{T}\int_0^T f(s) D_N(t-s)\,\mathrm{d}s$ <br><br> with $D_N(t-s) = \displaystyle\sum_{k=-N}^{N} e^{jk\omega_0(t-s)}, \quad j^2 = -1.$ |
| *Sine-Cosine form* | $f(t) = \dfrac{a_0}{2} + \displaystyle\sum_{k=1}^{N}\left(a_k \cos(k\omega_0 t) + b_k \sin(k\omega_0 t)\right).$ |
| *Conversion formulas* | $c_k = \dfrac{a_k - jb_k}{2}, \quad c_{-k} = \dfrac{a_k + jb_k}{2}$ <br> with $b_0 = 0, \quad k = 0, \dots, N;$ <br><br> $\dfrac{a_0}{2} = c_0, \quad a_k = c_k + c_{-k}, \quad b_k = j(c_k - c_{-k}).$ |
| *Orthonormality relations* | $\dfrac{1}{T}\int_0^T e^{jn\omega_0 t}\,\overline{e^{jk\omega_0 t}}\,\mathrm{d}t = \begin{cases} 0 & \text{for } k \neq n \\ 1 & \text{for } k = n. \end{cases}$ |
| *Computation of Fourier coefficients* | $c_k = \dfrac{1}{T}\int_0^T f(t) e^{-jk\omega_0 t}\,\mathrm{d}t$ <br><br> $\dfrac{a_0}{2} = \dfrac{1}{T}\int_0^T f(t)\mathrm{d}t, \quad a_k = \dfrac{2}{T}\int_0^T f(t)\cos(k\omega_0 t)\mathrm{d}t$ <br><br> $b_k = \dfrac{2}{T}\int_0^T f(t)\sin(k\omega_0 t)\mathrm{d}t.$ |
| *For real-valued $f$ hold* <br><br> *and the polar representation* | $c_k = \overline{c_{-k}}, \quad a_k = 2\Re(c_k), \quad b_k = -2\Im(c_k),$ <br><br> $f(t) = c_0 + 2\displaystyle\sum_{k=1}^{N} |c_k|\cos(k\omega_0 t + \arg(c_k)).$ |
| *Number of zeros per period* | maximally $2N$ zeros in $[0,T[$. |

*Visualization of the trigonometric polynomial $P(t) = 0.3\,j\,\sin(\omega_0 t) + j\,\sin(2\omega_0 t) - \cos(3\omega_0 t)$, $\omega_0 = \pi/2$, period $T = 4$, as a circular wave.*

see Fig. 2.2

**Fig. 2.2** Circular wave

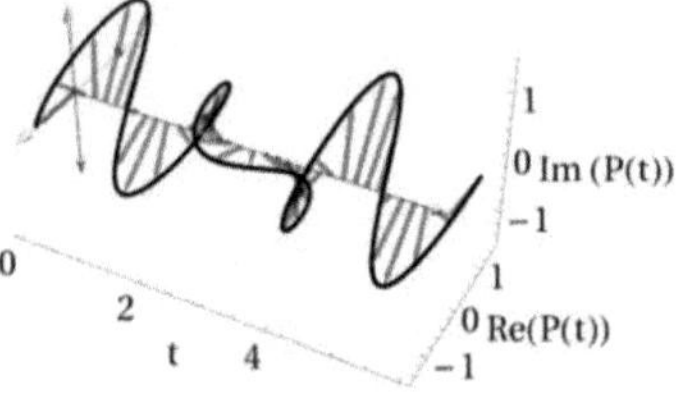

# Chapter 3
# Fourier Series

**Abstract** This chapter presents basic results on pointwise convergence of Fourier series. As a fundamental example, the Fourier series of the sawtooth function is studied. Properties of this series are deduced, such as pointwise convergence and uniform convergence in closed intervals, that do not contain a discontinuity point. The Gibbs phenomenon is worked out. The theorems of Dirichlet and Fejér are presented and discussed with examples. Their rigorous proofs are postponed until Chap. 7. Examples and exercises help the reader become familiar with the necessary calculations.

## 3.1 The First Fourier Series

By building the limit $N \to \infty$, $T$-periodic trigonometric polynomials evolve into trigonometric series

$$f_N(t) = \sum_{k=-N}^{N} c_k \, e^{jk\omega_0 t} \xrightarrow[N \to \infty]{} \sum_{k=-\infty}^{+\infty} c_k \, e^{jk\omega_0 t} \quad (\omega_0 = 2\pi/T).$$

However, as we have seen in the example of the Dirichlet kernels $\sum_{k=-N}^{N} e^{jkt}$, it may happen that no limit exists at any point.

If a limit function $f$ of a trigonometric series with $T$-periodic partial sums exists, then that function $f$ is also $T$-periodic. First of all, to develop an appropriate intuition for the behavior of trigonometric series, we study "the first Fourier series," mentioned for the first time by L. Euler as early as in 1744, i.e., we analyze the convergence of the series

$$\sum_{k=1}^{\infty} \frac{\sin(kt)}{k} = \sum_{\substack{k=-\infty \\ k \neq 0}}^{+\infty} \frac{1}{2kj} \, e^{jkt}$$

and work out a representation of its limit function by close inspection of the Dirichlet kernels. We start with a few preliminary remarks on functions with values in $\mathbb{R}$ or in $\mathbb{C}$.


## *Approximation Errors and Pointwise and Uniform Convergence*

In approximating a function $f$ using a sequence of functions $f_N$, $N \in \mathbb{N}$, the quality of the approximation (in other words the error $f_N - f$ for increasing $N$) plays a crucial role. For example, we can look at the error $f_N(t) - f(t)$ at single points $t$ in an interval $I$ in the domain of definition of the functions $f$ and $f_N$ or at the maximum error $\sup_{t \in I} |f_N(t) - f(t)|$ in $I$. It may indeed happen that $\lim_{N \to \infty} (f_N(t) - f(t)) = 0$ holds for any single $t \in I$, but the maximum error on $I$ nevertheless does not decrease when $N$ increases. Consider the example of the function sequence $(f_N)_{N \in \mathbb{N}}$ and $f$, defined through

$$f_N(t) = t^N \text{ auf } [0, 1] \quad \text{and} \quad f(t) = \begin{cases} 0 & \text{for } 0 \leqslant t < 1 \\ 1 & \text{for } t = 1. \end{cases}$$

For any $t \in [0, 1]$ we find $\lim_{N \to \infty} f_N(t) = f(t)$, but the maximum error on the interval $[0, 1]$ is $\sup_{0 \leqslant t \leqslant 1} |f_N(t) - f(t)| = \sup_{0 \leqslant t < 1} t^N = 1$ for all $N \in \mathbb{N}$. On the other hand, for the function sequence $(f_N)_{N \in \mathbb{N}}$ on the interval $[0, 1/2]$, we get $\lim_{N \to \infty} f_N(t) = f(t)$ for all $t \in [0, 1/2]$. The maximum error $(1/2)^N$ on that interval becomes arbitrarily small for increasing $N$. For the precise description, we remember the definitions of pointwise and uniform convergence of function sequences:

**Definition** A sequence of complex-valued functions $f_N : I \to \mathbb{C}$ converges pointwise to a function $f : I \to \mathbb{C}$, if $\lim_{N \to \infty} f_N(t) = f(t)$ for every $t \in I$. It converges to $f$ uniformly on $I$, if the maximum error $\sup_{t \in I} |f_N(t) - f(t)|$ on $I$ goes to zero for $N \to \infty$.

For readers whose acquaintance with these terms has yet to grow during further reading, the following important facts from first-year mathematical lectures are summarized again:

1. A pointwise convergent sequence $(f_N)_{N \in \mathbb{N}}$ can have a discontinuous limit $f$, even if all functions $f_N$ are continuous. This fact is shown by the example above. But, when a sequence of continuous functions $f_N$ converges uniformly to a function $f$, then $f$ is also continuous.

2. An important topic in the following studies is the uniform convergence of function series. Such a series $\sum_{k=1}^{\infty} f_k$ converges uniformly to a function $f$ in an interval $I$, if for $N \to \infty$ the sequence of its partial sums $\sum_{k=1}^{N} f_k$ converges to $f$ uniformly on $I$. A uniformly convergent series of continuous functions $f_k$ with limit $f$ on $I$ can be integrated term by term on any bounded subinterval $[a, b] \subset I$:

$$\int_a^b f(t)\, \mathrm{d}t = \int_a^b \sum_{k=1}^{\infty} f_k(t)\, \mathrm{d}t = \sum_{k=1}^{\infty} \int_a^b f_k(t)\, \mathrm{d}t.$$

If the $f_k$ are continuously differentiable and their series is pointwise convergent to $f$ and if the series of derivatives $f_k'$ converges uniformly on $I$, then the limit function $f$ is differentiable on $I$, and we have

$$f'(t) = \frac{\mathrm{d}}{\mathrm{d}t} \left( \sum_{k=1}^{\infty} f_k(t) \right) = \sum_{k=1}^{\infty} f_k'(t) \qquad (t \in I),$$

i.e., the series can be differentiated term by term.

3. The *Weierstrass M-Test* is a test for determining whether a series $\sum_{k=1}^{\infty} f_k$ of functions $f_k$ on an interval $I$ converges uniformly:
*If there is a sequence of positive numbers $(M_k)_{k \in \mathbb{N}}$ so that $\sup_{t \in I} |f_k(t)| \leqslant M_k$ for every $k \in \mathbb{N}$ and if $\sum_{k=1}^{\infty} M_k < \infty$, then the series $\sum_{k=1}^{\infty} f_k$ converges uniformly on $I$.*

As an example, the M-Test shows that the series $\sum_{k=1}^{\infty} \frac{\cos(kt)}{k^2}$ is uniformly convergent on $\mathbb{R}$, since $\sup_{t \in \mathbb{R}} \left| \frac{\cos(kt)}{k^2} \right| \leqslant \frac{1}{k^2} = M_k$ and $\sum_{k=1}^{\infty} \frac{1}{k^2} < \infty$ (see also p. 35).

## *An Initial Idea to Study the Series* $\sum_{k=1}^{\infty} \frac{\sin(kt)}{k}$

If you have ever felt daunted by the the "impulse function" $\delta(t)$ during a math or physics lecture, the following idea may be helpful: We have already seen that for large $N$ the Dirichlet kernels in the interval $[-\pi, \pi]$ behave like impulses. We can therefore recall a widely used introduction of the $\delta$-*Impulse* (see Fig. 3.1).

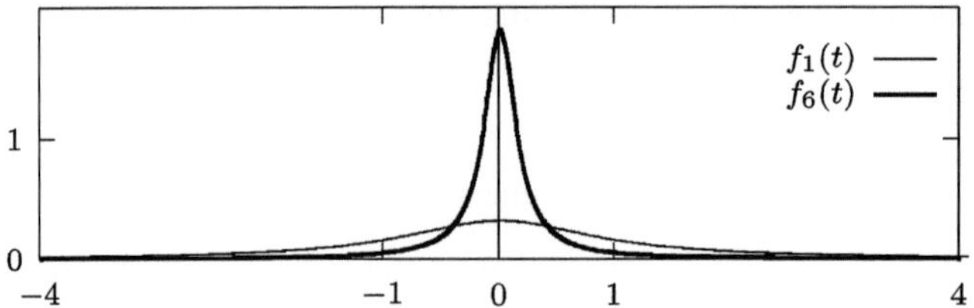

**Fig. 3.1** Functions $f_1$ and $f_6$ as elements of a $\delta$-sequence

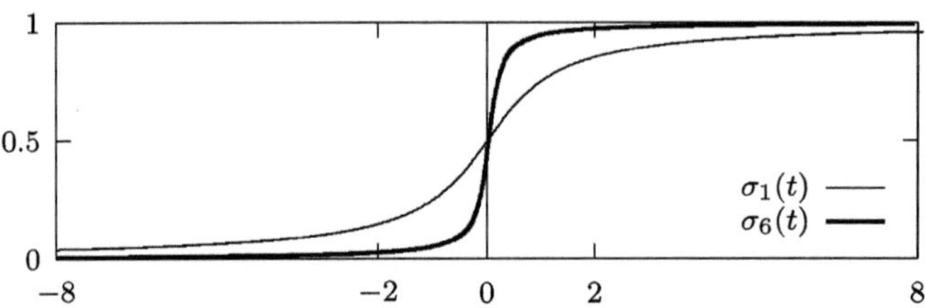

**Fig. 3.2** The primitives $\sigma_1$, $\sigma_6$ of $f_1$, $f_6$

The impulse $\delta(t)$ is often introduced as a limit of the sequence of functions

$$f_N(t) = \sigma_N'(t) = \frac{N}{\pi} \frac{1}{(1 + N^2 t^2)}.$$

The primitives $\sigma_N(t) = \dfrac{1}{2} + \dfrac{1}{\pi} \arctan(Nt)$ converge for $N \to \infty$ to the unit step function (see Fig. 3.2)

$$\sigma(t) = \begin{cases} 0 & \text{for } t < 0 \\[2mm] \dfrac{1}{2} & \text{for } t = 0 \\[2mm] 1 & \text{for } t > 0. \end{cases}$$

Looking at the convergence $\sigma_N \underset{N \to \infty}{\longrightarrow} \sigma$, we find:

1. $\sigma_N(t) \to \sigma(t)$ *pointwise everywhere* for $N \to \infty$, *i.e., for every* $t \in \mathbb{R}$.
2. *It holds the mean value property*

$$\sigma(t) = \frac{1}{2}\left(\sigma(t+) + \sigma(t-)\right) = \frac{1}{2}\left(\lim_{h \to 0+} \sigma(t + h) + \lim_{h \to 0+} \sigma(t - h)\right).$$

3. $\sigma(t)$ *is piecewise continuously differentiable.*

Therefore, we expect a similar behavior of the kernels $D_N(t)$—namely pointwise convergence of their primitives. These integrals are closely related with the series $\sum_{k=1}^{\infty} \dfrac{\sin(kt)}{k}$ because the derivatives of the series' partial sums yield—up to a factor and an additive constant—just the Dirichlet kernels.

## Study of the Series $\displaystyle\sum_{k=1}^{\infty} \frac{\sin(kt)}{k}$

Inexperienced readers might find the following calculations intricate at first sight. I therefore want to clarify that I have only assumed first-year course knowledge of differential and integral calculus in putting this section together. The goal of the following calculations is for the reader to achieve a basic estimation technique and a confident handling of trigonometric functions.

We build primitives of the Dirichlet kernels between two "impulse peaks" of $D_N$ (compare the figure on p. 16), i.e., we integrate for $t \in \,]0, 2\pi[$ from $\pi$ to $t$.

$$\int_{\pi}^{t} D_N(\tau)\,d\tau = \int_{\pi}^{t} \frac{\sin\left((N+\frac{1}{2})\tau\right)}{\sin(\frac{\tau}{2})}\,d\tau = \int_{\pi}^{t} (1 + 2\cos(\tau) + \ldots + 2\cos(N\tau))\,d\tau$$

$$= (t - \pi) + 2\left(\sin(t) + \frac{\sin(2t)}{2} + \frac{\sin(3t)}{3} + \ldots + \frac{\sin(Nt)}{N}\right) = I_N(t).$$

On the other hand, through integration by parts

$$\int_{\pi}^{t} u(\tau)v'(\tau)\,d\tau = u(\tau)v(\tau)\Big|_{\pi}^{t} - \int_{\pi}^{t} v(\tau)u'(\tau)\,d\tau\,,$$

$$u(\tau) = \frac{1}{\sin(\frac{\tau}{2})}, \quad v'(\tau) = \sin\left((N+\frac{1}{2})\tau\right), \quad v(\tau) = -\frac{\cos\left((N+\frac{1}{2})\tau\right)}{N+\frac{1}{2}},$$

we have

$$I_N(t) = -\frac{\cos\left((N+\frac{1}{2})t\right)}{(N+\frac{1}{2})\sin(\frac{t}{2})} + \frac{1}{N+\frac{1}{2}} \int_{\pi}^{t} \cos\left((N+\frac{1}{2})\tau\right)\left(\frac{1}{\sin(\frac{\tau}{2})}\right)' d\tau\,, \text{ thus}$$

$$
I_N(t) = -\frac{\cos\left((N+\frac{1}{2})t\right)}{(N+\frac{1}{2})\sin(\frac{t}{2})} + \frac{1}{N+\frac{1}{2}} \int_{\min(\pi,t)}^{\max(\pi,t)} \mathrm{sgn}(t-\pi)\cos\left((N+\frac{1}{2})\tau\right)
$$

$$
\times \left(\frac{1}{\sin(\frac{\tau}{2})}\right)' \, d\tau \; .
$$

Here, $\mathrm{sgn}(t - \pi)$ denotes the sign of $(t - \pi)$. Now, it holds $1/\sin(\tau/2) \geqslant 1$ for $0 < \tau < 2\pi$ and $\mathrm{sgn}(t - \pi)(1/\sin(\tau/2))' \geqslant 0$ for $\tau \in [\min(\pi, t), \max(\pi, t)]$. This is immediately apparent in the monotonicity properties of the function $1/\sin(\tau/2)$ in the interval $[\min(\pi, t), \max(\pi, t)]$. This function is strictly decreasing for $t < \pi$ while strictly increasing for $t > \pi$.

The cosine functions on the right-hand side are dominated by the constant $K = 1$ ($|\cos(x)| \leqslant 1$ everywhere). As a standard technique, application of the triangle inequality and increasing the terms at the right will yield the following estimation:

$$
|I_N(t)| \leqslant \frac{1}{(N + \frac{1}{2})\sin(\frac{t}{2})} + \frac{1}{N + \frac{1}{2}} \underbrace{\int_{\min(\pi,t)}^{\max(\pi,t)} \mathrm{sgn}(t - \pi)\left(\frac{1}{\sin(\frac{\tau}{2})}\right)' \, d\tau}_{\sin(t/2)^{-1}-1}
$$

$$
\leqslant \frac{2}{(N + \frac{1}{2})\sin(\frac{t}{2})} \; .
$$

Therefore, $I_N(t)$ disappears for increasing $N \to \infty$. For every fixed $t$ in $]0, 2\pi[$, we have the result

$$
(t - \pi) + 2\sum_{k=1}^{N} \frac{\sin(kt)}{k} \xrightarrow[N\to\infty]{} 0, \quad \text{i.e.,}
$$

$$
\sum_{k=1}^{\infty} \frac{\sin(kt)}{k} = \begin{cases} (\pi - t)/2 & \text{for } 0 < t < 2\pi \\ 0 & \text{for } t = 0. \end{cases}
$$

**Theorem 3.1**

1. *The series* $\displaystyle\sum_{k=1}^{\infty} \frac{\sin(kt)}{k}$ *represents the $2\pi$-periodic sawtooth function $S(t)$ (see Fig. 3.3)*

$$
S(t) = \begin{cases} (\pi - t)/2 & \textit{for } 0 < t < 2\pi \\ 0 & \textit{for } t = 0. \end{cases}
$$

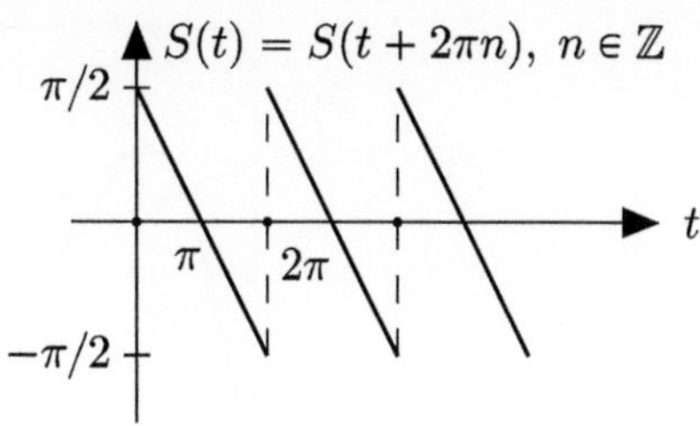


**Fig. 3.3** The sawtooth function

2. *The mean value property  is fulfilled for all $t \in \mathbb{R}$*

$$S(t) = \frac{1}{2}(S(t+) + S(t-)).$$

3. *$S(t)$ is piecewise continuously differentiable.*
4. *The trigonometric series representation for the sawtooth function converges uniformly on every closed interval which does not contain a jump discontinuity of $S(t)$. However, even though all partial sums are indefinitely often differentiable, the limit function is not continuous.*

Namely, for $h \leqslant t \leqslant 2\pi - h, h > 0$, we have $\sin(t/2) \geqslant \sin(h/2) > 0$, and thus

$$\left| S(t) - \sum_{k=1}^{N} \frac{\sin(kt)}{k} \right| = \frac{1}{2}|I_N(t)| \leqslant \frac{1}{(N + \frac{1}{2})\sin(\frac{t}{2})} \leqslant \frac{1}{(N + \frac{1}{2})\sin(\frac{h}{2})}$$

for all $t \in [h, 2\pi - h]$. The approximation of $I_N(t)$ to zero depends only on $N$, not on $t \in [h, 2\pi - h]$; in other words $\lim_{N \to \infty} I_N(t) = 0$ with uniform convergence in every closed interval $[h, 2\pi - h], h > 0$.

On the other hand, despite the uniform convergence of the partial sums $S_N(t)$ of $S(t)$ in every interval $[h, 2\pi - h], h > 0$, we will find wavelike overshoots of $S_N(t)$ over $S(t)$ in a small neighborhood of the jump discontinuities. These ripples move closer to the discontinuity points but do not die out as more terms are added to the sums; the deviation from $S(t)$ does not converge to zero. It turns out that the approximating partial sums $S_N(t)$ always overshoot $S(t)$ with about 9% of the jump height $S(0+) - S(0-)$. This property of the approximations $S_N(t)$ was discovered by J. W. Gibbs (1839–1903) and is therefore called the *Gibbs phenomenon*.

In the following illustration in Fig. 3.4, $\mathrm{Si}(\pi)$ is the value of the sine integral

$$\mathrm{Si}(t) = \int_0^t \frac{\sin(\tau)}{\tau}\, d\tau$$

at the point $t = \pi$.

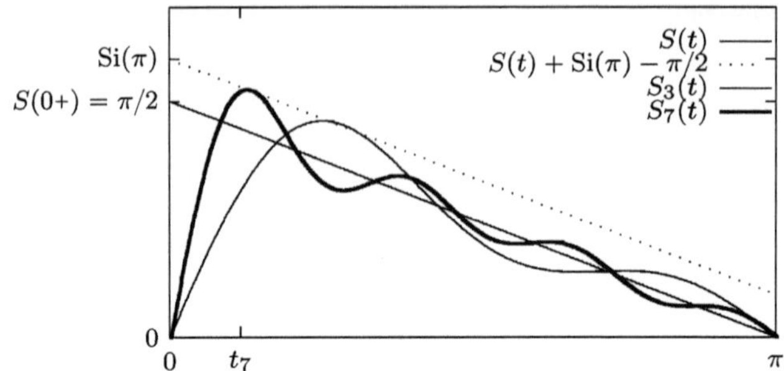



**Fig. 3.4**  Illustration of the Gibbs phenomenon

## *The Gibbs Phenomenon for the Sawtooth Function*

To prove the Gibbs phenomenon we consider the first positive extreme points at $t_N$ of the deviations $S_N(t) - S(t)$ between the sawtooth $S(t)$ and its partial sums $S_N(t)$ for $0 < t < \pi$. Since

$$S_N(t) - S(t) = \frac{1}{2} I_N(t) = \int_{\pi}^{t} \frac{\sin\left((N + \frac{1}{2})\tau\right)}{2\sin\left(\frac{\tau}{2}\right)}\, d\tau\,,$$

one obtains by piecewise integration and comparison with p. 23

$$S_N(t) - S(t) = \int_{0}^{t} \frac{\sin\left((N + \frac{1}{2})\tau\right)}{2\sin\left(\frac{\tau}{2}\right)}\, d\tau - \underbrace{\int_{0}^{\pi} \frac{\sin\left((N + \frac{1}{2})\tau\right)}{2\sin\left(\frac{\tau}{2}\right)}\, d\tau}_{S(0+)=\pi/2}\,.$$

Namely, the right integral can be written as $-\lim_{\varepsilon \to 0+} I_N(\varepsilon)/2 = \pi/2$.

The derivative $S'_N(t) - S'(t) = \dfrac{\sin\left((N + \frac{1}{2})t\right)}{2\sin\left(\frac{t}{2}\right)}$ shows the first positive zero as

$t_N = \dfrac{\pi}{N + \frac{1}{2}}$. To estimate the deviation at $t_N$

$$S_N(t_N) - S(t_N) = \int_{0}^{t_N} \frac{\sin\left((N + \frac{1}{2})\tau\right)}{2\sin\left(\frac{\tau}{2}\right)}\, d\tau - \frac{\pi}{2}\,,$$

we use the following transformation:

$$\frac{\sin\left((N+\frac{1}{2})\tau\right)}{2\sin\left(\frac{\tau}{2}\right)} = \frac{\sin\left((N+\frac{1}{2})\tau\right)}{\tau} + \frac{\tau - 2\sin\left(\frac{\tau}{2}\right)}{2\tau\sin\left(\frac{\tau}{2}\right)}\sin\left((N+\frac{1}{2})\tau\right).$$

Together with the substitution $t = (N+\frac{1}{2})\tau$ and $t_N = \dfrac{\pi}{N+\frac{1}{2}}$, we get

$$S_N(t_N) - S(t_N) = \int_0^{t_N} \frac{\sin\left((N+\frac{1}{2})\tau\right)}{\tau}\, d\tau - \frac{\pi}{2} +$$

$$+ \underbrace{\int_0^{t_N} \frac{\tau - 2\sin\left(\frac{\tau}{2}\right)}{2\tau\sin\left(\frac{\tau}{2}\right)}\sin\left((N+\frac{1}{2})\tau\right)\, d\tau}_{r_N(t_N)} = \int_0^{\pi} \frac{\sin(t)}{t}\, dt - \frac{\pi}{2} + r_N(t_N).$$

By $2\sin\left(\frac{\tau}{2}\right) < \tau$ and $\sin\left((N+\frac{1}{2})\tau\right) \geqslant 0$ for $0 < \tau < t_N$, it holds $r_N(t_N) \geqslant 0$.
Now, using the known value of the sine integral (Exercise)

$$\mathrm{Si}(\pi) = \int_0^{\pi} \frac{\sin(t)}{t}\, dt = 1.8519\ldots,$$

we accomplished

$$S_N(t_N) - S(t_N) \geqslant 0.28 + r_N(t_N) \geqslant 0,$$

i.e., $S_N(t)$ overshoots $S(t)$ at $t_N$.

Since the integrand of $r_N(t_N)$ has limit zero when $\tau \to 0+$ and $\lim\limits_{N\to\infty} t_N = 0$, it follows $\lim\limits_{N\to\infty} r_N(t_N) = 0$. Active readers can readily check this, for example, by power series expansion of the integrand or by application of L'Hospital's rule (Exercise).

***Result*** *Eventually, for increasing $N$ one obtains an overshoot of the partial sums $S_N(t)$ over the function $S(t)$ of about 9% of the jump height $S(0+) - S(0-) = \pi$, even when the $t_N$ move closer to the jump point at $t = 0$:*

$$\lim_{N\to\infty} (S_N(t_N) - S(t_N)) \approx 0.28 \approx 0.09\pi.$$

That result on the bad convergence of the series near jump points goes back to the work of Wilbraham (1848) and Gibbs (1898).

All so far considered properties for the particular sawtooth example are characteristic for many relevant trigonometric series in practice. Questions on the representability of other $T$-periodic functions as superpositions of harmonic oscillations are treated in the next section.

## 3.2  Basic Theorems on Fourier Series

*The Fourier series of a function* $f : [0, T] \to \mathbb{C}$ *is the series* $S_f(t) = \sum_{k=-\infty}^{+\infty} c_k\, e^{jk\omega_0 t}$ *with* $\omega_0 = \dfrac{2\pi}{T}$, *whose coefficients* $c_k$ *are defined by* $c_k = \dfrac{1}{T}\int_0^T f(t)\, e^{-jk\omega_0 t}\, \mathrm{d}t.$

An $n$th partial sum $S_n$ of $S_f$ is the sum $S_n(t) = \sum_{k=-n}^{+n} c_k\, e^{jk\omega_0 t}$, and we say that $S_f$ converges at $t \in \mathbb{R}$ if $\lim_{n \to \infty} S_n(t)$ exists.

In the following sections we restrict ourselves for the most part to assertions on piecewise continuous or piecewise continuously differentiable $T$-periodic functions $f$. A function $f$ is piecewise continuous when its real and imaginary parts are continuous except up to at most finitely many points in $]0, T[$. It is piecewise continuously differentiable when the same holds true for $f'$ instead of $f$. Further on, we postulate that all one-sided limits in $[0, T]$ of $f$ exist in $\mathbb{C}$ in the first case and of $f'$ in the second case. The right- and left-sided limits of $f$ at $t$ are denoted by $f(t+)$ and $f(t-)$, respectively.

Under the assumed conditions the functions $f$ and $f'$ are bounded, and the values at discontinuity points do not matter for definite integrals as in Fourier coefficients. We set $f(0) = f(T)$ and think $f$ as extended to a $T$-periodic function on $\mathbb{R}$, which is also denoted by $f$. Such functions build a sufficiently large class for many applications, and the following theorems on their representation as Fourier series can be shown with the knowledge of common first-year lectures in mathematics. The statements in the theorems go back to the work of Dirichlet (1829), Fejér (1904), Wilbraham (1848), and Gibbs (1898). The first theorem is a variant of the more general assertion, proven by Dirichlet, that periodic functions of bounded variation are representable by their Fourier series. We denote this variant as follows:

**Theorem 3.2 (Theorem of Dirichlet)** *If $f$ is piecewise continuously differentiable on $[0, T]$, then its Fourier series $S_f$ converges at every point $t$ to $\dfrac{1}{2}[f(t+) + f(t-)]$; hence it converges to $f(t)$ at every point $t$ of continuity. The Fourier series $S_f$ converges uniformly to $f$ in every closed interval which does not contain a discontinuity point of $f$.*

**Theorem 3.3 (Gibbs Phenomenon)** *At discontinuity points of piecewise continuously differentiable periodic functions $f$, the Gibbs phenomenon occurs. All $N$th partial sums of the Fourier series for the real or imaginary part of $f$ overshoot the respective jump for large $N$ with about 9% of the jump height.*

**Theorem 3.4 (Theorem of Fejér)**

1. *If $f$ is a continuous periodic function, then the arithmetic means*

$$\overline{S}_N = \frac{1}{N+1}(S_0 + S_1 + \cdots + S_N)$$

   *of the partial sums $S_n$, $n \in \mathbb{N}_0$, of $S_f$ converge uniformly to $f$ for $N \to \infty$.*
2. *If the Fourier series $S_f$ of a piecewise continuous function $f$ converges at a point $t_0$ at all, then it converges there to $\frac{1}{2}[f(t_0+) + f(t_0-)]$. If, in addition, $f$ is continuous at $t_0$, then it holds $S_f(t_0) = f(t_0)$.*

**Theorem 3.5 (Vanishing of the Gibbs phenomenon for Fejér Means)** *When one uses Fejér means $\dfrac{1}{N+1}(S_0 + S_1 + \cdots + S_N)$ of the partial sums of $S_f$ to approximate a piecewise continuously differentiable periodic function $f$, then the Gibbs phenomenon vanishes.*


## *First Explanations of the Theorems*

### Pointwise Convergence in the Theorem of Dirichlet

An impression about convergence of Fourier series at a single point $t$ is obtained from the already observed behavior of the Dirichlet kernels, here with period $2\pi/\omega_0$:

$$\sum_{k=-N}^{N} \left( \frac{1}{T} \int_0^T f(s)\, \mathrm{e}^{-jk\omega_0 s}\, \mathrm{d}s \right) \mathrm{e}^{jk\omega_0 t} = \frac{1}{T} \int_0^T f(s) \underbrace{\sum_{k=-N}^{N} \mathrm{e}^{jk\omega_0(t-s)}}_{D_N(t-s)}\, \mathrm{d}s = I(N,t).$$

With increasing $N$, the kernels $D_N(t - s)$ concentrate more and more around $t$ (see the figure below), while away from $t$ their oscillations grow with $N \to \infty$. On the other hand, $\dfrac{1}{T} \int_0^T D_N(t - s)\, \mathrm{d}s = 1$ for all $N$. The oscillating parts of $D_N(t - s)$ do not much contribute to that integral. Thus, the value of the integral $I(N,t)$ is approximately the mean value of $f$ in a small neighborhood $U(N,t)$ of $t$. $U(N,t)$ shrinks with the kernels $D_N(t - s)$ for $N \to \infty$ to the point $t$, and the mean of

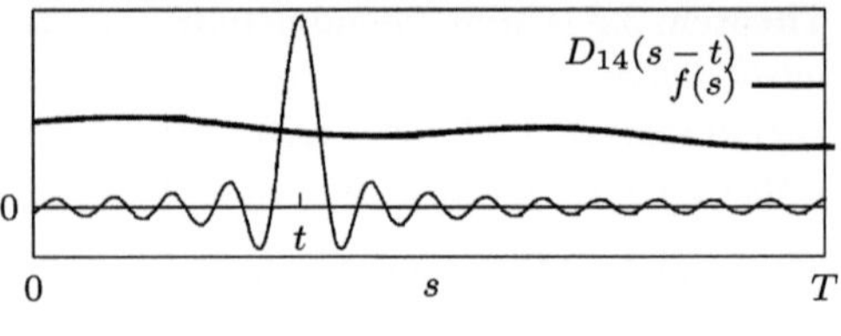

**Fig. 3.5** The integral of the product is largely canceled where $D_N$ highly oscillates

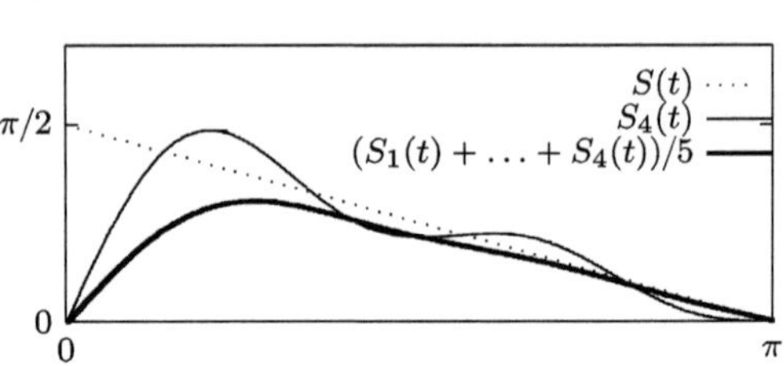

**Fig. 3.6** Averaging partial sums of the Fourier expansion for the sawtooth

$f$ on $U(N, t)$ converges to $\dfrac{1}{2}[f(t+) + f(t-)]$ (see Fig. 3.5). We will convert that impression into a mathematically precise proof later in Sect. 7.1.

## On Fejér's Theorem

P. Du Bois-Reymond (1831–1889) has shown that there are periodic continuous functions, whose Fourier series diverge on dense subsets of their domains of definition. Therefore, of special importance is the result of L. Fejér (1880–1959) that for Fourier series of continuous periodic functions $f$ the arithmetic means of the partial sums converge uniformly to $f$.

When we write the arithmetic mean of $S_n(t) = \sum\limits_{k=-n}^{n} c_k \, e^{jk\omega_0 t}$, $n = 0, \ldots, N$, in the form

$$\frac{1}{N+1}(S_0(t) + S_1(t) + \cdots + S_N(t)) = \sum_{k=-N}^{N} \left(1 - \frac{|k|}{N+1}\right) c_k \, e^{jk\omega_0 t},$$

we observe an attenuation of the higher frequency parts in that mean. Thus, we have a smoothing effect in the approximation of a function $f$ by averaging its Fourier expansion's partial sums. For a more detailed study of Fejér's theorem, we refer to the subsequent Chap. 7. The effect of the averaging is shown in Fig. 3.6 for the partial sum $S_4(t)$ of the sawtooth Fourier expansion and the corresponding arithmetic mean of its partial sums up to the order $N = 4$. $S(t)$ shows the sawtooth graph.

**On the Gibbs Phenomenon**

To obtain the Gibbs phenomenon at discontinuity points, it suffices to examine a $T$-periodic piecewise continuous real-valued function $f$ with a single jump discontinuity at $t_0$ in $[0, T]$ and the mean value property $f(t_0) = [f(t_0+) + f(t_0-)]/2$. Using the sawtooth function $S$ on page 24, we write $f$ in the form $f(t) = g(t)+r(t)$ with

$$g(t) = f(t) - \frac{1}{\pi}[f(t_0+) - f(t_0-)] \, S\left(\frac{2\pi}{T}(t - t_0)\right),$$

$$r(t) = \frac{1}{\pi}[f(t_0+) - f(t_0-)] \, S\left(\frac{2\pi}{T}(t - t_0)\right).$$

The function $g$ is continuous at $t_0$ with $g(t_0) = [f(t_0+) + f(t_0-)]/2$. By the Gibbs phenomenon for the sawtooth, the function $r$ shows the overshoot of the partial sums of its Fourier expansion, which amounts to about 9% of the jump height $f(t_0+) - f(t_0-)$ in the neighborhood of $t_0$.

The vanishing of the Gibbs phenomenon in Fejér means for the approximation comes from the fact, that the overshoot close to jump discontinuities, i.e., close to steep flanks, is caused by high-frequency parts in the approximating sum (see later p. 51). The amplitudes of the harmonic parts with higher frequencies, however, are heavily damped in the Fejér means. As a consequence, the approximations are smoothed and overshoots eliminated. On the other hand, this is paid by the price of less steep flanks in the approximations. For this compare the figures on p. 35 and p. 35 and Sect. 7.2, p. 136.

Fejér's averaging method for smoothing and convergence improvement corresponds to using the weight function (Fig. 3.7)

$$w_N(x) = \begin{cases} 1 - |x|/(N+1) & \text{for } |x| \leqslant N+1 \\ 0 & \text{otherwise} \end{cases}$$

at the discrete points $|k| = 0, \ldots, N$ (Fig. 3.7).

In engineering applications of Fourier analysis the technique of *smoothing by weight functions* is often called *windowing*. It plays an important role, for example, in signal analysis. The triangle window provides the amplitudes of the harmonic

**Fig. 3.7** The triangle window $w_3$ of Fejér's averaging

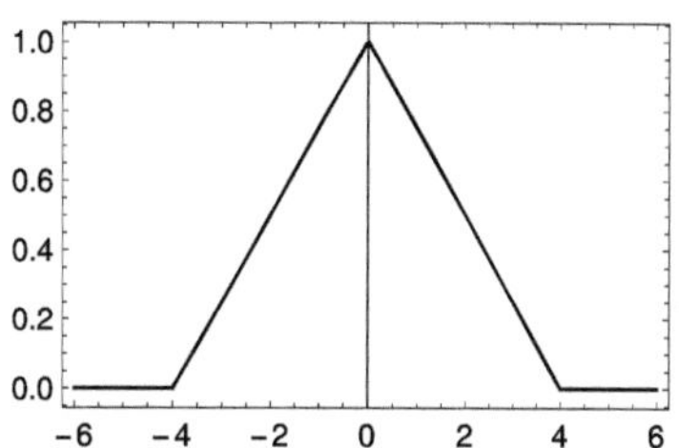

oscillations with angular frequencies $k\omega_0$, $|k| \leqslant N$, in a partial sum as above with the weights $1 - |k|/(N + 1)$. In signal processing, for instance, one often wants to work with much smaller tolerance limits than a 9% error by the Gibbs phenomenon at steep signal flanks. The theorem of Fejér shows a first mathematical method for improvement by windowing techniques without large additional effort.

The theorems show that one can represent a great many functions by their Fourier series. In contrast to Taylor series, which represent in their domain of convergence always infinitely often differentiable, i.e., very smooth, functions, Fourier series allow the representation of quite "irregular" functions by superposition of oscillations with increasing frequencies. Therefore, Fourier series offer much benefit in mathematics and its application fields. The whole thinking in spectral and frequency terms in many application areas goes back to the above theorems. For now we turn first to some application examples. That will provide us with sufficient motivation to study the theorems of Dirichlet and Fejér and their proofs in Chap. 7 in more detail.

## 3.3  The Spectrum of Periodic Functions

### *Significance of the Discrete Spectrum*

The sequence $(c_k)_{k\in\mathbb{Z}}$ of Fourier coefficients of a periodic function $f$ is called the *discrete spectrum of* $f$. For $T$-periodic, real "signals" $f : \mathbb{R} \to \mathbb{R}$, the *magnitude spectrum* $(|c_k|)_{k\in\mathbb{Z}}$ is symmetric because we have $c_k = \overline{c_{-k}}$ (Fig. 3.8).

Since

$$a_n = c_n + c_{-n}, \qquad b_n = j(c_n - c_{-n}),$$

we have

$$A_n = \sqrt{a_n^2 + b_n^2} = \sqrt{4c_n c_{-n}} = 2|c_n|$$

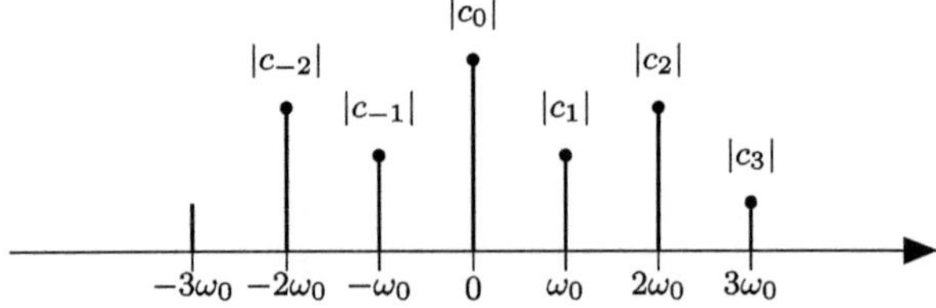


**Fig. 3.8**  Schematic diagram of a magnitude spectrum

for $n \neq 0$ and $\omega_0 = 2\pi/T$ as amplitude of the $n$th *order harmonic* of

$$f(t) = c_0 + \sum_{k=1}^{\infty} 2|c_k| \cos(k\omega_0 t + \arg(c_k)).$$

The sequence $(2|c_k|)_{k\in\mathbb{N}}$ is called *amplitude spectrum*; the sequence $(\arg(c_k))_{k\in\mathbb{N}}$ is the corresponding *phase spectrum* of $f$. $c_0 = a_0/2$ is the *DC part* in $f$, for example, the DC part in a periodic AC voltage $f$. The spectrum shows the amplitudes and phases of harmonics with specified angular frequencies $k\omega_0$, $k \in \mathbb{N}$, which build up as superposition a $2\pi/\omega_0$-periodic signal $f$. For real-valued periodic functions $f$,

the number $D = \sqrt{\sum_{k=2}^{\infty} |c_k|^2 \Big/ \sum_{k=1}^{\infty} |c_k|^2}$ is called *distortion factor*. It is a measure

for the amount of upper harmonics in $f$ and thus for the distortion compared with the pure fundamental oscillation. In Sect. 4.6 we will see how the distortion factor can be computed with the help of the normalized power of a periodic signal $f$.

## *Further Examples of Fourier Series*

### 1. Explicit Computation of a Fourier Series Representation

$$f(t) = \begin{cases} \dfrac{A}{2} & t = -\dfrac{T}{2} \\[2ex] -\dfrac{2A}{T}t & \text{for} \quad -\dfrac{T}{2} < t \leqslant 0 \\[2ex] 0 & 0 < t < \dfrac{T}{2} \end{cases}$$

$$f(t + T) = f(t), \ A > 0$$

(see the illustration in Fig. 3.9) $c_k, a_k, b_k$.

We compute the Fourier coefficients $c_k$, $a_k$, $b_k$. From the graph we see that $c_0 = \dfrac{a_0}{2} = \dfrac{A}{4}$.

For $k \in \mathbb{Z} \setminus \{0\}$ and $\omega_0 = \dfrac{2\pi}{T}$, we obtain

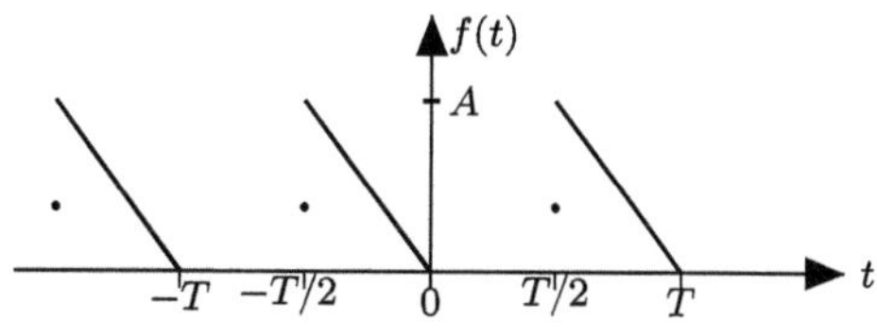

**Fig. 3.9** Sketch of the $T$-periodic function $f$

$$c_k = \frac{1}{T} \int_0^T f(t)\, e^{-jk\omega_0 t}\, dt = \frac{1}{T} \int_{T/2}^T -\frac{2A}{T}(t - T)\, e^{-jk\omega_0 t}\, dt \,.$$

With the substitution $u = t - T$ and $f(t) = 0$ on $[0, \frac{T}{2}[$, we have also

$$c_k = \frac{1}{T} \int_{-T/2}^0 -\frac{2A}{T} u\, e^{-jk\omega_0 u}\, du = -\frac{2A}{T^2} \underbrace{\left[ \frac{e^{-jk\omega_0 u}}{(-jk\omega_0)^2}(-jk\omega_0 u - 1) \right.}_{g(u)}{\left. \vphantom{\frac{e}{()}}\right]}_{u=-T/2}^{u=0}$$

$$\left( g'(u) = \frac{e^{-jk\omega_0 u}}{(-jk\omega_0)^2}(-jk\omega_0)\,[-jk\omega_0 u - 1] + \frac{e^{-jk\omega_0 u}}{(-jk\omega_0)^2}(-jk\omega_0) = u\, e^{-jk\omega_0 u} \right).$$

Inserting the limits of integration yields

$$c_k = -\frac{2A}{T^2} \cdot \left( -\frac{1}{k^2 \omega_0^2} \right) \left[ e^0(0 - 1) - e^{jk\omega_0 \frac{T}{2}} \left( jk\omega_0 \frac{T}{2} - 1 \right) \right]$$

$$= \frac{A}{2\pi^2 k^2} \cdot \left[ -1 - \underbrace{\underbrace{e^{jk\pi}}_{\underbrace{\cos(k\pi)}_{(-1)^k} + \underbrace{j\,\sin(k\pi)}_{0}}}\,(jk\pi - 1) \right]$$

$$= -\frac{A}{2\pi^2 k^2} \left[ 1 + (-1)^k (jk\pi - 1) \right].$$

With $c_k = \overline{c_{-k}}$, we obtain for $k \in \mathbb{N}$

$$a_k = c_k + c_{-k} = 2\,\Re(c_k) = -\frac{A}{\pi^2 k^2} \left[ 1 - (-1)^k \right] = \begin{cases} 0 & \text{for even } k \\ -\dfrac{2A}{k^2 \pi^2} & \text{for odd } k, \end{cases}$$

$$b_k = j(c_k - c_{-k}) = -2\,\Im(c_k) = (-1)^k \frac{A}{k\pi}.$$

With the spectral values $c_k$, respectively, $a_k$ and $b_k$ the harmonic "building blocks" of $f$ are known, and $f$ can approximately be reconstructed as a trigonometric polynomial from the partial sums of its Fourier series. The function $f$ possesses the Fourier series expansion

**Fig. 3.10** Approximation with a partial sum $S_7$ of the Fourier series expansion

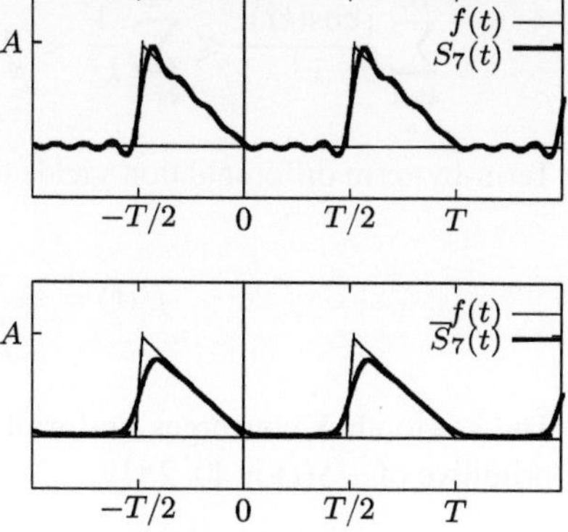

**Fig. 3.11** Approximation with the Fejér mean $\overline{S}_7$

$$f(t) = \frac{A}{4} + \sum_{k=1}^{\infty}\left(\frac{A}{k^2\pi^2}\left[(-1)^k - 1\right]\cos(k\omega_0 t) + \frac{A}{k\pi}(-1)^k\sin(k\omega_0 t)\right)$$

$$= \underbrace{\frac{A}{4}}_{\text{DC part}} \underbrace{-\frac{2A}{\pi^2}\cos(\omega_0 t) - \frac{A}{\pi}\sin(\omega_0 t)}_{\text{fundamental oscillation}} + \underbrace{\frac{A}{2\pi}\sin(2\omega_0 t)}_{\text{first harmonic}}$$

$$\underbrace{-\frac{2A}{9\pi^2}\cos(3\omega_0 t) - \frac{A}{3\pi}\sin(3\omega_0 t) + \ldots}_{\text{second harmonic}}$$

All partial sums are infinitely often differentiable, but their limit function is not continuous. Figure 3.10 with seven spectral values $c_k$ clearly shows the Gibbs phenomenon:

For comparison we look in Fig. 3.11 at a smoothed approximation by the corresponding Fejér mean $\overline{S}_7$ of the partial sums. The Gibbs phenomenon has disappeared. That improvement comes at the expense of a less steep slope in the jump neighborhood and thus a greater error at the edges of the graph.

### 2. Fourier Series Expansion with the Use of an Already Known Series

The series

$$f(t) = \sum_{k=1}^{\infty}\frac{\cos(kt)}{k^2} = \cos(t) + \frac{\cos(2t)}{4} + \frac{\cos(3t)}{9} + \ldots$$

is uniformly convergent (compare p. 20). By $\displaystyle\sum_{k=1}^{N}\frac{1}{k(k+1)} = \frac{N}{N+1}$ we have the estimate

$$\sum_{k=1}^{\infty} \frac{|\cos(kt)|}{k^2} \leqslant \sum_{k=1}^{\infty} \frac{1}{k^2} < \lim_{N \to \infty} \sum_{k=1}^{N} \frac{2}{k(k+1)} = \lim_{N \to \infty} \frac{2N}{N+1} = 2.$$

Term-by-term differentiation yields the sawtooth function

$$f'(t) = -\sum_{k=1}^{\infty} \frac{\sin(kt)}{k} = -S(t).$$

The sawtooth $S$ converges uniformly in $[h, 2\pi - h]$, $h > 0$. Therefore $f(t)$ is a primitive of $-S(t)$ in $]0, 2\pi[$:

$$f(t) = \frac{(t-\pi)^2}{4} + c \quad (t \in ]0, 2\pi[)\,.$$

To determine the constant $c$ we observe the DC value $c_0 = 0$ for $f$, i.e.,

$$\int_0^{2\pi} f(t)\,dt = \int_0^{2\pi} \left( \frac{(t-\pi)^2}{4} + c \right) dt = \frac{\pi^3}{6} + 2\pi c = 0, \quad \text{thus} \quad c = -\frac{\pi^2}{12}.$$

As an application one obtains for $t = 0$ the limit of the series $\displaystyle\sum_{k=1}^{\infty} \frac{1}{k^2} = \frac{\pi^2}{6}$.

## 3.4　Exercises

(**A1**)　(a)　Compute the Fourier coefficients $a_k$ and $b_k$ of the function given in
　　　　　　Fig. 3.12.
　　　　(b)　Give the Fourier expansion in trigonometric form and in polar form up to
　　　　　　the 5th harmonic.
　　　　(c)　What is the limit of the Fourier series at the point $x = 0$?

(**A2**)　Consider the $T$-periodic function $u(t)$ given by

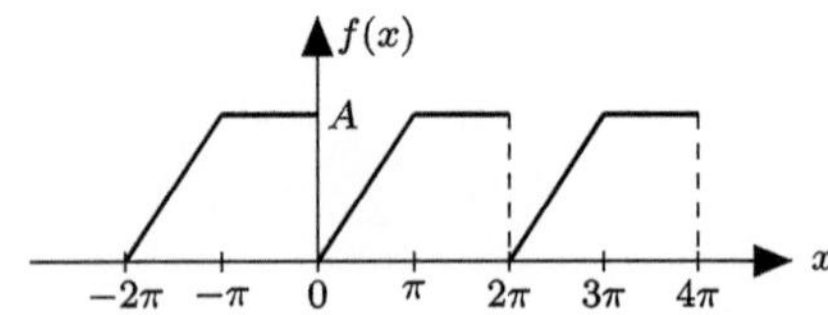


**Fig. 3.12**　Sketch of a periodic function, whose Fourier series shall be calculated

$$u(t) = \begin{cases} \hat{u}\,\sin(\omega_0 t) & \text{for} \quad 0 \leqslant t < \dfrac{T}{2} \\ 0 & \text{for} \quad \dfrac{T}{2} \leqslant t < T \end{cases} \qquad \left(\omega_0 = \dfrac{2\pi}{T}\right).$$

(a) Sketch a graph of $u(t)$ for $0 \leqslant t < 2T$.

(b) Compute the complex Fourier coefficients $c_k$ and the real Fourier coefficients $a_k$ and $b_k$.

(c) Write the Fourier series in complex and in trigonometric form, and generate graphically an approximation by a few trigonometric polynomials.

(**A3**) A function $f$ is given by

$$f(t) = \begin{cases} \left|\sin\left(\dfrac{\omega}{2}t\right)\right| & \text{for} \quad -\dfrac{T}{2} < t \leqslant 0 \\ 0 & \text{for} \quad 0 < t \leqslant \dfrac{T}{2} \end{cases}$$

$$f(t+T) = f(t), \quad t \in \mathbb{R}, \quad \omega = \dfrac{2\pi}{T}.$$

(a) What are the Fourier coefficients $a_k$, $b_k$, $c_k$ of $f$?

(b) What is the necessary degree of a Fourier partial sum so that it deviates not more than $0.5 \cdot 10^{-3}$ from the series limit at the point $t = T/2$?

(**A4**) Calculate for the sawtooth function $S(t) = \displaystyle\sum_{k=1}^{\infty} \dfrac{\sin(kt)}{k}$ and a given $h > 0$ a number $N \in \mathbb{N}$ so that for a fixed maximal error $\varepsilon > 0$ the deviation

$$\left| S(t) - \sum_{k=-N}^{N} \langle S(t) | \mathrm{e}^{jkt} \rangle \, \mathrm{e}^{jkt} \right|$$

in $[h, \pi]$ does not exceed $\varepsilon$.

(**A5**) (a) Calculate for $x, t \in \mathbb{R}$ the Fourier series expansion of $f(t) = \mathrm{e}^{jx\sin(t)}$.

(b) What are the Fourier series of $\cos(x\sin(t))$ and $\sin(x\sin(t))$ (see Fig. 3.13)?

Hint: Use the Bessel functions.

**Fig. 3.13** Illustration of a frequency modulated function

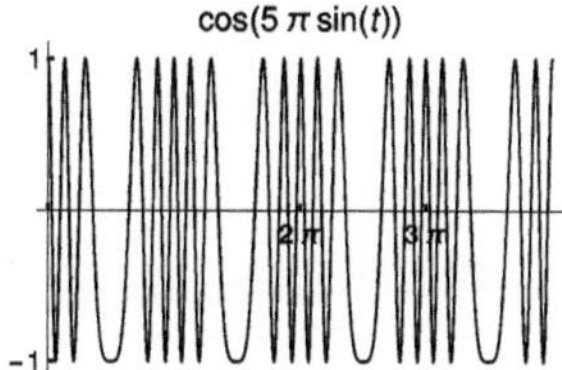

**(A6)** Let $a$ be real with $|a| > 1$. Find the sums of the series

$$f(t) = \sum_{k=0}^{\infty} \frac{\cos(kt)}{a^k} \quad \text{and} \quad g(t) = \sum_{k=1}^{\infty} \frac{\sin(kt)}{a^k}.$$

Hint: Use $F(z) = \sum_{k=0}^{\infty} \frac{z^k}{a^k}$ for $z \in \mathbb{C}$, $|z| < |a|$.

**(A7)** Calculate the Fourier series in $]0, \pi[$ of the function $f(t) = e^{at}$

(a) As a cosine series
(b) As a sine series

Realize the difference between the two cases, and plot the graphs of the $\pi$-periodically extended functions.

# Chapter 4
# Calculating with Fourier Series

**Abstract** General properties of Fourier series for piecewise continuously differentiable functions are worked out. This includes symmetry properties, amplitude modulation, derivatives and integrals of Fourier series, asymptotic decay of Fourier coefficients, spectrum, and the Parseval equation. An example of an everywhere convergent trigonometric series is given that cannot be the Fourier series of a classical function. Further examples and exercises on the contents complement the chapter.

## 4.1 Symmetry Properties, Linearity, and Similarity

In the following sections—if not otherwise stated—$f$ and $g$ are $\mathbb{C}$-valued piecewise continuously differentiable $T$-periodic functions with the mean value property. Therefore they can be represented by their Fourier series. The Fourier coefficients of $f$ are denoted by $c_k$ and those of $g$ by $d_k$, and we set $\omega_0 = 2\pi/T$. We often call the parameter $t$ a time parameter and $\omega$ a (angular) frequency parameter. We work out some important rules for the handling of trigonometric polynomials and Fourier series.

**Interval of Integration**
Since $f(t)\, \mathrm{e}^{-jk\omega_0 t}$ is $T$-periodic, one can integrate over an arbitrary interval of length $T$ to get the Fourier coefficients:

$$c_k = \frac{1}{T}\int_0^T f(t)\,\mathrm{e}^{-jk\omega_0 t}\,\mathrm{d}t = \frac{1}{T}\int_\alpha^{\alpha+T} f(t)\,\mathrm{e}^{-jk\omega_0 t}\,\mathrm{d}t \qquad (\alpha \in \mathbb{R}).$$

**Time Reversal**
Since we have $\dfrac{1}{T}\int_0^T f(-t)\,\mathrm{e}^{-jk\omega_0 t}\,\mathrm{d}t = \dfrac{1}{T}\int_{-T}^0 f(t)\,\mathrm{e}^{jk\omega_0 t}\,\mathrm{d}t = c_{-k}$, we obtain

$$f(-t) = \sum_{k=-\infty}^{+\infty} c_{-k}\, \mathrm{e}^{jk\omega_0 t}\,.$$

**Complex Conjugate Functions**

By

$$\frac{1}{T}\int_0^T \overline{f(t)}\, \mathrm{e}^{-jk\omega_0 t}\, \mathrm{d}t = \frac{1}{T}\overline{\int_0^T f(t)\, \mathrm{e}^{jk\omega_0 t}\, \mathrm{d}t} = \overline{c_{-k}},$$

we obtain

$$\overline{f(t)} = \sum_{k=-\infty}^{+\infty} \overline{c_{-k}}\, \mathrm{e}^{jk\omega_0 t}\,.$$

**Even Functions**

If $f$ is even, $f(t) = f(-t)$ for $t \in \mathbb{R}$, then by $c_k = c_{-k}$ and $b_k = j(c_k - c_{-k}) = 0$ all sine terms disappear, and it holds

$$a_k = c_k + c_{-k} = \frac{4}{T}\int_0^{T/2} f(t)\cos(k\omega_0 t)\, \mathrm{d}t \qquad (k \in \mathbb{N}_0).$$

The Fourier series of an even function is a cosine series.

**Odd Functions**

If $f$ is an odd function, $f(t) = -f(-t)$, $t \in \mathbb{R}$, then all terms $a_k = c_k + c_{-k} = 0$ disappear. Then, the Fourier series of $f$ is a sine series. Since $f(t)\sin(k\omega_0 t)$ is an even function, it holds

$$b_k = j(c_k - c_{-k}) = \frac{4}{T}\int_0^{T/2} f(t)\sin(k\omega_0 t)\, \mathrm{d}t\,, \qquad (k \in \mathbb{N}).$$

**Linearity**

$$\alpha f(t) + \beta g(t) = \sum_{k=-\infty}^{+\infty} (\alpha c_k + \beta d_k)\, \mathrm{e}^{jk\omega_0 t}, \qquad (\alpha,\ \beta \in \mathbb{C}).$$

**Similarity**

For $\alpha > 0$ the function $F(t) = f(\alpha t)$ has period $\dfrac{T}{\alpha}$. It possesses the same Fourier coefficients as $f(t)$:

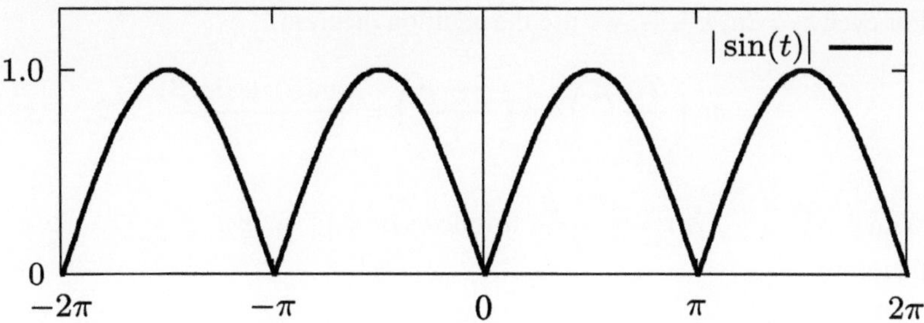

**Fig. 4.1**  Graph of $f(t) = |\sin(t)|$

$$F(t) = f(\alpha t) = \sum_{k=-\infty}^{+\infty} c_k \, e^{jk\alpha\omega_0 t} \,.$$

*A change of frequency does not change the amplitudes of the harmonics, but it changes the assignment of the Fourier coefficients $c_k$ to the angular frequencies $k\alpha\omega_0$.*

Namely, from $\tau = \alpha t,\, 0 \leqslant \tau \leqslant \dfrac{T}{\alpha}$, we get

$$\frac{\alpha}{T} \int\limits_0^{T/\alpha} f(\alpha t)\, e^{-jk\alpha\omega_0 t} \, \mathrm{d}t = \frac{1}{T} \int\limits_0^{T} f(\tau)\, e^{-jk\omega_0 \tau} \, \mathrm{d}\tau = c_k.$$

For known Fourier series of signals $f(t)$, this similarity property allows to immediately see the Fourier expansions of similar functions $f(\alpha t)$ without new computations.

**Examples**

1. $f(t) = |\sin(t)|$ (Fig. 4.1).

   For $T = 2\pi$, we have

   All $b_k = 0$, since $f$ is even; $\quad \dfrac{a_0}{2} = \dfrac{1}{\pi} \int\limits_0^{\pi} \sin(t)\, \mathrm{d}t = -\dfrac{1}{\pi} \cos(t) \Big|_0^{\pi} = \dfrac{2}{\pi}.$

   For odd $k$ we have $a_k = \dfrac{2}{\pi} \int\limits_0^{\pi} \sin(t) \cos(kt)\, \mathrm{d}t = 0$, because substitution $t = x + \pi/2$ leads with the addition theorem $\cos(k(x+\pi/2)) = -\sin(kx)\sin(k\pi/2)$ to an integral of an odd function over the interval $[-\pi/2, \pi/2]$; hence

$$a_k = -\frac{2}{\pi} \int\limits_{-\pi/2}^{\pi/2} \cos(x)\sin(kx)\sin\left(\frac{k\pi}{2}\right) \mathrm{d}x = 0.$$

For even $k = 2n$, $n \in \mathbb{N}$, we use the addition theorem

$$\sin\left(\frac{\alpha + \beta}{2}\right) \cos\left(\frac{\alpha - \beta}{2}\right) = \frac{\sin(\alpha) + \sin(\beta)}{2}.$$

With $t = \dfrac{\alpha + \beta}{2}$, $2nt = \dfrac{\alpha - \beta}{2}$, it follows $\alpha = (1 + 2n)t$, $\beta = (1 - 2n)t$, and therefore

$$a_{2n} = \frac{1}{\pi}\left(\int_0^\pi \sin((1 + 2n)t)\, dt + \int_0^\pi \sin((1 - 2n)t)\, dt\right)$$

$$= -\frac{1}{\pi}\left(\frac{1}{1 + 2n}\big[\underbrace{\cos((1 + 2n)\pi)}_{-1} - 1\big] + \frac{1}{1 - 2n}\big[\underbrace{\cos((1 - 2n)\pi)}_{-1} - 1\big]\right)$$

$$= \frac{2}{\pi}\left(\frac{1}{1 + 2n} + \frac{1}{1 - 2n}\right) = -\frac{4}{(2n + 1)(2n - 1)\pi}.$$

We obtain the result

$$f(t) = \frac{2}{\pi} - \frac{4}{\pi}\left(\frac{\cos(2t)}{3} + \frac{\cos(4t)}{3 \cdot 5} + \frac{\cos(6t)}{5 \cdot 7} + \ldots\right) = -\frac{2}{\pi}\sum_{k=-\infty}^{+\infty}\frac{e^{j2kt}}{4k^2 - 1}.$$

2. By the similarity theorem on p. 40 and the last example, the function $f(t) = |\sin(2t)|$ has the Fourier series expansion

$$f(t) = -\frac{2}{\pi}\sum_{k=-\infty}^{+\infty}\frac{e^{j4kt}}{4k^2 - 1} = \frac{2}{\pi} - \frac{4}{\pi}\left(\frac{\cos(4t)}{3} + \frac{\cos(8t)}{3 \cdot 5} + \frac{\cos(12t)}{5 \cdot 7} + \ldots\right).$$

**Remark** For a $T$-periodic $f$ and $\alpha = \dfrac{T}{2\pi}$, the function $f(\alpha t)$ is $2\pi$-periodic, and it holds

$$f(\alpha t) = \sum_{k=-\infty}^{+\infty} c_k\, e^{jkt} \quad \text{mit} \quad c_k = \frac{1}{2\pi}\int_0^{2\pi} f(\alpha t)\, e^{-jkt}\, dt.$$

Therefore in literature often only $2\pi$-periodic examples are treated.

3. When $f : {]0, T[} \to \mathbb{R}$ is given, then $f$ can be extended to a $2T$-periodic even or an odd function (see Fig. 4.2).
A $2T$-periodic extension then has—dependent on the chosen option— a pure cosine or a pure sine series representation.

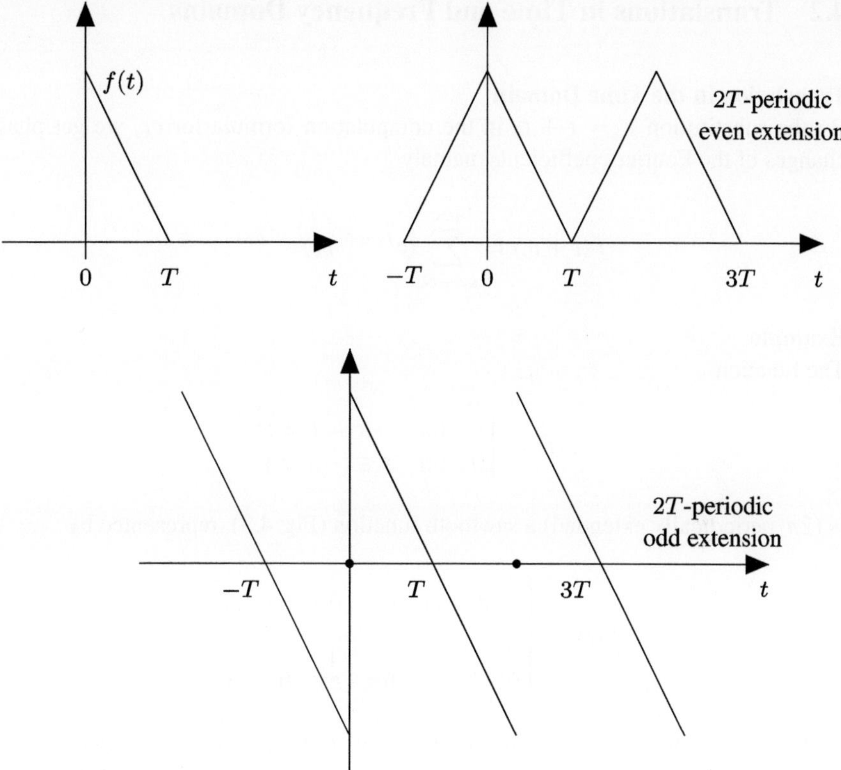

**Fig. 4.2** Even and odd $2T$-periodic extensions

*Cosine series for $f$ ($2T$-periodic evenly extended)*

$$f(t) = \frac{a_0}{2} + \sum_{k=1}^{\infty} a_k \cos\left(\frac{k\pi}{T} t\right) , \quad a_k = \frac{2}{T} \int_0^T f(t) \cos\left(\frac{k\pi}{T} t\right) dt .$$

*Sine series for $f$ ($2T$-periodic oddly extended)*

$$f(t) = \sum_{k=1}^{\infty} b_k \sin\left(\frac{k\pi}{T} t\right) , \quad b_k = \frac{2}{T} \int_0^T f(t) \sin\left(\frac{k\pi}{T} t\right) dt .$$

Both series represent the same function on the interval $]0, T[$. The sine form was already used when we treated the problem of a vibrating string in Sect. 1.2. There, the initial conditions $f(x)$ and $g(x)$ for the string vibration—$2l$-periodically extended to the real axis $\mathbb{R}$—had been odd functions on $\mathbb{R}$.

## 4.2  Translations in Time and Frequency Domains

**Translation in the Time Domain**

By the substitution $\tau = t + t_0$ in the computation formula for $c_k$, we get phase changes of the Fourier coefficients, namely

$$f(t + t_0) = \sum_{k=-\infty}^{+\infty} (e^{jk\omega_0 t_0} c_k)\, e^{jk\omega_0 t}\,.$$

**Example**

The function

$$f(t) = \begin{cases} t & \text{for} \quad -\pi < t < \pi \\ 0 & \text{for} \quad t \in \{-\pi, \pi\} \end{cases}$$

is ($2\pi$-periodically extended) a sawtooth function (Fig. 4.3), represented by

$$S(t) = \begin{cases} \dfrac{1}{2}(\pi - t) & \text{for} \quad 0 < t < 2\pi \\[2ex] 0 & \text{for} \quad t \in \{0, 2\pi\} \end{cases}$$

The amplitudes in the Fourier expansion of $f$ must be twice as large as in $S$, and with $e^{jk\pi} = (-1)^k$ we obtain from the Fourier series of $S(t)$ with the above principle (compare p. 19)

$$f(t) = -2 \sum_{\substack{k=-\infty \\ k\neq 0}}^{+\infty} \frac{e^{jk(t+\pi)}}{2kj} = -2 \sum_{k=1}^{\infty} (-1)^k \frac{\sin(kt)}{k}\,.$$

**Translation in the Frequency Domain and Amplitude Modulation**

From $\dfrac{1}{T} \displaystyle\int_0^T e^{jn\omega_0 t}\, f(t)\, e^{-jk\omega_0 t}\, dt = c_{k-n}$, we get $e^{jn\omega_0 t}\, f(t) = \displaystyle\sum_{k=-\infty}^{+\infty} c_{k-n}\, e^{jk\omega_0 t}\,.$

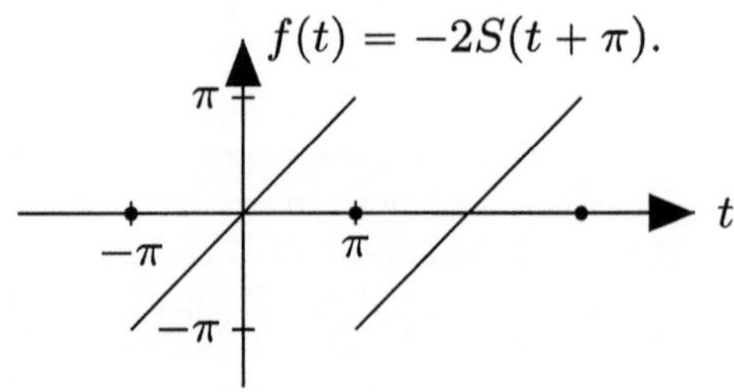


**Fig. 4.3** The graph of a scaled and shifted $2\pi$-periodic sawtooth

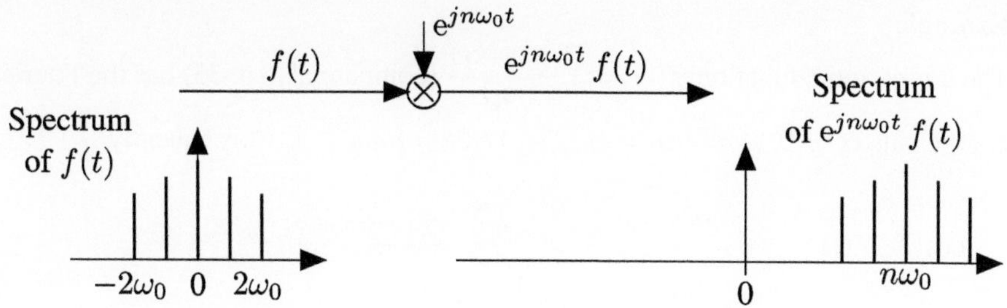

Fig. 4.4   Complex amplitude modulation causes a shift of the spectrum

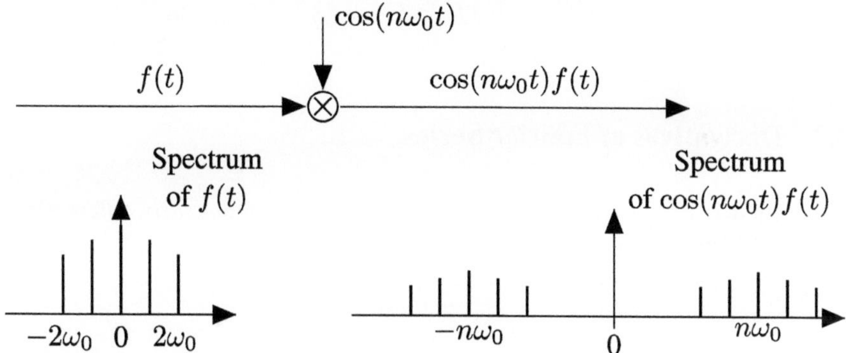

Fig. 4.5   Amplitude modulation with a cosine generates two sidebands in frequency domain

*Multiplication of a $2\pi/\omega_0$-periodic function $f(t)$ with the function $e^{jn\omega_0 t}$ produces a shift of the spectrum of $f(t)$ by $n\omega_0$ (Illustration Fig. 4.4).*

For better understanding of possible applications, we consider an amplitude modulation with $\cos(n\omega_0 t)f(t)$ and observe that $\cos(n\omega_0 t) = (e^{jn\omega_0 t} + e^{-jn\omega_0 t})/2$. The spectrum $(d_k)_{k\in\mathbb{Z}}$ of $\cos(n\omega_0 t)f(t)$ is then given by

$$d_k = \frac{1}{2}c_{k-n} + \frac{1}{2}c_{k+n},$$

i.e., the spectrum is shifted to the left and to the right by the angular frequency $n\omega_0$. We find two sidebands with halved amplitudes. In signal processing this property enables the shift of a signal spectrum to a freely selectable frequency band, for example, to transfer a speech signal spectrum into a non-audible frequency band and to bring it back to the audible band by repeated amplitude modulation (Illustration Fig. 4.5).

**Example**

The $2\pi$-periodic function $f(t) = \sum_{k=1}^{\infty} \dfrac{\cos(kt)}{k^2}$ (compare also p. 35) has the Fourier coefficients $c_0 = 0$, $c_k = c_{-k} = a_k/2 = 1/(2k^2)$ for $k \geqslant 1$. Consequently

$$\cos(2t) f(t) = f(t) \frac{e^{j2t} + e^{-j2t}}{2} = \sum_{k=-\infty}^{+\infty} \frac{c_{k+2} + c_{k-2}}{2} e^{jkt}$$

$$= \frac{a_2}{2} + \sum_{k=1}^{\infty} \frac{a_{k+2} + a_{|k-2|}}{2} \cos(kt)$$

$$= \frac{1}{8} + \frac{5}{9} \cos(t) + \frac{1}{32} \cos(2t) + \frac{13}{25} \cos(3t) + \frac{5}{36} \cos(4t) + \ldots$$

## 4.3   Derivatives of Fourier Series

**Theorem 4.1** *If $f$ is continuous on $\mathbb{R}$ and piecewise continuously differentiable, then the Fourier coefficients $c'_k$ of its derivative $f'$ are given by*

$$c'_k = jk\omega_0 c_k.$$

*The Fourier series of $f'$ is obtained through term-by-term differentiation of the Fourier series of $f$.*

Piecewise integration by parts between the points $0 = t_0 < t_1 < \ldots < t_m = T$, where $f'$ possibly does not exist, yields for $k \neq 0$

$$c_k = -\frac{1}{jk\omega_0 T} \sum_{\ell=1}^{m} \left[ f(t) \, e^{-jk\omega_0 t} \, \Big|_{t_{\ell-1}}^{t_\ell} - \int_{t_{\ell-1}}^{t_\ell} f'(t) \, e^{-jk\omega_0 t} \, \mathrm{d}t \right].$$

By continuity and $T$-periodicity of $f$, it follows

$$\frac{1}{T} \int_{0}^{T} f'(t) \, e^{-jk\omega_0 t} \, \mathrm{d}t = jk\omega_0 c_k.$$

For $k = 0$ we have $\dfrac{1}{T} \int_{0}^{T} f'(t) \, \mathrm{d}t = \sum_{\ell=1}^{m} (f(t_\ell) - f(t_{\ell-1})) = f(T) - f(0) = 0.$

**Example**

The function $f(t) = \sum\limits_{k=1}^{\infty} \dfrac{\cos(kt)}{k^2}$ is continuous and piecewise continuously differentiable. Term-by-term differentiation yields the sawtooth series $-S(t)$ (see p. 35).

**Remark**  If the derivative $f'$ is not again piecewise continuously differentiable, it is not ensured that the Fourier series of $f'$ converges at all. If, however, it converges at a point $t$, then by the theorem of Fejér it has the limit $(f'(t+) + f'(t-))/2$ there. Term-by-term differentiation of Fourier series of discontinuous functions generally leads to divergent series. For example, term-by-term differentiation of the sawtooth series $f(t) = \sum\limits_{k=1}^{\infty} \dfrac{\sin(kt)}{k}$ results in a series, which converges nowhere (see p. 15). G. Cantor and H. Lebesgue have shown that term-by-term differentiation of Fourier series of piecewise continuously differentiable functions with jump discontinuities leads to series, which converge at most in a null set (compare Exercise A7). Therefore, term-by-term differentiations of Fourier series, which are inserted into differential equations, for example (cf. Sect. 1.2), in general need clear mathematical arguments. We will only see in later chapters on distributions, in what sense we nevertheless can work successfully with such divergent series.

## 4.4  Integration of Fourier Series

For a piecewise continuous $T$-periodic function $f(t)$ with Fourier series expansion $\sum\limits_{k=-\infty}^{+\infty} c_k\, e^{jk\omega_0 t}$, we consider its integral function $F(t) = \int\limits_{0}^{t}(f(x) - c_0)\,\mathrm{d}x$. The function $F(t)$ is continuous and $T$-periodic with piecewise continuous derivative $f(t) - c_0$. Therefore, the integral function is representable by its Fourier series

$$F(t) = \sum_{k=-\infty}^{+\infty} F_k\, e^{jk\omega_0 t} .$$

Integration by parts and the fact $F(0) = F(T) = 0$ yield for $k \neq 0$

$$F_k = \frac{1}{T} \int\limits_{0}^{T} F(t)\, \frac{\mathrm{d}}{\mathrm{d}t}\left[\frac{e^{-jk\omega_0 t}}{-jk\omega_0}\right]\,\mathrm{d}t = \frac{1}{jk\omega_0 T} \int\limits_{0}^{T} (f(t) - c_0)\, e^{-jk\omega_0 t}\,\mathrm{d}t = \frac{c_k}{jk\omega_0} .$$

$$F_0 = \frac{1}{T} \int\limits_{0}^{T}\int\limits_{0}^{t} (f(x) - c_0)\,\mathrm{d}x\,\mathrm{d}t \ \text{ is the mean value of } F.$$

Hence $\displaystyle\int_0^t (f(x) - c_0)\,\mathrm{d}x = F_0 + \sum_{\substack{k=-\infty\\k\neq 0}}^{+\infty} \frac{c_k}{jk\omega_0}\, \mathrm{e}^{jk\omega_0 t}$, and

$$\int_0^t f(x)\,\mathrm{d}x = c_0 t + F_0 + \sum_{\substack{k=-\infty\\k\neq 0}}^{+\infty} \frac{c_k}{jk\omega_0}\, \mathrm{e}^{jk\omega_0 t}\,.$$

**Theorem 4.2** *The integral of a T-periodic, piecewise continuous function f is built of a T-periodic function, which oscillates around the ramp $c_0 t + F_0$. Definite integrals of f are obtained with term-by-term integration of the Fourier series of f:*

$$\int_\alpha^\beta f(t)\,\mathrm{d}t = \int_0^\beta f(t)\,\mathrm{d}t - \int_0^\alpha f(t)\,\mathrm{d}t =$$

$$c_0(\beta - \alpha) + \sum_{\substack{k=-\infty\\k\neq 0}}^{+\infty} \frac{c_k}{jk\omega_0}\left(\mathrm{e}^{jk\omega_0\beta} - \mathrm{e}^{jk\omega_0\alpha}\right) = \sum_{k=-\infty}^{+\infty} \int_\alpha^\beta c_k\, \mathrm{e}^{jk\omega_0 t}\,\mathrm{d}t\,.$$

**Comments** With more effort than above one can show that the integral function $F$ is representable by its (uniformly convergent) Fourier series, if $f$ is only absolutely integrable on $[0, T]$. $F$ is then absolutely continuous. This can be found, for example, in the textbook of Tolstov (1976).

**Example** The sawtooth series $f(t) = \displaystyle\sum_{k=1}^{\infty} \frac{\sin(kt)}{k}$ has no DC part. Integration from 0 to $t$ yields the $2\pi$-periodic function (see p. 35)

$$F(t) = \int_0^t f(x)\,\mathrm{d}x = \frac{\pi^2}{6} - \sum_{k=1}^{\infty} \frac{\cos(kt)}{k^2}$$

with DC part $F_0 = \dfrac{\pi^2}{6}$. For $0 \leqslant t \leqslant 2\pi$, we have $F(t) = \dfrac{2\pi t - t^2}{4}$.

## 4.5  Decrease of Fourier Coefficients and Riemann-Lebesgue Lemma

The objectives of this section are statements on the connection between smoothness properties of periodic functions and the qualitative behavior of their spectrum.

Thereby we will see that local properties of a function $f$ affect its entire spectrum. For a $T$-periodic function $f$, which is Riemann integrable on $[0, T]$ and $T$-periodic with the Fourier coefficients $a_k$, $b_k$, and $c_k$, respectively, we find the following important inequality. As usual again $\omega_0 = 2\pi/T$.

**The Bessel Inequality**   (F.W. Bessel, 1784–1846) is

$$\frac{|a_0|^2}{4} + \frac{1}{2} \sum_{k=1}^{\infty} (|a_k|^2 + |b_k|^2) = \sum_{k=-\infty}^{+\infty} |c_k|^2 \leqslant \frac{1}{T} \int_0^T |f(t)|^2 \, dt \, .$$

For the partial sums $f_N(t)$ of the Fourier series of $f$, it holds with $\omega_0 = 2\pi/T$:

$$\frac{1}{T} \int_0^T f(t) \overline{f_N(t)} \, dt = \sum_{k=-N}^{N} \overline{c_k} \frac{1}{T} \int_0^T f(t) \, e^{-jk\omega_0 t} \, dt = \sum_{k=-N}^{N} \overline{c_k} c_k = \sum_{k=-N}^{N} |c_k|^2.$$

By the orthonormality relations from 2.1, we also obtain for all $N \in \mathbb{N}$

$$\frac{1}{T} \int_0^T f_N(t) \overline{f_N(t)} \, dt = \sum_{k=-N}^{N} |c_k|^2.$$

Hence we have

$$0 \leqslant \frac{1}{T} \int_0^T |f(t) - f_N(t)|^2 \, dt = \frac{1}{T} \int_0^T (f(t) - f_N(t)) \overline{(f(t) - f_N(t))} \, dt$$

$$= \frac{1}{T} \left[ \int_0^T |f(t)|^2 \, dt - \int_0^T f(t) \overline{f_N(t)} \, dt - \int_0^T \overline{f(t)} f_N(t) \, dt + \int_0^T f_N(t) \overline{f_N(t)} \, dt \right]$$

$$= \frac{1}{T} \int_0^T |f(t)|^2 \, dt - \sum_{k=-N}^{N} |c_k|^2.$$

From this follows the right half of the Bessel inequality by the limit for $N \to \infty$. The left half follows from the conversions for the Fourier coefficients on p. 17:

$$|a_k|^2 + |b_k|^2 = (c_k + c_{-k})(\overline{c_k} + \overline{c_{-k}}) + j(c_k - c_{-k})(-j)(\overline{c_k} - \overline{c_{-k}})$$

$$= 2c_k \overline{c_k} + 2c_{-k} \overline{c_{-k}} \, ,$$

and therefore $|a_0|^2 = 4|c_0|^2$, $|a_k|^2+|b_k|^2 = 2(|c_k|^2+|c_{-k}|^2)$ for $k \geqslant 1$. In particular we always have

$$\frac{1}{T} \int_0^T |f(t) - f_N(t)|^2 \, dt \leqslant \frac{1}{T} \int_0^T |f(t)|^2 \, dt \, .$$

**Implication** *The Fourier coefficients of a Riemann integrable function $f$ on $[0, T]$ are square summable:*

$$\sum_{k=0}^{\infty} |a_k|^2 < \infty \, , \quad \sum_{k=1}^{\infty} |b_k|^2 < \infty \, , \quad \sum_{k=-\infty}^{+\infty} |c_k|^2 < \infty.$$

## The Riemann-Lebesgue Lemma

**Theorem 4.3 (Riemann-Lebesgue Lemma)**

1. *For the Fourier coefficients $a_k$, $b_k$, $c_k$ of a $T$-periodic function $f$, which is Riemann integrable on $[0, T]$, it holds*

$$\lim_{k \to \infty} a_k = \lim_{k \to \infty} b_k = \lim_{|k| \to \infty} c_k = 0.$$

2. *More general, for Lebesgue-integrable functions $f$, it holds for the function $\widehat{f}(\omega) = \int_{-\infty}^{+\infty} f(t)\, e^{-j\omega t}\, dt$, which is the Fourier transform of $f$ (cf. later Chap. 10)*

$$\lim_{|\omega| \to \infty} \widehat{f}(\omega) = \lim_{|\omega| \to \infty} \int_{-\infty}^{+\infty} f(t)\, e^{-j\omega t}\, dt = 0.$$

Thus, also integrals like $\int_0^T f(t)\sin(\omega t)\, dt$ are canceled with increasing $|\omega| \to \infty$ by the increasingly dense oscillations of the harmonics. For that, simply set $h(t) = f(t)$ in $[0, T]$, $h(t) = 0$ otherwise, and apply 2. to $h$. We will use the Riemann-Lebesgue Lemma repeatedly.

**Proof**

1. The assertion follows immediately from $|c_k|^2 \to 0$ for $|k| \to \infty$.
2. We use that a Lebesgue-integrable function can be approximated by a step function of the form $g = \sum_{k=0}^{n} a_k 1_{I_k}$ with bounded, pairwise disjoint intervals $I_k$

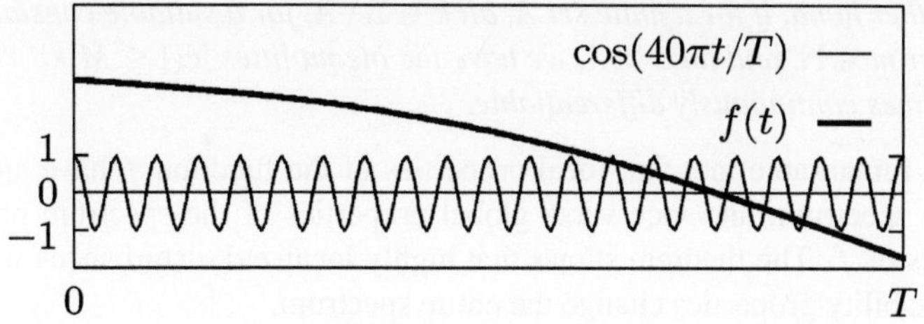


**Fig. 4.6** The more oscillations in the product, the less contribution to its integral

(cf. Appendix B). Choose $g$ such that $\int_{-\infty}^{\infty} |f(t) - g(t)|\, dt < \varepsilon$ for a given $\varepsilon > 0$.

For the Fourier transform $\widehat{g}$ of $g$ analogously as $\widehat{f}$ above, it holds $|\widehat{g}(\omega)| < \varepsilon$ for $|\omega|$ large enough (Exercise, left to the reader). Therefore, for large enough $|\omega|$,

$$|\widehat{f}(\omega)| \leqslant |\widehat{f}(\omega) - \widehat{g}(\omega)| + |\widehat{g}(\omega)| \leqslant \int_{-\infty}^{+\infty} |f(t) - g(t)|\, dt + |\widehat{g}(\omega)| \leqslant 2\varepsilon.$$

Since $\varepsilon$ is arbitrary, $\lim_{|\omega| \to \infty} \widehat{f}(\omega) = 0$.

The function $f$, illustrated in Fig. 4.6, varies only slowly compared with a fast oscillating harmonic factor. In the integral of their product the positive and negative parts cancel each other out more and more with increasing frequency of the harmonic factor. We will refer to the Riemann-Lebesgue Lemma later in the proof of the inversion theorem for the Fourier transform (cf. Theorem 10.1). The lemma also holds for $L^1$-functions of several variables.

## *Order of Magnitude of Fourier Coefficients and Smoothness of $f$*

For a $T$-periodic function $f$, which is integrable on $[0, T]$ and has Fourier coefficients $c_k$, we find the following relation between the magnitude of the Fourier coefficients $c_k$ and differentiability properties of $f$:

**Theorem 4.4** *If $f, f', \ldots, f^{(m-1)}$ are continuous on $\mathbb{R}$ and if $f^{(m)}$ is piecewise continuous, then it holds*

$$\sum_{k=-\infty}^{+\infty} |k^m c_k|^2 < \infty, \text{ in particular therefore } |k^m c_k| \longrightarrow 0 \text{ for } |k| \to \infty.$$

*On the other hand, if for a finite set A, all $k \in \mathbb{Z} \setminus A$, for a suitable constant $M > 0$, an integer $m \in \mathbb{N}$, and an $\alpha > 1$, we have the inequalities $|c_k| \leqslant M|k|^{-(m+\alpha)}$, then $f$ is m-times continuously differentiable.*

It is a remarkable fact that local properties of the function $f$ have an effect on its entire spectrum, and vice versa global properties of the spectrum reflect local properties of $f$. The theorem shows that highly localized disturbances with loss of differentiability properties change the entire spectrum.

***Proof*** For the Fourier coefficients $c_k$, $c_k'$, ..., $c_k^{(m)}$ of $f$, the results in Sect. 4.3 imply

$$c_k^{(m)} = (jk\omega_0)c_k^{(m-1)} = \ldots = (jk\omega_0)^m c_k.$$

The first part of the theorem is therefore obtained from Bessel's inequality and the Riemann-Lebesgue Lemma for $f^{(m)}$. With $\alpha > 1$ and the conditions in the second assertion, one finds for $n \leqslant m$ and sufficiently large $k_0 \in \mathbb{N}$

$$\sum_{|k|>k_0} |k^n c_k| \leqslant M \sum_{|k|>k_0} |k|^{-(m-n+\alpha)} \leqslant 2M \sum_{k>k_0} k^{-\alpha} < \infty.$$

Thus, the series $\sum_{k=-\infty}^{+\infty} (jk\omega_0)^n c_k\, e^{jk\omega_0 t}$ are uniformly convergent also for $n \leqslant m$ and represent the continuous functions $f^{(n)}$ by the theorem of Fejér on p. 29.

A more detailed discussion of the interrelations between smoothness properties of periodic functions and the magnitude of their Fourier coefficients can be found in Tolstov (1976) or Zygmund (2003).

**Examples**

1. The sawtooth function is piecewise continuous but not continuous. Its Fourier coefficients decrease like $1/|k|$ (case $m = 0$). The function $g$, defined by $g(t) = t^2$ for $t$ in $[-\pi, \pi]$, $g(t) = g(t + 2\pi k)$, $k \in \mathbb{Z}$, composed of parable arcs, has quadratically decreasing coefficients (Exercise):

$$g(t) = \frac{\pi^2}{3} - 4\left( \frac{\cos(t)}{1^2} - \frac{\cos(2t)}{2^2} + \frac{\cos(3t)}{3^2} \mp \ldots \right).$$

$g(t)$ is not continuously differentiable (case $m = 1$).

If a function $f(t) = \sum_{k=-\infty}^{+\infty} c_k\, e^{jk\omega_0 t}$ is continuously differentiable with a piecewise continuous second derivative (case $m = 2$), then the amplitudes of the harmonics with angular frequencies $k\omega_0$ decrease for increasing $|k| \in \mathbb{N}$ faster than $1/|k|^2$ (Illustration Fig. 4.7).

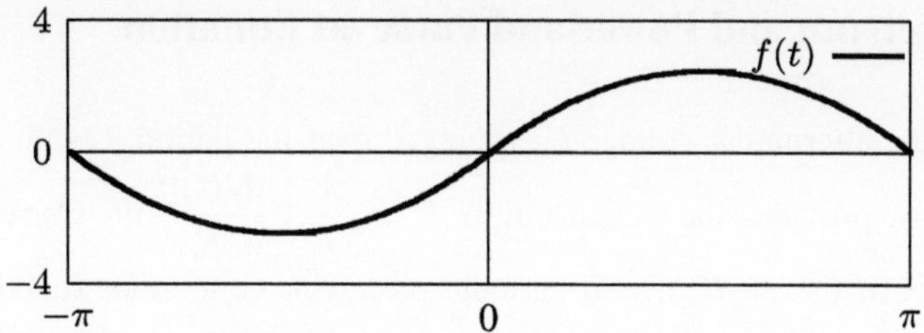

**Fig. 4.7** Graph of the twice piecewise continuously differentiable function $f$

For example, $f(t) = \begin{cases} t(\pi + t) & \text{for} \quad -\pi \leqslant t \leqslant 0, \\ t(\pi - t) & \text{for} \quad\;\; 0 \leqslant t \leqslant \pi, \end{cases}$

$$f(t) = \frac{8}{\pi}\left(\sin(t) + \frac{\sin(3t)}{3^3} + \frac{\sin(5t)}{5^3} + \cdots\right).$$

2. If $|c_k| \leqslant \dfrac{M}{|k|^3}$ for all $k \in \mathbb{Z} \setminus \{0\}$ and a suitable constant $M > 0$ as in the last example, then it follows

$$\sum_{k=-\infty}^{+\infty} |k|\,|c_k| = \sum_{k=1}^{\infty} k\,|c_k| + \sum_{k=1}^{\infty} k\,|c_{-k}| \leqslant 2M\sum_{k=1}^{\infty}\frac{1}{k^2} < \infty.$$

Then $f(t) = \displaystyle\sum_{k=-\infty}^{+\infty} c_k\, e^{jk\omega_0 t}$ is continuously differentiable.

3. When the Fourier coefficients $c_k$ of $f$ fulfill $\displaystyle\lim_{|k|\to\infty} |k|^m|c_k| = 0$ *for all* $m \in \mathbb{N}$, then $f$ is infinitely often differentiable. Namely in that case, for arbitrary $m \in \mathbb{N}$ the sequence $\left(|k|^{m+2}|c_k|\right)_{k\in\mathbb{Z}}$ is bounded, and with a suitable $M \in \mathbb{R}$ we find

$$\sum_{k=-\infty}^{+\infty} |k|^m|c_k| = \sum_{\substack{k=-\infty \\ k\neq 0}}^{+\infty} \frac{|k|^{m+2}}{|k|^2}|c_k| \leqslant M\sum_{\substack{k=-\infty \\ k\neq 0}}^{+\infty} \frac{1}{|k|^2} < \infty.$$

Hence it follows that $f$ is $m$-times differentiable for every $m \in \mathbb{N}$.

Intuitively spoken, the above asymptotic statements mean that for a good reproduction of periodic functions with a "kink" or jump discontinuities the amplitudes of the harmonics must not decrease too fast.

## 4.6  Spectrum and Power and Parseval Equation

A $T$-periodic alternating voltage $U$, averaged over the period $T$ with an effective resistance $R$, provides the output power $P = \dfrac{1}{T} \displaystyle\int_0^T \dfrac{|U(t)|^2}{R}\, dt$. Correspondingly, the *normalized power* $P$ of a $T$-periodic piecewise continuous function $f(t) = \displaystyle\sum_{k=-\infty}^{+\infty} c_k\, e^{jk\omega_0 t}$ is defined with the normalization $R = 1$

$$P = \frac{1}{T} \int_0^T |f(t)|^2\, dt\,.$$

$\sqrt{P}$ is called *effective value* or *root mean square* of $f$ (RMS). In mathematics the effective value is called the *norm* of $f$, and one writes $\|f\|_2 = \sqrt{P}$. The inner product of p. 12 is also defined for piecewise continuous functions, and it holds $\|f\|_2 = \sqrt{\langle f | f \rangle}$ for the norm of $f$.

For piecewise continuous $T$-periodic $f$ and $g$, the norm $\|f - g\|_2 = \langle f - g | f - g \rangle^{1/2}$ of their difference $f - g$ defines a *distance between the two functions*. In that context two functions are identified if they differ only on a null set. With this identification, the introduced inner product is positive definite. The functions $f$, more precisely, had to be replaced by their corresponding equivalence classes with this identification. Nevertheless, it is common to speak of functions further on instead of equivalence classes. The norm $\|f\|_2$ of $f$ is zero if and only if $f \neq 0$ at most in a null set, and for two functions $f$ and $g$ we have the triangle inequality $\|f \pm g\|_2 \leqslant \|f\|_2 + \|g\|_2$. In Sect. 5.1 we will come back to this distance and the related convergence concept for function sequences, i.e., to convergence in quadratic mean.

**Theorem 4.5 (Parseval Equation)** *The normalized power of $f$ can be expressed by the spectrum $(c_k)_{k \in \mathbb{Z}}$ of $f$:*

$$P = \|f\|_2^2 = \frac{1}{T} \int_0^T |f(t)|^2\, dt = \sum_{k=-\infty}^{+\infty} |c_k|^2.$$

This equation is called *Parseval equation* after M. A. Parseval (1755–1836). Because the normalized power of $f(t) = c_k\, e^{jk\omega_0 t}$ is just $|c_k|^2$, we can also formulate:

*The normalized power of $f$ is equal to the sum of the powers of all harmonic parts of $f$.* This important relation will be shown in Chap. 7 for piecewise continuous periodic functions. With theorems of Lebesgue's integration theory, it can be shown for all square Lebesgue integrable functions on $[0, T]$ (cf. Chap. 7, Exercise A8).

## 4.7  Exercises

The exercises with an asterisk $\star$ for this and the following chapters are mathematically more difficult than the others and are intended primarily for mathematicians among the readers.

(**A1**)  Let $f(t) = \cos(t)$ be given for $0 < t < \pi$.

Sketch a graph of the odd $2\pi$-periodic extension of $f$, and compute the Fourier series of this $2\pi$-periodic function.

(**A2**)  Let the $2\pi$-periodic function $h$ be given by

$$h(t) = \begin{cases} 1 \text{ for } 0 < t < \pi \\ 0 \text{ for } t = 0 \\ -1 \text{ for } -\pi < t < 0. \end{cases}$$

(a)  Compute its Fourier series (Fig. 4.8).

(b)  What is the Fourier series of the 4-periodic function $g(t)$ outlined in Fig. 4.9? Use the similarity theorem and the result on translations for this.

**Fig. 4.8**  One period of $h$

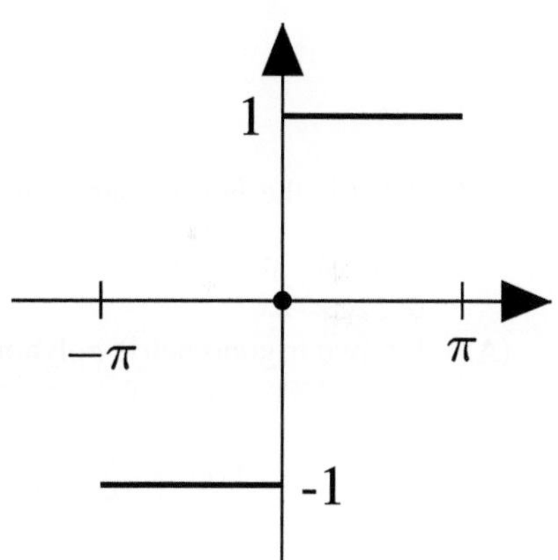


**Fig. 4.9**  One period of $g$

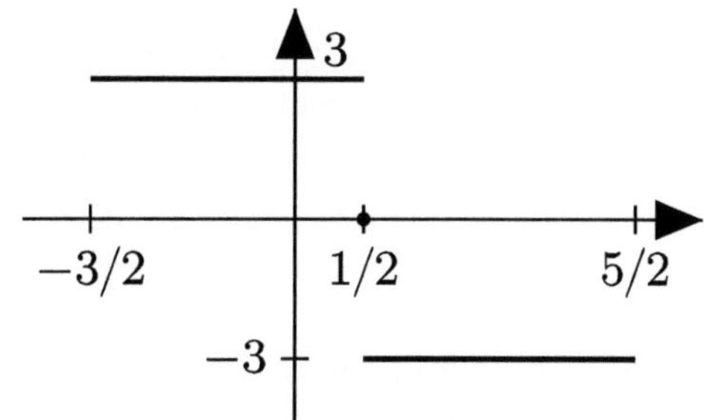

$$g(t) = \begin{cases} -3 \text{ for } \dfrac{1}{2} < t < \dfrac{5}{2} \\[2mm] 0 \text{ for } t = \dfrac{1}{2} \\[2mm] 3 \text{ for } -\dfrac{3}{2} < t < \dfrac{1}{2}. \end{cases}$$

**(A3)** (a)  For $|\sin(t)|$ give the trigonometric polynomial $f$ as an approximation with $\omega_0 = 2$

$$f(t) = \frac{a_0}{2} + \sum_{k=1}^{2} (a_k \cos(k\omega_0 t) + b_k \sin(k\omega_0 t)) \,.$$

     (b)  Sketch the amplitude spectrum of $f$.

     (c)  What is the amplitude spectrum of the modulated function $\cos(6t) f(t)$?

     (d)  Compute the Fourier series for $g(t) = \cos(6t) f(t)$.

     (e)  For $N \in \mathbb{N}$, what is generally the Fourier series of $\cos(N\omega_0 t) g(t)$ for

$$g(t) = \frac{a_0}{2} + \sum_{k=1}^{\infty} a_k \cos(k\omega_0 t) \,?$$

**(A4)**  What is the Fourier series of the $2\pi$-periodic function $\int_0^t f(x)\,\mathrm{d}x$, if $f(x)$ is the $2\pi$-periodic extension of the rectangle function $-\operatorname{sgn}(x - \pi)$ for $0 \leqslant x < 2\pi$?

**(A5)**  Let two trigonometric polynomials $f$ and $g$ be given by

$$f(t) = \sum_{k=-N}^{+N} c_k\, \mathrm{e}^{jkt} \quad \text{and} \quad g(t) = \sum_{k=-N}^{+N} d_k\, \mathrm{e}^{jkt} \,.$$

Prove that the product $f \cdot g$ has the Fourier coefficients $h_k = \sum_{n=-N}^{+N} c_n d_{k-n}$ for $-2N \leqslant k \leqslant +2N$ (set $d_m = 0$ for $|m| > N$).

**(A6)**  Compare continuity and differentiability properties of some Fourier series with previous examples here and in your formulary. Consider examples, whose Fourier coefficients decrease asymptotically like $\dfrac{1}{k}$, $\dfrac{1}{k^2}$, $\dfrac{1}{k^3}$, or $\dfrac{1}{k^2 - 1}$.

**(A7)**  Show that term-by-term differentiation of the Fourier series for the rectangle meander $f(t) = \operatorname{sgn}(t)$, $-\pi \leqslant t < \pi$, $f(t+2k\pi) = f(t)$, $k \in \mathbb{Z}$, leads to a series, which converges only at the points $t$ of the form $t = (2k-1)\pi/2$, $k \in \mathbb{Z}$.

(**A8**)  Show that the following series expansions hold:

(a)  $t^2 = \dfrac{\pi^2}{3} + 4 \sum\limits_{k=1}^{\infty} (-1)^k \dfrac{\cos(kt)}{k^2}$ for $-\pi \leqslant t \leqslant \pi$,

(b)  $t \cos(t) = -\dfrac{1}{2} \sin(t) + 2 \sum\limits_{k=2}^{\infty} (-1)^k \dfrac{k \sin(kt)}{k^2 - 1}$ for $-\pi < t < \pi$,

(c)  $t \sin(t) = 1 - \dfrac{1}{2} \cos(t) - 2 \sum\limits_{k=2}^{\infty} (-1)^k \dfrac{\cos(kt)}{k^2 - 1}$ for $-\pi \leqslant t \leqslant \pi$.

(**A9**)  Compute the limits of the series

$$\sum_{n=1}^{\infty} \frac{1}{4n^2 - 1}, \quad \sum_{n=1}^{\infty} (-1)^{n+1} \frac{1}{n^2}, \quad \sum_{n=1}^{\infty} (-1)^{n+1} \frac{1}{(2n-1)^3}.$$

Use Fourier series of periodic functions in your formulary.

(**A10**)$^\star$  The graphical display of the sums $\sum\limits_{k=1}^{n} \dfrac{\sin(kt)}{k}$ suggests for the sawtooth function that all partial sums in $]0, \pi[$ are strictly positive, in $]-\pi, 0[$ strictly negative, i.e., that they do not undershoot the sawtooth in $]0, \pi[$ and do not overshoot in $]-\pi, 0[$. Show this conjecture by induction. In which tolerance ranges can the sawtooth be approximated by such partial sums?

(**A11**)$^\star$  This exercise shall show that there are convergent trigonometric series, which are not Fourier series in the classical sense dealt with so far.

(a)  Prove that

$$\sum_{k=1}^{n} \sin(kt) = \frac{\sin((n+1)t/2) \sin(nt/2)}{\sin(t/2)}$$

(with continuous extension at the zeros of the denominator).
Hint: Use $1 - e^{j\varphi} = e^{j\varphi/2}(-2j) \sin(\varphi/2)$ and a similar computation as with the Dirichlet kernel on p. 15.

(b)  Show that the sine coefficients $b_k$ of the Fourier series of an integrable function $f$ on $[0, 2\pi]$—which does not necessarily represent the function $f$—fulfill $\sum\limits_{k=1}^{\infty} b_k/k < \infty$. For this use the comment on p. 48, and expand the integral function of $f$ into a Fourier series.

(c)  (Abel's Lemma) Show that for each two sequences $(a_k)_{k\in\mathbb{N}}$, $(b_k)_{k\in\mathbb{N}}$ and $s_n = \sum\limits_{k=1}^{n} a_k$, one has $\sum\limits_{k=1}^{n} a_k b_k = s_n b_{n+1} + \sum\limits_{k=1}^{n} s_k(b_k - b_{k+1})$.

(d)  (Abel-Dirichlet Test) Show with (c) the following generalization of the well-known Leibniz criterion for alternating series:

*For a sequence $(a_k)_{k\in\mathbb{N}}$, let $\left|\sum_{k=1}^{n} a_k\right| \leqslant M$ hold for all $n \in \mathbb{N}$ and a constant $M > 0$. Then the series $\sum_{k=1}^{\infty} a_k b_k$ converges for every monotonically decreasing sequence $(b_k)_{k\in\mathbb{N}}$ with $\lim_{k\to\infty} b_k = 0$.*

(e) Show using the comment on p. 48 that $\sum_{k=2}^{\infty} \dfrac{\sin(kt)}{\ln(k)}$ converges everywhere, for $h > 0$ in every interval $[h,\, 2\pi - h]$ even uniformly to a continuous function. But verify that this series cannot be the (classical) Fourier series of that function.

# Chapter 5
# Application Examples for Fourier Series

**Abstract** This chapter shows applications of classical Fourier series. The following topics are treated in respective sections: the best approximation in quadratic mean (RMS approximation), periodic convolution and its role in AC Circuit calculations, the boundary value problem for the 2D-potential equation on a circular disk with the Poisson integral formula, the classical solution for the vibrating string, the approximation theorem of K. Weierstrass, and the 1/f theorem of N. Wiener. Examples and exercises are provided. These include, for example, the inhomogeneous vibrating string, the homogeneous one-dimensional heat equation, periodic convolution of given periodic functions, and Kepler's equation.

## 5.1 Best Approximation in Quadratic Mean

The focus of the following considerations is no longer, as before, on pointwise or uniform approximations to periodic functions $f$ by trigonometric polynomials, but rather approximations, whose mean squared deviation from the considered function $f$ should be small.

A $T$-periodic piecewise continuous function $f : \mathbb{R} \to \mathbb{C}$ (or its restriction to $[0, T]$) shall be approximated by a trigonometric polynomial

$$P(t) = \sum_{k=-N}^{N} \alpha_k \, e^{jk\omega_0 t},$$

$\omega_0 = 2\pi/T$, so that the mean square error becomes minimal:

$$\frac{1}{T} \int_0^T |f(t) - P(t)|^2 \mathrm{d}t = \min !$$

If $f$ has the Fourier coefficients $c_k$, the mean square error can be written in the following form:

$$\frac{1}{T}\int_0^T |f(t) - P(t)|^2 \, dt$$

$$= \frac{1}{T}\int_0^T \left(f(t) - \sum_{k=-N}^{N} \alpha_k \, e^{jk\omega_0 t}\right)\left(\overline{f(t)} - \sum_{k=-N}^{N} \overline{\alpha_k} \, e^{-jk\omega_0 t}\right) dt$$

$$= \frac{1}{T}\int_0^T |f(t)|^2 dt - \sum_{k=-N}^{N} \alpha_k \overline{c_k} - \sum_{k=-N}^{N} \overline{\alpha_k} c_k + \sum_{k=-N}^{N} |\alpha_k|^2$$

$$= \underbrace{\frac{1}{T}\int_0^T |f(t)|^2 dt - \sum_{k=-N}^{N} |c_k|^2}_{\text{independent of the } \alpha_k} + \sum_{k=-N}^{N} |c_k - \alpha_k|^2 .$$

The above integral becomes minimal if and only if $\alpha_k = c_k$ for $|k| \leqslant N$. Thus, we obtain the following theorem on best approximation in quadratic mean:

**Theorem 5.1** *The best trigonometric approximation polynomial of degree at most $N$ for this purpose is the $N$th partial sum $P(t) = \sum_{k=-N}^{N} c_k \, e^{jk\omega_0 t}$ of the Fourier series of $f$.*

## *Geometric Interpretation*

When we look in a subspace $U$ of $\mathbb{R}^n$ for an approximating vector $\mathbf{y}$ to a given vector $\mathbf{x} = (x_1, \ldots, x_n) \in \mathbb{R}^n$ so that

$$|\mathbf{x} - \mathbf{y}|^2 = \sum_{i=1}^{n}(x_i - y_i)^2 = \min !$$

then, as is well known, $\mathbf{y}$ is the orthogonal projection of $\mathbf{x}$ into $U$. The inner products of $(\mathbf{x} - \mathbf{y})$ and vectors $\mathbf{u} \in U$ fulfill the orthogonality relation (Fig. 5.1)

$$(\mathbf{x} - \mathbf{y}) \cdot \mathbf{u} = 0 \text{ for all } \mathbf{u} \in U.$$

The same fact holds true in higher dimensional vector spaces with an inner product. In the same sense, the $N$th partial sum $f_N$ of the Fourier series of $f$ is the orthogonal projection of $f$ into the vector space $\mathcal{T}_N$ of $T$-periodic trigonometric polynomials

**Fig. 5.1** Orthogonal
projection into $U$

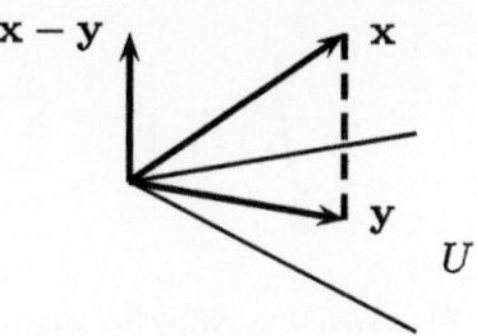


up to the degree $N$. The piecewise continuous $T$-periodic function $f$ has a unique
decomposition $f = f_N + f_N^\perp$ with $f_N \in \mathcal{T}_N$, and

$$\langle f_N^\perp | h \rangle = \langle f - f_N | h \rangle = 0 \text{ for all } h \in \mathcal{T}_N,$$

with the inner product on p. 12. For $h(t) = \sum_{k=-N}^{N} \alpha_k \, e^{jk\omega_0 t}$ we have

$$\langle f - f_N | h \rangle = \frac{1}{T} \int_0^T \left( f(s) - \sum_{k=-N}^{N} c_k \, e^{jk\omega_0 s} \right) \sum_{m=-N}^{N} \overline{\alpha_m} \, e^{-jm\omega_0 s} \, \mathrm{d}s$$

$$= \left( \sum_{k=-N}^{N} \overline{\alpha_k} c_k - \sum_{k=-N}^{N} c_k \overline{\alpha_k} \right) = 0.$$

The function $f$ is an element of the infinite dimensional vector space of all $T$-
periodic piecewise continuous functions. The orthogonal projection $f_N$ with the
integral representation

$$f_N(t) = \frac{1}{T} \int_0^T f(s) D_N(t - s) \mathrm{d}s$$

is an element of the $(2N + 1)$-dimensional subspace $\mathcal{T}_N$, where $D_N$ denotes the
Dirichlet kernel with degree $N$.

## Convergence in Quadratic Mean

**Theorem 5.2** *The Fourier series of a $T$-periodic piecewise continuous function $f$
converges to $f$ in quadratic mean.*

This theorem is equivalent to the validity of the Parseval equation. If $f$ has the
Fourier coefficients $c_k$, then we have with $\omega_0 = 2\pi/T$

$$\frac{1}{T}\int_0^T |f(t) - \sum_{k=-N}^{N} c_k\, \mathrm{e}^{jk\omega_0 t}\,|^2 \mathrm{d}t = \frac{1}{T}\int_0^T |f(t)|^2 \mathrm{d}t - \sum_{k=-N}^{N} |c_k|^2 \xrightarrow[N\to\infty]{} 0.$$

The notion of convergence in quadratic mean is often more important than pointwise convergence for technical and also theoretical purposes. By the theorem of Dirichlet (cf. p. 28) pointwise representation of periodic functions $f$ by their Fourier series, as presented in our context, is only ensured for piecewise continuously differentiable functions. In fact, there are examples of continuous periodic functions whose Fourier series diverge at infinitely many points. On the other hand, one knows from Carleson (1966) that Fourier series of continuous periodic functions $f$ converge almost everywhere (cf. Appendix B). Besides the remarkable theorem of Fejér (p. 29), the most important result for applications is probably the above formulated convergence in quadratic mean, which can be shown for the Fourier series of very general functions.

This result has been stated above for piecewise continuous functions and will be proven for this class of functions in Chap. 7. In functional analysis, using the Lebesgue integral (cf. Appendix B) for elements $f$ of the vector space $L^2([0, T])$ of all square integrable complex-valued functions, the following more general theorem is shown. Any function $f \in L^2([0, T])$, defined on the interval $[0, T]$, can be extended to a $T$-periodic function on $\mathbb{R}$ as usual.

**Theorem 5.3** *The Fourier series of a function $f \in L^2([0, T])$ converges to $f$ in quadratic mean.*

A proof of this theorem can be found, for instance, in Rudin (1991). Even if functions $f \in L^2([0, T])$ are not necessarily represented pointwise by their Fourier series, they can be approximated arbitrarily well in quadratic mean by partial sums $f_N$ of their Fourier series. In this sense, one also denotes $f$ by $f(t) = \sum_{k=-\infty}^{+\infty} c_k\, \mathrm{e}^{jk\omega_0 t}$. This means that the partial sums $f_N$ for $N \to \infty$ converge to $f$ with respect to the norm of p. 54: $\lim_{N\to\infty} \|f_N - f\|_2 = 0$. The distances between $f_N$ and $f$, measured with that norm, become arbitrarily small for increasing $N$, i.e., the mean square error $\|f_N - f\|_2^2$ converges to zero. Actually, pointwise convergence in general does not make sense for $L^2$-functions without additional conditions, since those represent entire equivalence classes of functions (cf. p. 54) whose elements are not determined at individual points.

**Computation of Distortion Factors**

By the Parseval equality we can compute distortion factors with the help of the normalized power $P$ (compare p. 33). The *distortion factor* $D$ for a real-valued signal $f$ with Fourier coefficients $c_k$ and normalized power $P$ is given by $D = \sqrt{Z/N}$, where $N = (P - |c_0|^2)/2$ and $Z = N - |c_1|^2$. For the example on p. 33, we find, for instance, a distortion factor $D$ of about $D = 0.56$.

## 5.2   Periodic Convolution and Application to Linear Systems

In this section, we apply Fourier series to solve linear differential equations with constant coefficients and periodic perturbation functions. Well-known examples are equations of motion in mechanics or differential equations describing $RLC$ networks in electrical engineering. For this we need the notion of periodic convolution.

**Definition** The $T$-*periodic convolution* for piecewise continuous $T$-periodic functions $f$ and $h$ is defined by

$$(f * h)_T(t) = \frac{1}{T} \int\limits_0^T f(u)h(t-u)\mathrm{d}u.$$

If $f$ and $h$ are piecewise continuous as assumed, then $(f * h)_T$ is continuous on $\mathbb{R}$ and $T$-periodic. This property will be demonstrated in Chap. 7 and used there to prove the Parseval theorem.

**Remark** With the *Lebesgue integration theory*, the periodic convolution can be defined more general. Thereby the convolution exists almost everywhere for functions $f$ and $h$, which are Lebesgue integrable on $[0, T]$. The continuity of $(f * h)_T$ can be shown for all $T$-periodic functions $f$ and $h$, which are square integrable on $[0, T]$, and then yields the Parseval equality also for these functions.

### *The Fourier Series of a Periodic Convolution*

Let the coefficients $c_k$ be the Fourier coefficients of $f$, the coefficients $h_k$ those of $h$, $f$ and $h$ piecewise continuous. For the $k$th Fourier coefficient of $(f * h)_T$, we obtain by interchanging the order of integration:

$$\frac{1}{T} \int\limits_0^T \frac{1}{T} \int_0^T f(u)h(t-u)\mathrm{d}u\, \mathrm{e}^{-jk\omega_0 t}\, \mathrm{d}t$$

$$= \frac{1}{T} \int\limits_0^T f(u)\, \underbrace{\frac{1}{T} \int_0^T h(t-u)\, \mathrm{e}^{-jk\omega_0 t}\, \mathrm{d}t}_{h_k \cdot \mathrm{e}^{-jk\omega_0 u}\ \text{by 4.2}}\, \mathrm{d}u = c_k h_k.$$

$$(f * h)_T(t) = \sum_{k=-\infty}^{+\infty} c_k h_k\, \mathrm{e}^{jk\omega_0 t}.$$

**Result** *The Fourier coefficients of the $T$-periodic convolution $(f * h)_T$ are the products of the corresponding Fourier coefficients of $f$ and $h$.*

Multiplying out the inequality $(|c_k| - |h_k|)^2 \geqslant 0$, we get $2|c_k h_k| \leqslant |c_k|^2 + |h_k|^2$. With the Bessel inequality we thereby obtain

$$\sum_{k=-\infty}^{+\infty} |c_k h_k| \leqslant \frac{1}{2} \sum_{k=-\infty}^{+\infty} (|c_k|^2 + |h_k|^2) < \infty.$$

Therefore the Fourier series of $(f * h)_T$ is uniformly convergent. According to Fejér's theorem (see later p. 134), the continuous function $(f * h)_T$ is actually represented pointwise by its Fourier series. The continuity of $(f * h)_T$ will be shown on p. 139.

## *Application to Asymptotically Stable Time-Invariant Linear Systems*

As application we consider linear differential equations $\sum_{k=0}^{n} a_k u^{(k)}(t) = f(t)$.

Here, the $u^{(k)}$ denote the $k$th derivatives of $u$, and the $a_k$ are constant real coefficients. Without loss of generality we set $a_n = 1$. Many differential equations in modeling physical or technical problems are of this type. For example, think about electrical networks built with resistors, capacitances, and inductances or about differential equations of oscillations in mechanics.

We now assume that the system *is asymptotically stable*, that is, for arbitrary initial values, the solution of the associated homogeneous differential equation vanishes for $t \to \infty$. This is exactly the case if all zeros of the characteristic polynomial have negative real parts. The characteristic polynomial must then be a *Hurwitz polynomial*, and all coefficients $a_k$ must be positive, i.e., with the same sign as $a_n > 0$.

Under these conditions, for right-hand sides of the form $f(t) = A\sin(\omega t + \phi)$ the *periodic solution* is uniquely determined. It has the same angular frequency $\omega$ but a different amplitude and phase than $f(t)$ (cf. the common methods for solving such differential equations).

For a right-hand side $f(t) = U_0 e^{j\omega t}$, the linear operator $L$, mapping $f$ to the uniquely determined periodic solution $L(f)$, describes a so-called *linear time-invariant system* (LTI system). We will analyze such systems after further mathematical preparations in more general terms in Chap. 11. Schematically, the facts presented here for harmonic oscillations $f$ are shown in Fig. 5.2:

The function $\hat{h}(\omega)$ expresses amplification or attenuation and phase shift in the transmission depending on the angular frequency $\omega$. We obtain $\hat{h}(\omega) = 1/P(j\omega)$ with the characteristic polynomial $P$ of the given differential equation.

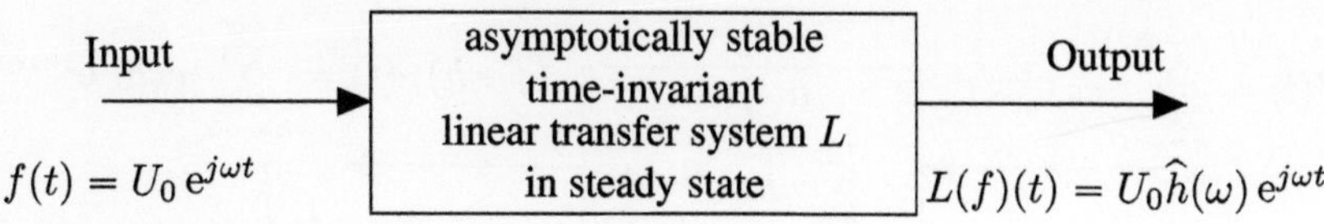


**Fig. 5.2**  Schematic figure of a stable LTI system in steady state

The question arises whether, in general, there is exactly one periodic solution for a periodic excitation. This solution would describe the long-term behavior after decay of transient solution parts. Under suitable constraints on the right-hand side $f$, we can prove the following theorem:

**Theorem 5.4** *An asymptotically stable linear ordinary differential equation of order $n$ with constant coefficients and with a continuous, continuously differentiable, $T$-periodic right-hand side $f$ has the uniquely determined $T$-periodic solution*

$$u(t) = \sum_{k=-\infty}^{+\infty} c_k h_k\, e^{jk\omega_0 t} \qquad \left(\omega_0 = \frac{2\pi}{T}\right).$$

*The coefficients $c_k$ are the Fourier coefficients of $f$ and $h_k = 1/P(jk\omega_0)$, where $P$ is the characteristic polynomial of the differential equation.*

***Proof*** Inserting the series into the differential equation shows this statement immediately. This series and all its term-by-term derivatives up to the order $n$ are uniformly convergent, since the Fourier series of $f$ converges uniformly. In particular, $u$ is an $n$-times continuously differentiable function. The uniqueness follows immediately from the fact that the difference of two $T$-periodic solutions is again $T$-periodic and must be a solution of the homogeneous equation. This can only be the zero function by the presupposed stability. $\qquad\Box$

We can show the estimate $|P(j\omega_0 k)|^{-1} \leqslant M\,|k|^{-3/2}$ for sufficiently large $|k|$, a suitable $M > 0$, and polynomial degree $n \geqslant 2$ (Exercise A4). Therefore the series

$$h(t) = \sum_{k=-\infty}^{+\infty} h_k\, e^{jk\omega_0 t}$$

represents a continuous function. For equations of order 1, we can see with some calculations, comparing with the sawtooth function, that $h(t) = \sum_{k=-\infty}^{+\infty} \frac{1}{jk+a_0}\, e^{jkt}$, $a_0 > 0$, is continuous except for the points $t = 2n\pi$, $n \in \mathbb{Z}$. At those points $h$ has right and left limits ($\pi(\coth(a_0\pi) \pm 1)$ for $T = 2\pi$). The series represents a piecewise continuously differentiable function (Exercise A5). This holds true also for other periods than $T = 2\pi$. Therefore, we obtain with the convolution property for Fourier series (Illustration Fig. 5.3):

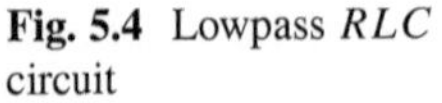

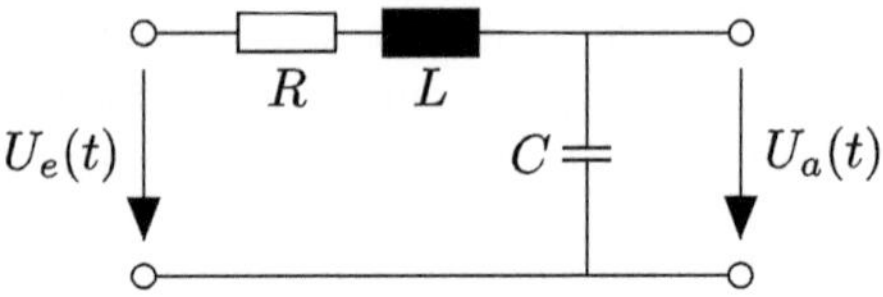

$$f(t) = \sum_{k=-\infty}^{+\infty} c_k\, e^{jk\omega_0 t} \quad\boxed{\begin{array}{c} h(t) = \\ \displaystyle\sum_{k=-\infty}^{+\infty} h_k\, e^{jk\omega_0 t} \end{array}}\quad (f*h)_T(t) = \sum_{k=-\infty}^{+\infty} c_k h_k\, e^{jk\omega_0 t}$$

$$= \frac{1}{T}\int_0^T f(s)h(t-s)\,\mathrm{d}s$$

**Fig. 5.3**  $T$-periodic convolution

**Fig. 5.4**  Lowpass $RLC$ circuit



*The Fourier coefficients of the solution u are the products of the corresponding Fourier coefficients of f and h, and u is the T-periodic convolution $u = (f*h)_T$.*

*The function h is called the T-periodic transfer function.*

**Remark**  If we want to treat more general right-hand sides like a sawtooth function or a rectangular meander as a model of switch-on and switch-off processes, a modification of the classical notion of a solution for differential equations is necessary. This can be done for the case with piecewise continuous right-hand sides $f$ by a modification of the common notion of a primitive function in the context of Riemannian integration theory, as, for instance, in Dieudonné (2006). With the results of the *Lebesgue integration theory, functional analysis,* and *distribution theory*, it is possible to introduce a new notion of a solution, thereby weakening the conditions on $f$ to a very large extent. It suffices, for example, that the coefficients $c_k$ of $f$ are square summable. All Fourier series treated above then converge in $L^2([0,T])$, and the function $u$ is the convolution $(f*h)_T$ of two functions in $L^2([0,T])$. With the concept of *generalized derivatives* (Sect. 8.5) in *distribution theory* and *generalized Fourier series f* and $h$ (cf. Sect. 9.1), finally $(f*h)_T$ can be interpreted as the solution of the differential equation, without additional continuity or differentiability conditions on $f$ as long as the Fourier coefficients of $f$ being of slow growth. This is a major progress in the treatment of many application problems. We consider, as already mentioned, *time-invariant linear systems* in more detail only after the necessary mathematical preparations in Chap. 11 and come back to the issues noted here in Sect. 11.5. The last theorem and its generalizations form the foundation of *complex AC circuit calculation* in electrical engineering.

**Example (AC Circuit Calculation)**  The $RLC$ lowpass filter shown in Fig. 5.4 with ohmic resistance $R$, inductivity $L$ and capacity $C$ due to Kirchhoff's law, is described by the differential equation

$$LC \frac{d^2 U_a}{dt^2}(t) + RC \frac{dU_a}{dt}(t) + U_a(t) = U_e(t).$$

The zeros of the characteristic polynomial $LC\lambda^2 + RC\lambda + 1$ are

$$\lambda_{1,2} = \begin{cases} -\frac{R}{2L} \pm \sqrt{\frac{R^2}{4L^2} - \frac{1}{LC}} & \text{for } \frac{R^2}{4L^2} \geqslant \frac{1}{LC} \\[2ex] -\frac{R}{2L} \pm j\sqrt{\frac{1}{LC} - \frac{R^2}{4L^2}} & \text{for } \frac{R^2}{4L^2} < \frac{1}{LC}. \end{cases}$$

They have negative real parts, and the given linear system is asymptotically stable. For $U_e(t) = U_0\, e^{jk\omega_0 t}$, $\omega_0 = 2\pi/T$, we obtain the $T$-periodic solution

$$U_a(t) = \frac{U_0}{1 + jk\omega_0 RC - k^2\omega_0^2 LC}\, e^{jk\omega_0 t}.$$

The continuous $T$-periodic transfer function is

$$\sum_{k=-\infty}^{+\infty} h_k\, e^{jk\omega_0 t} = \sum_{k=-\infty}^{+\infty} \frac{1}{1 + jk\omega_0 RC - k^2\omega_0^2 LC}\, e^{jk\omega_0 t}.$$

Thus, the convolution rule can be used to obtain Fourier series representations of the periodic system responses, if the Fourier expansions of (so far assumed continuous) periodic input signals and the corresponding periodic transfer functions are known.

**Remark**  The spectral sequence $(h_k)_{k\in\mathbb{Z}} = \left( \frac{1}{1 + jk\omega_0 RC - k^2\omega_0^2 LC} \right)_{k\in\mathbb{Z}}$ corresponds to samples of the function $\widehat{h}(\omega) = 1/(1 + j\omega RC - \omega^2 LC)$, which in electrical engineering is called *frequency response* of the filter. The lowpass effect of the circuit, i.e., the attenuation of high-frequency input parts, can be seen in the sequence $(h_k)_{k\in\mathbb{Z}}$ and in the frequency response $\widehat{h}(\omega)$ of the example.

## *Mechanical Systems of Second Order with Periodic Forces*

Analogously to the above example from electrical engineering, we obtain a solution for asymptotically stable mechanical systems of the form

$$m\ddot{x}(t) + k\dot{x}(t) + Dx(t) = K(t),$$

with the important case of periodic forces $K(t)$. The periodic solution is received by Fourier expansion of $K(t)$ and periodic convolution with the system's periodic transfer function (up to now again under the continuity condition on $K(t)$, later on with more generality in the Sects. 9.1 and 11.5). Replacing in analogy the coefficients of the preceding example by the constants $m$, $k$, and $D$ is left to the reader.

## 5.3   The Potential Equation on a Circular Disk

The *Laplace equation* $\Delta u = \frac{\partial^2 u}{\partial x^2} + \frac{\partial^2 u}{\partial y^2} + \frac{\partial^2 u}{\partial z^2} = 0$ occurs in many areas of mathematical physics. In the theory of heat conduction, $u$ *is the stationary temperature*, i.e., the temperature which is reached after some time. This temperature is obtained, if in the heat conduction equation $\frac{\partial u}{\partial t} = \alpha^2 \Delta u$ the left side is set to zero ($\alpha^2$ is the thermal diffusivity in m$^2$/s). In the theory of gravitation or electricity, the function $u$ represents a *gravitational potential* or an *electrical potential.* The equation $\Delta u = 0$ is therefore also called  *potential equation* .

The task to solve $\Delta u = 0$ within a domain $G$, where $u$ is given on the boundary of the domain $G$, is called a *Dirichlet boundary value problem.* It can be solved for functions of two variables on a circular disk by applying the convolution relation for Fourier series. In a circular disk around the zero point with radius $R$, we consider the problem

$$\Delta u = \frac{\partial^2 u}{\partial x^2} + \frac{\partial^2 u}{\partial y^2} = 0.$$

In polar coordinates this equation for $0 < r < R$ and $0 \leqslant \phi < 2\pi$ is given by

$$\Delta u = \frac{\partial^2 u}{\partial r^2} + \frac{1}{r} \frac{\partial u}{\partial r} + \frac{1}{r^2} \frac{\partial^2 u}{\partial \phi^2} = 0.$$

### *Solution by Fourier Series Expansion for Given Boundary Values*

Inserting

$$u_k = c_k \left( \frac{r}{R} \right)^k e^{jk\phi} \quad \text{and} \quad u_{-k} = c_{-k} \left( \frac{r}{R} \right)^k e^{-jk\phi}$$

into the equation, we prove that these functions are solutions of the potential equation for every $k \in \mathbb{N}_0$ and arbitrary constants $c_k$ and $c_{-k}$. With the superposition principle, we thereby obtain a solution of the form

$$u(r, \phi) = \sum_{k=-\infty}^{+\infty} c_k \left(\frac{r}{R}\right)^{|k|} e^{jk\phi},$$

provided that the series converges and represents a sufficiently smooth function. The constants $c_k$ have the physical unit of $u$. With given boundary values $U(\phi)$ on the circle $r = R$, the $c_k$ are just the Fourier coefficients of $U(\phi)$, $0 \leqslant \phi < 2\pi$,

$$u(R, \phi) = U(\phi) = \sum_{k=-\infty}^{+\infty} c_k\, e^{jk\phi}.$$

## *The Poisson Integral Formula*

For every $r \in [0, R[$ and $\phi \in [0, 2\pi[$, the following geometric series is absolutely convergent and represented as

$$\sum_{k=0}^{\infty} \left(\frac{r}{R} e^{j\phi}\right)^k = \frac{R}{R - r\,e^{j\phi}} = \frac{R^2 - Rr\cos(\phi) + jRr\sin(\phi)}{R^2 + r^2 - 2Rr\cos(\phi)}.$$

Hence it follows

$$\sum_{k=-\infty}^{+\infty} \left(\frac{r}{R}\right)^{|k|} e^{jk\phi} = \sum_{k=0}^{\infty} \left(\frac{r}{R}\right)^{k} e^{jk\phi} + \sum_{k=-\infty}^{-1} \left(\frac{r}{R}\right)^{-k} e^{jk\phi}$$

$$= \sum_{k=0}^{\infty} \left(\frac{r}{R}\right)^{k} e^{jk\phi} + \sum_{k=0}^{\infty} \left(\frac{r}{R}\right)^{k} e^{jk(-\phi)} - 1.$$

Thus, we can also write this series as

$$\sum_{k=-\infty}^{+\infty} \left(\frac{r}{R}\right)^{|k|} e^{jk\phi} = \frac{R^2 - Rr\cos(\phi) + jRr\sin(\phi) + R^2 - Rr\cos(\phi)}{R^2 + r^2 - 2Rr\cos(\phi)}$$

$$+ \frac{-jRr\sin(\phi) + 2Rr\cos(\phi) - R^2 - r^2}{R^2 + r^2 - 2Rr\cos(\phi)} = \frac{R^2 - r^2}{R^2 + r^2 - 2Rr\cos(\phi)}.$$

Because $u(r, \phi)$ is the *Fourier series of the $2\pi$-periodic convolution* between $U(\phi)$ and the function $g(r, \phi) = \frac{R^2 - r^2}{R^2 + r^2 - 2Rr\cos(\phi)}$, which is additionally dependent on $r$, we find from the convolution relation the resulting integral representation of the solution $u(r, \phi)$ for $0 \leqslant r < R$ and $0 \leqslant \phi < 2\pi$:

$$u(r, \phi) = \frac{1}{2\pi} \int_0^{2\pi} U(\psi) \frac{R^2 - r^2}{R^2 + r^2 - 2Rr \cos(\phi - \psi)} \, d\psi.$$

This solution formula for the potential equation with the boundary values $U(\phi)$ on the circle $r = R$ is known as *Poisson integral formula*. If the function $u(r, \phi)$ is, for example, a stationary temperature distribution, then the temperature for each point inside the circular disk is thus expressed by the (kept constant) temperature values $U(\psi), 0 \leqslant \psi < 2\pi$, on the boundary of the circular disk.

## *Smoothness and Uniqueness of the Solution and Maximum Principle*

In order to deepen our work with partial differential equations one more step, we briefly investigate the question of differentiability and uniqueness of the solution. For example, we require that the boundary condition $U(\phi)$ is a continuous, piecewise continuously differentiable, $2\pi$-periodic function. For every $m \in \mathbb{N}$ we get

$$\lim_{|k| \to \infty} |k|^m |c_k| \left(\frac{r}{R}\right)^{|k|} = \lim_{|k| \to \infty} |k|^m |c_k| \, e^{|k| \ln(r/R)} = 0.$$

By the results in Sect. 4.5 on summability properties of the series representation for the solution, we find that $u(r, \phi)$ is differentiable in both variables arbitrarily often, and we obtain for $\phi \in [0, 2\pi[$

$$\lim_{r \to R} u(r, \phi) = U(\phi).$$

The proof of uniqueness can be done with the help of the so-called *maximum principle for the potential equation*, which we formulate for more general domains as circular disks in the plane. A domain $G$ is a non-empty, open, and connected set.

**Theorem 5.5** *Let $G$ be an open bounded domain in the two-dimensional plane and $\partial G$ its boundary, and let a nonconstant function $u$ fulfill the potential equation $\Delta u = 0$ in $G$. If $u$ is continuous on $G \cup \partial G$, then it attains its maximum and its minimum on the boundary of the domain.*

**Proof** For $\varepsilon > 0$ we set $v(x, y) = u(x, y) + \varepsilon(x^2 + y^2)$. Then

$$\frac{\partial^2 v}{\partial x^2} + \frac{\partial^2 v}{\partial y^2} = 4\varepsilon > 0 \text{ in } G.$$

If $(x_0, y_0)$ is supposed to be an inner point of $G$ with

$$v(x_0, y_0) = \max\{v(x, y) \mid (x, y) \in G \cup \partial G\},$$

then, from well-known theorems on extreme values, necessarily follows

$$\frac{\partial^2 v}{\partial x^2}(x_0, y_0) \leqslant 0 \quad \text{and} \quad \frac{\partial^2 v}{\partial y^2}(x_0, y_0) \leqslant 0.$$

This is a contradiction to the above equation. Therefore, $v(x, y)$ attains its maximum on the boundary $\partial G$. By continuity of $u(x, y)$ and $u \leqslant v$, it follows immediately:

$$\max_{G \cup \partial G} u(x, y) \leqslant \max_{G \cup \partial G} v(x, y) = \max_{\partial G} v(x, y) \leqslant \max_{\partial G} u(x, y) + \varepsilon \max_{\partial G}(x^2 + y^2).$$

Since $\varepsilon > 0$ can be chosen arbitrarily small, we obtain the result

$$\max_{G \cup \partial G} u(x, y) = \max_{\partial G} u(x, y).$$

The same conclusion, applied to $-u(x, y)$, shows that $u$ attains also its minimum on the boundary. $\qquad\square$

The continuous functions $u$ on $G$ satisfying in $G$ the equation $\Delta u = 0$ are called *harmonic functions on $G$*. In a plausible simple meaning, the maximum principle for harmonic functions in a problem with constant temperature $T_0$ on the boundary of a bounded domain says that the stationary temperature in the interior of the region can be neither lower nor higher than at the boundary, i.e., after some time the temperature $T_0$ will be reached everywhere.

**Uniqueness of the Solution**

Supposed $\tilde{u}$ is a second solution, then we set $v(x, y) = u(x, y) - \tilde{u}(x, y)$. Hence, $v(x, y)$ fulfills the potential equation and has zero boundary values:

$$v(x, y) = 0 \quad \text{for all} \quad (x, y) \in \partial G.$$

Then, the maximum principle says that $v(x, y) = 0$ everywhere in $G \cup \partial G$, and this means

$$u(x, y) = \tilde{u}(x, y) \quad \text{everywhere in } G \cup \partial G.$$

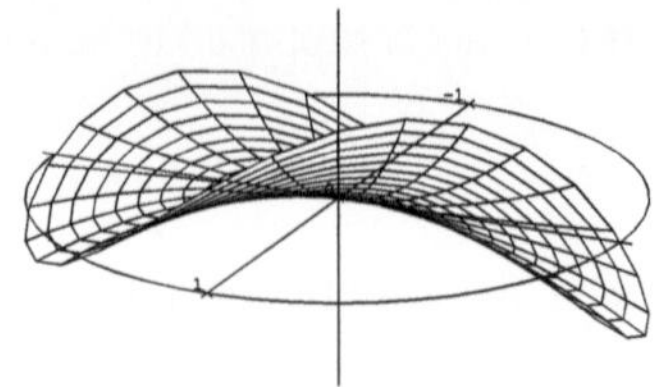

**Fig. 5.5** Potential in a circular disk for the given boundary function $U$

## *Illustration of the Solution for a Dirichlet Boundary Value Problem*

The illustration in Fig. 5.5 shows the solution of the Laplace equation on the unit circle disk for the given boundary function $U(\phi) = \cos(\phi) + \sin(2\phi)$. Maxima and minima of the solution lie on the boundary. Inside the circle there are no local extrema, but a saddle point. The solution represents the stationary temperature distribution or the electric potential inside of the circle for a given boundary temperature or boundary potential $U(\phi)$.

## 5.4  Solution for the Problem of the Force-Free Vibrating String

In our preceding work on the initial boundary value problem for the homogeneous force-free vibrating string

$$\frac{\partial^2 u}{\partial t^2} = c^2 \frac{\partial^2 u}{\partial x^2}, \qquad\qquad u(x,0) = f(x),$$

$$u(0,t) = u(l,t) = 0, \quad \lim_{t \to 0+} \frac{\partial u}{\partial t}(x,t) = g(x),$$

we had the solution approach

$$u(x,t) = \sum_{n=1}^{\infty} \sin\left(\frac{n\pi}{l}x\right)\left(a_n \cos\left(\frac{cn\pi}{l}t\right) + b_n \sin\left(\frac{cn\pi}{l}t\right)\right).$$

With term-by-term differentiation and interchanging the limit processes, we have

$$u(x,0) = \sum_{n=1}^{\infty} a_n \sin\left(\frac{n\pi}{l}x\right) = f(x),$$

$$\lim_{t \to 0+} \frac{\partial u}{\partial t}(x,t) = \sum_{n=1}^{\infty} \frac{cn\pi}{l} b_n \sin\left(\frac{n\pi}{l}x\right) = g(x).$$

For sufficiently smooth functions $f$ and $g$, therefore the solution can be determined with the Fourier coefficients of $f$ and $g$. The functions $f$ and $g$ are assumed to be $2l$-periodic and oddly extended. The coefficients $a_n$, $b_n$ are

$$a_n = \frac{2}{l} \int_0^l f(x) \sin\left(\frac{n\pi}{l} x\right) dx, \qquad b_n = \frac{2}{cn\pi} \int_0^l g(x) \sin\left(\frac{n\pi}{l} x\right) dx.$$

Thus, we have obtained a series representation for the solution.

## On Differentiability of the Solution

The question of which functions $f$ and $g$ are "sufficiently smooth" is answered by the following theorem:

**Theorem 5.6** *If $f$ is twice continuously differentiable on the entire axis $\mathbb{R}$ and $f'''$ is piecewise continuous and if $g$ is continuously differentiable on $\mathbb{R}$ and $g''$ is piecewise continuous, then the solution $u(x, t)$ is twice continuously partially differentiable. Differentiation with respect to $x$ or to $t$ twice results in convergent series, which represent continuous functions.*

**Proof** According to Sect. 4.3, we obtain the Fourier coefficients $f_n^{(3)}$ of $f'''$ by threefold term-by-term differentiation of the series $f(x) = \sum_{n=1}^{\infty} a_n \sin\left(\frac{n\pi}{l}x\right)$. Correspondingly we find the Fourier coefficients $g_n^{(2)}$ of $g''$ by differentiating term by term twice the series of the function $g$. For the coefficients $a_n$ of $f$ and $b_n$ of $g$, we have the relations

$$a_n = -\frac{l^3}{\pi^3 n^3} f_n^{(3)},$$

$$b_n = -\frac{l^3}{c\pi^3 n^3} g_n^{(2)}.$$

Thus, we can write $u(x, t)$ in the form

$$u(x, t) = -\left(\frac{l}{\pi}\right)^3 \sum_{n=1}^{\infty} \frac{1}{n^3} \sin\left(\frac{n\pi}{l} x\right) \left( f_n^{(3)} \cos\left(\frac{cn\pi}{l} t\right) + \frac{1}{c} g_n^{(2)} \sin\left(\frac{cn\pi}{l} t\right) \right).$$

With Bessel's inequality it is found that also after twofold term-by-term differentiation with respect to $x$ or $t$ the result is a uniformly convergent series:

$$\sum_{n=1}^{\infty} \frac{|f_n^{(3)}|}{n} \leq \frac{1}{2} \sum_{n=1}^{\infty} \left( \frac{1}{n^2} + |f_n^{(3)}|^2 \right) < \infty,$$

$$\sum_{n=1}^{\infty} \frac{|g_n^{(2)}|}{n} \leq \frac{1}{2} \sum_{n=1}^{\infty} \left( \frac{1}{n^2} + |g_n^{(2)}|^2 \right) < \infty.$$

$\square$

## *D'Alembert's Solution for the Force-Free Vibrating String*

We can also rewrite the solution into another form. Namely, if we set

$$a_n = A_n \sin(\phi_n) \quad \text{and} \quad b_n = A_n \cos(\phi_n),$$

then by the trigonometric addition theorems—again for sufficiently smooth functions $f$ and $g$—*D'Alembert's representation* of the solution is obtained (Exercise A7):

$$u(x,t) = \sum_{n=1}^{\infty} \frac{A_n}{2} \left( \cos \left( \frac{n\pi}{l}(x - ct) - \phi_n \right) - \cos \left( \frac{n\pi}{l}(x + ct) + \phi_n \right) \right)$$

$$= \frac{1}{2} \left( f(x - ct) + f(x + ct) + \frac{1}{c} \int_{x-ct}^{x+ct} g(\tau)\mathrm{d}\tau \right).$$

G. S. Ohm (1789–1854) concluded as an application of the series representations first basic principles of acoustics for the string vibration.

**Ohm's Law in Acoustics** *The sound of the string contains the tone pitch, determined by the fundamental frequency $c/(2l)$, and the overtones depending on the initial conditions with different amplitudes $A_n$. The sound perception depends on the ratio of these amplitudes $A_n = \sqrt{a_n^2 + b_n^2}$.*

**Remark** Transient vibrations, necessary for the recognition of a musical instrument, and phases, which are acoustically important for the localization of sound sources, are not taken into account in this characterization of timbre.

Concrete examples of string vibrations can be found in textbooks on mechanics or acoustics. Some initial boundary value problems of the discussed type and also string vibrations under the influence of constraining forces are dealt with in the exercises.

## *Uniqueness of the Solution*

A prerequisite that allows musicians among the readers to learn skills on a string instrument through regular practice, or even to risk an audition in front of an audience, is the fact that a string will always sound the same if the initial and boundary conditions are the same. Mathematically, this means that the solution to our initial boundary value problem must be unique. A standard method to show the uniqueness is to investigate the energy integral.

The *energy* of a twice continuously differentiable solution $u(x, t)$ is given at time $t \geqslant 0$ with tension $P$, mass density $\varrho$, and cross-sectional area $A$ of the string for small displacements by

$$E(t) = \int_0^l \underbrace{\frac{1}{2} \varrho A \left( \frac{\partial u}{\partial t} \right)^2}_{\text{kinetic}} + \underbrace{\frac{1}{2} P A \left( \frac{\partial u}{\partial x} \right)^2}_{\text{potential energy density}} dx.$$

By differentiation with respect to $t$ under the integral, observing the wave equation with $c^2 = P/\varrho$, and applying the chain and product rules when differentiating, we get

$$\frac{1}{A} \frac{dE}{dt}(t) = \int_0^l \left[ \varrho \frac{\partial u}{\partial t} \frac{\partial^2 u}{\partial t^2} + P \frac{\partial u}{\partial x} \frac{\partial^2 u}{\partial x \partial t} \right] dx = \int_0^l \left[ \varrho \frac{\partial u}{\partial t} \left( \frac{P}{\varrho} \frac{\partial^2 u}{\partial x^2} \right) + P \frac{\partial u}{\partial x} \frac{\partial^2 u}{\partial x \partial t} \right] dx$$

$$= P \int_0^l \frac{\partial}{\partial x} \left[ \frac{\partial u}{\partial t} \frac{\partial u}{\partial x} \right] dx = P \underbrace{\left[ \frac{\partial u}{\partial t} \frac{\partial u}{\partial x} \right]_{x=0}^{x=l}}_{=0 \text{ by the boundary condition}} = 0.$$

*This means that the law of energy conservation applies for the vibrating string:*

$$E(t) \text{ is constant.}$$

If now $\tilde{u}$ were a second solution, then we had for $v = u - \tilde{u}$

$$\frac{\partial^2 v}{\partial t^2} = c^2 \frac{\partial^2 v}{\partial x^2}, \quad v(x, 0) = 0 \text{ and } \lim_{t \to 0+} \frac{\partial v}{\partial t}(x, t) = 0 \text{ for all } x, \ v(0, t) = v(l, t) = 0$$

for all $t$; thus $\frac{\partial v}{\partial x}(x, 0) = 0$. Consequently we have $E(0) = 0$, and therefore we obtain by energy conservation $E(t) = 0$ for all $t$. Hence also $\frac{\partial v}{\partial x}(x, t) = 0$ for all $x$ and $t$, and finally

$$v(x, t) = v(0, t) + \int_0^x \frac{\partial v}{\partial x}(\tau, t)\mathrm{d}\tau = 0.$$

This says that the solution $u$ is unique: $u = \tilde{u}$. For the beautiful experience of being happy when listening to well-rehearsed music, we have to thank the law of energy conservation.

## *Meaning of the Solution*

For an initial velocity $g \equiv 0$ the solution $u(x, t)$ consists of the two waves $\frac{1}{2}f(x + ct)$ and $\frac{1}{2}f(x - ct)$, which move in opposite directions with velocity $c$ without changing their shape, superpose each other, and are reflected at the ends of the string with opposite phase. The influence of an initial velocity $g \neq 0$ is given by the additive component $\frac{1}{2c} \int_{x-ct}^{x+ct} g(\tau)\mathrm{d}\tau$ in the solution.

The tone pitch is determined by the fundamental frequency $c/(2l)$ in the Fourier series representation of the solution. Since $c^2 = P/\varrho$ is the quotient of the tension $P$ and the mass density $\varrho$ of the string, the influence of tension, mass, or length changes to the frequencies can be seen immediately in the series. Anyone who has ever manipulated or even tuned a string instrument probably knows these effects from experience.

Without regard to physical units, we illustrate the solution function $u(x, t)$ for $0 \leqslant x \leqslant 1$, $0 \leqslant t \leqslant 2$, $c = 1$, with concrete initial conditions $f$ and $g$ in two examples (Figs. 5.6 and 5.7). In the first example we set

$$f(x) = h(4x - 2) \text{ with } h(x) = \begin{cases} e^{-1/(1-x^2)} & \text{for} \quad -1 < x < 1, \\ 0 & \text{otherwise.} \end{cases}$$

In the second example we set $f(x) = \begin{cases} 2x & \text{for} \quad 0 \leqslant x \leqslant 1/4, \\ 2(1-x)/3 & \text{for} \quad 1/4 \leqslant x \leqslant 1. \end{cases}$

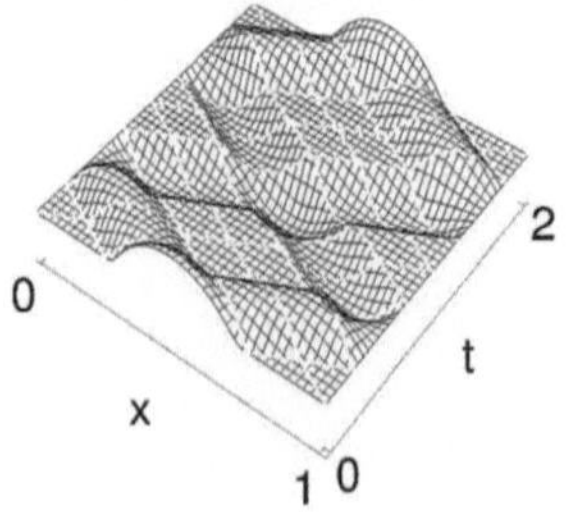


**Fig. 5.6** Smooth solution of the 1D wave equation

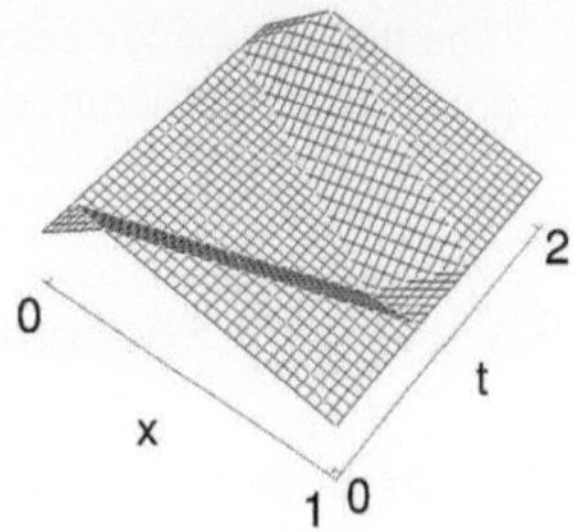

**Fig. 5.7** Non-differentiable solution of the 1D wave equation



In both cases we use $g = 0$. In the first example, the initial condition $f$ and the solution are infinitely often differentiable. In the second example, $f$ and thus also the solution $u$ are not differentiable (cf. later on Sect. 9.6).

## 5.5   The Approximation Theorem of Weierstrass

In the practice of mathematical modeling technical problems and also in many mathematical proofs, one replaces a continuous function $f : [a, b] \rightarrow \mathbb{C}$ approximately by a polynomial. A basis for this is the following theorem of K. Weierstrass (1815–1897).

**Theorem 5.7 (Theorem of Weierstrass)** *Every continuous $f : [a, b] \rightarrow \mathbb{C}$ on a closed bounded interval $[a, b]$ can be uniformly approximated by a polynomial.*

***Proof*** The function can be extended to a continuous $2(b - a)$-periodic function $\tilde{f}$ so that for any given $\varepsilon > 0$ there exists a trigonometric polynomial $P_n$ of the form $P_n = \frac{1}{n}(S_0 + S_1 + \cdots + S_{n-1})$, where $\sup_{t \in \mathbb{R}} |\tilde{f}(t) - P_n(t)| \leqslant \frac{\varepsilon}{2}$ (according to Fejér's theorem, p. 28). Here, the $S_k$ for $k \in \mathbb{N}_0$ are the $k$th partial sums of the Fourier series expansion of $\tilde{f}$. For each partial sum $S_k$ there is a Taylor polynomial $T_k$ such that $\sup_{t \in [a,b]} |S_k(t) - T_k(t)| \leqslant \frac{\varepsilon}{2}$. With $T = \frac{1}{n} \sum_{k=0}^{n-1} T_k$ it follows immediately that

$$\sup_{t \in [a,b]} |f(t) - T(t)| \leqslant \sup_{t \in [a,b]} |f(t) - P_n(t)| + \sup_{t \in [a,b]} |P_n(t) - T(t)| \leqslant \varepsilon.$$

$\square$

**Remark** We do not give any explicit examples here, because the method given in the proof for obtaining approximation polynomials is laborious and complicated. However, functions $f$ in practice often have additional smoothness properties. Then, one can find polynomials with less complicated methods, which interpolate the function at certain node points and which also have good overall approximation properties between the nodes (cf. p. 111).

The theorem of Weierstrass can be sharpened if the function $f$, which shall be approximated, has additional smoothness properties (cf. later on p. 142).

## 5.6  The 1/f-Theorem of Wiener

The subject of this section is an elementary proof of the famous $1/f$-Theorem of Wiener (1933). This theorem states that for a Fourier series $f$, which has no zero and has absolutely summable coefficients, also the reciprocal function $1/f$ has absolutely summable coefficients. When we discuss discrete linear filters in Sect. 11.6, an application of this theorem is shown to the reconstruction of filtered discrete data. The proof we follow was given by Newman (1975). Other proofs with the help of theorems on maximal ideals in normed algebras can be found in textbooks of functional analysis (e.g., in Rudin, 1991).

For a Fourier series $f(t) = \sum\limits_{k=-\infty}^{+\infty} c_k\, e^{jkt}$ with absolutely summable coefficients, the value

$$\|f\|_{\mathcal{A}} = \sum_{k=-\infty}^{+\infty} |c_k|$$

is a norm. Evidently, we have $\|f\|_\infty = \max |f(t)| \leqslant \|f\|_{\mathcal{A}}$. Two such series $f$ and $g$ each, now considered as continuous functions on $[0, 2\pi]$, fulfill the following inequalities:

$$\|f + g\|_{\mathcal{A}} \leqslant \|f\|_{\mathcal{A}} + \|g\|_{\mathcal{A}} \text{ and } \|f \cdot g\|_{\mathcal{A}} \leqslant \|f\|_{\mathcal{A}} \cdot \|g\|_{\mathcal{A}}.$$

With this norm, the vector space $\mathcal{A}$ formed by such Fourier series is a normed algebra with the function $f(t) = 1$ on $[0, 2\pi]$ as multiplicatively neutral element. One can show that this space $\mathcal{A}$ is complete, i.e., every Cauchy sequence in $\mathcal{A}$ converges to a function in $\mathcal{A}$ (for completeness see Rudin, 1991, for instance).

The first inequality is immediately obtained from the corresponding triangle inequality for the partial sums, and the second is seen as follows:

For $f_N(t) = \sum\limits_{k=-N}^{+N} c_k\, e^{jkt}$ and $g_N(t) = \sum\limits_{k=-N}^{+N} d_k\, e^{jkt}$, the Cauchy-Schwarz inequality and the Parseval equality of p. 54 yield the convergence of $f_N g_N$ to $fg$ in the norm $\|.\|_1$ of $L^1([0, 2\pi])$ (for the definition of $\|.\|_1$ see p. 500):

$$\frac{1}{2\pi} \int_0^{2\pi} |f_N(t)g_N(t) - f(t)g(t)|\mathrm{d}t \leqslant \|f_N - f\|_2\, \|g_N\|_2 + \|f\|_2\, \|g_N - g\|_2 \xrightarrow[N\to\infty]{} 0.$$

The Fourier coefficients $h_k(N)$ of $f_N g_N(t) = \sum\limits_{k=-2N}^{+2N} h_k(N)\,\mathrm{e}^{jkt}$ are (compare

Exercise A5, p. 56) $h_k(N) = \sum\limits_{n=-N}^{+N} c_n d_{k-n}$ (with $d_m = 0$ for $|m| > N$). They

converge for $N \to \infty$ to the Fourier coefficients $h_k$ of $fg$, since the $L^1$-convergence
of $f_N g_N$ implies

$$|h_k(N)-h_k| = \left| \frac{1}{2\pi} \int_0^{2\pi} (f_N(t)g_N(t) - f(t)g(t))\,\mathrm{e}^{-jkt}\,\mathrm{d}t \right| \leqslant \|f_N g_N - fg\|_1 \xrightarrow[N\to\infty]{} 0.$$

Thereby, we have $h_k = \sum\limits_{n=-\infty}^{+\infty} c_n d_{k-n}$, i.e., the Fourier coefficients of $f \cdot g$ are

obtained by *discrete convolution of the coefficient sequences of $f$ and $g$*. Thus, we
have for every $N \in \mathbb{N}$

$$\sum_{k=-N}^{+N} |h_k(N)| \leqslant \sum_{n=-N}^{+N} |c_n| \sum_{k=-N}^{+N} |d_k| \leqslant \|f\|_{\mathcal{A}} \cdot \|g\|_{\mathcal{A}},$$

and eventually from that the inequality $\|f \cdot g\|_{\mathcal{A}} \leqslant \|f\|_{\mathcal{A}} \cdot \|g\|_{\mathcal{A}}$.

Attentive readers are reminded of the absolute convergence of the Cauchy
product of power series shown in calculus, where one proceeds quite analogously.
We continue with another useful *preparatory inequality*: For $2\pi$-periodic and twice
continuously differentiable functions $f$ and their derivatives $f'$, the following
inequalities are valid:

$$\max_{t\in[0,2\pi]} |f(t)| \leqslant \|f\|_{\mathcal{A}} \leqslant \max_{t\in[0,2\pi]} |f(t)| + 2 \max_{t\in[0,2\pi]} |f'(t)|.$$

The first inequality is trivial. For the second one we estimate with the Cauchy-
Schwarz inequality (see also p. 35 and p. 51). For $f$ with Fourier coefficients $c_k$, we
have (compare also the Poincaré-Friedrichs inequality, p. 503)

$$\left( \sum_{k\in\mathbb{Z},k\neq 0} |c_k| \right)^2 \leqslant \sum_{k\in\mathbb{Z},k\neq 0} \frac{1}{k^2} \cdot \sum_{k\in\mathbb{Z},k\neq 0} k^2|c_k|^2 \leqslant \frac{\pi^2}{3} \cdot \frac{1}{2\pi} \int_0^{2\pi} |f'(t)|^2 \mathrm{d}t$$

$$\leqslant \frac{\pi^2}{3} \max_{t\in[0,2\pi]} |f'(t)|^2 \leqslant 4 \max_{t\in[0,2\pi]} |f'(t)|^2.$$

With $c_0 \leqslant \max_{t\in[0,2\pi]} |f(t)|$ now the upper bound for $\|f\|_{\mathcal{A}}$ in the inequality is
obtained.

**Theorem 5.8** (1/$f$-**Theorem of N. Wiener**) *For every Fourier series $f \in \mathcal{A}$, which has no zero, also $1/f$ belongs to $\mathcal{A}$.*

***Proof*** Let $f \in \mathcal{A}$ be given without a zero. We can assume that $|f(t)| \geqslant 1$ for all $t$. Then there exists a partial sum $P$ of the Fourier series of $f$, which has no zero so that $\|P - f\|_{\mathcal{A}} \leqslant 1/3$. Now we build the geometric series

$$S = \sum_{n=1}^{\infty} s_n = \sum_{n=1}^{\infty} \frac{(P-f)^{n-1}}{P^n}$$

and show that $S$ converges in $\mathcal{A}$ to $1/f$:
We have $|P(t) - f(t)| \leqslant 1/3$ for all $t$, and therefore by the triangle inequality also for all $t$

$$|P(t)| \geqslant |f(t)| - |P(t) - f(t)| \geqslant \frac{2}{3}.$$

From this, for $n \in \mathbb{N}$ we get the estimate

$$\max_{t \in [0, 2\pi]} \left| \frac{1}{P^n} \right| \leqslant \left( \frac{3}{2} \right)^n.$$

From $(1/P^n)' = -nP'/P^{n+1}$, it follows with $K = \max |P'|$ that

$$\max_{t \in [0, 2\pi]} \left| \left( \frac{1}{P^n} \right)' \right| \leqslant nK \left( \frac{3}{2} \right)^{n+1}.$$

Consequently, from the preceding preparatory inequality above, it follows that

$$\left\| \frac{1}{P^n} \right\|_{\mathcal{A}} \leqslant (3nK + 1) \left( \frac{3}{2} \right)^n.$$

Furthermore, by the norm inequality we have $\|(P-f)^{n-1}\|_{\mathcal{A}} \leqslant \|P - f\|_{\mathcal{A}}^{n-1} \leqslant \left( \frac{1}{3} \right)^{n-1}$, so that we now obtain for the summands $s_n$ of $S$, again with the norm inequality in $\mathcal{A}$,

$$\|s_n\|_{\mathcal{A}} = \left\| \frac{(P-f)^{n-1}}{P^n} \right\|_{\mathcal{A}} \leqslant \frac{9Kn + 3}{2^n}.$$

Therefore $\sum_{n=1}^{\infty} \|s_n\|_{\mathcal{A}} < \infty$, what can be seen, for example, by the well-known quotient criterion. Hence the series $S$ converges in the norm of $\mathcal{A}$, and by $\|s_n\|_{\infty} \leqslant \|s_n\|_{\mathcal{A}}$ it converges also uniformly (M-Test, p. 21). With $\|(P-f)/P\|_{\mathcal{A}}^n \to 0$ for $n \to \infty$ now, it follows for the geometric series

$$S = \frac{1}{P} \sum_{n=1}^{\infty} \left( \frac{P-f}{P} \right)^{n-1} = \frac{1}{f}.$$

To prove finally the absolute summability of the Fourier coefficients of $1/f$, we denote the $k$th Fourier coefficient of $1/f$ by $c_k(1/f)$ and those of $s_n = (P - f)^{n-1}/P^n$ analogously by $c_k(s_n)$.

The Fourier coefficients of $S$ can be computed with term-by-term integration, because the series converges uniformly. We obtain with $s_n \in \mathcal{A}$ and interchanging the order of summation in the following absolutely convergent double sequence:

$$\left\| \frac{1}{f} \right\|_{\mathcal{A}} = \sum_{k=-\infty}^{+\infty} |c_k(1/f)| = \sum_{k=-\infty}^{+\infty} \left| \sum_{n=1}^{\infty} c_k(s_n) \right| \leqslant \sum_{k=-\infty}^{+\infty} \sum_{n=1}^{\infty} |c_k(s_n)|$$

$$= \sum_{n=1}^{\infty} \sum_{k=-\infty}^{+\infty} |c_k(s_n)| = \sum_{n=1}^{\infty} \|s_n\|_{\mathcal{A}} \leqslant \sum_{n=1}^{\infty} \frac{9Kn+3}{2^n} < \infty.$$

Thus, the assertion of the $1/f$-Theorem is proven. $\qquad\qquad\qquad\qquad\square$

**Remark** The $1/f$-Theorem is also valid for Fourier series of several variables. See Rudin (1991) for that. An analogous result of this type for so-called Dirichlet series can be found in Goodman and Newman (1984).

## 5.7  Exercises

**(A1)** (a) What is the distortion factor of the odd $2\pi$-periodic extension of the function $f(t) = \cos(t)$ for $0 < t < \pi$ ?

    (b) What is the distortion factor of the $2\pi$-periodic rectangle signal $r(t)$ (Fig. 5.8)?

**(A2)** Let $f$ and $g$ be given by Fig. 5.9.

**Fig. 5.8** $2\pi$-periodic rectangle signal

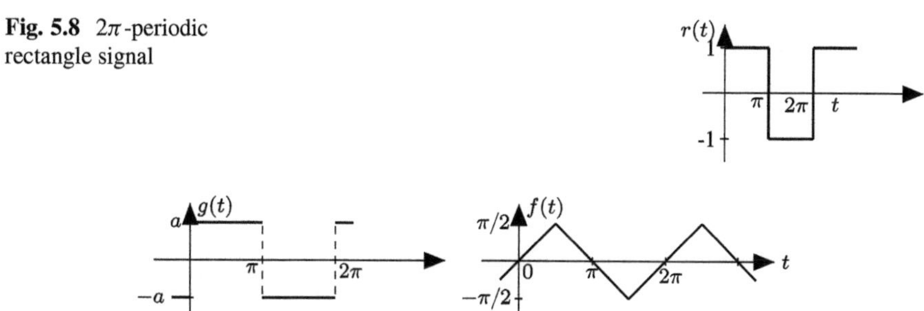

**Fig. 5.9** A rectangle and a triangle signal that shall be convolved

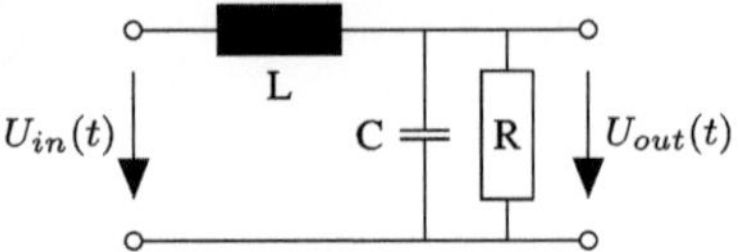

**Fig. 5.10** *RLC* circuit whose asymptotic output voltage for the given input shall be calculated

What is the Fourier series of $(f * g)_{2\pi}$? Is $(f * g)_{2\pi}$ a differentiable function?

**(A3)** What is the $T$-periodic transfer function for the circuit in Fig. 5.10 with inductance $L$, capacitance $C$, and ohmic resistance $R$?

What is for $t \to \infty$ the output voltage $U_{out}(t)$ for $U_{in}(t) = U_0 |\sin(\omega_0 t)|$?

**(A4)** Set $h_k = 1/P(j2\pi k/T)$, $k \in \mathbb{Z}$, for an asymptotically stable differential equation $\sum_{k=0}^{n} a_k u^{(k)} = f$ with the characteristic polynomial $P$.

Show for $n \geqslant 2$ that the coefficients $h_k$ of the $T$-periodic transfer function satisfy for sufficiently large $|k|$ the inequality

$$|h_k| \leqslant M|k|^{-3/2}$$

with a suitable constant $M > 0$ (cf. p. 65).

**(A5)** Show that the $2\pi$-periodic transfer function

$$h(t) = \sum_{k=-\infty}^{+\infty} \frac{e^{jkt}}{jk + a_0}$$

of the equation $u'(t) + a_0 u(t) = f(t)$, $a_0 > 0$, is a $2\pi$-periodic extension of the function $2\pi \, e^{-a_0 t} (1 - e^{-2\pi a_0})^{-1}$ on $]0, 2\pi[$.

Remark: Since $h$ solves the equation for the $2\pi$-periodic impulse sequence

$$f(t) = 2\pi \sum_{k=-\infty}^{+\infty} \delta(t - 2k\pi),$$

which vanishes between two impulses (cf. later Sect. 9.1), the solution in $]0, 2\pi[$ must coincide with a solution of the homogeneous differential equation. The series of $h$ converges uniformly on every closed subinterval of $]0, 2\pi[$ and the one-sided limits for $t \to 0+$ and $t \to 2\pi-$ exist. $h$ is piecewise continuously differentiable. Calculate these limits and the jump height at $t = 0$ (see also p. 65).

**(A6)** What is the Fourier series representation for the potential $u(r, \phi)$ in a circular disk around zero with radius $R$, if the potential on the boundary of the disk is given by

$$u(R, \phi) = \begin{cases} -\frac{U_0}{\pi}(\phi - \pi) & \text{for } 0 < \phi \leqslant \pi \\ \frac{U_0}{\pi}(\phi - \pi) & \text{for } \pi < \phi \leqslant 2\pi \end{cases} \; ?$$

**(A7)** Using the trigonometric addition theorems, transform the Fourier series solution for the problem of the force-free vibrating string, with sufficiently smooth functions $f$ and $g$, into D'Alembert's form

$$u(x, t) = \frac{1}{2}(f(x + ct) + f(x - ct)) + \frac{1}{2c} \int\limits_{x-ct}^{x+ct} g(\tau)d\tau$$

and to

$$u(x, t) = \sum_{n=1}^{\infty} \frac{A_n}{2} \left( \cos \left( \frac{n\pi}{l}(x - ct) - \phi_n \right) - \cos \left( \frac{n\pi}{l}(x + ct) + \phi_n \right) \right).$$

**(A8)** Solve—as Fourier did in 1807—the one-dimensional heat equation from page 1 with thermal diffusivity $k$

$$\frac{\partial u}{\partial t}(x, t) = k \frac{\partial^2 u}{\partial x^2}(x, t) \quad \text{(no external energy input)}$$

$$u(x, 0) = f(x) \qquad \text{(initial temperature distribution } f)$$

$$u(0, t) = u(l, t) = 0 \quad \text{(the bar ends are chilled with ice)}$$

by a separation of the variables approach as for the vibrating string. Assume that all occurring series converge uniformly.

**(A9)**$^\star$ For the *inhomogeneous one-dimensional wave equation*

$$\frac{\partial^2 u}{\partial t^2} = c^2 \frac{\partial^2 u}{\partial x^2} + F(x, t), \qquad u(x, 0) = f(x),$$
$$u(0, t) = u(l, t) = 0, \qquad \lim_{t \to 0+} \frac{\partial u}{\partial t}(x, t) = g(x),$$

one searches for a solution of the form $u = v + w$, where $w$ is the solution of the homogeneous initial boundary value problem as in Sect. 1.2, and $v$ is a solution of the inhomogeneous equation. The function $v$ shall satisfy the boundary conditions $v(0, t) = v(l, t) = 0$, the initial conditions $v(x, 0) = 0$, and $\lim_{t \to 0+} \frac{\partial v}{\partial t}(x, t) = 0$.

Solve the task by the approach $v(x, t) = \sum_{k=1}^{\infty} v_k(t) \sin \left( \frac{k\pi x}{l} \right)$. Use term-by-term differentiation and coefficient comparison with the Fourier series of the inhomogeneous part $F(x, t) = \sum_{k=1}^{\infty} F_k(t) \sin \left( \frac{k\pi x}{l} \right)$,

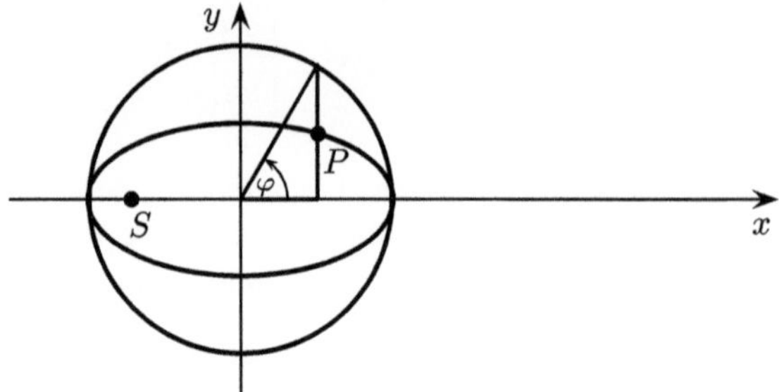

**Fig. 5.11** Sketch of the orbit of planet $P$ around its Sun $S$

$F_k(t) = \frac{2}{l} \int_0^l F(x,t) \sin\left(\frac{k\pi x}{l}\right) dx$. Assume that all occurring series converge uniformly.

**(A10)** Solve the homogeneous equation

$$\frac{\partial^2 u}{\partial t^2} = c^2 \frac{\partial^2 u}{\partial x^2} - 2\kappa \frac{\partial u}{\partial t} \quad \text{with } 0 < \kappa < \frac{\pi c}{l},$$

for the damped vibration of a string with the same initial and boundary conditions as in the preceding exercise.

**(A11)** *Kepler's equation* (J. Kepler 1571–1630) for the elliptic orbit of a planet $P$ is

$$\varphi(t) - \varepsilon \sin(\varphi(t)) = \omega t.$$

Here, $\omega = 2\pi/T$ is the angular frequency with orbital period $T$, $0 \leqslant \varepsilon < 1$ the eccentricity of the ellipse, and $\varphi(t)$ the eccentric anomaly at time $t$ (see Fig. 5.11).

For all $t \in \mathbb{R}$ the following is valid:

$$\dot{\varphi}(t) = \frac{d\varphi}{dt}(t) = \frac{\omega}{1 - \varepsilon \cos(\varphi(t))} > 0.$$

Furthermore $\varphi(0) = 0$ and $\varphi(T) = 2\pi$. Therefore, $\varphi(t)$ is monotonically increasing with $t$ and $\sin(\varphi(t))$ must be an odd function of $t$, due to the motion's symmetry. This motivates the solution approach $\varphi(t) = \omega t + \sum_{k=1}^{\infty} b_k \sin(k\omega t)$. Find the solution, which goes back to J. L. Lagrange and F. W. Bessel, by calculating the Fourier coefficients $b_k$.

# Chapter 6
# Discrete Fourier Transforms, First Applications

**Abstract** This chapter presents the discrete Fourier transform DFT with applications and examples. The alias effect is studied in detail with its disadvantages, but also with its great advantages for low-cost signal processing. The connection of the DFT with interpolation by Chebyshev polynomials is deduced. Further applications worked out are: trigonometric interpolation and interpolation with Chebyshev polynomials. The use of the discrete cosine transform DCT in numerical Clenshaw-Curtis integration is shown as well as the 2D-Cosine transform in image processing like JPEG. The principle of the Fast Fourier Transform FFT is demonstrated with a programmable algorithm. The exercises treat approximation error estimates of trigonometric interpolations, dependent on the number of nodes, DFT frequency assignments, low-cost subsampling, comparison of interpolations on an interval with equidistant nodes versus Chebyshev abscissae. As practice tasks, a Chebyshev lowpass filter can be designed with the help of the Joukowsky transformation, and characteristic values like DC gain, distortion, or RMS value for a transmitter in emitter circuit can be computed with a DFT.

## 6.1 Finite Discrete Fourier Transform (DFT)

The task of approximating signals $f : [0, T] \rightarrow \mathbb{C}$ of finite duration $T$ by superposition of harmonic oscillations is solved by Fourier series expansion. In signal processing practice or numerical integration, however, the signal $f(t)$, $t \in [0, T]$, is often not given as a continuous curve, but only by values $f(t_n)$ at certain equidistant sampling times $t_n = n\Delta t$. From those, a trigonometric polynomial has to be found by which approximate values of $f(t)$ shall be computed for times $t \neq t_n$ as well as approximations for the spectral values $c_k$ of the signal.

We assume that $f : [0, T[ \rightarrow \mathbb{C}$ is continuous and piecewise continuously differentiable and that the limit $f(T-)$ exists for $t \rightarrow T$, $t \in [0, T[$. For $f$ on $[0, T[$, let

$$\mathbf{y} = (y_0, y_1, \ldots, y_{N-1}), \ y_n = f(n\Delta t),$$

be a given sampling vector with $\Delta t > 0$, $n = 0, 1, \ldots, N-1$ and $T = N\Delta t$. The signal $f$ has a piecewise continuously differentiable $T$-periodic extension $f_p$, which then has the sampling sequence $(y_n)_{n \in \mathbb{Z}} = (\ldots, y_{-2}, y_{-1}, y_0, y_1, \ldots, y_{N-1}, y_N, \ldots)$. This sequence is $N$-periodic, i.e., $y_{n+mN} = y_n$, $m \in \mathbb{Z}$, $n = 0, 1, \ldots, N-1$.

*As approximation $\widehat{c}_k$ for the kth Fourier coefficient $c_k$ of $f_p$,*

$$c_k = \frac{1}{T} \int_0^T f_p(t)\, \mathrm{e}^{-jk\omega_0 t}\, \mathrm{d}t \qquad (k \in \mathbb{Z},\ \omega_0 = 2\pi/T),$$

*the following Riemannian sum $\widehat{c}_k$ is chosen with the available samples:*

$$\widehat{c}_k = \frac{1}{T} \sum_{n=0}^{N-1} f(n\Delta t)\, \mathrm{e}^{-jkn\omega_0 \Delta t}\, \Delta t = \frac{1}{N} \sum_{n=0}^{N-1} y_n\, \mathrm{e}^{-jkn2\pi/N}.$$

However, when using these approximations $\widehat{c}_k$ for the spectral values $c_k$, we have to take the following aspects into account:

1. The periodicity of the complex exponential function implies

$$\widehat{c}_k = \widehat{c}_l \quad \text{for all } l = k + mN,\ m \in \mathbb{Z},$$

because for $m \in \mathbb{Z}$ one has:   $\mathrm{e}^{-jkn2\pi/N} = \mathrm{e}^{-j(k+mN)n2\pi/N}$.

Thus, the resulting sequence $(\widehat{c}_k)_{k \in \mathbb{Z}}$ is $N$-periodic. On the other hand, for the Fourier coefficients $c_k$ of $f_p$, we know that $\lim_{|k| \to \infty} c_k = 0$ (cf. Sect. 4.5).

Therefore, we can use at most a segment of length $N$ of this sequence for approximation of $N$ spectral values of the function $f_p$. For the DFT coefficients $\widehat{c}_k$ of real-valued functions $f$, we have $\widehat{c}_k = \overline{\widehat{c}_{N-k}}$, $1 \leqslant k \leqslant N/2$, i.e., for even $N$ the coefficient $\widehat{c}_{N/2}$ is real.

2. With $f_p(n\Delta t) = f(n\Delta t) = \sum_{l=-\infty}^{+\infty} c_l\, \mathrm{e}^{jln2\pi/N}$ for $n \neq 0$ and (cf. Theorem of

Dirichlet, p. 28) $\sum_{l=-\infty}^{+\infty} c_l = \frac{f_p(0)+f_p(T-)}{2}$ we obtain for $k \in \mathbb{Z}$, applying term-by-term summation of the convergent series,

$$\widehat{c}_k = \frac{1}{N}\left( f_p(0) + \sum_{n=1}^{N-1} \sum_{l=-\infty}^{+\infty} c_l\, \mathrm{e}^{jln2\pi/N}\, \mathrm{e}^{-jkn2\pi/N} \right)$$

$$= \sum_{l=-\infty}^{+\infty} c_l \frac{1}{N} \sum_{n=0}^{N-1} \mathrm{e}^{-j(k-l)n2\pi/N} - \frac{1}{N} \sum_{l=-\infty}^{+\infty} c_l + \frac{f_p(0)}{N}.$$

The finite geometric series $\frac{1}{N} \sum_{n=0}^{N-1} e^{-j(k-l)n2\pi/N}$ yields

$$\frac{1}{N} \sum_{n=0}^{N-1} e^{-j(k-l)n2\pi/N} = \begin{cases} 1 & \text{for } l = k + mN, \ m \in \mathbb{Z}, \\ 0 & \text{for } l \neq k + mN, \ m \in \mathbb{Z}. \end{cases}$$

Summarizing, we refer to the result as **Alias Formula**:

$$\widehat{c}_k = \sum_{m=-\infty}^{+\infty} c_{k+mN} + \frac{f_p(0) - f_p(T-)}{2N}.$$

The coefficient $\widehat{c}_k$ contains the sum of all exact Fourier coefficients $c_{k+mN}$ of $f_p$, $m \in \mathbb{Z}$. The corresponding circular oscillations $e^{j(k+mN)\omega_0 t}$ with angular frequencies $(k + mN)\omega_0$ cannot be distinguished on the basis of the samples $f(nT/N)$, because all functions $e^{j(k+mN)\omega_0 t}$ match at all points $nT/N$:

$$e^{jkn2\pi/N} = e^{j(k+mN)n2\pi/N} \quad \text{for all } m \in \mathbb{Z} \text{ and all } n = 0, 1, \ldots, N-1.$$

*This fact is called the "alias effect." The complex amplitudes of all oscillations with angular frequencies $(k + mN)\,\omega_0, m \in \mathbb{Z}$ arbitrary are represented in $\widehat{c}_k$ as sum. If $f_p$ has a jump discontinuity at $t = T$, then the term $(f_p(0) - f_p(T-))/(2N)$ is added to all DFT coefficients. This term vanishes, if one changes the value $f(0)$ to the mean value $(f_p(0+) + f_p(T-))/2$.*

**Example** For $N = 10$ and $\omega_0 = 2\pi/T$, $T = 1$ s, the following Fig. 6.1 shows as example that the oscillations $f_1(t) = \sin(4\omega_0 t)$ and $f_2(t) = \sin(14\omega_0 t)$ are indistinguishable on the basis of 10 samples at the times $t_k = kT/N$, $k = 0, \ldots, 9$. A DFT of $f_1 + f_2$ with these samples results in $\widehat{c}_4 = -j$ and $\widehat{c}_6 = j$ as if the DFT were of $2f_1$.

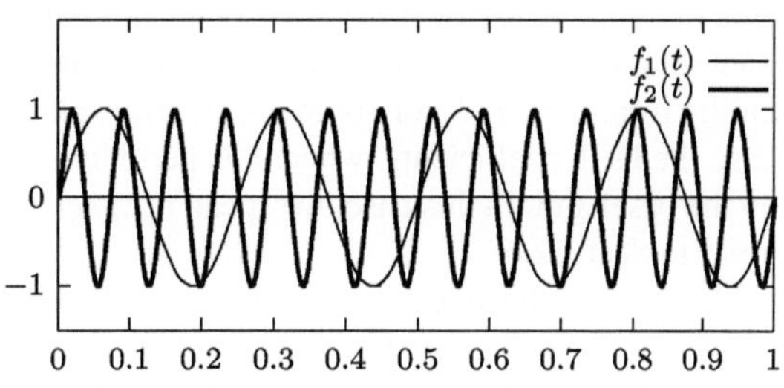


**Fig. 6.1** Illustration on the alias effect

## *Consequences for Applications of the DFT*

**Frequency Assignment in the Baseband** By assigning the DFT coefficients $\widehat{c}_k$ and $\widehat{c}_{N-k}$ to oscillations in the *baseband* with frequencies $k/T$, $-N/(2T) \leqslant k/T \leqslant N/(2T)$, and *bandwidth* $B = N/(2T)$, one chooses the frequencies of smallest magnitude that are possible according to the alias relation. For real-valued signals, the symmetry of the frequency band in both semiaxes makes sense, due to $\widehat{c}_k = \overline{\widehat{c}_{N-k}}$.

**Example** For $N = 15$ and $T$-periodic $f_p$ with Fourier coefficients $c_k$, $k \in \mathbb{Z}$,

$(\widehat{c}_0, \widehat{c}_1, \ldots, \widehat{c}_7)$ serves in the baseband as approximation for $(c_0, c_1, \ldots, c_7)$,

$(\widehat{c}_8, \widehat{c}_9, \ldots, \widehat{c}_{14})$ serves in the baseband as approximation for $(c_{-7}, c_{-6}, \ldots, c_{-1})$.

For $N = 14$ and $T$-periodic $f_p$ with Fourier coefficients $c_k$, $k \in \mathbb{Z}$,

$(\widehat{c}_0, \widehat{c}_1, \ldots, \widehat{c}_6)$ serves in the baseband as approximation for $(c_0, c_1, \ldots, c_6)$,

$(\widehat{c}_8, \widehat{c}_9, \ldots, \widehat{c}_{13})$ serves in the baseband as approximation for $(c_{-6}, c_{-5} \ldots, c_{-1})$.

Since $N$ is even, the coefficient $\widehat{c}_7 = \widehat{c}_{N/2}$ serves as approximation for the amplitude of the oscillation $\cos(2\pi t N/(2T))$. For real-valued $f_p$, the coefficient $\widehat{c}_{N/2}$ is real and otherwise $\widehat{c}_k = \overline{\widehat{c}_{N-k}}$.

Frequency mappings in the baseband as above are convenient for $T$-periodic signals, which are several times differentiable, if their Fourier coefficients $c_k$ decay rapidly at higher angular frequencies $|(k + mN)\omega_0|$ $(\omega_0 = 2\pi/T)$ (cf. Sect. 4.5 on the asymptotic behavior of $c_k$). Then one can take $\widehat{c}_k$ with $|k| \leqslant N/2$ as useful approximation for $c_k$. For $T$-periodic *baseband signals*, i.e., with frequencies only in the baseband, the DFT coefficients reveal the exact Fourier coefficients. Such signals are trigonometric polynomials and can be exactly reconstructed with the DFT. Signal frequencies outside the baseband produce alias effects. A disadvantage of the alias effect for detecting periodic oscillations of a frequency $\nu$ in a specified baseband is the requirement that the sampling frequency must be at least $2\nu$. In practice, one uses lowpass filters with the band limit as cutoff frequency before sampling to mitigate alias effects. Signal components of $f$ with frequencies $\nu$ not of the form $\nu = n/T$ for some $n \in \mathbb{Z}$ affect all Fourier coefficients of $f_p$, and thus also the DFT coefficients, when $f$ is periodized to $f_p$, even if such frequencies are in the baseband. In this case, we speak of *spectral leakage*. For more on that, please refer to Sect. 12.6.

The better the frequency resolution $1/T$ of a DFT should be, the longer the sampling time $T$ must be. The number $N$ of the DFT samples determines the bandwidth $B$ for a given $T$. The frequency band $[0, B]$ is also called *Nyquist interval*, and $B$ also *Nyquist frequency*.

Our human senses likewise make an assignment in the specified baseband in the case of visual impressions. The human eye can perceive image sequences in a video with 24 frames per second—a frame rate in movies—as a continuous process. Optical illusions, such as rotating wheels appearing to rotate slowly in the opposite direction of motion in a film, are alias effects: We assign the alias frequency with the lowest magnitude matching the sampled values. This, associated with a phase change, can cause the impression of a slower motion opposite to the actual rotation (*Stroboscope effect*, see Example 2 on p. 93).

**Remark on DFT Scaling Factors**  Since different scaling factors are used in software for a DFT, it should be noted that the correct values of the complex amplitudes of analyzed harmonic oscillations are obtained only with the prefactor $1/N$ in the definition of the DFT coefficients $\widehat{c_k}$.

We can visualize the alias effect and the possible mappings between DFT coefficients and spectral values of $T$-periodic functions, if we extend the DFT spectrum $N$-periodically. The following figure shows two possible correspondences in the baseband of bandwidth $B = N/(2T)$. $T$ is the duration of an $N$-point DFT of a real-valued function with spectral values at frequencies $\nu$, $|\nu| \in {]}3B\,,\,4B{[}$. There is a *phase reversal* when the frequency of an alias oscillation changes its sign. This is the case when an alias frequency lies in a half-band of the form $[mB, (m+1)B]$ or $[-(m+1)B, -mB]$ for odd $m \in \mathbb{N}$.

**Example**  A DFT is performed for the 8 Hz oscillation

$$f(t) = 2\sin(32\,\omega_0 t)$$

with $T = 4\,\mathrm{s}$, $N = 20$ and $\omega_0 = 2\pi/T$. With $32 - N = 12$, $-32 + 2N = 8$, the DFT coefficients $\widehat{c}_8$ and $\widehat{c}_{12}$ are nonzero. We find

$$\widehat{c}_8 = j, \ \widehat{c}_{12} = -j = \overline{\widehat{c}_8}.$$

In the baseband $[-2.5\,\mathrm{Hz},\ 2.5\,\mathrm{Hz}]$, $f$ corresponds to $f_a(t) = -2\sin(8\omega_0 t)$ as alias with frequency 2 Hz and *phase reversal* compared to $f$ (cf. the following Fig. 6.2 with 8 Hz in $[\,3B,\ 4B\,]$).

**Remark**  We will later see in Sect. 12.2 that signal sampling always generates a periodic spectrum. Therefore the observations on aliasing here apply to every sampling scenario accordingly.


## *Alias Effect and Frequency Assignment with Undersampling*

We have seen that the bandwidth of a segment of the spectrum of $f_p$ is determined by $N$ and $T$ and thus a segment whose spectrum is representable by a DFT without aliasing. Not a priori determined is the position of such a spectral part on the frequency axis. Its position can be determined from a priori knowledge

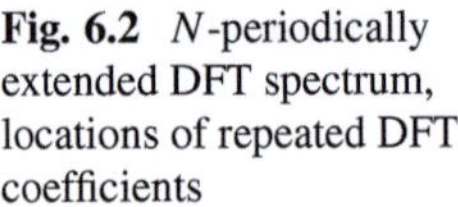

**Fig. 6.2** $N$-periodically extended DFT spectrum, locations of repeated DFT coefficients

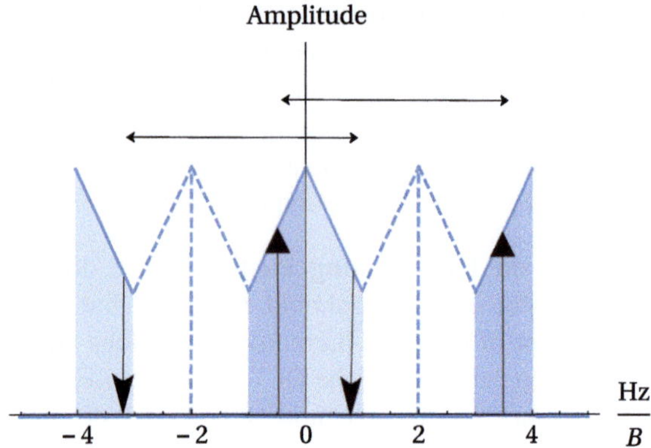


or deliberately. This has disadvantages in observing unknown signals, but also enormous advantages in signal processing for technical systems as for example in communications engineering, because there the signals and the allocation of signal frequencies in the spectrum can be chosen intentionally.

Again, we assume that a DFT with $N$ values and sampling frequency $N/T = 2B$ is given for a function $f$ as at the beginning of the section, and $f$ is $T$-periodically extended to $f_p$.

While the above alias formula shows how amplitudes of real signals with frequencies above N/(2T) appear as aliases in the DFT, in frequency detection problems we want to detect signal components $c_l\, e^{j2\pi lt/T}$ in frequency bands of the form $\frac{mN}{2T} \leqslant \frac{l}{T} \leqslant \frac{(m+1)N}{2T}$ or $-\frac{(m+1)N}{2T} \leqslant \frac{l}{T} \leqslant -\frac{mN}{2T}$ with $m \geqslant 1$.

The assignment of DFT coefficients to frequencies in a band $[-(m+1)B\,,\, -mB\,]$ $\cup\ [\,mB\,,\, (m+1)B\,]$, $B = N/(2T)$, $m \in \mathbb{N}$, is useful if you know that the sampled signal has frequencies only in the selected spectral range. If the signal is also $T$-periodic, then it is a trigonometric polynomial, which can be reconstructed exactly from the DFT, although the DFT—measured with the Nyquist frequency of the baseband—has a too low sampling rate. This is called *subsampling* or *undersampling in a passband* or *bandpass sampling*.

In practice a one-to-one mapping of signal components in those half-bands to DFT coefficients and their respective signal frequencies could be done by amplitude modulation (cf. p. 45) and subsequent sampling or equivalently by an appropriate undersampling (cf. Examples 3 and 4 on p. 93). Therefore, we ask for the correspondence between DFT coefficients $\widehat{c}_k$ to unique circular waves $c_l\, e^{j2\pi lt/T}$ in these spectral bands.

In other words, we want to know the mapping of these half-bands to the baseband through rephrasing the alias relations. The correspondences $\widehat{c}_k \longleftrightarrow c_l$ are given in the following theorem, whose statements follow directly from the alias formula.

**Theorem 6.1**

*1. For each $m \in \mathbb{N}_0$ and each $k$, $1 \leqslant k < N/2$, there is a unique circular wave $c_l\, e^{j2\pi lt/T}$ in the Fourier series of $f_p$ with $\frac{mN}{2} < l < \frac{(m+1)N}{2}$, whose complex* *amplitude corresponds to $\widehat{c}_k$ for even $m$ or $\widehat{c}_{N-k}$ for odd $m$ according to the alias formula.*

2. *There is a unique oscillation $c_l\, e^{j2\pi l t/T}$ with $-\frac{(m+1)N}{2} < l < -\frac{mN}{2}$, whose amplitude corresponds to $\widehat{c}_k$ for odd $m$ or $\widehat{c}_{N-k}$ for even $m$.*

*For a selected $m$ the correspondences between $\widehat{c}_k$, $\widehat{c}_{N-k}$, and $c_l$ are as follows:*

$$m \text{ even}:$$
$$\widehat{c}_k \longleftrightarrow c_l \text{ with } l = k + \tfrac{mN}{2}, \qquad \text{and } \widehat{c}_{N-k} \longleftrightarrow c_l \text{ with } l = -k - \tfrac{mN}{2}.$$
$$m \text{ odd}:$$
$$\widehat{c}_k \longleftrightarrow c_l \text{ with } l = k - \tfrac{(m+1)N}{2}, \text{ and } \widehat{c}_{N-k} \longleftrightarrow c_l \text{ with } l = -k + \tfrac{(m+1)N}{2}.$$

3. *For $m \in \mathbb{N}_0$, $k = 0$ and $k = N/2$ at even $N$, and for the band-edge frequencies $mN/(2T)$ and $(m+1)N/(2T)$, the following statements are valid with $\omega_0 = 2\pi/T$:*

(a) *If $m$ is even, then the complex amplitudes of $e^{j\omega_0 t mN/2}$ and $e^{-j\omega_0 t mN/2}$ as parts of $f_p$ are added to $\widehat{c}_0$. If $N$ is also even, then the complex amplitudes of $e^{j\omega_0 t(m+1)N/2}$ and $e^{-j\omega_0 t(m+1)N/2}$ as parts of $f_p$ are added to $\widehat{c}_{N/2}$.*

(b) *If $m$ is odd, then the complex amplitudes of $e^{j\omega_0 t(m+1)N/2}$ and of $e^{-j\omega_0 t(m+1)N/2}$ as parts of $f_p$ are added to $\widehat{c}_0$. If $m$ is odd and $N$ is even, then the complex amplitudes of $e^{j\omega_0 t mN/2}$ and $e^{-j\omega_0 t mN/2}$ are added to $\widehat{c}_{N/2}$.*

Since $\sin(\omega_0 t mN/2)$ always yields zero at the sampling points of the considered DFT for all $m \in \mathbb{Z}$, only the complex amplitudes of the cosine parts in the circular waves $e^{\pm j\omega_0 t mN/2} = \cos(\omega_0 t mN/2) \pm j\sin(\omega_0 t mN/2)$, as considered in 2.a) and 2.b) above, contribute to $\widehat{c}_0$ or $\widehat{c}_{N/2}$. If $m$ and $N$ are both odd, then $mN/(2T) \neq n/T$ for all $n \in \mathbb{Z}$, i.e., these are not frequencies in the Fourier series of the $T$-periodic extension $f_p$ of $f$. The effect of such frequencies in the signal $f$ on indeed all Fourier coefficients of $f_p$, and thus also on the DFT, is discussed later under the keyword "*leakage effect*" in Sect. 12.6.

The following Fig. 6.3 schematically illustrates the correspondences of DFT coefficients to high-frequency signal parts as given in the theorem before with $m = 3$.

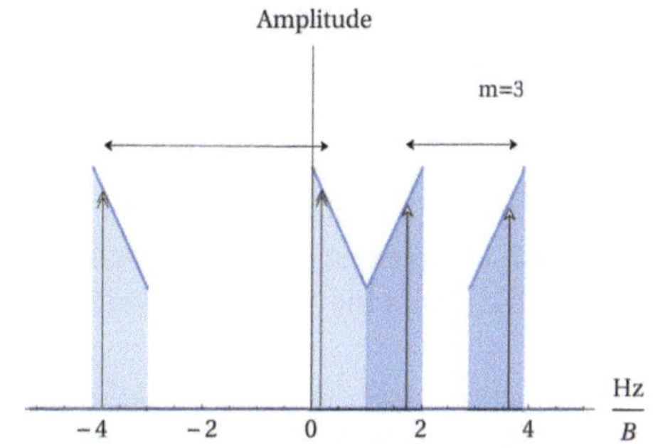


**Fig. 6.3** Undersampling shifts by aliasing high-frequency bands into the range of the DFT spectrum

**Important Observation** The theorem shows an enormous advantage of the alias effect for applications in signal processing. For the purpose of detecting signal components in a high-frequency band, the signal to be observed is subjected to bandpass filtering before sampling, i.e., only signal components with frequencies in the selected subband are permitted to pass the filter. This permits to detect signal components with very high frequencies by a DFT with only a few samples and short observation times $T$ (cf. Example 3, p. 93). That is a key feature in modern transmissions of high-frequency signals.

In digital communications, high transmission frequencies—in the range of several GHz for WLAN or LTE—are used with much lower CPU clocks of the digital end-user devices and limited memory capacity in real-time operation. The transmitted signal consists of good approximation of time segments of trigonometric polynomials with mutually orthogonal components and known constant frequency spacings in a given high-frequency band. The complex amplitudes of the oscillations carry the encoded user information. The associated oscillations are therefore also called *carriers*. Examples are OFDM transmissions (*Orthogonal Frequency Division Multiplexing*, see later Sect. 12.2), applying, e.g., 64-QAM modulation and 40 MHz bandwidth for WLAN according to IEEE802.11n.

A DFT with bandpass sampling makes it possible to reconstruct the amplitudes, required for the user information at the receiver, with a low sampling rate— depending only on the bandwidth and the spacing of the carrier frequencies. Trigonometric polynomials in a passband (with large $m$ in the last theorem) are mapped by subsampling to a signal of the same bandwidth and the same amplitude distribution which lies in another low-frequency alias band (small $m$ in the last theorem). One simply reads the DFT coefficients as amplitudes of carriers with alias frequencies in the chosen band. For carrier frequencies in a high-frequency passband, the analog-to-digital conversion (ADC) thus can save enormous cost and energy, compared with alternative amplitude modulations using mixers, by undersampling matching the bandwidth and carrier frequencies[1] (cf. Example 4, p. 93). Frequency assignments as in the theorem on p. 90 can be understood as an amplitude modulation into the baseband, by a simple mathematical operation without additional hardware in practice. Readers with interest in digital communications should consult the textbooks of Proakis and Salehi (2013) or Tietze and Schenk (2008).

**Examples  (Frequency Assignments)**

1. *Alias Effect in the Baseband.* Let us assume that we observe a real signal which is a superposition of oscillations with frequencies up to a bandwidth of 400 Hz. A DFT performed over $T = 2$ seconds with $N = 512$ points is assumed to show only the DFT coefficients $\widehat{c}_{80}$ and $\widehat{c}_{432}$ as nonzero.

   All complex amplitudes of signal components with frequencies $|k + mN|/T$, $m \in \mathbb{Z}$, within the given bandwidth up to $B = 400$ Hz are added in the two DFT

---

[1] Check, for example, the specification of the ADC12DL040/65 of Texas Instruments.

coefficients $\widehat{c}_k$ and $\widehat{c}_{-k+N}$. This means that an oscillation with $|k + mN|/T$ Hz can contribute to these coefficients as long as $|k + mN| \leqslant BT$. With $k = 80$, $(80 + 512)/2 = 296$, and $|(80 - 512)/2| = 216$ in the example, besides the oscillation with 40 Hz, also 216 Hz and 296 Hz are possible signal frequencies affecting $\widehat{c}_{80}$ and $\widehat{c}_{432}$.

All functions with the same samples have the same DFT spectrum. The true spectra of this variety of conceivable functions can be very different but are indistinguishable from the DFT coefficients without additional information. In signal processing one tries to avoid unwanted alias effects by using bandpass filters and weight functions, the so-called window functions. For more details on this, we refer again to the later Sect. 12.6 in Chap. 12.

2. *Stroboscope Effect.* We consider the DFT of the two complex-valued functions

$$f_1(t) = e^{j20\omega_0 t} \text{ and } f_2(t) = e^{-j4\omega_0 t} \text{ with } N = 24, T = 1\,\text{s and } \omega_0 = 2\pi/T.$$

They represent opposite circular motions. The DFT of $f_1$ yields $\widehat{c}_{20} = 1, \widehat{c}_k = 0$ for $k \neq 20$. In the baseband with bandwidth $B = 12$ Hz, the alias for $f_1$ is the slower opposite rotation $e^{-j4\omega_0 t} = f_2(t)$.

Since for the Fourier coefficients $c_k$ of complex-valued periodic functions, the equations $c_k = \overline{c_{-k}}$ need not be true, we can dispense with the symmetry of the baseband in both semiaxes, if we know that only complex circular waves with positive frequencies are sampled. Choosing the interval $[0, N/T[$ for frequency assignments in the example and assigning $\widehat{c}_{20}$ to $\widehat{c}_{20} e^{j20\omega_0 t}$, we obtain the observed rotation $f_1$, which in turn appears as an alias of $f_2$. Rotations $e^{jnN\omega_0 t}$, $n \in \mathbb{Z}$, result with the above DFT in the point 1 at rest as alias. The function $f_3(t) = e^{j25\omega_0 t}$ in turn has the slower rotation $e^{j\omega_0 t}$ of the same direction as alias.

3. *High-Frequency Detection with Undersampling.* Let us assume that a DFT with $N = 512$ points, duration $T = 0.256 \times 10^{-3}$s is sampling a real signal $f$ in the frequency band ]1GHz, 1GHz + 1MHz[, and yields only the DFT coefficients $\widehat{c}_{160} = \widehat{c}_{352}$ and $\widehat{c}_{164} = \widehat{c}_{348}$ as nonzero.

With $m = 1000$, $mN/(2T) = 1$ GHz, $N/(2T) = 1$ MHz, $\nu_1 = 160/T = 625$ kHz, and $\nu_2 = 164/T = 640625$ Hz, it must be true that $f(t) = 2\widehat{c}_{160}\cos(2\pi(1\,\text{GHz} + \nu_1)t) + 2\widehat{c}_{164}\cos(2\pi(1\,\text{GHz} + \nu_2)t)$. The trigonometric alias polynomial in the baseband is $2\widehat{c}_{160}\cos(2\pi\nu_1 t) + 2\widehat{c}_{164}\cos(2\pi\nu_2 t)$.

4. *Gain in Computational Effort by Undersampling in the Radio-Frequency Band.* Assume we have signals in the radio-frequency band $FM$ from 87.5 MHz to 108 MHz], which shall be digitally processed. Sampling with 216 MHz according to the Nyquist frequency would require an anti-alias lowpass filter with cutoff frequency 108 MHz and yield a data stream of $2 \cdot 216 = 432$ MB/s from a 16 Bit ADC to the signal processing unit. Undersampling with sampling frequency $f_s = 43.5$ MHz shifts the signal spectrum to [0.5 MHz, 21 MHz]. This would result in a data stream of only 87 MB/s for further signal processing, which is a gain of about 80 % in computation time, compared to 432 MB/s, without the need of an (costly) analogue mixer.

5. *Delayed Sampling, Correction in the Spectrum of Trigonometric Polynomials.*
   We consider a sampling of $f(t) = 2\,e^{j\omega_0 t} + (1+j)\,e^{j2\omega_0 t} + (1-j)\,e^{j3\omega_0 t}$ with
   $T = 1\,\text{s}$, $\omega_0 = 2\pi/T$. Assume that the sampling times are $t_n = nT/N + 0.1\,\text{s}$
   with $N = 4$ and $n = 0, \dots, N-1$. The DFT spectrum of the delayed sampling
   with the "synchronization error" $\Delta t = 0.1\,\text{s}$ is

$$(\tilde{c}_0, \dots, \tilde{c}_3) = (0,\ 1.6180 + 1.1755j,\ -0.6420 + 1.2600j,\ 0.6420 + 1.2600j).$$

Since $f$ is a $T$-periodic trigonometric polynomial with frequencies only in the
passband $[0,\ N/T[$, the DFT coefficients $\widehat{c}_k$ of $f$ are simply phase-shifted
toward $\tilde{c}_k = \widehat{c}_k z^k$, $z = e^{j\omega_0 \Delta t}$, due to the delayed sampling (cf. p. 44).

If the amplitude $A$ of a "pilot carrier" is known, here for example $A = 2$
for the carrier frequency 1 Hz, one can recognize the phase shifts from the
obtained DFT coefficient of this carrier and correct the entire DFT spectrum.
In the example, the products $\tilde{c}_k z^{-k}$ with $z = \tilde{c}_1/A$, $k = 0, \dots, 3$ yield the true
spectrum $(0,\ 2,\ 1+j,\ 1-j)$ of $f$ in the frequency band up to 3 Hz.

Of course, from $z = e^{j\omega_0 \Delta t}$ the time delay $\Delta t = \arg(z)/\omega_0$ can be
calculated. In a transmit-receive scenario, where the amplitudes of trigonometric
polynomials represent the encoded information in a suitably chosen frequency
band, the use of known amplitudes on known carriers (preambles and pilot
symbols) is standard in transmissions such as DAB, DVB-T, DSL, WLAN, LTE,
etc. They are used for synchronization and generally for channel estimation.

6. *Limits of Special Series.* We had already seen in example 2 on p. 35 that
   sometimes limits of series can be found if the series elements are Fourier
   coefficients of a known periodic function. Also the alias relation for a coefficient
   $\widehat{c}_k$ of a DFT of length $N$ permits this, if the coefficients of a series $\sum\limits_{m=1}^{\infty} a_m$ are of
   the form $a_m = c_{k+mN} + c_{k-mN}$ and the Fourier coefficients $c_{k\pm mN}$ result from a
   function with known necessary samples and spectrum.

   We choose as an example the Fourier series of the 2-periodic extension of
   $f(t) = -t+1$ on $[0, 1]$ and $f(t) = 0$ on $[1, 2[$. We know from example 1 on p. 33
   the Fourier coefficients $c_0 = 1/4$ and $c_k = ((-1)^{k+1} + 1)/(2\pi^2 k^2) - j/(2\pi k)$
   for $k \neq 0$. There, the Fourier coefficients must be multiplied by $(-1)^k$, due to
   the right shift.
   A 3-point DFT with $T = 2$ yields $\widehat{c}_0 = 4/9$ and $\widehat{c}_1 = 5/18 - j\sqrt{3}/18$. With the
   alias effect for the coefficient $\widehat{c}_0$ one immediately calculates

$$\sum_{k=1}^{\infty} \frac{1}{(6k-3)^2} = \left(\widehat{c}_0 - c_0 - \frac{1}{6}\right)\frac{\pi^2}{2} = \frac{\pi^2}{72}.$$

From $\Re(\widehat{c}_1) = \frac{1}{6} + \sum\limits_{k=1}^{\infty} \Re(c_{6k-5} + c_{1-6k})$ and the according equation for the
imaginary part, left to the reader, we obtain the two limits

$$\sum_{k=1}^{\infty} \left[ \frac{1}{(6k-5)^2} + \frac{1}{(1-6k)^2} \right] = \frac{\pi^2}{9}, \quad \sum_{k=1}^{\infty} \frac{1}{(3k-2)(1-3k)} = -\frac{\sqrt{3}\pi}{9}.$$

A systematic theory for the calculation of series limits uses the residue theorem, hypergeometric summation, or special functions like the polygamma function $\Psi(n, z) = \frac{d^{n+1}}{dz^{n+1}} \ln \Gamma(z)$. We find

$$\sum_{k=1}^{\infty} \left[ \frac{1}{(6k-5)^2} + \frac{1}{(1-6k)^2} \right] = \frac{\Psi(1, 1/6) + \Psi(1, 5/6)}{36} = \frac{\pi^2}{9}.$$

More details can be found in the work of Grosjean (1984) and of Choi and Cvijović (2010) on specific values of the polygamma function. Summation algorithms in computer algebra systems are discussed in the textbook of Koepf (1998).

We now show that the DFT has an inverse, which is called IDFT.

**Definition**  The linear transform

$$(y_0, y_1, \ldots, y_{N-1}) \longrightarrow (\widehat{c}_0, \widehat{c}_1, \ldots, \widehat{c}_{N-1}),$$

$$\widehat{c}_k = \frac{1}{N} \sum_{n=0}^{N-1} y_n \, e^{-jkn2\pi/N}$$

is called finite discrete Fourier transform or in short **DFT**. The coefficients $\widehat{c}_k$ are uniquely determined for $k = 0, 1, \ldots, N-1$ by the samples $y_0, \ldots, y_{N-1}$ and are called DFT coefficients of $\mathbf{y} = (y_0, \ldots, y_{N-1})$.

## *The Inverse Discrete Fourier Transform (IDFT)*

Conversely, by the vector $(\widehat{c}_0, \widehat{c}_1, \ldots, \widehat{c}_{N-1})$ exactly one vector $(y_0, y_1, \ldots, y_{N-1})$ is determined, whose DFT coefficients are the $\widehat{c}_k$:

For $n = 0, \ldots, N-1$ and $k = 0, \ldots, N-1$ we obtain with $y_n = \sum_{k=0}^{N-1} \widehat{c}_k \, e^{jkn2\pi/N}$

$$\frac{1}{N} \sum_{n=0}^{N-1} y_n \, e^{-jkn2\pi/N} = \frac{1}{N} \sum_{n=0}^{N-1} \sum_{l=0}^{N-1} \widehat{c}_l \, e^{jln2\pi/N} \, e^{-jkn2\pi/N}$$

$$= \sum_{l=0}^{N-1} \widehat{c}_l \underbrace{\frac{1}{N} \sum_{n=0}^{N-1} e^{-j(k-l)n2\pi/N}}_{1 \text{ for } l=k, \ 0 \text{ otherwise (cf. 2. on p. 87)}} = \widehat{c}_k.$$

**Theorem 6.2** *The mapping* $(y_0, y_1, \ldots, y_{N-1}) \longrightarrow^{\mathrm{DFT}} (\widehat{c}_0, \widehat{c}_1, \ldots, \widehat{c}_{N-1})$ *is linearly one to one on* $\mathbb{C}^N$. *Its inverse mapping is called inverse finite discrete Fourier transform, in short* IDFT.

$$\mathbf{DFT}: \quad \widehat{c}_k = \frac{1}{N} \sum_{n=0}^{N-1} y_n \, \mathrm{e}^{-jkn2\pi/N} \qquad (k = 0, \ldots, N-1)$$

$$\mathbf{IDFT}: \quad y_n = \sum_{k=0}^{N-1} \widehat{c}_k \, \mathrm{e}^{jkn2\pi/N} \qquad (n = 0, \ldots, N-1).$$

## *Properties and Calculation Rules for the Discrete Fourier Transform*

Now, let $\mathbf{y}, \mathbf{x} \in \mathbb{C}^N$ be given vectors and $\mathbf{c}, \mathbf{d} \in \mathbb{C}^N$ the vectors of their corresponding DFT coefficients. For computational purposes, one extends these vectors in $\mathbb{C}^N$ to $N$-periodic sequences so that $y_{n+mN} = y_n$ for $m \in \mathbb{Z}$, $n = 0, 1, \ldots, N-1$. Then the DFT is a bijective *linear* map on the vector space of all $N$-periodic complex sequences with

$$\widehat{c}_k = \frac{1}{N} \sum_{n=0}^{N-1} y_n z^{-kn}, \quad k = 0, \ldots, (N-1), \quad z = \mathrm{e}^{j2\pi/N} .$$

We obtain analogous calculation rules as for Fourier series. The most important rules are summarized in the following Table 6.1.

Here, we only prove the convolution relation as an example. With $z = \mathrm{e}^{j2\pi/N}$ it follows for the $m$-th DFT coefficient of the cyclic convolution $(y * x)_{n \in \mathbb{Z}}$—in this text with the same prefactor $1/N$ as in the DFT—by interchanging the order of summation:

$$\frac{1}{N} \sum_{n=0}^{N-1} \left( \frac{1}{N} \sum_{k=0}^{N-1} x_k y_{n-k} \right) z^{-nm} = \frac{1}{N} \sum_{k=0}^{N-1} \left( \frac{1}{N} \sum_{n=0}^{N-1} x_k y_{n-k} z^{-nm} \right)$$

$$= \left( \frac{1}{N} \sum_{k=0}^{N-1} x_k z^{-mk} \right) \left( \frac{1}{N} \sum_{n=0}^{N-1} y_{n-k} z^{-m(n-k)} \right) = \widehat{d}_m \widehat{c}_m .$$

In the last line it is used that the sequence $(y_n)_{n \in \mathbb{Z}}$ is $N$-periodic. Note again the alias effect. Readers are encouraged to prove the remaining relations of the table and Exercise A21 by themselves.

**Table 6.1** Properties of Fourier series and the DFT compared

| Fourier series | DFT | |
| --- | --- | --- |
| Time-domain function | Samples | Spectral values |
| $f(t)$  $T$-periodic $$f(t) = \sum_{k=-\infty}^{+\infty} c_k\, \mathrm{e}^{jk\omega_0 t}, \quad \omega_0 = \tfrac{2\pi}{T}$$ $$c_k = \tfrac{1}{T} \int_0^T f(t)\, \mathrm{e}^{-jk\omega_0 t}\, dt$$ | $y_n = f(nT/N)$   $N$-periodic $$y_n = \sum_{k=0}^{N-1} \widehat{c}_k\, \mathrm{e}^{jkn2\pi/N}$$ | $\widehat{c}_k =$ $$\tfrac{1}{N} \sum_{n=0}^{N-1} y_n\, \mathrm{e}^{-jkn2\pi/N}$$ |
| Similarity $$f(\alpha t) = \sum_{k=-\infty}^{+\infty} c_k\, \mathrm{e}^{jk\alpha\omega_0 t}$$ $T/\alpha$-periodic, $\alpha > 0$ | $y_n = f(nT/(\alpha N))$ otherwise as above | as above |
| Translations, amplitude modulation $$f(t + t_0) = \sum_{k=-\infty}^{+\infty} c_k\, \mathrm{e}^{jk\omega_0(t+t_0)}$$ $$\mathrm{e}^{jm\omega_0 t}\, f(t) = \sum_{k=-\infty}^{+\infty} c_{k-m}\, \mathrm{e}^{jk\omega_0 t}$$ | $(y_{n+m})_{n\in\mathbb{Z}}, \quad (m \in \mathbb{Z})$ $(z^{nm} y_n)_{n\in\mathbb{Z}}$ | $(z^{km}\widehat{c}_k)_{k\in\mathbb{Z}}, z = \mathrm{e}^{j2\pi/N}$ $(\widehat{c}_{k-m})_{k\in\mathbb{Z}}$ |
| $T$-periodic convolution For $$f(t) = \sum_{k=-\infty}^{+\infty} c_k\, \mathrm{e}^{jk\omega_0 t}$$ $$g(t) = \sum_{k=-\infty}^{+\infty} d_k\, \mathrm{e}^{jk\omega_0 t}$$ $$(f * g)_T(t) = \sum_{k=-\infty}^{+\infty} c_k d_k\, \mathrm{e}^{jk\omega_0 t}$$ | $N$-periodic convolution For $y_n = f(nT/N)$ $x_n = g(nT/N)$ $$(y * x)_{n\in\mathbb{Z}} = \tfrac{1}{N} \sum_{m=0}^{N-1} x_m y_{n-m}$$ | with DFT coefficients $(\widehat{c}_k)_{k\in\mathbb{Z}}$ $(\widehat{d}_k)_{k\in\mathbb{Z}}$ $(\widehat{c}_k\widehat{d}_k)_{k\in\mathbb{Z}}$ |
| Parseval equality $$\|f\|^2 = \sum_{k=-\infty}^{+\infty} |c_k|^2$$ | $$\tfrac{1}{N} \sum_{n=0}^{N-1} |y_n|^2 = \sum_{k=0}^{N-1} |\widehat{c}_k|^2$$ | |

## 6.2  Trigonometric Interpolation

Let be given a sample vector $(y_0, y_1, \ldots, y_{N-1})$ of a continuous function $f$ on $[0, T]$, $T > 0$, $y_n = f(n\Delta t)$, $N\Delta t = T$. We ask for a trigonometric polynomial $P(t) = \sum_{k=-m}^{m} \alpha_k\, \mathrm{e}^{jk\omega_0 t}$ with $\omega_0 = 2\pi/T$ of degree at most $m \leqslant N/2$ so that $P(n\Delta t) = f(n\Delta t)$ at the equidistant sampling points $n\Delta t$ for $n = 0, \ldots, N - 1$. To accomplish this we must find $2m + 1$ coefficients for $P$.

*Trigonometric Interpolation with an Odd Number of Samples*
If the number $N$ of the interpolation points is odd, $N = 2m + 1$, then the trigonometric interpolation polynomial is uniquely determined. Because a trigonometric polynomial $P(t) \neq 0$ of degree $m$ has at most $2m$ zeros per period (cf. p. 14), each two such polynomials which coincide at $2m + 1$ points per period are already identical.

By construction of the DFT and the IDFT, $Q(t) = \sum\limits_{k=0}^{N-1} \widehat{c}_k \, e^{jk\omega_0 t}$ with the DFT coefficients $\widehat{c}_k$ of $(y_0, \ldots, y_{N-1})$ is a trigonometric interpolation polynomial. Since the functions $e^{jk\omega_0 t}$ and $e^{j(k+N)\omega_0 t}$ for $k \in \mathbb{Z}$ match at all points $n\Delta t$ and the sequence $(\widehat{c}_k)_{k\in\mathbb{Z}}$ is $N$-periodic, it follows immediately

$$Q\left(n\frac{T}{N}\right) = \sum_{k=0}^{2m} \widehat{c}_k \, e^{jkn\frac{2\pi}{N}} = \sum_{k=0}^{m} \widehat{c}_k \, e^{jkn\frac{2\pi}{N}} + \sum_{k=-m}^{-1} \widehat{c}_{k+N} \, e^{j(k+N)n\frac{2\pi}{N}} = \sum_{k=-m}^{m} \widehat{c}_k \, e^{jkn\frac{2\pi}{N}}.$$

**Theorem 6.3** *The uniquely determined trigonometric interpolation polynomial $P$ with a degree of at most $m = (N-1)/2$ results from the DFT of $(y_0, \ldots, y_{N-1})$ with $\omega_0 = 2\pi/T$:*

$$P(t) = \sum_{k=-m}^{m} \widehat{c}_k \, e^{jk\omega_0 t} \qquad (N = 2m+1).$$

The trigonometric form of $P$ follows with the values $P(n\Delta t) = y_n$ for $n = 0, \ldots, N-1$ as in Sect. 2.1:

$$P(t) = \frac{\widehat{a}_0}{2} + \sum_{k=1}^{m} (\widehat{a}_k \cos(k\omega_0 t) + \widehat{b}_k \sin(k\omega_0 t)),$$

$$\widehat{a}_k = \widehat{c}_k + \widehat{c}_{-k} = \frac{2}{N} \sum_{n=0}^{N-1} y_n \cos\left(nk\frac{2\pi}{N}\right),$$

$$\widehat{b}_k = j(\widehat{c}_k - \widehat{c}_{-k}) = \frac{2}{N} \sum_{n=1}^{N-1} y_n \sin\left(nk\frac{2\pi}{N}\right).$$

If the samples are real, then $P(t)$ is also real-valued. In particular, $P = f$ if $f$ is a $T$-periodic trigonometric polynomial of degree at most $m$.

*Trigonometric Interpolation with an Even Number of Samples*
If $N = 2m$ is even, then the interpolation problem is not uniquely solvable; the trigonometric polynomial $P(t) = \sum\limits_{k=-m}^{m} \alpha_k \, e^{jk\omega_0 t}$ has $N+1$ coefficients. The DFT of $(y_0, y_1, \ldots y_{N-1})$ yields $N$ coefficients $(\widehat{c}_0, \ldots, \widehat{c}_{m-1}, \widehat{c}_m, \ldots, \widehat{c}_{N-1})$.

If we map the coefficient $\widehat{c}_m = \widehat{c}_{N/2}$ as amplitude to the oscillation $e^{-jm\omega_0 t}$ and set $\alpha_m = 0$, i.e.,

$$(\alpha_{-m}, \ldots, \alpha_0, \ldots, \alpha_m) = (\widehat{c}_m, \ldots, \widehat{c}_{N-1}, \widehat{c}_0, \ldots, \widehat{c}_{m-1}, 0),$$

then we obtain the trigonometric interpolation polynomial

$$P_1(t) = \sum_{k=-m}^{m-1} \widehat{c}_k \, e^{jk\omega_0 t} \, .$$

However, in general $P_1(t)$ is not real-valued. Though, by aliasing we can also use the assignment

$$(\alpha_{-m}, \ldots, \alpha_0, \ldots, \alpha_m) = \left( \frac{1}{2}\widehat{c}_m, \ldots, \widehat{c}_{N-1}, \widehat{c}_0, \ldots, \widehat{c}_{m-1}, \frac{1}{2}\widehat{c}_m \right),$$

and then obtain as trigonometric interpolation polynomial with $\widehat{c}_m = \sum\limits_{n=0}^{N-1} y_n (-1)^n / N$

$$P_2(t) = \sum_{k=-m+1}^{m-1} \widehat{c}_k \, e^{jk\omega_0 t} + \widehat{c}_m \cos(m\omega_0 t).$$

$P_2$ is real-valued for given real samples $y_n$. We now denote by $V_m$ the real vector space spanned by the functions $\cos(k\omega_0 t)$, $\sin(k\omega_0 t)$ for $k = 1, \ldots, m - 1$, the constant one and the function $\cos(m\omega_0 t)$. Then we can formulate the following theorem:

**Theorem 6.4** *Let the number $N = 2m$ of nodes $t_n = nT/N$ $(n = 0, \ldots, N - 1)$ be even and $y_n = f(t_n)$ be samples of a real-valued function $f$ on $[0, T]$. Then $\widehat{c}_m$ is real, and with $\widehat{a}_k$, $\widehat{b}_k$ as above, $\omega_0 = 2\pi/T$, the function*

$$P_2(t) = \frac{\widehat{a}_0}{2} + \sum_{k=1}^{m-1} (\widehat{a}_k \cos(k\omega_0 t) + \widehat{b}_k \sin(k\omega_0 t)) + \frac{\widehat{a}_m}{2} \cos(m\omega_0 t), \quad P_2(t_n) = f(t_n),$$

*is the uniquely determined real-valued trigonometric interpolation polynomial in the vector space $V_m$. If $f$ can be extended to a $T$-periodic even function, then all coefficients $\widehat{b}_k = 0$. If an odd $T$-periodic extension is possible, then all $\widehat{a}_k = 0$.*

***Proof*** Let $P$ be in $V_m$ and have the samples $y_0, \ldots, y_{2m-1}$ of $P_2$ and their DFT coefficients $\widehat{c}_k$. Since $P(0) = P(T-)$ by continuity, it holds true (cf. the Alias Formula on p. 87)

$$\widehat{c}_k = \sum_{n=-\infty}^{+\infty} p_{k+2nm} \quad \text{for } k = -m + 1, \ldots, m.$$

Since $p_n = 0$ for $|n| > m$ for $P$ in $V_m$ and $p_m = p_{-m}$ are real, it follows $p_k = \widehat{c}_k$ for $|k| \leqslant m-1$ and $p_m = \widehat{c}_m/2 = p_{-m}$. Thus, having the same Fourier coefficients, $P$ and $P_2$ coincide. If $f$ can be extended $T$-periodically to an even function, then by symmetry it must be $y_n = y_{2m-n}$, and thus $\widehat{c}_k = \widehat{c}_{2m-k} = \widehat{c}_{-k}$. Then all $\widehat{b}_k = j(\widehat{c}_k - \widehat{c}_{-k}) = 0$ and $P_2$ is even. For an odd symmetry in a $T$-periodic extension of $f$, we have correspondingly $y_n = -y_{2m-n}$, $\widehat{c}_k = -\widehat{c}_{-k}$, i.e., all $\widehat{a}_k$ are zero and $P_2$ is odd. $\hfill\square$

**Example** For $N = 4$, $t_n = n\pi/2$, $T = 2\pi$, and with samples $y_0 = 1$, $y_1 = 2$, $y_2 = 1$, and $y_3 = 3$, we compute $P_2$ as above and obtain

$$P_2(t) = \frac{7}{4} - \frac{1}{2}\sin(t) - \frac{3}{4}\cos(2t).$$

Also $P(t) = P_2(t) + \alpha\sin(2t)$ with arbitrary real $\alpha$ is a trigonometric interpolation polynomial of degree 2, since $\sin(2t_n)$ always yields zero. However, such a function $P$ is not in $V_2$ for $\alpha \neq 0$.

The given interpolations $P_1$ and $P_2$ are trigonometric polynomials in the baseband to a DFT. For bandpass signals $f$, trigonometric interpolation polynomials in the corresponding passband can also be given with the help of a DFT and bandpass sampling. In particular, trigonometric polynomials in a passband can be reconstructed exactly with a DFT. The formulation of this is left to the readers.

## 6.3   The Discrete Cosine Transform DCT I

The interpolation formula gives reason to introduce a real-valued discrete Fourier transform for real-valued functions, which is known in the literature as *Discrete Cosine Transform of Type I* or DCT I for short. To do this, we assume a continuous, piecewise continuously differentiable real-valued function $f$ on $[0, T]$, which we think of as being extended to an *even $2T$-periodic function* $f_p$, and consider samples $y_n$ of $f_p$ with the symmetry $y_n = y_{-n}$.

With $N = 2m$ samples $y_n = f_p(nT/m)$ for $n = -m + 1, \ldots, m$, we obtain for the DFT coefficients $\widehat{c}_k$ of $f_p$ and $0 \leqslant k \leqslant 2m - 1$, due to the symmetry $y_n = y_{n\pm 2m} = f_p(nT/m \pm 2T)$ and the relation $\mathrm{e}^{-j\pi kn/m} = \mathrm{e}^{-j\pi k(n\pm 2m)/m}$,

$$\widehat{c}_k = \frac{1}{2m}\left(\sum_{n=0}^{m} y_n\, \mathrm{e}^{-j\pi\frac{kn}{m}} + \sum_{n=m+1}^{2m-1} y_{n-2m}\, \mathrm{e}^{-j\pi\frac{kn}{m}}\right) = \frac{1}{2m}\sum_{n=-m+1}^{m} y_n\, \mathrm{e}^{-j\pi\frac{kn}{m}}$$

$$= \frac{1}{m}\left(\frac{y_0}{2} + \sum_{n=1}^{m-1} y_n \cos\left(\frac{\pi kn}{m}\right) + \frac{y_m}{2}\cos(k\pi)\right).$$

Since we have also $\widehat{c}_k = \widehat{c}_{-k} = \widehat{c}_{2m-k}$, by virtue of the symmetry requirement, it is sufficient to calculate the coefficients for $k = 0, \ldots, m$ from the samples $y_0, \ldots, y_m$. One defines the DCT I of $y_k$ for $k = 0, \ldots, m$ by $\widehat{c}_k$ as above.

This transform is invertible like the DFT, due to the interpolation property of the trigonometric polynomial $P_2$ from above, and the inverse transform is directly readable from $P_2$, because of $y_n = P_2(nT/m)$ with $\omega_0 = \pi/T$. We set $\widehat{a}_k = \widehat{c}_k + \widehat{c}_{-k} = 2\widehat{c}_k$ and obtain the discrete cosine transform DCT I and its inverse:

$$\textbf{DCTI}: \ \widehat{a}_k = \tfrac{2}{m}\left(\tfrac{y_0}{2} + \sum_{n=1}^{m-1} y_n \cos\left(\tfrac{\pi kn}{m}\right) + \tfrac{y_m}{2}\cos(k\pi)\right), \ (k=0,\ldots,m),$$

$$\textbf{IDCTI}: \ y_n = \tfrac{\widehat{a}_0}{2} + \sum_{k=1}^{m-1} \widehat{a}_k \cos\left(\tfrac{\pi kn}{m}\right) + \tfrac{\widehat{a}_m}{2}\cos(n\pi), \qquad (n=0,\ldots,m).$$

Before showing applications of the DCT, we turn to another option and consider interpolation with a shifted set of nodes in comparison. This case results in the variant known as DCT II, which is particularly widespread in DCT applications. One reason for this is the optimality statement no. 4 of the later following theorem on page 112.

## 6.4   Shifted Nodes, Discrete Cosine Transform DCT II

As before, we assume a given continuous, piecewise continuously differentiable real-valued even $2T$-periodic function $f_p$. However, we now choose the following shifted set of nodes, at which the samples are taken:

$$t_n = \frac{2n+1}{2m}T \ \text{ for } \ 0 \leqslant n \leqslant 2m-1, \ m \in \mathbb{N}.$$

We set $y_n = f_p(t_n)$ and obtain by symmetry of $f_p$ for $n = 0, \ldots, m-1$

$$y_n = f_p(t_n) = f_p(2T - t_n) = f_p(t_{2m-1-n}) = y_{2m-1-n}.$$

To use the previous result, we define the function $g$ on $[0, 2T]$ by $g(t) = f_p(t + T/(2m))$. In general, the function $g$ is not even, but $g_n = g(nT/m) = f_p(nT/m + T/(2m)) = y_n$ is true for $n = 0, \ldots 2m-1$. Now, as before, we interpolate this function $g$ on the interval $[0, 2T]$ with $P_2$ as above, where $\omega_0 = \pi/T$. The DFT coefficients $\widehat{c}_k$, $k = -m+1, \ldots, m$, for the samples $g_n = y_n$ are

$$\widehat{c}_k = \frac{1}{2m}\sum_{n=0}^{2m-1} g_n\, e^{-j\pi kn/m} = \frac{1}{2m}\sum_{n=0}^{2m-1} y_n\, e^{-j\pi kn/m}.$$

By $e^{-j\pi kn/m} = e^{j\pi k/(2m)} \, e^{-j\pi k(2n+1)/(2m)}$, $e^{-j\pi k(2M+1)/(2m)} = e^{+j\pi k(2n+1)/(2m)}$
for $M = 2m - 1 - n$, we get from the symmetry $y_n = y_{2m-1-n}$

$$\widehat{c}_k = e^{j\pi k/(2m)} \frac{1}{m} \sum_{n=0}^{m-1} y_n \cos\left(\frac{\pi k(2n+1)}{2m}\right).$$

We have $\widehat{c}_m = 0$, since the cosine terms in the sum are zero for $k = m$.

Eventually, we obtain for the shift $f_p(t) = g(t - T/(2m))$ and its DFT coefficients
$\widehat{c(f_p)}_k = \widehat{c}_k \, e^{-j\pi k/(2m)}$ (compare p. 44) with $\omega_0 = \pi/T$ the corresponding real-valued *trigonometric interpolation polynomial* $P_3$ from the formula for $P_2$ on p. 99:

$$P_3(t) = \sum_{k=-m+1}^{m-1} \widehat{c}_k \, e^{-j\pi k/(2m)} \, e^{jk\omega_0 t} = \frac{\widehat{a}_0}{2} + \sum_{k=1}^{m-1} \widehat{a}_k \cos(k\omega_0 t),$$

$$\widehat{a}_k = \frac{2}{m} \sum_{n=0}^{m-1} y_n \cos\left(\frac{\pi k(2n+1)}{2m}\right) \quad \text{for } k = 0, \ldots, m - 1.$$

As before, the map $(y_0, y_1, \ldots, y_{m-1}) \longrightarrow (\widehat{a}_0, \widehat{a}_1, \ldots, \widehat{a}_{m-1})$ is invertible and the inverse can be read directly from the interpolation formula. This map is called DCT II, its inverse correspondingly IDCT II. We denote the DCT II and the IDCT II with $m$ samples $y_n = f((2n + 1)T/(2m))$, $n = 0, \ldots, m - 1$ by

**DCTII :** $\quad \widehat{a}_k = \frac{2}{m} \sum_{n=0}^{m-1} y_n \cos\left(\frac{\pi k(2n+1)}{2m}\right) \qquad (k = 0, \ldots, m - 1),$

**IDCTII :** $\quad y_n = \frac{\widehat{a}_0}{2} + \sum_{k=1}^{m-1} \widehat{a}_k \cos\left(\frac{\pi k(2n+1)}{2m}\right) \qquad (n = 0, \ldots, m - 1).$

Thus, with the same number of samples $y_0, \ldots, y_m$, it is possible to exactly represent even real $2T$-periodic trigonometric polynomials $f$ up to the degree $m$ by a DCT I or a DCT II with the associated trigonometric interpolation polynomials $P_2$ and $P_3$ with $\omega_0 = \pi/T$. For $P_2$ and the DCT I the samples are $y_n = f(nT/m)$; for the DCT II accordingly the samples are $y_n = f((2n + 1)T/(2m + 2))$ with $n = 0, \ldots, m$. In the above formulas of the DCT II and $P_3$, then $m$ has to be replaced by $m + 1$.

**Remarks**

(a) The coefficient $\widehat{a}_0/2$ is the DC component ( e.g., the DC gain of an alternating voltage $f$). In the literature and in software implementations of DFT and DCT variants (as for example in Matlab, Mathematica or Maple) different scaling factors are in use. Also the indexing often starts there with one instead of zero.

Applying the DFT or DCT, attention has to be paid to such differences in the definitions.

(b) As we have already seen, the decay of the spectral values $|c_k|$ of a periodic function $f$ for growing $|k|$ depends on smoothness properties of $f$. For functions with pointwise representation by their Fourier series as considered last, the alias relation $\widehat{c}_k = \sum\limits_{l=-\infty}^{+\infty} c_{k+lN}$ permits estimates for the approximation errors of the trigonometric interpolations, and for the decay of the DFT coefficients depending on the number $N$ of samples. You can find such estimates for instance in the textbooks of Briggs and Van Emden Henson (1995) or Kincaid and Cheney (2002).

## 6.5   Numerical Integration by Clenshaw-Curtis Quadrature

A first application of the discrete cosine transform, shown here, is numerical integration of a function on a bounded interval.

Let be given a continuous $2T$-periodic real-valued even function $f$. The $k$-th coefficient $\widehat{a}_k$ of the DCT I with samples $y_n = f(nT/m)$, $m \in \mathbb{N}$, $n = 0, \ldots, m$ is an approximation for the integral $\frac{2}{T} \int_0^T f(t) \cos(k\pi t/T)\mathrm{d}t$ with the trapezoidal rule. If $f$ belongs to the vector space $V_m$ introduced before, this quadrature with the trapezoidal rule yields the exact Fourier coefficients of $f$ according to the interpolation theorem of p. 99.

Let us now find an approximation for the integral

$$I = \int_a^b g(t)\mathrm{d}t$$

of a function $g$ assuming it is continuous and piecewise continuously differentiable on $[a, b]$. Mapping the interval $[-1, 1]$ to $[a, b]$ with $u(x) = \frac{b-a}{2} x + \frac{a+b}{2}$ and defining $f$ by $f(x) = g(u(x))u'(x)$, we obtain

$$I = \int_{-1}^{+1} f(x)\mathrm{d}x.$$

Substituting $x = \cos(\varphi)$, $\varphi \in [0, \pi]$, we have with the notation $I = I(f)$:

$$I(f) = \int_0^\pi f(\cos(\varphi)) \sin(\varphi)\mathrm{d}\varphi.$$

The function $f(\cos(\varphi))$ can be extended to a continuous even $2\pi$-periodic function represented by its Fourier series pointwise. We obtain from the Fourier series

$$f(\cos(\varphi)) = \frac{a_0}{2} + \sum_{k=1}^{\infty} a_k \cos(k\varphi),$$

therefore through integration by parts (cf. p. 48) a representation of the integral as a series

$$I(f) = a_0 + \sum_{k=1}^{\infty} a_k \int_0^{\pi} \cos(k\varphi)\sin(\varphi)\mathrm{d}\varphi = a_0 + \sum_{k=1}^{\infty} a_k \frac{(-1)^k + 1}{1 - k^2},$$

$$I(f) = a_0 + \sum_{k=1}^{\infty} \frac{2a_{2k}}{1 - (2k)^2}.$$

With *quadrature after Clenshaw and Curtis* (1960), one approximates the function $f(\cos(\varphi))$ in $[0, 2\pi]$ by the even trigonometric interpolation polynomial

$$P(\varphi) = \frac{\widehat{a_0}}{2} + \sum_{k=1}^{N-1} \widehat{a_k} \cos(k\varphi) + \frac{\widehat{a_N}}{2} \cos(N\varphi)$$

*with the $2N$ samples $P(n\pi/N) = f(\cos(n\pi/N))$ for $n = 0, \ldots, 2N - 1$.* It is uniquely determined by the theorem of p. 99. We require now that $N = 2m$ *is even.* Then the corresponding approximation $S_N(f) = \int_0^{\pi} P(\varphi) \sin(\varphi)\mathrm{d}\varphi$ for the integral $I(f)$ is

$$S_N(f) = \widehat{a_0} + \sum_{k=1}^{m-1} \frac{2\widehat{a_{2k}}}{1 - 4k^2} + \frac{\widehat{a_{2m}}}{1 - 4m^2}.$$

The approximations $\widehat{a_k}$ for the Fourier coefficients $a_k$ of $f(\cos(\varphi))$ are now computed with the $N + 1 = 2m + 1$ samples $f\left(\cos\left(\frac{n\pi}{N}\right)\right)$ for $0 \leqslant n, k \leqslant N = 2m$, according to the trapezoidal rule with the DCT I:

$$\widehat{a_k} = \frac{2}{2m}\left(\frac{f(1)}{2} + \sum_{n=1}^{N-1} f\left(\cos\left(\frac{n\pi}{N}\right)\right)\cos\left(\frac{\pi k n}{N}\right) + \frac{f(-1)}{2}(-1)^k\right).$$

The necessary coefficients $\widehat{a_{2k}}$ can be obtained by $\cos\left(\frac{(N-n)\pi}{N}\right) = -\cos\left(\frac{n\pi}{N}\right)$ already from a DCT I with $m + 1$ summands. For $0 \leqslant n, k \leqslant m$ and

$$y_n = f\left(\cos\left(\frac{n\pi}{N}\right)\right) + f\left(-\cos\left(\frac{n\pi}{N}\right)\right)$$

we find

$$\widehat{a}_{2k} = \frac{1}{m}\left(\frac{y_0}{2} + \sum_{n=1}^{m-1} y_n \cos\left(\frac{\pi k n}{m}\right) + \frac{y_m}{2}(-1)^k\right).$$

With $x_n = \cos(n\pi/N)$, the obtained quadrature rule is usually written in the form

$$S_N(f) = \sum_{n=0}^{m} w_n(f(x_n) + f(-x_n)) = \mathbf{w}^T\mathbf{y}.$$

With precomputed weights $w_n$, various functions $f$ can quickly be integrated numerically by inserting their samples at the nodes $\pm x_n$.

To specify the weights $w_n$, we write the quadrature formula in vector notation with the matrix $D$ belonging to the DCT I as follows: $D = (d_{kn})_{0\leq k,n\leq m}$ is the DCT I matrix with row index $k$ and column index $n$

$$d_{kn} = \begin{cases} \frac{1}{2m}\cos\left(\frac{\pi nk}{m}\right) & \text{for } n = 0, n = m \\[2mm] \frac{1}{m}\cos\left(\frac{\pi nk}{m}\right) & \text{otherwise.} \end{cases}$$

With the vector of the necessary samples $\mathbf{y} = (y_0, \ldots, y_m)^T$, $\mathbf{w} = (w_0, \ldots, w_m)^T$, with $\mathbf{a} = (\widehat{a}_0, \widehat{a}_2, \widehat{a}_4, \ldots, \widehat{a}_{2m})^T$—$T$ stands for the transposition as usual— and the column vector

$$\mathbf{b} = (\beta_k/(1 - 4k^2))_{0\leq k\leq m}, \ \beta_0 = \beta_m = 1, \ \beta_k = 2 \text{ otherwise,}$$

we get

$$S_N(f) = \mathbf{b}^T\mathbf{a} = \mathbf{b}^T(D\mathbf{y}) = (D^T\mathbf{b})^T\mathbf{y} = \mathbf{w}^T\mathbf{y}, \ \text{and thus } \mathbf{w} = D^T\mathbf{b}.$$

$D^T\mathbf{b}$ can also be regarded as a DCT I with only slightly different normalization factors.

*All the weights $w_n$ are positive and their sum is one.*

To see this, we consider for $n = 0, \ldots, m$ and $\alpha_0 = 2m = \alpha_m$, $\alpha_n = m$ otherwise,

$$\alpha_n w_n = 1 - \left(\sum_{k=1}^{m-1} \frac{2}{4k^2 - 1}\cos\left(\frac{\pi nk}{m}\right) + \frac{1}{4m^2 - 1}\cos(\pi n)\right) = 1 - s.$$

We estimate the bracketed sum $s$

$$|s| \leq \sum_{k=1}^{m-1} \frac{2}{4k^2 - 1} + \frac{1}{4m^2 - 1} = 1 - \frac{2m}{4m^2 - 1} < 1.$$

The last equation is readily seen from $\frac{2}{4k^2-1} = \frac{1}{2k-1} - \frac{1}{2k+1}$ (exercise for readers). Thus, the positivity of the weights $w_n$ follows. With the constant function $f = 1$, it follows from $I(f) = S_N(f)$ that their sum is one.

As an alternative to the approximation of the Fourier coefficients by a DCT I, you could also use a DCT II with the shifted sampling points considered before. Such a quadrature was already given by L. Fejér. From the interpolation theorem of p. 99, it follows that the Clenshaw-Curtis quadrature with the $N+1$ given nodes is exact for trigonometric polynomials $f(\cos(\varphi))$ and thus, according to known addition theorems for trigonometric functions, also for polynomials $f$ up to degree $N$. For this, it is necessary that the weights $w_n$ above are positive with the sum equal to one. The same assertions are valid for quadrature according to Fejér with the $N + 1$ nodes as given before for the DCT II.

Error estimates can be obtained from the known estimate for the trapezoidal rule, according to which the DCT I integrates. Literature with such estimates was already referred to before. Convergence of the approximations $S_N(f)$ to $I(f)$ for $N \to \infty$ follows from the convergence of the Fourier series of $f(\cos(\varphi))$. Since we had used $2N$ samples above, the alias relation between the Fourier coefficients $c_k$ of the periodic function $f(\cos(\varphi))$ and its DFT coefficients is $\widehat{c}_k = \sum\limits_{n=-\infty}^{+\infty} c_{k+2Nn}$. We see that the smoother the integrand is, the better the method converges.

It is remarkable that the Clenshaw-Curtis method is a so-called *universal quadrature method*, i.e., for any $k \in \mathbb{N}$ and all $k$-times continuously differentiable functions $f$ the maximum errors of $S_N(f)$ are of order $N^{-k}$ for $N \geqslant k - 1$. To clarify this statement, in the following we denote by $F^k$ the set

$$F^k = \{\, f \in C^k([-1, 1]) \ : \ \|f^{(k)}\|_\infty \leqslant 1 \,\}.$$

Here $\|f^{(k)}\|_\infty$ denotes the supremum norm of the $k$-th derivative of $f$, $C^k([-1, 1])$ the vector space of $k$-times continuously differentiable real-valued functions on $[-1, 1]$. By $\mathbb{P}_N$ we denote the set of all polynomials of degree $\leqslant N$. The above statement about $S_N(f)$ then follows by a theorem of Jackson (1912) from approximation theory:

**Theorem 6.5 (Theorem of Jackson)** *For every $k \in \mathbb{N}$ there exists a constant $\alpha_k$ so that for all $N \in \mathbb{N}$ with $N \geqslant k - 1$ and all $f \in C^k([-1, 1])$ the following inequality is true:*

$$E_N(f) = \inf_{P \in \mathbb{P}_N} \|f - P\|_\infty \leqslant \alpha_k N^{-k} \|f^{(k)}\|_\infty.$$

A proof with a sharper upper bound can be found in Rivlin (1974, 2010).

**Theorem 6.6** *The maximum errors of the Clenshaw-Curtis quadrature $S_N(f)$ on the sets $F^k$ fulfill for any $k \in \mathbb{N}$ and for all $N = 2m \geqslant k - 1$ the following inequality with a constant $\gamma_k$ depending only on $k$:*

$$\sup\{\, |I(f) - S_N(f)| \ : \ f \in F^k \,\} \leqslant \gamma_k N^{-k}.$$

***Proof*** We consider as before the quadrature formula $S_N(f) = \sum\limits_{n=0}^{m} w_n(f(x_n) + f(-x_n))$ with nodes $x_n = \cos(n\pi/N)$ for a function $f$ on $[-1, 1]$ with $N = 2m$, $m \in \mathbb{N}$. Since the weights $w_0, \ldots, w_m$ are positive and their sum is one, we obtain for all $f \in F^k$ and all polynomials $P$ from $\mathbb{P}_N$, due to $I(P) = S_N(P)$, the inequality

$$|I(f) - S_N(f)| = |I(f-P) - S_N(f-P)| \leqslant |I(f-P)| + |S_N(f-P)| \leqslant 4\|f - P\|_\infty.$$

With this, for $N \geqslant k-1$ the assertion follows from Jackson's theorem with $\gamma_k = 4\alpha_k$.

$$\square$$

An analysis and comparison of the Clenshaw-Curtis quadrature with the Gaussian quadrature can be found in Trefethen (2008), a fast algorithm for computing the weights $w_n$ of the method in Waldvogel (2003). For a detailed reading about quadrature methods the textbook of Brass and Petras (2011) is recommended.

The Clenshaw-Curtis method is also used in the construction of interpolatory algorithms for numerical integration of functions on high-dimensional cuboids. For an introduction to the topic, interested readers are referred to the work of Novak et al. (1999) and further sources mentioned there. However, for large dimensions $d \gg 10$ one will rather prefer Monte Carlo methods for numerical integration. A reference to these methods is the textbook of Leobacher and Pillichshammer (2014).

## 6.6  Approximation and Interpolation by Chebyshev Polynomials

To conclude our first excursion into numerical mathematics, it should be pointed out the close relation of the Clenshaw-Curtis quadrature to interpolation and approximation of a considered function $f$ on $[-1, 1]$ by *Chebyshev*[2] *polynomials*. This gives us examples for approximations of functions on bounded intervals with a system of orthogonal functions different from the trigonometric functions considered so far.

The Chebyshev polynomials of the first kind are defined on $[-1, 1]$ for $n \geqslant 0$ by

$$T_n(x) = \cos(n \arccos x).$$

With addition theorems for the cosine function, we obtain for $n \in \mathbb{N}$, with $T_0(x) = 1$ and $T_1(x) = x$, the recursion equation (Exercise: Substitute $\arccos(x) = \varphi \in [0, \pi]$)

$$T_{n+1}(x) = 2x T_n(x) - T_{n-1}(x).$$

---

[2] Russian mathematician Chebyshev is written in Cyrillic as Пафнутий Чебышёв (1821–1894).

Thus $T_n$ is a polynomial of degree $n$, also defined on all of $\mathbb{R}$ and $\mathbb{C}$. $T_n$ is even for even $n$, for odd $n$ an odd polynomial of degree $n$. $T_n$ has $n$ different zeros in $[-1, 1]$ and always satisfies $|T_n| \leqslant 1$ on $[-1, 1]$. The coefficient $a_n$ in $a_n x^n$ with the largest power of $x$ in $T_n$ is $a_n = 2^{n-1}$. The Chebyshev polynomials form an orthogonal system over the interval $[-1, 1]$ with respect to the inner product with the weight function $w(x) = 1/\sqrt{1 - x^2}$,

$$\langle T_n, T_m \rangle_w = \int_{-1}^{+1} T_n(x) T_m(x) \frac{1}{\sqrt{1 - x^2}} dx$$

$$= \begin{cases} 0 & \text{for} \quad n \neq m \\ \pi & \text{for} \quad n = m = 0 \\ \pi/2 & \text{for} \quad n = m \neq 0. \end{cases}$$

In the exercises at the end of the chapter, readers can work out on their own these and some other properties of Chebyshev polynomials.

In our context, because of $T_n(\cos(\varphi)) = \cos(n\varphi)$, we see that the Fourier series expansion of $f(\cos(\varphi))$ on p. 104, with the substitution $x = \cos(\varphi)$, corresponds to a *series expansion of $f$ with respect to the orthogonal system of Chebyshev polynomials*. For continuous, piecewise continuously differentiable functions $f$ on $[-1, 1]$, this series converges uniformly to $f$. When we normalize the Chebyshev polynomials $T_k$ to $\widetilde{T}_k$ with respect to the introduced inner product, i.e., $\langle \widetilde{T}_k, \widetilde{T}_k \rangle_w = 1$ for all $k$, we obtain for $x \in [-1, 1]$

$$f(x) = \frac{a_0}{2} T_0(x) + \sum_{k=1}^{\infty} a_k T_k(x) = \sum_{k=0}^{\infty} \langle f, \widetilde{T}_k \rangle_w \widetilde{T}_k(x).$$

The coefficients $a_k$ are $a_k = \alpha_k \langle f, \widetilde{T}_k \rangle_w$ with the orthonormalized Chebyshev polynomials $\widetilde{T}_k$, $\alpha_0 = 2/\sqrt{\pi}$, and $\alpha_k = \sqrt{2/\pi}$ for $k \neq 0$, and they are the Fourier coefficients of the cosine series representation of $f(\cos(\varphi))$, due to the choice of the weight function $w = 1/\sqrt{1 - x^2}$ in the inner product.

If functions that differ only on a null set (cf. p. 54) are identified, then the inner product is positive definite. We obtain series representations as above for all functions $f : [-1, 1] \to \mathbb{R}$ with $\langle f, f \rangle_w < \infty$. In general, these series no longer converge pointwise, but in the norm generated by the inner product. We denote the real vector space of the functions $f$ with $\langle f, f \rangle_w < \infty$ by $L_w^2([-1, 1])$. . For $f \in L_w^2([-1, 1])$ this norm is given by

$$\|f\|_w = \sqrt{\langle f, f \rangle_w}.$$

Analogous to the theorem on p. 62, we get the following theorem that states the completeness of the orthogonal system of Chebyshev polynomials in this vector space (cf. Mason & Handscomb, 2002):

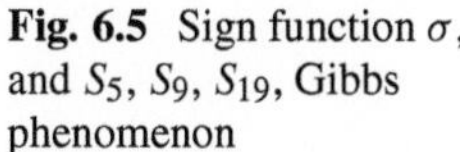

**Fig. 6.4** The Chebyshev polynomials $T_1, T_2, T_3$, and $T_4$

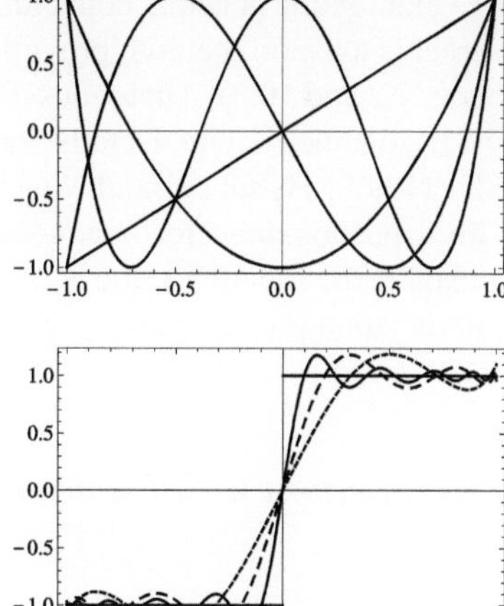

**Fig. 6.5** Sign function $\sigma$, and $S_5, S_9, S_{19}$, Gibbs phenomenon

**Theorem 6.7** *Each function $f \in L_w^2([-1, 1])$ has the series representation*

$$f = \frac{a_0}{2} T_0 + \sum_{k=1}^{\infty} a_k T_k,$$

*with coefficients $a_k = \alpha_k \langle f, \widetilde{T}_k \rangle_w$ as above. The series converges in the norm of $L_w^2([-1, 1])$ to $f$. The coefficients $a_k$ are the Fourier coefficients $a_k = c_k + c_{-k}$, $k \in \mathbb{N}_0$, of $f(\cos(\varphi))$.*

The smoother the function $f$ is, the faster the above series converges (cf. p. 51). An algorithm going back to Clenshaw and Curtis (1960) permits a fast computation of partial sums of that series representation of $f$.

For series expansions of functions $f$ on $[-1, 1]$ with Chebyshev polynomials, because of their relation to the corresponding Fourier series expansion of $f(\cos(\varphi))$, we obtain analogous statements on pointwise convergence, convergence of arithmetic means as in Fejér's theorem, the Gibbs phenomenon, and on approximations by partial sums as in Chap. 3, in Sect. 5.1 or in the following Chap. 6.

The following Figs. 6.4 and 6.5 show the Chebyshev polynomials $T_1$ to $T_4$ on $[-1, 1]$ and three approximations to the sign function $\sigma$ by partial sums $S_n$ of the series expansion as in the last theorem with polynomial degrees $n = 5$, $n = 9$, and $n = 19$. As in Fourier series expansions, we also observe the Gibbs phenomenon, i.e., an overshoot in the neighborhood of the jump discontinuity at $x = 0$.

**Remark** The Gibbs phenomenon also occurs in approximations with other orthogonal systems. Examples are the systems of Legendre, Hermite, or Laguerre polynomials, and for several variables the spherical harmonics. For improvement of approximations by Fourier series expansion in the neighborhood of jump discontinuities of a function $f$, we had used arithmetic means of partial sums according to Fejér. More general, convolution of a function $f$ with a suitable summation kernel is a useful method, providing weights to the spectrum of the function $f$ (cf. Sects. 7.2 and 10.3). Analogous methods are used for orthogonal systems other than the trigonometric one. Details and further references can be found in Gottlieb and Shu (1997). Helmberg and Wagner (1997) have shown a method to mitigate the Gibbs phenomenon for trigonometric interpolation polynomials by appropriately changing the function value at a jump discontinuity, when this point is a node of the interpolation polynomial.

## *Interpolation with Chebyshev Polynomials*

We consider real-valued continuous, piecewise continuously differentiable functions $f$ on $[-1, 1]$. Interpolation of $f(\cos(\varphi))$ with the trigonometric polynomials $P_2$ and $P_3$ of p. 99 and p. 102 directly implies corresponding formulas for the interpolation of $f$ with Chebyshev polynomials. Sampling $f$ so that $y_n = f(x_n)$, $x_n = \cos(\pi n/m))$ for the DCT I, alternatively $x_n = \cos(\pi(2n + 1)/(2m + 2))$ for the DCT II with $n = 0, \ldots, m$, we obtain the following polynomials $P_{2,T}$ and $P_{3,T}$ interpolating $f$ at the points $x_n$:

$$\text{with DCT I coefficients } \widehat{a}_k : \quad P_{2,T}(x) = \frac{\widehat{a_0}}{2} + \sum_{k=1}^{m-1} \widehat{a}_k T_k(x) + \frac{\widehat{a_m}}{2} T_m(x),$$

$$\text{with DCT II coefficients } \widehat{a}_k : \quad P_{3,T}(x) = \frac{\widehat{a_0}}{2} + \sum_{k=1}^{m} \widehat{a}_k T_k(x).$$

Approximation errors of the trigonometric interpolation of a periodic function $f$ can be seen as alias effects, i.e., amplitudes of oscillatory components of $f$ with frequencies higher than the maximum frequency of the interpolation polynomial are added in the amplitudes of the approximation due to aliasing. The error of the interpolation with Chebyshev polynomials can be described correspondingly. For example, let us consider the case of interpolation with the DCT II and the $m + 1$ nodes $x_n = \cos(\pi(2n + 1)/(2m + 2))$, which are the zeros of the Chebyshev polynomial $T_{m+1}$ ($0 \leqslant n \leqslant m$). These nodes are also called *Chebyshev abscissae*. We see from $T_k(\cos(x_n)) = \cos(kx_n)$, using the trigonometric addition theorems, that the polynomials $T_k$ and $(-1)^l T_{l(2m+2)\pm k}$ are indistinguishable at the nodes $x_n$. Thus, for $k = 0, \ldots, m$, for $n = 0, \ldots, m$, and for arbitrary $l \in \mathbb{N}$ we find the relation (to be proven as an exercise for readers):

$$T_k(x_n) = (-1)^l T_{l(2m+2)\pm k}(x_n).$$

For the coefficients $\widehat{a}_k$ of the interpolation polynomial $P_{3,T}$, this means:

**DCT II Alias Formula**   *The coefficients $a_k$ of $f = \frac{a_0}{2}T_0 + \sum\limits_{k=1}^{\infty} a_k T_k$ and the DCT II coefficients $\widehat{a}_k$, $k = 0, \ldots, m$, with samples of $f$ at the Chebyshev abscissae, are related by*

$$\widehat{a}_k = a_k + \sum_{l=1}^{\infty} (-1)^l (a_{l(2m+2)+k} + a_{l(2m+2)-k}).$$

Readers can themselves find an analogous alias relation for the case of interpolation with a DCT I and the nodes $x_n = \cos(\pi n/m)$ $(0 \leqslant n \leqslant m)$. In particular, we have to pay attention to such effects when, for instance, in nonlinear problems a term of the form $f(x)^3$ shall be approximated by Chebyshev interpolation of $f$.

To illustrate this (Fig. 6.6), we interpolate the function $f = 2T_{10} + T_{20}$ with 5 Chebyshev abscissae as nodes. The coefficients $a_{10} = 2$ and $a_{20} = 1$ yield the interpolation polynomial $P_{3,T} = -T_0 = -1$, due to the alias effect with $m = 4$.

There are many studies on convergence of interpolation polynomials for the increasing numbers of nodes, depending on the norm used to measure the approximation errors. In the following some statements are given, and hints to details in the literature. A reference is, for example, Rivlin (2010). In the following theorem, $C([-1, 1])$ and $C^n([-1, 1])$ are the spaces of continuous and of $n$-times continuously differentiable real-valued functions on $[-1, 1]$, provided with the norm $\|.\|_\infty$ of uniform convergence. By $S_n(f)$ we denote the interpolation polynomial of degree $\leqslant n - 1$ with the $n$ zeros of the Chebyshev polynomial $T_n$ as nodes.

**Theorem 6.8**

1. *For every function $f \in C([-1, 1])$, the polynomials $S_n(f)$ converge to $f$ for $n \to \infty$ with respect to the norm of $L_w^2([-1, 1])$.*
2. *For any array of interpolation nodes $x_k^{(n)}$ in $[-1, 1]$, $-1 \leqslant x_1^{(n)} < \ldots < x_n^{(n)} \leqslant 1$, $n \in \mathbb{N}$, $k = 1, \ldots, n$, there exists a function $f \in C([-1, 1])$ so that the sequence of the associated interpolation polynomials $P_n$ does not uniformly converge to $f$ for $n \to \infty$.*
3. *For Lipschitz continuous functions $f$ on $[-1, 1]$, i.e., functions $f$ satisfying*

$$|f(x) - f(y)| \leqslant L|x - y|$$

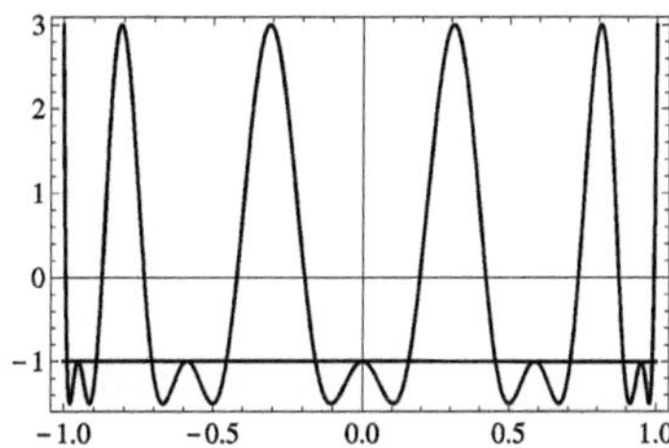

**Fig. 6.6**  Alias effect with Chebyshev polynomials $f$ and $P_{3,T}$ have the same values at the Chebyshev abscissae $x_n = \cos(\pi(2n + 1))/10$ for $n = 0, \ldots, 4$. The Chebyshev abscissae are approximately $-0.951, -0.588, 0, 0.588,$ and $0.951$

*for x, y in* $[-1, 1]$, *and a suitable constant L, in particular for continuously differentiable functions f, the sequence of* $S_n(f)$ *converges uniformly to f for* $n \to \infty$.

4. *The polynomials* $S_n(f)$ *with the Chebyshev abscissae as nodes have the following minimax property: For all* $f \in C^n([-1, 1])$ *with n-th derivative* $f^{(n)}$ *the following estimate is true:*

$$- S_n(f)\infty \leqslant \frac{2^{1-n}}{n!} \|f^{(n)}\|_\infty.$$

*For any other choice of nodes for interpolation polynomials, this bound for the maximum error of the interpolation on* $C^n([-1, 1])$ *is exceeded, i.e., the choice of the Chebyshev abscissae as nodes is the optimal choice with respect to the worst case error on* $C^n([-1, 1])$.

The first statement is proven in Erdös and Turàn (1937), the second in Faber (1912), the others in the already given literature on approximation theory. Comparing different strategies for node selection, we note that the third statement of the theorem for equidistant nodes does not hold true even for analytic functions $f$. A well-known example for this by Runge (1901) is the function $f(x) = 1/(1 + 25x^2)$ on $[-1, 1]$ (cf. Exercise A23).

Different choices of nodes can also be compared by the norms of the operators $A_n$ on $C([-1, 1])$ that map for $n$ nodes a function $f$ to the interpolation polynomial $A_n(f)$ of degree $\leqslant n - 1$. These norms $\|A_n\| = \sup\{\|A_n(f)\|_\infty : \|f\|_\infty \leqslant 1\}$ can be shown to grow with $n$ like $\log(n)$ when Chebyshev abscissae are chosen as nodes, but grow exponentially for equidistant nodes. Details on this can be found in Rivlin (2010).

Altogether it results from the theorem that an interpolation with polynomials of high degree, without known smoothness properties of the interpolated function, in general is not reasonable. Therefore it should be pointed out that by piecewise polynomial interpolation with the so-called *splines* uniform convergence can be accomplished under relatively mild smoothness requirements for the interpolated function. Interpolation with splines is also indicated if the nodes cannot be chosen deliberately but are predefined. For more, please see the literature on approximation theory.

**An Extremal Property of Chebyshev Polynomials Useful in Filter Design**

Finally, we review an extremal property of Chebyshev polynomials which explains why these polynomials are often used in electrical engineering in lowpass filter design. In transmission systems, such filters should be able to filter signal components up to a cutoff frequency as undistorted as possible and to attenuate as well as possible signal components with frequencies higher than the cutoff frequency.

**Theorem 6.9**

1. *For any $x_0 \notin [-1, 1]$ and among all polynomials $P$ of degree $n$ with $P(x_0) = 1$, the polynomial $P_n = T_n / T_n(x_0)$ has minimum supremum norm on $[-1, 1]$.*
2. *Compared to all polynomials $P$ of degree $n$ with $|P(x)| \leqslant 1$ on $[-1, 1]$, the Chebyshev polynomial $T_n$ grows fastest outside $[-1, 1]$, i.e., for $x \notin [-1, 1]$, we have*

$$|T_n(x)| \geqslant |P(x)|.$$

*Proof*

1. Choose $x_0 \notin [-1, 1]$. Since all zeros of $T_n$ are in $[-1, 1]$, $T_n(x_0) \neq 0$ for $x_0 \notin [-1, 1]$. At the $n + 1$ points $t_k = \cos(k\pi/n)$ for $k = 0, \ldots, n$, by definition $T_n$ successively has the alternating extremal values $\pm 1$ beginning with $+ 1$.

   If we assume that there is a polynomial $P$ of degree $n$ with $P(x_0) = 1$, and with smaller norm $\|P\|_\infty$ on $[-1, 1]$ than $T_n / T_n(x_0)$, then $|P(t_k)| < |T_n(t_k)/T_n(x_0)|$ would also be true at all points $t_k$. Therefore $T_n / T_n(x_0) - P$ would have at least $n$ sign changes and thus zeros: For example, for $T_n(x_0) > 0$ it would follow that

$$P(t_0) < T_n(t_0)/T_n(x_0) = 1/T_n(x_0),$$

$$P(t_1) > T_n(t_1)/T_n(x_0) = -1/T_n(x_0) \text{ and so on.}$$

The difference polynomial would have another zero in $x_0$. Contrary to the assumption, $P$ and $T_n / T_n(x_0)$ would then be equal.

2. For any $x_0$ outside $[-1, 1]$ and for polynomials $P$ with $|P(x)| \leqslant 1$ on $[-1, 1]$, again with the supremum norm on $[-1, 1]$, it follows from the first part of the theorem

$$\frac{1}{|P(x_0)|} \geqslant \left\| \frac{P}{P(x_0)} \right\|_\infty \geqslant \left\| \frac{T_n}{T_n(x_0)} \right\|_\infty = \frac{1}{|T_n(x_0)|},$$

if $P(x_0) \neq 0$. There is nothing to prove for the case $P(x_0) = 0$.

$\square$

## *Chebyshev Lowpass Filters*

*Chebyshev lowpass filters* in electrical engineering possess frequency responses $\widehat{h}$, which fulfill the equation

$$|\widehat{h}(\omega)| = K / \sqrt{1 + \varepsilon^2 T_n^2(\omega/\omega_c)}.$$

Such filters let pass signal components with low frequencies largely undistorted and strongly attenuate components with angular frequencies $\omega > \omega_c$. The cutoff frequency $\omega_c/(2\pi)$, the order $n$ of the filter, and the constants $\varepsilon$ and $K$ are chosen according to an attenuation plan (compare again the lowpass example on p. 66, Exercise A15 at the end of the chapter and subsequently Chap. 11).

The statement of the last Theorem 6.9 shows that Chebyshev lowpass filters have advantages in attenuation for $\omega > \omega_c$ compared to other lowpass filters with rational frequency responses. This advantage comes at the expense of a distortion in the passband of the filter due to the ripple of the Chebyshev polynomials $T_n$ with $n$ zeros in $[-1, 1]$.

For a discussion of further applications of Chebyshev polynomials, we refer to the textbook of Mason and Handscomb (2002) devoted entirely to these polynomials. With the Chebyshev polynomials we have acquired, besides the trigonometric functions, a second system of orthogonal functions with respect to an inner product with a weight function, and we have seen series expansions of functions by orthogonal projections onto subspaces of $L_w^2([-1, 1])$ spanned by these polynomials.

By rescaling, all results can be applied to functions over intervals other than $[-1, 1]$ (cf. Exercise A15 at the end of the chapter). The same concept permits representations of functions with many other families of functions which form a complete orthogonal system in a space with inner product. We will discuss this aspect later on in Chap. 14. Readers who want to deepen their knowledge of this are recommended to study the book of Folland (1992) and the rich literature on functional analysis.

## 6.7   Further Application Examples for the DFT

From the abundance of technical applications, which are not possible without the DFT or its closely related methods, only a few further examples shall be outlined here with some appropriate references for further reading. Every prospective engineer or scientist will get to know such applications in his or her studies.

### *Discrete Linear Filters*

A discrete, causal linear filter with a rational transfer function ( cf. later Chap. 11) processes a sequence of input values $x_k,\, k \geqslant 0$, into a sequence of output values $y_n$, $n \geqslant 0$, according to the formula

$$y_n = \sum_{k=0}^{N} a_k x_{n-k} + \sum_{l=1}^{M} b_l y_{n-l},$$

with $x_k, y_k = 0$ for $k < 0$, $N \geqslant 0$, $M \geqslant 1$.

Such filters occur in linear transmission systems in electrical engineering, and indeed in discretization of linear systems in any other scientific field.

The $N+1$ coefficients $a_k$ and the $M$ coefficients $b_l$ have constant values for time-invariant systems and determine the filter response depending on the input values. If the values $x_k$ are samples of a time-dependent function $f$ at time interval $T$, $x_k = f(kT)$, $k = 0, 1, 2, \ldots$, then the output value $y_n$ at time $nT$ is calculated from $x_n$, the $N$ previous values $x_{n-N}, \ldots, x_{n-1}$, and from the $M$ previous output values $y_{n-M}, \ldots, y_{n-1}$. This is called a *causal filter* because the current $y_n$ is only determined by $x_n$ and backward values $x_0, \ldots, x_{n-1}$. It is assumed that the system is initially at rest.

When the second sum in the formula is omitted, such filters are called *non-recursive filters*. When $M \geqslant 1$ and some of the coefficients $b_l$ are nonzero, the filter is called a *recursive filter*. The *frequency response* of the filter (cf. Chap. 11 for details) is defined as the function

$$\widehat{h}(\omega) = \frac{\displaystyle\sum_{k=0}^{N} a_k\, e^{-jk\omega T}}{1 - \displaystyle\sum_{l=1}^{M} b_l\, e^{-jl\omega T}}.$$

In stationary state, the frequency response shows the amplitude and phase changes in samples of an oscillation, sampled at times $kT$, $k \in \mathbb{N}_0$, when passing through the filter as a function of the angular frequency $\omega$ of the oscillation ($|\omega| < \pi/T$). An example for of a non-recursive filter is the following scheme with hold elements delaying the propagation of values by one time step $T$.

With hold time $T$, the filter has the $2\pi/T$-periodic frequency response

$$\widehat{h}(\omega) = \sum_{k=0}^{N} a_k\, e^{-jk\omega T} \quad \text{(Block diagram Fig. 6.7).}$$

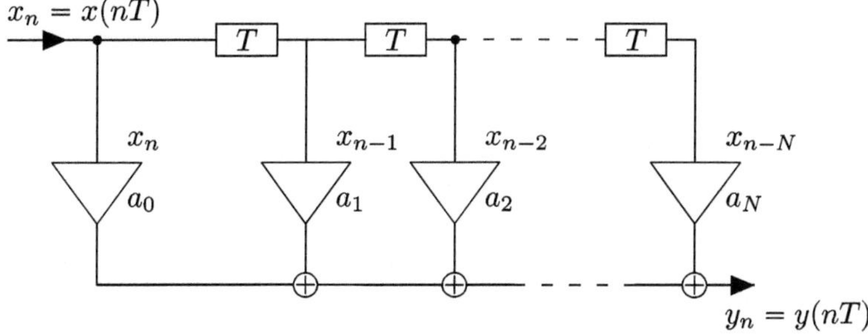


**Fig. 6.7**  Schematic non-recursive discrete linear filter

For example, if the input values of the filter are $x_k = \cos(k\omega T)$ for $k \geqslant 0$, $x_k = 0$ for $k < 0$, then for real coefficients $a_k$ in stationary state ($n > N$) the output value $y_n$ is given by $y_n = |\widehat{h}(\omega)| \cos(n\omega T + \phi)$, $\phi = \arg(\widehat{h}(\omega))$ (Exercise A1).

We note that the filter response is a *convolution* of the filter coefficients with the input values and in the recursive case also with the output values from behind. This requires per time step $N + 1$ or $N + 1 + M$ multiplications. It can be shown that the convolution theorem for the DFT (cf. Table on p. 97) in connection with fast algorithms for the computation of a DFT and IDFT—which we will come to in the following— yields a considerable reduction of the number of necessary multiplications. This is important in real-time applications with very high sampling frequencies and high filter degrees $N$. Examples for the use of such discrete filters are again DAB, DVB-T, DSL, WLAN, mobile broadcasting, etc.

The DFT is applied also in filter design for an intended frequency response $\widehat{h}$. The filter coefficients are then calculated from prescribed samples of $\widehat{h}$ so that the result yields a close approximation of the desired response. Such DFT calculations are embedded in iteration procedures for the stepwise optimization of the filter coefficients.

Details of this, the so-called Remez-Parks-McClellan Exchange Algorithm, often used in the design of non-recursive filters, can be found in the textbook on discrete time signal processing by Oppenheim and Schafer (2013). We discuss other filter design techniques, both for recursive and for non-recursive linear filters, later in Sect. 11.6.


## *Time Series Analysis*


The samples $f(nT/N)$ of a function $f$ are also called a *time series*. In wireless transmissions, also in the medicine in electroencephalograms or in seismography, one often has to deal with randomly noisy signals. *Spectrum estimation* can provide essential information about such signals. An electroencephalogram can be used in medicine to detect brain damage. For example, electroencephalograms of patients with epilepsy or Alzheimer's disease show increased amplitudes compared to healthy patients in certain frequency ranges.

In safety engineering of power plants, one uses wide-ranging systems of vibration detectors for monitoring. On the basis of the spectrum of mechanical vibrations, periodic components with certain frequencies can be detected. By analyzing amplitude and phase spectra with the aid of the DFT, loose vibrating components can be located and safety risks eliminated.

Another field of DFT applications is, for example, radiation measurement in the high-frequency range for testing electromagnetic compatibility (EMC). The signals are typically in the range of up to 1 GHz. In order to avoid baseband sampling with more than 2 GS/s (GS=Gigasamples) and unacceptable data throughputs with required long measurement times, the observed frequency range is divided into segments by bandpass filters, and the spectra are calculated part by part with the

DFT. The results are combined to form the overall image. In the segment passbands one can work with undersampling, few sampling points and short observation times (cf. Example 3 on p. 93).

A key technology in modern communication systems such as WLAN networks, digital radio, and digital television (DAB, DVB-T), but also in wired techniques such as DSL or Powerline, is *multicarrier transmission*. The currently most widespread transmission method is based on the discrete Fourier transform and is known as OFDM (Orthogonal Frequency Division Multiplexing) in the context of communication applications, cf. later Sect. 12.1. In wired transmission it is also called DMT (Discrete Multitone). The transmission method takes its name from the basic idea of transmitting the user information as amplitudes of a trigonometric polynomial with orthogonal carriers. Besides the DFT also linear filters are used in combination with sampling, coding, and estimation algorithms.

Details on communications engineering can be found in the textbook of Proakis and Salehi (2013). Multidimensional DFT variants are essential in medical imaging or in the generation of satellite images from SAR data (Synthetic Aperture Radar). The textbooks of Salditt et al. (2017) and Cumming and Wong (2005) are appropriate references for this. These and numerous other examples such as MP3 players, mobile phones, etc., clearly show that much in today's households and our technical society would not exist at all without the methods of Fourier analysis.

In measuring applications and in filtering or prediction of noisy random signals $f$ the autocorrelation function $r(t) = \lim\limits_{T \to \infty} \frac{1}{2T} \int\limits_{-T}^{T} f(\tau)f(t + \tau)\mathrm{d}\tau$ plays a fundamental role. With the help of DFT and IDFT, one obtains fast algorithms for numerical computation of the autocorrelation from the convolution rule, by multiplication of the Fourier coefficients belonging to $f$.

*Numerical Solution of Integral and Differential Equations*
The discrete Fourier transform can also be extended to the case of functions with several variables. It opens up possibilities for the numerical differentiation of analytic functions, for the numerical inversion of Laplace transforms as well as for the treatment of integral and differential equations. For example, use of the DFT in potential problems of the form $\Delta u = f$ is one of the fastest numerical solution methods in rectangular domains. Thereby, the discretization of the potential equation yields difference equations, which can be solved with the DFT.

We do not go into any of the mentioned application fields in detail. As a recommendation, however, interested students should read the paper *Fast Fourier Methods In Computational Complex Analysis* by Henrici (1979), which deals with a part of the mentioned topics, or the textbook of Briggs and Van Emden Henson (1995).

For all applications one needs an algorithm to calculate the DFT and IDFT. Naive calculation of a DFT for a sample vector $(y_0, y_1, \ldots, y_{N-1})$ requires $N^2$ operations (1 operation $=$ 1 complex multiplication $+$ addition). However, fast algorithms for the computation of the DFT by exploiting its symmetries can considerably reduce the number of necessary operations. Therefore, in the following the basic principle of such algorithms will be briefly presented.

## 6.8   The Basic Principle of the Fast Fourier Transform (FFT)

The history of fast algorithms for computing trigonometric series goes back to Gauss, who used the same approach, as early as 1805 even before Fourier's work, as Cooley and Tukey (1965) in their famous article *An Algorithm For The Machine Calculation Of Complex Fourier Series*. Thereby, the number of operations for a DFT of length $N = 2^n$ can be reduced from $N^2$ to $N \log_2(N)$. An overview is given in the paper *Fast Fourier Transforms: A Tutorial Review and the State of the Art* by Duhamel and Vetterli (1990). We follow a presentation in Nussbaumer (1982).

The basic idea of all FFT algorithms is to compute a DFT of length $N$ by a factorization $N = n_1 n_2 \cdots n_k$ $(n_1, n_2, \ldots, n_k \in \mathbb{N})$, so that it is recursively computed by DFTs of smaller lengths $n_1, n_2, \ldots, n_k$, symbolically

$$\mathrm{DFT}_N = \mathrm{DFT}_{n_k}(\mathrm{DFT}_{n_{k-1}}(\ldots(\mathrm{DFT}_{n_1})\ldots)).$$

For the case $k = 2$, $N = r \cdot s$, $r, s \in \mathbb{N}$, the procedure shall be shown by example. We use the following notations here:

$$\mathrm{e}(n) = \mathrm{e}^{-j2\pi n}$$
$$u = p_0 + p_1 r, \qquad p_0 = 0, \ldots, r - 1$$
$$p_1 = 0, \ldots, s - 1$$
$$v = q_1 + q_0 s, \qquad q_0 = 0, \ldots, r - 1$$
$$q_1 = 0, \ldots, s - 1.$$

We have $0 \leqslant u \leqslant rs - 1 = N - 1$, $0 \leqslant v \leqslant N - 1$, and

$$\mathrm{e}(n + m) = \mathrm{e}(n)\,\mathrm{e}(m)$$
$$\mathrm{e}(n) = 1 \quad \text{for } n, m \in \mathbb{Z}.$$

For a given vector of samples $(y_0, y_1, \ldots, y_{N-1})$, $N = r \cdot s \in \mathbb{N}$ and its corresponding vector of DFT coefficients $(\widehat{c}_0, \widehat{c}_1, \ldots, \widehat{c}_{N-1})$, we write $C(u)$ for $\widehat{c}_u$ and $Y(v)$ for $\frac{1}{N} y_v$.

*Thereby, we find the following representation for the Fourier coefficients $C(u)$:*

$$C(u) = \sum_{v=0}^{rs-1} Y(v)\,\mathrm{e}\left(\frac{uv}{N}\right) = \sum_{q_1+q_0 s=0}^{rs-1} Y(q_1 + q_0 s)\,\mathrm{e}\left(\frac{(p_0 + p_1 r)(q_1 + q_0 s)}{N}\right)$$

$$= \sum_{q_1=0}^{s-1} \sum_{q_0=0}^{r-1} Y(q_1 + q_0 s)\,\mathrm{e}\left(\frac{(p_0 + p_1 r)(q_1 + q_0 s)}{N}\right).$$

Observing $N = r \cdot s$ and $\mathrm{e}\left(\frac{p_1 r q_0 s}{N}\right) = \mathrm{e}(p_1 q_0) = 1$, we see

$$C(p_0 + p_1 r) = \sum_{q_1=0}^{s-1} \sum_{q_0=0}^{r-1} Y(q_1 + q_0 s) \, \mathrm{e}\left(\frac{p_0 q_0}{r}\right) \mathrm{e}\left(\frac{p_0 q_1}{rs}\right) \mathrm{e}\left(\frac{p_1 q_1}{s}\right)$$

$$= \sum_{q_1=0}^{s-1} \mathrm{e}\left(\frac{p_1 q_1}{s}\right) \underbrace{\left( \mathrm{e}\left(\frac{p_0 q_1}{rs}\right) \underbrace{\sum_{q_0=0}^{r-1} Y(q_1 + q_0 s) \, \mathrm{e}\left(\frac{p_0 q_0}{r}\right)}_{\text{``DFT''}_r} \right)}_{\text{``DFT''}_s}.$$

With this, we can formulate as algorithm:

## *FFT Algorithm*

1. For $p_0 = 0, \ldots, r - 1$ and $q_1 = 0, \ldots, s - 1$, compute

$$C(p_0, q_1) = \mathrm{e}\left(\frac{p_0 q_1}{rs}\right) \sum_{q_0=0}^{r-1} Y(q_1 + q_0 s) \, \mathrm{e}\left(\frac{p_0 q_0}{r}\right).$$

2. For $p_0 = 0, \ldots, r-1$ and $p_1 = 0, \ldots, s-1$, i.e., for $u = 0, \ldots, N-1$, compute

$$C(u) = C(p_0 + p_1 r) = \sum_{q_1=0}^{s-1} \mathrm{e}\left(\frac{p_1 q_1}{s}\right) C(p_0, q_1).$$

The necessary values of the exponential function are calculated and stored in advance for a given $N = r \cdot s$. Then the algorithm requires $rs(r + 1)$ operations for the first step and $rs^2$ operations for the second step.

**Example** For $N = 10^3 = 10 \cdot 100$, naive calculation of the DFT needed $N^2 = 1{,}000{,}000$ operations. With $r = 10$, $s = 100$ the above algorithm needed only $N(r + s + 1) = 111{,}000$ operations, i.e., we were approximately 9 times faster.

As mentioned, the algorithm can be extended to the case of $k$ factors, $N = n_1 n_2 \ldots n_k$, and thus the computational effort can be further reduced. The choice of $N$ as a power of two proves to be particularly advantageous for practical applications. We leave with the presented exemplary introduction of the basic idea. Concerning questions about optimal factorization of $N$, questions about error analysis (fast FFT algorithms are also more accurate than naive methods) etc., we refer to the given literature. If one counts a complex multiplication and addition together as one complex operation, then the following reductions of the computational effort can be achieved:

*If $N = n_1 n_2 \cdots n_k$, then an FFT does not require more than $2N \sum\limits_{i=1}^{k} (n_i - 1)$ complex operations. If $N$ is a power of two, then $N \log_2(N)$ operations are sufficient.*

Also for calculations of discrete cosine transforms, there are fast algorithms. See for example Rao and Hwang (1996). In numerical software, such as Matlab, or computer algebra systems such as Mathematica or Maple, there are already implemented powerful fast algorithms for DFT and DCT computations.

At the end of this short excursion into numerical mathematics, a few examples are given below to illustrate and to encourage for deepening:

**Examples**

1. Let a signal $f$ be given whose signal frequencies $\nu$ are in the range $0 \leqslant \nu < 50$ Hertz. Assume that it is superposed with random noise. When sampling over a time interval $[0, T[$ with a sampling frequency $\nu_a = N/T > 100$ Hz, for example with $T = 2$ s and $N = 256$, an interpolation of the signal by a polygon does not show which oscillations the signal is composed of (see Fig. 6.8).

   Application of DFT and computation of the magnitude spectrum $|\widehat{c}_k|$, $0 \leqslant k \leqslant N - 1$ (right figure below), show "*Peaks*" for $k = 60$ and $k = 90$. From this we conclude that the signal essentially is a superposition of two noisy oscillations with frequencies $\nu_1 = 60/T = 30$ Hz and $\nu_2 = 45$ Hz, the first with amplitude of about one, and the second with about 0.7 as amplitude. The symmetry of the shown magnitude spectrum is a consequence of the alias effect. Looking at the signal shape, the noise appears also to have amplitudes up to about one, which were not detected by the DFT but can be explained by superposition effects of numerous noise components with small amplitudes. In the example, in fact it was used $f(t) = \cos(60\pi t) + 0.7 \sin(90\pi t)$ and additive noise with random values in the interval $[-0.7, 0.7]$ (cf. Figs. 6.8 and 6.9).

   The clear detection of the frequencies in the example is due to the fact that they coincidentally agree with those computed in the DFT spectrum. Such luck will be rare in real practice. At this point we do not go into lack of luck and other practical problems but will focus on the discrete Fourier transform again in a later Chapter (Sect. 12.6). There, we will discuss some aspects which are important for applications in practice.

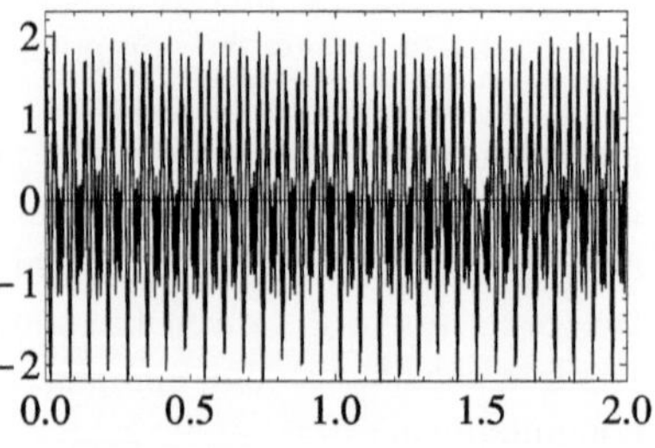

**Fig. 6.8** Signal progression in 2s

**Fig. 6.9**  DFT magnitude
spectrum

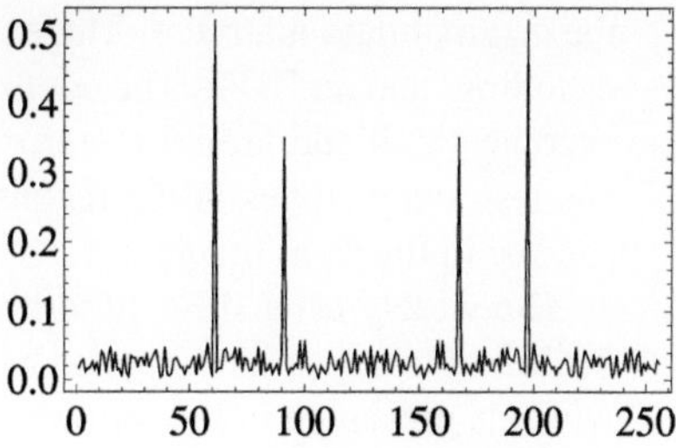

**Fig. 6.10**  Original image of
G. S. Ohm

**Fig. 6.11**  Soft-focussed
Ohm, 90% of pixels set to
zero

2. The next example shows the possibility of using Fourier methods for image data
   compression. The first Fig. 6.10 is a scanned image of G. S. Ohm (1789–1854).
   Each pixel of it was assigned to an integer gray value in the interval $[0, 255]$.
   These image data were stored in the associated *grayscale matrix A*.

   This matrix, conceived as a discrete signal of gray values, was subjected to a
   DFT (in two variables, see next example). Finally, all DFT coefficients, whose
   magnitude was smaller than $2M \cdot 10^{-3}$, $M$ being the maximum magnitude of
   all occurring DFT coefficients, were set to zero. About 90% of all coefficients
   of the example were thus replaced by zero. The second Fig. 6.11 shows the
   reconstruction with an inverse DFT applied to the modified coefficients.

   Since the modified DFT coefficients contain many zeros, their storage with an
   entropy encoding (e.g., Huffman encoding) requires significantly less space than

the original data matrix $A$. The expense for this is a DFT, entropy encoding and decoding, and an IDFT. The coefficients belonging to higher frequencies rapidly become small and are set to zero with this kind of compression. Therefore, one has less steep slopes in the reconstruction and thus a blurring effect, which we can see in the right image.

The widely used JPEG algorithms for image data compression (JPEG stands for Joint Photographic Experts Group) use the DCT II in a 2-dimensional variant, which is presented in the following example. In JPEG compression, pixel blocks of size $8 \times 8$ or $16 \times 16$ are transformed with the DCT and the results per block are quantized. This is done so that the according DCT coefficients of such blocks of a grayscale matrix, depending on their position in the coefficient matrix, are divided by accordingly positioned values of a so-called *luminance table* and rounded to integers. The values of the luminance table depend on the desired compression ratio. In the following example such a luminance table is shown and used.

Since the DCT coefficients for higher frequency components usually decrease rapidly and the divisors of the table for such coefficients increase, one mostly gets many zeros in the high frequencies as a result of quantization. These quantized spectral data can be stored or transmitted in compressed form by entropy encoding. When transmitting a JPEG image, the used encoding method (e.g., Huffman table, not uniquely determined) is specified in the file header as necessary information for decoding. At the viewer, the data stream is decoded back into the DCT matrix and subjected to IDCT block by block. As a rule, the IDCT data for the image must also be rendered again if there are values, which do not belong to $\mathbb{N}_0 \cap [0, 255]$.

In color images, the color information is quantized analogously with chrominance tables. The modified quantization can lead to undesired artifacts in the neighborhood of edges in combination with the Gibbs phenomenon, since the IDCT after compression usually yields a trigonometric interpolation polynomial different from that of the original DCT data. This can be quickly verified by zooming in on the edges in a JPEG image. Current standards in image data, audio and also video encoding can be found in Rao and Hwang (1996). Modern and sometimes more powerful mathematical methods for signal compression will be introduced in the final Sect. 14.2 about *Wavelets*. Such wavelet procedures— see for example Taubman and Marcellin (2001)—are used with the newer JPEG 2000 standard or also in the file format DjVu.

## 6.9   DCT-2D

As announced in the example before, we first specify the 2-dimensional variant of DCT II, which we use below. For an $(M \times N)$-matrix $A$ with components $A_{mn}$, the DCT-2D of $A$ is the $(M \times N)$-matrix $B$ with components $B_{pq}$ for $0 \leqslant p \leqslant M - 1$, $0 \leqslant q \leqslant N - 1$, defined by

$$\textbf{DCT} - \textbf{2D}: \quad B_{pq} = \alpha_p \alpha_q \sum_{m=0}^{M-1} \sum_{n=0}^{N-1} A_{mn} \cos\left(\frac{\pi(2m+1)p}{2M}\right) \cos\left(\frac{\pi(2n+1)q}{2N}\right)$$

$$\textbf{IDCT} - \textbf{2D}: A_{mn} = \sum_{p=0}^{M-1} \sum_{q=0}^{N-1} \alpha_p \alpha_q B_{pq} \cos\left(\frac{\pi(2m+1)p}{2M}\right) \cos\left(\frac{\pi(2n+1)q}{2N}\right) \ .$$

Thereby, the normalization factors are those of the numerics software Matlab given by

$$\alpha_p = \begin{cases} 1/\sqrt{M} & \text{if } p = 0, \\ \sqrt{2/M} & \text{if } 1 \leqslant p \leqslant M - 1, \end{cases} \qquad \alpha_q = \begin{cases} 1/\sqrt{N} & \text{if } q = 0, \\ \sqrt{2/N} & \text{if } 1 \leqslant q \leqslant N - 1. \end{cases}$$

The DCT-2D is the concatenation of a DCT over the rows of the matrix $A$, followed by a DCT over the columns of the preceding transformation result. The larger $p+q$, the higher frequency components in the signal the coefficient $B_{pq}$ is assigned to. Detailed information about the geometrical aspects of the DCT-2D can be found for example in the textbook of Briggs and Van Emden Henson (1995).

For our demonstration example, the word "Geheimnis" (German for "Secret") was initially stored as a suitably scaled black and white image. The values 0 for black, 1 for white pixels were sequentially encoded into the DCT matrix of the subsequent image so that one pixel value of information was stored per $8 \times 8$ block. To identify the pixels, white pixels were encoded so that the relation $B_{32} > B_{41}$ was fulfilled. Where appropriate, these values were interchanged to achieve this result. Accordingly, an information pixel was encoded and identified as black by $B_{32} < B_{41}$. If both values were equal, the block was skipped.

The difference between the two values was increased by a threshold value to achieve better stability against attacks by noise and data compression. The coefficients $B_{32}$ and $B_{41}$ were chosen because they belong to the middle frequencies in the block and are quantized equally according to the following luminance table for JPEG. This lets expect a certain robustness against JPEG compression if both values are scaled equally.

The example does not show a professional algorithm for watermark generation, but it demonstrates in a simple way how the frequency domain can be used to store information. Since a watermark constructed in this way is not stable against geometric attacks such as scaling or rotation of the image, different methods are used for professional purposes. One example is the *spread spectrum method*, in which the information is spread over the entire spectrum. These methods originated in radio transmission technology and were also used in military communications since about 1950. As with images, digital watermarks can also be introduced into audio and video data. Figs. 6.12, 6.13, and 6.14 illustrate the experiment.

**Fig. 6.12** $8 \times 8$—luminance table

| $p \backslash q$ | 0 | 1 | 2 | 3 | 4 | 5 | 6 | 7 |
|---|---|---|---|---|---|---|---|---|
| 0 | 16 | 11 | 10 | 16 | 24 | 40 | 51 | 61 |
| 1 | 12 | 12 | 14 | 19 | 26 | 58 | 60 | 55 |
| 2 | 14 | 13 | 16 | 24 | 40 | 57 | 69 | 56 |
| 3 | 14 | 17 | **22** | 29 | 51 | 87 | 80 | 62 |
| 4 | 18 | **22** | 37 | 56 | 68 | 109 | 103 | 77 |
| 5 | 24 | 35 | 55 | 64 | 81 | 104 | 113 | 92 |
| 6 | 49 | 64 | 78 | 87 | 193 | 121 | 120 | 101 |
| 7 | 72 | 92 | 95 | 98 | 112 | 100 | 103 | 99 |

**Fig. 6.13** Image with watermark (197 KB)

The hidden watermark



The identified watermark
in the noisy and compressed
image with pixel errors,
but well recognizable

**Fig. 6.14** JPEG compressed with additional noise (30 KB)

We see the luminance table and the result of the described example[3] in the following images reconstructed from the modified DCT spectra:

There is extensive literature on the subject of digital watermarking and steganography in the context of Digital Rights Management (DRM). Interested readers are referred to the specialized literature on this subject. A reference is the textbook by Cox et al. (2008). Also, by simply searching the Internet for the above-mentioned keywords, you can very quickly find numerous sources.

---

[3] Readers are encouraged to create their own analog examples using a DCT with appropriate software.

## 6.10  Exercises

**(A16)** For a non-recursive discrete filter with hold time $T$, real coefficients $a_k$, and $2\pi/T$-periodic frequency response $\widehat{h}(\omega) = \sum_{k=0}^{N} a_k\, e^{-jk\omega T}$ (cf. p. 116), let the input values be given by $x_k = \cos(k\omega T)$ for $k \geqslant 0$, $x_k = 0$ for $k < 0$. Compute the output values $y_n$ for $n > N$.

**(A2)** Let be given the *modified Dirichlet kernel*

$$W_m(t) = \sum_{k=-m}^{m} e^{jk\omega_0 t} - \cos(m\omega_0 t) \text{ for } \omega_0 = \frac{2\pi}{T}.$$

Prove that for $N = 2m$ nodes the $T$-periodic trigonometric interpolation polynomial $P_2$ in the theorem on p. 99 can also be written in the form

$$P(t) = \frac{1}{2m} \sum_{k=0}^{2m-1} f(t_k)W_m(t-t_k) = P_2(t), \quad t_k = \frac{kT}{2m}\ (k = 0, \ldots, 2m-1).$$

**(A3)** For the function $f(t) = \cos(t)$, $0 \leqslant t < \pi$, carry out a DFT with $N = 15$ samples $t_n = n\pi/N$, $n = 0, \ldots, N-1$. Verify the alias formula of p. 87 on the DFT coefficients $\widehat{c}_0$ and $\widehat{c}_1$. Why is $\sum_{k=-7}^{7} \widehat{c}_k\, e^{j2kt}$—with $N$-periodically extended $\widehat{c}_k$—not odd? Set $f(0) = 0$, use it to repeat the DFT, and again form the associated trigonometric interpolation polynomial. What do you find?

**(A4)** *Approximation Quality of Trigonometric Interpolation Polynomials.*
Let be given a continuous $2\pi$-periodic function $f : \mathbb{R} \to \mathbb{C}$ with absolutely summable Fourier coefficients $c_k$, $k \in \mathbb{Z}$. For $N = 2m$ let $P_2$ be the interpolation polynomial of p. 99. Show that the following error estimate holds true for all $t \in \mathbb{R}$:

$$|P_2(t) - f(t)| \leqslant 2 \sum_{|k|\geqslant N/2}{}'' |c_k|.$$

The symbol $\sum''$ means that the summands with indices $N/2$ and $-N/2$ are to be multiplied by the factor $1/2$.

**(A5)** Consider a signal whose signal frequencies $\nu$ are in the range $0 \leqslant \nu < 80$ Hz. The signal is analyzed with the DFT. The observation time $T$ is $T = 2$ seconds. How large the number $N$ of samples must be at least, in order to avoid adverse alias effects in the spectral analysis?

**(A6)** (a) For the oscillation $\cos(2\pi\nu t)$, $\nu = 6$ Hz, which values are nonzero in the amplitude spectrum of a DFT with $N = 128$ points in the period $0 \leqslant t \leqslant 4$ seconds?

(b) For $\cos(2\pi \nu t)$, $\nu = 100$ Hz, let a DFT be given with $N = 128$ samples from a time window of one second. Which coefficients $\widehat{c}_k$, $0 \leqslant k \leqslant 127$, in the DFT spectrum are nonzero and why?

**(A7)** For a real periodic signal $f$ in the frequency band up to 600 Hz, a DFT is given with $N = 512$ samples from a time window of 2 seconds. Let be nonzero only the DFT coefficients $\widehat{c}_{90}$ and $\widehat{c}_{422}$. Which signal frequencies can produce this DFT spectrum?

**(A8)** Signal amplitudes in the frequency band $F = ]2\text{GHz}, 2\text{GHz} + 1\text{MHz}[$ are to be detected by a DFT. Let the frequency resolution be $1/T = 5$ kHz. Determine $m$ and the sampling frequency $N/T$, so that $F \subset ]mN/(2T), (m+1)N/(2T)[$. Which DFT coefficients of a DFT, performed with these values, belong to $f(t) = \sin(2\pi \nu t)$ with $\nu = 2000150$ kHz? (cf. Example 3, p. 93)

**(A9)** A DFT is performed with the samples $t_n = n/8 + \Delta t$, $n = 0, \ldots, 7$, and $\Delta t = 0.05$ of the function $f(t) = 6\cos(2\pi t) + 3\sin(4\pi t) - 4\sin(6\pi t) + 5\cos(8\pi t)$. How can the DFT spectrum be corrected, only with knowledge of the amplitude $A = 6$ of the "pilot carrier" $\cos(2\pi t)$, so as to obtain the spectrum of the real-valued function $f$? (Cf. Example 5, p. 94)

**(A10)** Show that the DFT of $1/N \cdot (x_n y_n)_{0 \leqslant n \leqslant N-1}$ is the N-periodic convolution of the DFT coefficients of $(x_n)_{0 \leqslant n \leqslant N-1}$ and $(y_n)_{0 \leqslant n \leqslant N-1}$.

**(A11)** Program the Clenshaw-Curtis method for numerical integration and test your program on polynomials and in a comparison with the trapezoidal rule, using as example the integral $\int_0^1 \frac{1}{1+x} dx = \ln(2)$.

Compare the relative errors with an increasing number of nodes.

**(A12)** *Runge's Example.* Write a program to interpolate the function

$$f(x) = 1/(1 + 25x^2)$$

on $[-1, 1]$ (example from Runge, 1901) with $n$ equidistant nodes and with the Chebyshev abscissae as nodes for $n = 8$, $n = 13$, and $n = 17$.

Generate graphs of the results and discuss the quality of the polynomials obtained as approximations to the function $f$ on $[-1, 1]$.

**(A13)** Show the orthogonality of the Chebyshev polynomials $T_k$ on $[-1, 1]$ with respect to the inner product $\langle ., . \rangle_w$. Calculate the alias relation for the coefficients of the polynomials $T_k$ when interpolating with the nodes $x_n = \cos(n\pi/m)$, $n = 0, \ldots, m$. Test your result by interpolating the function $f = T_{11} + T_{13} - 2T_{23}$ with seven such nodes with a DCT I as on p. 110.

**(A14)** *Interpolation with Chebyshev polynomials on intervals other than* $[-1, 1]$. Let the function $f$ be given by $f : [-3, 7] \to \mathbb{R}$,

$$f(t) = \frac{1}{1 + (t - 2)^2}.$$

Map the interval $[-1, 1]$ with an affine mapping $L$ to $[-3, 7]$, and compute with a DCT II the interpolation polynomial $P$ of degree $m = 12$ for

$g(t) = f(L(t))$, with the Chebyshev abscissae $x_n$ $(n = 0, \ldots, m)$ as nodes. Plot the function $f$, the interpolation polynomial $P \circ L^{-1}$ with nodes $t_n = L(x_n)$, and the interpolation error $f - P \circ L^{-1}$. Use your program of exercise A12.

**(A15)** *Complex Chebyshev polynomials, design of Chebyshev lowpass filters.*

(a)$^\star$ If you are familiar with complex functions, then verify that for variable $z \in \mathbb{C}$ the $n$-th Chebyshev polynomial is $T_n(\cos(z)) = \cos(nz)$. If you are not sufficiently familiar with complex functions to solve (a)–(c), please take (a)–(c) for granted and solve (d).

(b)$^\star$ The *Joukowsky transformation* $z = z(w) = (w + w^{-1})/2$ maps the complement of the unit circle invertibly to $\mathbb{C} \setminus [-1, 1]$.

When the principal values

$$\sqrt{1 - 1/z^2} = \sum_{n=0}^{\infty} \binom{1/2}{n} \frac{(-1)^n}{z^{2n}}$$

$$\sqrt{1 - z^2} = \sum_{n=0}^{\infty} \binom{1/2}{n} (-1)^n z^{2n}$$

are chosen for the roots, show that the inverse mapping is explicitly given by

$$w(z) = \begin{cases} z + z\sqrt{1 - 1/z^2} & |z| > 1, \\ z + j\sqrt{1 - z^2} & \text{for } |z| \leqslant 1, \, \Im(z) > 0, \\ z - j\sqrt{1 - z^2} & |z| \leqslant 1, \, \Im(z) < 0. \end{cases}$$

Prove that $T_n(z) = \frac{w^n + w^{-n}}{2}$.

(c)$^\star$ Using the approach

$$Q(z) = H(z)H(-z) = 1/(1 + \varepsilon^2 T_n^2(z/(j\omega_c))),$$

find the poles of $Q$ with negative real part. Set $z/(j\omega_c) = \cos(x + jy)$, and show that for $\varepsilon > 0$ and $\omega_c > 0$ these poles are given by

$$z_k = \omega_c \sin(x_k) \sinh(y) + j\omega_c \cos(x_k) \cosh(y) \quad (k = 0, \ldots, n-1),$$

$$x_k = \frac{(2k + 1)\pi}{2n}, \quad y = -\frac{1}{n} \operatorname{arsinh}\left(\frac{1}{\varepsilon}\right).$$

Use (b), trigonometric addition theorems, and the relations of complex trigonometric functions and complex hyperbolic functions: $\cos(jz) = \cosh(z)$ and $\sin(jz) = j\sinh(z)$.

(d) Write a program to solve the following problem. With the poles $z_k$ from (c), $k = 0, \ldots, n - 1$, the frequency response $\widehat{h}(\omega) = H(j\omega)$ of a Chebyshev lowpass filter of order $n$ with

$$|\widehat{h}(\omega)|^2 = \frac{1}{1 + \varepsilon^2 T_n^2(\omega/\omega_c)}$$

is given by

$$\widehat{h}(\omega) = \widehat{h}(0) \prod_{k=0}^{n-1} \frac{-z_k}{j\omega - z_k}.$$

The DC gain is $\widehat{h}(0) = 1$ for odd $n$ and $\widehat{h}(0) = 1/\sqrt{1 + \varepsilon^2}$ for even $n$.

With $\omega_c/(2\pi)$ as cutoff frequency, the attenuation in dB (decibel) is

$$A(\omega) = 10 \log_{10}(1 + \varepsilon^2 T_n^2(\omega/\omega_c)).$$

Now, calculate a *Chebyshev lowpass filter* with lowest possible order $n$ according to the following specification:

Let the cutoff angular frequency $\omega_c$ be given by $\omega_c = 2\pi \cdot 1000$ Hz; let the stopband edge be $\omega_s = 2\pi \cdot 2500$ Hz. The maximum attenuation at the passband edge $\omega_c$ shall be $A_{max} = 0.2$ dB, the minimum attenuation at the stopband edge $A_{min} = 40$ dB.

First calculate $\varepsilon$ and the necessary filter order $n \in \mathbb{N}$ by substituting the given attenuations at $\omega_c$ and $\omega_s$ in $A(\omega)$. Then calculate the poles with negative real part as in (c) and build the frequency response of the filter.

Plot the amplitude response $|\widehat{h}|$, the phase response $\Phi(\omega) = \arg(\widehat{h}(\omega))$, the delay $- \Phi(\omega)/\omega$, and the group delay $D(\omega) = -\mathrm{d}\Phi(\omega)/\mathrm{d}\omega$.

See also later in Sect. 11.3 the design of other analog filter types with rational frequency responses—such as *Butterworth lowpass filters*—and in Sect. 11.6 corresponding *discrete filter variants*.

**(A16)** *Transistor in emitter circuit.* Suppose the collector current $i_C(t)$ is given by

$$i_C(t) = e^{1.1 + 0.75 \sin(\omega_0 t)} - 1 \ [\text{mA}] \quad (\omega_0 = 1\text{rad/s}).$$

Compute a DFT with sampling points $i_C(2\pi k/16)$, $k = 0, \ldots, 15$ and estimate the DC gain, the RMS value, and the distortion factor (cf. p. 33).

# Chapter 7
# Convergence of Fourier Series

**Abstract** This chapter is devoted to the proofs of the previously given theorems on pointwise convergence of Fourier series. The theorems of Dirichlet and Fejér with their implications are proven. The mitigation of the Gibbs phenomenon by summation kernels is shown as well as the Parseval equation for piecewise continuous periodic functions. As a summary of the acquired knowledge at that point, mathematical results are discussed, which have historically led from classical Fourier analysis and integration theory to the Lebesgue integral and distributions. The theoretical foundations of distribution theory and its countless practical applications are developed in the following chapters.

In the last chapters we have learned about first examples and applications of Fourier series. In this chapter the central statements of convergence in the theorems of P. L. Dirichlet and L. Fejér from Sect. 3.2, and the Parseval equation are studied in more detail. The proofs are presented in such a way that the emphasis is not so much on their mathematical "justification character," but rather that the reader learns a good portion of arithmetic technique in dealing with trigonometric functions, sums, and integrals.

The common basic principle in studying approximations $f_N$ to $T$-periodic functions $f$ in the following sections is the representation of the approximations in the form

$$f_N(t) = \frac{1}{T} \int_0^T f(s) K_N(t-s) \, \mathrm{d}s \,,$$

with suitable integral kernels $K_N$. From the properties of the function $f$ and the convolution kernels $K_N$ result the properties of the approximating functions $f_N$. With regard to convergence of the approximations we distinguish between pointwise convergence, uniform convergence, and convergence in quadratic mean.

© The Author(s), under exclusive license to Springer Nature Switzerland AG 2025      129
R. Brigola, *Fourier Analysis and Distributions*, Texts in Applied Mathematics 79,
https://doi.org/10.1007/978-3-031-81311-5_7

## 7.1    The Theorem of Dirichlet

The subject of this section is the proof of Dirichlet's theorem on pointwise representation of piecewise continuously differentiable periodic functions by their Fourier series (cf. S. 28). A detailed discussion of pointwise convergence of Fourier series of more general functions, for example monotone functions and functions of bounded variation, can be found for instance in Zygmund (2003). As a typical example of such a function, which is not piecewise continuously differentiable, only the famous Cantor function, also called devil's staircase, may be mentioned here. Further examples for Fourier series expansions of functions, which are not piecewise continuously differentiable, are given in Exercise A6.

**Theorem 7.1** *The Fourier series of a piecewise continuously differentiable periodic function $f : \mathbb{R} \to \mathbb{C}$ converges at each point $t$ to $(f(t+) + f(t-))/2$.*

**Proof** For the proof, assume that the function $f$ is $T$-periodic with $T = 1$ and piecewise continuously differentiable. The partial sum of degree $N$ of the Fourier series of $f$ is denoted by $f_N$. The proof of the theorem is carried out in four steps:

1. According to $T = 1$, we use the 1-periodic Dirichlet kernels $D_N$, given by

$$
D_N(t) = \sum_{k=-N}^{N} e^{j2\pi kt} =
\begin{cases}
\dfrac{\sin((2N+1)\pi t)}{\sin(\pi t)} & \text{for} \quad t \notin \mathbb{Z}, \\
2N+1 & \text{for} \quad t \in \mathbb{Z},
\end{cases}
$$

   and prove immediately

$$
\int_{0}^{1} D_N(t)\, \mathrm{d}t = \int_{-1/2}^{1/2} D_N(t)\, \mathrm{d}t = 2 \int_{0}^{1/2} D_N(t)\, \mathrm{d}t = 1.
$$

2. $D_N$ and $f$ are 1-periodic. As already shown in 3.2 it holds

$$
f_N(t) = \sum_{k=-N}^{N} \int_{0}^{1} f(s)\, e^{-j2\pi ks}\, \mathrm{d}s\; e^{j2\pi kt} = \int_{0}^{1} D_N(t-s) f(s)\, \mathrm{d}s
$$

$$
= \int_{-1/2}^{1/2} D_N(s) f(t-s)\, \mathrm{d}s = \int_{-1/2}^{1/2} \frac{\sin((2N+1)\pi s)}{\sin(\pi s)} f(t-s)\, \mathrm{d}s \, .
$$

3. We can write the last integral in the form:

$$f_N(t) = \underbrace{\int_{-1/2}^{0} \sin((2N+1)\pi s)\,\frac{f(t-s)-f(t+)}{\sin(\pi s)}\,ds + f(t+)\int_{-1/2}^{0} D_N(s)\,ds}_{I_1(N,t)}$$

$$+ \underbrace{\int_{0}^{1/2} \sin((2N+1)\pi s)\,\frac{f(t-s)-f(t-)}{\sin(\pi s)}\,ds + f(t-)\int_{0}^{1/2} D_N(s)\,ds}_{I_2(N,t)}$$

$$= \frac{1}{2}[f(t+)+f(t-)] + I_1(N,t) + I_2(N,t).$$

4. Since $f$ is piecewise continuously differentiable, the right- and left-sided derivatives of $f$ at $t$ exist, and thus also the limits (cf. Exercise A8):

$$\lim_{s\to 0-}\frac{f(t-s)-f(t+)}{\sin(\pi s)} = -\frac{f'(t+)}{\pi} \quad\text{and}\quad \lim_{s\to 0+}\frac{f(t-s)-f(t-)}{\sin(\pi s)} = -\frac{f'(t-)}{\pi}.$$

Therefore, both functions $\dfrac{f(t-s)-f(t\pm)}{\sin(\pi s)}$ can be continuously extended from their integration intervals to $s = 0$ by these limits. Hence, by the Riemann-Lebesgue Lemma in 4.5, p. 50, we eventually obtain $\lim_{N\to\infty} I_1(N,t) = \lim_{N\to\infty} I_2(N,t) = 0$ for every $t$.

Conclusion: *The Fourier series of $f$ converges everywhere to $\frac{1}{2}[f(t+)+f(t-)]$.*

*If $f$ has the mean value property, then* $f(t) = \lim\limits_{N\to\infty}\displaystyle\int_{0}^{1} f(s)D_N(t-s)\,ds$ *for*

*every $t \in \mathbb{R}$.*

$\square$


## 7.2   The Theorem of Fejér, Convergence by Smoothing

### *Uniform Convergence of Fejér Means for Continuous Functions*

After P. Du Bois-Reymond (1831–1889) had shown that there are periodic functions $f$ whose Fourier series diverge on a dense set in their domain of definition, L. Fejér succeeded in 1904, only as late as 100 years after Fourier's work, to show the following theorem, already stated on p. 29:

**Fig. 7.1**  The Fejér kernels
$F_4$ and $F_{10}$

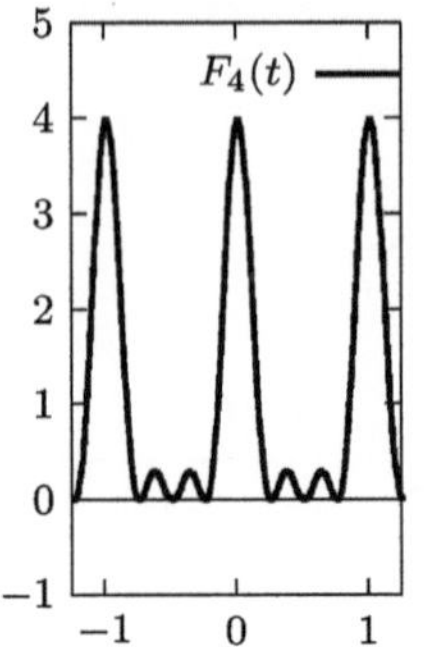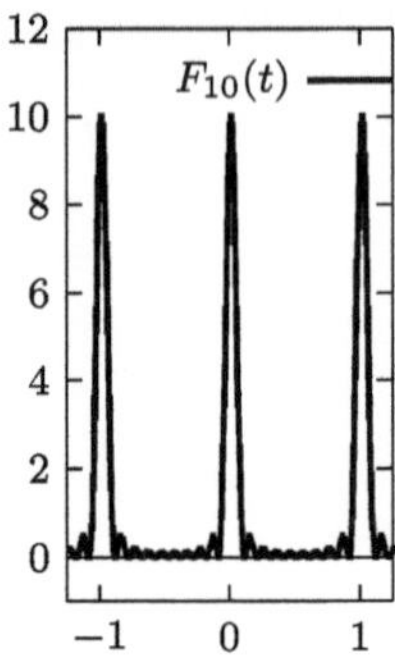

**Theorem 7.2 (Theorem of Fejér)** *Let $f : \mathbb{R} \to \mathbb{C}$ be a periodic continuous function. Then the arithmetic means of the partial sums $S_k$ of the Fourier series of $f$ converge uniformly to $f$.*

Again, we assume $T = 1$ as period. Instead of the Dirichlet kernels $D_n$ we use for the proof of this result the Fejér kernels (see Fig. 7.1)

$$F_n(t) = \frac{1}{n}(D_0 + \ldots + D_{n-1}).$$

With period $T = 1$ and $t \in \mathbb{R} \setminus \mathbb{Z}$ we have for $n \in \mathbb{N}$:

$$F_n(t) = \frac{1}{n}\sum_{k=0}^{n-1} \frac{\sin((2k+1)\pi t)}{\sin(\pi t)} \qquad \text{for} \quad t \notin \mathbb{Z}.$$

The $F_n$ are even functions, and $\displaystyle\int_{-1/2}^{1/2} D_n(t)\,\mathrm{d}t = 1$ implies

$$\int_{-1/2}^{1/2} F_n(t)\,\mathrm{d}t = 2\int_{0}^{1/2} F_n(t)\,\mathrm{d}t = 1.$$

With $\sin(x) = \Im(e^{jx})$ and with the formula for the sum of a finite geometric series, we obtain (Exercise)

$$F_n(t) = \begin{cases} \dfrac{1}{n}\dfrac{\sin^2(n\pi t)}{\sin^2(\pi t)} & \text{for} \quad t \in \mathbb{R}\setminus\mathbb{Z}, \\[2ex] n & \text{for} \quad t \in \mathbb{Z}. \end{cases}$$

Therefore, we find that all $F_n \geqslant 0$. The inequality $F_n(t) \leqslant \dfrac{1}{n \sin^2(\pi t)}$ for all $t \in \mathbb{R} \setminus \mathbb{Z}$ shows that the function sequence $(F_n)_{n \in \mathbb{N}}$ converges uniformly to zero in each interval $[\delta, 1/2]$, $0 < \delta < 1/2$.

***Proof of Fejér's Theorem***   After these remarks we now obtain for periodic continuous functions $f$ with period $T = 1$ and arbitrary $t \in \mathbb{R}$

$$\frac{1}{n} \sum_{k=0}^{n-1} S_k(t) - f(t) = \frac{1}{n} \sum_{k=0}^{n-1} \int_{-1/2}^{1/2} D_k(s) f(t - s) \, \mathrm{d}s - f(t)$$

$$= \int_{-1/2}^{1/2} F_n(s)(f(t - s) - f(t)) \, \mathrm{d}s \, .$$

The function $g(s) = f(t - s) - f(t)$ is continuous with $g(0) = 0$. Then for every $\varepsilon > 0$ there is a $\delta \in \,]0, \frac{1}{2}]$, so that $|g(s)| \leqslant \varepsilon$ for $|s| < \delta$.

Hence, from the inequality $\left| \dfrac{\sin^2(n\pi s)}{\sin^2(\pi s)} \right| \leqslant \dfrac{1}{\sin^2(\pi \delta)}$ for all $s \in [\delta, \frac{1}{2}]$ we obtain

$$\left| \frac{1}{n} \sum_{k=0}^{n-1} S_k(t) - f(t) \right| \leqslant \int_{|s|<\delta} F_n(s)|g(s)| \, \mathrm{d}s + \int_{\delta \leqslant |s| \leqslant 1/2} F_n(s)|f(t - s) - f(t)| \, \mathrm{d}s$$

$$\leqslant \varepsilon \int_{|s|<\delta} F_n(s) \, \mathrm{d}s + 2 \max_{0 \leqslant t \leqslant 1} |f(t)| \cdot \frac{2}{n} \int_{\delta}^{1/2} \frac{\sin^2(n\pi s)}{\sin^2(\pi s)} \, \mathrm{d}s$$

$$\leqslant \varepsilon + \frac{2}{n \sin^2(\pi \delta)} \cdot \max_{0 \leqslant t \leqslant 1} |f(t)| \, .$$

This yields the uniform convergence of the Fejér means $\dfrac{1}{n}(S_0 + \ldots + S_{n-1})$ to $f$, since the right-hand side becomes arbitrarily small for $n \to \infty$ independent of $t$.

$$\square$$

## Convergence of Fejér Means for Piecewise Continuous Functions

**Theorem 7.3**   *For piecewise continuous periodic functions $f : \mathbb{R} \to \mathbb{C}$, the Fejér means converge to $(f(t_0+) + f(t_0-))/2$ at any point $t_0$.*

***Proof*** Analogous to the proof of Dirichlet's theorem, we write the Fejér averaging in the form of a convolution integral. As period we assume again $T = 1$. At a point $t_0$ we have

$$\int_{-1/2}^{1/2} F_n(s)f(t_0 - s)\,\mathrm{d}s = \underbrace{\int_{-1/2}^{0} F_n(s)(f(t_0 - s) - f(t_0+))\,\mathrm{d}s + f(t_0+)\int_{-1/2}^{0} F_n(s)\,\mathrm{d}s}_{I_1(n,t_0)}$$

$$+ \underbrace{\int_{0}^{1/2} F_n(s)(f(t_0 - s) - f(t_0-))\,\mathrm{d}s + f(t_0-)\int_{0}^{1/2} F_n(s)\,\mathrm{d}s}_{I_2(n,t_0)} .$$

Setting $g(s) = f(t_0 - s) - f(t_0+)$ in $[-1/2, 0[$, we have $\lim_{s \to 0-} g(s) = 0$, and therefore with an estimation as in the proof before $\lim_{n \to \infty} I_1(n, t_0) = 0$. Analogously follows $\lim_{n \to \infty} I_2(n, t_0) = 0$, and thus the assertion. $\qquad\square$

An already discussed consequence of Fejér's theorem was the approximation theorem of Weierstrass. As further applications we now show the remaining statements from Sect. 3.2 on pointwise convergence of Fourier series.


## Convergence of Fourier Series of Piecewise Continuous Functions

The theorem on pointwise convergence of the Fejér means has an important consequence for Fourier series of piecewise continuous periodic functions $f : \mathbb{R} \to \mathbb{C}$.

**Theorem 7.4** *If the Fourier series $S_f$ of a piecewise continuous periodic function converges at a point $t_0$ at all, then $S_f(t_0) = (f(t_0+) + f(t_0-))/2$. Moreover, if $f$ is continuous at $t_0$, then $S_f(t_0) = f(t_0)$.*

***Proof*** The assertion follows directly from the fact that the Fejér means at $t_0$ converge to the limit $(f(t_0+) + f(t_0-))/2$, and from the fact that each convergent sequence of numbers and its arithmetic means[1] converge to the same limit. $\qquad\square$

---

[1] Arithmetic means in mathematical literature are also called *Cesàro means*.

## Completeness of the Trigonometric System

The following theorem is called the *completeness theorem* for the trigonometric function system.

**Theorem 7.5**  *A piecewise continuous periodic function $f : \mathbb{R} \to \mathbb{C}$ with the mean value property, whose Fourier coefficients $c_k$ all vanish, is the zero function.*

**Proof**  If there were point $t_0 \in \mathbb{R}$ at which such a $T$-periodic function $f$ were continuous with $f(t_0) \neq 0$, then $f$ would satisfy $f(t) \neq 0$ on a suitable interval around $t_0$. Then there were a $T$-periodic Fejér kernel $F_n$ such that

$$\frac{1}{T} \int\limits_{-T/2}^{T/2} f(t) F_n(t_0 - t)\, \mathrm{d}t \neq 0.$$

But this is a contradiction to the assumption: From $c_k = 0$ for all $k \in \mathbb{Z}$ we get

$$\int\limits_{-T/2}^{T/2} f(t) P(t)\, \mathrm{d}t = 0$$ for each $T$-periodic trigonometric polynomial, thus also for
$F_n(t_0 - t)$. Therefore $f$ must be zero under the conditions of the theorem.  □

Applying the completeness theorem to the difference $f - g$ of two piecewise continuous periodic functions $f$ and $g$, we immediately obtain the following theorem about uniqueness of Fourier series:

**Theorem 7.6**  *Two piecewise continuous periodic functions $f$ and $g$ with the mean value property and the same Fourier coefficients are equal.*

## Fourier Series of Piecewise Continuously Differentiable Functions

**Theorem 7.7**  *If a piecewise continuously differentiable periodic $f : \mathbb{R} \to \mathbb{C}$ is continuous, then its Fourier series converges uniformly to $f$.*

**Proof**  If $f$ is $T$-periodic and satisfies the assumptions, then we have for the Fourier coefficients $c_k$ of $f$ and $c'_k$ of $f'$

$$c'_k = jk\frac{2\pi}{T} c_k \quad \text{and} \quad c_k = \frac{T}{j2\pi k} c'_k \quad \text{for } k \neq 0 \,.$$

Since for complex numbers $a, b$ we have $0 \leqslant (|a| - |b|)^2 = |a|^2 + |b|^2 - 2|a||b|$, we obtain for $k \neq 0$ the estimate

$$|c_k| \leqslant \frac{1}{2} \left( \frac{T^2}{4\pi^2 |k|^2} + |c'_k|^2 \right).$$

From the convergence of $\sum_{k=1}^{\infty} \frac{1}{k^2}$ and from the Bessel inequality for $f'$ follows therefore that $\sum_{k=-\infty}^{+\infty} |c_k| < \infty$, i.e., the Fourier series of $f$ converges absolutely and uniformly to a continuous function $g$. According to the uniqueness theorem, then $g = f$. $\qquad\square$

**Theorem 7.8** *If a periodic piecewise continuously differentiable $f : \mathbb{R} \to \mathbb{C}$ has discontinuities, the uniform convergence of the Fourier sequence $S_f$ still follows on any closed interval, which does not contain a discontinuity point of $f$.*

**Proof** It suffices to consider $T$-periodic, piecewise continuously differentiable functions $f$ with a single discontinuity at $t_0$ in $[0, T]$. We assume that $f$ is continuous otherwise and possesses the mean value property $f(t_0) = (f(t_0+) + f(t_0-))/2$. We write $f$ in the form $f(t) = g(t) + r(t)$ with

$$g(t) = f(t) - \frac{1}{\pi}[f(t_0+) - f(t_0-)] \, S\left( \frac{2\pi}{T}(t - t_0) \right),$$

$$r(t) = \frac{1}{\pi}[f(t_0+) - f(t_0-)] \, S\left( \frac{2\pi}{T}(t - t_0) \right),$$

where $S(2\pi t/T)$ is the $T$-periodic sawtooth function of p. 24. Then the Fourier series of $g$ uniformly converges to the continuous function $g$. The Fourier series of $r$ converges uniformly on each closed interval not containing $t_0$, as we have shown in Sect. 3.1. From this follows the convergence of $S_f = S_g + S_r$ as claimed. $\qquad\square$

## *Vanishing of the Gibbs Phenomenon in Fejér Means*

For periodic piecewise continuously differentiable functions $f(t) = \sum_{k=-\infty}^{+\infty} c_k \, e^{jk\omega_0 t}$, we had seen in Sect. 3.2 that using partial sums to approximate $f$ causes the Gibbs phenomenon. Using such a partial sum

$$S_n(t) = \sum_{k=-n}^{n} c_k \, e^{jk\omega_0 t} = \sum_{k=-n}^{n} w_n(k) c_k \, e^{jk\omega_0 t}$$

with

$$w_n(t) = \begin{cases} 1 & \text{for} \quad |t| \leqslant n \\ 0 & \text{for} \quad |t| > n \end{cases}$$

corresponds to weighting the spectral values $c_k$ with the *rectangular window function* $w_n$. At all jump discontinuities of the real or imaginary part of $f$, there is an overshoot of approximately 9% of the respective jump height. We now show that the Fejér means of $f$ do not exhibit the Gibbs phenomenon anymore. The Fejér means

$$\frac{1}{n+1}\sum_{k=0}^{n} S_k(t) = \sum_{k=-n}^{n} \left(1 - \frac{|k|}{n+1}\right) c_k\, e^{jk\omega_0 t}$$

prevent the Gibbs phenomenon through weighting the spectrum with the *triangle window.*

$$w_n(t) = \begin{cases} 1 - |t|/(n+1) & \text{for} \quad |t| \leqslant n+1 \\ 0 & \text{for} \quad |t| > n+1. \end{cases}$$

***Proof*** It suffices to consider real-valued functions. We note that for any $\varepsilon > 0$, any $0 < \delta < 1/2$, for 1-periodic Fejér kernels $F_n$ and 1-periodic, piecewise continuously differentiable real-valued functions $f$, we have the inequalities

$$-\varepsilon < \int_{-1/2}^{-\delta} F_n(s)f(t-s)\,\mathrm{d}s + \int_{\delta}^{1/2} F_n(s)f(t-s)\,\mathrm{d}s < \varepsilon,$$

if $n$ is greater than a suitably chosen $n_0 \in \mathbb{N}$.

Now, if $m \leqslant f(t) \leqslant M$ for $t \in [a, b]$ and $0 < \delta < \min\left\{\frac{1}{2}, \frac{b-a}{2}\right\}$, then to give $\varepsilon > 0$ there is a $n_0 \in \mathbb{N}$, so that we obtain for $n \geqslant n_0$ and $t \in [a+\delta, b-\delta]$ the estimate

$$\frac{1}{n+1}\sum_{k=0}^{n} S_k(t) = \int_{-1/2}^{1/2} F_{n+1}(s)f(t-s)\,\mathrm{d}s \leqslant \int_{-\delta}^{\delta} F_{n+1}(s)f(t-s)\,\mathrm{d}s + \varepsilon \leqslant M + \varepsilon,$$

because $(t - s) \in [a, b]$ for $|s| \leqslant \delta$, thus $f(t-s) \leqslant M$ and $F_{n+1} \geqslant 0$, and

$$\int_{-\delta}^{\delta} F_{n+1}(s)\,\mathrm{d}s \leqslant 1.$$

Analogously one obtains $m - \varepsilon \leqslant \dfrac{1}{n+1}\sum_{k=0}^{n} S_k(t)$ for $t \in [a+\delta, b-\delta]$.   $\square$

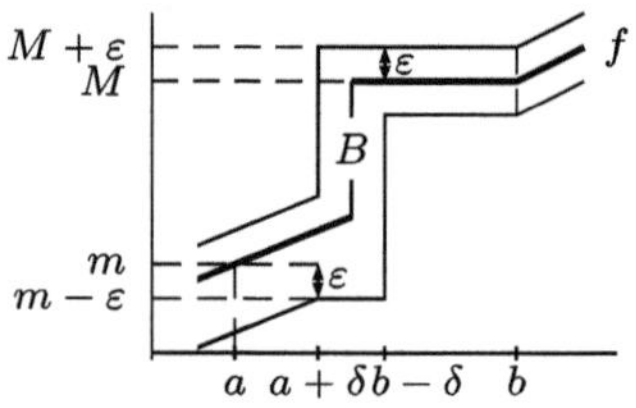

**Fig. 7.2** Disappearance of the Gibbs phenomenon

These inequalities show that the Fejér means can be kept in a specified tolerance zone $B = B(a, b, \varepsilon, \delta)$ with arbitrarily small $\varepsilon$ for sufficiently large $n \in \mathbb{N}$, even in the neighborhood of discontinuities. Therefore, the Gibbs phenomenon cannot occur. To illustrate this, consider the following Fig. 7.2 and again Fig. 3.11 on p. 35.

When partial sums of Fourier series are used for approximation and reconstruction of $T$-periodic functions from spectral values $c_k$, the Gibbs phenomenon, more generally speaking, the oscillatory behavior of the approximation plays an important role. The preceding considerations show that the approximations can be smoothed and kept within certain tolerance ranges if, for example, the Fejér means are used for the approximation. The proofs also show that other weight functions can be used instead of the spectral triangle window. A triangle window as a weighting function in the spectrum corresponds to convolution with a Fejér kernel $F_n$ in time domain.

In all theorems the Fejér kernels $F_n$ can be replaced by arbitrary kernels $K_n$, if these convolution kernels, here related to 1-periodic functions, are nonnegative continuous even functions with $\displaystyle\int_0^1 K_n(t)\,\mathrm{d}t = 1$, and if $\displaystyle\lim_{n\to\infty} K_n(t) = 0$ uniformly in each interval $[\delta, 1/2]$, $0 < \delta < 1/2$. Such kernels are called *summation kernels*.

This finding is the starting point for the construction of other window functions and related kernels, which—depending on the purpose of application—produce more advantageous approximations than the Fejér means. Such advantages can be, for example, steeper slopes at jump discontinuities, thereby also higher power of the approximation, less smoothing—technically speaking a higher resolution—and much more. See also the later Sects. 12.5 and 12.6 on windowed Fourier transforms. The mentioned conditions on the sequence $K_n$ of convolution kernels can still be weakened, so that it is not necessary to require $K_n \geq 0$. An example is the *de la Vallée Poussin kernel* $V_{2n} = 2F_{2n+1} - F_n$ with the Fejér kernel $F_n$. For this and other convolution kernels, refer to Walker (1988) and further references cited there.

## 7.3  The Parseval Equation

We first show that periodic convolutions of piecewise continuous $T$-periodic functions $f : \mathbb{R} \to \mathbb{C}$ are continuous, and then we deduce the Parseval equation for such functions:

$$\textit{If f has Fourier coefficients } c_k, \textit{ then } \quad \frac{1}{T}\int_0^T |f(t)|^2\,\mathrm{d}t = \sum_{k=-\infty}^{+\infty} |c_k|^2.$$

## *Continuity of Periodic Convolutions of Piecewise Continuous Functions*

It is sufficient to consider $2\pi$-periodic functions $f$ and $h$, each with only one jump discontinuity at $t_0$ and $t_1$ in $[0, 2\pi]$, respectively, and otherwise continuous. They then have the form $f = g_1 + r_1$ and $h = g_2 + r_2$,

$$f(t) = g_1(t) + \frac{1}{\pi}(f(t_0+) - f(t_0-))S(t - t_0)\,,$$

$$h(t) = g_2(t) + \frac{1}{\pi}(h(t_1+) - h(t_1-))S(t - t_1)\,,$$

where $g_1$ and $g_2$ are continuous on $\mathbb{R}$ and $S$ is the $2\pi$-periodic sawtooth function.

The uniform continuity of $g_1$ and $g_2$ implies immediately the continuity of the convolutions $(g_1 * g_2)_{2\pi}$, $(g_1 * r_2)_{2\pi}$, and $(g_2 * r_1)_{2\pi}$, by deductions like, for example,

$$\left|\int_0^{2\pi} r_1(s)(g_2(t + \delta - s) - g_2(t - s))\,\mathrm{d}s\right| \leqslant \int_0^{2\pi} |r_1(s)||g_2(t + \delta - s) - g_2(t - s)|\,\mathrm{d}s$$

$$< 2\pi\varepsilon \sup_{0 \leqslant s \leqslant 2\pi} |r_1(s)|.$$

The second term in the integrand of the right-hand side integral becomes for any $s$ in $[0, 2\pi]$ smaller than any $\varepsilon > 0$, if only $\delta > 0$ is chosen small enough. So it remains to prove the continuity of the $2\pi$-periodic convolution of two sawtooth functions $S_{t_0}(t) = S(t - t_0)$ and $S_{t_1}(t) = S(t - t_1)$, $t_0, t_1 \in [0, 2\pi]$. In the equation

$$\int_0^{2\pi} S(x - t_0)S(t - t_1 - x)\,\mathrm{d}x = \int_0^{2\pi} S(u)S(t - t_0 - t_1 - u)\,\mathrm{d}u$$

we observe $(S_{t_0} * S_{t_1})_{2\pi}(t) = (S * S)_{2\pi}(t - t_0 - t_1)$. Therefore, we have to show the continuity of $(S * S)_{2\pi}$. However, with $S(t) = \begin{cases} (\pi - t)/2 & \text{for} \quad 0 < t < 2\pi, \\ (-t - \pi)/2 & \text{for} \quad -2\pi < t < 0, \end{cases}$ we simply calculate for $t \in \,]0, 2\pi[$:

$$(S * S)_{2\pi}(t) = \frac{1}{8\pi}\left(\int_0^t (\pi - s)(\pi - t + s)\,\mathrm{d}s + \int_t^{2\pi} (\pi - s)(-t - \pi + s)\,\mathrm{d}s\right)$$

$$= -\frac{1}{8}t^2 + \frac{\pi}{4}t - \frac{\pi^2}{12}.$$

The continuity of the $2\pi$-periodic extension then results from $(S * S)_{2\pi}(0+) = (S * S)_{2\pi}(2\pi-)$.

## The Parseval Equation for Piecewise Continuous Periodic Functions

If $f$ is piecewise continuous on $[0, T]$ with Fourier coefficients $c_k$, $k \in \mathbb{Z}$, then the function $\overline{f(-t)}$ possesses the Fourier coefficients $\overline{c_k}$, $k \in \mathbb{Z}$, according to Sect. 4.1.

Hence, we have for the $T$-periodic convolution $g(t) = \dfrac{1}{T}\displaystyle\int_0^T f(u)\overline{f(u - t)}\,\mathrm{d}u =$

$\displaystyle\sum_{k=-\infty}^{+\infty} |c_k|^2\, e^{jk\omega_0 t}$ with $\omega_0 = 2\pi/T$ (cf. p. 64).

Now, the function $g$ is continuous, and its Fourier series converges uniformly to $g$ by the completeness theorem on p. 135. In particular, the *Parseval equation* holds.

$$g(0) = \frac{1}{T}\int_0^T |f(u)|^2\,\mathrm{d}u = \sum_{k=-\infty}^{+\infty} |c_k|^2.$$

This also shows that the Fourier series of $f$ converges to $f$ in quadratic mean

**Remark** As already remarked on p. 63, one can prove the continuity of the convolution $(f * h)_T$ also for $T$-periodic functions $f$ and $h$, which are square-Lebesgue-integrable on $[0, T]$. Hence, the Parseval equation and the convergence of their Fourier series in quadratic mean follow also for such functions $f$ or $h$ (cf. Zygmund (2003) and Exercise A8).

## 7.4  Fourier Series for Functions of Several Variables

With Fourier series expansions one can also represent many functions of several variables, which are defined on cubic domains $Q$ in $\mathbb{R}^n$. One obtains quite analogous results as in the one-dimensional case. We restrict ourselves to some statements for the case of two variables without proofs.

In general, the theory for Fourier series of functions with several variables is far more complex and complicated than that for one-dimensional domains of definition. Think for example of different geometrical shapes as domains of definition, which are not axis parallel. If you are interested, you can find details of this field, e.g., in Tolstov (1976), Stein and Weiss (1971), or Zygmund (2003).

**Theorem 7.9** *If $f : Q \to \mathbb{C}$ is square-integrable on $Q =]-\pi, \pi[\times]-\pi, \pi[$, then Parseval's equation*

$$\frac{1}{4\pi^2} \int\limits_{-\pi}^{\pi} \int_{-\pi}^{\pi} |f(x, y)|^2 \, dx \, dy = \sum_{l,m=-\infty}^{+\infty} |c_{lm}|^2$$

*is valid with the Fourier coefficients* $c_{lm} = \dfrac{1}{4\pi^2} \int\limits_{-\pi}^{\pi} \int_{-\pi}^{\pi} f(x, y) \, e^{-j(lx+my)} \, dx \, dy$ .

*The Fourier series* $\displaystyle\sum_{l,m=-\infty}^{+\infty} c_{lm} \, e^{j(lx+my)}$ *converges in quadratic mean to $f$, i.e., for $N_1, N_2 \to \infty$ simultaneously, the error*

$$\int\limits_{-\pi}^{\pi} \int_{-\pi}^{\pi} \left| f(x, y) - \sum_{|l| \leqslant N_1} \sum_{|m| \leqslant N_2} c_{lm} \, e^{j(lx+my)} \right|^2 dx \, dy$$

*becomes arbitrarily small. In that sense, $f$ has a Fourier series expansion.*

*If $f$ is twice continuously differentiable with support $\mathrm{supp}(f) \subset Q$, then the Fourier series of $f$ converges uniformly and represents $f$ pointwise.*

The *support* $\mathrm{supp}\,(f)$ is the closure of the set $\{(x, y) \in Q \mid f(x, y) \neq 0\}$ in $\mathbb{R}^2$.

It holds true the following extension of *Fejér's theorem*:

**Theorem 7.10** *If $f : Q \to \mathbb{C}$ is continuous in $Q =]-\pi, \pi[\times]-\pi, \pi[$, and $\mathrm{supp}(f) \subset Q$, then the Fejér means*

$$M_{N_1, N_2}(x, y) = \frac{1}{(N_1 + 1)(N_2 + 1)} \sum_{k_1=0}^{N_1} \sum_{k_2=0}^{N_2} S_{k_1, k_2}(x, y)$$

*converge uniformly to $f$, when simultaneously $N_1 \to \infty$ and $N_2 \to \infty$.*

Here, the partial sums $S_{k_1, k_2}(x, y)$ are defined by

$$S_{k_1, k_2}(x, y) = \sum_{|l| \leqslant k_1} \sum_{|m| \leqslant k_2} c_{lm} \, e^{j(lx+my)} \, .$$

For the proof one uses the representation

$$M_{N_1,N_2}(x, y) = \frac{1}{4\pi^2} \int_{-\pi}^{\pi} \int_{-\pi}^{\pi} f(u, v) F_{N_1,N_2}(x - u, y - v) \, du \, dv \, ,$$

where $F_{N_1,N_2}$ is the product of the $2\pi$-periodic Fejér kernels $F_{N_1}(x)$ and $F_{N_2}(y)$. Then, the proof follows completely the line of the proof in the one-dimensional case (cf. 7.2). In particular, we obtain the following variant of Weierstrass' approximation theorem:

**Theorem 7.11 (Theorem of Weierstrass)** *If $f$ is continuous on $Q =] - \pi, \pi[^2$ and* $\mathrm{supp}(f) \subset Q$, *then there exists for each $\varepsilon > 0$ a polynomial $P(x, y)$ such that the following inequality is valid for all $(x, y) \in Q$:*

$$\left| f(x, y) - P(x, y) \right| < \varepsilon.$$

*Thus, the function $f$ can be uniformly approximated by polynomials.*

All theorems can be rephrased for rectangles other than $Q$ as above and are also valid for more than two variables. As an example we consider a square-integrable function $f$ on the rectangle $Q =]0, L_1[\times]0, L_2[$, which can be expanded into a double sine series:

**Theorem 7.12** *If $f$ is square-integrable on $Q =]0, L_1[\times]0, L_2[$, then the series*

$$\sum_{n,m=1}^{\infty} b_{n,m} \sin\left(\frac{n\pi x}{L_1}\right) \sin\left(\frac{m\pi y}{L_2}\right),$$

$$b_{n,m} = \frac{4}{L_1 L_2} \int_{0}^{L_2} \int_{0}^{L_1} f(x, y) \sin\left(\frac{n\pi x}{L_1}\right) \sin\left(\frac{m\pi y}{L_2}\right) dx \, dy \, ,$$

*converges to $f$ in quadratic mean.*

To give an idea for the theorem, without an exact proof, we expand $f(x, y)$ for fixed $y$ into a sine series (cf. p. 43)

$$f(x, y) = \sum_{n=1}^{\infty} b_n \sin\left(\frac{n\pi x}{L_1}\right) \quad \text{with} \quad b_n = \frac{2}{L_1} \int_{0}^{L_1} f(x, y) \sin\left(\frac{n\pi x}{L_1}\right) dx \, .$$

If we consider $b_n$ as a function of $y$, which in turn can be expanded into a sine series, then

$$b_n = \sum_{m=1}^{\infty} b_{n,m} \sin\left(\frac{m\pi y}{L_2}\right) \quad \text{with} \quad b_{n,m} = \frac{2}{L_2} \int_0^{L_2} b_n \sin\left(\frac{n\pi y}{L_2}\right) dy,$$

and thereby

$$f(x, y) = \sum_{n=1}^{\infty} \sum_{m=1}^{\infty} b_{n,m} \sin\left(\frac{n\pi x}{L_1}\right) \sin\left(\frac{m\pi y}{L_2}\right),$$

$$b_{n,m} = \frac{4}{L_1 L_2} \int_0^{L_2} \int_0^{L_1} f(x, y) \sin\left(\frac{n\pi x}{L_1}\right) \sin\left(\frac{m\pi y}{L_2}\right) dx\, dy.$$

Applications of Fourier series of several variables arise in linear partial differential equations with constant coefficients in cubic domains. One can then try to solve such equations with a separation of variables approach, analogous to the procedure for the string vibration. For such problems the series expansions had just been introduced by Bernoulli and Fourier. First applications were the solution of heat conduction problems and also the treatment of vibrating membranes. Here, the *eigensolutions* (cf. p. 5) lead to the trigonometric function system. We consider an example to which we refer later on in Sect. 9.5 in more detail.

### *A Dirichlet Boundary Value Problem for a Rectangle Membrane*

Let an elastic membrane be fixed at the boundary of the rectangle $Q = [0, L] \times [0, L]$ in the plane. Load by a force, perpendicular to the plane, causes a displacement of the membrane. Let the tension, which is exerted by the fastening, be isotropic, so that it is described by a scalar $k$ (of physical dimension N/m). If $f$ is the area density of the force, then small displacements $u$ in equilibrium state are described approximately by the differential equation

$$-k\Delta u = -k\left(\frac{\partial^2 u}{\partial x^2} + \frac{\partial^2 u}{\partial y^2}\right) = f \text{ in } Q \setminus \partial Q, \quad u = 0 \text{ on the boundary } \partial Q \text{ of } Q.$$

This is called a Dirichlet boundary value problem. The functions

$$u_{n,m} = \sin\left(\frac{n\pi x}{L}\right) \sin\left(\frac{m\pi y}{L}\right)$$

are *eigenfunctions* of $-\Delta$ for the *eigenvalues* $\lambda_{n,m} = \left(\frac{n\pi}{L}\right)^2 + \left(\frac{m\pi}{L}\right)^2$, i.e., for all $n, m \in \mathbb{N}$ holds

$$- \Delta u_{n,m} = \lambda_{n,m} u_{n,m}.$$

If the force density $f$ is a linear combination of these eigenfunctions $u_{n,m}$,

$$f(x, y) = \sum_{n=1}^{N} \sum_{m=1}^{M} b_{n,m} \sin\left(\frac{n\pi x}{L}\right) \sin\left(\frac{m\pi y}{L}\right),$$

then the displacement $u$ with $u(x, y) = 0$ on the boundary $\partial Q$ is given by

$$u(x, y) = \frac{1}{k} \sum_{n=1}^{N} \sum_{m=1}^{M} \frac{b_{n,m}}{\lambda_{n,m}} \sin\left(\frac{n\pi x}{L}\right) \sin\left(\frac{m\pi y}{L}\right),$$

verified by inserting it into the differential equation.

Right-hand sides $f$ in the differential equation, which are of the form of a trigonometric polynomial as above, are best understood as approximations for the exact physical force action. The solution $u$ is then an approximation of the real membrane displacement. For this, cf. exercise A7 at the end of this section and later on Sect. 9.5, p. 253.

In order to obtain good approximations for different physical situations one would like to have a solution theory for right-hand sides $f$ being as general as possible. This is achievable with Fourier series expansions of $f$ and $u$. The higher the order of the partial sums of these series expansions is, the better approximations can be expected. For square -integrable $f$ on $Q$ with the Fourier series representation

$$f(x, y) = \sum_{n,m=1}^{\infty} b_{n,m} \sin\left(\frac{n\pi x}{L}\right) \sin\left(\frac{m\pi y}{L}\right),$$

we obtain by the approach

$$u(x, y) = \frac{1}{k} \sum_{n,m=1}^{\infty} \frac{b_{n,m}}{\lambda_{n,m}} \sin\left(\frac{n\pi x}{L}\right) \sin\left(\frac{m\pi y}{L}\right)$$

with term-by-term differentiation of the Fourier series of $u$

$$- k\Delta u = - \sum_{n,m=1}^{\infty} \frac{b_{n,m}}{\lambda_{n,m}} \Delta\left(\sin\left(\frac{n\pi x}{L}\right) \sin\left(\frac{m\pi y}{L}\right)\right) = f(x, y).$$

Thus $u$ is the desired solution. The method is elegant, but it requires a mathematically exact reasoning: The Fourier series of $f$ does not converge pointwise in general, but in quadratic mean to $f$, the term-by-term differentiation of the Fourier series of $u$ is a questionable procedure (cf. Sect. 4.3), and the question arises in which sense this series converges and takes on the zero boundary values. Satisfying answers are provided by the distribution theory, in literature also called the theory of generalized functions. Therefore, in the next chapter we work out some basics of distribution theory and take up again the Dirichlet problem of the membrane in Chap. 9.

A detailed discussion of the cited theorems about Fourier series of several variables and their application to partial differential equations can be found, e.g., in Shapiro (2019). Analogously to the just given example, questions of mathematical physics often lead to function systems arising as eigensolutions, having similar properties as the trigonometric system. Then, one can find solutions for such problems through replacing the trigonometric system by *series expansions according to the eigenfunction system*. This more general concept, which has its roots in classical Fourier analysis, is outlined in Chap. 14.

## *A Warning Example*

As a warning against a purely formal approach, consider the following example: Given is the differential equation $y''(x) = x - \pi/2$ on $[0, \pi]$ with the boundary conditions $y^{(3)}(0) = y^{(3)}(\pi) = 0$. We try to find a trigonometric series solution, i.e., we assume $y(x) = a_0/2 + \sum_{k=1}^{\infty} (a_k \cos(kx) + b_k \sin(kx))$. By differentiating twice term by term, we obtain with the uniformly convergent Fourier cosine series of $x - \pi/2$ on the right-hand side

$$-\sum_{k=1}^{\infty} (a_k k^2) \cos(kx) + b_k \sin(kx)) = -\frac{4}{\pi} \sum_{k=0}^{\infty} \frac{\cos((2k+1)x)}{(2k+1)^2}.$$

Comparing the coefficients yields: $a_k = 0$ for even $k$, $b_k = 0$ for all $k$, and $a_{2k+1} = 4/(\pi(2k+1)^4)$ for $k \in \mathbb{N}$. The coefficient $a_0$ can be arbitrary. We thus obtain

$$y(x) = \frac{a_0}{2} + \frac{4}{\pi} \sum_{k=0}^{\infty} \frac{\cos((2k+1)x)}{(2k+1)^4}.$$

This also satisfies the boundary conditions, since $y^{(3)}(x) = \frac{4}{\pi} \sum_{k=0}^{\infty} \frac{\sin((2k+1)x)}{(2k+1)}$, which is zero at the endpoints $0$ and $\pi$. However, with this formal procedure we did not pay attention whether or not all operations made sense. Namely, the procedure must involve a serious mistake, because the problem does not have a solution at all. Actually, differentiation of $y''(x) = x - \pi/2$ implies $y^{(3)}(x) = 1$ for all $x \in [0, \pi]$, and thus the boundary conditions cannot be fulfilled by any solution. The point is that the series given for $y^{(3)}$ is the Fourier series of the *sgn* function. It is zero at

$x = 0$ and $x = \pi$, while the limits from the left and right at those points are one. The series itself is not any first derivative, and thus not a third derivative of any function on the entire interval $[0, \pi]$.


## 7.5  Reasons for the Transition to Distributions

### *A Reviewing Summary*

We have seen that Fourier series are appropriate to solve wave and heat equations and potential problems in suitable domains. They also permitted to describe time-invariant linear systems in electrical engineering or mechanics for periodic inputs or forces. At studying such examples, we have found an approach to frequency and spectral concepts, which are fundamental for many fields of physics and technology. We have learned the most important calculation rules in dealing with trigonometric sums and series and have worked with the discrete Fourier transform. With DFT, DCT, Chebyshev polynomials, interpolation, and Clenshaw-Curtis quadrature, we have also made first steps in numerical applications, and we have studied some essential convergence properties of Fourier series.

Most of these results can be applied to more general classes of functions than piecewise continuous or continuously differentiable functions. For example, the Parseval equation or the continuity of periodic convolutions can still be proven for all $T$-periodic functions $f$ with $\displaystyle\int_0^T |f(t)|^2 \, dt < \infty$. The transition to such more general functions, connected with the transition from the traditional Riemann integral to the more modern Lebesgue integral, which is more flexible in decisive points, and finally to distribution theory is not only a mathematical pastime. In fact it goes back to the objections against Fourier's approach in 1807, and to requirements, which arose from real-world application problems and necessitate such a further development of the mathematical tools. This can easily be explained by some examples with the knowledge we have acquired so far.


### *Transition to Distributions and Lebesgue Integral*

For example, let us choose initial conditions $f(x)$ of the following forms for the vibrating string problem to study a computationally straightforward mathematical model for a plucked string. Then, we immediately recognize at D'Alembert's form of the solution (cf. p. 74) that the formally calculated solution $u(x, t)$ is by no means a twice differentiable function. But what should it then mean, since we want to use it in a second-order differential equation? (see Fig. 7.3 for simple initial conditions).

**Fig. 7.3** Non-differentiable
initial conditions

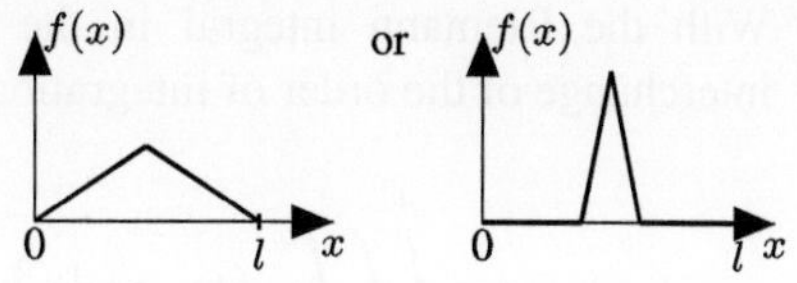


We have also seen that term-by-term derivatives of the Fourier series of periodic functions with jump discontinuities usually no longer converge. Nevertheless, the partial sums of the Dirichlet kernels can be well understood as approximations of a periodic impulse sequence (cf. p. 16). Couldn't we then consider the corresponding Fourier series as an ideal impulse sequence and also calculate mathematically correct with it, although it diverges everywhere? Answers to such questions, and also reliable methods of calculation, result from the theory of distributions. Moreover, these methods permit much easier computations compared with classical differential calculus.

In the treatment of stable time-invariant linear systems given by ordinary differential equations, in Sect. 5.2 for periodic right-hand sides $f$ of such equations, it was assumed that $f$ should be continuous and piecewise continuously differentiable. For many applications this is a very restrictive condition. An example would be a periodic switch-on and switch-off process, described by a discontinuous rectangle meandering function $f$. The reason for this restriction was the Riemann integral used with the traditional notion of primitive functions.

The Bessel inequality and the Parseval equation have been shown without theory effort only for piecewise continuous periodic functions. The Parseval equation for functions $f$ from $L^2([0, T])$ (cf. S. 62) removes this restriction and opens up practicable application of Fourier series in the study of linear systems (cf. also the remark on p. 66).

In these generalizations, the Riemann integral is replaced by the Lebesgue integral. The Lebesgue integral completes the set of integrable functions in a similar way as the real numbers complete the set of rational numbers. There are examples where a sequence $(f_n)_{n \in \mathbb{N}}$ of Riemann integrable functions on $[a, b] \subset \mathbb{R}$ for $n \to \infty$ converges to a function $f$ which is no longer Riemann integrable. In particular, then *it does not hold*

$$\lim_{n \to \infty} \int_a^b f_n(t)\, \mathrm{d}t = \int_a^b \lim_{n \to \infty} f_n(t)\, \mathrm{d}t \,.$$

V. Volterra (1860–1940) gave an example of a differentiable function $f$ on $[0, 1]$ whose derivative $f'$ is bounded but not Riemannian integrable. In particular, *it does not hold*

$$\int_0^1 f'(t)\, \mathrm{d}t = f(1) - f(0).$$

With the Riemann integral in the case of functions of several variables, the interchange of the order of integration and the equality

$$
\int\limits_c^d \left( \int_a^b f(x, y)\, dx \right) dy = \int\limits_a^b \left( \int_c^d f(x, y)\, dy \right) dx
$$

in general is only true if, besides the existence of the integrals

$$
\int\limits_a^b f(x, y)\, dx \quad \text{and} \quad \int\limits_c^d f(x, y)\, dy \, ,
$$

it is required that $f$ is bounded. The interchangeability of the integration order is therefore not ensured for improper Riemann integrals without additional conditions. Moreover, the existence of the improper Riemann integral of a function $f$ is not equivalent to the existence of the corresponding integral of the function $|f|$.

Now, a practitioner wants to work on his actual problem without having to worry about convergence problems all the time. In fact, he would like to differentiate and integrate series term by term, convolve, interchange limit processes in integrals and series—and usually does so without too much concern. Why such questionable procedures, nevertheless, and often just because of this, produce meaningful results will be discussed in the next chapter on distribution theory.

From now on we use the integration theory, established in 1902 by H. Lebesgue (1875–1941), which is already taught today in beginners' lectures and which is more efficient with respect to interchange of integrals with limits and therefore computationally simpler than the Riemann integral.

For application-oriented readers, there will be no additional difficulties in the following chapters compared to the usually acquired integral calculus. Mathematically interested readers will find the used theorems from integration theory with corresponding literature references in Appendix B.

## 7.6   Exercises

**(A1)** Prove that $\displaystyle\sum_{k=0}^{n-1} \sin((2k+1)\pi t) = \frac{\sin^2(n\pi t)}{\sin(\pi t)}$.

**(A2)** Compute approximately by Taylor series expansion $\displaystyle\mathrm{Si}(\pi) = \int\limits_0^\pi \frac{\sin(t)}{t}\, dt$.

**(A3)**   Compute the gradient of the Fejér means at the point $t_0 = 0$ for the sawtooth function $S(t) = (\pi - t)/2$ in $]0, 2\pi[$, $S(t + 2k\pi) = S(t)$, $k \in \mathbb{Z}$. Compare the corresponding gradients of the partial sums of the Fourier series of $S$.

**(A4)**   Show that the approximations $\dfrac{1}{T} \displaystyle\int_{-T/2}^{T/2} f(t - s) K_n(s)\, ds$ converge uniformly to $f$ for continuous $T$-periodic functions $f$ and $T$-periodic summation kernels $K_n$, which were introduced on p. 138.

**(A5)**   Show that the Fourier series of the $2\pi$-periodic function

$$f(t) = \begin{cases} -1 & \text{for } t \in\, ]-\pi\,,\, 0], \\ +1 & \text{for } t \in\, ]0\,,\, \pi], \end{cases}$$

has strictly positive partial sums in $]0, \pi[$, and strictly negative partial sums in $]-\pi, 0[$. Consider for sufficiently large $n \in \mathbb{N}$ the Fejér means and the tolerance region around the graph of $f$, where they can be restricted to.

**(A6)**★   *A Fourier Series Representation of an Unbounded Periodic Function.*

(a)   Show that the Fourier series of a $2\pi$-periodic function, which is absolutely integrable on $[0, 2\pi]$, converges to $f(t_0)$ provided $f$ is differentiable at $t_0$.

(b)   Show that $f(t) = \ln \left| 2 \sin\left(\dfrac{t}{2}\right) \right|$, $t \neq 2k\pi$, $k \in \mathbb{Z}$, is absolutely integrable on $[0, 2\pi]$.

(c)   Show that $f(t) = -\displaystyle\sum_{n=1}^{\infty} \dfrac{\cos(nt)}{n}$ for $t \neq 2k\pi$, $k \in \mathbb{Z}$, and $f$ from (b).

(d)   Show $\ln \left| 2 \cos\left(\dfrac{t}{2}\right) \right| = \displaystyle\sum_{n=1}^{\infty} (-1)^{n+1} \dfrac{\cos(nt)}{n}$ for $t \neq (2k+1)\pi$, $k \in \mathbb{Z}$.

Hint: Examine the proof of Dirichlet's theorem on p. 130 and use $S(0+) = \pi/2$ for the sawtooth function $S$ (cf. p.. 26).

**(A7)**★   Let a square $Q = [0, L]^2$ be given, at the boundary of which a loaded elastic membrane is fixed. The side length is $L = 1\,\text{m}$, the tension $k = 2\,\text{N/m}$. The area density of the external force is constantly $f(x, y) = 1\,\text{N/m}^2$.
Calculate an approximation of the displacement $u(x, y)$ of the membrane in the equilibrium state, i.e., solve $-k\Delta u = f$ in $Q$, $u = 0$ on the boundary of $Q$, replacing the function $f$ by the partial sum

$$\sum_{n,m=1}^{3} b_{n,m} \sin\left(\frac{n\pi x}{L}\right) \sin\left(\frac{m\pi y}{L}\right)$$

of its Fourier series expansion. What is the calculated displacement at the point with coordinates $x = y = L/2$? Generate a graphical representation of this approximate solution and compare it with the figure on p. 258.

**(A8)**$^\star$  Show that the periodic convolution of $f, g \in L^2[-\pi, \pi]$, is continuous. Use Hölder's inequality (cf. Appendix B) and $\lim_{h \to 0} \|f(. + h) - f(.)\|_2 = 0$. Implication: *Parseval's equation in* $L^2[-\pi, \pi]$! Elementary proofs of the given assertions can be found in Zygmund (2003).

# Chapter 8
# Fundamentals of Distribution Theory

**Abstract** The fundamentals of distribution theory are developed. The Dirac impulse is introduced motivated with a circuit that causes a derivation of an input signal. Starting from this example, the space of distributions is defined and examples of its elements are given. Such elements are, for example, all locally integrable functions, the principal value, and other pseudofunctions like rational functions or $1/|t|$. The calculus of distributions is developed to the extent as necessary in the further text. This includes generalized derivatives and convolution of distributions. The results are generalized for multidimensional parameters and test functions over the complex scalar field. Examples for every topic and exercises complete the chapter.

## 8.1 Characterizing Functions by Their Means

In basic mathematics or physics lectures we have learned to describe, for example, oscillations or voltages, current, etc., by functions $f(t)$, $t \in \mathbb{R}$, and to calculate with them. The idea associated with such a mathematical model is that the values of physical quantities of interest, for a time parameter $t$, are known exactly at any time.

But this is an *idealized approach*. In real practice, physical quantities are known mainly from measurements. If, for example, $f(t) = v(t) = \dot{x}(t)$ is the velocity of a train, then it is common to estimate the instantaneous velocity $v(t_0)$ at a time $t_0$ by the average velocity in a certain time interval $[t_0 - \varepsilon, t_0 + \varepsilon]$:

$$v(t_0) \approx \int_{-\infty}^{+\infty} v(t)\varphi_\varepsilon(t)\mathrm{d}t = \frac{1}{2\varepsilon}(x(t_0 + \varepsilon) - x(t_0 - \varepsilon)),$$



R. Brigola, *Fourier Analysis and Distributions*, Texts in Applied Mathematics 79,
https://doi.org/10.1007/978-3-031-81311-5_8

with $\varphi_\varepsilon(t) = \dfrac{1}{2\varepsilon}$ for $|t - t_0| \leqslant \varepsilon$, $\varphi_\varepsilon(t) = 0$ for $|t - t_0| > \varepsilon$. With $\varepsilon = 1/n$ follows

$$v(t_0) = \lim_{n \to \infty} \int_{-\infty}^{+\infty} v(t)\varphi_{1/n}(t)\,\mathrm{d}t.$$

*The instantaneous velocity is a (ideal) limit of mean values, and in practice a velocity is never pointwise accessible at all.* More generally, an ideal measurement of the value $f(t_0)$ of a continuous function $f$ at a time $t_0$ can be described schematically by Fig. 8.1.

However, a realistic measuring device, e.g., an electrical circuit, will show a rise and fall output during this sampling. A real measurement will therefore never exactly yield the sampled value $f(t_0)$, but a weighted average $\displaystyle\int_{-\infty}^{+\infty} f(t)\varphi(t)\,\mathrm{d}t$ of the function $f$ with a weight function $\varphi$ characteristic for the measuring device. This is schematically shown in Fig. 8.2:

However, mathematically we can show that any continuous function $f$ can also be reconstructed pointwise by its weighted means $\displaystyle\int_{-\infty}^{+\infty} f(t)\varphi(t)\,\mathrm{d}t$ with sufficiently many weight functions $\varphi$.

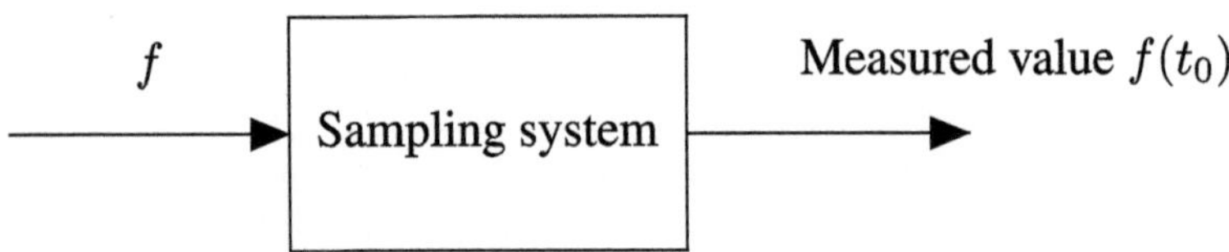


**Fig. 8.1**  Schematic sampling

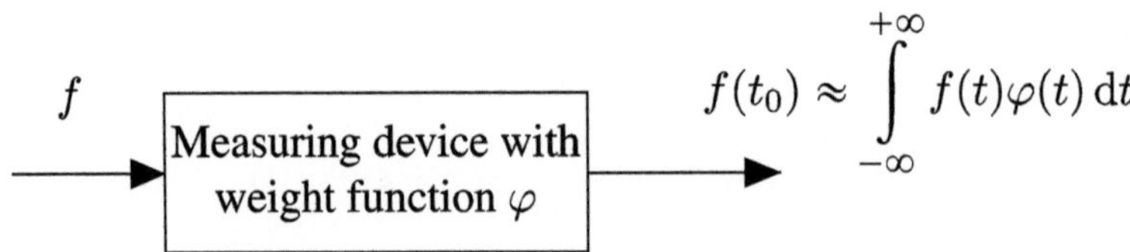


**Fig. 8.2**  Schematic measurement

## *Pointwise Reconstruction of Continuous Functions by Means*

We consider the following smooth weight function (see Figs. 8.3 and 8.4):

$$\varphi(t) = \begin{cases} c \cdot \mathrm{e}^{-1/(1-t^2)} & \text{for} \quad |t| < 1 \\ 0 & \text{for} \quad |t| \geqslant 1 \,, \end{cases}$$

where the constant $c$ is chosen so that $\displaystyle\int_{-\infty}^{+\infty} \varphi(t)\mathrm{d}t = 1$.

With this infinitely often differentiable function $\varphi$, we define for $t_0 \in \mathbb{R}$ and $n \in \mathbb{N}$

$$\varphi_{t_0,n}(t) = n\varphi(n(t - t_0)).$$

We then obtain $\varphi_{t_0,n}(t) = 0$ for $|t - t_0| \geqslant \dfrac{1}{n}$ and $\displaystyle\int_{-\infty}^{+\infty} \varphi_{t_0,n}(t)\mathrm{d}t = 1$ for all $n \in \mathbb{N}$.

($\varphi_{t_0,n}$ concentrates for increasing $n$ more and more around $t_0$.)

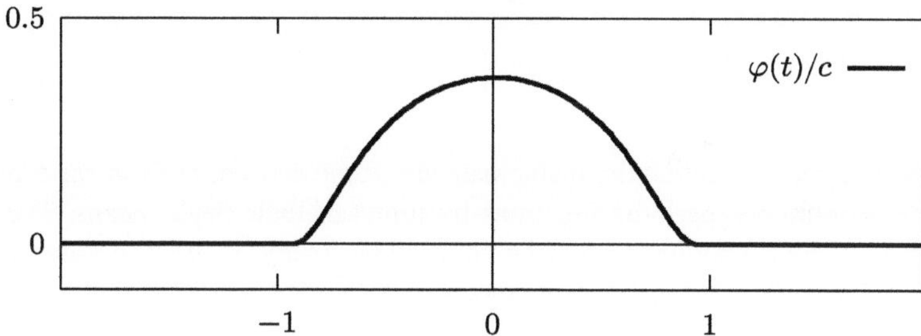

Fig. 8.3 A smooth weight function, called Sobolev's mollifier

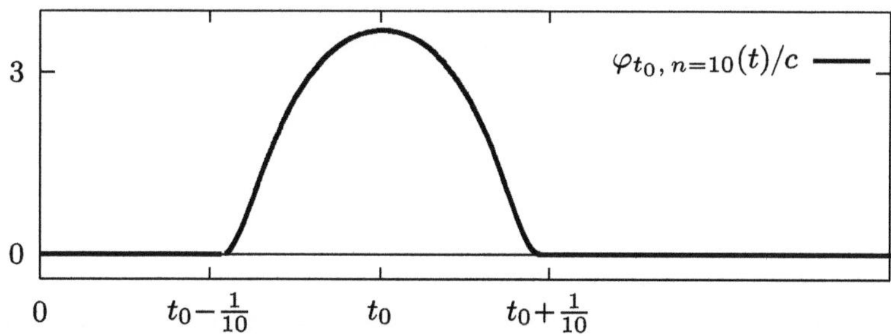

Fig. 8.4 A scaled version of Sobolev's mollifier

For a continuous function $f$, we therefore observe that

$$\left| \int_{-\infty}^{+\infty} f(t)\varphi_{t_0,n}(t)\mathrm{d}t - f(t_0) \right| = \left| \int_{t_0-1/n}^{t_0+1/n} (f(t) - f(t_0))\varphi_{t_0,n}(t)\mathrm{d}t \right|$$

$$\leqslant \sup_{|t-t_0|\leqslant 1/n} |f(t) - f(t_0)| \underbrace{\int_{t_0-1/n}^{t_0+1/n} \varphi_{t_0,n}\mathrm{d}t}_{=1} \xrightarrow[n\to\infty]{} 0.$$

*Thus, the function value of $f$ at any point $t_0$ can be recovered from the weighted means* $\int_{-\infty}^{+\infty} f(t)\varphi_{t_0,n}(t)\mathrm{d}t$:

$$f(t_0) = \lim_{n\to\infty} \int_{-\infty}^{+\infty} f(t)\varphi_{t_0,n}(t)\mathrm{d}t.$$

The term "mean" is appropriate according to the mean value theorem of integral calculus, since $\int_{-\infty}^{+\infty} f(t)\varphi_{t_0,n}(t)\,\mathrm{d}t = f(t_n)$ for a certain point $t_n$ close to $t_0$ with distance $|t_0 - t_n| \leqslant 1/n$.

Attentive readers notice the mathematically same procedure as in the representation of continuous periodic functions by limits of their Fejér means. The Fejér kernels as weight functions have been replaced here only by the smoothing kernels $\varphi_{t_0,n}$ (cf. Chap. 7).

**Summary** We normally obtain information on a physical function $f$ by measurements, i.e., the object $f$ is determined by certain weighted mean values of $f$. Instead of giving physical functions $f(t)$, $t \in \mathbb{R}$, point by point, we can describe them by their means: Each weight function $\varphi$ from a suitable vector space $\mathcal{D}$ is mapped to the mean value $\int_{-\infty}^{+\infty} f(t)\varphi(t)\mathrm{d}t$. If the set $\mathcal{D}$ of weight functions $\varphi$ is rich enough, then we can find continuous functions $f$ also pointwise by the linear mapping $T_f : \mathcal{D} \to \mathbb{R}$, $T_f(\varphi) = \int_{-\infty}^{+\infty} f(t)\varphi(t)\mathrm{d}t$.

This is one of the basic ideas of distribution theory. In the following section we introduce a suitable, i.e., a sufficiently large, set of infinitely often differentiable weight functions. Instead of weight functions we speak of *test functions*. All functions are assumed to be real-valued until further notice.

## 8.2   The Space of Test Functions

We consider functions $\varphi : \mathbb{R} \to \mathbb{R}$ which are arbitrarily often differentiable and zero outside a bounded interval $[a, b]$ (depending on $\varphi$). The *support* $\mathrm{supp}(\varphi)$ *of* $\varphi$ is the closure of the set $\{\, t \in \mathbb{R} \mid \varphi(t) \neq 0 \,\}$. A bounded support of $\varphi$ is *compact*, i.e., a closed and bounded set in $\mathbb{R}$.

**Definition** The set $\mathcal{D}$ of all functions $\varphi$ having a compact support is called the space of test functions.

We immediately see that $\mathcal{D}$ is a *vector space* over $\mathbb{R}$, i.e., $\lambda \varphi_1 \in \mathcal{D}$ and $\varphi_1 + \varphi_2 \in \mathcal{D}$ for any $\varphi_1, \varphi_2 \in \mathcal{D}$, $\lambda \in \mathbb{R}$. The space $\mathcal{D}$ contains very many functions: Examples are the functions $\varphi(t)$ and $\varphi_{t_0,n}(t) = n\varphi(n(t - t_0))$ used in the last section; the support $\mathrm{supp}(\varphi_{t_0,n})$ of $\varphi_{t_0,n}$ is the closed interval $[t_0 - 1/n, t_0 + 1/n]$. Also products of these functions with arbitrary, infinitely often differentiable functions generate again test functions in $\mathcal{D}$.

### *Convergence of Test Functions*

Two weight functions $\varphi_1$ and $\varphi_2$ in $\mathcal{D}$ are only "slightly different," if besides $\varphi_1$ and $\varphi_2$ also all their derivatives $\varphi_1^{(k)}$ and $\varphi_2^{(k)}$, $k \in \mathbb{N}$, differ only slightly. The experience shows that approximately the same measuring devices, i.e., those with only slightly different weight functions $\varphi_1$ and $\varphi_2$ at measurement of $f$, yield only slightly different measured values $\displaystyle\int_{-\infty}^{+\infty} f(t)\varphi_1(t)\mathrm{d}t$ and $\displaystyle\int_{-\infty}^{+\infty} f(t)\varphi_2(t)\mathrm{d}t$. This observation finds its mathematical equivalent in a continuity requirement for the mapping $T_f(\varphi) = \displaystyle\int_{-\infty}^{+\infty} f(t)\varphi(t)\mathrm{d}t$. For this we need an appropriate definition of convergence in $\mathcal{D}$, which expresses what "only slightly different" test functions are.

**Definition** A sequence $(\varphi_n)_{n\in\mathbb{N}}$ of test functions converges against $\varphi$ in $\mathcal{D}$ if there is a bounded interval containing the supports of all $\varphi_n$ and $\varphi$, and if furthermore the $\varphi_n$ converges uniformly to $\varphi$, and all derivatives of $\varphi_n^{(k)}$ converge uniformly to $\varphi^{(k)}$, $k \in \mathbb{N}$, in other words if for all $n \in \mathbb{N}$ and a suitable $r > 0$ hold true

$$\varphi_n(t) = 0 \quad \text{and} \quad \varphi(t) = 0 \quad \text{for } |t| \geqslant r \,,$$

and if for all $k \in \mathbb{N}_0$

$$\sup_{t\in\mathbb{R}} \left| \varphi_n^{(k)}(t) - \varphi^{(k)}(t) \right| \xrightarrow[n\to\infty]{} 0.$$

We then denote $\varphi = \mathcal{D}\text{-}\lim\limits_{n\to\infty} \varphi_n$ to clearly distinguish this convergence definition from other types of convergence.

**Example** The function $\varphi(t) = \begin{cases} c \cdot e^{-1/(1-t^2)} & \text{for} \quad |t| < 1 \\ 0 & \text{for} \quad |t| \geqslant 1 \end{cases}$ is infinitely often differentiable and zero for $|t| \geqslant 1$. The support of $\varphi$ is the interval $[-1, 1]$. This is valid also for all derivatives $\varphi^{(k)}$.

*The sequence $\varphi_n = \dfrac{1}{n}\varphi$ converges in $\mathcal{D}$ to the null function:* $\quad \mathcal{D}\text{-}\lim\limits_{n\to\infty} \varphi_n = 0.$

Since all derivatives $\varphi^{(k)}$ are bounded, it follows

$$\varphi_n^{(k)}(t) = \frac{1}{n}\,\varphi^{(k)}(t) \xrightarrow[n\to\infty]{} 0 \text{ uniformly.}$$

In contrast, the sequence $\psi_n(t) = \frac{1}{n}\varphi\left(\frac{t}{n}\right) = \begin{cases} \frac{c}{n}\,e^{-n^2/(n^2-t^2)} & \text{for} \quad |t| < n \\ 0 & \text{for} \quad |t| \geqslant n \end{cases}$ and all

the derivatives $\psi_n^{(k)}$ likewise converge uniformly to zero, but *this sequence does not converge in $\mathcal{D}$,* because there is no bounded interval containing jointly the supports of all the functions $\psi_n$.

The *distribution theory* involves the study of linear, continuous mappings on the vector space $\mathcal{D}$ of test functions, thus the study of physical objects by means of weighted averages. This theory goes back to P. Dirac (1902–1984) and was developed about 1935 by S. Ł. Sobolev (1908–1989), in the years 1945–1950 by L. Schwartz (1915–2002) and others. It makes possible, for example, a mathematical model for *impulses* and a *differentiability notion also for functions with discontinuities.*

## 8.3  The Dirac Impulse

### *Impulses in Electrical Engineering*

In electrical engineering, there are circuits that have a differentiating effect (Fig. 8.5):

An *ideal operational amplifier* in the above circuit yields for the currents $I_n = I_p = 0$ and for the voltages $U_n = U_p$. From Kirchhoff's law for the currents and voltages, we have the following nodal equations:

$$K_1: \qquad \frac{U_a - U_n}{R} - C\frac{dU_n}{dt} = 0,$$

**Fig. 8.5** A circuit that differentiates the input voltage $U_e$ to output voltage $U_a$

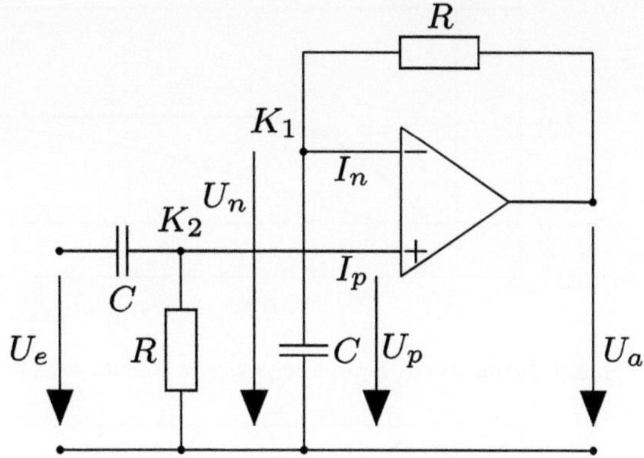


$$K_2: \qquad C\frac{\mathrm{d}(U_e - U_p)}{\mathrm{d}t} - \frac{U_p}{R} = 0.$$

With $U_n = U_p$ we obtain by equating the left sides

$$U_a = RC\frac{\mathrm{d}U_e}{\mathrm{d}t}.$$

The circuit realizes a (approximately ideal) *differentiator*.

As input $U_e(t)$ we now choose a DC voltage $U_0$ beginning at $t = 0$:

$$U_e(t) = U_0 s(t), \quad s(t) = \begin{cases} 0 & \text{for} \quad t \leqslant 0 \\ 1 & \text{for} \quad t > 0. \end{cases}$$

$U_e(t)$ is not differentiable at $t = 0$. This voltage function is again a simplified model with an *ideal switch*, which raises the problem how $U_a(t) = RC\dfrac{\mathrm{d}U_e(t)}{\mathrm{d}t}$ is to be understood. We approach the answer by considering the step function $U_e(t)$ as a limit of a sequence of smooth (more realistic) voltage functions $U_n(t)$ with increasingly steep slopes:

$$U_e(t) = \lim_{n\to\infty} U_n(t) \quad \text{for} \quad t \in \mathbb{R}.$$

As a model, we could start with the smooth function

$$\psi(t) = \begin{cases} e^{-1/(1-t^2)} & \text{for} \quad |t| < 1 \\ 0 & \text{for} \quad |t| \geqslant 1 \end{cases}$$

and build smooth voltage functions $U_n(t)$ (see Fig. 8.6)

$$U_n(t) = \begin{cases} 0 & \text{for} & kt \leqslant 0 \\ U_0 \cdot e \cdot \psi(n(kt - 1/n)) & \text{for} \quad 0 < kt < 1/n \\ U_0 & \text{for} & kt \geqslant 1/n \end{cases}$$

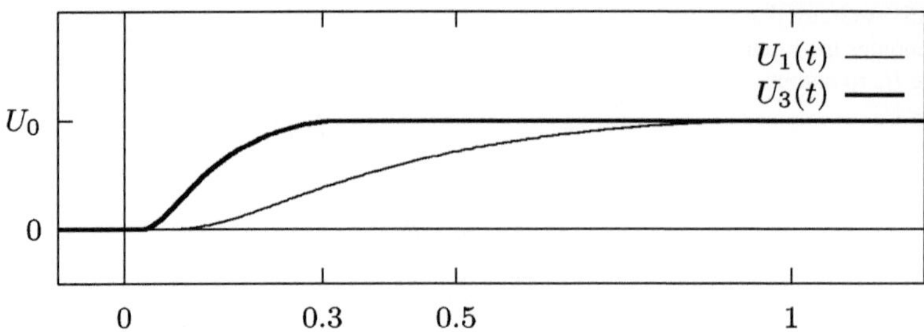

**Fig. 8.6** Elements $U_1$, $U_3$ of a sequence of smooth voltages converging to $U_0 s(t)$

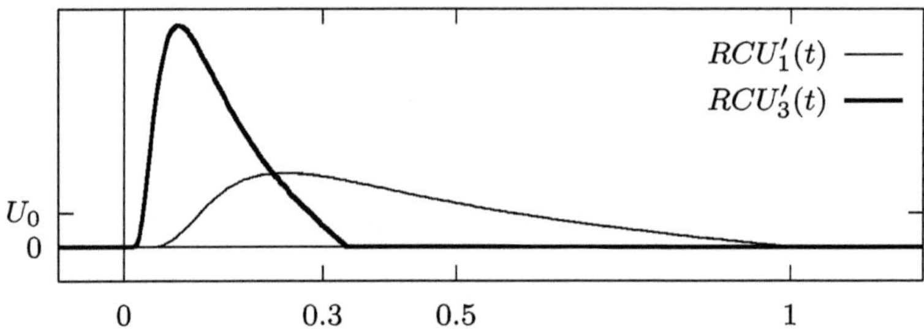

**Fig. 8.7** The smooth output voltages $RCU_1'$, $RCU_3'$ of the circuit

( $e = e^1$ is Euler's number, $kt$ physically dimensionless with the value of t).

We would expect that the output voltages $RCU_0 f_n(t) = RCU_n'(t)$, associated with inputs $U_n(t)$, approximate for increasing $n \in \mathbb{N}$ the response $U_a(t)$ of the differentiator to the step function input $U_e(t)$. We illustrate the functions $RCU_n'(t)$ for $R = 1\,\Omega$, $C = 1\,\mathrm{F}$ in Fig. 8.7:

For fixed $n \in \mathbb{N}$ and input $U_n(t)$, we thus find as the differentiator's output the voltage surge $RCU_0 f_n(t) = RCU_n'(t)$, approximating a voltage impulse for increasing $n$. We always have

$$\int_{-\infty}^{+\infty} RCU_0 f_n(t)\mathrm{d}t = RC(U_n(1/n) - U_n(0)) = RCU_0.$$

On the other hand, since $f_n(t) = U_n'(t)/U_0 = 0$ for $kt \leqslant 0$ and $kt \geqslant 1/n$, the following holds true in the sense of pointwise convergence, since we get $kt \geqslant 1/n$ for every $t > 0$ with sufficiently large $n$:

$$\lim_{n \to \infty} f_n(t) = 0 \quad \textit{for all } t \in \mathbb{R}.$$

If we would use in the idealized limit case $U_a(t) = \lim_{n \to \infty} RCU_0 f_n(t)$, then we would have $U_a(t) = 0$ *for each t*, whereas interchanging the limiting process with integration we would find:

$$\int\limits_{-\infty}^{+\infty} U_a(t)\mathrm{d}t = \int\limits_{-\infty}^{+\infty} \lim_{n\to\infty} RCU_0 f_n(t)\mathrm{d}t = \lim_{n\to\infty} \int\limits_{-\infty}^{+\infty} RCU_0 f_n(t)\mathrm{d}t = RCU_0.$$

*Such a function $U_a(t)$ cannot exist in classical sense.* Mathematically, the situation is as follows: Given is a sequence of infinitely often differentiable functions $f_n(t)$ such that

$$\lim_{n\to\infty} f_n(t) = 0 \text{ for all } t \in \mathbb{R} \text{ and } \int\limits_{-\infty}^{+\infty} f_n(t)\mathrm{d}t = 1 \text{ for all } n \in \mathbb{N}.$$

## *Definition of $\delta$-Impulses*

There is no classical function $\delta(t)$ so that pointwise $\delta(t) = \lim_{n\to\infty} f_n(t)$, $f_n(t)$ as above, and $\int\limits_{-\infty}^{+\infty} \delta(t)\mathrm{d}t = \lim_{n\to\infty} \int\limits_{-\infty}^{+\infty} f_n(t)\mathrm{d}t = 1$.

Although $\delta(t)$ as a function of $t \in \mathbb{R}$ cannot be defined, it is however quite reasonable, to build the limit of the integrals $\int\limits_{-\infty}^{+\infty} f_n(t)\varphi(t)\,\mathrm{d}t$ for $n \to \infty$ and each test function $\varphi$. Therefore, we do not define the $\delta$-impulse pointwise for $t \in \mathbb{R}$, but by integral values with test functions $\varphi \in \mathcal{D}$. The functions $f_n$ in our example are given by $f_n(t) = U_n'(t)/U_0$.

**Definition**  The $\delta$-impulse is defined by the mapping

$$\varphi \in \mathcal{D} \to \delta(\varphi) = \lim_{n\to\infty} \int\limits_{-\infty}^{+\infty} f_n(t)\varphi(t)\mathrm{d}t.$$

In many cases the notation $\delta(t) = \lim_{n\to\infty} f_n(t)$ is used in literature and $\delta(\varphi)$ is denoted by an *integral symbol*:

$$\delta(\varphi) = \int\limits_{-\infty}^{+\infty} \delta(t)\varphi(t)\mathrm{d}t = \lim_{n\to\infty} \int\limits_{-\infty}^{+\infty} f_n(t)\varphi(t)\mathrm{d}t.$$

The included argument $t$ in the notation $\delta(t)$ for the $\delta$-impulse only serves as a reference to the parameter of the right side and does not mean that function values can be assigned at individual points $t$. *The integral on the left side is not an integral in the common sense, but merely a symbol,* whose meaning is determined by the right side. If one writes symbolically $\delta(t) = \lim\limits_{n \to \infty} f_n(t)$, then the above corresponds formally to an interchange of the limit with the integration. An integral by definition is also the result of a limiting process. This interchange of limit processes leads to contradictions in the sense of classical functions. $\delta(t)$ is not a function of $t$ in common sense but becomes a *generalized function* or synonymously a *distribution*. This distribution is also called *Dirac distribution, Dirac impulse*, or briefly $\delta$-*impulse*.

We also use the mentioned notations and learn how to work correctly with generalized functions.

### Evaluation of Dirac Impulses, $\delta$ as Sampling Functional

Despite the still common notation $\delta(t)$, this generalized function itself *has no value at any single point t*. The use of $\delta(t)$ is always understood in the sense that only applying

$$\delta(\varphi) = \int\limits_{-\infty}^{+\infty} \delta(t)\varphi(t)\mathrm{d}t$$

with a test function $\varphi$ yields a numerical value. We want to calculate this value and show that it results in

$$\delta(\varphi) = \int\limits_{-\infty}^{+\infty} \delta(t)\varphi(t)\mathrm{d}t = \varphi(0).$$

We had in our example $f_n(t) = U_n'(t)/U_0$ with

$$U_n(t) = \begin{cases} 0 & \text{for} & kt \leqslant 0 \\ U_0 \cdot \mathrm{e} \cdot \psi(n(kt - 1/n)) & \text{for} & 0 < kt < 1/n \\ U_0 & \text{for} & kt \geqslant 1/n \end{cases}$$

and

$$\psi(t) = \begin{cases} \mathrm{e}^{-1/(1-t^2)} & \text{for} & |t| < 1 \\ 0 & \text{for} & |t| \geqslant 1. \end{cases}$$

For every $n \in \mathbb{N}$, we have $f_n \geqslant 0$; $f_n$ is infinitely often differentiable with support

$$\mathrm{supp}(f) = [0, 1/n], \text{ and } \int_{-\infty}^{+\infty} f_n(t)\mathrm{d}t = \int_0^{1/n} f_n(t)\mathrm{d}t = 1. \text{ For } \varphi \in \mathcal{D} \text{ now follows}$$

$$\left| \int_{-\infty}^{+\infty} f_n(t)\varphi(t)\mathrm{d}t - \varphi(0) \right| \leqslant \sup_{0 \leqslant t \leqslant 1/n} |\varphi(t) - \varphi(0)| \int_0^{1/n} f_n(t)\mathrm{d}t \xrightarrow[n \to \infty]{} 0.$$

The assertion is thus already shown:

$$\left| \int_{-\infty}^{+\infty} \delta(t)\varphi(t)\mathrm{d}t - \varphi(0) \right| = \left| \lim_{n \to \infty} \int_{-\infty}^{+\infty} f_n(t)\varphi(t)\,\mathrm{d}t - \varphi(0) \right| = 0.$$

If now $\delta(t - t_0)$ denotes the $\delta$-impulse shifted by $t_0$, defined for $\varphi \in \mathcal{D}$ through

$$\int_{-\infty}^{+\infty} \delta(t - t_0)\varphi(t)\,\mathrm{d}t = \lim_{n \to \infty} \int_{-\infty}^{+\infty} f_n(t - t_0)\varphi(t)\,\mathrm{d}t ,$$

it holds true correspondingly that

$$\int_{-\infty}^{+\infty} \delta(t - t_0)\varphi(t)\mathrm{d}t = \int_{-\infty}^{+\infty} \delta(t)\varphi(t + t_0)\mathrm{d}t = \varphi(t_0).$$

Such a shifted impulse appears as output at our (ideal) differentiating circuit, if the input voltage $U_0 s(t)$ is shifted to $U_0 s(t - t_0)$.

The introduction of $\delta$ by the above chosen sequence $f_n$ shows that we obtain not only for test functions, but for arbitrary continuous functions $f : \mathbb{R} \to \mathbb{R}$ and every $t_0 \in \mathbb{R}$ the outcome

$$\int_{-\infty}^{+\infty} \delta(t - t_0)f(t)\mathrm{d}t = f(t_0).$$

**Result** *Applying $\delta$-impulses is an appropriate mathematical model in describing pointwise evaluation or sampling processes of continuous functions. We also say that $\delta(t)$ is a sampling functional.*

### *Dirac Distributions as Generalized Density Functions*

In physics (generalized) $\delta$-functions are often used to specify discrete distributions, for example of masses or charges. One uses for example $\varrho(x) = \sum_{i=1}^{n} m_i \delta(x - x_i)$ as *generalized mass density* to denote $n$ point masses $m_i$ at the points $x_i$ (on the real axis). Then, the same formulas can be used for calculations with continuous and discrete distributions. For example, the center of gravity $S$ of the $n$ point masses $m_i$ in $x_i$ fulfills with that notation

$$
S = \frac{\int\limits_{-\infty}^{+\infty} x\varrho(x)\mathrm{d}x}{\int\limits_{-\infty}^{+\infty} \varrho(x)\mathrm{d}x} = \frac{\int\limits_{-\infty}^{+\infty} x \sum_{i=1}^{n} m_i \delta(x - x_i)\mathrm{d}x}{\int\limits_{-\infty}^{+\infty} \sum_{i=1}^{n} m_i \delta(x - x_i)\mathrm{d}x}
$$

$$
= \frac{\sum_{i=1}^{n} m_i \int\limits_{-\infty}^{+\infty} x\delta(x - x_i)\mathrm{d}x}{\sum_{i=1}^{n} m_i \int\limits_{-\infty}^{+\infty} \delta(x - x_i)\mathrm{d}x} = \frac{\sum_{i=1}^{n} m_i x_i}{\sum_{i=1}^{n} m_i}.
$$

**Remark** Comparing Appendix B, we recognize that integrals of the form

$$
\int\limits_{-\infty}^{+\infty} f(x)\varrho(x)\,\mathrm{d}x = \sum_{i=1}^{n} m_i f(x_i)
$$

define the discrete measure $m$, which gives the mass $m(I) = \sum_{x_i \in I} m_i$ to an interval $I$ in $\mathbb{R}$. Thereby, $m(I)$ measures the mass distributed in $I$. Often also

$$
\mathrm{d}m = \varrho(x)\,\mathrm{d}x = \sum_{i=1}^{n} m_i \delta(x - x_i)\,\mathrm{d}x
$$

is denoted as description of the measure $m$ with generalized density $\varrho$. The introduced term "distribution" is deduced from this aspect.

### *The $\delta$-Impulse as Derivative of the Unit Step Function*

For our example with the simplified input voltage $U_e(t) = U_0 s(t)$, the distribution $RCU_0\delta(t)$ means the (ideal) *impulse*, which appears at the time $t = 0$ as output of the treated differentiator with the *impulse strength* $RCU_0$ (Fig. 8.8):

**Fig. 8.8** Schematic differentiator

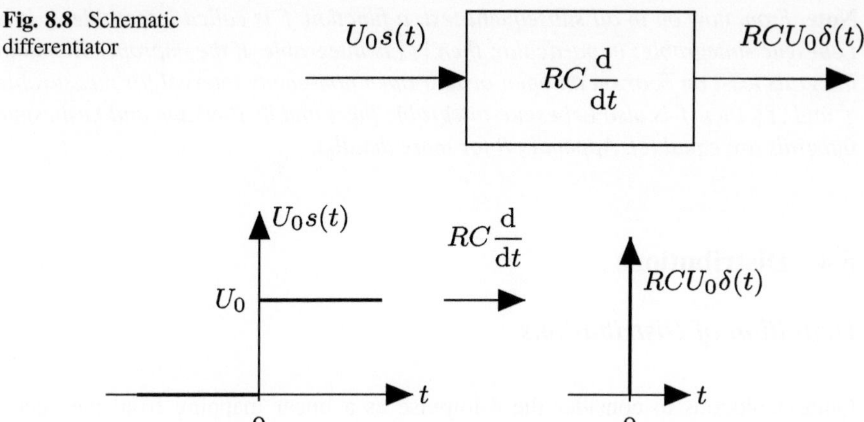



**Fig. 8.9** Graphical representation of an impulse with strength $RCU_0$

*Therefore $\delta(t)$ can be seen as the generalized derivative of the unit step function.*

We can illustrate $\delta(t)$ by an arrow in Fig. 8.9, whose height corresponds to the impulse strength.

**Remark** If $U_0$ is a voltage in V, and $s(t)$ is a physically dimensionless function of time $t$ in seconds, then $\delta(t)$ can be given the physical unit 1/s. The impulse strength then has the unit Vs, and the impulse $RCU_0\delta(t)$, appearing as output of our ideal differentiator, has consistently again the voltage unit V.

**Summary** We recognize that the map

$$\varphi \in \mathcal{D} \to \delta(\varphi) = \int\limits_{-\infty}^{+\infty} \delta(t)\varphi(t)\,\mathrm{d}t = \varphi(0)$$

is a *linear continuous operator* from the space $\mathcal{D}$ of test functions into the reals. Thus, our first distribution $\delta(t)$ is an example of the concept outlined in Sect. 8.1 of describing physical quantities—in this case an impulse—by mean values. The impulse $\delta(t)$ cannot be directly measured at any time, but averaging with weight functions $\varphi \in \mathcal{D}$ provides numerical values. As some readers probably already know from basic lectures, it is possible to describe linear time-invariant systems in a simple way by means of its *impulse response*. In this context, the Dirac distribution can be interpreted as a right-hand side and possibly as a solution part of linear ordinary differential equations with constant coefficients. We will discuss this in more examples later.

These few remarks alone indicate a variety of possible applications. In the following sections the concept of distributions and their use will be explained in more detail.

**Note** *From now on in all subsequent text, a function $f$ is called integrable, if it is Lebesgue -integrable; in particular, then $|f|$ is integrable. If the improper Riemann integrals exist on $\mathbb{R}$ or on an open or half-open non-empty interval for measurable $f$ and $|f|$, then $f$ is also Lebesgue-integrable there and its Riemann and Lesbesgue integrals are equal (cf. Appendix B for more details).*


## 8.4  Distributions

### *Definition of Distributions*


Quite analogous to consider the $\delta$-impulse as a linear mapping from the vector space $\mathcal{D}$ of test functions to the real numbers, we define distributions in general. *Distributions* are also called synonymously *generalized functions.*

**Definition** A distribution $T$ is a linear continuous map $T : \mathcal{D} \to \mathbb{R}$, i.e., for $a, b$ in $\mathbb{R}$, $\varphi_1, \varphi_2 \in \mathcal{D}$ and $\varphi = \mathcal{D}\text{-}\lim_{n \to \infty} \varphi_n$ in $\mathcal{D}$ hold true: $T(a\varphi_1 + b\varphi_2) = aT(\varphi_1) + bT(\varphi_2)$ and $T(\varphi) = \lim_{n \to \infty} T(\varphi_n)$. The set of all distributions is denoted by $\mathcal{D}'$.

**Remarks**

1. It is immediately seen from the definition that $\mathcal{D}'$ *is a real vector space.*
2. For the value $T(\varphi)$ of a distribution $T$ with a given test function $\varphi$, in literature also the following notations are found:

$$T(\varphi) = \langle T, \varphi \rangle = \langle T(t), \varphi(t) \rangle = \int\limits_{-\infty}^{+\infty} T(t)\varphi(t)\mathrm{d}t.$$

   We will use them as well. The motivation for these notations are based on the subsequent theorem and the following examples of distributions. We write $T(t)$ instead of $T$, if we want to indicate the variable of the underlying parameter space, even if $T(t)$ is not to be understood in the sense of a function value at a point $t$.
3. For concretely given linear mappings $T : \mathcal{D} \to \mathbb{R}$, it is usually easy to show the required continuity. There are also no physical linear mappings $T : \mathcal{D} \to \mathbb{R}$ known, which are not continuous on $\mathcal{D}$. Readers interested in the topological structure of $\mathcal{D}$ and $\mathcal{D}'$ are referred to the books of Schwartz (1957) or Rudin (1991).

The $\delta$-distribution can be represented as a limit of a sequence of infinitely many differentiable functions $f_n$ as we had seen before. The following is likewise true in general:

**Theorem 8.1** *For each distribution $T$ there is a sequence $(f_n)_{n\in\mathbb{N}}$ of infinitely often differentiable functions, so that for each test function $\varphi \in \mathcal{D}$ holds*

$$T(\varphi) = \lim_{n\to\infty} \int\limits_{-\infty}^{+\infty} f_n(t)\varphi(t)\mathrm{d}t.$$

*The functions $f_n$ can be chosen so that they have bounded support, i.e., that they are test functions.*

**Notation** *We write $T = \mathcal{D}'\text{-}\lim\limits_{n\to\infty} f_n$ and call $T$ the distributional limit of $f_n$ or the weak limit of them, in other words, the $f_n$ are weakly convergent to $T$.*

A distribution $T$ is thus the limit of a sequence of classical functions $(f_n)_{n\in\mathbb{N}}$. In general such a limit is not a limit in the sense of pointwise convergence of the function sequence, as we have already seen for the $\delta$ impulse. However, according to the above theorem: *The weighted means of $f_n$ with a test function $\varphi$ converge to a real number for $n \to \infty$.* This number $T(\varphi)$ can be obtained in arbitrarily good approximation by an integral $\int f_n(t)\varphi(t)\mathrm{d}t$ with an approximating function $f_n$ for the distribution $T$, if only $n$ is sufficiently large. We prove this fact later (p. 201) by theorems on convolutions and turn to examples first.

### Basic Examples of Distributions

All functions in this and the next section are assumed to be real-valued. The extension of definitions and examples to the case of complex-valued functions and distributions is given in Sect. 8.6:

1. *The impulse $\delta : \mathcal{D} \to \mathbb{R}$ is a distribution in the sense of the definition above.*
   With $c_1, c_2 \in \mathbb{R}$, $\varphi$, and $\varphi_n$ in $\mathcal{D}$ for $n \in \mathbb{N}$, $\varphi = \mathcal{D}\text{-}\lim\limits_{n\to\infty} \varphi_n$ we have

$$\langle \delta, c_1\varphi_1 + c_2\varphi_2 \rangle = c_1\varphi_1(0) + c_2\varphi_2(0) = c_1\langle \delta, \varphi_1 \rangle + c_2\langle \delta, \varphi_2 \rangle\,,$$

$$\lim_{n\to\infty} \langle \delta, \varphi_n \rangle = \lim_{n\to\infty} \varphi_n(0) = \varphi(0) = \langle \delta, \mathcal{D}\text{-}\lim_{n\to\infty} \varphi_n \rangle.$$

   Therefore, $\delta$ is linear and continuous on $\mathcal{D}$. The same holds true for a shift of $\delta$ by $t_0$ to $\delta(t - t_0)$.
2. *Every locally integrable function $f$* (i.e., $f$ and $|f|$ are integrable on every bounded interval) *can be considered as a distribution $T_f$ by*

$$T_f(\varphi) = \langle f, \varphi \rangle = \int\limits_{-\infty}^{+\infty} f(t)\varphi(t)\mathrm{d}t \qquad (\varphi \in \mathcal{D}).$$

This also motivates the notation, mentioned in the previous remark, for distributions in general. The definition shows immediately that $T_f$ is linear on $\mathcal{D}$. If $\varphi = \mathcal{D}\text{-}\lim\limits_{n\to\infty} \varphi_n$ and $[a, b]$ an interval containing the supports of all $\varphi_n$ and $\varphi$, then we obtain

$$|T_f(\varphi_n) - T_f(\varphi)| \leqslant \sup_{t\in[a,b]} |\varphi_n(t) - \varphi(t)| \int\limits_{[a,b]} |f(t)|\mathrm{d}t \xrightarrow[n\to\infty]{} 0.$$

Therefore, the continuity of $T_f$ on $\mathcal{D}$ follows by the uniform convergence of the $\varphi_n$ to $\varphi$ on $[a, b]$. Thus, with $f$ a distribution is given through

$$T_f(\varphi) = \int\limits_{-\infty}^{+\infty} f(t)\varphi(t)\mathrm{d}t.$$

For example, the function $\ln(|t|)$ can be considered as a distribution. The aspect, to consider locally integrable functions now also as distributions, corresponds exactly to the concept, presented in Sect. 8.1, that a function can be represented by its mean values with weight functions $\varphi$ from $\mathcal{D}$. Thus, we already know a very large set of distributions.

Distributions which are such classical, locally integrable functions are called *regular*. Distributions which are not locally integrable functions, e.g., the $\delta$-distribution, are called *singular*. For regular distributions $T_f$ it is common to write again only $f$ instead of $T_f$, and to specify their values for $\varphi \in \mathcal{D}$ by the common alternative notations

$$T_f(\varphi) = \langle T_f, \varphi\rangle = \langle f, \varphi\rangle = \int\limits_{-\infty}^{+\infty} f(t)\varphi(t)\mathrm{d}t.$$

Two functions $f$ and $g$ on $\mathbb{R}$ are equal if and only if $f(t) = g(t)$ holds true for all $t$ in $\mathbb{R}$. Equivalently, two distributions $T$ and $G$ are equal if and only if $T(\varphi) = G(\varphi)$ for all test functions $\varphi \in \mathcal{D}$. For two regular distributions $T_f$ and $T_g$ and any test function $\varphi$ all integral values $T_f(\varphi)$ and $T_g(\varphi)$ are equal, if $f$ and $g$ differ, e.g., only at finitely many points, in general at most on a null set (cf. Appendix B). In such a case we have $T_f = T_g$, i.e., these both distributions are identified.

3. *Principal Values, Pseudofunctions, Regularization of Divergent Integrals.*

   In addition to regular distributions there are, besides the $\delta$ distribution, many *singular distributions*. Typical examples arise with divergent integrals of functions with singularities such as rational functions. Rational functions and their Fourier transforms play a major role in linear systems theory and circuit design (for applications cf. Chap. 11). We therefore consider such examples of singular distributions:

(a) *The Cauchy Principal Value.* Starting point is the function $f(t) = 1/t, t \neq 0$, which is not locally integrable on $\mathbb{R}$. The improper integral $\int_{-\varepsilon}^{\varepsilon} f(t) dt, \varepsilon > 0$, is divergent. On the other hand, for $a > 0$ there exists the limit

$$\lim_{\varepsilon \to 0+} \left[ \int_{-a}^{-\varepsilon} f(t) dt + \int_{\varepsilon}^{+a} f(t) dt \right] = 0.$$

The *Cauchy principal value* $\mathrm{vp}(f)$ of $f$ is defined by

$$\mathrm{vp}(f)(\varphi) = \lim_{\varepsilon \to 0+} \left[ \int_{-\infty}^{-\varepsilon} f(t)\varphi(t) dt + \int_{\varepsilon}^{+\infty} f(t)\varphi(t) dt \right].$$

To prove that this is a distribution, we observe that we have for $\varphi \in \mathcal{D}$ with $\mathrm{supp}(\varphi) \subset [-a, a], a > 0$,

$$\mathrm{vp}(f)(\varphi) = \lim_{\varepsilon \to 0+} \left[ \int_{-a}^{-\varepsilon} \frac{\varphi(t) - \varphi(0)}{t} dt + \int_{-a}^{-\varepsilon} \frac{\varphi(0)}{t} dt \right.$$

$$\left. + \int_{\varepsilon}^{a} \frac{\varphi(0)}{t} dt + \int_{\varepsilon}^{a} \frac{\varphi(t) - \varphi(0)}{t} dt \right].$$

By the mean value theorem $|\varphi(t) - \varphi(0)| \leqslant |t| \max_{-a \leqslant t \leqslant a} |\varphi'(t)|$, the limits of the two integrals at the left and right exist for $\varepsilon \to 0, \varepsilon > 0$. The two integrals between add up to zero. From this it follows that $\mathrm{vp}(f)(\varphi)$ is defined for all $\varphi \in \mathcal{D}$.

Linearity of $\mathrm{vp}(f)$ on $\mathcal{D}$ follows immediately; its continuity on $\mathcal{D}$ is implied by the last inequality: It suffices to prove continuity for $\mathcal{D}\text{-}\lim_{n \to \infty} \varphi_n \to 0$, by virtue of linearity. The continuity of $\mathrm{vp}(f)$ is thus seen by the estimate $|\mathrm{vp}(f)(\varphi_n)| \leqslant 2a \max_{-a \leqslant t \leqslant a} |\varphi_n'(t)|$ for $\mathrm{supp}(\varphi_n) \subset [-a, a], n \in \mathbb{N}$.

The principal value $\mathrm{vp}(f)$ is a singular distribution (*valeur principale*, in English literature also denoted by $\mathrm{pv}(f)$ for *principal value*). It is also called a *regularization of the divergent integral of* $1/t$. For $\varphi \in \mathcal{D}$ with $0 \notin \mathrm{supp}(\varphi)$, $\mathrm{vp}(f)(\varphi)$ is simply the (convergent) integral of $f(t)\varphi(t)$ over $\mathbb{R}$:

$$\mathrm{vp}(f)(\varphi) = \int_{-\infty}^{\infty} \frac{\varphi(t)}{t} dt = \int_{0}^{\infty} \frac{\varphi(t) - \varphi(-t)}{t} dt.$$

Analogously, the principal values $\mathrm{vp}(f)$ are defined for other functions with singularities like $f(t) = \tan(t)$ or $f(t) = \cot(t)$, provided the involved integrals are convergent (cf. pp. 213 and 299).

(b) *The Pseudofunctions* $\mathrm{pf}(t^{-m})$, $m \in \mathbb{N}$. For the functions $t^{-m}$ with $m \geqslant 2$ the

integrals $\displaystyle\int_{\varepsilon}^{\infty} t^{-m}\left(\varphi(t) + (-1)^m \varphi(-t)\right) \mathrm{d}t$ are in general divergent for $\varepsilon \to 0$;

the principal values as in a) therefore do not exist.

In order to compensate the singularity at $t = 0$, one can subtract in the regularization from each test function $\varphi$ its Taylor polynomial $T_{m-1}\varphi$ of degree $m - 1$ about the singularity $t = 0$ and substitute for $\varphi$ the Taylor remainder $R_m\varphi = \varphi - T_{m-1}\varphi$. The divergent part of the integral is thereby thrown away. The resulting distribution is called a *pseudofunction* and is denoted by $\mathrm{pf}(t^{-m})$.

According to Hadamard (1932), this is the finite part (*partie finie*) of the divergent integral. The pseudofunction $\mathrm{pf}(t^{-m})$ is thus defined for $\varphi \in \mathcal{D}$, $m \geqslant 1$, with the Taylor remainder $R_m\varphi(t) = \varphi(t) - \displaystyle\sum_{k=0}^{m-1} \frac{\varphi^{(k)}(0)}{k!} t^k$, by

$$\mathrm{pf}(t^{-m})(\varphi) = \int_0^{\infty} t^{-m}\left(R_m\varphi(t) + (-1)^m R_m\varphi(-t)\right) \mathrm{d}t.$$

By Taylor's formula we have $|R_m\varphi(t)| \leqslant \frac{|t|^m}{m!} \max\{|\varphi^{(m)}(t)| : t \in \mathrm{supp}(\varphi)\}$. The improper integral $\mathrm{pf}(t^{-m})(\varphi)$ is therefore convergent for all $\varphi \in \mathcal{D}$. Linearity and continuity of $\mathrm{pf}(t^{-m})$ on $\mathcal{D}$ follow immediately. For $m = 1$ we have $\mathrm{pf}(t^{-1}) = \mathrm{vp}(t^{-1})$. An advantage of this regularization according to Hadamard is that the considered Taylor polynomial about $t = 0$ vanishes for $\varphi \in \mathcal{D}$ with $0 \notin \mathrm{supp}(\varphi)$. Then $R_m\varphi = \varphi$ and $\mathrm{pf}(t^{-m})(\varphi)$ coincides with the convergent integral of $t^{-m}\varphi(t)$:

$$\mathrm{pf}(t^{-m})(\varphi) = \int_{-\infty}^{+\infty} \frac{\varphi(t)}{t^m} \mathrm{d}t = \int_0^{\infty} \frac{\varphi(t) + (-1)^m \varphi(-t)}{t^m} \mathrm{d}t \quad (0 \notin \mathrm{supp}(\varphi)).$$

As an explicit example, we have $\mathrm{pf}(t^{-2})(\varphi) = \displaystyle\int_0^{\infty} \frac{\varphi(t) - 2\varphi(0) + \varphi(-t)}{t^2} \mathrm{d}t$.

(c) *The Pseudofunctions* $\mathrm{pf}(t_+^{-m})$, $\mathrm{pf}(t_-^{-m})$, *and* $\mathrm{pf}(|t|^{-m})$ *for* $m \in \mathbb{N}$.
For $t_+^{-m} = s(t)t^{-m}$, $s(t)$ the unit step function (Heaviside function), $\varphi \in \mathcal{D}$ and $R_m$ the Taylor remainder as before, we first consider

$$\int\limits_{\varepsilon}^{\infty} t^{-m} R_m \varphi(t)\mathrm{d}t = \int\limits_{\varepsilon}^{1} t^{-m} R_m \varphi(t)\mathrm{d}t + \int\limits_{1}^{\infty} t^{-m} R_{m-1}\varphi(t)\mathrm{d}t - \int_{1}^{\infty} \frac{\varphi^{(m-1)}(0)}{(m-1)t}\,\mathrm{d}t,$$

where $R_0\varphi(t) = \varphi(t)$. The first two integrals of the right-hand side are convergent for all $\varphi \in \mathcal{D}$ with $\varepsilon \to 0$. The third integral of the right-hand side is divergent.

A possibility for a regularization of $t_{+}^{-m}$ is therefore to replace this divergent part by zero, and to define the pseudofunction $\mathrm{pf}(t_{+}^{-m})$ by:

$$\mathrm{pf}(t_{+}^{-m})(\varphi) = \int\limits_{0}^{1} t^{-m} R_m \varphi(t)\mathrm{d}t + \int_{1}^{\infty} t^{-m} R_{m-1}\varphi(t)\mathrm{d}t$$

$$= \int\limits_{0}^{\infty} t^{-m}\left( R_{m-1}\varphi(t) - \frac{\varphi^{(m-1)}(0)}{(m-1)!} t^{m-1} s(1-t) \right)\mathrm{d}t.$$

Linearity and continuity on $\mathcal{D}$ are easily seen and left to the reader. Analogously we can define the pseudofunctions $\mathrm{pf}(t_{-}^{-m})$ and $\mathrm{pf}(|t|^{-m})$ with $t_{-}^{-m} = s(-t)t^{-m}$:

$$\mathrm{pf}(t_{-}^{-m})(\varphi) = \int\limits_{0}^{\infty} t^{-m}\left( (-1)^m R_{m-1}\varphi(-t) + \frac{\varphi^{(m-1)}(0)}{(m-1)!} t^{m-1} s(1-t) \right)\mathrm{d}t$$

$$\mathrm{pf}(|t|^{-m}) = \mathrm{pf}(t_{+}^{-m}) + (-1)^m \mathrm{pf}(t_{-}^{-m}).$$

From this we get $\mathrm{pf}(t^{-m}) = \mathrm{pf}(t_{+}^{-m}) + \mathrm{pf}(t_{-}^{-m})$, and for even $m \in \mathbb{N}$ we also obtain $\mathrm{pf}(|t|^{-m}) = \mathrm{pf}(t^{-m})$.

Regularizations as above are extensions of bounded linear functionals on the subspace of test functions with a support not containing a singularity like $t = 0$ in the examples. Therefore, such extensions are not uniquely determined. Think of a linear functional $T$ of the form $T(x) = a_1 x_1 + a_2 x_2$ for a vector $x = (x_1, x_2) \in \mathbb{R}^2 \subset \mathbb{R}^3$. Even in the finite-dimensional case you have infinitely many possible extensions of $T$ to a functional on $\mathbb{R}^3$. There is extensive literature on regularizations of divergent integrals, in particular for the case of several variables in physics. For more it is referred, e.g., to Schwartz (1957), Gel'fand et al. (1964), Horváth (1966), or Zemanian (2010).

4. For a *measure m* on $\mathbb{R}$ with a density function $\varrho$, often denoted by $\mathrm{d}m = \varrho(x)\,\mathrm{d}x$, a distribution

$$T(\varphi) = \int \varphi\,\mathrm{d}m = \int_{-\infty}^{+\infty} \varphi(x)\varrho(x)\,\mathrm{d}x \qquad (\varphi \in \mathcal{D})$$

is defined. Generally, any measure $m$ (cf. Appendix B) defines also a distribution by $T(\varphi) = \int \varphi\,dm$ $(\varphi \in \mathcal{D})$. *Distributions therefore are generalizations of functions and of measures as well.*

The measure $m$ above, with an integrable density function $\varrho$, corresponds to the regular distribution $T_\varrho$, and the discrete measure $\widetilde{m} = \sum_{i=1}^{n} m_i \delta(x - x_i)$ (cf. p. 162) is identified with the singular distribution $S$

$$S(\varphi) = \langle \sum_{i=1}^{n} m_i \delta(x - x_i), \varphi(x) \rangle = \int \varphi\,d\widetilde{m} = \sum_{i=1}^{n} m_i \varphi(x_i).$$

**Remark** In application problems also expressions of the form $T(\varphi)$ must be evaluated, where $\varphi$ is not necessarily a test function from $\mathcal{D}$. However, given distributions $T$ can often be defined as continuous linear functionals on a larger class of functions than $\mathcal{D}$. The space $\mathcal{D}$ is a set of functions on which all such linear functionals $T$ operate together. For example, the Dirac distribution $\delta(\varphi)$ can be defined for all $\varphi$ which are continuous around zero; regular distributions $T_f$ can be extended to all $\varphi$, for which the product $f\varphi$ is integrable. The pseudofunctions $\mathrm{pf}(t_\pm^{-m})$ can be applied to all sufficiently fast decaying, arbitrarily often differentiable functions $\varphi$, etc. Restricting the functionals $T$ to the common domain $\mathcal{D}$, the distribution theory provides a calculus that can be used for all such functionals $T$.

In later chapters about the Fourier transform we will take up this remark, and work also with another common test function space which is larger than $\mathcal{D}$ (cf. Chap. 10, p. 288).

**Summary** Besides the classical concept of functions, it seems reasonable to the engineer or scientist understanding weighted averages from measurements of an object $T$ as a distribution. The object of interest is the distribution $T$ with its properties. Its values $T(\varphi)$ on test functions are numerical values, which are single weighted means from single measurements of $T$.

An engineer can, as in the preceding example 2, straightforwardly consider locally integrable functions $f$ as the corresponding distributions $T_f$. He or she knows from experience that a periodic rectangular function $f$ in practice is realized approximately by a finite superposition of harmonic oscillations. Thereby, the *ideal* rectangular function $f$ is to be regarded as a distribution, namely as the distributional limit of the infinitely often differentiable partial sums of the Fourier series of $f$. The approximation can be so close that, e.g., power differences in comparison to the ideal function $f$ (these are also integral means) become arbitrarily small; and calculating with $f$ as a simple model is much easier than calculating with a Fourier expansion or a possibly more realistic smooth function as in our differentiator example from the beginning, which is complicated to describe analytically.

Analogously, we can imagine every distribution as such a limit of classical functions, limit in the sense of existence of limits for weighted means and not necessarily pointwise. Often enough in practice only such averages of physical quantities are of interest. Attentive reading of technical literature shows that in many cases it is calculated "distributionally," without this being explicitly noted.

The advantages of this concept regarding differentiation and other mathematical limit processes will become apparent in the following sections, where calculating with distributions will be explained.

## 8.5   Calculating with Distributions

Distributions are characterized by the fact that you can much easier compute with than with conventional functions. For this it is necessary to introduce some operations in $\mathcal{D}'$. For the derivative of a differentiable function $f$, we use the notation $f'$. For the following introduced generalized derivative of a distribution $T(t)$, we use the notation $\dot{T}(t)$, later again also $T'(t)$.

### *Differentiation of Distributions*

Distributions can be differentiated as often as you like without any restrictions. To see this we consider a distribution $T = \mathcal{D}'\text{-}\lim_{n\to\infty} f_n$, all $f_n$ arbitrarily often differentiable, and $\varphi \in \mathcal{D}$ with support $\mathrm{supp}(\varphi) \subset [a, b]$. Then it follows through *integration by parts* that the following limit exists:

$$\lim_{n\to\infty} \int_{-\infty}^{+\infty} f_n'(t)\varphi(t)\mathrm{d}t = \lim_{n\to\infty} \left[ \underbrace{f_n(t)\varphi(t)\Big|_a^b}_{=0} - \int_a^b f_n(t)\varphi'(t)\mathrm{d}t \right].$$

You also have

$$\lim_{n\to\infty} \int_{-\infty}^{+\infty} f_n'(t)\varphi(t)\mathrm{d}t = - \lim_{n\to\infty} \int_{-\infty}^{+\infty} f_n(t)\varphi'(t)\mathrm{d}t = -T(\varphi').$$

Therefore, the derivative $\dot{T}$ of a distribution $T$ can be introduced as follows:

**Definition** The derivative $\dot{T}$ of a distribution $T = \mathcal{D}'\text{-}\lim_{n\to\infty} f_n$ is defined by

$$\dot{T} = \mathcal{D}'\text{-}\lim_{n\to\infty} f_n'.$$

For $\varphi \in \mathcal{D}$ we have

$$\dot{T}(\varphi) = -T(\varphi').$$

For regular distributions $T_f$ we also write again $\dot{f}$ instead of $\dot{T}_f$. Linearity and continuity of $\dot{T}$ on $\mathcal{D}$ are easily proven. For $\varphi = \mathcal{D}\text{-}\lim\limits_{m \to \infty} \varphi_m$ and $a, b \in \mathbb{R}$ it holds true that

$$\lim_{m \to \infty} \dot{T}(\varphi_m) = -\lim_{m \to \infty} T(\varphi'_m) = -T(\varphi') = \dot{T}(\varphi)$$

$$\dot{T}(a\varphi_1 + b\varphi_2) = -T(a\varphi'_1 + b\varphi'_2) = a\dot{T}(\varphi_1) + b\dot{T}(\varphi_2).$$

*Higher derivatives* of order $k$ are analogously defined by

$$T^{(k)} = \mathcal{D}'\text{-}\lim_{n \to \infty} f_n^{(k)}.$$

Applying to $\varphi \in \mathcal{D}$, this means for the $k$-th derivative of $T$

$$T^{(k)}(\varphi) = (-1)^k T(\varphi^{(k)}).$$

**Example**  For the step function

$$\sigma(t) = \begin{cases} 0 & \text{for } t < 0, \\ \frac{1}{2} & \text{for } t = 0, \\ 1 & \text{for } t > 0 \end{cases}$$

(considered as a regular distribution) and arbitrary $\varphi \in \mathcal{D}$, we obtain with the notations $T_\sigma = \sigma$ and $\dot{T}_\sigma = \dot{\sigma}$

$$\langle \dot{T}_\sigma, \varphi \rangle = \langle \dot{\sigma}, \varphi \rangle = -\langle \sigma, \varphi' \rangle = -\int_0^{+\infty} \varphi'(t)\,\mathrm{d}t = \varphi(0) = \langle \delta, \varphi \rangle.$$

$\dot{\sigma}(\varphi)$ and $\delta(\varphi)$ thus yield for any $\varphi \in \mathcal{D}$ the same value; we therefore have as result the *equation* $\dot{\sigma} = \delta$ *in* $\mathcal{D}'$, also denoted by $\dot{\sigma}(t) = \delta(t)$, if we still want to indicate the initial function variable $t$.

Thus, we can now differentiate a discontinuous step function. This has not been possible within the framework of classical analysis. Correspondingly, we have for a translation $\dot{\sigma}(t - t_0) = \delta(t - t_0)$.

Changing $\sigma(t)$ at $t = 0$ to $s(t) = \begin{cases} 0 & \text{for } t \leqslant 0 \\ 1 & \text{for } t > 0 \end{cases}$ does not have an effect on integrals of $\sigma(t)$ or $s(t)$, i.e., $\sigma(t)$ and $s(t)$ are the same distribution: $T_\sigma = T_s$, and we have the equation $\dot{s} = \delta$ in $\mathcal{D}'$. This is also found symbolically denoted by

$$\int_{-\infty}^{t} \delta(\tau)\mathrm{d}\tau = s(t) \ \text{(Fig. 8.10)}.$$

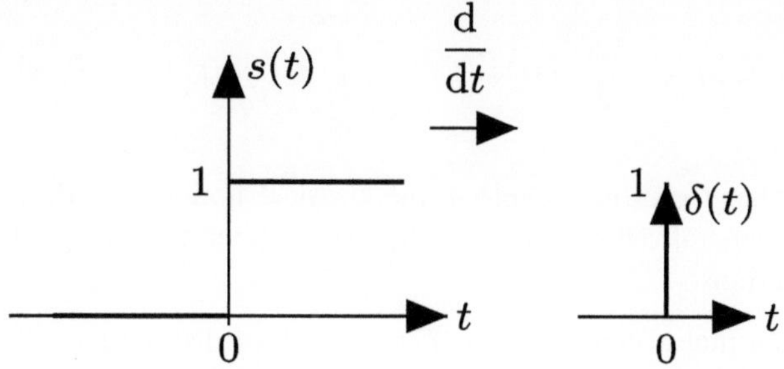

**Fig. 8.10** Derivative of a unit step

*We easily verify the following facts*:

1. $\mathcal{D}'$ *is a vector space.*
2. *For $T \in \mathcal{D}'$ and infinitely often differentiable functions $f$, the product $f \cdot T$ is a distribution, defined for $\varphi \in \mathcal{D}$ by $\langle f \cdot T, \varphi \rangle = \langle T, f \cdot \varphi \rangle$.*

**Further Rules for Derivatives**

*The derivative is linear, and the product rule is valid for products with infinitely often differentiable functions:*

$$(cT)^{(k)} = cT^{(k)}, \quad (S+T)^{(k)} = S^{(k)} + T^{(k)},$$

$$(fT)^{(k)} = \sum_{n=0}^{k} \binom{k}{n} f^{(n)} T^{(k-n)}$$

*for $k \in \mathbb{N}_0, c \in \mathbb{R}, S, T \in \mathcal{D}'$ and infinitely often differentiable functions $f$.*
To confirm these relations, we observe for $\varphi \in \mathcal{D}$ and $k = 1$:

$$\langle (cT)^{(1)}, \varphi \rangle = -c\langle T, \varphi' \rangle = \langle c\dot{T}, \varphi \rangle,$$

$$\langle (S+T)^{(1)}, \varphi \rangle = -\langle S, \varphi' \rangle - \langle T, \varphi' \rangle = \langle \dot{S}, \varphi \rangle + \langle \dot{T}, \varphi \rangle,$$

$$\langle (fT)^{(1)}, \varphi \rangle = -\langle fT, \varphi' \rangle = -\langle T, f\varphi' \rangle = -\langle T, f\varphi' + f'\varphi \rangle + \langle T, f'\varphi \rangle$$

$$= \langle \dot{T}, f\varphi \rangle + \langle T, f'\varphi \rangle = \langle f\dot{T} + f'T, \varphi \rangle.$$

For derivatives of higher order $k > 1$, we then obtain the rules by induction.

Understanding now locally integrable functions as distributions, we can differentiate them without any restriction, even if they are discontinuous. To indicate differentiation in this distributional sense, one speaks of *generalized derivatives*. For a regular distribution $T_f$ belonging to a differentiable function $f$, we have

$$\langle \dot{T}_f, \varphi \rangle = \langle \dot{f}, \varphi \rangle = - \int\limits_{-\infty}^{+\infty} f(t)\varphi'(t)\mathrm{d}t = \int\limits_{-\infty}^{+\infty} f'(t)\varphi(t)\mathrm{d}t = \langle f', \varphi \rangle = \langle T_{f'}, \varphi \rangle.$$

This means that for differentiable $f$ the classical notion of a derivative and the notion of the generalized derivative are equivalent in terms of integrating derivatives with test functions.

**Remark** Note that a multiplication $T \cdot G$ in $\mathcal{D}'$ of distributions $T$ and $G$ is not defined in general. For example, the locally integrable function $f(t) = 1/\sqrt{|t|}$ is a regular distribution; on the other hand $f^2(t) = 1/|t|$ is not locally integrable, and the product $f^2$ can only be interpreted as a distribution by the regularization $\mathrm{pf}(|t|^{-1})$. Also not defined in $\mathcal{D}'$ are expressions like $\delta(t) \cdot \dot{\delta}(t)$, $\delta^2(t)$, or $f(t)\delta(t)$ for functions $f$ which are not infinitely often differentiable. However, it should be noted that products like $s(t)\delta(t)$ or $\delta^2(t)$ can be explained on an extended class of generalized functions containing $\mathcal{D}'$. Such an extension of the distribution theory has fundamental importance in studying nonlinear equations between generalized functions. For this it is referred to Oberguggenberger (1992) and further references cited there.

## *Further Examples*

1. *For infinitely often differentiable functions f, the following important relation is true*:

$$f(t)\delta(t - t_0) = f(t_0)\delta(t - t_0).$$

Again we have to show that the equation is true applying both sides to an arbitrary test function $\varphi \in \mathcal{D}$:

$$\langle f(t)\delta(t - t_0), \varphi(t) \rangle = \langle \delta(t - t_0), f(t)\varphi(t) \rangle$$
$$= f(t_0)\varphi(t_0) = \langle f(t_0)\delta(t - t_0), \varphi(t) \rangle.$$

A remarkable consequence of this relation is that the equation $tT(t) = 1$ in $\mathcal{D}'$ has the solutions $T(t) = \mathrm{vp}(1/t) + k\delta(t)$ with arbitrary constants $k$, by virtue of $kt\delta(t) = 0$. In the following example 8 we show that there are no further solutions.

2. For $f(t) = |t|$ we get

$$\dot{f}(t) = \mathrm{sgn}(t) = \begin{cases} -1 & \text{for } t < 0 \\ 0 & \text{for } t = 0 \\ 1 & \text{for } t > 0. \end{cases}$$

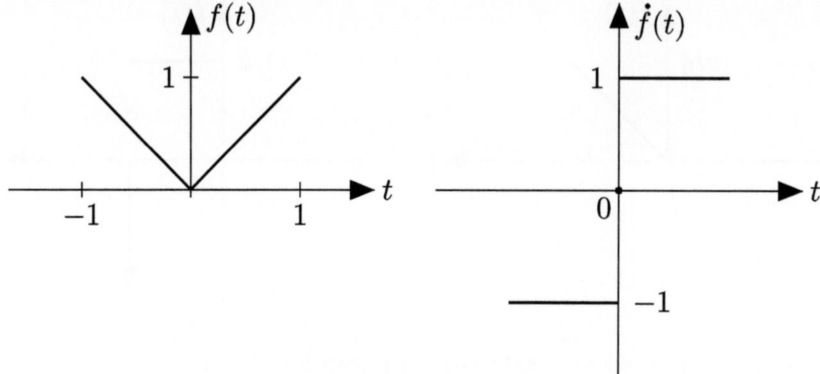

**Fig. 8.11**  A function with a kink and its generalized derivative

Namely, for $\varphi \in \mathcal{D}$ we obtain through integration by parts

$$\langle \dot{f}, \varphi \rangle = -\langle f, \varphi' \rangle = -\int_{-\infty}^{+\infty} |t|\varphi'(t)\mathrm{d}t = \int_{-\infty}^{0} t\varphi'(t)\mathrm{d}t - \int_{0}^{+\infty} t\varphi'(t)\mathrm{d}t$$

$$= -\int_{-\infty}^{0} \varphi(t)\mathrm{d}t + \int_{0}^{+\infty} \varphi(t)\mathrm{d}t = \int_{-\infty}^{+\infty} \mathrm{sgn}(t)\varphi(t)\mathrm{d}t = \langle \mathrm{sgn}, \varphi \rangle.$$

Illustratively in Fig. 8.11
We thus can differentiate functions with "kinks" considering them as distributions. The examples demonstrate the following:

**Rule for Generalized Derivatives.** *If a function $f(t)$ has a kink at $t_0$, a jump at $t_1$, and is otherwise differentiable, the generalized derivative $\dot{f}(t)$ results in a jump from $f'(t_0-)$ to $f'(t_0+)$ at $t_0$ and in a $\delta$-impulse of strength $f(t_1+) - f(t_1-)$ at $t_1$.*

3. Let $f$ be given by

$$f(t) = \begin{cases} 0 & \text{for} & t < 0 \\ at & \text{for} & 0 \leqslant t < t_0 \\ \frac{at_0}{2} & \text{for} & t = t_0 \\ 0 & \text{for} & t > t_0. \end{cases}$$

$f(t) = at[\sigma(t) - \sigma(t - t_0)]$, considered as a distribution (see Fig. 8.12), yields

$$\dot{f}(t) = a[\sigma(t) - \sigma(t - t_0)] + at[\dot{\sigma}(t) - \dot{\sigma}(t - t_0)]$$

$$= a[\sigma(t) - \sigma(t - t_0)] + a \cdot 0 \cdot \delta(t) - at_0\delta(t - t_0).$$

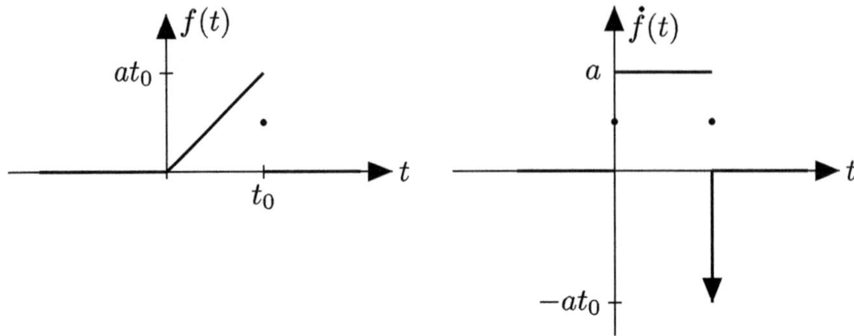

**Fig. 8.12** A function with a kink and a jump and its generalized derivative

Therefore we have $\dot{f}(t) = a[\sigma(t) - \sigma(t - t_0)] - at_0\delta(t - t_0)$.

4. With $f(t) = t$ we have for $m \in \mathbb{N}$ and $\varphi \in \mathcal{D}$ by the product rule for derivatives

$$\langle f\delta^{(m)}, \varphi \rangle = (-1)^m \langle \delta, (f\varphi)^{(m)} \rangle = (-1)^m m \varphi^{(m-1)}(0), \text{ thus}$$

$$f\delta^{(m)} = -m\delta^{(m-1)}.$$

5. *The Derivative of* $\ln(|t|)$. We show that the derivative of $T(t) = \ln(|t|)$ is the principal value (cf. p. 167). Through integration by parts, it follows for $\varphi \in \mathcal{D}$ with support $\operatorname{supp}(\varphi) \subset [-a, a]$, $a > 0$,

$$\left\langle \operatorname{vp}\left(\frac{1}{t}\right), \varphi \right\rangle = \lim_{\varepsilon \to 0+} \left[ \int_{-a}^{-\varepsilon} \frac{\varphi(t)}{t}\, dt + \int_{\varepsilon}^{a} \frac{\varphi(t)}{t}\, dt \right]$$

$$= \lim_{\varepsilon \to 0+} \left[ (\varphi(-\varepsilon) - \varphi(\varepsilon))\ln(\varepsilon) + \underbrace{(\varphi(a) - \varphi(-a))\ln(a)}_{=0} \right.$$

$$\left. - \int_{-a}^{-\varepsilon} \ln(|t|)\varphi'(t)\, dt - \int_{\varepsilon}^{a} \ln(|t|)\varphi'(t)\, dt \right].$$

By the mean value theorem, we have $\varphi(-\varepsilon) - \varphi(\varepsilon) = 2\varepsilon\varphi'(x)$, $x \in [-\varepsilon, \varepsilon]$ suitable; thus we obtain $\lim_{\varepsilon \to 0+}(\varphi(-\varepsilon) - \varphi(\varepsilon))\ln(\varepsilon) = 0$. Since $\ln(|t|)$ is locally integrable, it turns out the result

$$\left\langle \operatorname{vp}\left(\frac{1}{t}\right), \varphi \right\rangle = -\int_{-a}^{a} \ln(|t|)\varphi'(t)\, dt = -\int_{-\infty}^{+\infty} \ln(|t|)\varphi'(t)\, dt = \langle \dot{T}, \varphi \rangle.$$

The distribution $T(t) = \ln(|t|)$ is regular, and its derivative $\dot{T}(t) = \mathrm{vp}(1/t)$ is a singular distribution.

6. *Derivatives of the Pseudofunctions* $\mathrm{pf}(t_+^{-m})$ *and* $\mathrm{pf}(t_-^{-m})$, $m \in \mathbb{N}$.

We compute the generalized derivative of $\mathrm{pf}(t_+^{-m})$, denoting it by $\mathrm{pf}(t_+^{-m})'$.

At first we observe that the derivative of $R_m \varphi(t) = \varphi(t) - \sum_{k=0}^{m-1} \dfrac{\varphi^{(k)}(0)}{k!} t^k$, $\varphi \in \mathcal{D}$,

is

$$(R_m \varphi(t))' = R_m \varphi'(t) + \frac{\varphi^{(m)}(0)}{(m-1)!} t^{m-1} \quad (\varphi \in \mathcal{D}).$$

$R_m \varphi(t) - \frac{\varphi^{(m)}(0)}{m!} t^m = R_{m+1} \varphi(t)$ is therefore a primitive of $R_m \varphi'(t)$. This primitive with the constant $K = -\varphi(0)$ permits integration by parts of the following improper integrals. Using integration by parts, we compute with some patience for $\varphi \in \mathcal{D}$ (cf. p. 168 and use $\int_a^b u(t)v'(t)\mathrm{d}t = uv\big|_a^b - \int_a^b u'(t)v(t)\mathrm{d}t$ choosing $u(t) = t^{-m}$)

$$\mathrm{pf}\left(t_+^{-m}\right)'(\varphi) = -\mathrm{pf}(t_+^{-m})(\varphi') = -\int_0^1 t^{-m} R_m \varphi'(t)\mathrm{d}t - \int_1^\infty t^{-m} R_{m-1}\varphi'(t)\mathrm{d}t.$$

With $(\star)$ $\lim_{\varepsilon \to 0+}\left(\varepsilon^{-m} R_{m+1}\varphi(\varepsilon)\right) = 0$, we obtain for any $\varphi \in \mathcal{D}$:

$$-\ \mathrm{pf}(t_+^{-m})(\varphi') = -\left( t^{-m} R_{m+1}\varphi(t)\Big|_0^1 - \int_0^1 -mt^{-(m+1)} R_{m+1}\varphi(t)\mathrm{d}t \right.$$

$$\left. +\ t^{-m} R_m \varphi(t)\Big|_1^\infty - \int_1^\infty -mt^{-(m+1)} R_m \varphi(t)\mathrm{d}t \right)$$

$$\underset{\mathrm{by}(\star)}{=} \frac{\varphi^{(m)}(0)}{m!} - \int_0^1 mt^{-(m+1)} R_{m+1}\varphi(t)\mathrm{d}t - \int_1^\infty mt^{-(m+1)} R_m \varphi(t)\mathrm{d}t.$$

*Observing* $\dfrac{\varphi^{(m)}(0)}{m!} = \dfrac{(-1)^m}{m!} \langle \delta^{(m)}(t), \varphi(t) \rangle$, *we have found the result*:

$$\mathrm{pf}\left(t_+^{-m}\right)' = \mathrm{pf}\left(-mt_+^{-m-1}\right) + \frac{(-1)^m}{m!} \delta^{(m)}(t).$$

*Completely analogous we also obtain the following generalized derivatives:*

$$\mathrm{pf}\left(t_-^{-m}\right)' = \mathrm{pf}\left(-mt_-^{-m-1}\right) - \frac{(-1)^m}{m!}\delta^{(m)}(t),$$

$$\mathrm{pf}\left(t^{-m}\right)' = \mathrm{pf}\left(-mt_+^{-m-1}\right) + \mathrm{pf}\left(-mt_-^{-m-1}\right) = \mathrm{pf}\left(-mt^{-m-1}\right),$$

$$\mathrm{pf}\left(|t|^{-m}\right)' = \mathrm{pf}\left(-mt_+^{-m-1}\right) + (-1)^m\mathrm{pf}\left(-mt_-^{-m-1}\right) + \frac{((-1)^m-1)}{m!}\delta^{(m)}(t).$$

7. **The Equation $t^n T(t) = 0$ in $\mathcal{D}'$.**
   In subsequent chapters (Chaps. 10 and 11) related to calculating Fourier transforms, we have to solve equations of the form $t^n T(t) = u(t)$ with $u(t) = 1$, $u(t) = s(t)$, or $u(t) = \mathrm{sgn}(t)$. To find their general solutions in $\mathcal{D}'$ we first determine the general solution of the homogeneous equation:

**Theorem 8.2**  *The general solution of $t^n T(t) = 0$ in $\mathcal{D}'$ for $n \in \mathbb{N}$ is given by*

$$T = \sum_{k=0}^{n-1} c_k \delta^{(k)}$$

*with arbitrary constants $c_k$, $k = 0, \ldots n - 1$.*

***Proof*** From example 1 on p. 174 it follows $t^n \delta^{(k)}(t) = 0$ for $0 \leqslant k < n$; thus $T = \sum_{k=0}^{n-1} c_k \delta^{(k)}$ is a solution of $t^n T(t) = 0$ for arbitrary constants $c_k$. We now show that conversely any solution of this equation in $\mathcal{D}'$ is a linear combination of the distributions $\delta^{(k)}$, $k = 0 \ldots n - 1$:

Let $T$ fulfill $t^n T(t) = 0$ for $n \in \mathbb{N}$. By Taylor's formula, we have for $\varphi \in \mathcal{D}$ :

$$\varphi(t) = \sum_{k=0}^{n-1} \frac{\varphi^{(k)}(0)}{k!} t^k + t^n \varrho(t)$$

with $\varrho(t) = \dfrac{1}{(n-1)!} \displaystyle\int_0^1 (1-x)^{n-1} \varphi^{(n)}(xt)\,\mathrm{d}x$.

The function $\varrho(t)$ is infinitely often differentiable (differentiation under the integral is possible). Now let $\alpha$ be a function in $\mathcal{D}$ so that $\alpha(t) = 1$ in an open interval $U$ around zero. We define with the Taylor polynomial $T_{n-1}\varphi$ of $\varphi$ with degree up to $n - 1$

$$\psi(t) = \frac{1}{t^n}\left(\varphi(t) - \alpha(t)T_{n-1}\varphi(t)\right).$$

Then both $\psi(t)$ and $t^n\psi(t)$ are again test functions in $\mathcal{D}$, since $\psi$ coincides with $\varrho$ in $U$. Since $\psi(t)$ is in particular infinitely often differentiable at $t = 0$, we get for arbitrary $\varphi \in \mathcal{D}$

$$\langle T, \varphi \rangle = \langle T, \alpha T_{n-1}\varphi \rangle + \langle T, t^n\psi \rangle.$$

The last addend is zero, due to $t^n T = 0$. With $\langle \delta^{(k)}, \varphi \rangle = (-1)^k \varphi^{(k)}(0)$ and the constants $c_k = (-1)^k \dfrac{\langle T, t^k\alpha \rangle}{k!}$ now follows the claimed assertion for $T$:

$$\langle T, \varphi \rangle = \sum_{k=0}^{n-1} \frac{\varphi^{(k)}(0)}{k!}\langle T, t^k\alpha \rangle = \sum_{k=0}^{n-1} c_k \langle \delta^{(k)}, \varphi \rangle.$$

$\square$

8. **The Equation $t^{nT(t)} = 1$ in $\mathcal{D}'$**

From the preceding results now follows for $n \in \mathbb{N}$:

**Theorem 8.3**  *The equation $t^n T(t) = 1$ has in $\mathcal{D}'$ the general solution*

$$T(t) = \mathrm{pf}(t^{-n}) + \sum_{k=0}^{n-1} c_k \delta^{(k)}(t)$$

*with arbitrary constants $c_k$.*

Correspondingly we obtain:

**Theorem 8.4**  *The equation $t^n T(t) = s(t)$ has in $\mathcal{D}'$ the general solution*

$$T(t) = \mathrm{pf}(t_+^{-n}) + \sum_{k=0}^{n-1} c_k \delta^{(k)}(t).$$

*The equation $t^n T(t) = s(-t)$ has in $\mathcal{D}'$ the general solution*

$$T(t) = \mathrm{pf}(t_-^{-n}) + \sum_{k=0}^{n-1} c_k \delta^{(k)}(t).$$

*The equation $t^n T(t) = \mathrm{sgn}(t)$ has in $\mathcal{D}'$ the general solution*

$$T(t) = \mathrm{pf}(t_+^{-n}) - \mathrm{pf}(t_-^{-n}) + \sum_{k=0}^{n-1} c_k \delta^{(k)}(t).$$

With the pseudofunctions $\mathrm{pf}(t_\pm^{-n})$ we have got to know first examples for regularizations of functions with singularities. Historically, the work of J. Hadamard on regularizations of divergent integrals, which appear in solutions of hyperbolic partial differential equations, contributed significantly to the development of the distribution theory. Readers who want to learn more about this are referred to the references Gel'fand et al. (1964) or Ortner and Wagner (2015).

## *Primitives of Distributions*

The goal of this section is to show that the differential equation $\dot{T} = G$ has a solution $T \in \mathcal{D}'$ for every $G \in \mathcal{D}'$ and that two solutions differ by at most a constant $c$. We call each such solution a *primitive* or synonymously an *indefinite integral of $G$*.

First we observe that the integral $\displaystyle\int_{-\infty}^{+\infty} \varphi'(t)\mathrm{d}t = 0$ for all test functions $\varphi \in \mathcal{D}$.

Conversely, it follows for all $\psi \in \mathcal{D}$ with $\displaystyle\int_{-\infty}^{+\infty} \psi(t)\mathrm{d}t = 0$ that $\varphi(t) = \displaystyle\int_{-\infty}^{t} \psi(x)\mathrm{d}x$

belongs to $\mathcal{D}$ and is a primitive of $\psi$. Therefore it holds true

$$\mathcal{D}_0 = \{\varphi' : \varphi \in \mathcal{D}\} = \{\psi \in \mathcal{D} : \int_{-\infty}^{+\infty} \psi(t)\mathrm{d}t = 0\}.$$

For the following two proofs, let $\alpha$ be a test function in $\mathcal{D}$ so that $\displaystyle\int_{-\infty}^{+\infty} \alpha(t)\mathrm{d}t = 1$,

and for $\varphi \in \mathcal{D}$ set $P\varphi = \varphi - \alpha I(\varphi)$ with $I(\varphi) = \displaystyle\int_{-\infty}^{+\infty} \varphi(t)\mathrm{d}t$. Then $P\varphi \in \mathcal{D}_0$ and

$P\varphi^{(k)} = \varphi^{(k)}$ for all $k \in \mathbb{N}$.

**Theorem 8.5** *For $T \in \mathcal{D}'$, the equation $\dot{T} = 0$ is true if and only if $T = c$ with a constant $c$.*

**Proof** Evidently, $\dot{T} = 0$ for $T = c$, $c$ constant. Conversely assume $\dot{T} = 0$. Then we have $\langle T, \psi \rangle = 0$ for all $\psi \in \mathcal{D}_0$. Since $\varphi(t) = P\varphi(t) + \alpha(t)I(\varphi)$, it follows with $c = \langle T, \alpha \rangle$ and $\langle T, P\varphi \rangle = 0$ that $\langle T, \varphi \rangle = \langle T, \alpha I(\varphi) \rangle = \displaystyle\int_{-\infty}^{+\infty} c\,\varphi(t)\mathrm{d}t$, hence $T = c$. $\qquad\qquad\square$

**Theorem 8.6** *Every distribution $G$ has a primitive $T \in \mathcal{D}'$, and every distribution $S$ with $\dot{S} = G$ fulfills $S = T + c$ with a constant $c$.*

***Proof*** For two primitives $T$ and $S$ of $G$ holds $\dot{T} - \dot{S} = 0$, so $T - S = c$ with a constant $c$ according to the previous theorem. To determine a distribution $T$ with

$$\dot{T} = G, \text{ we define } F\varphi(t) = \int_{-\infty}^{t} P\varphi(x)\mathrm{d}x \text{ for } \varphi \in \mathcal{D}. \text{ Then } F\varphi \text{ is a primitive of } P\varphi,$$

and we observe that $(F\varphi)' \in \mathcal{D}_0$ and $F\varphi \in \mathcal{D}$, since $P\varphi \in \mathcal{D}$ and $\displaystyle\int_{t}^{+\infty} P\varphi(t)\mathrm{d}t = 0$

for sufficiently large $t$.

We define for a given distribution $G$

$$\langle T, \varphi \rangle = -\langle G, F\varphi \rangle.$$

$T$ is linear and also continuous on $\mathcal{D}$: Let $\varphi_n$ be a sequence in $\mathcal{D}$ with $\mathcal{D}\text{-}\lim_{n\to\infty} \varphi_n = 0$. Due to $P\varphi^{(k)} = \varphi^{(k)}$ and $\lim_{n\to\infty} I(\varphi_n^{(k)}) = 0$ for all $k \in \mathbb{N}$, we get $\mathcal{D}\text{-}\lim_{n\to\infty} P\varphi_n = 0$.

Now, let $[a, b]$ be an interval containing the supports of $\alpha$ and of all $P\varphi_n$. Then $[a, b]$ contains also all $\operatorname{supp}(F\varphi_n), n \in \mathbb{N}$, and we have the estimate

$$\sup_{t\in\mathbb{R}} |F\varphi_n(t)| \leqslant \int_{-\infty}^{+\infty} |P\varphi_n(t)|\mathrm{d}t \leqslant (b - a) \sup_{t\in\mathbb{R}} |P\varphi_n(t)|.$$

This implies $F\varphi_n \to 0$ uniformly. Furthermore we have for $k \geqslant 1 : (F\varphi_n)^{(k)} = (P\varphi_n)^{(k-1)}$ and $(P\varphi_n)^{(k-1)} \to 0$ uniformly for $n \to \infty$. Thus, $\mathcal{D}\text{-}\lim_{n\to\infty} F\varphi_n = 0$ and eventually $\lim_{n\to\infty} \langle T, \varphi_n \rangle = - \lim_{n\to\infty} \langle G, F\varphi_n \rangle = 0$. Therefore, $T$ is continuous on $\mathcal{D}$, i.e., it is a distribution. The distribution $T$ is a primitive of $G$: In fact, with $F\varphi' = \varphi$ we have accomplished

$$\langle \dot{T}, \varphi \rangle = -\langle T, \varphi' \rangle = \langle G, F\varphi' \rangle = \langle G, \varphi \rangle.$$

$\qquad\qquad\qquad\qquad\qquad\qquad\qquad\qquad\qquad\qquad\qquad\qquad\qquad\qquad\qquad\qquad\qquad\square$

An important consequence of the two theorems is the conclusion that a homogeneous linear differential equation, whose coefficients are constant or infinitely often differentiable functions, considered as an equation in $\mathcal{D}'$, has no further solutions in $\mathcal{D}'$ than the known classical solutions (Exercise A20).

## *Convergence of Sequences of Distributions*

Assume two approximately equal physical quantities are represented by distributions $T_1$ and $T_2$. Experience shows that, when measuring with the same weight function $\varphi \in \mathcal{D}$, the values of $\langle T_1, \varphi \rangle$ and $\langle T_2, \varphi \rangle$ are approximately the same. This experience corresponds to the notion of convergence for distributions.

**Definition** A sequence $(T_n)_{n \in \mathbb{N}}$ of distributions in $\mathcal{D}'$ converges to a distribution $T \in \mathcal{D}'$, if $\lim_{n \to \infty} \langle T_n, \varphi \rangle = \langle T, \varphi \rangle$ holds true for all $\varphi \in \mathcal{D}$. We then denote this by $T = \mathcal{D}'\text{-}\lim_{n \to \infty} T_n$.

**Remark** It can be shown that for every sequence $T_n \in \mathcal{D}'$, whose limits $\lim_{n \to \infty} \langle T_n, \varphi \rangle$ exist for every $\varphi \in \mathcal{D}$, there is indeed a linear continuous functional $T$ on $\mathcal{D}$ defined by $\langle T, \varphi \rangle = \lim_{n \to \infty} \langle T_n, \varphi \rangle$. A proof for this *completeness property* of $\mathcal{D}'$ can be found, e.g., in Schwartz (1957), Zemanian (2010), or Vladimirov (2002).

**Examples**

1. For an arbitrary integrable function $f$, define $f_n(t) = nf(nt)$, $n \in \mathbb{N}$. For $\varphi \in \mathcal{D}$, we then get by substituting $x = nt$

$$\int_{-\infty}^{+\infty} f_n(t)\varphi(t)\, dt = \int_{-\infty}^{+\infty} nf(nt)\varphi(t)\, dt = \int_{-\infty}^{+\infty} f(x)\varphi(x/n)\, dx \ .$$

If $\int_{-\infty}^{+\infty} f(t)\, dt = 1$, then we get by $|f(x)\varphi(x/n)| \leqslant |f(x)| \max_{x \in \mathbb{R}} |\varphi(x)|$ with

interchange of limiting $n \to \infty$ and the integration (possible by the dominated convergence theorem of Lebesgue, cf. p. 496 in Appendix B)

$$\lim_{n \to \infty} \int_{-\infty}^{+\infty} f_n(t)\varphi(t)\, dt = \int_{-\infty}^{+\infty} f(x) \lim_{n \to \infty} \varphi(x/n)\, dx = \varphi(0) = \langle \delta, \varphi \rangle.$$

Therefore $\mathcal{D}'\text{-}\lim_{n \to \infty} f_n = \delta$. Such a sequence of functions is called a $\delta$-*sequence*. The previously used notation is thus compatible with the defined notion of convergence. In particular, all approximating functions $f_n$ themselves can be considered as distributions. As concrete examples, consider the functions $f(t) = \frac{1}{\varepsilon}(s(t + \varepsilon/2) - s(t - \varepsilon/2))$ or $f(t) = \frac{1}{\pi(1+t^2)}$, $s(t)$ the unit step function. These functions are often used to introduce the $\delta$-distribution as a limit of function sequences in the sense of the above defined convergence in $\mathcal{D}'$ (cf. p. 22 and p. 152). There are also $\delta$-sequences constructed from non-integrable functions as is seen in the following examples.

2. We consider the functions $f_n(t) = n \sin(nt)s(t)$, $s(t)$ the unit step function, $n \in \mathbb{N}$. Intuitively it is not obvious whether the sequence $(f_n)_{n \in \mathbb{N}}$ converges in any sense. However, for $\varphi \in \mathcal{D}$ we get with integration by parts:

$$\int_{-\infty}^{+\infty} f_n(t)\varphi(t)\mathrm{d}t \;=\; \int_{0}^{+\infty} n\sin(nt)\varphi(t)\mathrm{d}t \;=\; -\cos(nt)\varphi(t)\Big|_{0}^{+\infty} + \int_{0}^{+\infty}\cos(nt)\varphi'(t)\mathrm{d}t$$

$$= \varphi(0) + \Bigg[\; \underbrace{\frac{1}{n}\sin(nt)\varphi'(t)\Big|_{0}^{\infty}}_{=0} \;-\; \frac{1}{n}\int_{0}^{+\infty}\sin(nt)\varphi''(t)\mathrm{d}t \;\Bigg].$$

By $\left|\displaystyle\int_{0}^{+\infty}\sin(nt)\varphi''(t)\mathrm{d}t\right| \;\leqslant\; \displaystyle\int_{0}^{+\infty}|\varphi''(t)|\mathrm{d}t \;<\; \infty$ follows

$$\lim_{n\to\infty}\int_{-\infty}^{+\infty} f_n(t)\varphi(t)\mathrm{d}t = \varphi(0)\,,$$

i.e., we obtain the result: $\quad \mathcal{D}'\text{-}\lim_{n\to\infty} f_n = \delta.$

On the other hand, the sequence $g_n(t) = n\cos(nt)s(t)$ yields $\mathcal{D}'\text{-}\lim\limits_{n\to\infty} g_n = 0.$

3. For the functions $\frac{\sin(nt)}{\pi t}$, $n \in \mathbb{N}$, $\varphi \in \mathcal{D}$, we prove

$$\lim_{n\to\infty}\int_{-\infty}^{+\infty}\frac{\sin(nt)}{\pi t}\varphi(t)\mathrm{d}t = \varphi(0).$$

The functions $\sin(nt)/t$ are continuously extended to $t = 0$ with respective value $n$. The integrals converge, since $\varphi$ has a bounded support. Substituting $x = nt$, we obtain

$$\int_{-\infty}^{+\infty}\frac{\sin(nt)}{\pi t}\varphi(t)\mathrm{d}t = \int_{-\infty}^{+\infty}\frac{\sin(x)}{\pi x}\varphi\left(\frac{x}{n}\right)\mathrm{d}x.$$

Substituting now $x = \left(n + \frac{1}{2}\right)t$, we consider

$$I_n = \int_{-(n+\frac{1}{2})\pi}^{+(n+\frac{1}{2})\pi}\frac{\sin(x)}{\pi x}\varphi\left(\frac{x}{n}\right)\mathrm{d}x = \int_{-\pi}^{+\pi}\frac{\sin\!\left(\left(n+\frac{1}{2}\right)t\right)}{\pi t}\varphi\left((1 + 1/(2n))\,t\right)\mathrm{d}t$$

$$= \int_{-\pi}^{+\pi} D_n(t)\frac{\sin(t/2)}{t/2}\varphi((1 + 1/(2n))t)\mathrm{d}t$$

with the $2\pi$-periodic Dirichlet kernel

$$D_n(t) = \frac{1}{2\pi} \sum_{k=-n}^{+n} e^{jkt} = \begin{cases} \dfrac{1}{2\pi} \dfrac{\sin((n+\frac{1}{2})t)}{\sin(t/2)} & \text{for} \quad t \notin 2\pi\mathbb{Z}, \\[2ex] \frac{1}{2\pi}(2n+1) & \text{for} \quad t \in 2\pi\mathbb{Z}. \end{cases}$$

For all $n \in \mathbb{N}$ it holds $\displaystyle\int_{-\pi}^{+\pi} D_n(t)\mathrm{d}t = 1$, and for piecewise continuously differen-tiable functions $f : [-\pi, \pi] \to \mathbb{C}$, we have $\displaystyle\int_{-\pi}^{+\pi} D_n(t)f(t)\mathrm{d}t \xrightarrow[n\to\infty]{} \frac{1}{2}(f(0-) + f(0+))$, by virtue of Dirichlet's theorem. We thus obtain $I_n \xrightarrow[n\to\infty]{} \varphi(0)$, i.e.,

$$\mathcal{D}'\text{-}\lim_{n\to\infty} \frac{\sin(nt)}{\pi t} = \delta(t).$$

From $\dfrac{1}{2\pi} \displaystyle\int_{-n}^{n} e^{-j\omega t}\,\mathrm{d}\omega = \dfrac{\sin(nt)}{\pi t}$ we get $\mathcal{D}'\text{-}\lim\limits_{n\to\infty} \dfrac{1}{2\pi} \displaystyle\int_{-n}^{n} e^{-j\omega t}\,\mathrm{d}\omega = \delta(t)$.

In the subsequent Chap. 10, this relation will show us that the constant function $f = 1$ has $2\pi\delta$ as Fourier transform (cf. p. 292).

We note that the approximation $f_1(t) = 100/(\pi(1 + 10000t^2))$ from the previous example 1 corresponds most closely to the usual idea of an impulse function, whereas function sequences like in examples 2 and 3 have little in common with the idea that a $\delta$-sequence converges to infinity at $t = 0$ and to zero otherwise. Consider the following Fig. 8.13 with illustrations of $f_1(t) = 100/(\pi(1 + 10000t^2))$, $f_2(t) = 100\sin(100t)s(t)$, and $f_3(t) = \sin(100t)/(\pi t)$.

However, comparing the sampling properties of $f_1$, $f_2$ and $f_3$, for example with the function $\varphi(t) = e^{-1/(1-t^2)}$ for $|t| \leqslant 1$, zero otherwise, shows that

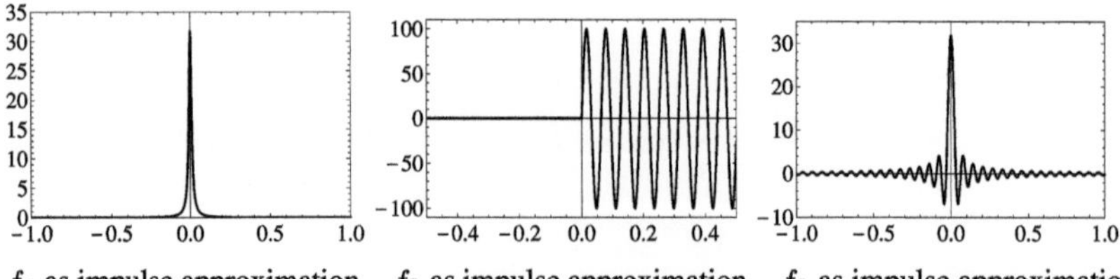

$f_1$ as impulse approximation    $f_2$ as impulse approximation    $f_3$ as impulse approximation

**Fig. 8.13** Three different functions as impulse approximations

the functions $f_2$ and $f_3$ give by far better approximations for the sample $\langle \delta, \varphi \rangle$ than the more impulse-like appearing function $f_1$: For $\varphi(0) = 1/\mathrm{e} \approx 0.36788$, numerical calculation of the integrals with a computer algebra system yields $\langle f_k, \varphi \rangle$, $k = 1, 2, 3$, as follows:

$$\langle f_1, \varphi \rangle \approx 0.36293, \quad \langle f_2, \varphi \rangle \approx 0.36807, \quad \langle f_3, \varphi \rangle \approx 0.36788.$$

4. We consider the distributions $T_n = \displaystyle\sum_{k=-n}^{n} \delta(t - k)$, $n \in \mathbb{N}$. For every $\varphi \in \mathcal{D}$ exists

$$\lim_{n \to \infty} \langle T_n, \varphi \rangle = \lim_{n \to \infty} \sum_{k=-n}^{n} \langle \delta(t - k), \varphi \rangle = \lim_{n \to \infty} \sum_{k=-n}^{n} \varphi(k) = \sum_{k=-\infty}^{+\infty} \varphi(k),$$

because the series at the right is indeed a finite sum, due to the bounded support of $\varphi$. Therefore, by

$$T = \sum_{k=-\infty}^{+\infty} \delta(t - k) = \mathcal{D}'\text{-}\lim_{n \to \infty} \sum_{k=-n}^{n} \delta(t - k)$$

a distribution is defined. Linearity and continuity on $\mathcal{D}$ are immediately seen. The series $T = \sum_{k=-\infty}^{+\infty} \delta(t - k)$ is convergent in $\mathcal{D}'$ with $T = \mathcal{D}'\text{-}\lim_{n \to \infty} T_n$.

5. For $T_n, T \in \mathcal{D}'$ with $T = \mathcal{D}'\text{-}\lim_{n \to \infty} T_n$ we get $\dot{T} = \mathcal{D}'\text{-}\lim_{n \to \infty} \dot{T}_n$, because we have for $\varphi \in \mathcal{D}$

$$\langle \dot{T}_n, \varphi \rangle = \langle T_n, -\varphi' \rangle \xrightarrow[n \to \infty]{} \langle T, -\varphi' \rangle = \langle \dot{T}, \varphi \rangle.$$

Analogously we obtain for distribution series:

$$\text{If } \sum_{n=-\infty}^{+\infty} T_n = T, \text{ then } \sum_{n=-\infty}^{+\infty} \dot{T}_n = \dot{T}.$$

**Result** *Every distribution series, which converges in $\mathcal{D}'$, can be differentiated without any restriction. Differentiation can be carried out term by term.*

Such a result is extremely practical and not achievable in classical analysis. It means that differentiation, with the introduced notion of convergence, is a continuous operation on the vector space $\mathcal{D}'$ of the distributions.

## *Coordinate Transformations for Distributions*

Substituting $t = ax + b$, $a \neq 0$ in an integral of the form $\displaystyle\int_{-\infty}^{+\infty} f(ax + b)\varphi(x)\mathrm{d}x$, $\varphi \in \mathcal{D}$, $f$ a locally integrable function, yields that

$$\int_{-\infty}^{+\infty} f(ax + b)\varphi(x)\mathrm{d}x = \int_{-\infty}^{+\infty} f(t)\frac{1}{|a|}\varphi\left(\frac{t - b}{a}\right)\mathrm{d}t.$$

For non-regular distributions $T$ we define analogously the transformed distribution $T_A$ with a transform $A(t) = at + b$, $a \neq 0$, and $\varphi \in \mathcal{D}$ by

$$\langle T_A, \varphi \rangle = \langle T, |(A^{-1})'| \, \varphi \circ A^{-1} \rangle,$$

and use also the notation $\langle T(at + b), \varphi(t) \rangle = \frac{1}{|a|}\left\langle T(t), \varphi\left(\frac{t-b}{a}\right)\right\rangle$.

Here, $\varphi \circ A^{-1}$ is the composition of the two mappings $A^{-1}$ and $\varphi$.

**Example**  For $a \neq 0$ we have:

$$\delta(at + b) = \frac{1}{|a|}\delta\left(t + \frac{b}{a}\right),$$

because for $\varphi \in \mathcal{D}$

$$\langle \delta(at + b), \varphi(t) \rangle = \frac{1}{|a|}\left\langle \delta(t), \varphi\left(\frac{t - b}{a}\right)\right\rangle = \frac{1}{|a|}\varphi\left(\frac{-b}{a}\right).$$

By the above definition, we can describe some symmetry properties for distributions in the same way as for functions:

$T$ is *even*, if $T(t) = T(-t)$, i.e., if $T = T_A$ with $A(t) = -t$.

$T$ is *odd*, if $T(t) = -T(-t)$, i.e., if $T = -T_A$ with $A(t) = -t$.

$T$ is *periodic with period $p > 0$*, if $T(t + p) = T(t)$, i.e., if $T = T_A$ with $A(t) = t + p$.

For example, $\delta(t)$ is an even distribution. Classical Fourier series can be regarded as periodic distributions. A consequence of the introduced transformations is for example that in case of symmetries, translations or frequency changes of Fourier series can be computed in the usual way, even if these represent non-regular distributions. Such examples will be discussed in the next chapter (cf. Sect. 9.1).

It can be shown that in the sense of the introduced rules for the derivative $\dot{T}$ of a distribution $T$ holds

$$\dot{T} = \lim_{h \to 0} \frac{T(t+h) - T(t)}{h} \quad \text{(Exercise)}.$$

For the derivative of a distribution $S(t) = T(at+b)$, $a \neq 0$, we then have again the *chain rule*: $\dot{S}(t) = a\dot{T}(at+b)$, because with the chain rule for $\varphi(\frac{t-b}{a}) \in \mathcal{D}$ we obtain

$$\langle \dot{S}(t), \varphi(t) \rangle = -\langle T(at+b), \varphi'(t) \rangle = -\frac{1}{|a|} \left\langle T(t), \varphi'\left(\frac{t-b}{a}\right) \right\rangle$$

$$= \frac{a}{|a|} \left\langle \dot{T}(t), \varphi\left(\frac{t-b}{a}\right) \right\rangle = \langle a\dot{T}(at+b), \varphi(t) \rangle.$$

In the following section we also introduce coordinate transformations more generally for distributions on multidimensional parameter sets.


## 8.6   Test Functions and Distributions with Several Variables

All discussed terms can be applied to functions with several variables. Readers who have experience with functions of several variables will quickly recognize the analogies. We sketch here only a few corresponding basic notions  and work out a more familiar way of dealing with them in the following section on convolutions and in the next chapter with application examples of distributions.

Before we define test functions and distributions for multidimensional parameter sets, it is useful to introduce a compact notation for partial differential operators. For a multi-index $k = (k_1, \ldots, k_n) \in \mathbb{N}_0^n$ and $\mathbf{x} \in \mathbb{R}^n$ one defines

$$|k| = k_1 + k_2 + \ldots k_n \quad \text{and} \quad \mathbf{x}^k = x_1^{k_1} x_2^{k_2} \cdots x_n^{k_n}.$$

Then, the partial differential operators $\partial_i$, $\partial_i^{k_i}$, and $\partial^k$ are defined for $1 \leqslant i \leqslant n$, $k_i \in \mathbb{N}$, and a multi-index $k$ by

$$\partial_i = \frac{\partial}{\partial x_i}, \quad \partial_i^{k_i} = \frac{\partial^{k_i}}{\partial x_i^{k_i}} \quad \text{und} \quad \partial^k = \frac{\partial^{k_1+k_2+\ldots+k_n}}{\partial x_1^{k_1} \ldots \partial x_n^{k_n}} = \partial_1^{k_1} \partial_2^{k_2} \cdots \partial_n^{k_n}.$$

The *space of test functions* $\mathcal{D}(\Omega)$ for a domain $\Omega \subset \mathbb{R}^n$ is the set of all those functions $\varphi : \Omega \to \mathbb{R}$, which are arbitrarily differentiable and have a bounded support $\mathrm{supp}(\varphi)$ in $\Omega$. The support of $\varphi$ in $\mathbb{R}^n$ is the closure of the set of all points $\mathbf{x} \in \mathbb{R}^n$, where $\varphi(\mathbf{x}) \neq 0$.

A sequence $(\varphi_m)_{m \in \mathbb{N}}$ converges in $\mathcal{D}(\Omega)$ to the zero function, if and only if there is a compact, i.e., a closed and bounded subset of $\Omega$, which contains all the supports of the $\varphi_m$ and if all derivatives of the $\varphi_m$ converge uniformly to zero, i.e., if for arbitrary $k = (k_1, \ldots, k_n) \in \mathbb{N}_0^n$ holds:

$$\sup_{\mathbf{x} \in \Omega} \left| \partial^k \varphi_m(\mathbf{x}) \right| \xrightarrow[m \to \infty]{} 0.$$

Now, we define the *vector space $\mathcal{D}'(\Omega)$ of distributions on $\Omega$* by

$$\mathcal{D}'(\Omega) = \{ T : \mathcal{D}(\Omega) \to \mathbb{R} \mid T \text{ linear and continuous} \}.$$

A characterization of continuity for linear functionals on $\mathcal{D}$ by estimates of $|\langle T, \varphi \rangle|$, $\varphi \in \mathcal{D}$, is shown at the end of the section.

*Partial derivatives $\partial^k T$ of distributions $T \in \mathcal{D}'(\Omega)$ for a multi-index $k$* are defined by

$$\langle \partial^k T, \varphi \rangle = (-1)^{|k|} \langle T, \partial^k \varphi \rangle.$$

The order of differentiations for distributions can always be chosen arbitrarily. According to the well-known theorem Schwarz, in general this is not the case for classical functions.

*Convergence in $\mathcal{D}'(\Omega)$* is defined as before: For $T, T_m \in \mathcal{D}'(\Omega)$ it holds

$$T = \mathcal{D}'\text{-}\lim_{m \to \infty} T_m, \quad \text{if} \quad \lim_{m \to \infty} \langle T_m, \varphi \rangle = \langle T, \varphi \rangle \text{ for all } \varphi \in \mathcal{D}(\Omega).$$

All terms can be defined analogously for *complex-valued test functions and distributions*. A distribution $T$ has the form $T = T_1 + jT_2$ with $T_1, T_2 \in \mathcal{D}'(\Omega)$. Application to a complex-valued test function $\varphi = \varphi_1 + j\varphi_2$, $\varphi_1, \varphi_2 \in \mathcal{D}(\Omega)$, is defined by

$$\langle T, \varphi \rangle = (\langle T_1, \varphi_1 \rangle - \langle T_2, \varphi_2 \rangle) + j \left( \langle T_1, \varphi_2 \rangle + \langle T_2, \varphi_1 \rangle \right).$$

*Thus from now on, we can use complex-valued test functions and distributions with the scalar field $\mathbb{C}$.*

For *coordinate transformations $A$* and distributions $T$ on $\mathbb{R}^n$, one defines the distribution $T_A$ by generalization of the substitution rule for integrals (cf. p. 497). The inverse transformation $A^{-1}$ is assumed to be an infinitely often differentiable bijective mapping, whose Jacobian determinant $\det \partial A^{-1} \neq 0$ in $\mathbb{R}^n$ (cf. p. 497). For $T \in \mathcal{D}'(\mathbb{R}^n)$ and a test function $\varphi \in \mathcal{D}(\mathbb{R}^n)$ the distribution $T_A \in \mathcal{D}'(\mathbb{R}^n)$ is defined by

$$\langle T_A, \varphi \rangle = \langle T, |\det \partial A^{-1}| \, \varphi \circ A^{-1} \rangle.$$

With that definition we can formulate symmetry properties, for example rotational invariance, for distributions and can make coordinate transformations. For example, a distribution $T$ on $\mathbb{R}^n$ is *rotationally invariant*, if for all orthogonal $(n \times n)$-matrices $A$ with $\det A = 1$ we have $T_A = T$. For those matrices $A$ hold $A^{-1} = A^*$, $A^*$ the transposed matrix and $\det \partial A^{-1} = 1$. For any test function $\varphi$ and rotationally invariant $T$ we obtain $\langle T_A, \varphi \rangle = \langle T(\mathbf{x}), \varphi(A^{-1}\mathbf{x}) \rangle = \langle T, \varphi \rangle$. More examples can be found in the following sections.

**Examples**

1. The function $h(\mathbf{x}) = \frac{1}{|\mathbf{x}|} = (x^2 + y^2 + z^2)^{-1/2}$ for $\mathbf{x} = (x, y, z)$ is locally integrable in $\mathbb{R}^3$, since the integrals

$$\int\limits_{0<\varepsilon \leqslant |\mathbf{x}| \leqslant R} h(\mathbf{x})\, d\lambda^3(\mathbf{x}) = \int\limits_0^{2\pi} \int\limits_0^{\pi} \int\limits_\varepsilon^{R} r \sin(\theta)\, dr\, d\theta\, d\phi$$

converge for $\varepsilon \to 0$ to $2\pi R^2$ ($d\lambda^3(\mathbf{x}) = dx\, dy\, dz$ *denotes the differential volume element on* $\mathbb{R}^3$). Thus, the function $h$ can be considered as a regular distribution on $\mathbb{R}^3$.

2. Generalized $\delta$-functions with three variables, defined for $\varphi$ in $\mathcal{D}(\mathbb{R}^3)$ and $\mathbf{x}_0$ in $\mathbb{R}^3$ by $\langle \delta(\mathbf{x} - \mathbf{x}_0), \varphi(\mathbf{x}) \rangle = \varphi(\mathbf{x}_0)$, can be used as generalized density functions, for example to describe spatially discrete distributions. For example, $\varrho(\mathbf{x}) = \sum\limits_{i=1}^{n} q_i \delta(\mathbf{x} - \mathbf{x}_i)$ can be a generalized density function for $n$ electric charges $q_i$ at the points $\mathbf{x}_i \in \mathbb{R}^3$. The distribution $\varrho$ can be extended to a discrete measure in $\mathbb{R}^3$ (see remark on p. 162). Conversely, each measure $m$ in $\mathbb{R}^3$ defines a distribution $T$ by $\langle T, \varphi \rangle = \int \varphi\, dm$, $\varphi \in \mathcal{D}(\mathbb{R}^3)$ (cf. Appendix B).

3. With $\varphi \in \mathcal{D}(\mathbb{R}^3)$, $R > 0$, with the usual surface measure $do$ on a sphere of radius $R$ around zero (cf. Appendix B) and a smooth density function $\varrho$ there, a distribution $T$ is defined by the surface integral $\langle T, \varphi \rangle = \int\limits_{|\mathbf{x}|=R} \varrho \varphi\, do$. That distribution is also denoted by $T(\mathbf{x}) = \varrho(\mathbf{x})\delta(|\mathbf{x}| - R)$.

## *Characterization of Continuity of Distributions*

The following theorem describes continuity of linear functionals on $\mathcal{D}(\Omega)$ by estimates of $|\langle T, \varphi \rangle|$ for $\varphi \in \mathcal{D}(\Omega)$. There, $\|\varphi\|_p$ for an integer $p \geqslant 0$ is the maximum norm

$$\|\varphi\|_p = \max\{|\partial^k \varphi(\mathbf{x})| \ : \ \mathbf{x} \in \Omega, |k| \leqslant p\}$$

of $\varphi$ in the space of $p$-times continuously differentiable functions with compact support in the domain $\Omega$.

**Theorem 8.7** *A linear functional $T$ on $\mathcal{D}(\Omega)$ is continuous if and only if for every compact set $K \subset \Omega$ there are a constant $C > 0$ and an integer $p \geqslant 0$ such that for every test function $\varphi$ with support in $K$ the following estimation applies:*

$$|\langle T, \varphi \rangle| \leqslant C \|\varphi\|_p.$$

***Proof***

(a) Let $T \in \mathcal{D}'(\Omega)$ and a compact set $K \subset \Omega$ be given. To prove the necessity of the given condition, we assume that it is wrong. Then for each natural number $p$ there is a test function $\varphi_p$ with support in $K$ such that

$$|\langle T, \varphi_p \rangle| > p \|\varphi_p\|_p.$$

The functions $\psi_p = \varphi_p / (p \|\varphi_p\|_p)$ have their support in $K$ and converge to zero in $\mathcal{D}(\Omega)$ for $p \to \infty$, since for each fixed $k \in \mathbb{N}_0^n$ and all $p \geqslant |k|$ and $\mathbf{x} \in \Omega$

$$|\partial^k \psi_p(\mathbf{x})| \leqslant \|\psi_p\|_p = \frac{1}{p}$$

is fulfilled. It then follows from the continuity of $T$ that for $p \to \infty$ the numbers $|\langle T, \psi_p \rangle|$ also converge to zero. This contradicts the conclusion $|\langle T, \psi_p \rangle| > 1$ for all $p$ according to the assumption made above.

(b) For every null sequence $\varphi_m$ in $\mathcal{D}(\Omega)$ there is a compact set $K \subset \Omega$ that contains all supports of $\varphi_m$, and the following holds $\lim\limits_{m \to \infty} \|\varphi_m\|_p = 0$ even for all $p \in \mathbb{N}_0$. From the given condition for a linear functional $T$ on $\mathcal{D}(\Omega)$ it therefore follows $\lim_{m \to \infty} \langle T, \varphi_m \rangle = 0$, i.e., the continuity of $T$.  $\square$

## 8.7   Tensor Product and Convolution

The aim of this section is the introduction of convolutions for distributions. Therewith the next chapter provides a basic solution procedure for inhomogeneous linear differential equations with constant coefficients. This central result, which allows the calculation of particular solutions of such equations, requires some preparations. Convolutions are also a fundamental theoretical tool for linear systems theory and its applications. We will go into this in more detail in Chap. 11.

## *The Tensor Product of Distributions*

For locally integrable functions $f$ and $g$ on $\mathbb{R}$ the function

$$f \otimes g(x, y) = f(x)g(y)$$

is a locally integrable function on $\mathbb{R}^2$. Then by $f \otimes g$ a distribution on the test functions $\varphi$ in $\mathcal{D}(\mathbb{R} \times \mathbb{R})$ is defined by

$$\langle f \otimes g, \varphi \rangle = \iint_{\mathbb{R}\ \mathbb{R}} f(x)g(y)\varphi(x, y)\, \mathrm{d}y\, \mathrm{d}x .$$

With regard to applications of distributions, we transfer all the following considerations immediately to functions with several variables. In the following we denote with $X = \mathbb{R}^n$, $Y = \mathbb{R}^m$, $Z = \mathbb{R}^p$ and with $\mathrm{d}\lambda^p(\mathbf{z})$ the differential volume element in $\mathbb{R}^p$. For locally integrable functions $f$ on $X$, $g$ on $Y$, and a test function $\varphi$ in $\mathcal{D}(X \times Y)$, a regular distribution $f \otimes g \in \mathcal{D}'(X \times Y)$ is defined by

$$\langle f \otimes g, \varphi \rangle = \iint_{X\ Y} f(\mathbf{x})g(\mathbf{y})\varphi(\mathbf{x}, \mathbf{y})\, \mathrm{d}\lambda^m(\mathbf{y})\, \mathrm{d}\lambda^n(\mathbf{x}) = \langle f(\mathbf{x}), \langle g(\mathbf{y}), \varphi(\mathbf{x}, \mathbf{y}) \rangle \rangle.$$

The distribution $f \otimes g$ is called *the tensor product of $f$ and $g$.* The factor $f$ only affects the parameters from $X$, and the factor $g$ affects the parameters from $Y$. Instead of $f \otimes g$, we also write $f(\mathbf{x}) \otimes g(\mathbf{y})$, if we want to specify the integration variables for better orientation. Exchanging the order of integration corresponds to the permutation of the tensor product, i.e., the following holds: $f(\mathbf{x}) \otimes g(\mathbf{y}) = g(\mathbf{y}) \otimes f(\mathbf{x})$.

**Definition** For two distributions $T \in \mathcal{D}'(X)$ and $G \in \mathcal{D}'(Y)$ the tensor product $T \otimes G$ is defined analogously with $\varphi \in \mathcal{D}(X \times Y)$ by

$$\langle T(\mathbf{x}) \otimes G(\mathbf{y}), \varphi(\mathbf{x}, \mathbf{y}) \rangle = \langle T(\mathbf{x}), \langle G(\mathbf{y}), \varphi(\mathbf{x}, \mathbf{y}) \rangle \rangle.$$

The definition is meaningful because for every fixed $\mathbf{x} \in X$ the function $\varphi_{\mathbf{x}}(\mathbf{y}) = \varphi(\mathbf{x}, \mathbf{y})$ belongs to $\mathcal{D}(Y)$, and the function $\psi(\mathbf{x}) = \langle G, \varphi_{\mathbf{x}} \rangle$ can be shown to be a test function in $\mathcal{D}(X)$. The tensor product is linear and continuous on $\mathcal{D}(X \times Y)$, i.e., $T \otimes G$ *is a distribution from* $\mathcal{D}'(X \times Y)$. At this point and in the following statements about convolutions we refrain from very technical, detailed proofs and focus on the essential aspects for the applications. Interested readers will find a more detailed presentation of the contents of this and the following section on convolutions with proofs in Schwartz (1957), Vladimirov (2002), or Zemanian (2010).

For test functions $\varphi \in \mathcal{D}(X \times Y)$ of the special form $\varphi(\mathbf{x}, \mathbf{y}) = \varphi_1(\mathbf{x})\varphi_2(\mathbf{y})$ with $\varphi_1 \in \mathcal{D}(X)$ and $\varphi_2 \in \mathcal{D}(Y)$ it follows from the definition of the tensor product

$$\langle T(\mathbf{x}) \otimes G(\mathbf{y}), \varphi(\mathbf{x}, \mathbf{y}) \rangle = \langle T(\mathbf{x}), \varphi_1(\mathbf{x}) \rangle \langle G(\mathbf{y}), \varphi_2(\mathbf{y}) \rangle$$
$$= \langle G(\mathbf{y}) \otimes T(\mathbf{x}), \varphi(\mathbf{x}, \mathbf{y}) \rangle.$$

With the theorem of Weierstrass (cf. p. 142) it can be shown that every test function $\varphi$ can be approximated by linear combinations of the special form

$$\sum_{k=1}^{n} \varphi_{1,k}(\mathbf{x})\varphi_{2,k}(\mathbf{y})$$

in $\mathcal{D}(X \times Y)$. From the linearity and the continuity of $T \otimes G$ the *commutativity of the tensor product on whole* $\mathcal{D}(X \times Y)$ follows:

$$T(\mathbf{x}) \otimes G(\mathbf{y}) = G(\mathbf{y}) \otimes T(\mathbf{x}).$$

In an analogous way, one also obtains the *associativity* of the tensor product of distributions $T$, $G$, and $H$ on $X$, $Y$, and $Z$ respectively:

$$T(\mathbf{x}) \otimes G(\mathbf{y}) \otimes H(\mathbf{z}) = (T(\mathbf{x}) \otimes G(\mathbf{y})) \otimes H(\mathbf{z}) = T(\mathbf{x}) \otimes (G(\mathbf{y}) \otimes H(\mathbf{z})).$$

**Examples**

1. The $\delta$-distribution $\delta(\mathbf{x}) = \delta(x, y, z)$ in $\mathbb{R}^3$ for $\mathbf{x} = (x, y, z)$ is simply the tensor product $\delta(x) \otimes \delta(y) \otimes \delta(z)$. Because for $\varphi \in \mathcal{D}(\mathbb{R}^3)$ it holds

$$\langle\, \delta(x, y, z), \varphi(x, y, z) \rangle = \langle \delta(x), \langle \delta(y), \langle \delta(z), \varphi(x, y, z) \rangle \rangle \rangle$$
$$= \langle \delta(x), \langle \delta(y), \varphi(x, y, 0) \rangle \rangle = \langle \delta(x), \varphi(x, 0, 0) \rangle = \varphi(0, 0, 0).$$

2. If $g$ is locally integrable on $Y$, then for $\varphi \in \mathcal{D}(X \times Y)$

$$\langle \delta(\mathbf{x}) \otimes g(\mathbf{y}), \varphi(\mathbf{x}, \mathbf{y}) \rangle = \langle g(\mathbf{y}) \otimes \delta(\mathbf{x}), \varphi(\mathbf{x}, \mathbf{y}) \rangle$$
$$= \int_Y g(\mathbf{y})\varphi(\mathbf{0}, \mathbf{y})\, d\lambda^m(\mathbf{y})\,.$$

3. A mass or charge density $\varrho(z)$ on a thin rod of the length $2l$, which is idealized by the (degenerate) interval $\{0\} \times \{0\} \times [-l, l]$ in $\mathbb{R}^3$, is described with the help of the unit step function $s(z)$ by the tensor product

$$\delta(x) \otimes \delta(y) \otimes \varrho(z)[s(z+l) - s(z-l)].$$

## The Support of a Distribution

In the following, we will describe what is meant by the support of a distribution. The support of a continuous function $f$ contains all points at which $f$ does not vanish. Since for distributions $T$ it does not make sense to speak of values at individual points, we say *T vanishes on an open set A, $T = 0$ in A*, if $\langle T, \varphi \rangle = 0$ for all test functions $\varphi$ with $\operatorname{supp}(\varphi) \subset A$. For example, $\delta(\mathbf{x}) = 0$ on any open set $A$ that does not contain the zero point.

Conversely, a point $\mathbf{x}$ *is an essential point for T,* if for every open neighborhood $U$ of $\mathbf{x}$ there is a test function $\varphi$ with $\operatorname{supp}(\varphi) \subset U$ and $\langle T, \varphi \rangle \neq 0$. For example, the zero point is the only essential point for $\delta(\mathbf{x})$.

**Definition** The support $\operatorname{supp}(T)$ of a distribution T is the closed set of all essential points for T.

The support $\operatorname{supp}(T)$ of a distribution $T$ is therefore the smallest closed set on whose complement it holds $T = 0$.

**Examples**

1. For regular distributions $T_f$ with continuous $f$, $\operatorname{supp}(T_f) = \operatorname{supp}(f)$. The support of locally integrable functions $f$ is defined as the support of the distribution $T_f$. The support of $\delta(\mathbf{x} - \mathbf{x}_0)$ is the set $\{\mathbf{x}_0\}$, the support of $\delta(|\mathbf{x}| - R)$ is the spherical surface $|\mathbf{x}| = R$, and the support of $\delta(x) \otimes \delta(y) \otimes [s(z+l) - s(z-l)]$ is the set $\{0\} \times \{0\} \times [-l, l]$ in $\mathbb{R}^3$.
2. *For a distribution T and a test function $\varphi$, $\langle T, \varphi \rangle$ is only dependent on the values of $\varphi$ on the support of $T$.* Namely, if $\varphi$ is changed outside of a neighborhood $U$ of $\operatorname{supp}(T)$ so that again a test function $\psi$ is obtained, then $\psi = \varphi + h$ with a test function $h$ that vanishes on $U$. Because $\langle T, h \rangle = 0$, it follows that $\langle T, \psi \rangle = \langle T, \varphi + h \rangle = \langle T, \varphi \rangle$.

## The Convolution of Distributions

The convolution $f * g$ of two integrable functions $f$ and $g$ on $\mathbb{R}^n$ is defined by

$$(f * g)(\mathbf{x}) = \int f(\mathbf{x} - \mathbf{y}) g(\mathbf{y}) \, d\lambda^n(\mathbf{y}) .$$

By the theorem of Fubini-Tonelli (cf. Appendix B), the convolution $f * g$ is again an integrable function (Exercise A14, Chap. 9). If one looks at it as a regular distribution, then for each test function $\varphi$ it follows with the substitution rule for integrals

$$\langle f * g, \varphi \rangle = \int \int f(\mathbf{x} - \mathbf{y}) g(\mathbf{y}) \varphi(\mathbf{x}) \, d\lambda^n(\mathbf{y}) \, d\lambda^n(\mathbf{x})$$

$$= \int \int f(\mathbf{x}) g(\mathbf{y}) \varphi(\mathbf{x} + \mathbf{y}) \, d\lambda^n(\mathbf{y}) \, d\lambda^n(\mathbf{x}) .$$

With the help of the tensor product $f(\mathbf{x}) \otimes g(\mathbf{y})$, the convolution of $f$ and $g$ can then be described by the formula

$$\langle f * g, \varphi \rangle = \langle f(\mathbf{x}) \otimes g(\mathbf{y}), \varphi(\mathbf{x} + \mathbf{y}) \rangle = \langle f(\mathbf{x}), \langle g(\mathbf{y}), \varphi(\mathbf{x} + \mathbf{y}) \rangle \rangle.$$

With the *reflection* $\check{\varphi}(\mathbf{x}) = \varphi(-\mathbf{x})$ of $\varphi$ one equivalently obtains

$$\langle f * g, \varphi \rangle = \langle f, \check{g} * \varphi \rangle.$$

As usual, the regular distribution $T_{f*g}$ is identified with the function $f * g$:

$$T_{f*g} = f * g.$$

**Definition** For two arbitrary, not necessarily regular distributions $T$ and $G$, the convolution $T * G$ is defined by the same approach:

$$\langle T * G, \varphi \rangle = \langle T(\mathbf{x}) \otimes G(\mathbf{y}), \varphi(\mathbf{x} + \mathbf{y}) \rangle = \langle T(\mathbf{x}), \langle G(\mathbf{y}), \varphi(\mathbf{x} + \mathbf{y}) \rangle \rangle.$$

With the *reflection* $\check{T}(\varphi) = \langle T, \check{\varphi} \rangle = \langle T(\mathbf{x}), \varphi(-\mathbf{x}) \rangle$ of $T$, the convolution can also be written as above by

$$\langle T * G, \varphi \rangle = \langle T, \check{G} * \varphi \rangle.$$

However, it should be noted in the definition that $\varphi(\mathbf{x} + \mathbf{y})$ generally does not have a bounded support. The defining formula therefore usually *only makes sense under additional assumptions.* If the convolution $T * G$ of two distributions exists, then $T * G$ is again a distribution, and from the commutativity of the tensor product the commutativity of the convolution follows: $T * G = G * T$.

The following theorem specifies conditions under which the convolution of distributions exists. Afterward the most important properties for calculations with convolutions are summarized. Some facts used in previous chapters about convolutions of classical functions are collected in Appendix B. Further statements about convolutions that are needed in connection with the Fourier transform are discussed in Chap. 10. For detailed proofs it is again referred to Zemanian (1995) or Vladimirov (2002).

## *Sufficient Conditions for the Existence of Convolutions*

**Theorem 8.8 (Existence of Convolutions)** *The convolution $T * G$ of two distributions $T$ and $G$ on $\mathbb{R}^n$ is meaningfully defined for all $\varphi \in \mathcal{D}(\mathbb{R}^n)$ by $\langle T * G, \varphi \rangle = \langle T(\mathbf{x}) \otimes G(\mathbf{y}), \varphi(\mathbf{x} + \mathbf{y}) \rangle$ under each of the following conditions:*

1. *T or G has a bounded support.*
2. *The supports of T and G are both contained in a "quadrant"*

$$Q_+^c = \{\mathbf{x} = (x_1, \ldots, x_n) | x_i \geqslant c, i = 1, \ldots, n\}$$

*or both in one "quadrant"*

$$Q_-^c = \{\mathbf{x} = (x_1, \ldots, x_n) | x_i \leqslant c, i = 1, \ldots, n\},$$

*$c$ in $\mathbb{R}$ suitable.*

*Thus, in the one-dimensional case $n = 1$, $T * G$ exists if the supports of $T$ and $G$ are bounded on the same side. This is the case, for example, if $\mathrm{supp}(T) \subset [0, \infty[$ and $\mathrm{supp}(G) \subset [0, \infty[.$*

***Proof*** In the first case, for example, let $G$ have a bounded support. Then the infinitely often differentiable function $\psi(\mathbf{x}) = \langle G(\mathbf{y}), \varphi(\mathbf{x} + \mathbf{y}) \rangle$ disappears, if $|\mathbf{x}|$ is so large such that the supports $\mathrm{supp}\,(G)$ and $\mathrm{supp}\,(\varphi_{\mathbf{x}})$, $\varphi_{\mathbf{x}}(\mathbf{y}) = \varphi(\mathbf{x} + \mathbf{y})$ no longer intersect. So $\psi$ in that case is a test function, and $T$ can be applied to $\psi$, i.e., the convolution $T * G$ is possible (see Exercise 14).

For the second condition, we consider the illustrative case $n = 1$ and distributions $T$ and $G$ with supports in $[0, \infty[$. For growing $x$ the support of $\varphi_x$ shifts to the left, so that at some point $\mathrm{supp}\,(\varphi_x) \cap \mathrm{supp}\,(G) = \emptyset$, i.e., the support of $\psi(x) = \langle G(y), \varphi(x + y) \rangle$ is bounded to the right. Consequently, the intersection $\mathrm{supp}\,(T) \cap \mathrm{supp}\,(\psi)$ is bounded. If one chooses a test function $\alpha$ to be constantly $\alpha = 1$ on this intersection, then the test function $\alpha\psi$ coincides with $\psi$ on the support of $T$. The convolution $T * G$ exists and it holds:

$$\langle T * G, \varphi \rangle = \langle T, \alpha\psi \rangle.$$

These arguments can be applied under the "quadrant conditions" from Point 2 also to the multidimensional case.                                                                    $\square$

Further theorems on the existence of convolutions, their properties, and detailed proofs can be found in Schwartz (1957) and Vladimirov (2002). Before we come to examples of convolutions, some basic properties of convolutions are listed below.

## *Properties of Convolutions*

1. *Distributivity of Convolutions.* If the convolution of a distribution $T$ with two distributions $G$ and $S$ exists, then the convolution is distributive, i.e., the following applies for any constants $\alpha$ and $\beta$

$$T * (\alpha G + \beta S) = \alpha(T * G) + \beta(T * S).$$

The following applies $\mathrm{supp}(\alpha G + \beta S) \subset \mathrm{supp}(G) \cup \mathrm{supp}(S)$.

2. *Commutativity, associativity of convolutions* Since tensor products are commutative, this also applies to convolutions. If three distributions $T$, $G$, and $S$ fulfill the quadrant condition with a common quadrant $Q_+^c$ or $Q_-^c$ or at least two of them have a bounded support, then the convolution is associative, i.e., the following holds

$$T * (G * S) = (T * G) * S.$$

3. *Convolution with the Dirac distribution.* For all distributions $T$ there exists $T * \delta$ and for each test function $\varphi$ it holds

$$\langle T * \delta, \varphi \rangle = \langle T(\mathbf{x}), \langle \delta(\mathbf{y}), \varphi(\mathbf{x} + \mathbf{y}) \rangle \rangle = \langle T(\mathbf{x}), \varphi(\mathbf{x}) \rangle.$$

From the definition of convolution and commutativity, it follows that $\delta$ acts like the one in multiplication:

$$T * \delta = \delta * T = T.$$

4. *Differentiation and translation of convolutions.* If the convolution $T * G$ exists, for a differential operator $\partial_i = \frac{\partial}{\partial x_i}$ and the derivative $\partial_i(T * G)$ of the convolution it holds

$$\partial_i(T * G) = T * \partial_i G = \partial_i T * G.$$

We exemplarily show the first equation: For each test function $\varphi$ it holds

$$\langle \partial_i(T * G), \varphi \rangle = \langle T * G, -\partial_i \varphi \rangle = \langle T(\mathbf{x}), \langle \partial_i G(\mathbf{y}), \varphi(\mathbf{x} + \mathbf{y}) \rangle \rangle = \langle T * \partial_i G, \varphi \rangle.$$

Specifically, for all distributions $T$ we have that $\partial_i(\delta * T) = \partial_i(T * \delta) = \partial_i T$.
In an analogous way, for a shift $A(\mathbf{x}) = \mathbf{x} + \mathbf{a}$ and the translation $(T * G)_A$ of the convolution (cf. p. 186), we obtain $(T * G)_A = T_A * G = T * G_A$.
This follows with $(T * G)_A = (T * G) * \delta_A$ from the commutativity and associativity of convolutions. Therefore, when differentiating and translating a convolution product, one can arbitrarily choose among the factors.
5. *Convolution with infinitely often differentiable functions.* For every distribution $T$ and every test function $g$, the convolution $T * g$ is an infinitely often differentiable function. It holds $(T * g)(\mathbf{x}) = \langle T(\mathbf{y}), g(\mathbf{x} - \mathbf{y}) \rangle$. The same applies if $g$ is infinitely often differentiable and has no bounded support but $T$ has a bounded support. As derivative $\partial_i(T * g)$ of the convolution $T * g$ for $\partial_i = \partial/\partial x_i$, we obtain $T * \partial_i g$ (see Point 4 and Exercise A14).
6. *The support of convolutions.* For the support of a convolution $T * G$, the relation
$\mathrm{supp}(T * G) \subset \mathrm{supp}(T) + \mathrm{supp}(G) = \{\mathbf{x} + \mathbf{y} \mid \mathbf{x} \in \mathrm{supp}(T), \mathbf{y} \in \mathrm{supp}(G)\}$ holds true.

7. *Convergence of convolutions.* For distributions $G$ and $T = \mathcal{D}'\text{-}\lim_{k\to\infty} T_k$ it holds

$$T * G = \mathcal{D}'\text{-}\lim_{k\to\infty} (T_k * G),$$

if one of the following conditions is fulfilled:

(a)  There is a bounded set that contains all supports of the $T_k$.
(b)  The support of $G$ is bounded.
(c)  The supports of all $T_k$ and the support of $G$ lie together in a "quadrant" $Q_+^c$ or $Q_-^c$, for suitable $c \in \mathbb{R}$ (cf. p. 195).

The proof is obtained analogously as in Theorem 8.8 by the imposed support conditions (Exercise).

## *Examples of Convolutions*

1. According to Property 4, for any distributions $T$ and $G$ whose convolution exists, it holds $T(t - a) * G(t - b) = T(t) * G(t - (a + b))$. For example, the following relations apply to the Dirac distribution $\delta(t)$ and the unit step function $s(t)$:

$$\delta(t - a) * \delta(t - b) = \delta(t - (a + b)),$$

$$\delta(t - a) * s(t - b) = \delta(t) * s(t - (a + b)) = s(t - (a + b)).$$

2. For integrable functions $f(t)$ and the unit step function $s(t)$ it is useful to remember

$$(f * s)(t) = \int_{-\infty}^{+\infty} f(u)s(t - u)\,\mathrm{d}u = \int_{-\infty}^{t} f(u)\,\mathrm{d}u\,,$$

$$((fs) * s)(t) = \int_{-\infty}^{t} f(u)s(u)\,\mathrm{d}u = s(t) \int_{0}^{t} f(u)\,\mathrm{d}u\,.$$

The convolution $s * r_n$ of the unit step function with the rectangular function $r_n$, defined by $r_n(t) = s(t) - s(t - n)$, results in $(s * r_n)(t) = \min(t, n)s(t)$. With $\mathcal{D}'\text{-}\lim_{n\to\infty} r_n = s$, it follows

$$(s * s)(t) = ts(t).$$

3. *The convolution as a continuous superposition.* The potential $U$ vanishing at infinity of a spatially bounded charge distribution with the density function $\varrho$ is given by the Poisson formula (see later Sect. 9.4) ( $\mathrm{supp}\,(\varrho) \subset B$, $B$ bounded in $\mathbb{R}^3$):

$$U(\mathbf{x}) = \frac{1}{4\pi\varepsilon_0} \int_B \frac{\varrho(\mathbf{y})}{|\mathbf{x} - \mathbf{y}|}\, d\lambda^3(\mathbf{y})\,.$$

The potential $U$ is the convolution of $\varrho$ with the regular distribution $\frac{1}{4\pi\varepsilon_0|\mathbf{x}|}$. It is thus obtained by continuous superposition of the influences of all spatially distributed charges.

4. If the convolution $T*G$ of two distributions $T$ and $G$ on $\mathbb{R}$ exists, then for any linear differential operator $L = \sum_{k=0}^{m} c_k \frac{d^k}{dt^k}$, with constant coefficients $c_k$ the equation $L(T*G) = LT*G = T*LG$ holds true. The same applies for several variables and corresponding partial differential operators. These equations follow immediately from Property 4 due to the linearity of convolutions. They represent a distributional analog of the interchange of differentiation and integration, which is not possible with classical functions without further assumptions.

5. The support of a convolution $T*G$ is often enlarged compared to the supports of $T$ and $G$. As a simple example consider the indicator function $f = 1_{[-1,1]}$ of the interval $[-1, 1]$. It has the value one in $[-1, 1]$ and zero otherwise. It is a regular distribution with $\mathrm{supp}(f) = [-1, 1]$ and

$$(f * f)(t) = \int_{-\infty}^{+\infty} 1_{[-1,1]}(x)1_{[-1,1]}(t - x)\, dx = \int_{-1}^{1} 1_{[-1,1]}(t - x)\, dx\,.$$

With $1_{[-1,1]}(t - x) = 1_{[-1,\infty[}(t - x) - 1_{]1,\infty[}(t - x) = 1_{]-\infty,t+1]}(x) - 1_{]-\infty,t-1[}(x)$ one immediately computes

$$(f*f)(t) = \int_{-1}^{1} 1_{]-\infty,t+1]}(x)\, dx - \int_{-1}^{+1} 1_{]-\infty,t-1[}(x)\, dx = \begin{cases} 0 & \text{for } t \leq -2 \\ t + 2 & \text{for } -2 < t \leq 0 \\ 2 - t & \text{for } 0 \leq t \leq 2 \\ 0 & \text{for } t > 2. \end{cases}$$

Therefore, $\mathrm{supp}(f * f) = \mathrm{supp}(f) + \mathrm{supp}(f) = [-2, 2]$.

From Property 6 it follows that for bounded supports of $T$ and $G$, $\mathrm{supp}(T * G)$ is also bounded. For distributions $T$ and $G$ on $\mathbb{R}$, whose support lie in $[0, \infty[$, the support of $T * G$ is also contained in $[0, \infty[$.

6. *Convolutions of impulse sequences.* The *discrete convolution* plays a central role in discrete signal processing, the basics of which we explain in Chap. 11.

There, discrete time signals are modeled by an impulse sequence of the form $x = \sum_{k=-\infty}^{+\infty} x_k \delta_k$. We use $x_k \delta_k$ to denote impulses $\delta_k = \delta(t - ka)$ of the strength $x_k$ at time $ka$, where $1/a > 0$ is the sampling frequency, with which the discrete system operates. For discrete linear filters with the impulse response $h = \sum_{k=-\infty}^{+\infty} h_k \delta_k$ and discrete input signals $x$ the convolution relation $y = x * h$ then applies to the output signals $y$ (cf. Chap. 11 later).

The convolution of the impulse response $h$ of the filter with the allowed input signals $x$ must be possible for this relationship to make sense. We calculate the discrete convolution for two special filters, which we will return to in Chap. 11. In the first case, we assume that $h$ has a bounded support, and in the second case, that the supports of $h$ and $x$ are bounded below. Then the convolution $y = h * x$ exists in each case, and we show that $y$ is of the form $y = \sum_{m=-\infty}^{+\infty} y_m \delta_m$, where

$$y_m = \sum_{k=-\infty}^{+\infty} h_k x_{m-k} = \sum_{k=-\infty}^{+\infty} h_{m-k} x_k.$$

(a) In the case $h_k = 0$ for $|k| > M$ and suitable $M \in \mathbb{N}$, it follows from the distributivity of the convolution with the index transformation $k + n = m$

$$h * x = \left( \sum_{k \in \mathrm{supp}(h)} h_k \delta_k \right) * \left( \sum_{n=-\infty}^{+\infty} x_n \delta_n \right)$$

$$= \sum_{n=-\infty}^{+\infty} \sum_{k \in \mathrm{supp}(h)} x_n h_k \delta_{k+n} = \sum_{m=-\infty}^{+\infty} \sum_{k \in \mathrm{supp}(h)} x_{m-k} h_k \delta_m$$

and thus the representation of $y_m$ claimed above.

**Example** Let $x = \sum_{n=-\infty}^{+\infty} x_n \delta_n$ and $h = \sum_{k=-2}^{1} h_k \delta_k$ with support $\{-2, -1, 0, 1\}$ be given. For example, to obtain $y_{-2}$ in $x * h = \sum_{m=-\infty}^{+\infty} y_m \delta_m$, write the mirrored sequence of $h_k$ over the sequence of coefficients $x_n$, so that $h_0$ is above the $x_{-2}$ associated with $\delta_{-2}$. Then multiply all overlapping coefficients and add them together.

So $y_{-2}$ is calculated in the example as follows:

$$\ldots 0 \quad h_1 \quad h_0 \quad h_{-1} \ h_{-2} \ 0 \ 0 \ 0 \ \ldots$$
$$\ldots x_{-4} \ x_{-3} \ x_{-2} \ x_{-1} \ x_0 \quad x_1 \ x_2 \ x_3 \ \ldots$$
$$\uparrow$$

$$y_{-2} = x_{-3}h_1 + x_{-2}h_0 + x_{-1}h_{-1} + x_0 h_{-2}.$$

Note that the convolution $y = h * x$ usually has a larger support than the convolution factors involved, even if $h$ and $x$ have finite supports.

(b)  In the second example, we assume that $h = \sum\limits_{k=k_0}^{+\infty} h_k \delta_k$ and $x = \sum\limits_{n=n_0}^{+\infty} x_n \delta_n$ have bounded below supports. Then the following equations hold for any test function $\varphi$

$$\langle h, \varphi \rangle = \sum_{k=k_0}^{+\infty} h_k \varphi(ka) \quad \text{and} \quad \langle x, \varphi \rangle = \sum_{n=n_0}^{+\infty} x_n \varphi(na)$$

and both sums have only finitely many nonzero summands due to the bounded support of $\varphi$. Therefore it follows from

$$\langle h * x, \varphi \rangle = \sum_{k=k_0}^{+\infty} h_k \left( \sum_{n=n_0}^{+\infty} x_n \varphi((n+k)a) \right)$$

by exchanging the summation order and index transformation $m = n + k$

$$\langle h * x, \varphi \rangle = \sum_{n=n_0}^{+\infty} \sum_{k=k_0}^{+\infty} h_k x_n \varphi((n+k)a)$$
$$= \sum_{m=k_0+n_0}^{+\infty} \left( \sum_{k=k_0}^{+\infty} h_k x_{m-k} \right) \varphi(ma).$$

In this case, too, the coefficient $y_m$ of $y = h * x = \sum\limits_{m=-\infty}^{+\infty} y_m \delta_m$ is given by the following convolution formula (with $h_k = 0$ for $k < k_0$, $x_n = 0$ for $n < n_0$)

$$y_m = \sum_{k=-\infty}^{+\infty} h_k x_{m-k} = \sum_{k=-\infty}^{+\infty} h_{m-k} x_k.$$

The coefficient $y_m$ is zero if $m < k_0 + n_0$, and the series for $y_m$ is de facto a finite sum. Further examples and concrete applications of discrete convolutions will be elaborated in Sect. 10.5 and in the already mentioned Chap. 11 about linear filters.

### *Approximations of Distributions by Smooth Functions*

The above properties of convolutions include the theorem that every distribution $T$ of $\mathcal{D}'(\mathbb{R}^n)$ is the $\mathcal{D}'$-limit of a sequence of smooth functions $f_k$, which can even be chosen as test functions. It is also said that $\mathcal{D}$ *is dense in* $\mathcal{D}'$. To see this, choose a sequence of test functions $\varphi_k$ with $\mathcal{D}'\text{-}\lim_{k\to\infty} \varphi_k = \delta$, as we did with the introduction of the $\delta$-distribution. Such a sequence is called a *smoothing sequence*. You can start from any test function $\varphi$ with $\int \varphi(\mathbf{x})\, d\lambda^n(\mathbf{x}) = 1$, and define

$$\varphi_k(\mathbf{x}) = k^n \varphi(k\mathbf{x})$$

(cf. Ex. 1, p. 182). Then according to the previously mentioned Properties 3 and 7, it holds (p. 196)

$$\mathcal{D}'\text{-}\lim_{k\to\infty} (T * \varphi_k) = T * \delta = T.$$

The sequence of the $\varphi_k$ is called smoothing sequence, because the convolutions $T * \varphi_k$ are infinitely often differentiable functions according to Property 5 of p. 196.

Now choose an arbitrary test function $g$ with $g(\mathbf{x}) = 1$ in a neighborhood of zero, and define for $k \in \mathbb{N}$ the function $g_k(\mathbf{x}) = g(\mathbf{x}/k)$. For an arbitrary test function $\psi$ and sufficiently large $k$ is then $g_k(\mathbf{x}) = 1$ for all $\mathbf{x} \in \operatorname{supp}(\psi)$, and thus

$$\lim_{k\to\infty} \langle g_k(T * \varphi_k), \psi \rangle = \langle T, \psi \rangle.$$

The functions $f_k = g_k(T * \varphi_k)$ are the sought test functions for the approximation of $T$: $\quad \mathcal{D}'\text{-}\lim_{k\to\infty} f_k = T$. Therefore it holds:

**Theorem 8.9** *Each distribution $T$ is the $\mathcal{D}'$-times of a sequence of test functions.*

**Remark** For specific given functions or distributions, it can be quite difficult to "calculate" convolutions. As an important tool we will get to know the Fourier transform in Chap. 10. In practical, numerical applications, convolutions of functions are often approximately calculated using the discrete Fourier transform.

## *The Spaces $\mathcal{E}'$ and $\mathcal{D}'_R$, Continuity of Convolution Operators*

To conclude this section, an aspect of property 7 of p. 197 on the convergence of convolutions will be emphasized. For this purpose, we introduce two new vector spaces of distributions and define convergence terms for sequences in these spaces, which we will come back to later in the application examples in Chap. 11.

### Definition

1. $\mathcal{E}'$ denotes the space of distributions $T \in \mathcal{D}'(\mathbb{R})$ with compact support. A sequence $T_n$ converges in $\mathcal{E}'$ to a distribution $T$ if $T = \mathcal{D}'\text{-}\lim_{n\to\infty} T_n$ and all $T_n$ and $T$ have their supports in a common compact set $K$.
2. With $\mathcal{D}'_r$ we denote the space of all distributions $T \in \mathcal{D}'(\mathbb{R})$ whose supports lie in the interval $[r, \infty[$ for $r \in \mathbb{R}$. A sequence of distributions $T_n$ converges to a distribution $T$ in $\mathcal{D}'_r$ if $T = \mathcal{D}'\text{-}\lim_{n\to\infty} T_n$ and all $T_n$ and $T$ have their support in $[r, \infty[$.

   The space $\mathcal{D}'_R = \bigcup_{r\in\mathbb{R}} \mathcal{D}'_r$ is called the space of causal distributions. A sequence of distributions $T_n$ converges in $\mathcal{D}'_R$ toward a distribution $T$ if $T = \mathcal{D}'\text{-}\lim_{n\to\infty} T_n$ and, additionally, all supports of the $T_n$ and $T$ for a suitable $r \in \mathbb{R}$ lie in the interval $[r, \infty[$. Causal distributions are also called right-sided distributions.

These definitions can also be applied to the case of multidimensional parameter sets in an obvious way. We restrict ourselves to the one-dimensional case and again briefly note only $\mathcal{D}'$ for $\mathcal{D}'(\mathbb{R})$ below.

If we now consider for a distribution $G$ the *convolution operator* $L_G(T) = G * T$ for distributions $T$ in spaces that always allow the convolutions $G * T$, then the properties of convolutions from No. 7 of p. 197 are continuity statements for such convolution operators. We will refer to the following statements in Chap. 11:

**Theorem 8.10** *For a distribution $G$, the convolution operator $L_G \colon \mathcal{Z} \to \mathcal{A}$ defined by $L_G(T) = G * T$ is a linear translation-invariant continuous operator in the following cases:*

1. $\mathcal{Z} = \mathcal{E}'$, $\mathcal{A} = \mathcal{D}'$, and $G \in \mathcal{D}'$.
2. $\mathcal{Z} = \mathcal{D}' = \mathcal{A}$ and $G \in \mathcal{E}'$.
3. $\mathcal{Z} = \mathcal{D}'_R = \mathcal{A}$ and $G \in \mathcal{D}'_R$.

The translation invariance means that the convolution operators $L_G$ for translations $A$ of the parameter set (on $\mathcal{D}'$ defined by $AT = T_A$), can be interchanged with the translation, i.e., if for all $T \in \mathcal{Z}$ and $L_G(T) = G * T$ it holds:

$$L_G(AT) = A(L_G(T)) = (G * T)_A.$$

The theorem is only a reformulation of No. 4 and No. 7 on p. 197, if the convergence notions introduced above for $\mathcal{E}'$ and $\mathcal{D}'_R$ are taken into account. In Chap. 11 we address the question, which translation-invariant linear operators $L : \mathcal{Z} \to \mathcal{A}$ can be represented as convolution operators for certain given distribution spaces $\mathcal{Z}$ and $\mathcal{A}$. There are also examples of convolution operators that are not continuous and examples of continuous translation-invariant operators, which are not convolution operators (p. 327 and p. 355). Preliminarily, we already formulate a fundamental result here, which goes back to Schwartz (1957). There one can also find the proof.

**Theorem 8.11 (Theorem of L. Schwartz)** *Every continuous translation-invariant linear operator $L$ from $\mathcal{E}'$ to $\mathcal{D}'$ is interchangeable with convolutions between elements of $\mathcal{E}'$ and can be represented as a convolution with $h = L\delta$, i.e., $L(T) = L(\delta * T) = L(\delta) * T = h * T$ for all $T \in \mathcal{E}'$.*

A detailed discussion of variants of this theorem for operators also on distribution spaces other than $\mathcal{E}'$ can be found in Zemanian (2010) and Albrecht and Neumann (1979).

## 8.8 Exercises

**(A1)** Which of the following functionals on $\mathcal{D}$ are distributions?

$$T(\varphi) = -\int_{0}^{+\infty} \varphi'(t)\,\mathrm{d}t\,, \quad G(\varphi) = \max_{t \in \mathbb{R}} \varphi(t), \quad H(\varphi) = \int_{-\infty}^{+\infty} |\varphi(t)|\,\mathrm{d}t\,,$$

$$R(\varphi) = |\varphi(0)|, \quad S(\varphi) = \sum_{k=0}^{\infty} \varphi^{(k)}(0), \quad U(\varphi) = \sum_{|k|\leqslant p} \partial^k \varphi(0) \text{ for } \varphi \in \mathcal{D}(\mathbb{R}^n),$$

$(p \in \mathbb{N})$.

**(A2)** Calculate $t\dot{\delta}(t)$, $t^2\dot{\delta}(t)$ and $t\ddot{\delta}(t)$, $t^2\ddot{\delta}(t)$.

**(A3)** Prove $t\,\mathrm{pf}(t^{-2}) = \mathrm{pf}(t^{-1})$, $t\,\mathrm{pf}(t_+^{-2}) = \mathrm{pf}(t_+^{-1})$, $t\delta^{(m)}(t) = -m\delta^{(m-1)}(t)$ for $m \in \mathbb{N}$.

**(A4)** Show that the principal value $\mathrm{vp}(t^{-1})$ can be represented in the form $\mathrm{vp}(t^{-1}) = T(t) + f(t)$, where $T$ is a distribution with bounded support and $f(t)$ is a square-integrable function.

**(A5)**★ Show: For every $\varphi \in \mathcal{D}$ it holds $\displaystyle\lim_{\varepsilon \to 0} \int_{-\infty}^{+\infty} \varphi(t)\,\frac{t}{t^2 + \varepsilon^2}\,\mathrm{d}t = \mathrm{vp}(t^{-1})(\varphi),$

i.e., the regular distributions $T_\varepsilon(t) = \dfrac{t}{t^2 + \varepsilon^2}$ converge for $\varepsilon \to 0$ to the principal value $\mathrm{vp}(t^{-1})$.

**(A6)**$^\star$  Show that for $-1 < \lambda < 0$ by

$$T(\varphi) = \mathrm{pf}(\lambda t_+^{\lambda-1})(\varphi) = \int\limits_0^\infty \lambda t^{\lambda-1}(\varphi(t) - \varphi(0))\mathrm{d}t$$

a singular distribution is defined which regularizes the derivative of $t_+^\lambda = t^\lambda s(t)$, i.e., the generalized derivative of the regular distribution $t_+^\lambda$ is $T$ (cf. p. 168). As usual, $s(t)$ is the unit step function.

Example: $\left(t_+^{-1/2}\right)' = -\tfrac{1}{2}\mathrm{pf}\left(t_+^{-3/2}\right).$

**(A7)**  What is the generalized second derivative $\ddot{f}$ of

$$f(t) = (\sin(t) + \alpha)[s(t) - s(t - \tfrac{\pi}{2})] + ((t - \tfrac{\pi}{2})^2 + \alpha + 1)s(t - \tfrac{\pi}{2}),$$

$s(t)$ the unit step function, $\alpha > 0$?

**(A8)**$^\star$  Show with the use of the transformation rules and the notion of convergence for distributions that the generalized derivative $\dot{T}(t)$ of a distribution $T(t)$ reads as:

$$\dot{T}(t) = \lim_{\Delta t \to 0} \frac{T(t + \Delta t) - T(t)}{\Delta t}.$$

**(A9)** (a)  Show for the improper integral $\displaystyle\int\limits_{-\infty}^{+\infty} \frac{\sin(nt)}{\pi t}\mathrm{d}t = 1.$

(b)  Show for the $2\pi$-periodic Dirichlet-Kernels $D_n(t) = \dfrac{1}{2\pi}\displaystyle\sum_{k=-n}^{+n} e^{jkt}$ and piecewise continuously differentiable $f$ that

$$\int\limits_{-\pi}^{+\pi} D_n(t)f(t)\mathrm{d}t \xrightarrow[n\to\infty]{} \frac{1}{2}(f(0-) + f(0+)).$$

**(A10)**  Check that $\sum_{k=1}^\infty k \sin(kt)$ converges in $\mathcal{D}'$, and calculate

$$\left\langle \sum_{k=1}^\infty k \sin(kt), \varphi(t - \tfrac{1}{2})\right\rangle \quad \text{for} \quad \varphi(t) = \begin{cases} e^{-1/(1-t^2)} & \text{for } |t| < 1 \\ 0 & \text{otherwise.} \end{cases}$$

**(A11)**  Calculate for the Dirac distribution $\delta(t)$, the unit step function $s(t)$, an integrable function $f(t)$, and real numbers $a$ and $b$ the convolutions

$$\dot{\delta}(t-a) * s(t-b), \quad \dot{\delta}(t-a) * \dot{\delta}(t-b), \quad s(t-a) * f(t),$$

$$s(t-a) * s(t-b), \quad \text{and the convolution } s(t) * [\ln(t+1)s(t+1)].$$

**(A12)** For Gaussian functions of the form $G_\sigma^m(x) = \frac{1}{\sigma\sqrt{2\pi}}\,e^{-(x-m)^2/(2\sigma^2)}$ with $m \in \mathbb{R}$ and $\sigma > 0$, verify that the convolution again results in a Gaussian function:

$$G_\sigma^{m_1} * G_\tau^{m_2} = G_{\sqrt{\sigma^2+\tau^2}}^{m_1+m_2}.$$

Note: To do this, write the exponent of the integrand in the convolution integral $G_\sigma^{m_1} * G_\tau^{m_2}(x)$ in the form $u^2 + v^2$ with $v = (x - (m_1 + m_2))/\sqrt{\sigma^2 + \tau^2}$ and substitute the integration variable with $u$.

**Remark** The relationship $G_\sigma^{m_1} * G_\tau^{m_2} = G_{\sqrt{\sigma^2+\tau^2}}^{m_1+m_2}$ plays an important role in probability theory. $G_\sigma^m$ is the *probability density of the so-called* $(m, \sigma)$ *normal distribution with expected value m and standard deviation* $\sigma$. The standard deviation of the sum of $n$ independently measured values of $(m, \sigma)$-normally distributed random variables is then $\sigma\sqrt{n}$, and that of the arithmetic mean of the measured values is therefore $\sigma/\sqrt{n}$. Averaging from several measured values thus reduces the variance, and the arithmetic mean of the measured values provides a useful estimate of the expected value $m$.

**(A13)** For integrable functions $f$ and $g$, prove $\operatorname{supp}(f * g) \subset \operatorname{supp}(f) + \operatorname{supp}(g)$.

**(A14)**$^\star$ Prove that the convolution $T * \varphi$ of a distribution $T \in \mathcal{D}'$ with a test function $\varphi \in \mathcal{D}$ is an infinitely often differentiable function.

**(A15)** Show that the support of the derivative $\dot{T}$ of a distribution $T \in \mathcal{D}'(\mathbb{R})$ is contained in the support of $T$.

**(A16)**$^\star$ Give an example with three distributions whose convolution is not associative.

**(A17)** What is the general solution of the equation $tT(t) = G(t)$ for $G \in \mathcal{D}'$ with $0 \notin \operatorname{supp}(G)$?

**(A18)** Specify a distribution $T$ for $n \in \mathbb{N}$ with $\operatorname{supp}(T) = \{0\}$ and $t^n T(t) \neq 0$.

**(A19)** The space $\mathcal{D}'_+$ of all distributions with support in $\mathbb{R}_0^+$ is a convolution algebra.

Determine the convolution inverses $T$ with $T * G = \delta$ for the distributions $G = \dot{\delta}$, $G = \dot{\delta} - \alpha\delta$ and the unit step $G(t) = s(t)$ in $\mathcal{D}'_+(\mathbb{R})$.

**(A20)**$^\star$ Show that every distribution $T \in \mathcal{D}'_+$ has exactly one indefinite integral $S \in \mathcal{D}'_+$ and that $S = s * T$ holds ($s$ is the unit step function).

**(A21)**$^\star$ Show that a homogeneous linear differential equation of $n$-th order

$$\sum_{k=0}^{n} a_k(t)x^{(k)}(t) = 0$$

with infinitely often differentiable coefficients $a_k(t)$, taken as equation in $\mathcal{D}'$, does not have any further solutions than the regular classical solutions in $\mathcal{D}'$.

Show that this also applies to a homogeneous first-order system

$$\mathbf{x}'(t) = A(t)\mathbf{x}(t)$$

with a componentwise infinitely often differentiable matrix $A$.

# Chapter 9
# Application Examples for Distributions

**Abstract** The chapter is devoted to practice applications of distributions in various fields. This includes generalized Fourier series, fundamental solutions for linear differential equations with constant coefficients in a myriad of applications. Linear circuits and input-output relations by convolution of input signals with the impulse or step response form general examples. For 3D problems, the fundamental solution for the potential equation is calculated and applied in examples. Furthermore, as one of the most important practice applications, the method of finite elements (FEM) is treated, and its solution in a suitable Sobolev space by the Lax-Milgram theorem is shown and applied in examples, e.g., the Poisson boundary value problem. The problem of the vibrating string from the beginning in Chap. 1 is now solved in the sense of distributions, i.e., weak solutions are obtained as in the FEM problems.

## 9.1  Periodic Distributions are Generalized Fourier Series

For $p > 0$, a distribution $T$ of $\mathcal{D}'$ is called $p$-periodic, if for all $k \in \mathbb{Z}$ it holds in the sense of translations for distributions:

$$T(t + p) = T(t).$$

For $\varphi \in \mathcal{D}$ and $k \in \mathbb{Z}$, it then follows

$$\langle T(t + kp), \varphi(t) \rangle = \langle T(t), \varphi(t - kp) \rangle = \langle T, \varphi \rangle.$$

### *Fourier Series as Distributions*

**Theorem 9.1** *Every trigonometric series* $\displaystyle\sum_{k=-\infty}^{+\infty} c_k\, e^{jkt}$, *whose coefficients $c_k$ are polynomially bounded, i.e.,* $|k|^{-n}|c_k| \xrightarrow[|k|\to\infty]{} 0$ *for suitable $n \in \mathbb{N}$, converges in the distributional sense to a $2\pi$-periodic distribution $T$.*

***Proof*** For polynomially bounded coefficients, the series $\displaystyle\sum_{\substack{k=-\infty \\ k\neq 0}}^{+\infty} \frac{c_k}{(jk)^n}\, \mathrm{e}^{jkt}$ is uni-

formly convergent for sufficiently large $n \in \mathbb{N}$ and thus represents a continuous, $2\pi$-periodic function $f$. With the $n$-th generalized derivative $f^{(n)}$ of $f$, it then follows

$$T(t) = \sum_{k=-\infty}^{+\infty} c_k\, \mathrm{e}^{jkt} = c_0 + f^{(n)}(t).$$

Since for $\varphi \in \mathcal{D}$ and $k \in \mathbb{Z}$, it always

$$\int_{-\infty}^{+\infty} \mathrm{e}^{jkt}\, \varphi(t)\mathrm{d}t = \int_{-\infty}^{+\infty} \mathrm{e}^{jkt}\, \varphi(t - 2\pi)\mathrm{d}t$$

holds, $T$ is a $2\pi$-periodic distribution. $\qquad\square$

An analogous statement also applies to other periods $p \neq 2\pi$. If a classical Fourier series is to converge pointwise almost everywhere or in the quadratic mean, its Fourier coefficients $c_k$ must necessarily form a zero sequence. In the distributional sense, however, convergence is even achieved if the $c_k$ only do not grow too fast. All these series now considered as distributions, to which we refer as *generalized Fourier series*, may also be differentiated term by term, which is not the case in classical analysis (cf. Sect. 4.3).

**Example (Periodic Impulse Sequences)** The $2\pi$-periodic Dirichlet kernels

$$D_n(t) = \sum_{k=-n}^{n} \mathrm{e}^{jkt}$$

converge for no $t \in \mathbb{R}$ pointwise. For increasing $n$ these functions oscillate on $\mathbb{R}$ more and more (cf. p. 14). However, we now obtain convergence for $n \to \infty$ in the distribution sense:

**Theorem 9.2** *The Dirichlet kernels converge in the distributional sense toward the periodic distribution*

$$T(t) = \sum_{k=-\infty}^{+\infty} \mathrm{e}^{jkt} = 2\pi \sum_{k=-\infty}^{+\infty} \delta(t - 2\pi k).$$

***Proof*** It holds

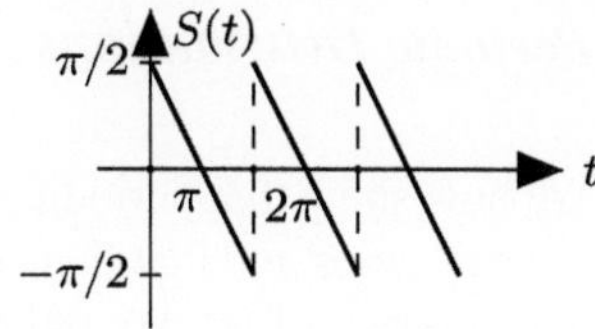

**Fig. 9.1** The periodic sawtooth function

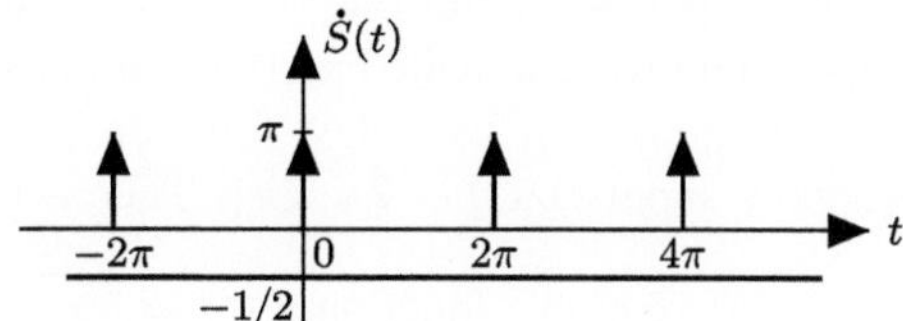

**Fig. 9.2** The derivative of the sawtooth

$$\mathcal{D}'\text{-}\lim_{n\to\infty} D_n(t) = \sum_{k=-\infty}^{+\infty} e^{jkt} = 1 + 2\sum_{k=1}^{\infty} \cos(kt) = 1 + 2\frac{d}{dt}\sum_{k=1}^{\infty}\frac{\sin(kt)}{k}\ ,$$

and

$$\sum_{k=1}^{\infty}\frac{\sin(kt)}{k} = \begin{cases} (\pi - t)/2 & \text{for} \quad 0 < t < 2\pi \\ 0 & \text{for} \qquad t = 0 \end{cases}$$

represents the sawtooth function $S(t)$, $S(t + 2k\pi) = S(t)$ (Fig. 9.1).
Distributional (term by term) differentiation yields (Fig. 9.2)

$$\sum_{k=1}^{\infty} \cos(kt) = -\frac{1}{2} + \pi \sum_{k=-\infty}^{+\infty} \delta(t - 2\pi k).$$

So one obtains for the Dirichlet kernels $D_n(t)$:

$$D_n(t) = \sum_{k=-n}^{n} e^{jkt} = 1 + 2\sum_{k=1}^{n} \cos(kt) \xrightarrow[n\to\infty]{\mathcal{D}'} 2\pi \sum_{k=-\infty}^{+\infty} \delta(t - 2\pi k).$$

This confirms the heuristic impression we had already gained from the Dirichlet kernels (cf. p. 14 and p. 50): The Dirichlet kernels converge to a periodic impulse sequence. This impulse sequence is a singular distribution.  □

**Theorem 9.3** *For p-periodic impulse sequences ($p > 0$, $\omega_0 = 2\pi/p$) apply accordingly*

$$\sum_{k=-\infty}^{+\infty} \delta(t-kp) = \frac{1}{p}\sum_{k=-\infty}^{+\infty} e^{jk\omega_0 t} \quad and \quad \sum_{k=-\infty}^{+\infty} \dot{\delta}(t-kp) = \frac{1}{p}\sum_{k=-\infty}^{+\infty} jk\omega_0\, e^{jk\omega_0 t}.$$

## *Periodic Distributions Are Generalized Fourier Series*

We now show that periodic distributions can always be represented by generalized Fourier series and that their Fourier coefficients $c_k$ for $|k| \to \infty$ do not grow faster than a power of $|k|$. We only consider $2\pi$-periodic distributions. The reformulation for other period lengths may serve as an exercise for the reader. A standard proof of this result uses a so-called partition of unity. We construct such a partition:

We choose any even test function $\varphi \geqslant 0$ with $\varphi(t) \geqslant 1/2$ in $[-\pi, \pi]$ and support $\mathrm{supp}(\varphi) \subset ]-2\pi, 2\pi[$. Therewith, we define the $2\pi$-periodic, infinitely often differentiable function $\Phi(t) = \sum\limits_{k=-\infty}^{+\infty} \varphi(t + 2k\pi)$. The periodicity is obvious, the smoothness follows from the fact that the above series has only finitely many nonzero summands in each bounded interval, all of which are infinitely often differentiable. It always holds $\Phi(t) \geqslant 1/2$. We introduce the following functions:

$$h(t) = \varphi(t)/\Phi(t) \in \mathcal{D} \text{ and } H(t) = \sum_{k=-\infty}^{+\infty} h(t + 2k\pi).$$

Then $H(t) = 1$ for all $t \in \mathbb{R}$ and $h(0) = 1$. The representation of $H$ is called a *partition of unity by means of a test function $h$*. To illustrate these functions and to understand the calculation afterward, consider the following graphs, for which we choose as a concrete example $\varphi$ as follows (see Figs. 9.3 and 9.4):

With $\psi(t) = \mathrm{e}^{-t^2/(1-t^2)}$ for $|t| \leqslant 1$, $\psi(t) = 0$ otherwise, we define $\varphi$ by

$$\varphi(t) = \begin{cases} \psi\left(\frac{4t+1}{19}\right) & \text{for } -\infty < t < -1/4 \\ 1 & \text{for } -1/4 \leqslant t \leqslant 1/4 \\ \psi\left(\frac{4t-1}{19}\right) & \text{for } 1/4 < t < +\infty. \end{cases}$$

The image on the right shows $h(t + 2\pi)$, $h(t)$, $h(t - 2\pi)$ and the sum $H = 1$ on the interval $[-2\pi, 2\pi]$. The support of $\varphi$ and $h$ is $[-5.5]$.

One verifies that the Fourier coefficients $f_k$ of a $2\pi$-periodic function $f$ that is integrable on $[0, 2\pi]$ are given by applying $f$ to the test function $h(t)\,\mathrm{e}^{-jkt}$:

**Fig. 9.3**
$\varphi(t + 2\pi), \varphi(t), \varphi(t - 2\pi), \Phi$

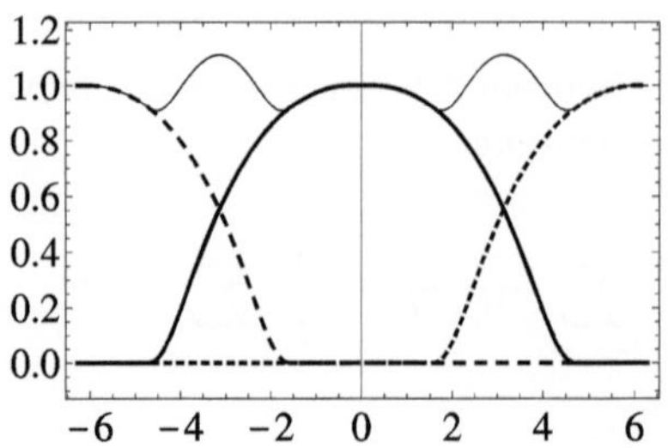

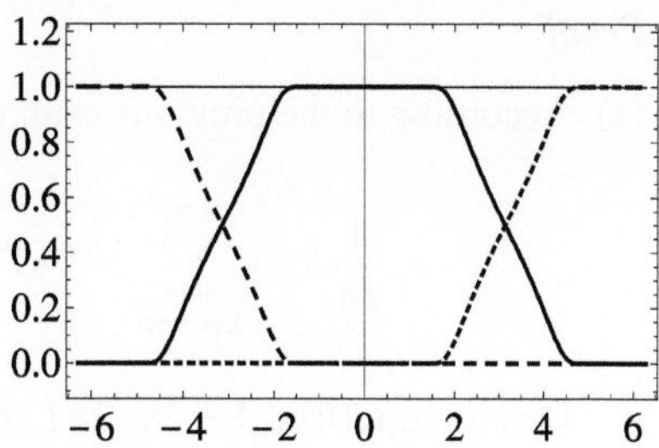

**Fig. 9.4** Generated partition of unity

$$\langle f(t)\,,\,h(t)\,\mathrm{e}^{-jkt}\rangle = \int\limits_{-\infty}^{+\infty} f(t)h(t)\,\mathrm{e}^{-jkt}\,\mathrm{d}t = \sum_{k=-\infty}^{+\infty}\ \int\limits_{2k\pi}^{2(k+1)\pi} f(t)h(t)\,\mathrm{e}^{-jkt}\,\mathrm{d}t$$

$$= \sum_{k=-\infty}^{+\infty}\int\limits_{0}^{2\pi} f(x)h(x+2k\pi)\,\mathrm{e}^{-jkx}\,\mathrm{d}x = \int_{0}^{2\pi} f(x)H(x)\,\mathrm{e}^{-jkx}\,\mathrm{d}x = 2\pi f_k.$$

Because $h$ has a bounded support, the resulting series are de facto finite sums and may be interchanged with the integral. This relationship is independent of the specific choice we have made for $h$ and thus $H$, and applies to any analogous partition of unity. If we replace $f$ with a $2\pi$-periodic distribution $T$, the quantity $\langle T(t)\,,\,h(t)\,\mathrm{e}^{-jkt}\rangle$ is also independent of the choice of $h$, since any such $T$ can be represented with the help of a smoothing sequence $(f_n)_{n\in\mathbb{N}}$ as on p. 201 as a distributional limit of regular and again $2\pi$-periodic distributions such as $f$. From this it follows that the Fourier coefficients $f_k^n$ of $f_n$ defined with any $h$ as above always have the same limit $\langle T(t)\,,\,h(t)\,\mathrm{e}^{-jkt}\rangle/(2\pi)$ for $n \to \infty$. This expression can therefore be used to define the Fourier coefficients of periodic distributions in general. Therewith, we obtain the desired theorem on the representation of periodic distributions as generalized Fourier series.

**Definition** The Fourier coefficients $c_k$ of a $2\pi$-periodic distribution $T$ are—with $h$ constructed as above—for $k \in \mathbb{Z}$ the complex numbers $c_k = \dfrac{1}{2\pi}\langle T(t)\,,\,h(t)\,\mathrm{e}^{-jkt}\rangle$.

**Theorem 9.4** *Every $2\pi$-periodic distribution $T$ has the representation*

$$T(t) = \sum_{k=-\infty}^{+\infty} c_k e^{jkt}$$

*with the Fourier coefficients $c_k$ given in the definition. The Fourier coefficients $c_k$ are polynomially bounded, i.e., there exists a constant $C$ and a natural number $n$ such that for all $k \in \mathbb{Z} \setminus \{0\}$ the inequality $|c_k| \leqslant C|k|^n$ holds.*

***Proof***

(a)  According to the previous example, we have

$$\frac{1}{2\pi}h(t)\sum_{k=-\infty}^{+\infty}\mathrm{e}^{jkt}*T(t)=h(t)\sum_{k=-\infty}^{+\infty}\delta(t-2k\pi)*T(t).$$

Due to $\mathrm{supp}(h)\subset\,]-2\pi,2\pi[$ and $h(0)=1$, we obtain the representation for $T$

$$T(t)=T(t)*h(t)\delta(t)=h(t)\sum_{k=-\infty}^{+\infty}\delta(t-2k\pi)*T(t).$$

(b)  Because for every $t$, $\widetilde{h}(s)=h(t-s)$ gives a partition of unity $\widetilde{H}$ as above as well, $c_k$ can also be calculated with $\widetilde{h}$ instead of $h$. Thus, it holds that

$$T(t)*h(t)\,\mathrm{e}^{jkt}=\langle T(s)\,,\,\widetilde{h}(s)\,\mathrm{e}^{jk(t-s)}\rangle=2\pi c_k\,\mathrm{e}^{jkt}\,.$$

(c)  From (a) and (b), the desired representation of $T$ as a generalized Fourier series follows:

$$T(t)=T(t)*\frac{1}{2\pi}h(t)\sum_{k=-\infty}^{+\infty}\mathrm{e}^{jkt}=\frac{1}{2\pi}\sum_{k=-\infty}^{+\infty}\left(h(t)\,\mathrm{e}^{jkt}*T(t)\right)=\sum_{k=-\infty}^{+\infty}c_k\,\mathrm{e}^{jkt}\,.$$

The second equation uses the convergence of convolutions (cf. p. 197), since $h\in\mathcal{D}$.

(d)  If $I$ is an open interval containing the support of $h$, then according to the characterization of the continuity of $T$ on p. 190, there exists a $K>0$ and $n\in\mathbb{N}_0$, such that for all $\varphi\in\mathcal{D}$ with support in $I$, the inequality $|\langle T\,,\,\varphi\rangle|\leqslant K$ holds. Specifically, for $k\neq 0$ with Leibniz rule for derivatives of $h(t)\,\mathrm{e}^{-jkt}$ up to order $n$, we obtain (for $\|h\|_n$ see p. 189)

$$2\pi|c_k|=|\langle T(t)\,,\,h(t)\,\mathrm{e}^{-jkt}\rangle|\leqslant K\|h(t)\,\mathrm{e}^{-jkt}\|_n\leqslant 2^n K\|h\|_n|k|^n.$$

Thus, the Fourier coefficients $c_k$ are polynomially bounded.

$$\square$$

From the theorem, it particularly follows that two generalized Fourier series

$$T(t)=\sum_{k=-\infty}^{+\infty}c_k\,\mathrm{e}^{jkt}\quad\text{and}\quad S(t)=\sum_{k=-\infty}^{+\infty}d_k\,\mathrm{e}^{jkt}\quad\text{are equal if and only if }c_k=d_k$$

for all $k\in\mathbb{Z}$, i.e., *comparison of coefficients is possible.*

**Remark** Also for distributions over a multidimensional parameter space, which are $2\pi$-periodic in each variable, analogous representations as generalized Fourier series can be shown. Proofs can be found in Schwartz (1957) or Vladimirov (2002).

**Example (Generalized Fourier Series of the Tangent Function)** With

$$T(t) = -\ln|2\cos(t)| = \sum_{k=1}^{\infty}(-1)^k \frac{\cos(2kt)}{k}$$

for $t \neq (2k+1)\pi/2$, $k \in \mathbb{Z}$ (Exercise 6 of Chap. 7), differentiation yields the generalized Fourier series for the regularized tangent function $T_{\tan} = \dot{T}$ considered as a periodic distribution. It represents as a weak limit the principal value $\mathrm{vp}(\tan(t))$.

$$T_{\tan}(t) = \mathrm{vp}(\tan(t)) = 2\sum_{k=1}^{\infty}(-1)^{k+1}\sin(2kt), \quad \langle T_{\tan}, \varphi \rangle = -\langle T, \varphi' \rangle \text{ for } \varphi \in \mathcal{D}.$$

**Remark** Regularizations of functions with singularities, such as the tangent, are generally not possible in only one way. Therefore, it should be noted that the abovementioned distribution $\dot{T}$ as a regularization of the tangent is the uniquely determined so-called *canonical regularization* in the sense of Gel'fand et al. (1964), Vol. I.

The calculation rules for Fourier series treated in Sects. 4.1–4.3 apply to generalized Fourier series as well due to the transformation rules for distributions. For example, a translation of the $p$-periodic impulse sequence $\sum_{k=-\infty}^{+\infty}\delta(t - kp)$ to the right by $t_0 > 0$ results in

$$\sum_{k=-\infty}^{+\infty}\delta((t - t_0) - kp) = \frac{1}{p}\sum_{k=-\infty}^{+\infty}e^{jk\omega_0(t-t_0)} \qquad (\omega_0 = 2\pi/p).$$

Two $p$-periodic distributions generally cannot be convolved in the way defined in Sect. 8.7. None of the conditions for the supports given there are met. However, the $p$-periodic convolution $(T * S)_p$ of

$$T(t) = \sum_{k=-\infty}^{+\infty}c_k\, e^{jk\omega_0 t} \quad \text{and} \quad S(t) = \sum_{k=-\infty}^{+\infty}d_k\, e^{jk\omega_0 t} \qquad (\omega_0 = 2\pi/p),$$

can be introduced analogously as in Sect. 5.2, p. 63 by $(T * S)_p(t) = \sum_{k=-\infty}^{+\infty}c_k d_k\, e^{jk\omega_0 t}$. If, for instance, $T = f^{(n)}$ and $S = g^{(m)}$ are generalized derivatives of order $n$ and $m$, respectively, of two continuous $p$-periodic functions $f$ and $g$ (cf. p. 208), then $(T * S)_p = (f * g)_p^{(n+m)}$, i.e., the $p$-periodic convolution of $T$ and $S$ is the generalized derivative of order $n + m$ of $(f * g)_p$.

**Application in Asymptotically Stable Differential Equations as in Sect. 5.2**
Therewith, the long-term behavior of the solutions of asymptotically stable ordinary
linear differential equations with constant coefficients under periodic excitations can
be described, as on p. 66, by the periodic transfer function. Continuity assumptions,
as required there, can be dispensed with if periodic input signals, transfer functions,
and output signals are considered as periodic distributions with generalized Fourier
series representations. In particular, all Fourier series, regarded as distributions
$T$, can always be differentiated term by term, and the result then represents the
generalized derivative $\dot{T}$.

## *The Impulse Method for Calculating Fourier Coefficients*

Understanding Fourier series as distributions can be used, for example, to calculate the Fourier coefficients for " simple" piecewise continuously differentiable
functions not through integration, but through differentiation and comparison of
coefficients with series that represent impulse sequences with known coefficients.
We exemplarily show this with the following example:

We choose, for instance, $f(t) = \cos(\omega_0 t)$ for $0 \leqslant t < \frac{p}{2}$, $\omega_0 = \frac{2\pi}{p}$,
$f\left(t + k\frac{p}{2}\right) = f(t), k \in \mathbb{Z}$ (Fig. 9.5).

For $0 \leqslant t < \frac{p}{2}$ it holds

$$\dot{f}(t) = 2\delta(t) - \omega_0 \sin(\omega_0 t)$$

$$\ddot{f}(t) = 2\dot{\delta}(t) - \omega_0^2 \cos(\omega_0 t) = 2\dot{\delta}(t) - \omega_0^2 f(t). \tag{9.1}$$

On the other hand, $f$ can be represented as a Fourier series, and term-by-term
differentiation yields

$$f(t) = \sum_{k=-\infty}^{+\infty} c_k\, e^{j2k\omega_0 t}$$

$$\ddot{f}(t) = -\sum_{k=-\infty}^{+\infty} (2k\omega_0)^2 c_k\, e^{j2k\omega_0 t}. \tag{9.2}$$

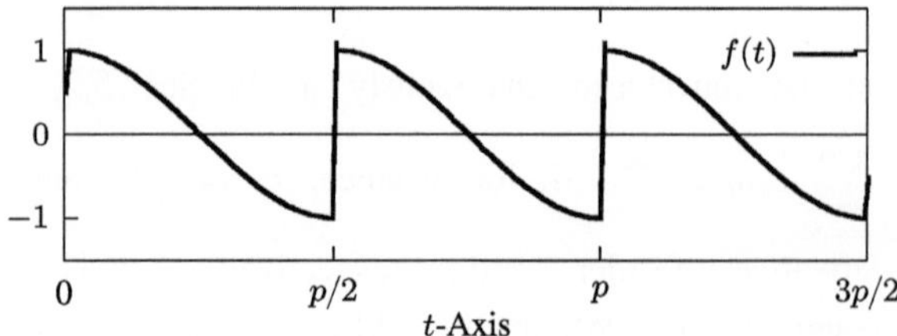


**Fig. 9.5** The $p/2$-periodic $f(t) = \cos(\omega_0 t)$

By periodic extension on $\mathbb{R}$, it follows from (9.1) and (9.2) by substituting the Fourier series of $f$ into (9.1):

$$2 \sum_{k=-\infty}^{+\infty} \dot{\delta}(t - k\frac{p}{2}) - \omega_0^2 \sum_{k=-\infty}^{+\infty} c_k \, e^{j2k\omega_0 t} = - \sum_{k=-\infty}^{+\infty} (2k\omega_0)^2 c_k \, e^{j2k\omega_0 t} . \qquad (9.3)$$

*With the series representation* $2 \sum_{k=-\infty}^{+\infty} \dot{\delta}(t - k\frac{p}{2}) = \frac{4}{p} \sum_{k=-\infty}^{+\infty} j2k\omega_0 \, e^{j2k\omega_0 t}$ *one immediately obtains from (9.3) the sought Fourier coefficients $c_k$ by a comparison of coefficients:*

$$\frac{4 \cdot 2jk\omega_0}{p} = (\omega_0^2 - (2k\omega_0)^2)c_k , \qquad c_k = \frac{4 \cdot 2jk\omega_0}{p(1 - (2k)^2)\omega_0^2} = \frac{4k}{j\pi((2k)^2 - 1)}.$$

Thus, $f$ has the representation

$$f(t) = \sum_{k=-\infty}^{+\infty} \frac{4k}{j\pi((2k)^2 - 1)} \, e^{j2k\omega_0 t} .$$

If the series is written so that only positive $k$ appears,

$$f(t) = \frac{2}{j\pi} \sum_{k=1}^{\infty} \left( \frac{2k}{(2k)^2 - 1} e^{j2k\omega_0 t} - \frac{2k}{(2k)^2 - 1} e^{-j2k\omega_0 t} \right),$$

one obtains with $n = 2k$ the *series representation with sine functions*

$$f(t) = \frac{4}{\pi} \sum_{\substack{n=2 \\ n\,\text{even}}}^{\infty} \frac{n}{n^2 - 1} \sin(n\omega_0 t) = \frac{4}{\pi} \left( \frac{2\sin(2\omega_0 t)}{3} + \frac{4\sin(4\omega_0 t)}{15} + \dots \right).$$

## 9.2 Linear Differential Equations with Constant Coefficients

### *Fundamental Solutions*

Consider a differential equation with constant coefficients $a_k$

$$Au(t) = f(t), \quad A = \sum_{k=0}^{n} a_k \frac{\mathrm{d}^k}{\mathrm{d}t^k} \qquad (n > 0, \, a_n \neq 0).$$

A well-known example is the equation for the displacement $u$ of a spring pendulum:

$$m\frac{\mathrm{d}^2 u}{\mathrm{d}t^2}(t) + k\frac{\mathrm{d}u}{\mathrm{d}t}(t) + Du(t) = K(t).$$

The coefficients denote the mass $m$ of the pendulum, the damping coefficient $k$, and the spring constant $D$. The external force acting on the pendulum is $K(t)$.

The differential operator $A$ in this example is

$$A = m\frac{\mathrm{d}^2}{\mathrm{d}t^2} + k\frac{\mathrm{d}}{\mathrm{d}t} + D.$$

Every student is familiar with other equations of this type, whether from mechanics, electricity, or other fields of application. In agreement with typical examples, we denote the variable $t$ as a time parameter. For a continuous perturbation $f(t)$ and vanishing initial values, the solution is regular and uniquely determined, and for a time translation of the right-hand side to $f(t+t_0)$, the corresponding shifted solution $u$ of $Au = f$ gives the solution $v(t) = u(t+t_0)$ of $Av(t) = f(t+t_0)$. The equation then describes a so-called time-invariant linear system.

The central importance of convolutions lies in the following method for finding particular solutions of such differential equations, which we now regard as equations between distributions.

**Fundamental Solution Method** *First, one determines a fundamental solution $g$, i.e., a distribution $g \in \mathcal{D}'$, such that*

$$Ag = \delta.$$

*If $f$ is a distribution for which the convolution $g * f$ exists, then one obtains a distributional particular solution $u$ of the equation $Au = f$ through $u = g * f$.*

Then $A(g * f) = Ag * f = \delta * f = f$. The convolution $g * f$ certainly exists if $f$ has a bounded support (is "bounded in time"), or if the supports of $g$ and $f$ are both semi-bounded in the same direction, for example, if they lie in $[0, \infty[$.

**Remarks**

1. From the theorem on indefinite integrals of distributions (see p. 181), it follows that the *homogeneous* differential equation $Au = 0$ has no additional distributional solutions besides the classical infinitely often differentiable solutions. This is immediately plausible because any solution $u$ on the left side of the equation $Au = 0$ is differentiated by $A$, so it must be an infinitely often differentiable function as the right side.
2. In the case of *initial value problems*, one usually requires additional conditions to ensure that the solution $u$ of $Au = f$ can be discussed in terms of initial values. We will address this later (p. 220).

3. The fundamental solution $g$ introduced above is in general not a solution in the physical sense for technical-physical equations because the physical unit of $g$ typically does not match the dimension of such a solution. The distribution $g$ should be understood as a functional whose convolution $u = g * f$ with a right-hand side $f$ yields a solution to the equation $Au = f$. The solution $u = g * f$ then possesses the correct physical dimension along with $f$ and the coefficients of the equation. The same applies to the impulse response introduced below. See also the subsequent examples in the next section.

## *The Causal Fundamental Solution*

If the differential equation $Au = f$ describes a physical system, then the solutions $u$ are possible system responses to the excitation of the system mathematically modeled by $f$. Each particular solution $u$ depends not only on $f$ but also on the initial conditions of the system. We now assume that the system is in a resting state without energy up to an initial time $t_0$, i.e., all initial values are zero, and we consider an excitation $f$ from this time $t_0$, i.e., $\operatorname{supp}(f) \subset [t_0, \infty[$. Assuming the *causality* of the system, the system response cannot be present before the excitation. Therefore, we seek a solution $u$ of $Au = f$ whose support is also contained in $[t_0, \infty[$. If such a causal solution exists, then it is *uniquely determined*; because for any other causal solution $\tilde{u}$, we have $A(u - \tilde{u}) = 0$, and $u - \tilde{u}$ vanishes for $t < t_0$. Due to the uniqueness of this solution of the homogeneous equation, $u - \tilde{u}$ must then be the zero function. To not violate the causality principle, no nontrivial solutions of the homogeneous equation can be added to $u$.

We call a distribution *causal*, if its support is contained in a half-axis $[t_0, \infty[$. We now show that there is exactly one causal fundamental solution $g \in \mathcal{D}'$. Its support lies in $[0, \infty[$, and it is the response of the system described by $Au = f$ to an input impulse at time $t = 0$. The convolution $g * f$ with the causal fundamental solution then exists for every right-hand side $f \in \mathcal{D}'$ whose support is bounded to the left, and results in the sought causal solution of $Au = f$ due to $\operatorname{supp}(g * f) \subset \operatorname{supp}(g) + \operatorname{supp}(f)$.

**Theorem 9.5 (Causal Fundamental Solutions)** *The causal fundamental solution $g$ is given by $g(t) = v(t)s(t)$. The support of $g$ is contained in $[0, \infty[$. Here, $s(t)$ is the unit step function and $v(t)$ is the solution of the homogeneous equation $Au = 0$ of n-th order that satisfies the following initial conditions:*
For $n = 1 :$ $v(0) = 1/a_1$,
for $n \geqslant 2 :$ $v^{(k)}(0) = 0$ *for* $k = 0, \ldots, n - 2$, $v^{(n-1)}(0) = 1/a_n$.

One can therefore determine $g$ using the zeros of the characteristic polynomial.

***Proof*** For $n = 1$, $\mathrm{e}^{-a_0 t/a_1} s(t)/a_1$ is the sought fundamental solution. For $n \geqslant 2$ and for $k = 1, \ldots, n$, it holds based on the initial condition with an arbitrary test function $\varphi$

$$\langle (vs)^{(k)}, \varphi \rangle = \langle v^{(k)}s + v^{(k-1)}\delta, \varphi \rangle.$$

For $k = 1$ this is immediately evident, for $1 \leqslant k \leqslant n - 1$, it results by induction and the initial condition

$$\langle (vs)^{(k+1)}, \varphi \rangle = -\langle (vs)^{(k)}, \varphi' \rangle = -\langle v^{(k)}s + v^{(k-1)}\delta, \varphi' \rangle$$
$$= -\langle v^{(k)}s, \varphi' \rangle = \langle v^{(k+1)}s + v^{(k)}\delta, \varphi \rangle.$$

Therewith, it then follows from applying the initial condition again with $Av = 0$

$$\langle Ag, \varphi \rangle = \left\langle \sum_{k=0}^{n} a_k (vs)^{(k)}, \varphi \right\rangle = \left\langle \left( \sum_{k=0}^{n} a_k v^{(k)} \right) s + \delta, \varphi \right\rangle = \langle \delta, \varphi \rangle.$$

The uniqueness of the causal fundamental solution has already been established above.                                                                              $\square$

**Example** The differential equation $u^{(3)} + \dot{u} = 0$ has the characteristic polynomial $P(\lambda) = \lambda^3 + \lambda$ with the zeros $\lambda_1 = 0$, $\lambda_2 = j$, and $\lambda_3 = -j$. Thus, the general solution is $u(t) = c_1 + c_2 \sin(t) + c_3 \cos(t)$ with parameters $c_1, c_2, c_3$ from $\mathbb{R}$. From the initial condition follows the particular solution $v(t) = 1 - \cos(t)$ and the causal fundamental solution

$$g(t) = s(t) - \cos(t)s(t).$$

**Remark** If $v(t)s(t)$ is the causal fundamental solution of $Au = f$, then the function $-v(t)s(-t)$ is also a fundamental solution. By convex combinations of $v(t)s(t)$ and $-v(t)s(-t)$, one obtains infinitely many noncausal fundamental solutions. The causal fundamental solution $g$ is also called the *Green's function* after G. Green (1793–1841).

## *Impulse Response, Step Response of Time-Invariant Linear Systems*

Many causal, time-invariant linear systems in technical applications are described by differential equations of the form

$$A_1 u = A_2 f,$$

where $A_1$ and $A_2$ are two linear differential operators with constant coefficients. Derivatives also appear on the right-hand side. We continue to use $t_0 = 0$ as the initial time for disturbances $f$ and understand a distribution $f$ with $\text{supp}(f) \subset [0, \infty[$ as the input quantity, and the uniquely determined distributional solution $u$

with $\text{supp}(u) \subset [0, \infty[$ as the sought output quantity of the linear system. The second example in the next section is of this type. If $g$ is the causal fundamental solution of $A_1 g = \delta$, then one obtains the *causal impulse response* $h$ of the system $A_1 u = A_2 f$ through

$$h = g * A_2 \delta = A_2 g.$$

For then $A_1 h = A_1 (g * A_2 \delta) = A_1 g * A_2 \delta = \delta * A_2 \delta = A_2 \delta$. The support of $h$ is contained in $[0, \infty[$. Since the impulse response $h$ can be convolved with any distribution $f$ whose support is bounded below, the following relationship exists between such input quantities $f$, the impulse response $h$, and the system response $u$:

$$A_1 (h * f) = A_1 h * f = A_2 \delta * f = \delta * A_2 f = A_2 f.$$

**Theorem 9.6** *The causal solution $u$ for an excitation $f$ with* $\text{supp}(f) \subset [0, \infty[$ *is obtained by the convolution $h * f$ of $f$ with the causal impulse response $h$:*

$$u = h * f.$$

*The support of $u$ is contained in $[0, \infty[$.*

Because of this relationship, the impulse response $h$ is used in system theory to characterize the transfer behavior of causal, time-invariant linear systems of the form $A_1 u = A_2 f$. This characterization of the transfer behavior applies more generally also to causal time-invariant linear systems that cannot be described by differential equations (for example, delay elements or integrators in electrical engineering). We will return to the basics of linear system theory in more detail in Chap. 11.

Easier to measure than the impulse response $h$ is usually the causal step response $a$, i.e., the reaction to the unit step function $s$. From $A_1 a = A_2 s$, however, it immediately follows $A_1 \dot{a} = A_2 \dot{s} = A_2 \delta$.

*The impulse response $h$ is obtained by differentiating the step response $a$, so causal, time-invariant linear systems $A_1 u = A_2 f$ can be characterized equally well by their step response $a$.*

**Explication of the Convolution** Physically realizable, time-invariant linear systems of the form $Au = x$ are causal (circuits, controllers, etc.). The system response to a regular excitation $x(\tau)$ with $x(\tau) = 0$ for $\tau < 0$ at time $t \geqslant 0$ is given by the convolution integral

$$u(t) = \int_0^t x(\tau) h(t - \tau) d\tau.$$

The convolution $u(t)$ is therefore a *continuous superposition* of values of the excitation $x$, namely of values with which the excitation began at all ($x(\tau)$ for $\tau = 0$) over the entire course of time of the " signal" $x$ up to the " present" $t$.

The *strength* with which the values $x(\tau)$ enter this superposition is controlled by the factors $h(t-\tau)$. The present value $x(t)$ is weighted with the factor $h(0)$, the most distant value $x(0)$ with the factor $h(t)$. One could therefore say that the impulse response $h$ contains the *"physical memory"* of the system (an electrical circuit, a controller, etc.): The entire time course from 0 to $t$ enters the system response $h * x$ at time $t$ through the superposition of signal values $x(\tau)$, $0 \leqslant \tau \leqslant t$, with the weights $h(t - \tau)$. In the weights $h(t - \tau)$ lies the information about the strength with which past events in the system due to its construction still affect the present.

Compare this in particular with the computation of the impulse response $h(t)$ from the general solution of the homogeneous system (see p. 217). From this, one can clearly see how the *eigensolutions* and the *decay behavior in transient processes* and thus how the roots of the characteristic polynomial uniquely determine the impulse response $h(t)$ up to a factor. If we also consider that typical impulse responses of *asymptotically stable*, causal time-invariant systems described by linear differential equations usually tend to zero very fast for $t \to \infty$, then we recognize that in the superposition $h * x$ values of $x(\tau)$ typically have stronger weight the closer $\tau$ is to the " present" $t$, and that the weights $h(t - \tau)$ decrease all the more rapidly the further one goes with $\tau \to 0$ into the " past." When $h$ has a bounded support, then the convolution $x * h$ is a weighted moving average of $x$.

## *Linear Initial Value Problems of n-th Order with Constant Coefficients*

We continue to consider equations with constant coefficients of the form

$$P(D)u = Q(D)f$$

for $D = \mathrm{d}/\mathrm{d}t$, polynomials $P(\lambda) = \sum_{k=0}^{n} a_k \lambda^k (n > 0,\ a_n \neq 0)$, and $Q(\lambda) = \sum_{k=0}^{m} b_k \lambda^k$.

The notation $P(D)$ of the differential operator means

$$P(D) = \sum_{k=0}^{n} a_k \frac{\mathrm{d}^k}{\mathrm{d}t^k}.$$

For equations $P(D)u = Q(D)f$ on $\mathcal{D}'$, it is not clear what initial values mean for distributional right-hand sides without additional conditions. For example, the equation $\dot{u} = \dot{\delta}$ has the general solution $u = \delta + c$ with arbitrary constants $c$, but an initial value $u(0) = a$ does not make sense.

We want to explain how an initial value problem should be understood in the following. For this purpose, we assume that the differential equation describes a transmission system that transforms a given time-dependent input signal $f$ into a corresponding output signal $u$. In order for the differential equation of this output signal $u$ to have a unique solution, further conditions need to be formulated regarding the nature of the system and the type of the input signal $f$.

To this end, we assume that our system is *causal*, i.e., a disturbance of the system at rest by an excitation with support in $[t_0, \infty[$ generates a system response with support in $[t_0, \infty[$. We assume that the right-hand side $f$ can be decomposed as $f = f_r + f_g$ with $f_r \in C^m(\mathbb{R})$ and $f_g \in \mathcal{D}'_+$. Here, $C^m(\mathbb{R})$ denotes the space of $m$-times continuously differentiable functions on $\mathbb{R}$, and $\mathcal{D}'_+$ denotes the space of distributions with support in $[0, \infty[$. We choose the initial time $t_0 = 0$ and prescribe initial conditions of the form $u^{(k)}(0-) = \lim\limits_{t \to 0, t<0} u^{(k)}(t) = c_k$ for $k = 0, \ldots, n-1$.

Thus, the initial state of the system at time $t_0 = 0$ originates from the past $t < 0$ of the system. In order for the left-sided limits $c_k$ to exist, regularity conditions on the right-hand side $f$ in a left-sided neighborhood of $t_0$ are necessary. For simplicity, we use signals $f$ that are $m$-times continuously differentiable for $t < 0$. This is sufficiently general for all applications covered in the following chapters. We now extend the classical definition of initial value problems for the specified framework as follows.

**Definition**  A causal initial value problem for the equation $P(D)u = Q(D)f$ on $\mathcal{D}'$ with $f = f_r + f_g$, $f_r \in C^m(\mathbb{R})$, $f_g \in \mathcal{D}'_+$, is to find a distribution $u \in \mathcal{D}'$, such that:

1. The distribution $u$ solves the equation $P(D)u = Q(D)f$ on $\mathcal{D}'$.
2. The distribution $u$ for $t < 0$ coincides with the classical solution $z$ of $P(D)u = Q(D)f_r$, which satisfies $z^{(k)}(0) = c_k$ for the $k$-th derivatives $z^{(k)}$, $k = 0, \ldots, n-1$.

**Comment**  A distribution $T$ and a locally integrable function $f$ coincide on an open set $G$ if and only if $\langle T, \varphi \rangle = \langle f, \varphi \rangle$ for all test functions $\varphi$ with $\mathrm{supp}(\varphi) \subset G$.

**Examples**

1. For initial conditions $\ddot{u}(0-) = \dot{u}(0-) = u(0-) = 0$, the differential equation $u^{(3)} + \dot{u} = \delta$ in $\mathcal{D}'$ has the unique causal solution $u(t) = (1 - \cos(t))s(t)$ with the unit step function $s$. The noncausal solution $w(t) = u(t) + \cos(t) - 1$ satisfies $w(0+) = \dot{w}(0+) = \ddot{w}(0+) = 0$ and $\ddot{w}(0-) = -1$ (cf. p. 218). Condition 2 in the above definition yields the uniquely determined, causal solution $u$ as the system output signal for initial conditions from the signal's past $t < 0$.

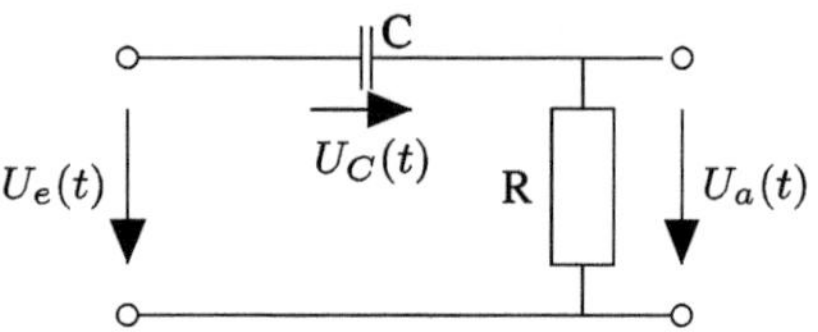



**Fig. 9.6** Circuit that realizes a first-order highpass filter

2. *Impact Forces in Mechanics.* For the linear motion $x(t)$ of an initially force-free particle of mass $m$, let $x(0-) = x_0$, $\dot{x}(0-) = v_0$. At time $t = 0$, an impact force (e.g., by a hammer blow) acts, which is mathematically modeled by $Z\delta$ with the Dirac distribution $\delta$. The constant $Z$ represents the change in momentum in Ns. The equation of motion for the desired solution $x$ reads

$$m\ddot{x} = Z\delta, \quad x(0-) = x_0, \quad \dot{x}(0-) = v_0.$$

3. *Schematic Highpass Filter.* We consider the following *RC* circuit—neglecting induction (see Fig. 9.6):
   Let $U_e(t) = U_0 - U_0 s(t)$ be the input voltage, and let $U_a(t)$ be the output voltage to be determined. At time $t_0 = 0$, the capacitor is charged with $U_C(0-) = U_0$. The output voltage at time $t_0$ is then $U_a(0-) = 0$. Using Kirchhoff's laws and Ohm's law, the problem can be described by the initial value problem

$$\dot{U}_a + \frac{1}{RC}U_a = \dot{U}_e, \quad U_a(0-) = 0.$$

For $U_e(t) = U_0 - U_0 s(t) = U_0 s(-t)$, the right-hand side becomes $\dot{U}_e = -U_0\delta$.

There exists exactly one classical solution $z$ of the equation $P(D)u = Q(D)f_r$ for given values $z(0) = c_0$, $\dot{z}(0) = c_1, \ldots, z^{(n-1)}(0) = c_{n-1}$. For $t < 0$, the solution $u$ of $P(D)u = Q(D)f$ coincides with $z$. A sudden disturbance occurring at time $t_0 = 0$ such as an impulse or its derivatives can cause a sudden change in the function $u$ which describes the temporal evolution of the system state. The temporal evolution from $t_0$ is determined by the final state that the system has reached in the past up to time $t_0$, and by the effect of the disturbance $f_g$ from the initial time $t_0$. Due to the already proven results, we obtain the following result about the solution of the stated initial value problem:

**Theorem 9.7** *The causal initial value problem for the equation $P(D)u = Q(D)f$ in $\mathcal{D}'$ with $f = f_r + f_g$, $f_r \in C^m(\mathbb{R})$, $f_g \in \mathcal{D}'_+$, with given initial values $c_k$, $k = 0, \ldots, n-1$, has the unique solution*

$$u = g * Q(D)f_g + z.$$

*Here, $g$ is the causal fundamental solution of $P(D)u = \delta$, $g * Q(D)f_g$ is the convolution of $g$ with the distribution $Q(D)f_g$, and $z$ is the classical solution of* the equation $P(D)u = Q(D)f_r$ which satisfies the conditions $z^{(k)}(0) = c_k$. Then $u^{(k)}(0-) = c_k (k \leqslant n - 1)$.

**Proof** Since the difference of two solutions solves the homogeneous equation with vanishing initial conditions, the solution is unique and independent of the representation of the superposition $f = f_r + f_g$. The convolution of the causal fundamental solution $g$ with $Q(D)f_g$ is the unique solution for the input signal $f_g$ with support in $[0, \infty[$ with vanishing initial conditions. Due to linearity and the regularity conditions on $f_r$, the unique solution $z$ of $P(D)u = Q(D)f_r$ is added, so that the required initial conditions for the overall solution $u$ are satisfied.     $\square$

**Example** Consider $f(t) = 3s(t) - 1$. The differential equation

$$\frac{\mathrm{d}u}{\mathrm{d}t} + 2u = 3\frac{\mathrm{d}f}{\mathrm{d}t} + 5f$$

with initial value $u(0-) = -5/2$. Its causal fundamental solution is $g(t) = \mathrm{e}^{-2t} s(t)$.

The solution on $\mathbb{R}$ is then $u(t) = g * (9\delta + 15s)(t) - \frac{5}{2} = \left(\frac{3}{2}\mathrm{e}^{-2t} + \frac{15}{2}\right) s(t) - \frac{5}{2}$. We will solve the previous examples following the next section.


## *Initial Value Problems on Half-Lines, Suppression of the Past*

In initial value problems, one is often only interested in the evolution of a system for $t \geqslant t_0$, where the system state at time $t_0$ is given. With the system description by the equation $P(D)u = Q(D)f$, we often consider disturbances from $t_0$ under "suppression of the past" only for the half-line $t \geqslant t_0$ and do not ask how the initial values in a real system could have come about. Mathematically, we can assume that they have been imposed on the present system by a suitable solution $z$ of the homogeneous differential equation. We set $t_0 = 0$ and look for a distribution $T$, which has its support in $[0, \infty[$ and for $t > 0$ and sufficiently smooth $f$ agrees with the classical solution of the initial value problem. The following statement holds:

**Theorem 9.8 (Initial Value Problems)**

*1. For $f \in \mathcal{D}'_+$, the distribution $T$, defined by $T = g * Q(D)f + zs$, is the unique causal solution of the distributional equation*

$$P(D)u = Q(D)f + \sum_{k=1}^{n} a_k \left( \sum_{p=0}^{k-1} c_p \delta^{(k-1-p)} \right).$$

*Here, $P(\lambda) = \sum_{k=0}^{n} a_k \lambda^k$, $g$ is the causal fundamental solution of $P(D)u = \delta$, $s$ is the unit step function, and $z$ is the classical solution of the homogeneous equation*

$P(D)u = 0$, *which satisfies the initial conditions* $z^{(k)}(0) = c_k$, $k = 0, \ldots, n-1$. *The solution is the convolution of $g$ with the right side of the above differential equation.*

2. *For $f \in C^m(\mathbb{R})$ with support in $[0, \infty[$, $T = g * Q(D)f + zs$ is regular and for $t > 0$ agrees with the classical solution $u$ of the initial value problem with the initial values $u^{(k)}(0-) = c_k$, $k = 0, \ldots, n-1$.*

***Proof*** The differential equation has a unique solution in $\mathcal{D}'_+$. Substituting $T$ into the equation shows the assertion, as for $k = 1, \ldots, n-1$

$$(zs)^{(k)} = z^{(k)}s + \sum_{p=0}^{k-1} c_p \delta^{(k-1-p)}.$$

For $f \in C^m(\mathbb{R})$ with support in $[0, \infty[$, the classical solution of the initial value problem is $u = g * Q(D)f + z$. It agrees for $t > 0$ with the regular distribution $T$.

□

**Example** The solution of the initial value problem from the preceding Example 9.2 on page 222 is $x(t) = \dfrac{Z}{m}ts(t) + v_0 t + x_0$. It is continuous, but as a result of the impact force at $t = 0$ it shows an abrupt change in velocity. The solution of the initial value problem in $\mathcal{D}'_+$, i.e., the modified differential equation $m\ddot{u} = Z\delta + m(x_0\dot{\delta} + v_0\delta)$ "with the past excluded" from $t = 0$ is $T(t) = \dfrac{Z}{m}ts(t) + v_0 ts(t) + x_0 s(t)$, thus $T = xs$ (see Figs. 9.7 and 9.8).

The images show $x(t)$ and $T(t)$ for $m = 1\,\text{kg}$, $x_0 = 2\,\text{m}$, $v_0 = 0.4\,\text{m/s}$, $Z = 5\,\text{Ns}$. The third example from page 222 will be dealt with in the next section.

**Fig. 9.7** Solution $x(t)$ of Example 9.2, p. 222

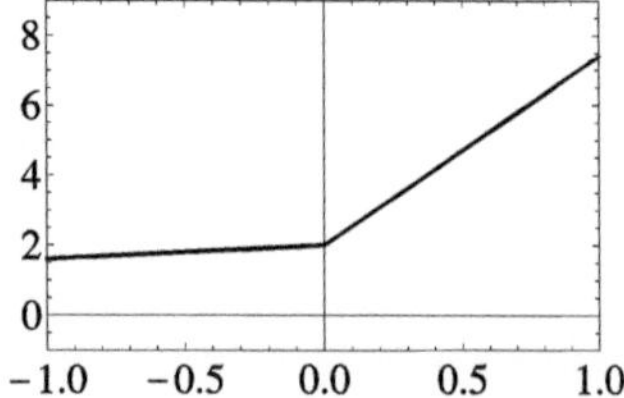

**Fig. 9.8** Solution $T(t)$ in $\mathcal{D}'_+$

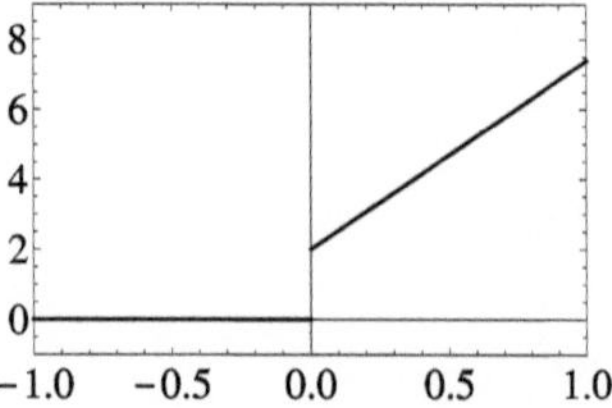

**Remarks**

1. The differential equation in the first part of the last theorem for initial value problems in $\mathcal{D}'_+$ is modified compared to $P(D)u = Q(D)f$ and goes back to Zemanian (2010), Schwartz (1957), and Shilov (1968). This equation directly includes the initial values, and their influence precisely yields the effect of the initial state on the solution for $t > 0$. It does not matter how the initial state was "really" achieved (cf. also Example 9.2 in the next section). An advantage of this formulation for problems where only the solution from $t_0 = 0$ is of interest is the following: The initial value problem is formulated in the *convolution algebra* $\mathcal{D}'_+$, and *with the modified equation* one can then solve such problems with distributional right-hand sides also with the *Laplace transform* often used by engineers, which operates precisely in $\mathcal{D}'_+$. See Schwartz (1966) or Zemanian (2010) for this. Under suitable additional conditions, which are often fulfilled in practical applications, the *Fourier transform* can also be used to solve the problem. We will address this in Sect. 12.7.

2. In a similar way, initial value problems in half-spaces for partial differential equations with distributional right-hand sides can be treated. Interested readers are referred to Schwartz (1957), Shilov (1968), or Triebel (1986).

3. The proofs of the last two theorems show that only sufficient smoothness of the right-hand side in a left-sided neighborhood of $t_0 = 0$ is necessary to obtain the results. One can then also calculate the solution of the initial value problem for the half-line $t < 0$ with the help of the parameter transformation $t \to -t$ similarly to the last theorem and also use the right-sided Laplace transform for this. For further details, refer to the previously cited literature.

**Theorem 9.9** *For $t < 0$, the solution of the initial value problem with given initial values $u^{(k)}(0-) = c_k$ $(k = 0, \ldots, n-1)$ is the reflection $u(t) = y(-t)$ of the solution $y \in \mathcal{D}'_+$ of*

$$P(-D)y = (Q(-D)\check{f}_r)s + \sum_{k=1}^{n}(-1)^k a_k \left( \sum_{p=0}^{k-1}(-1)^p c_p \delta^{(k-1-p)} \right).$$

*Here, $\check{f}_r(t) = f_r(-t)$, $f = f_r + f_g$ are as in the theorem on p. 222.*

***Proof*** Consider the causal fundamental solution $g = vs$ as on p. 217, $\check{v}(t) = v(-t)$, and the solution $z$ of $P(D)u = 0$ with initial values $z^{(k)}(0) = c_k$. Then $-\check{v}s$ is the causal fundamental solution of the reflected equation $P(-D)y = \delta$ in $\mathcal{D}'_+$. The convolution of $-\check{v}s$ with the right-hand side of the reflected equation

$$P(-D)y = (Q(-D)\check{f}_r)s + \sum_{k=1}^{n}(-1)^k a_k \left( \sum_{p=0}^{k-1}(-1)^p c_p \delta^{(k-1-p)} \right)$$

yields the reflection $y \in \mathcal{D}'_+$ of the solution $u$ of our initial value problem for $t < 0$. Due to the regularity properties of $v$ and $f_r$, the continuous convolutions $\left(-\breve{v}s * (Q(-D)\widetilde{f_r})s\right)^{(k)}$ vanish for $k = 0, \ldots, n-1$ and $t \to 0+$, while the convolution of $(-\breve{v}s)^{(k)}$ with the singular term on the right-hand side matches $(\breve{z}s)^{(k)}$, i.e., converges to $\breve{z}^{(k)}(0) = (-1)^k c_k$ as $t \to 0+$. The reflection $u$ of $y$ then yields the required initial values $c_k = u^{(k)}(0-)$, $k = 0, \ldots, n-1$.                              $\square$

**Example** As illustration, consider the equation $\ddot{u} + u = \dot{f}$ with initial conditions $u(0-) = 0$, $\dot{u}(0-) = 1$, and $f(t) = s(t+1) - s(t) + \delta(t)$.

The causal fundamental solution is $g(t) = v(t)s(t) = \sin(t)s(t)$. The initial value problem for $t \geq 0$ in $\mathcal{D}'_+$ with $f_g(t) = -s(t) + \delta(t)$ is $\ddot{x} + x = \dot{\delta}$. It has the unique causal solution $x(t) = \cos(t)s(t)$. For $t < 0$, we proceed as in the theorem shown above:

With $\breve{f_r}(t) = s(-t+1)$, $P(\lambda) = \lambda^2 + 1$, $Q(\lambda) = \lambda$, and $-\breve{v}s = g$, one solves the equation $P(-D)y(t) = \delta(t-1) - \delta(t)$ as in the previous remark in $\mathcal{D}'_+$ and obtains

$$y(t) = -\sin(t)s(t) + \sin(t-1)s(t-1).$$

Thus, we have $u(t) = x(t) + \breve{y}(t)$ as the overall solution, which satisfies the initial conditions $u(0-) = 0$ and $\dot{u}(0-) = 1$:

$$u(t) = \cos(t)s(t) + \sin(t)s(-t) - \sin(t+1)s(-t-1) \text{ for } t \in \mathbb{R}.$$

## *Causal Linear First-Order Systems with Constant Coefficients*

We still consider initial value problems on the half-line $t \geq 0$ for linear first-order systems of the form

$$\dot{X} = AX + F$$

with a constant $(n \times n)$-matrix $A$ and vector distributions $X$ and $F$. The components of $F$ are assumed to have support in $[0, \infty[$. We use the following notation:

**Definition** For $(n \times n)$ matrices $G$ with components $g_{km} \in \mathcal{D}'_+$ and vector distributions $F$ with $n$ components $f_m \in \mathcal{D}'_+$, i.e., $F \in \mathcal{D}'^n_+$, the convolution $G * F = (c_k)_{1 \leq k \leq n}$ is defined by

$$c_k = \sum_{m=1}^{n} g_{km} * f_m.$$

One easily verifies that for the componentwise generalized derivative $(G * F)'$, as in the one-dimensional case, $(G * F)' = G' * F = G * F'$ holds. For constant matrices $A$, $G$, and $B$ as above, and a vector distribution $F$, we have $(AG+B)*F = A(G * F) + B * F$.

With the well-known fundamental matrix $\mathrm{e}^{At}$ of the system, we can now formulate the following theorem, where $s$ is again the unit step function:

**Theorem 9.10**

1. *The system $\dot{X} = AX + F$ has the uniquely determined causal solution*

$$U = G * F$$

*for $F \in \mathcal{D}'^{n}_{+}$ with $G(t) = \mathrm{e}^{At} s(t)$.*
2. *For $F \in \mathcal{D}'^{n}_{+}$ and a given vector $\mathbf{x_0} \in \mathbb{R}^n$, the vector distribution $T$, defined by $T = G*(F+\mathbf{x_0}\delta) = G*F+G\mathbf{x_0}$, is the uniquely determined causal solution of*

$$\dot{X} = AX + F + \mathbf{x_0}\,\delta.$$

*For continuous disturbances $F$, $T$ is regular, and for $t > 0$, it coincides with the classical solution $X$ of the initial value problem $\dot{X} = AX + F$, $X(0) = \mathbf{x_0}$.*

$$X(t) = \mathrm{e}^{At}\,\mathbf{x_0} + \int\limits_{0}^{t} \mathrm{e}^{A(t-\tau)}\, F(\tau)\mathrm{d}\tau.$$

*Proof*

1. With the identity matrix $E$, it follows from

$$\dot{U} = \dot{G} * F = (AG + E\delta) * F = AU + F\,,$$

that $U = G * F$ is a solution. $U$ is causal, and thus uniquely determined.
2. For the distribution $T = G * F + G\mathbf{x_0}$,

$$\dot{T} = \dot{G} * F + \dot{G}\mathbf{x_0}$$

$$= (AG + E\delta) * F + (AG + E\delta)\mathbf{x_0}$$

$$= AT + F + \mathbf{x_0}\delta.$$

Thus, $T$ is the uniquely determined causal solution of the posed problem. With continuous $F$, $T$ is also regular, continuous, and obviously for $t > 0$, it coincides with the known classical solution $X$ of the initial value problem with $X(0) = \mathbf{x_0}$.

$\square$

**Remarks**

1. *What entries are in the matrix* $e^{At}$ *?*
   The matrix $e^{At} s(t)$ here plays the role that the causal fundamental solution had in the one-dimensional case. The $k$-th column of $e^{At} s(t)$ shows the response of all state variables to a $\delta$-disturbance of the state variables $x_k$. The response of the system to the disturbance $F = 1_n \delta$ is $e^{At} 1_n s(t)$ with the vector $1_n$, whose components are all one. It describes the reaction of each component of the state vector $X$ to simultaneous $\delta$-disturbances of these components.

   The elements of the matrix $e^{At}$ are therefore linear combinations of functions of the form $t^m \sin(\alpha t) e^{\beta t}$ and $t^n \cos(\gamma t) e^{\lambda t}$ with $m, n \in \mathbb{N}_0$ and $\alpha, \beta, \gamma, \lambda \in \mathbb{R}$, depending on the roots of the characteristic polynomial of the matrix $A$.

2. The equation $\dot{X} = AX + F + \mathbf{x_0} \delta$ is the adequate mathematical formulation for an initial value problem that one wants to solve in $\mathcal{D}'_+ \times \cdots \times \mathcal{D}'_+$. For the explicit determination of the fundamental matrix $e^{At}$, one can work well with a computer algebra system or with the Laplace transform already mentioned for small matrices in simple application examples. An analysis of numerical algorithms for the computation of $e^{At}$ for larger matrices $A$ can be found in the recommendable work of Moler and Van Loan (2003).

## *The Malgrange-Ehrenpreis Theorem*

The method of fundamental solutions remains valid for linear partial differential equations with constant coefficients (see p. 196). The existence of fundamental solutions is guaranteed by the theorem of B. Malgrange (1956) and Ehrenpreis (1954). Interested readers can find proofs of this important result on the solvability of linear differential equations, for example, in Hörmander (2003) or W. Rudin (1991). In Chap. 12, we follow a proof as an application of the Fourier transform, which is based on the work of Ortner and Wagner (1994), and Wagner (2009), providing an explicit formula for the representation of fundamental solutions.

## 9.3  Application to Linear Electrical Networks

We test the method of fundamental solutions on four simple application examples for electrical networks. The analysis of the behavior of such oscillating circuits under different excitations is part of the basic education in a physics degree and, of course, in electrical engineering.

**Example 9.1**  Given is the following "$RL$-network" with electrical resistance $R$ and inductance $L$ (Fig. 9.9):

**Fig. 9.9** Schematic *RL* circuit

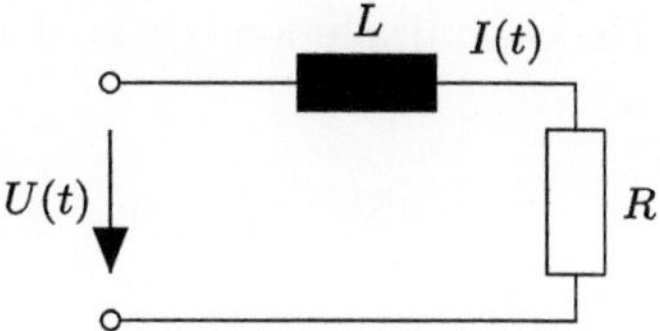

For $U(t) = \begin{cases} 0 & \text{for } t < 0 \\ U_0 \sin(\omega t) & \text{for } t \geqslant 0 \end{cases}$ the current $I(t)$ with $I(t) = 0$ for $t < 0$ is sought from the corresponding differential equation

$$L\frac{dI(t)}{dt} + RI(t) = U(t).$$

The causal impulse response is given by

$$h(t) = \frac{1}{L}\,e^{-(R/L)t}\,s(t)\,, \quad s(t) \text{ the unit step function}\,,$$

because

$$L\dot{h}(t) + Rh(t) = -\frac{R}{L}\,e^{-(R/L)t}\,s(t) + e^{-(R/L)t}\,\delta(t) + \frac{R}{L}\,e^{-(R/L)t}\,s(t) = \delta(t).$$

The solution $I(t)$ follows from this immediately through the convolution of the regular distributions $h(t)$ and $U(t)$. Both distributions have their supports in $[0, \infty[$.
   For $t \geqslant 0$ it is

$$I(t) = (h * U)(t) = \int_0^t h(t - \tau)U(\tau)d\tau.$$

Carrying out the integration yields for $t \geqslant 0$, $I(t) = 0$ for $t < 0$

$$I(t) = \underbrace{\frac{U_0}{R^2 + (\omega L)^2}\,(R\sin(\omega t) - \omega L\cos(\omega t))}_{\substack{\text{stationary}\\\text{part}}} + \underbrace{\frac{U_0\omega L}{R^2 + (\omega L)^2}\,e^{-(R/L)t}}_{\substack{\text{decaying}\\\text{transient}}}.$$

**Example 9.2 (Schematic Highpass Filter)** In the highpass filter of Example 9.3 on p. 222   assume the input voltage to be $U_e(t) = U_0 - U_0 s(t)$. The circuit is described with a charged capacitor by the initial value problem

$$\dot{U}_a + \frac{1}{RC}U_a = -U_0\delta \quad \text{and} \quad U_a(0-) = 0.$$

The general solution $U_H(t)$ of the corresponding homogeneous differential equation is

$$U_H(t) = K\,\mathrm{e}^{-t/(RC)}, \qquad (K \in \mathbb{R}\,\text{arbitrary}).$$

As a causal fundamental solution, we obtain

$$g(t) = \mathrm{e}^{-t/(RC)}\,s(t).$$

The corresponding impulse response $h$ is determined with the differential operator $\frac{\mathrm{d}}{\mathrm{d}t}$ on the right side of the output equation by differentiating the fundamental solution $g$:

$$h(t) = \dot{g}(t) = \delta(t) - \frac{1}{RC}\,\mathrm{e}^{-t/(RC)}\,s(t).$$

For the input voltage $U_e(t) = U_0 s(-t)$, the solution is

$$U_a(t) = -U_0(g * \delta)(t) = -U_0\,\mathrm{e}^{-t/(RC)}\,s(t).$$

If the excitation is a (ideal) voltage pulse $U_e(t) = U_0 R_1 C_1 \delta(t)$ with the impulse strength $U_0 R_1 C_1$ in the physical unit Vs (cf. p. 163), produced, for example, with an (ideal) upstream differentiator, and $U_a(0-) = 0$, then the output voltage $U_a$ for this input impulse is obtained by convolving with the causal impulse response $h$:

$$U_a(t) = h(t) * U_0 R_1 C_1 \delta(t) = U_0 R_1 C_1 \delta(t) - \frac{U_0 R_1 C_1}{RC}\,\mathrm{e}^{-t/(RC)}\,s(t).$$

**Example 9.3**  Given is the depicted $RLC$ circuit with resistance $R$, capacitance $C$, and inductance $L$ (Fig. 9.10).

The differential equation

$$\ddot{U}_a + \frac{2}{\sqrt{LC}}\dot{U}_a + \frac{1}{LC}U_a = U_1 \dot{\delta} \quad\text{and}\quad U_a(0-) = U_0\,,\ \dot{U}_a(0-) = 0$$

describes the circuit under critical damping $(R^2{=}4L/C)$ with input voltage $U_e(t) = U_1 s(t)$ and given initial values. The solution is the voltage at the inductance.

The causal fundamental solution is



**Fig. 9.10**  Schematic $RLC$ circuit

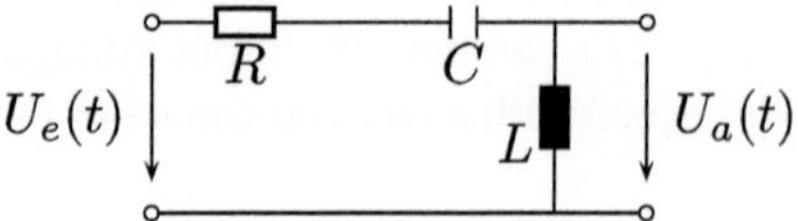

$$g(t) = t\,e^{-t/\sqrt{LC}}\,s(t).$$

The solution according to our theorem from p. 222 is

$$U_a(t) = g * (U_1\dot{\delta}) + z,$$

$$U_a(t) = \left(U_1 - \frac{U_1 t}{\sqrt{LC}}\right) e^{-t/\sqrt{LC}}\,s(t) + \left(U_0 + \frac{U_0 t}{\sqrt{LC}}\right) e^{-t/\sqrt{LC}}.$$

It satisfies $U_a(0-) = U_0$ and $U_a(0+) = U_0 + U_1$. For large negative values $t$, $U_a(t)$ certainly does not represent a realistic voltage, and in general, the "real" voltage characteristic over the entire past $t < 0$ remains probably unknown. Therefore, we consider the initial value problem only on the half-line $t \geqslant 0$, and disregarding the past we obtain the solution $T \in \mathcal{D}'_+$ with the distributional equation

$$\ddot{T} + \frac{2}{\sqrt{LC}}\dot{T} + \frac{1}{LC}T = U_1\dot{\delta} + \frac{2U_0}{\sqrt{LC}}\delta + U_0\dot{\delta}$$

from the theorem on p. 223, namely $T = U_a s$,

$$T(t) = \left(U_0 + U_1 + \frac{(U_0 - U_1)t}{\sqrt{LC}}\right) e^{-t/\sqrt{LC}}\,s(t).$$

**Example 9.4**  We consider the sketched block diagram of adders, integrators, and multipliers in Fig. 9.11. Its components (operational amplifiers, resistors, capacitors) can be realized in such a way that for voltage inputs $f$, all occurring state variables $x_0, \ldots x_{m-1}$ and $\dot{x}_0, \ldots \dot{x}_{m-1}$ are again voltages. We will see later that a large class of linear transmission systems, possessing a specified frequency response, can be constructed in this way (Sect. 11.2).

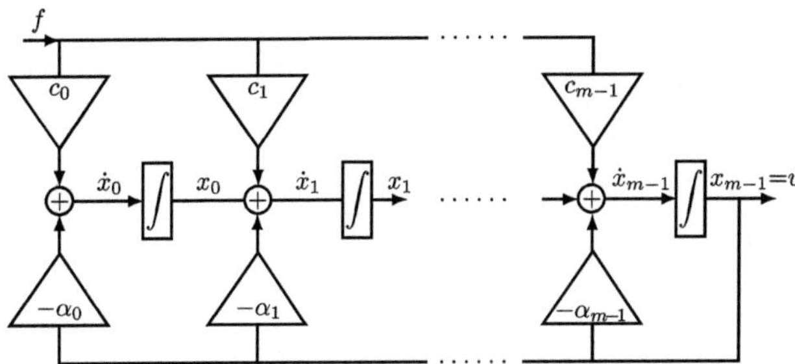


**Fig. 9.11**  Block diagram for a realization of a first-order system

The state description of this network by a first-order system with vanishing initial values is

$$\dot{\mathbf{x}}(t) = \mathbf{A}\mathbf{x}(t) + \mathbf{c}f(t)$$

with $\mathbf{x}(t) = (x_0(t), x_1(t), \ldots, x_{m-1}(t))^T$, $\quad \mathbf{c} = (c_0, c_1, \ldots, c_{m-1})^T$ and

$$\mathbf{A} = \begin{pmatrix} 0 \cdots\cdots\cdots\cdots & -\alpha_0 \\ 1 \;\; 0 \;\cdots & \cdots \;\; -\alpha_1 \\ \quad 1 \;\; \ddots & \vdots \\ \quad\quad \ddots\;\ddots & \vdots \\ \quad\quad\quad \ddots\;\ddots & \vdots \\ 0 \quad\quad 1 \;\; 0 & -\alpha_{m-2} \\ \quad\quad\quad\quad 1 & -\alpha_{m-1} \end{pmatrix}.$$

To calculate concretely, we choose

$$m = 3, \;\; c_0 = \Omega^3, \;\; c_1 = c_2 = 0, \;\; \alpha_0 = \Omega^3, \;\; \alpha_1 = 2\Omega^2, \;\; \alpha_2 = 2\Omega.$$

Later in Sect. 11.2 it will be shown that this describes a *Butterworth lowpass filter* of order 3 with cutoff frequency $\Omega/(2\pi)$. A representation of the corresponding frequency response can be found on p. 335.

We choose as the input signal a voltage pulse $f = k\delta$ ($k$ in the unit Vs) and ask, for example, for the output voltage $x_2(t) = v(t) \in \mathcal{D}'_+$ as the corresponding impulse response. The state description then reads:

$$\begin{pmatrix} \dot{x}_0 \\ \dot{x}_1 \\ \dot{x}_2 \end{pmatrix} = \begin{pmatrix} 0 & 0 & -\Omega^3 \\ 1 & 0 & -2\Omega^2 \\ 0 & 1 & -2\Omega \end{pmatrix} \begin{pmatrix} x_0 \\ x_1 \\ x_2 \end{pmatrix} + \begin{pmatrix} k\Omega^3\delta \\ 0 \\ 0 \end{pmatrix}.$$

Usually, it is quite tedious to calculate the fundamental matrix $\mathrm{e}^{At}$. As already mentioned, the Laplace transform or computer algebra systems like Mathematica, Maple, or Matlab offer good support in this process. In the present case, the required component $g_{31}$ of $G(t) = \mathrm{e}^{At}\, s(t)$ results in:

$$g_{31}(t) = \frac{1}{3\Omega^2}\left(3\,\mathrm{e}^{-\Omega t} - 3\,\mathrm{e}^{-\Omega t/2}\cos\left(\frac{\sqrt{3}\Omega t}{2}\right) + \sqrt{3}\,\mathrm{e}^{-\Omega t/2}\sin\left(\frac{\sqrt{3}\Omega t}{2}\right)\right)s(t).$$

The desired output voltage for the input signal $k\delta$ with $k = 1$ Vs, i.e., *the causal impulse response of the given Butterworth filter, is then* $x_2(t) = k\,\Omega^3 g_{31}(t)$.

Readers are encouraged to determine the complete matrix $\mathrm{e}^{At}$ as an exercise using the methods of linear algebra (e.g., with support of a suitable computer algebra system).

## 9.4  3D Potential Problems

We will now begin with applications of distribution methods to spatial problems. It is assumed that readers have already studied functions with several variables in their basic mathematics courses and can use the most important theorems of vector analysis, particularly Gauss's divergence theorem and Green's formulas. These theorems are also collected in Appendix B.

In many physical problems, the task is to calculate a force field from its divergences and vortices. Such tasks can often be formulated as potential problems. For example, if a spatially bounded charge distribution in a vacuum is given by the charge density $\varrho$, the fact that the charges are the sources or sinks of the generated electric field has been described by J. C. Maxwell (1831–1879) through the equation $\operatorname{div} \mathbf{E} = \varrho/\varepsilon_0$, where $\varepsilon_0$ denotes the electric field constant. With $\mathbf{E} = -\operatorname{grad} u$, the electrostatic field $\mathbf{E}$ in space can be calculated from a solution $u$ of the *potential equation*:

$$\operatorname{div} \operatorname{grad} u = \Delta u = -\frac{\varrho}{\varepsilon_0}.$$

In Cartesian coordinates, the Laplace operator is given by $\Delta = \frac{\partial^2}{\partial x^2} + \frac{\partial^2}{\partial y^2} + \frac{\partial^2}{\partial z^2}$. A solution $u$ is a Coulomb potential corresponding to $\varrho$. For $u : \mathbb{R}^3 \to \mathbb{R}$ and $\mathbf{E} = (E_1, E_2, E_3)$ with scalar components $E_1$, $E_2$, and $E_3$, we have

$$\operatorname{grad} u = \left( \frac{\partial u}{\partial x}, \frac{\partial u}{\partial y}, \frac{\partial u}{\partial z} \right) \quad \text{and} \quad \operatorname{div} \mathbf{E} = \frac{\partial E_1}{\partial x} + \frac{\partial E_2}{\partial y} + \frac{\partial E_3}{\partial z}.$$

Similarly, in the case of a stationary $\mathbf{E}$ field and a given current density $\mathbf{j}$, Maxwell's equations $\operatorname{rot} \mathbf{B} = \mu_0 \mathbf{j}$ and $\operatorname{div} \mathbf{B} = 0$ for the generated magnetic vortex field in a vacuum with $\mathbf{B} = \operatorname{rot} \mathbf{A}$ can be transformed into an *equation for the vector potential* $\mathbf{A}$:

$$-\operatorname{rot} \operatorname{rot} \mathbf{A} = \Delta \mathbf{A} = -\mu_0 \mathbf{j}.$$

For $\mathbf{A} = (A_1, A_2, A_3)$, $\Delta \mathbf{A} = (\Delta A_1, \Delta A_2, \Delta A_3)$, $\mu_0$ is the magnetic field constant, and

$$\operatorname{rot} \mathbf{A} = \left( \frac{\partial A_3}{\partial y} - \frac{\partial A_2}{\partial z}, \frac{\partial A_1}{\partial z} - \frac{\partial A_3}{\partial x}, \frac{\partial A_2}{\partial x} - \frac{\partial A_1}{\partial y} \right).$$

Corresponding potential problems arise when calculating gravitational fields from given mass distributions or in the mechanics of fluids and gases.

Since all the concepts and results from the previous sections can be translated to functions with more than one variable (see Sect. 8.7),  one can use the fundamental solution method to solve partial linear differential equations with constant coefficients. We calculate, as an example, a potential $u$ generated by an electric charge distribution in a bounded spatial region. We treat the charge density $\varrho$ as a distribution with bounded support. The corresponding partial differential equation is

$$\Delta u = -\frac{\varrho}{\varepsilon_0}.$$

First, we verify that the function $g(x, y, z) = \frac{1}{\sqrt{x^2+y^2+z^2}} = \frac{1}{r}$ provides a fundamental solution. Here, $r^2 = |\mathbf{r}|^2 = x^2 + y^2 + z^2$ for $\mathbf{r} = (x, y, z)$, and the function $g$ is considered as a regular distribution (see page 189):

1. $\Delta g(x, y, z) = 0$ in any region that does not include the origin. The reader should verify this through appropriate differentiation.
2. For test functions $\varphi = \varphi(x, y, z) = \varphi(\mathbf{r})$, we have

$$\langle \Delta g, \varphi \rangle = \langle g, \Delta\varphi \rangle$$

$$= \iiint\limits_{\mathbb{R}^3} g(x, y, z) \Delta\varphi(x, y, z)\mathrm{d}x\mathrm{d}y\mathrm{d}z$$

$$= \lim_{\varepsilon \to 0} \iiint\limits_{r \geqslant \varepsilon > 0} g(x, y, z) \Delta\varphi(x, y, z)\mathrm{d}x\mathrm{d}y\mathrm{d}z.$$

We use *Green's second formula from vector analysis* (see Appendix B on page 499):

$$\int\limits_{G} (g\Delta\varphi - \varphi\Delta g)\mathrm{d}\lambda^3 = \int\limits_{\partial G} (g\,\mathrm{grad}\varphi - \varphi\,\mathrm{grad}g) \cdot \mathbf{n}\mathrm{d}o,$$

where $\mathbf{n}$ denotes the unit outward normal, $\mathrm{d}\lambda^3$ is the volume element, $\mathrm{d}o$ is the surface element, i.e., the surface measure on the boundary $\partial G$ of $G$, and $G$ is the spherical shell

$$G = \{\mathbf{r} \in \mathbb{R}^3 \mid 0 < \varepsilon \leqslant |\mathbf{r}| \leqslant a\}.$$

The outer radius $a$ is chosen such that for $r = |\mathbf{r}| \geqslant a$ the test function $\varphi(\mathbf{r})$ vanishes:

$$\varphi(x, y, z) = 0 \quad \text{for} \quad \sqrt{x^2 + y^2 + z^2} \geqslant a.$$

Then, with $\Delta g = 0$ for $r \geqslant \varepsilon$ and the directional derivative $\frac{\mathrm{d}\varphi}{\mathrm{d}\mathbf{n}} = \mathrm{grad}\varphi \cdot \mathbf{n}$ :

$$\int\limits_{r \geqslant \varepsilon} g\Delta\varphi\mathrm{d}\lambda^3 = \int\limits_{r=\varepsilon} g\,\frac{\mathrm{d}\varphi}{\mathrm{d}\mathbf{n}}\,\mathrm{d}o - \int\limits_{r=\varepsilon} \varphi\,\frac{\mathrm{d}g}{\mathrm{d}\mathbf{n}}\,\mathrm{d}o.$$

Now, $g(\mathbf{r}) = \frac{1}{\varepsilon}$ for all $\mathbf{r}$ with $|\mathbf{r}| = \varepsilon$. From the boundedness of $\frac{d\varphi}{d\mathbf{n}}$, it follows:

$$\left| \int_{r=\varepsilon} g \frac{d\varphi}{d\mathbf{n}}\, do \right| = \frac{1}{\varepsilon} \left| \int_{r=\varepsilon} \frac{d\varphi}{d\mathbf{n}}\, do \right| \leqslant \frac{1}{\varepsilon} \cdot 4\pi \varepsilon^2 \cdot K \quad \text{for suitable } K \in \mathbb{R}.$$

Hence, for $\varepsilon \to 0$, the first surface integral on the right-hand side vanishes:

$$\lim_{\varepsilon \to 0} \int_{r=\varepsilon} g \frac{d\varphi}{d\mathbf{n}}\, do = 0.$$

Because on the inner sphere $r = \varepsilon$, the normal vector $\mathbf{n}$ points toward the origin, we there have

$$\frac{dg}{d\mathbf{n}}(\mathbf{r}) = \frac{1}{\varepsilon^2}.$$

Consequently, for the second surface integral, we get:

$$-\int_{r=\varepsilon} \varphi \frac{dg}{d\mathbf{n}}\, do = -\frac{1}{\varepsilon^2} \int_{r=\varepsilon} \varphi\, do = -4\pi M_\varepsilon(\varphi).$$

Here, $M_\varepsilon(\varphi)$ denotes the mean value of $\varphi$ on the sphere with radius $\varepsilon$:

$$M_\varepsilon(\varphi) = \frac{1}{4\pi \varepsilon^2} \int_{r=\varepsilon} \varphi\, do.$$

For $\varepsilon \to 0$, $M_\varepsilon(\varphi) \xrightarrow[\varepsilon \to 0]{} \varphi(0)$, and thus, we obtain:

$$\langle \Delta g, \varphi \rangle = \lim_{\varepsilon \to 0} \int_{r \geqslant \varepsilon} \frac{\Delta \varphi}{r}\, d\lambda^3 = -4\pi \varphi(0) = -4\pi \langle \delta, \varphi \rangle.$$

**Theorem 9.11 (Coulomb Potential)** *The function* $-\dfrac{1}{4\pi} g(\mathbf{r}) = -\dfrac{1}{4\pi r}$ *is a fundamental solution of the potential equation in* $\mathbb{R}^3$. *By convolution, a particular solution for the equation* $\Delta u = -\dfrac{\varrho}{\varepsilon_0}$ *is obtained as follows:*

$$u = \frac{\varrho}{\varepsilon_0} * \frac{1}{4\pi r}. \tag{9.4}$$

*Here, $\varrho$ can be any distribution with bounded support. For regular charge densities $\varrho$, this is the Poisson integral formula in* $\mathbb{R}^3$

$$u(x, y, z) = \frac{1}{4\pi\varepsilon_0} \iiint\limits_{\mathbb{R}^3} \frac{\varrho(u, v, w)}{\sqrt{(x-u)^2 + (y-v)^2 + (z-w)^2}} \, du\,dv\,dw.$$

Two Coulomb potentials of the same charge distribution $\varrho$ differ at most by a constant due to the relationship $\nabla \cdot \mathbf{E} = \varrho/\varepsilon_0$ for the electric field strength. The regular distribution $q/(4\pi\varepsilon_0 r)$ is the potential of a charge $q$ at the origin, which vanishes at infinity. Accordingly, $q/(4\pi\varepsilon_0(r - r_0))$ is the contribution to the potential at the point $\mathbf{r}$ by such a charge at the point $\mathbf{r}_0$. Through the above convolution integral with the charge density $\varrho(\mathbf{r})$, the contributions of all points in space to the total potential are summed. For *"simple"* charge densities $\varrho$, the potential $u$ can be directly calculated from the Poisson convolution integral.

## *Examples*

1. In space, let there be a thin rod $S = \{ (x, y, z) \in \mathbb{R}^3 : x = y = 0, |z| \leqslant l\}$ of length $2l$ with homogeneous mass density $\varrho_0$ (with the unit kg/m) given. We describe this spatial mass distribution with the indicator function $1_{[-l,l]}(z)$ through the distributional tensor product (see p. 191)

$$\varrho(x, y, z) = \delta(x) \otimes \delta(y) \otimes \varrho_0 1_{[-l,l]}(z).$$

From the potential equation $\Delta u = 4\pi\gamma\varrho$ with the gravitational constant $\gamma$, the gravitational potential vanishing at infinity follows

$$u = -\gamma\varrho * \frac{1}{r}.$$

For $(x, y, z) \notin S$, the potential $u$ is obtained through integration:

$$u(x, y, z) = -\gamma\varrho_0 \int_{-l}^{l} \frac{1}{\sqrt{x^2 + y^2 + (z-w)^2}} \, dw$$

$$= -\gamma\varrho_0 \ln\left( 2\sqrt{x^2 + y^2 + (z-w)^2} + 2w - 2z \right)\Bigg|_{w=-l}^{w=+l}$$

$$= -\gamma\varrho_0 \ln\left( \frac{\sqrt{x^2 + y^2 + (z-l)^2} + l - z}{\sqrt{x^2 + y^2 + (z+l)^2} - l - z} \right).$$

With $\mathbf{F} = -\operatorname{grad} u$ one then finds the corresponding gravitational field $\mathbf{F}$.

**Fig. 9.12** Equipotential surface of two charges $q$ and $-2q$ in the half-space $y > 0$

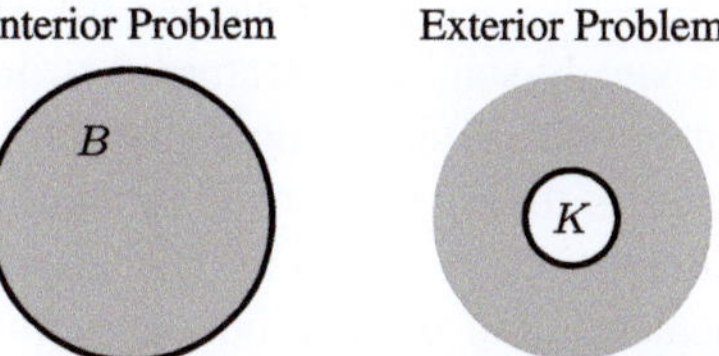

**Fig. 9.13** Illustration of the interior and the exterior Dirichlet problem



2. If $\sigma$ is a regular surface charge density on the surface of the sphere with radius $R$ around the origin, denoted by the singular distribution $\varrho(\mathbf{x}) = \sigma(\mathbf{x})\delta(|\mathbf{x}| - R)$ (see p. 189), then from $u = \varrho * 1/(4\pi\varepsilon_0 r)$, one obtains *Coulomb's formula* for the potential vanishing at infinity

$$u(\mathbf{x_0}) = \frac{1}{4\pi\varepsilon_0} \int\limits_{|\mathbf{x}|=R} \frac{\sigma(\mathbf{x})}{|\mathbf{x} - \mathbf{x_0}|} \, do(\mathbf{x}) .$$

3. Figure 9.12 illustrates an equipotential surface of the Coulomb field generated by two charges $q$ and $-2q$ in the half-space $y > 0$:

## *Approaches for Solving Boundary Value Problems*

Poisson's formula can be used for potential computation when a known charge distribution is available. However, many problems in electrostatics are of the form where the potential $u$ is given on certain surfaces without knowing the generating charge distribution, or where, given a charge distribution in a region, the values of the potential on the boundary of the region are additionally prescribed. Two typical associated *boundary value problems* are (cf. Fig. 9.13):

1. *The interior Dirichlet problem:*   For a bounded region $B$ with boundary $\partial B$, a distribution $v$ with $\operatorname{supp}(v) \subset B$ and a function $f$ on $\partial B$ are given. Determine $u$ such that

$$\Delta u = v \text{ in } B \quad \text{and} \quad u = f \text{ on } \partial B.$$

2. *The exterior Dirichlet problem*:   For a bounded region $K$ with boundary $\partial K$, a distribution $v$ with $\text{supp}(v) \subset \mathbb{R}^3 \setminus (K \cup \partial K)$ and a function $f$ on $\partial K$ are given. Determine $u$ such that

$$\Delta u = v \text{ in } \mathbb{R}^3 \setminus (K \cup \partial K) \quad \text{and} \quad u = f \text{ on } \partial K.$$

Examples of the first type include the question of the potential of an electric charge distribution $\varrho = -v\varepsilon_0$ within an electrically shielded area (i.e., $f = 0$) or the question of the steady-state temperature distribution ($v = 0$) in a body $B$ with a time-constant boundary temperature distribution $f$. An example of an exterior Dirichlet problem is the question of the potential $u$ that results when a grounded electrical conductor (i.e., $f = 0$) is in the electric field of a charge $\varrho$.

The analytical solution of such boundary value problems generally requires a considerable amount of mathematical effort and additional smoothness properties of the data $\partial B$, $\partial K$, $v$, and $f$. To get a general impression of a classical solution method, we consider the boundary value problems under simplifying assumptions:

For $\mathbf{x_0} \in \mathbb{R}^3$ we set $g(\mathbf{x}, \mathbf{x_0}) = 1/(4\pi |\mathbf{x} - \mathbf{x_0}|)$. Then it holds that $\Delta_{\mathbf{x}} g = 0$ in $\mathbb{R}^3 \setminus \{\mathbf{x_0}\}$. $\Delta_{\mathbf{x}}$ means that the Laplace operator is to be applied to the variable $\mathbf{x}$. As region $B$ we consider $B = \{\mathbf{x} \in \mathbb{R}^3 \mid 0 \leqslant r < |\mathbf{x}| < R\}$, i.e., a *sphere* or a *spherical shell*. $K$ is the sphere around the origin with radius $r$. If $\mathbf{x_0} \in B$, then let the set $B_\varepsilon(\mathbf{x_0}) \subset B$ denote a small closed sphere with radius $\varepsilon$ around this singularity $\mathbf{x_0}$ of $g$ (see Fig. 9.14).

For a sufficiently smooth function $u$ on $B$, Green's second formula holds, and since $\Delta_{\mathbf{x}} g = 0$ in $B \setminus \{\mathbf{x_0}\}$, it follows with the outward normal vectors $\mathbf{n}$ of $\partial B$ respectively $\partial B_\varepsilon(\mathbf{x_0})$

$$\int_{B \setminus B_\varepsilon(\mathbf{x_0})} g(\mathbf{x}, \mathbf{x_0}) \Delta u(\mathbf{x}) \, d\lambda^3(\mathbf{x}) = \int_{\partial B} \left( g(\mathbf{x}, \mathbf{x_0}) \frac{du}{d\mathbf{n}}(\mathbf{x}) - u(\mathbf{x}) \frac{dg}{d\mathbf{n}}(\mathbf{x}, \mathbf{x_0}) \right) do(\mathbf{x})$$

$$- \int_{\partial B_\varepsilon(\mathbf{x_0})} \left( g(\mathbf{x}, \mathbf{x_0}) \frac{du}{d\mathbf{n}}(\mathbf{x}) - u(\mathbf{x}) \frac{dg}{d\mathbf{n}}(\mathbf{x}, \mathbf{x_0}) \right) do(\mathbf{x}).$$

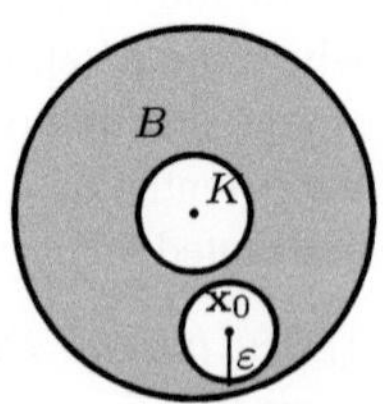


**Fig. 9.14** Illustration with a small sphere around a singularity $\mathbf{x_0}$

On p. 235 we showed that the second surface integral converges to $u(\mathbf{x_0})$ for $\varepsilon \to 0$. Thus, it follows that

$$u(\mathbf{x_0}) = \int_{\partial B} \left( g(\mathbf{x}, \mathbf{x_0}) \frac{du}{d\mathbf{n}}(\mathbf{x}) - u(\mathbf{x}) \frac{dg}{d\mathbf{n}}(\mathbf{x}, \mathbf{x_0}) \right) do(\mathbf{x}) - \int_B g(\mathbf{x}, \mathbf{x_0}) \Delta u(\mathbf{x}) \, d\lambda^3(\mathbf{x}) .$$

$$(9.5)$$

The value of a sufficiently smooth solution $u$ of the interior Dirichlet problem at a "starting point" $\mathbf{x_0} \in B$ is thus determined by the values of the inhomogeneity $\Delta u$ in $B$ and by the values of the potential $u$ and its normal derivative (from the inside) on the boundary $\partial B$. The potential $u$ and therefore also its normal derivative are already *uniquely determined* by specifying $\Delta u$ in $B$ and $u$ on the boundary $\partial B$: If $u_1$ and $u_2$ are two (sufficiently smooth) solutions, then $\varphi = u_1 - u_2$ satisfies the equations $\Delta \varphi = 0$ in $B$ and $\varphi = 0$ on $\partial B$. Substituting into the Green's first formula (p. 499) shows

$$\int_B (\varphi \Delta \varphi + \operatorname{grad} \varphi \cdot \operatorname{grad} \varphi) \, d\lambda^3 = \int_{\partial B} \varphi \frac{d\varphi}{d\mathbf{n}} \, do ,$$

and thus because $\Delta \varphi = 0$,

$$\int_B \operatorname{grad} \varphi \cdot \operatorname{grad} \varphi \, d\lambda^3 = 0,$$

i.e., $\operatorname{grad} \varphi = \mathbf{0}$. Then $\varphi$ is constant, and from the boundary values, it follows $\varphi = 0$, $u_1 = u_2$ everywhere.

If it is possible, in the above representation formula (9.5) for all $\mathbf{x_0} \in B$, to replace the function $g(\mathbf{x}, \mathbf{x_0})$ by a function $G(\mathbf{x}, \mathbf{x_0})$, which still satisfies $\Delta_{\mathbf{x}} G(\mathbf{x}, \mathbf{x_0}) = 0$ for all $\mathbf{x}$ in $B \setminus \{\mathbf{x_0}\}$ and also vanishes on the boundary of $B$ and leaves the volume integral unchanged, then it follows from (9.5) the following:

**Approach to the Solution u of the Interior Dirichlet Problem at a Point $\mathbf{x_0} \in \mathbf{B}$**

$$u(\mathbf{x_0}) = - \int_{\partial B} f(\mathbf{x}) \frac{dG}{d\mathbf{n}}(\mathbf{x}, \mathbf{x_0}) \, do(\mathbf{x}) - \int_B v(\mathbf{x}) G(\mathbf{x}, \mathbf{x_0}) \, d\lambda^3(\mathbf{x}) . \qquad (9.6)$$

The desired function $G$ is called *Green's function*. It is defined by

$$G(\mathbf{x}, \mathbf{x_0}) = \frac{1}{4\pi |\mathbf{x} - \mathbf{x_0}|} + F(\mathbf{x}, \mathbf{x_0}) \qquad (\mathbf{x} \neq \mathbf{x_0},\ \mathbf{x_0} \in B,\ \mathbf{x} \in B \cup \partial B).$$

According to the above considerations, the following properties are imposed on $F$:

1. *For each $\mathbf{x_0} \in B$, $F(\mathbf{x}, \mathbf{x_0})$ as a function of $\mathbf{x}$ is twice continuously differentiable in $B$ with first partial derivatives that can be continuously extended to the boundary $\partial B$. For each $\mathbf{x_0} \in B$, $F$ is harmonic in $B$, i.e., for all $\mathbf{x}, \mathbf{x_0}$ in $B$, it holds that $\Delta_\mathbf{x} F(\mathbf{x}, \mathbf{x_0}) = 0$.*
2. *For each $\mathbf{x_0} \in B$ and each $\mathbf{x} \in \partial B$, it holds that $F(\mathbf{x}, \mathbf{x_0}) = -\frac{1}{4\pi |\mathbf{x}-\mathbf{x_0}|}$.*

Analogous to the uniqueness proof for the solution $u$, it can be seen that there is at most one function $F$ with the required properties, i.e., if there exists a Green's function $G$ at all, then it is uniquely determined.

The part $g(\mathbf{x}, \mathbf{x_0}) = 1/(4\pi |\mathbf{x} - \mathbf{x_0}|)$ of Green's function $G$ is up to a factor a potential of a point charge at the location $\mathbf{x_0}$ in $B$. The *physical meaning of* $F(\mathbf{x}, \mathbf{x_0})$ is therefore, because $\Delta_\mathbf{x} F(\mathbf{x}, \mathbf{x_0}) = 0$ for $\mathbf{x} \in B$, the potential of a charge distribution outside the region $B$. The size and location of this external charge distribution must be such that the superposition of its potential with that of the point charge at $\mathbf{x_0}$ is zero on the boundary $\partial B$. This consideration forms the basis of the *method of image charges*, which can be used to find Green's function $G$ for simple regions such as spheres. Before we compute an example, let us turn to the *exterior Dirichlet problem for the sphere $K$* (see Fig. 9.13 on p. 237):

Coulomb potentials $u$, generated by spatially bounded, continuous charge distributions $\varrho$ and vanishing at infinity, decay according to Poisson's formula for $|\mathbf{x}| \to \infty$ like $1/|\mathbf{x}|$, their normal derivatives on spherical surfaces $|\mathbf{x}| = R$ for $R \to \infty$ like $1/R^2$: If $\alpha > 0$ is large enough so that the existing total charge lies within the sphere around the origin with radius $\alpha$, then for all $\beta$ with $0 < \beta < 1$ and $\mathbf{x}$ with $|\mathbf{x}| \geqslant \alpha/\beta$, the inequality $1 - \alpha/|\mathbf{x}| \geqslant 1 - \beta$ holds and thus (see Fig. 9.15 for illustration)

$$|u(\mathbf{x})| = \left| \frac{1}{4\pi\varepsilon_0} \int\limits_{|\mathbf{y}|\leqslant\alpha} \frac{\varrho(\mathbf{y})}{|\mathbf{x}-\mathbf{y}|}\,\mathrm{d}\lambda^3(\mathbf{y}) \right| \leqslant \frac{1}{4\pi\varepsilon_0} \int\limits_{|\mathbf{y}|\leqslant\alpha} \frac{|\varrho(\mathbf{y})|}{|\mathbf{x}-\frac{\alpha}{|\mathbf{x}|}\mathbf{x}|}\,\mathrm{d}\lambda^3(\mathbf{y})$$

$$= \frac{1}{4\pi\varepsilon_0(1-\frac{\alpha}{|\mathbf{x}|})|\mathbf{x}|} \int\limits_{|\mathbf{y}|\leqslant\alpha} |\varrho(\mathbf{y})|\,\mathrm{d}\lambda^3(\mathbf{y})$$

$$\leqslant \frac{1}{4\pi\varepsilon_0(1-\beta)|\mathbf{x}|} \int\limits_{|\mathbf{y}|\leqslant\alpha} |\varrho(\mathbf{y})|\,\mathrm{d}\lambda^3(\mathbf{y}).$$

Accordingly, one shows that for $u$ the normal derivative on spherical surfaces $|\mathbf{x}| = R$ decays for $R \to \infty$ uniformly in all directions like $1/R^2$ (exercise).

We now notice in Green's formula (9.5) on p. 239, that under the above assumptions on $u$ at a spherical shell $B = \{\mathbf{x} \in \mathbb{R}^3 \mid r < |\mathbf{x}| < R\}$ the surface integral over the outer surface $|\mathbf{x}| = R$ for $R \to \infty$ vanishes: The integrand decays for $R \to \infty$ like $1/R^3$, while the sphere surface grows like $R^2$. Also, the volume integral in (9.5) remains bounded, if spatially bounded, continuous charge

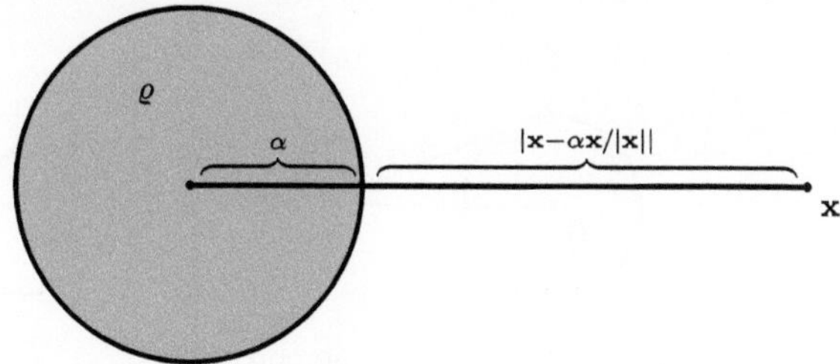


**Fig. 9.15** Illustration for the estimation of $|u|$

densities $\varrho$ are present. Following the same considerations as in the inner Dirichlet problem, formula (9.6) under the above physically motivated assumptions about the solution $u$ also provides an approach for the outer Dirichlet problem at a point $x_0 \notin K \cup \partial K$.

**Approach for the Outer Dirichlet Problem at a Point $x_0 \notin K \cup \partial K$**

$$u(\mathbf{x_0}) = - \int\limits_{|\mathbf{x}|=r} f(\mathbf{x}) \frac{\mathrm{d}G}{\mathrm{d}\mathbf{n}}(\mathbf{x}, \mathbf{x_0}) \, \mathrm{d}o(\mathbf{x}) - \int\limits_{|\mathbf{x}|>r} v(\mathbf{x})G(\mathbf{x}, \mathbf{x_0}) \, \mathrm{d}\lambda^3(\mathbf{x}) \tag{9.7}$$

for $|\mathbf{x_0}| > r$ with the *inward normal* $\mathbf{n}$ to the spherical surface $|\mathbf{x}| = r$ and Green's function $G$ for the sphere $K$ around the origin with radius $r$.

Since $\delta$-distributions can be approximated by smooth functions (cf. p. 201), formula (9.6) can also be used if the charge density $\varrho$ represents a point charge $q\delta(\mathbf{x} - \mathbf{x_1})$. With boundary values $f$ on the spherical surface $|\mathbf{x}| = r$, $|\mathbf{x_1}| > r$, and $v(\mathbf{x}) = -q\delta(\mathbf{x} - \mathbf{x_1})/\varepsilon_0$, the approach for the potential at a point $\mathbf{x_0} \neq \mathbf{x_1}$, $|\mathbf{x_0}| > r$ using the inward normal $\mathbf{n}$ to the spherical surface is

$$u(\mathbf{x_0}) = \frac{q}{\varepsilon_0} G(\mathbf{x_1}, \mathbf{x_0}) - \int\limits_{|\mathbf{x}|=r} f(\mathbf{x}) \frac{\mathrm{d}G}{\mathrm{d}\mathbf{n}}(\mathbf{x}, \mathbf{x_0}) \, \mathrm{d}o(\mathbf{x}) . \tag{9.8}$$

**Example (The Grounded, Conductive Sphere in the Field of a Point Charge)**
Given is an electrically conductive sphere $K$ around the origin with radius $r$. It is at ground potential $u = 0$. Outside this sphere, there is a point charge $q$ at the point $\mathbf{x_1}$, $|\mathbf{x_1}| > r$. The corresponding outer Dirichlet problem reads as:
The potential $u$ vanishing at infinity is sought such that

$$\Delta u(\mathbf{x}) = -\frac{q}{\varepsilon_0}\delta(\mathbf{x} - \mathbf{x_1}) \quad \text{for } |\mathbf{x}| > r, \mathbf{x} \neq \mathbf{x_1},$$
$$u(\mathbf{x}) = 0 \quad\quad\quad\quad \text{for } |\mathbf{x}| = r.$$

According to the method of image charges, an approach is made for Green's function $G$ with a second charge $q'$ inside the sphere $K$. The location $\mathbf{x_2}$ and magnitude of $q'$ are to be determined such that $G(\mathbf{x}, \mathbf{x_0}) = 0$ for all $\mathbf{x}$ and $\mathbf{x_0}$ with

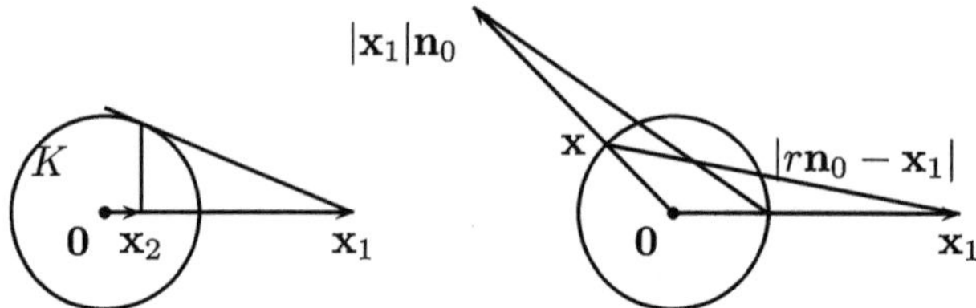

**Fig. 9.16**  Illustration for $\mathbf{x}_1$ and its mirror point $\mathbf{x}_2$

$|\mathbf{x}_0| > r$ and $|\mathbf{x}| = r$. Using (9.8), the approach for the Coulomb potential vanishing at infinity of the two point charges $q$ and $q'$ at a point $\mathbf{x}_0 \neq \mathbf{x}_1$ is

$$u(\mathbf{x}_0) = \frac{q}{\varepsilon_0} G(\mathbf{x}_1, \mathbf{x}_0) = \frac{q}{4\pi\varepsilon_0 |\mathbf{x}_0 - \mathbf{x}_1|} + \frac{q'}{4\pi\varepsilon_0 |\mathbf{x}_0 - \mathbf{x}_2|}.$$

The boundary condition is satisfied if $\mathbf{x}_2$ is the *mirror point* of $\mathbf{x}_1$ at $\partial K$, i.e., if $\mathbf{x}_2 = r^2 \mathbf{x}_1 / |\mathbf{x}_1|^2$, and if $q' = -rq/|\mathbf{x}_1|$ (cf. Fig. 9.16 left).

For arbitrary boundary points $\mathbf{x} = r\mathbf{n}_0$, $|\mathbf{n}_0| = 1$, it holds (see Fig. 9.16 right)

$$\frac{q'}{|r\mathbf{n}_0 - \mathbf{x}_2|} = -\frac{q}{\left|\, |\mathbf{x}_1|\mathbf{n}_0 - \frac{r}{|\mathbf{x}_1|}\mathbf{x}_1 \,\right|} = -\frac{q}{|r\mathbf{n}_0 - \mathbf{x}_1|} \quad \text{and thus } u(\mathbf{x}) = 0.$$

By substituting into the problem formulation, it is verified that the potential

$$u(\mathbf{x}) = \frac{q}{4\pi\varepsilon_0}\left( \frac{1}{|\mathbf{x} - \mathbf{x}_1|} - \frac{r}{|\mathbf{x}_1|\,\left|\mathbf{x} - \frac{r^2}{|\mathbf{x}_1|^2}\mathbf{x}_1\right|} \right)$$

is the unique solution to the given boundary value problem. Green's function $G(\mathbf{x}, \mathbf{y})$ is symmetric in $\mathbf{x}$ and $\mathbf{y}$ (calculation exercise with the cosine rule for triangles) and in this example reads in the complement of the sphere as

$$G(\mathbf{x}, \mathbf{y}) = \frac{1}{4\pi}\left( \frac{1}{|\mathbf{x} - \mathbf{y}|} - \frac{r}{|\mathbf{y}|\,\left|\mathbf{x} - \frac{r^2}{|\mathbf{y}|^2}\mathbf{y}\right|} \right).$$

The presented method can be translated to all domains for which Gauss's divergence theorem applies (see p. 499). In mathematical potential theory, it is shown under which—as minimal as possible—conditions on the data $\partial B$, $v$, and $f$ the approaches with the formulas (9.6) and (9.7) actually yield solutions to the given boundary value problems. For more complicated domains than spheres, however, it is generally difficult to determine Green's functions or to calculate the occurring convolution integrals. Therefore, other solution methods are also of great importance. Mentioned here are methods from the calculus of variations and functional analysis, or for problems in the plane, the methods of function theory. In the specified literature—one might study the textbooks by Courant and Hilbert (1993), Cassel (2013) et al.—one finds, in addition to the further theoretical treatment of boundary value problems, also concrete applications of the approaches dealt with here to questions of mathematical physics and electrical engineering.

For practical problems, numerical methods for approximate solutions are very important. The *Finite Element Method* has gained particular significance, the basic idea of which is to be explained in the next section against the background of distributional viewpoints. The following pages should therefore be understood as an invitation to learn these indispensable methods in engineering mathematics through further literature.

## 9.5   The Basic Idea of Finite Elements

The method of finite elements goes back to techniques used in 1908 by W. Ritz, 1915, by B. Galerkin, and 1943 by R. Courant for solving variational problems. It has been developed in the engineering disciplines since the 1950s with the use of electronic computing devices into a standard tool, for example, for solving elasticity problems of deformable bodies with complicated geometry or problems in fluid mechanics. Systematic presentations of this method can be found, for example, in Dautray and Lions (1992) or Braess (1992). A first insight will be elaborated using a boundary value problem for the Poisson equation as an example.

**Example (Equilibrium State of a Loaded Membrane)** Consider a bounded domain $\Omega$ in the plane with a piecewise linear boundary $\partial\Omega$, where an elastic membrane is fixed. Under the influence of an external force acting perpendicular to the plane, the membrane deflects (Fig. 9.17). The tension due to the fixing is isotropic, so it is described by a scalar quantity $k$ (with the dimension N/m). If $f$ denotes the surface density of the force, then for small displacements $u$ in the equilibrium state

$$-k\Delta u = f \text{ in } \Omega, \quad u = 0 \text{ on } \partial\Omega. \tag{9.9}$$

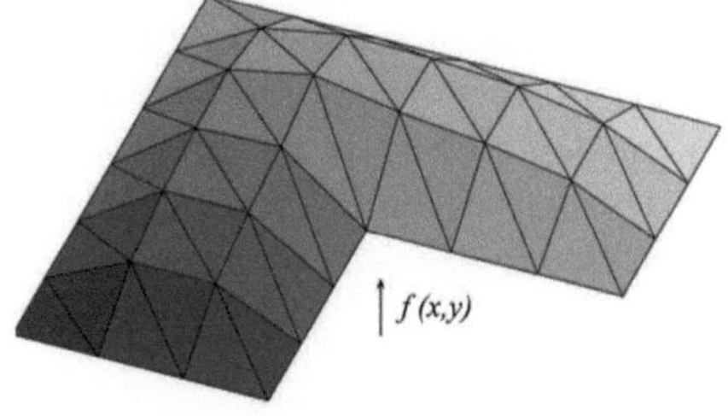


**Fig. 9.17** *L*-shaped membrane under external force

Here, $\Delta$ denotes the Laplace operator. Thus, the equilibrium position is the solution of a *Dirichlet boundary value problem* (see also the previous section). The following considerations can also be translated to electrical potential problems or stationary heat conduction problems. A derivation of the above fact from Hooke's law can be found in works such as Courant and Hilbert (1993) et al.

If the boundary $\partial\Omega$ has a complicated shape, it will not be possible to calculate a solution using the classical analytical methods discussed so far. However, a practical solution approach is opened up by distributional considerations.

The equation $-k\Delta u = f$ is interpreted as an equation between distributions, i.e., one seeks a distribution $u$, so that using Cartesian coordinates for all test functions $\varphi \in \mathcal{D}(\Omega)$ it holds

$$- k\langle\Delta u, \varphi\rangle = k \int_{\Omega} \operatorname{grad} u(x, y) \cdot \operatorname{grad} \varphi(x, y)\,\mathrm{d}x\,\mathrm{d}y = \langle f, \varphi\rangle. \tag{9.10}$$

The equality with the middle expression in (9.10) follows from the definition of the generalized derivatives of $u$ (cf. p. 188). Since one now has to solve a boundary value problem, one seeks a distribution solution $u$ that is *regular* and also allows for speaking about boundary values $u = 0$ of this distribution $u$ on $\partial\Omega$.

According to S. L. Sobolev (1908–1989), one seeks the solution $u$ among those functions $v$ that are square-integrable along with their partial generalized derivatives on $\Omega$ and vanish on the boundary $\partial\Omega$. The set of all such functions $v$ forms a *function vector space* $V$ over $\mathbb{R}$, which is denoted as $V = H_0^1$.

Even for complicated domains whose boundary has only minimal regularity properties, this vector space can be introduced in such a way that it is possible to meaningfully speak of boundary values of its elements. This is assumed for $\Omega$ and $V$ in the following.[1] It holds that $\mathcal{D}(\Omega) \subset V = H_0^1 \subset \mathcal{D}'(\Omega)$, and two functions in $V$ are identified if they differ only on a null set. The space $V$ is an example of a function vector space called a *Sobolev space*. More generally, *Sobolev spaces* are vector spaces of regular distributions of $\mathcal{D}'(\Omega)$ whose partial derivatives up to a certain order are also regular. Details about Sobolev spaces and their applications in partial differential equations can be found, for example, in Dautray and Lions (1992), Triebel (1986), or Atkinson and Han (2005). The needed properties of $V = H_0^1$ in the following are to be found in Appendix B.

A look at the second equation in (9.10) shows that the test functions $\varphi$ from $\mathcal{D}(\Omega)$ in (9.10) can be replaced by functions of $V$, because the product of two square-integrable functions is integrable again. The basis for statements on the solvability of problem (9.10) and also for the construction of numerical approximate solutions in $V$ using the finite element method is then the following reformulation of the problem:

---

[1] Mathematically precise: $\Omega$ is a bounded *Lipschitz domain* (see Appendix B, p. 502). Simplified: Consider the domain as the L-shaped figure or the figures on pp. 265 and 432.

*The force density f is square-integrable on $\Omega$, and V is the above-described Sobolev space. We seek a function $u \in V$, such that for all $v \in V$ the following holds:*

$$a(u, v) = l(v) \tag{9.11}$$

$$\textit{with} \quad a(u, v) = k \int_{\Omega} \operatorname{grad} u \cdot \operatorname{grad} v \, dx \, dy = k \int_{\Omega} \left( \frac{\partial u}{\partial x} \frac{\partial v}{\partial x} + \frac{\partial u}{\partial y} \frac{\partial v}{\partial y} \right) dx \, dy$$

$$l(v) = \langle f, v \rangle = \int_{\Omega} f(x, y) v(x, y) \, dx \, dy \, .$$

Due to the assumptions, $a(u, v)$ and $l(v)$ are well defined for all $u, v \in V$. The derivatives involved are to be understood as generalized derivatives. The boundary condition is included in the problem formulation by seeking the solution $u$ in the vector space $V$, whose elements are functions that vanish on the boundary $\partial \Omega$. The solution $u$ is—if it exists—to be understood as a distributional solution of (9.9) and is also called a *weak solution*.

**Potential Energy and Energy Functional of the Membrane**

The current task (9.11) is closely related to the physical consideration that the equilibrium state of the membrane adjusts so that the *total potential energy is minimal*.

Assuming a linear elastic material behavior according to Hooke's law, the deformation energy is proportional to the change in area. The total potential energy $E(v)$ of the membrane is then given for a displacement $v$ by

$$E(v) = k \left( \int_{\Omega} (1 + |\operatorname{grad} v(x, y)|^2)^{1/2} \, dx \, dy - \int_{\Omega} dx \, dy \right) - \int_{\Omega} f(x, y) v(x, y) \, dx \, dy \, .$$

For small displacements, one obtains with

$$(1 + |\operatorname{grad} v(x, y)|^2)^{1/2} - 1 \approx \frac{1}{2} |\operatorname{grad} v(x, y)|^2$$

the approximation

$$E(v) \approx J(v) = \frac{k}{2} \int_{\Omega} |\operatorname{grad} v(x, y)|^2 \, dx \, dy - \int_{\Omega} f(x, y) v(x, y) \, dx \, dy \, .$$

$J$ is called the *energy functional* of the membrane. If there is a function $u$ for which $J(u)$ is *minimal*, then $u$ approximately describes the equilibrium position of the membrane.

We note that $J(v)$ can be written in the form

$$J(v) = \frac{1}{2}a(v, v) - l(v)$$

with $a$ and $l$ as above in (9.11). The connection of the posed boundary value problem with the variational problem of minimizing the functional $J$ is established by the following version of a theorem by P. Lax and A. Milgram (see, for example, Dautray and Lions (1992)). The theorem shows that both problems have a common solution in the Sobolev space $V$.

**Theorem 9.12 (Theorem of Lax-Milgram)**

1. *For a function $u \in V$, the equation $a(u, v) = l(v)$ holds for all $v \in V$ if and only if $J(u) = \inf\{J(v)|v \in V\}$, i.e., if $u$ minimizes the energy functional $J$.*
2. *Under the given conditions in (9.11) the energy functional $J$ is bounded below on $V$, and there is a uniquely determined function $u \in V$ that minimizes $J$. This function $u$ is thus also the desired distributional solution of the boundary value problem (9.11).*

It is worth studying the justifications of this fundamental solvability statement more closely, because it teaches us that not only our exemplary problem (9.11), but also other problems of the same type can be solved in the same way. Many boundary value problems can be formulated such that one seeks a function $u$ in a function space $V$ adapted to the respective task, so that an equation of the form $a(u, v) = l(v)$ holds for all $v \in V$. The statements of the theorem then also apply to all such problems for which the essential properties of the vector space $V$ and the (problem-dependent) functionals $a$ and $l$ are satisfied. Two main steps in the following proof are indicated by *italics* and use arguments found in Appendix B.

*Proof*

1. The mapping $a : V \times V \to \mathbb{R}$ defines a symmetric bilinear form ($a(u, v) = a(v, u)$ for $u, v \in V$). The functional $l$ is linear on $V$. For $u, v \in V$ and $t \in \mathbb{R}$, it follows that

$$J(u + tv) = \frac{1}{2}a(u + tv, u + tv) - l(u + tv)$$

$$= J(u) + t(a(u, v) - l(v)) + \frac{1}{2}t^2 a(v, v).$$

If $u$ satisfies the equation $a(u, v) = l(v)$ for all $v \in V$, then we get with $t = 1$

$$J(u + v) = J(u) + \frac{1}{2}a(v, v) \geqslant J(u)$$

and equality holds only for $v = 0$. Thus, $u$ is the uniquely determined minimum point of $J$. The term $a(v, v)/2$ describes the increase in potential energy when the displacement $u$ is replaced by $u + v$.

Conversely, if $J$ has a minimum at $u$, then for every $v$ the derivative of the differentiable function $t \mapsto J(u + tv)$ at $t = 0$ must vanish. This derivative is precisely $a(u, v) - l(v)$. Then it follows that $a(u, v) = l(v)$ for every $v \in V$.

*Proof of* 2.   The bilinear form $a$ is positive definite on $V \times V$ ($a(v, v) > 0$ for $v \neq 0$). This follows from the *Poincaré-Friedrichs inequality* (Appendix B, p. 503): There exists a constant $c > 0$ such that for all $v \neq 0$ in $V$ we have

$$0 < \langle v, v \rangle = \int_{\Omega} v^2 \, dx \, dy \leqslant c \, a(v, v) = c \, k \int_{\Omega} |\operatorname{grad} v|^2 \, dx \, dy \,.$$

Therefore, $a$ defines an inner product and a norm so that the vector space $V$ is a Hilbert space. For each $v \in V$, $\|v\|_a = \sqrt{a(v, v)}$ is called the *energy norm* of $v$ according to its physical meaning (see also p. 54 and p. 62). The distance between two functions $v, w \in V$ is given in this norm by $\|v - w\|_a$. Now, $V$ is *the completion of* $\mathcal{D}(\Omega)$ *in the energy norm*, i.e., for each $v \in V$ there is a sequence $(\varphi_n)_{n \in \mathbb{N}}$ in $\mathcal{D}(\Omega)$ such that $\lim_{n \to \infty} \|v - \varphi_n\|_a = 0$, and in $V$ every Cauchy sequence $(v_n)_{n \in \mathbb{N}}$ with respect to this norm ($\|v_n - v_m\|_a \to 0$ for $n, m \to \infty$) converges to a function $v \in V$. If $u$ is a minimum point of $J$, then $l(v) = a(u, v)$ holds for all $v \in V$. According to the Cauchy-Schwarz inequality for inner products, the linear functional $l$ must then necessarily be continuous with respect to the energy norm: For each $v \in V$ we have

$$|l(v)|^2 = |a(u, v)|^2 \leqslant a(u, u) a(v, v) = \|u\|_a^2 \, \|v\|_a^2 \,,$$

in particular $\lim_{n \to \infty} l(v_n) = l(v)$ if $\lim_{n \to \infty} \|v_n - v\|_a = 0$ for $v_n, v \in V$. For a square-integrable force density $f$ in our example (9.11), the functional $l(v) = \langle f, v \rangle$ satisfies this necessary condition. This again follows from the *Poincaré-Friedrichs inequality* $\langle v, v \rangle \leqslant c \|v\|_a^2$ by the estimation

$$|l(v)|^2 \leqslant \langle f, f \rangle \langle v, v \rangle \leqslant c \, \langle f, f \rangle \|v\|_a^2.$$

Thus, $|l(v)| \leqslant C \|v\|_a$ with $C = \sqrt{c \, \langle f, f \rangle}$. This then gives

$$J(v) = \frac{1}{2} \|v\|_a^2 - l(v) \geqslant \frac{1}{2} (\|v\|_a - C)^2 - \frac{1}{2} C^2 \geqslant -\frac{1}{2} C^2$$

for all $v \in V$, i.e., the energy functional $J$ is bounded below on $V$. Therefore, there is now a minimizing sequence $(u_n)_{n \in \mathbb{N}}$ for $J$ in $V$,

$$\lim_{n \to \infty} J(u_n) = \inf\{J(v) \mid v \in V\}.$$

In all vector spaces where the norm of vectors is given by an inner product, the parallelogram law known from elementary geometry holds, so it follows for elements $u_n$ and $u_m$ of this minimizing sequence in $V$ that

$$2\|u_n\|_a^2 + 2\|u_m\|_a^2 = \|u_n - u_m\|_a^2 + \|u_n + u_m\|_a^2.$$

Subtracting $4l(u_n + u_m)$ from both sides, we get the inequality

$$4J(u_n) + 4J(u_m) = 8J\left(\frac{u_n + u_m}{2}\right) + \|u_n - u_m\|_a^2 \geqslant 8 \inf_{v \in V} J(v) + \|u_n - u_m\|_a^2.$$

Since $J(u_n) \to \inf_{v \in V} J(v)$ and $J(u_m) \to \inf_{v \in V} J(v)$ for $n, m \to \infty$, we obtain

$$\|u_n - u_m\|_a^2 \to 0 \text{ for } n, m \to \infty.$$

Thus, the minimizing sequence is a Cauchy sequence with respect to the energy norm. Therefore, from the *completeness* of $V$ and the continuity of $J$, again with respect to the energy norm, it follows that there is a function $u = \lim_{n \to \infty} u_n$ in $V$ for which $J(u) = \inf\{J(v) \mid v \in V\}$ holds. The uniqueness of the minimum point of $J$ has already been shown in the first part of the proof.                     $\square$

**Summary** From the proof, we learn that *any* problem of the form $a(u, v) = l(v)$ in a function space $V$ has a solution, if $a$ defines an inner product on $V$, such that firstly $V$ is complete with the energy norm associated with $a$, and secondly the functional $l$ is continuous in this norm. For many practical boundary value problems, the Sobolev concept of constructing a suitable vector space $V$ is therefore the starting point first for a theoretical solvability statement and then for the development of methods to determine approximate solutions.

**Example (Stationary Temperature Distribution in a Rod)**   We formulate as another example, for which the readers can later calculate an approximate solution using finite elements themselves, the following one-dimensional boundary value problem, which describes the stationary temperature distribution $T$ in a rod of length $L$ with piecewise constant thermal diffusivity $k$ and prescribed temperature values at the rod ends:

$$\left(k(x)T'(x)\right)' = 0, \quad T(0) = T_0, \quad T(L) = T_1 \tag{9.12}$$

$$k(x) = k_1 > 0 \text{ for } 0 \leqslant x \leqslant \frac{L}{3}, \quad k(x) = k_2 > 0 \text{ for } \frac{L}{3} < x \leqslant L.$$

First, by substituting $u(x) = T(x) - T_0 + \frac{x}{L}(T_0 - T_1)$, the problem is transformed into a differential equation for $u$ with homogeneous boundary values:

$$\left(k(x)\left(u'(x) - \frac{1}{L}(T_0 - T_1)\right)\right)' = 0, \quad u(0) = u(L) = 0, \quad k(x) \text{ as above.}$$

$$\tag{9.13}$$

For this task, $V$ is chosen as the vector space of all absolutely continuous functions $v$ on $[0, L]$, which have a square-integrable derivative $v'$ and satisfy $v(0) = v(L) = 0$. From the absolute continuity, it follows that all $v$ are continuous and differentiable outside of a null set $N \subset [0, L]$, and that for them the fundamental theorem of calculus $\int_0^x v'(x)\,\mathrm{d}x = v(x)$ holds. From the problem statement

$$\int_0^L k(x)(u'(x) - \frac{1}{L}(T_0 - T_1))v'(x)\,\mathrm{d}x = 0 \text{ for all } v \in V,$$

one obtains after a short calculation the following formulation of the problem (9.13), which is completely analogous to (9.11):

  *A function $u \in V$ is sought such that for all $v \in V$*

$$a(u, v) = l(v) \text{ with } a(u, v) = \int_0^L k(x)u'(x)v'(x)\,\mathrm{d}x \tag{9.14}$$

$$l(v) = \frac{T_0 - T_1}{L}\int_0^L k(x)v'(x)\,\mathrm{d}x = \frac{T_0 - T_1}{L}(k_1 - k_2)\,v\!\left(\frac{L}{3}\right).$$

The *Poincaré-Friedrichs inequality* $\langle v, v\rangle \leqslant ca(v, v)$ with $c > 0$ holds analogously for this one-dimensional problem. As in the previous example, $a$ thus defines an inner product on $V$. It can be shown that $V$ is *complete with respect to the energy norm* $\|v\|_a = \sqrt{a(v, v)}$. With the Cauchy-Schwarz inequality, one sees that $l$ is continuous on $V$ with respect to this norm: Writing $|l(v)|$ for $v \in V$ in the form

$$|l(v)| = \left| \frac{T_0 - T_1}{L} \int_0^L \sqrt{k(x)}\sqrt{k(x)}v'(x)\,\mathrm{d}x \right|,$$

$$\text{then } |l(v)| \leqslant \left| \frac{T_0 - T_1}{L} \right| \left( \int_0^L k(x)\,\mathrm{d}x \right)^{1/2} \left( \int_0^L k(x)v'(x)^2\,\mathrm{d}x \right)^{1/2},$$

$$|l(v)| \leqslant C\sqrt{a(v, v)} = C\|v\|_a \text{ with } C = \left| \frac{T_0 - T_1}{L} \right| \left( \int_0^L k(x)\,\mathrm{d}x \right)^{1/2}.$$

Thus, the proof of the solvability theorem can be adopted verbatim, and it follows that the problem (9.14) has exactly one solution $u$ in $V$. From $u$,

$$T(x) = u(x) + T_0 - \frac{x}{L}(T_0 - T_1)$$

immediately follows as the solution of the given heat conduction problem (9.12). Based on this guaranteed solvability statement, it now makes sense to consider algorithms for constructing approximations for $u$ or $T$.

## *The Ritz-Galerkin Method*

With the work done so far in this section, we have learned how to formulate boundary value problems according to (9.11) and that their (distributional) solution is to be sought in a vector space $V$ that has the inner product $a(u, v)$ for $u, v \in V$ and the energy norm $\|u\|_a$. This now makes it easy to describe the basics of approximation methods according to Ritz and Galerkin and later as a special case the finite element principle.

In all vector spaces $V$ where a norm $\|f\|$ of elements $f \in V$ is given by an inner product, one obtains an approximation in a sub-vector space $U$ of $V$ by orthogonal projection of $f$ onto $U$ (see also p. 12, Sects. 5.1 and 14.1, p. 449). The concept of orthogonality is directly related to the inner product: $f, g$ from $V$ are orthogonal if and only if their inner product is zero. The orthogonal projection $f_U$ of $f$ onto $U$ is an optimal approximation for $f \in V$ by an element $g \in U$ in the following sense:

$$\|f - f_U\| = \langle f - f_U, f - f_U \rangle^{1/2} = \inf_{g \in U} \|f - g\|,$$

i.e., the norm of the error $f - g$ is minimal among all $g \in U$ for $g = f_U$.

The exemplary problem (9.11) discussed on p. 245 now has a (unknown) solution $u$ in the infinitely dimensional function vector space $V$ described there. In this vector space $V$, the *bilinear form $a(u, v)$* belonging to the problem defines an inner product and the norm $\|u\|_a$ for all $u, v \in V$. According to Ritz-Galerkin, one constructs a *finite-dimensional sub-vector space $V_N$* of $V$ and calculates the *orthogonal projection $u_N$ of $u$ onto $V_N$ with the inner product given by $a$* as an approximation for the sought solution of the posed boundary value problem. Even if the function $u$ remains unknown, its orthogonal projection $u_N$ can be determined from the specification of $V_N$ and from the equation $a(u, v) = l(v)$ valid for every $v \in V$. $u_N$ is called the *Ritz-Galerkin solution* belonging to $V_N$. The choice of $V_N$ and hence how well a function $u \in V$ can be approximated by functions from $V_N$ is crucial for the error $\|u - u_N\|_a$ of the approximation.

To achieve satisfactory numerical results, the specification of the subspace $V_N$ and its approximation properties is the key to the construction of approximate solutions.

## *The Linear System of Equations for a Ritz-Galerkin Solution*

By specifying $N$ linearly independent functions $v_1, v_2, \ldots, v_N$ in $V$, a *basis* of an $N$-dimensional subspace $V_N$ of $V$ is determined. The vector space $V_N$ is the set of all linear combinations of the $v_k$, $1 \leqslant k \leqslant N$; thus the Ritz-Galerkin solution $u_N$ in $V_N$ has a representation of the form $u_N = \sum_{k=1}^{N} u_{N,k} v_k$ with uniquely determined real coefficients $u_{N,k}$. The orthogonality relations $a(u - u_N, v_i) = 0$ and the equations $a(u, v_i) = l(v_i)$ yield

$$a(u_N, v_i) = l(v_i) \text{ for } 1 \leqslant i \leqslant N.$$

With the linearity of $a$ and the above representation of $u_N$, one obtains the linear system of equations

$$\sum_{k=1}^{N} u_{N,k} a(v_k, v_i) = l(v_i)$$

for the sought coefficients $u_{N,1}, \ldots, u_{N,N}$. In matrix form, with column vectors $\mathbf{u}$ and $\mathbf{l}$, the task is thus as follows:

**Task**   *Determine* $\mathbf{u} \in \mathbb{R}^N$, *so that* $\mathbf{A}\mathbf{u} = \mathbf{l}$ *is satisfied for*

$$A = (\alpha_{i,k})_{\substack{1 \leqslant i \leqslant N \\ 1 \leqslant k \leqslant N}}, \qquad \alpha_{i,k} = a(v_k, v_i),$$

$$\mathbf{l} = (l_i)_{1 \leqslant i \leqslant N}, \qquad\quad l_i = l(v_i).$$

The quantities $\alpha_{i,k}$ and $l_i$ can be calculated from the given functionals $a$ and $l$ and the chosen basis functions $v_i$. The matrix $A$ is symmetric and positive definite, particularly regular. For $\mathbf{x} = (x_1, \ldots, x_N) \neq \mathbf{0}$ from $\mathbb{R}^N$, due to the *Poincaré-Friedrichs inequality* for our example (9.11), i.e., because $a$ is positive definite:

$$A\mathbf{x} \cdot \mathbf{x} = \sum_{i,k=1}^{N} x_i \alpha_{i,k} x_k = a\left( \sum_{k=1}^{N} x_k v_k, \sum_{i=1}^{N} x_i v_i \right) > 0.$$

In elasticity problems, $A$ is called the *stiffness matrix*. The uniquely determined solution $\mathbf{u} = (u_{N,1}, \ldots, u_{N,N})$ of the system of equations yields the desired approximate solution $u_N = \sum_{k=1}^{N} u_{N,k} v_k$ of the original problem $a(u, v) = l(v)$ for elements $v \in V$.

**Example (A Ritz-Galerkin Solution for a Loaded Membrane)** As an application, we calculate a Ritz-Galerkin solution for the boundary value problem (9.11).

Here, the domain $\Omega = ]0, L[\times]0, L[$ is the square with side length $L$, and $f(x, y) = F$ is a force density constant over $\Omega$. To calculate an approximate solution, we choose the following four linearly independent functions $v_1, \ldots, v_4$:

$$v_1(x, y) = L \sin\left(\tfrac{\pi}{L}x\right) \sin\left(\tfrac{\pi}{L}y\right), \qquad v_2(x, y) = L \sin\left(\tfrac{3\pi}{L}x\right) \sin\left(\tfrac{\pi}{L}y\right),$$
$$v_3(x, y) = L \sin\left(\tfrac{\pi}{L}x\right) \sin\left(\tfrac{3\pi}{L}y\right), \qquad v_4(x, y) = L \sin\left(\tfrac{3\pi}{L}x\right) \sin\left(\tfrac{3\pi}{L}y\right).$$

They vanish on the boundary $\partial\Omega$ of $\Omega$, lie in the Sobolev space $V$ of the problem (9.11), and form a basis of the subspace $V_4 \subset V$ ($N = 4$) generated by them. For the elements $\alpha_{i,k}$ of the stiffness matrix $A$, the following holds (exercise)

$$\alpha_{1,1} = a(v_1, v_1) = k \int_0^L \int_0^L |\operatorname{grad} v_1(x, y)|^2 \, dx \, dy$$

$$= k\pi^2 \int_0^L \int_0^L \left( \cos^2\left(\frac{\pi}{L}x\right) \sin^2\left(\frac{\pi}{L}y\right) + \sin^2\left(\frac{\pi}{L}x\right) \cos^2\left(\frac{\pi}{L}y\right) \right) dx \, dy = \frac{kL^2\pi^2}{2},$$

$$\alpha_{2,2} = \alpha_{3,3} = \frac{5kL^2\pi^2}{2}, \qquad \alpha_{4,4} = \frac{9kL^2\pi^2}{2}.$$

All off-diagonal elements $\alpha_{i,k}$, $i \neq k$, of $A$ are zero because of the orthogonality relations for the trigonometric functions (cf. p. 12). The coefficients $l_i$ of the vector $\mathbf{l}$ on the right-hand side are calculated as

$$l_1 = F \int_0^L \int_0^L v_1(x, y) \, dx \, dy = \frac{4FL^3}{\pi^2}, \qquad l_2 = l_3 = \frac{4FL^3}{3\pi^2}, \qquad l_4 = \frac{4FL^3}{9\pi^2}.$$

As the solution $\mathbf{u} = (u_{4,1}, u_{4,2}, u_{4,3}, u_{4,4})$ of $A\mathbf{u} = \mathbf{l}$ we get

$$u_{4,1} = \frac{8FL}{k\pi^4}, \qquad u_{4,2} = u_{4,3} = \frac{8FL}{15k\pi^4}, \qquad u_{4,4} = \frac{8FL}{81k\pi^4}.$$

*The Ritz-Galerkin solution in $V_4$ for the deflection of the membrane under the given conditions is thus*

$$u_4(x, y) = \frac{8FL^2}{k\pi^4}\left[ \sin\left(\frac{\pi}{L}x\right) \sin\left(\frac{\pi}{L}y\right) + \frac{1}{15} \sin\left(\frac{3\pi}{L}x\right) \sin\left(\frac{\pi}{L}y\right) \right.$$

$$\left. + \frac{1}{15} \sin\left(\frac{\pi}{L}x\right) \sin\left(\frac{3\pi}{L}y\right) + \frac{1}{81} \sin\left(\frac{3\pi}{L}x\right) \sin\left(\frac{3\pi}{L}y\right) \right].$$

For $L = 1\,\mathrm{m}$, $F = 1\,\mathrm{N/m}^2$, $k = 2\,\mathrm{N/m}$ the deflection at the point $x = y = L/2$ is approximately $u_4\,(L/2, L/2) = 2848FL^2/(405k\pi^4) \approx 0.0361\,\mathrm{m}$.

**Remark** Since the domain $\Omega$ in the example is a rectangle, one can obtain a Fourier series representation for the solution $u$ through a separation approach. Our approximate solution is, due to the choice of basis functions $v_1, \ldots, v_4$, exactly the Fourier expansion of the solution to Exercise A7 in Chap. 7 (cf. also Sect. 7.4, p. 143). The example thus makes it clear that Ritz-Galerkin solutions and consequently the approximate solutions with finite elements are *generalizations of Fourier series expansions*. Orthogonal projection of the solution onto a subspace $V_N$, generated by trigonometric functions, allows for a representation of the approximate solution as a trigonometric polynomial, and other basis functions of $V_N$ yield correspondingly different expansion coefficients $u_{N,1}, \ldots, u_{N,N}$ of the approximation $u_N$.

## *Finite Elements*

We continue to study our membrane problem (9.11) for explanation. When choosing a basis arbitrarily for an $N$-dimensional subspace $V_N$ of $V$, the stiffness matrix $A$ is generally fully populated, and $N^2$ integrations are required to determine its elements exactly or approximately. The final solution of the equation system $A\mathbf{u} = \mathbf{l}$ would require a number of computational operations that grow with $N$ as $N^3$. If one wants good approximations $u_N$ for the solution $u$ of the given problem in a high-dimensional subspace $V_N$, the general Ritz-Galerkin method proves to be impractical.

The principal idea of the finite element method is now to choose the basis functions $v_1, \ldots, v_N$ so that as many of the $v_k$ as possible have disjoint supports. If $\mathrm{supp}(v_i) \cap \mathrm{supp}(v_k)$ is empty for $i \neq k$ or if the supports only meet parts of their boundaries, then for problems like those in our examples, $a(v_i, v_k) = 0$. If this is true for many $i$ and $k$, then the stiffness matrix is sparse, i.e., it has zeros in many places. If the $v_k$ are all of the same type, one can also use existing symmetry properties and thus save considerable computational effort.

### Triangulation of the Domain, Choice of Basis Functions, Linear Elements

To implement the described idea, the domain $\Omega$ is divided into small pieces. To illustrate, we consider a domain $\Omega$ in the plane, bounded by axis-parallel lines (with respect to a Cartesian coordinate system). By subdividing with a rectangular grid and then subdividing each rectangle into two triangles, a *triangulation $\mathcal{T}$ of $\Omega$* is created as shown in the following Fig. 9.18.

The vertices of a triangle lying within $\Omega$ are called the *internal nodes* of the triangulation $\mathcal{T}$. We denote by $N$ the number of internal nodes. Consider them

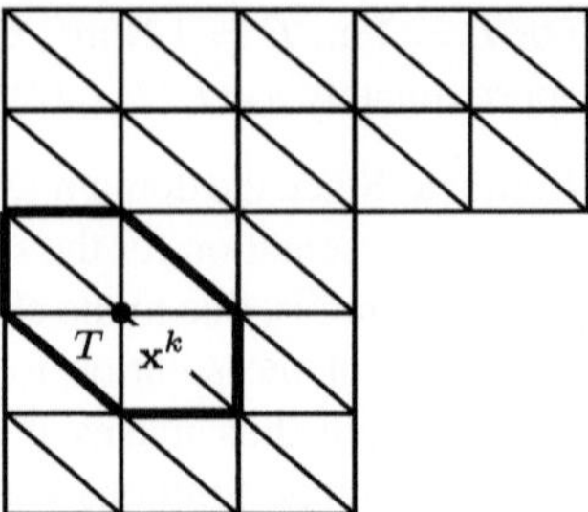

**Fig. 9.18**  A triangulation of
the $L$-shaped membrane

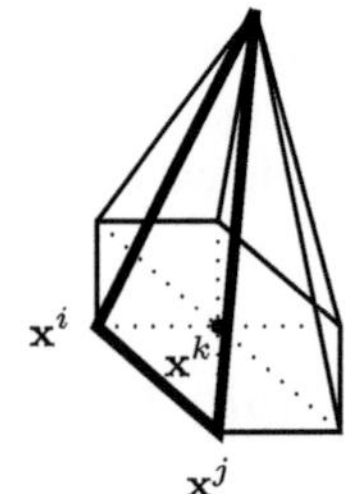

**Fig. 9.19**  A hat function $v_k$
over the triangulation

numbered in a sequence $\mathbf{x}^1, \ldots, \mathbf{x}^N$. A polygon $P_k$ is formed by the triangles that
have an internal node $\mathbf{x}^k$ as a common vertex (e.g., the hexagon with thick borders
in the picture above left). For each internal node $\mathbf{x}^k$, $1 \leqslant k \leqslant N$, we define a
continuous function $v_k \in V$ with $v_k(\mathbf{x}^k) \neq 0$, which has precisely the closed
polygon $P_k$ as its support. A particularly simple example is the choice of continuous
functions $v_k$ such that all $v_k$ are affine-linear on each triangle $T$ of the triangulation,
i.e., of the form $v_k(x, y) = a + bx + cy$, with

$$v_k(\mathbf{x}^k) = H > 0 \text{ and } v_i(\mathbf{x}^k) = 0 \text{ for all } i \neq k \quad (1 \leqslant i, k \leqslant N).$$

The graph of each function $v_k$, $1 \leqslant k \leqslant N$, then looks like a *tent roof* over the
corresponding polygon $P_k$, supported by a stake of height $H$ at the point $\mathbf{x}^k$ and
attached to the ground at the neighboring nodes (Fig. 9.19). If $T$ is a triangle of
the triangulation with vertices $\mathbf{x}^k = (x_k, y_k)$, $\mathbf{x}^i = (x_i, y_i)$, and $\mathbf{x}^j = (x_j, y_j)$, it
is quickly calculated that $v_k$ on $T$ is given by the following formula (compare the
thick-bordered "tent surface" in the picture above right):

$$v_k(x, y) = H \frac{(x - x_i)(y_j - y_i) - (y - y_i)(x_j - x_i)}{(x_k - x_i)(y_j - y_i) - (y_k - y_i)(x_j - x_i)} \quad \text{for } (x, y) \in T.$$

$$(9.15)$$

The functions $v_1, \ldots, v_N$ are linearly independent: If one successively sets $\mathbf{x} =
\mathbf{x}^1, \ldots, \mathbf{x}^N$ in the equation $\sum_{k=1}^{N} \alpha_k v_k(\mathbf{x}) = 0$, it follows that $H\alpha_1 = 0, \ldots, H\alpha_N = 0$.
Therefore, the functions $v_1, \ldots, v_N$ span an $N$-dimensional subspace $V_N$ of $V$.

Once the choice of basis functions is made and their form over a triangle $T$ of the triangulation is established, $T$ is referred to as a *finite element*. In our example, due to the linearity of the basis functions $v_k$ on each triangle $T$, we speak of *linear triangular elements*. With the specification of the functions $v_1, \ldots, v_N$, the stiffness matrix $A$ and the right-hand side $\mathbf{l}$ of the equation system $A\mathbf{u} = \mathbf{l}$ are determined.

The approximate solution $u_N = \displaystyle\sum_{k=1}^{N} u_{N,k} v_k$ is uniquely determined by the components $u_{N,k}$ of $\mathbf{u} = A^{-1}\mathbf{l}$.

The described method of triangulation of $\Omega$ can be used for all domains *Omega* in the plane that are bounded by a polygon. A (not necessarily regular) decomposition of $\Omega$ into triangles is then a *permissible triangulation* if any two different triangles $T_i$ and $T_k$ that meet boundaries either have only one common vertex and no other boundary points in common or share a common side with two common vertices.

For example,  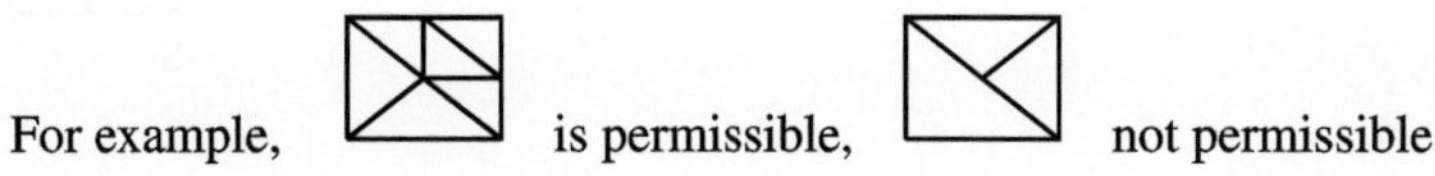 is permissible,  not permissible.

Piecewise linear basis functions $v_k$ can then be defined exactly as above, and each function $u$ from the generated vector space $V_N$ is uniquely determined by its values on the internal nodes.

For spatial problems with polyhedral domains $\Omega$ in $\mathbb{R}^3$, the principle described here can also be used with corresponding admissibility conditions for the decomposition. For example, tetrahedra, plates, or prisms are used as finite elements, and suitable polynomials or wavelets (see Sect. 14.2) are used as basis functions. We will calculate an approximation for the deflection of our loaded square membrane as a concrete, simple example using finite elements.

**Approximate Solution with Finite Elements for a Loaded Membrane**

As already mentioned on p. 251, let $\Omega$ be the square with side length $L$ and $f(x, y) = F$ a force density constant on $\Omega$. For the approximate solution of the problem (9.11) using finite elements, we first divide $\Omega$ into $(p + 1)^2$, $p \in \mathbb{N}$, squares and then each resulting square into two triangles as shown in the picture. With $p = 3$, this decomposition has $N = p^2 = 9$ internal nodes. Each triangle has an area of $h^2/2$ with $h = L/(p+1)$. As basis functions we use the piecewise linear functions $v_k$ described in (9.15), $1 \leqslant k \leqslant N$ (Illustration Figs. 9.20 and 9.21).

**Setting up the Linear System of Equations**

We now denote a selected inner node by $Z$, its neighboring nodes according to the cardinal directions by $S, SE, \ldots, W$ and the triangular areas between these nodes by Roman numerals $I, \ldots, VI$ as in the picture above. If all nodes $S, SE, \ldots, W$

**Fig. 9.20** Mesh with
$h = L/(p + 1)$

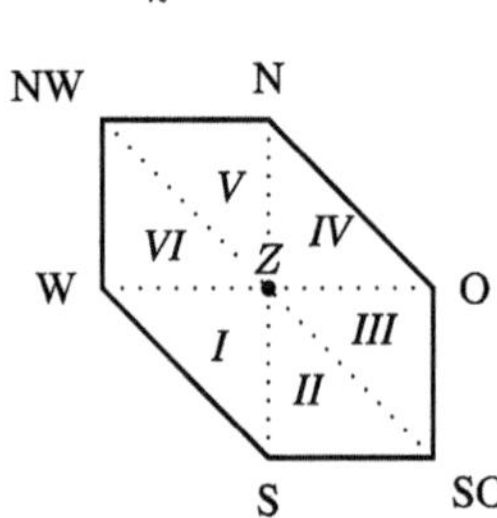

**Fig. 9.21** Six permissible
triangles with numbering and
orientation

are inner nodes, then the quantities $a(v_Z, v_Z), a(v_Z, v_S), \ldots, a(v_Z, v_W)$ for the corresponding basis functions $v_Z, v_S, \ldots, v_W$ are to be calculated. If some of the nodes $S, SE, \ldots W$ lie on the boundary $\partial\Omega$, then the respective basis functions $v_S, v_{SE} \ldots v_W$ and accordingly the respective quantities $a(v_Z, v_S), a(v_Z, v_{SE})$ or $a(v_Z, v_W)$ are omitted.

From the graph of $v_Z$ or using formula (9.15), the partial derivatives of $v_Z$ on the triangles $I, \ldots, VI$ can be immediately seen. Since all basis functions have the same shape, the derivatives of $v_S, \ldots, v_W$ on these triangles follow from this. Summarized in a table, we get

|  | $I$ | $II$ | $III$ | $IV$ | $V$ | $VI$ |
|---|---|---|---|---|---|---|
| $\dfrac{\partial v_Z}{\partial x}$ | $\dfrac{H}{h}$ | $0$ | $-\dfrac{H}{h}$ | $-\dfrac{H}{h}$ | $0$ | $\dfrac{H}{h}$ |
| $\dfrac{\partial v_Z}{\partial y}$ | $\dfrac{H}{h}$ | $\dfrac{H}{h}$ | $0$ | $-\dfrac{H}{h}$ | $-\dfrac{H}{h}$ | $0.$ |

With the notation $I \ldots VI$ for the union of the areas $I$ to $VI$, one gets

$$a(v_Z, v_Z) = k \int_{I\ldots VI} |\operatorname{grad} v_Z|^2 \, dx \, dy = k \int_{I\ldots VI} \left(\frac{\partial v_Z}{\partial x}\right)^2 + \left(\frac{\partial v_Z}{\partial y}\right)^2 dx \, dy$$

$$= k\frac{H^2}{h^2}\left(2\frac{h^2}{2} + \frac{h^2}{2} + \frac{h^2}{2} + 2\frac{h^2}{2} + \frac{h^2}{2} + \frac{h^2}{2}\right) = 4kH^2,$$

$$a(v_S, v_Z) = k \int_{I\cup II} \left(\frac{\partial v_Z}{\partial x}\frac{\partial v_S}{\partial x} + \frac{\partial v_Z}{\partial y}\frac{\partial v_S}{\partial y}\right) dx \, dy = k\frac{H^2}{h^2}\left(-\frac{h^2}{2} - \frac{h^2}{2}\right) = -kH^2.$$

**Fig. 9.22** Numbering of nine inner nodes for the example with $p = 3$

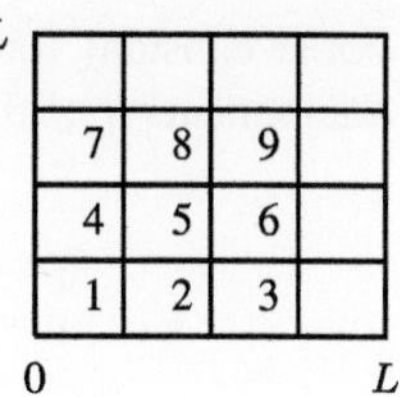

Similarly, one finds

$$a(v_N, v_Z) = a(v_W, v_Z) = a(v_E, v_Z) = -kH^2, \; a(v_{SE}, v_Z) = a(v_{NW}, v_Z) = 0.$$

By exploiting the existing symmetries, the six calculated integrals are sufficient to set up the stiffness matrix $A$. For this purpose, we number the inner nodes as in the following picture for our example with $p = 3$ (Fig. 9.22):
Each inner node now successively takes on the role of $Z$, and its inner neighboring nodes according to the cardinal directions of their position relative to $Z$ take on the roles of the nodes $S$, $SE$, etc. Starting with the first node, it has the inner neighboring nodes numbered 2 and 4. The first row of $A$ is then populated at the positions $\alpha_{1,1}$, $\alpha_{1,2}$ and $\alpha_{1,4}$. All other elements of the first row are zero. Proceeding accordingly, the matrix $A$ is given by

$$
A = kH^2
\left[
\begin{array}{ccc|ccc|ccc}
4 & -1 &    & -1 &    &    &    &    &    \\
-1 & 4 & -1 &    & -1 &    &    &    &    \\
   & -1 & 4 &    &    & -1 &    &    &    \\ \hline
-1 &    &    & 4 & -1 &    & -1 &    &    \\
   & -1 &    & -1 & 4 & -1 &    & -1 &    \\
   &    & -1 &    & -1 & 4 &    &    & -1 \\ \hline
   &    &    & -1 &    &    & 4 & -1 &    \\
   &    &    &    & -1 &    & -1 & 4 & -1 \\
   &    &    &    &    & -1 &    & -1 & 4
\end{array}
\right].
$$

The empty spaces are filled with zeros. The matrix is sparse with $5N - 4\sqrt{N}$ (here with 33) nonzero coefficients. For a square grid decomposition with a large number $N = p^2$ of inner nodes, one obtains the coefficients $\alpha_{i,k}$ of $A$ by the corresponding numbering analogously

$$
\begin{aligned}
\alpha_{i,i} \;\; &= 4kH^2 \;\; (1 \leqslant i \leqslant N) \\
\alpha_{i,i+1} &= -kH^2 \;\; (1 \leqslant i \leqslant N - 1, \; i \bmod p \neq 0, \\
&\qquad\qquad\quad \text{that is, when } i \text{ is not divisible by } p) \\
\alpha_{i,i-1} &= -kH^2 \;\; (2 \leqslant i \leqslant N, \; (i-1) \bmod p \neq 0) \\
\alpha_{i,i+p} &= -kH^2 \;\; (1 \leqslant i \leqslant N - p) \\
\alpha_{i,i-p} &= -kH^2 \;\; (p + 1 \leqslant i \leqslant N) \\
\alpha_{i,k} \;\; &= 0 \qquad\quad \text{otherwise.}
\end{aligned}
$$

For a constant force density $f(x, y) = F$ on $\Omega$, for reasons of symmetry, all components $l_i$ of the right-hand side of $A\mathbf{u} = \mathbf{l}$ are given by

$$l_i = F \int_{I-VI} v_Z(x, y)\, \mathrm{d}x\, \mathrm{d}y = F H h^2 \text{ for } 1 \leqslant i \leqslant N,$$

if $Z$ is again an inner node as in the picture on p. 256, because above the hexagon around $Z$, the graph of $v_Z$ forms a pyramid with a volume of $H h^2$.

**Graphical Representation of an Approximate Solution**

The approximate solution $u_N$ for the deflection of the membrane calculated with the discussed finite elements for $p = 7$, $N = 49$, $L = 1\,\text{m}$, $k = 2\,\text{N/m}$, $F = 5\,\text{N/m}^2$ is shown in Fig. 9.23. One can clearly see that the approximation is not differentiable at the edges of the triangulation, and thus it can only be meaningfully interpreted as a distributional approximate solution for the original boundary value problem (9.9). Active readers can enjoy the pleasure of carrying out the calculation and graphics themselves on a rainy weekend and as an exercise, similarly treat an L-shaped membrane as on p. 243 and the heat conduction problem from p. 248 with finite elements (cf. Exercise A10). The maximum deflection of the square membrane calculated here is approximately $0.182\,\text{m}$.

Similar approaches to those in the treated examples can also be developed for other differential equations with different types of boundary conditions and for the 3D case as well. A 3D example is given later in Chap. 12 on p. 432. At the end of this section, which aims to provide a first insight into the importance of distribution methods for the theoretical and practical solution of boundary value problems, important tasks of numerical mathematics in the use of the described methods will

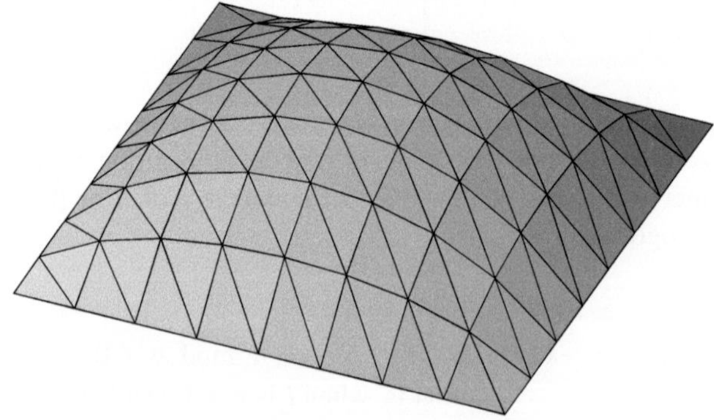

**Fig. 9.23** Graphical result of the FEM example

be briefly mentioned: Practical tasks include, besides the mathematical modeling of given problems, for example, the development of triangulation routines even for areas or bodies with complex boundaries, the selection of suitable numerical integration methods for the calculation of the stiffness matrix, the choice of suitable equation solvers for the resulting large linear systems of equations, a useful graphics-related post-processing to obtain meaningful results from "data cemeteries",[2] and much more.

Generally, one expects that with refinements of the triangulation, the approximate solutions $u_N$ converge in $V$ to the solution $u$ with increasing $N$. Only with convergence studies and error estimates for the approximations can one obtain efficient and useful numerical results, stopping criteria for computational programs that iteratively calculate approximations by refining the triangulations, reliable criteria for the quality of the calculated approximate solutions. The framework for this is provided by approximation and distribution theory. Considering that the safety of nuclear power plants, chemical plants, electronic systems in aircraft, etc., depends on the quality of numerically calculated approximate solutions— for example, for deformations and loads on components under mechanical and thermal influences—it becomes clear that even with exactly solved systems of equations, reliable estimates for the error $u_N - u$ are necessary to protect against high risks. Modern technology requires a high degree of equally modern mathematics. Engineers and scientists should not hesitate to seek collaboration with competent mathematicians. For those interested in deepening their understanding and in error estimates for the numerical methods sketched here, reference is made once again to the literature given at the beginning of the section on p. 244. As a newer source for error estimates, the work of Dahlke et al. (2010) is particularly recommended.

## 9.6   Distributional Solution of the 1D Wave Equation

We briefly reconsider the initial boundary value problem for the force-free vibrating string as an example. A twice continuously differentiable function $u(x, t)$ was sought, such that

---

[2] In the 1970s with the NASTRAN FEM Software one had to create the mesh for the structure with several thousands of punch cards, and the computation results were simply long listings with again thousands of numerical values, which had to be analyzed by looking at them. There existed no graphics output at that time. This is of course much better now. NASTRAN (Nasa Structural Analysis System) is still a commercial FEM standard. With MYSTRAN a largely compatible FEM software is available free of charge.

**Fig. 9.24** Non-differentiable initial conditions

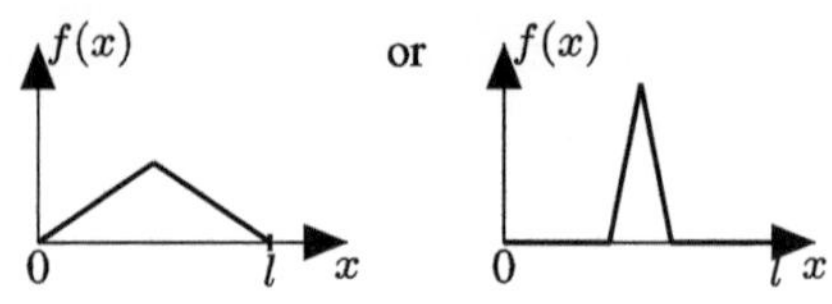

$$\frac{\partial^2 u}{\partial t^2} - c^2 \frac{\partial^2 u}{\partial x^2} = 0 \qquad \text{for } 0 < x < l \,,\ t > 0 \,,$$

$$u(0,t) = u(l,t) = 0 \qquad \text{for } t \geqslant 0 \,,$$

$$u(x,0) = f(x) \quad \text{for } 0 \leqslant x \leqslant l \,,\ f(0) = f(l) = 0 \,,$$

$$\lim_{t \to 0+} \frac{\partial u}{\partial t}(x,t) = g(x) \quad \text{for } 0 \leqslant x \leqslant l \,,\ g(0) = g(l) = 0.$$

For solvability within the framework of classical theory, smoothness conditions on $f$ and $g$ were necessary. Now, if one wants to treat initial conditions $f$ for a guitar string of the following forms as in Fig. 9.24, the problem is considered as a differential equation for distributions under the following additional assumptions on the initial conditions $f$ and $g$:

$f : [0, l] \to \mathbb{R}$ *should be continuous and piecewise continuously differentiable with* $f(0) = f(l) = 0$, *and* $g : [0, l] \to \mathbb{R}$ *should be piecewise continuously differentiable,* $g(0) = g(l) = 0$ . The odd, $2l$-periodic extensions of $f$ and $g$ are again called $f$ and $g$, respectively.

The Fourier series of $f$ converges uniformly, and in the quadratic mean to $f$, the Fourier series of $g$ converges in the quadratic mean to $g$ and pointwise except at possible points of discontinuity of $g$. The calculated series solution $u(x,t)$ from Sect. 5.4, p. 72, is interpreted as a **distributional solution with the following properties**:

1. *The Fourier series represents the following distribution for* $0 < x < l, t > 0$:

$$u(x,t) = \frac{1}{2}[f(x+ct) + f(x-ct)] + \frac{1}{2c} \int_{x-ct}^{x+ct} g(\tau)\, d\tau \,.$$

*This distribution is regular,* $u(x,t)$ *is continuously extendable on* $[0, l] \times \mathbb{R}_0^+$, *but not necessarily differentiable in* $]0, l[\times\mathbb{R}^+$. *Its continuous extension on* $[0, l] \times \mathbb{R}_0^+$ *is again denoted by* $u(x,t)$.

2. $u(x,t)$ *solves the wave equation for* $0 < x < l, t > 0$ *in the distributional sense, i.e., for every test function* $\varphi(x,t)$ *with support in* $]0, l[\times\mathbb{R}^+$ *and for the generalized derivatives of* $u(x,t)$, *it holds that* $\left\langle \frac{\partial^2 u}{\partial t^2} - c^2 \frac{\partial^2 u}{\partial x^2}, \varphi \right\rangle = 0.$

3. *For the boundary values, it holds that* $u(0,t) = u(l,t) = 0 \quad (t \geqslant 0)$.

4. *By considering the limit $t \rightarrow 0+$ and (generalized) term-by-term differentiation of the Fourier series of $u(x, t)$, one obtains for the initial values*:

    (a) $\lim\limits_{t \to 0+} u(x, t) = f(x)$ *for* $0 \leqslant x \leqslant l$ *pointwise.*

    (b) $\lim\limits_{t \to 0+} \dfrac{\partial u}{\partial t}(x, t) = g(x)$ *for* $0 \leqslant x \leqslant l$ *pointwise except at points of discontinuity.*

5. *The solution $u(x, t)$ is uniquely determined by the initial conditions $f$ and $g$.*

One may calculate distributionally with the series approach as we have done, and obtain a distributional solution in $]0, l[ \times \mathbb{R}^+$. All partial sums of the series for $u$ satisfy the wave equation in the classical and thus also in the distributional sense. For their distributional limit $u$, the above statement 2 immediately follows from the continuity of derivatives on $\mathcal{D}'(]0, l[ \times \mathbb{R}^+)$ (see p. 185). Since distributions generally do not have pointwise values, the question of the meaning of initial and boundary values arises. In the present case, however, the solution is regular, so an ordinary function. The attainment of the initial and boundary values results here from the continuity of $u(x, t)$ on $[0, l] \times \mathbb{R}_0^+$, from the assumptions about $f$ and $g$, and from the convergence properties of the Fourier series representing $f$ and $g$. With sufficient smoothness of the initial conditions, the distributional solution, as seen in the d'Alembert form of the solution and in the asymptotic behavior of the Fourier coefficients, is a correspondingly smooth, thus classical solution.

Thus, the concept of distributions provides a solid foundation for the old approaches of Bernoulli and Fourier, and largely frees one from the concerns and limitations one had to consider within the classical framework regarding convergence, term-by-term differentiability of series, etc. In a similar manner to the solution of the string problem, inhomogeneous wave, heat conduction, and potential problems can also be treated using separation approaches that lead to series representations for distributional solutions. Fundamental questions that always need to be resolved are questions about the regularity of such solutions and in what sense the initial and boundary values of the distributional solutions can be discussed. For this purpose, methods based on the works of S. L. Sobolev (1964) are used (see also Sect. 9.5). These methods can be studied in depth with the provided references to further textbooks on partial differential equations and their applications.

## 9.7  Summary

To conclude this chapter, we once again present some important concepts and facts from classical differential calculus in comparison with the corresponding concepts and statements of distribution theory. The comparison shows why distribution methods have become a frequently used mathematical tool, especially in engineering disciplines. These methods provide a calculus that allows for correct and easy calculations in a manner that often was only heuristic or even incorrect from the perspective of classical analysis.

Compare the statements in Sect. 7.5 and the theorems known from analysis with the properties of the distribution calculus that we developed in Chap. 8 using the following Table 9.1.

**Table 9.1** Properties of functions and distributions compared

| Classical analysis | Distribution theory |
| --- | --- |
| • One studies pointwise defined functions $f : \mathbb{R} \to \mathbb{R}$ | • One studies linear mappings $T : \mathcal{D} \to \mathbb{R}$, $\mathcal{D}$ the space of test functions |
| • Values: To each $t \in \mathbb{R}$ is assigned the function value $f(t)$ | • Values: To each test function $\varphi \in \mathcal{D}$ is assigned the value $T(\varphi)$. Despite the often common notation $T = T(t)$, distributions generally have no values for individual $t \in \mathbb{R}$, but only "averages" $T(\varphi)$ are defined |
| | • For each distribution $T$ there is a sequence of infinitely differentiable functions $f_n$, such that $T = \mathcal{D}'\text{-}\lim_{n\to\infty} f_n$, i.e., for each $\varphi \in \mathcal{D}$, $$T(\varphi) = \lim_{n\to\infty} \int_{-\infty}^{+\infty} f_n(t)\varphi(t)\,\mathrm{d}t$$ |
| | • Each classical, locally integrable function $f : \mathbb{R} \to \mathbb{R}$ is also a distribution through $$f(\varphi) = \langle f, \varphi \rangle = \int_{-\infty}^{+\infty} f(t)\varphi(t)\,\mathrm{d}t \quad (\varphi \in \mathcal{D})$$ |
| • Many functions $f$ are not continuous or not differentiable | • All distributions are continuous on $\mathcal{D}$. By a more general derivative concept, all distributions are also arbitrarily often differentiable. Linearity, chain rule, and product rule hold in the sense of distributions |
| • There is no ideal impulse function $\delta(t) = \frac{\mathrm{d}}{\mathrm{d}t} s(t)$, $s(t)$ the unit step function | • The $\delta$ impulse $\delta(\varphi) = \varphi(0)$, $\varphi \in \mathcal{D}$, is the generalized derivative of $s(t)$: $\dot{s} = \delta$, $$\delta(\varphi) = -\int_{-\infty}^{+\infty} s(t)\varphi'(t)\,\mathrm{d}t$$ |
| • A sequence of functions $f_n : \mathbb{R} \to \mathbb{R}$ converges pointwise to $f : \mathbb{R} \to \mathbb{R}$ if for each $t \in \mathbb{R}$: $\lim_{n\to\infty} f_n(t) = f(t)$ | • A sequence of distributions $T_n : \mathcal{D} \to \mathbb{R}$ converges to $T : \mathcal{D} \to \mathbb{R}$ if for each $\varphi \in \mathcal{D}$: $\lim_{n\to\infty} T_n(\varphi) = T(\varphi)$ |
| • For a pointwise convergent sequence of functions $f_n$ with $\lim_{n\to\infty} f_n(t) = f(t)$, it generally does not hold that $\lim_{n\to\infty} f_n'(t) = f'(t)$ | • For a convergent sequence of distributions $T_n$ with $\lim_{n\to\infty} T_n = T$ it always holds that $\lim_{n\to\infty} \dot{T}_n = \dot{T}$ |

(continued)

**Table 9.1**  (continued)

| Classical analysis | Distribution theory |
|---|---|
| • For pointwise convergent series of functions $\displaystyle\sum_{n=0}^{\infty} f_n(t) = f(t)$, it generally does not hold that $\displaystyle\sum_{n=0}^{\infty} f_n'(t) = f'(t)$ | • For convergent series of distributions $\displaystyle\sum_{n=0}^{\infty} T_n = T$ it always holds that $\displaystyle\sum_{n=0}^{\infty} \dot{T}_n = \dot{T}$ |
| • Classical Fourier series $\displaystyle\sum_{k=-\infty}^{+\infty} c_k\, e^{jkt}$ converge at most to an integrable function on $[0, 2\pi]$, if $c_k \to 0$ for $|k| \to \infty$ | • Generalized Fourier series $\displaystyle\sum_{k=-\infty}^{+\infty} c_k\, e^{jkt}$ converge in the distributional sense even if the $|c_k|$ grow polynomially, i.e., if with suitable $n \in \mathbb{N}$ it holds: $|k|^{-n}|c_k| \to 0$ for $|k| \to \infty$ |
| • The convolution $f * g$ generally exists only under suitable integrability conditions on $f$ and $g$. The order of differentiation and convolution can be interchanged only under differentiability properties of $f$ or $g$ | • The convolution $T * G$ of two distributions $T$ and $G$ on $\mathbb{R}$ exists if one of the distributions has a bounded support, or if both supports are semi-bounded on the same side, e.g., if $\operatorname{supp}(T) \subset [0, \infty[$ and $\operatorname{supp}(G) \subset [0, \infty[$ are defined |
| | • For distributions in $\mathbb{R}^n$, the convolution exists under the conditions specified on p. 194. Differentiation and convolution are interchangeable. This allows, together with the existence of the $\delta$ distribution, the simple description of many time-invariant, linear systems through their impulse response (see later Chap. 11) |
| • Classical solutions for many initial boundary value problems often exist only under strong smoothness conditions on the initial and boundary conditions | • Distributions allow for a solution concept that often permits simple, not necessarily smooth initial and boundary conditions. This can greatly facilitate solving practical problems with simple mathematical models |

## 9.8  Exercises

**(A1)** Calculate the Fourier series of the following function using the impulse method from p. 214:

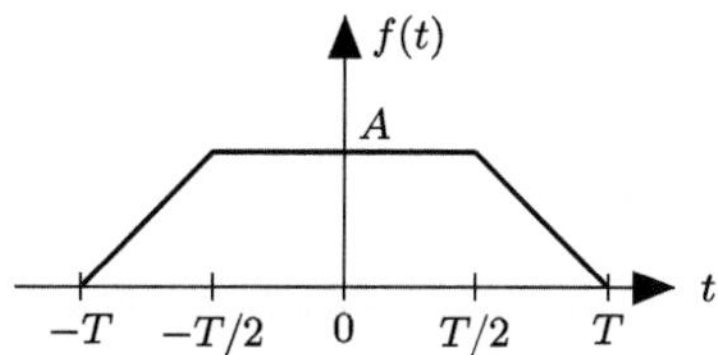

**(A2)** Calculate the step response and impulse response for the $RLC$ circuit $PT_2$ element from p. 66. Determine the system response $U_a(t)$ in the steady state, i.e., for $t \to \infty$, if:

(a) $U_e(t) = U_0 1_{[0,T]}(t)$ is a rectangular signal of duration $T$.

(b) The excitation is $U_e(t) = U_0 \sin(\omega t)$.

(c) Transform the differential equation from (a) into a first-order system with system matrix $A$, calculate the matrix $e^{At}$ for various values of $R$, $C$, and $L$ using a computer algebra system of your choice, and identify the impulse response as an element in $e^{At} s(t)$.

**(A3)** Let a time-invariant linear transfer system be described by a differential equation. Assume that the zeros of the characteristic polynomial $P$ of the differential equation are $z_1 = -1 + j$, $z_2 = -1 - j$, $z_3 = -2$. Assume that $P(0) = 4$:

(a) State the third-order differential equation that describes the system.

(b) What are the eigenfunctions of the system and the general solution of the homogeneous equation?

(c) Determine the causal impulse response and note its determination by the eigenfunctions.

(d) Describe the same problem using a first-order system. Provide a fundamental matrix for it. What is the solution of the inhomogeneous system when all initial values are set to zero?

**(A4)** Calculate the potential $u$ of the rod from the example on p. 236 at the points $(0, 0, z)$ and $(0, 0, -z)$ for $z > l$ and determine the shape of the equipotential surfaces of the solution $u$.

**(A5)** A hemispherical shell $H$ in a vacuum with radius $R$ is described in spherical coordinates by $H = \{(R, \theta, \phi) : 0 \leqslant \theta \leqslant \pi/2, \ 0 \leqslant \phi < 2\pi\}$. It carries a surface charge density $\sigma(\theta, \phi) = \sigma_0 \cos(\theta)$. Calculate the value of the potential vanishing at infinity for the center of the shell with the following values: $R = 2$ m, $\sigma_0 = 3\,\mu C/m^2$, $\varepsilon_0 = 8.85 \cdot 10^{-12}$ As/Vm.

**(A6)**$^\star$ Calculate the normal derivative of Green's function for a spherical surface around the origin with radius $R$. Use this to express the values of a harmonic function inside the sphere based on its boundary values on the sphere surface and compare it with Poisson's formula on p. 70.

**(A7)**$^\star$ (a) Using polar coordinates, verify that

$$g(r, \phi) = \frac{1}{2\pi} \ln(r)$$

is a *fundamental solution of the potential equation in the plane*.

(b) Find Green's function for the circle $K$ around the origin of radius $R$.

(c) Calculate the normal derivative of Green's function on this circle. Use formula (9.6) on p. 239 to solve the Dirichlet problem $\Delta u = 0$ in $K$, $u = f$ on $\partial K$. Compare your result with Poisson's formula on p. 70.

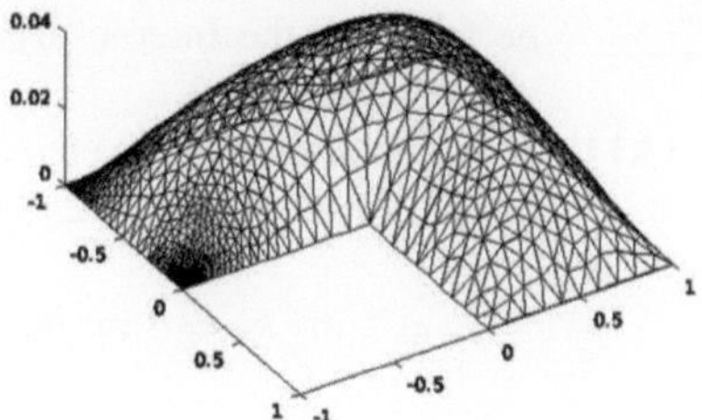

**Fig. 9.25**  Matlab's FEM solution for the $L$-shaped membrane

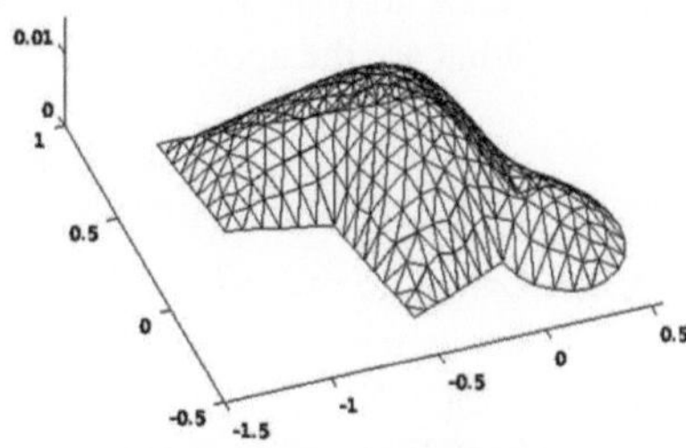

**Fig. 9.26**  FEM solution for a different region with a Lipschitz boundary

**(A8)**  Solve the boundary value problem using an image charge:

$$\Delta u(\mathbf{x}) = -\frac{q}{\varepsilon_0}\delta(\mathbf{x} - \mathbf{x_0}) \text{ for } \mathbf{x_0} = (2,0,0),\ \mathbf{x} = (x,y,z) \text{ with } x > 0,$$
$$u(\mathbf{x}) = 0 \qquad\qquad \text{for } \mathbf{x} = (x,y,z) \text{ with } x = 0.$$

**(A9)**  Using the values from the example on p. 258 and the linear finite triangular elements used there, calculate the approximate solution $u_N$, $N = p^2 = 81$, for the displacement of the membrane and its value at the point $x = y = L/2$.

**(A10)**  Use a numerical software to solve the problem with the L-shaped membrane of page 244 by the finite element method with a finer triangulation than before. With Fig. 9.25 you see the solution, which was computed for a finer mesh, force density $f = 0.5$ N/$m^2$ and $k = 2$ N/$m$ with Matlab, which is widely used in engineering disciplines, where FEM methods[3] are "State of the Art." Fig. 9.26 shows the FEM solution with the same input data on a different region with a Lipschitz boundary (cf. pp. 244 and 502). For a 3D problem, see p. 432.

Of course, given the importance of the method, there are numerous textbooks about it and also numerous special purpose FEM software depending on the application field. When you will be working in an engineering discipline, you will fairly soon have to learn more about FEM. If you are interested, you can use the FEM Software Elmer for free that can easily

---

[3] Worth reading is the article "Clough R. W. And E. L. Wilson, Early finite element research at Berkeley, Fifth U.S. National Conference on Computational Mechanics, (1999)", which can be found in the Internet.

be found in the Internet by searching for "Elmer FEM." I used it for the example on p. 432.

**(A11)** Using analogous finite elements, calculate the stationary temperature distribution for the rod on p. 248 with the values

$$L = 1 \text{ m}, \ k_1 = 2 \text{ m}^2/\text{s}, \ k_2 = 3\text{m}^2/\text{s}, \ T_0 = 273.2\,^\circ\text{K}, \ T_1 = 283.2\,^\circ\text{K}$$

with a subdivision of the interval $[0, L]$ into $N + 1 = 12$ equal subintervals, i.e., with $N = 11$ internal nodes.

What are the approximate values for the stationary temperature at the points

$$x_1 = L/6, \ x_2 = L/3, \ x_3 = 2L/3 \,?$$

**(A12)** Calculate the amplitudes $A_k$ up to the 5th overtone for the freely vibrating string with initial displacements $f_1$ and $f_2$ and initial velocities zero for $x \in [0, l]$,
$l = 200h = 1\text{m}, \ n \in \mathbb{N}, \ a > 0, \ h > 0$ with

$$f_1(x) = \begin{cases} ax & \text{for } 0 \leqslant x \leqslant \frac{l}{2} \\ a(l - x) & \text{for } \frac{l}{2} \leqslant x \leqslant l \end{cases}$$

and

$$f_2(x) = \begin{cases} \frac{h}{l - \frac{l}{n}} x & \text{for } 0 \leqslant x \leqslant l - \frac{l}{n} \\ \frac{nh}{l}(l - x) & \text{for } l - \frac{l}{n} \leqslant x \leqslant l. \end{cases}$$

Compare the amplitude ratios with your experiences about the timbre differences of the two models. The first function models a displacement in the middle of the string, and the second for large $n$ a displacement near the end of the string. Are the solutions of the wave equation to the above initial conditions $f_1$ and $f_2$ classical solutions, or can they only be understood in the distributional sense?

**(A13)** *On convolution equations, numerical solutions, Tikhonov regularization.* In applications, convolution equations $h * f = x$ often occur, in which $x$ and $h$ are known and $f$ is sought. The following examples should illustrate that such equations are the so-called *ill-posed problems*, i.e., if the right-hand side $x$ (the data) has even only small data errors, a naively calculated numerical solution can be far from the actual solution of the problem. To make approximate solutions less sensitive to data errors, regularization methods are common. Suitable literature includes the reference Engl and Groetsch (1987). In this exercise, the so-called Tikhonov regularization will be tested for two examples:

(a) Calculate the impulse response of the equation $f(t) = x''(t)$. In the corresponding inverse problem, let

$$x(t) = (\sin(t) - t)s(t)$$

($s$ the unit step function) be given. With the impulse response $h$, it holds that $h * f = x$. Calculate $f$. Now suppose only equidistantly taken sample values of $x$ are known, which are arbitrarily corrupted by small errors, for example

$$x_i = x(t_i)(1 + (-1)^i \, 10^{-3}).$$

Discretize the convolution equation into a system of equations of the form $A\mathbf{f} = \mathbf{x}$ with 100 data values over $0 \leqslant t \leqslant 10$ and solve it "naively" with the help of a computer algebra system. Compare the numerical solution with the analytical solution $f$ using a graph. Convince yourself by comparing $A\mathbf{f}$ with $\mathbf{x}$ that your equation solver is very good and that the large deviations of this numerical solution lie in the nature of the problem. Consider $\det(A)$ and think about the effects of data errors, for example, when considering Cramer's rule for solving the equation. The problem corresponds to the calculation of an acceleration based on an observed motion.

(b) Now find an approximate solution with the error-affected data using *Tikhonov regularization*, i.e., solve

$$(A^T A + \alpha E)\mathbf{f} = A^T \mathbf{x}$$

instead of $A\mathbf{f} = \mathbf{x}$ with a regularization parameter $\alpha \ll 1$, for example, $\alpha = 10^{-3}$ ($E$ the identity matrix). Compare your result again with the analytical solution without data errors. Test with different parameters $\alpha$.

(c) *On the reception problem in a transmission.* In an analogous way, solve the problem $h * f = x$ with $f(t) = U_0 \sin(\omega_0 t)s(t)$ and the impulse response $h$ of the following transmission system (RC lowpass filter of the 2nd order). Neglect the coupling impedance, and set for the test $\omega_0 = 2\,\text{rad/s}$, $R = 1\,\Omega$, $C = 1\,\text{F}$, $U_0 = 1\,\text{V}$. Use error-affected values of $\mathbf{x}$ corresponding to slightly disturbed reception data in the transmission. Recalculate $f$ from this (see also later p. 401):

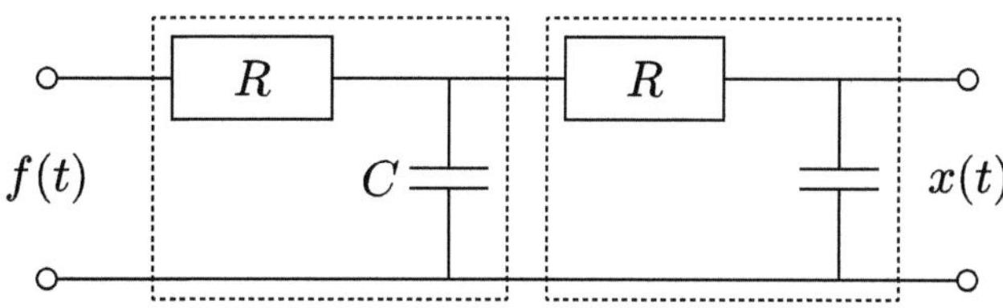

**(A14)*** Prove for integrable functions $f$ and $g$ and

$$h_\alpha(t) = (4\pi\alpha)^{-1/2}\, e^{-t^2/(4\alpha)}$$

that the convolution $f * g$ is integrable, and that the pointwise limit is

$$\lim_{\alpha \to 0+} f * h_\alpha(t) = f(t),$$

if $f$ is continuous.

# Chapter 10
# The Fourier Transform

**Abstract** The Fourier transform is introduced, and a pointwise inversion theorem for classical functions is proven. The Fourier transform of tempered distributions is then established. Calculations with Fourier transforms are derived with the corresponding rules regarding symmetries, derivatives, integrals, and convolutions. Typical application examples are generalized Fourier series and impulse sequences, polynomials, and pseudofunctions such as rational functions. Important examples for discrete signal processing are also convolutions of impulse sequences with suitable growth properties of their pulse strengths. The Fourier transforms for square-integrable functions and for functions or distributions with several variables are dealt with in separate sections. Fraunhofer diffraction on rectangular and circular apertures is one of the examples. Further examples on all topics can be found in the text and in the exercises of the chapter.

## 10.1  Representation of Functions by Harmonic Oscillations

In the preceding chapters, it was shown that many periodic functions and periodic distributions can be represented by their Fourier series as superpositions of harmonic functions. These representations were the key to solving initial boundary value problems such as the vibrating string and some boundary value problems for the potential equation. They were also convenient for describing time-invariant linear systems with periodic excitations. If an analogous harmonic analysis can be achieved for nonperiodic functions and distributions, i.e., a representation as a superposition of harmonic functions, useful applications can be expected even in problems where periodic phenomena do not play a dominant role.

To find an approach, we consider a function $f$ on the real axis that is integrable and piecewise continuously differentiable on every bounded interval. On each interval $]-T/2, T/2[$, $T > 0$, $f$ has a Fourier series representation with the mean value property (cf. p. 24):

© The Author(s), under exclusive license to Springer Nature Switzerland AG 2025     269
R. Brigola, *Fourier Analysis and Distributions*, Texts in Applied Mathematics 79,
https://doi.org/10.1007/978-3-031-81311-5_10

$$\frac{f(t+) + f(t-)}{2} = \sum_{k=-\infty}^{+\infty} c_k e^{2\pi j k t / T} \quad \text{with} \quad c_k = \frac{1}{T} \int_{-T/2}^{+T/2} f(s) e^{-2\pi j k s / T} \, ds.$$

$$(10.1)$$

Following the idea of enlarging the considered domain and finally taking the limit $T \to \infty$, one sees that all $c_k$ would vanish. To circumvent this difficulty, we use the products $c_k T$, set $\Delta\omega = 2\pi/T$, and define a function $\widehat{f}$ at all points $k\Delta\omega$, $k \in \mathbb{Z}$, by

$$\widehat{f}(k\Delta\omega) = c_k T = \int_{-T/2}^{+T/2} f(s) e^{-js(k\Delta\omega)} \, ds. \qquad (10.2)$$

Then for $t \in \, ] - T/2, T/2 \, [$,

$$\frac{f(t+) + f(t-)}{2} = \lim_{N \to \infty} \frac{1}{2\pi} \sum_{k=-N}^{N} \widehat{f}(k\Delta\omega) e^{jt(k\Delta\omega)} \Delta\omega. \qquad (10.3)$$

Now consider the quantities $\widehat{f}(k\Delta\omega)$ as sample values of a function $\widehat{f}(\omega)$, and take the limit as $T \to \infty$. Then from (10.2) we obtain

$$\widehat{f}(\omega) = \int_{-\infty}^{+\infty} f(s) e^{-j\omega s} \, ds.$$

The right side of (10.3) could then be understood as a discrete approximation for

$$\frac{f(t+) + f(t-)}{2} = \frac{1}{2\pi} \int_{-\infty}^{+\infty} \widehat{f}(\omega) e^{j\omega t} \, d\omega. \qquad (10.4)$$

At all continuity points, $f(t) = \dfrac{1}{2\pi} \displaystyle\int_{-\infty}^{+\infty} \widehat{f}(\omega) e^{j\omega t} \, d\omega$ would be represented in this way by a continuous superposition of all functions of the form $e^{j\omega t}$ with respective amplitude $\dfrac{1}{2\pi} |\widehat{f}(\omega)|$ and phase $\arg(\widehat{f}(\omega))$ with $\omega \in \mathbb{R}$.

**Definition** The mapping $\mathcal{F} : f \to \widehat{f}$ is called the Fourier transform. The Fourier transform $\mathcal{F}(f) = \widehat{f}$ of $f$ is the function

$$\widehat{f}(\omega) = \int_{-\infty}^{+\infty} f(t) e^{-j\omega t} \, dt.$$

The above consideration of the representation of $f$ by (10.4) is mathematically not quite exact because of the questionable interchange of the limits $T \to \infty$ and $N \to \infty$, but it at least shows which formula is to be expected. In the following, we consider integrable functions $f$ on $\mathbb{R}$. Integrability means that with $f$, also $|f|$ is integrable over $\mathbb{R}$. A function $f$ is piecewise continuous or piecewise continuously differentiable if $f$ is piecewise continuous or piecewise continuously differentiable on every bounded interval and all one-sided limits of $f$ and $f'$ exist.

## *The Fourier Inversion Theorem for Piecewise Continuously Differentiable Functions*

**Theorem 10.1** *For an integrable, piecewise continuously differentiable function $f$ on $\mathbb{R}$ and its Fourier transform $\widehat{f}$, the following Fourier inversion formula holds at every point $t \in \mathbb{R}$:*

$$\frac{f(t+) + f(t-)}{2} = \lim_{\Omega \to \infty} \frac{1}{2\pi} \int_{-\Omega}^{+\Omega} \widehat{f}(\omega) e^{j\omega t} \, d\omega.$$

For a better understanding, we denote by $f_\Omega$ the function

$$f_\Omega(t) = \frac{1}{2\pi} \int_{-\Omega}^{+\Omega} \widehat{f}(\omega) e^{j\omega t} \, d\omega = \frac{1}{2\pi} \int_{-\Omega}^{+\Omega} \int_{-\infty}^{+\infty} f(s) e^{j\omega(t-s)} \, ds \, d\omega.$$

It is then to be shown that $\displaystyle\lim_{\Omega \to \infty} f_\Omega(t) = \frac{f(t+) + f(t-)}{2}$. For fixed $\Omega > 0$, it follows first by interchanging the order of integration

$$f_\Omega(t) = \int_{-\infty}^{+\infty} f(s) \int_{-\Omega}^{+\Omega} \frac{1}{2\pi} e^{j\omega(t-s)} \, d\omega \, ds = \int_{-\infty}^{+\infty} f(s) \frac{\sin(\Omega(t-s))}{\pi(t-s)} \, ds. \qquad (10.5)$$

As we have already seen earlier (cf. p. 183), the sequence of functions $\dfrac{\sin(\Omega(t-s))}{\pi(t-s)}$ converges in distributional sense to the $\delta$-impulse $\delta(t-s)$ as $\Omega \to \infty$, so that the proof of the inversion formula becomes intuitively clear. The convolution kernel $\sin(\Omega(t-s))/(\pi(t-s))$ here plays the role that the periodic Dirichlet kernel played in the Fourier series expansion of periodic functions. The function $\sin(\Omega t)/(\pi t)$ is referred to as the *Fourier kernel* or also as the *Dirichlet kernel*.

***Proof*** To carry out the proof, choose an arbitrary number $\varepsilon > 0$, and divide the integration range of the integral on the right side of (10.5).

$$f_\Omega(t) = \int\limits_{|s-t|>\varepsilon} f(s)\frac{\sin(\Omega(t-s))}{\pi(t-s)}\,ds + \int\limits_{t}^{t+\varepsilon} f(s)\frac{\sin(\Omega(t-s))}{\pi(t-s)}\,ds$$

$$+ \int\limits_{t-\varepsilon}^{t} f(s)\frac{\sin(\Omega(t-s))}{\pi(t-s)}\,ds.$$

Since $\dfrac{f(s)}{\pi(t-s)}$ is integrable with respect to $s$ for $|s-t| > \varepsilon$, it follows from the Riemann-Lebesgue Lemma (Theorem 4.3, p. 50) that for increasing $\Omega \to \infty$, the first integral vanishes due to the damping effect of the increasing oscillations of $\sin(\Omega(t-s))$. The second integral is written exactly as in the proceeding on p. 130:

$$\int\limits_{t}^{t+\varepsilon} f(s)\frac{\sin(\Omega(t-s))}{\pi(t-s)}\,ds = \int\limits_{-\varepsilon}^{0}(f(t-u)-f(t+))\frac{\sin(\Omega u)}{\pi u}\,du$$

$$+ \int\limits_{-\varepsilon}^{0} f(t+)\frac{\sin(\Omega u)}{\pi u}\,du. \qquad (10.6)$$

Since $(f(t-u)-f(t+))/(\pi u)$ remains bounded for $-\varepsilon < u < 0$, the first integral on the right side vanishes again for increasing $\Omega$. The second integral converges for $\Omega \to \infty$ to $f(t+)/2$ (substitute $\Omega u = x$ and use Exercise 9 in Chap. 8, p. 204).

Similarly, the third integral in (10.6) converges to $f(t-)/2$ as $\Omega$ increases. Thus, the stated Fourier inversion formula is shown.  $\square$

***Remark*** After possibly modifying $f$ at points of discontinuity such that everywhere the mean value property $f(t) = \dfrac{f(t+) + f(t-)}{2}$ is satisfied, the inversion formula is often written briefly as

$$f(t) = \frac{1}{2\pi}\int\limits_{-\infty}^{+\infty} \widehat{f}(\omega)e^{j\omega t}\,d\omega.$$

The integral is then understood in the sense of the shown theorem as a *Cauchy principal value integral*, i.e., as the limit of the integrals over $[-\Omega, \Omega]$ for $\Omega \to \infty$. This is to be noted because integrability of $f$ does not imply integrability of $\widehat{f}$ (cf. Examples 1 and 3 in the next section). The conditions assumed about $f$ in the theorem are sufficient but not necessary for the validity of the Fourier inversion formula. We will see later that the Fourier transform and the inversion formula can also be introduced in a sense appropriate for applications for a large class of distributions.

## 10.2   Fourier Transform of Real-Valued Functions

The Fourier transform $\widehat{f}$ of a function $f$ is also called the *spectral function* of $f$. It has the same meaning as the discrete spectral values for periodic functions (cf. p. 32), i.e., it indicates for each angular frequency $\omega$ how amplitude and phase of the corresponding oscillation contribute to the composition of the "signal" $f$.

We consider real-valued functions $f$ for which the Fourier inversion formula holds. Then, because $e^{-j\omega t} = \cos(\omega t) - j\sin(\omega t)$, the real part $R(\omega)$ and the imaginary part $X(\omega)$ of the spectral function $\widehat{f}(\omega)$ are given by

$$R(\omega) = \int_{-\infty}^{+\infty} f(t)\cos(\omega t)\,dt\,, \quad X(\omega) = -\int_{-\infty}^{+\infty} f(t)\sin(\omega t)\,dt.$$

From this it follows that $R$ is an even function and $X$ is an odd function:

$$R(\omega) = R(-\omega)\,, \quad X(\omega) = -X(-\omega).$$

For the spectral function of a real-valued function $f$ and any $\omega \in \mathbb{R}$, it holds that

$$\widehat{f}(-\omega) = \overline{\widehat{f}(\omega)} \quad \text{and} \quad |\widehat{f}(-\omega)| = |\widehat{f}(\omega)|.$$

Conversely, these symmetry properties of the spectral function are sufficient to ensure that $f$ is real-valued:

$$\Im f(t) = \frac{1}{2\pi} \int_{-\infty}^{+\infty} (R(\omega)\sin(\omega t) + X(\omega)\cos(\omega t))\,d\omega = 0.$$

For a real-valued $f$, one obtains the representation (see also p. 13)

$$f(t) = \Re f(t) = \Re\left(\frac{1}{2\pi} \int_{-\infty}^{+\infty} \widehat{f}(\omega)e^{j\omega t}\,d\omega\right) = \frac{1}{2\pi} \int_{-\infty}^{+\infty} \Re(\widehat{f}(\omega)e^{j\omega t})\,d\omega$$

$$= \frac{1}{\pi} \int_{0}^{\infty} |\widehat{f}(\omega)|\cos(\omega t + \arg(\widehat{f}(\omega)))\,d\omega.$$

Intuitively, $f$ is composed of cosine oscillations of all angular frequencies $\omega \geqslant 0$ with respective amplitude $\frac{1}{\pi}|\widehat{f}(\omega)|$ and phase $\Phi(\omega) = \arg(\widehat{f}(\omega))$.

*Symmetry Properties*  For real *even* functions $f$, $X = 0$, so $\widehat{f}$ is real-valued and

$$f(t) = \frac{1}{\pi} \int_0^\infty R(\omega) \cos(\omega t)\, d\omega.$$

For real *odd* functions $f$, $R = 0$, so $\widehat{f}$ is purely imaginary and

$$f(t) = -\frac{1}{\pi} \int_0^\infty X(\omega) \sin(\omega t)\, d\omega.$$

For the *even part* $f_g(t) = (f(t) + f(-t))/2$ of a real function $f$, it holds that

$$\widehat{f_g}(\omega) = R(\omega) = 2 \int_0^\infty f_g(t) \cos(\omega t)\, dt.$$

For the *odd part* $f_u(t) = (f(t) - f(-t))/2$, it similarly holds that

$$\widehat{f_u}(\omega) = jX(\omega) = -2j \int_0^\infty f_u(t) \sin(\omega t)\, dt.$$

*Causal Functions*  For real causal functions $f$, i.e., $f(t) = 0$ for $t < 0$, it holds that $f(-t) = 0$ for $t > 0$, so $f(t) = 2f_g(t) = 2f_u(t)$ for $t > 0$. Thus, for $t > 0$,

$$f(t) = \frac{2}{\pi} \int_0^\infty R(\omega) \cos(\omega t)\, d\omega = -\frac{2}{\pi} \int_0^\infty X(\omega) \sin(\omega t)\, d\omega.$$

These relationships imply in particular that the real part $R$ and the imaginary part $X$ of the spectral function $\widehat{f}$ of a causal function $f$ are *not independent* of each other. By substituting these representations of $f$ into the first two equations of the section, one finds

$$R(\omega) = -\frac{2}{\pi} \int_0^\infty \int_0^\infty X(z) \sin(zt) \cos(\omega t)\, dz\, dt,$$

$$X(\omega) = -\frac{2}{\pi} \int_0^\infty \int_0^\infty R(z) \cos(zt) \sin(\omega t)\, dz\, dt.$$

This fact plays an important role in the design of causal filters in system theory.

## *Examples of Spectral Functions*

1. For the *rectangular function* $r_T(t) = 1_{[-T,T]}(t) = \begin{cases} 1 & \text{for} & -T \leqslant t \leqslant T \\ 0 & \text{for} & |t| > T \end{cases}$ is

$$\widehat{r_T}(\omega) = \int\limits_{-T}^{+T} e^{-j\omega s}\, ds = -\frac{1}{j\omega}\left(e^{-j\omega T} - e^{j\omega T}\right) = 2T\frac{\sin(\omega T)}{\omega T}.$$

$r_T$ is integrable, but the spectral function $\widehat{r_T}$ is not absolutely integrable, although its improper Riemann integral exists (see Fig. 10.1). If the frequency components for angular frequencies above $\pi/T$ are neglected and $\pi/T$ is referred to as the bandwidth of the rectangular signal, it is found: *The shorter the duration of the rectangular signal, the greater its bandwidth.* We will later observe a similar relationship between duration and bandwidth for other signals as well (Sect. 12.4).

2. For the *triangular function* $f(t) = \begin{cases} 1 - |t|/T & \text{for} & |t| \leqslant T \\ 0 & \text{for} & |t| > T \end{cases}$, one obtains with integration by parts the Fourier transform (Illustration Fig. 10.2)

$$\widehat{f}(\omega) = \int\limits_{0}^{T} \left(1 - \frac{s}{T}\right)\left(e^{-j\omega s} + e^{j\omega s}\right) ds = 2\int\limits_{0}^{T}\left(1 - \frac{s}{T}\right)\cos(\omega s)\, ds$$

$$= \frac{2}{T}\int\limits_{0}^{T}\frac{\sin(\omega s)}{\omega}\, ds = \frac{2}{T\omega^2}(-\cos(\omega T) + 1) = \frac{4}{T\omega^2}\sin^2\left(\frac{T\omega}{2}\right).$$

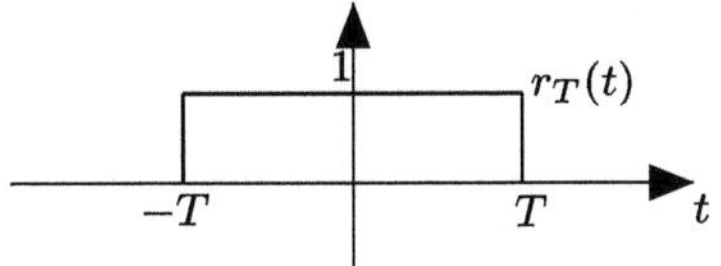


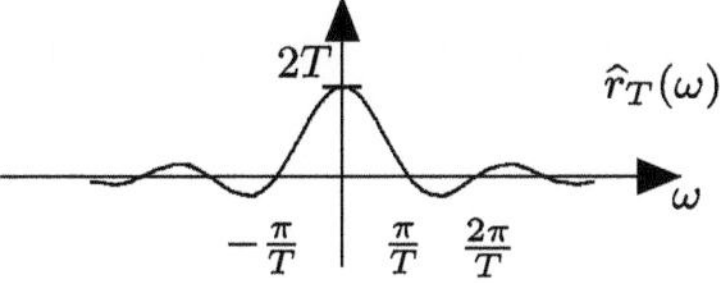


**Fig. 10.1**  A rectangle function and its Fourier transform

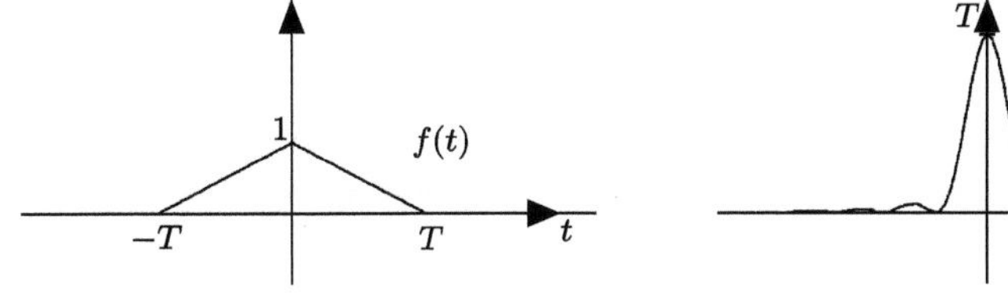


**Fig. 10.2**  A triangle function and its Fourier transform

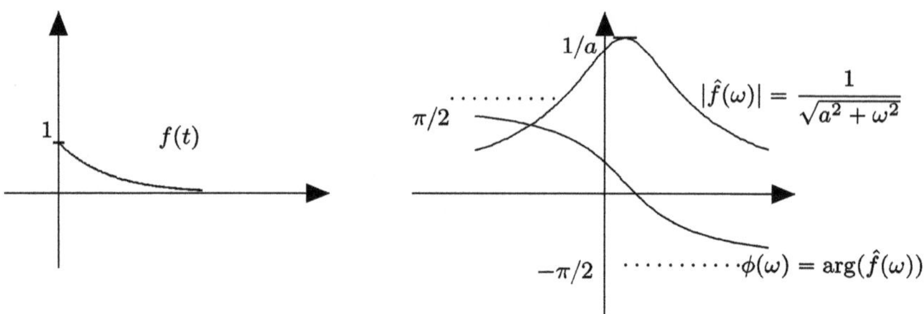

**Fig. 10.3** The function $f(t) = \mathrm{e}^{-at}s(t)$, magnitude and phase of its Fourier transform

3. The spectral function of $f(t) = \mathrm{e}^{-at}s(t)$, $s(t)$ the unit step function and $a > 0$, is

$$
\widehat{f}(\omega) = \int_0^\infty \mathrm{e}^{-at}\mathrm{e}^{-j\omega t}\,\mathrm{d}t = \lim_{R\to\infty}\left[\frac{\mathrm{e}^{-(a+j\omega)t}}{-(a+j\omega)}\right]_{t=0}^{t=R} = \frac{1}{a+j\omega}.
$$

Also in this example $f$ is absolutely integrable, but $\widehat{f}$ is not (Illustration Fig. 10.3).

4. *The Gaussian function.* To compute the Fourier transform

$$
\widehat{f}(\omega) = \int_{-\infty}^{+\infty} \mathrm{e}^{-t^2/2}\mathrm{e}^{-j\omega t}\,\mathrm{d}t
$$

for the Gaussian function $f(t) = \mathrm{e}^{-t^2/2}$, we differentiate $\widehat{f}$.
Interchanging differentiation and integration is allowed because for the function $g(t, \omega) = \mathrm{e}^{-t^2/2}\mathrm{e}^{-j\omega t}$ it holds that $\left|\dfrac{\partial g}{\partial \omega}(t, \omega)\right| \leqslant |t|\mathrm{e}^{-t^2/2}$ and the right side of this inequality is integrable over $t$. Therefore,

$$
\frac{\mathrm{d}}{\mathrm{d}\omega}\widehat{f}(\omega) = -j\int_{-\infty}^{+\infty} t\,\mathrm{e}^{-t^2/2}\mathrm{e}^{-j\omega t}\,\mathrm{d}t.
$$

Integration by parts yields for $R > 0$

$$
\int_{-R}^{+R} t\,\mathrm{e}^{-t^2/2}\mathrm{e}^{-j\omega t}\,\mathrm{d}t = -\mathrm{e}^{-t^2/2}\mathrm{e}^{-j\omega t}\Big|_{t=-R}^{t=+R} - j\omega\int_{-R}^{+R}\mathrm{e}^{-t^2/2}\mathrm{e}^{-j\omega t}\,\mathrm{d}t,
$$

and thus $\dfrac{\mathrm{d}\widehat{f}}{\mathrm{d}\omega}(\omega) = -\lim_{R\to\infty}\omega\int_{-R}^{+R}\mathrm{e}^{-t^2/2}\mathrm{e}^{-j\omega t}\,\mathrm{d}t = -\omega\widehat{f}(\omega).$

The Fourier transform $\widehat{f}$ therefore satisfies the initial value problem

$$\dot{y}(\omega) = -\omega y(\omega) \quad \text{with} \quad y(0) = \widehat{f}(0).$$

The unique solution to this problem is $y(\omega) = \widehat{f}(\omega) = \widehat{f}(0)\mathrm{e}^{-\omega^2/2}$. With the known value of the integral $\widehat{f}(0) = \displaystyle\int_{-\infty}^{+\infty} \mathrm{e}^{-t^2/2}\,\mathrm{d}t = \sqrt{2\pi}$ (see also Example 2 on p. 286), it therefore follows

$$\widehat{f}(\omega) = \sqrt{2\pi}\,\mathrm{e}^{-\omega^2/2}.$$

We note that the Fourier transform maps the Gaussian function $f(t) = \mathrm{e}^{-t^2/2}$ onto itself up to the factor $\sqrt{2\pi}$. In other words, the Gaussian function is an *eigenfunction of the Fourier transform with the eigenvalue* $\sqrt{2\pi}$.

## 10.3   Gibbs Phenomenon and Smoothing

In the Fourier inversion formula on p. 271, we obtained $f$ as the pointwise limit of the functions $f_\Omega$ for $\Omega \to \infty$:

$$\lim_{\Omega\to\infty} f_\Omega(t) = \frac{f(t+) + f(t-)}{2},$$

where $f_\Omega$ is defined by $f_\Omega(t) = \dfrac{1}{2\pi}\displaystyle\int_{-\Omega}^{+\Omega} \widehat{f}(\omega)\mathrm{e}^{j\omega t}\,\mathrm{d}\omega = \dfrac{1}{2\pi}\displaystyle\int_{-\infty}^{+\infty} \widehat{f}(\omega)r_\Omega(\omega)\mathrm{e}^{j\omega t}\,\mathrm{d}\omega.$

As with partial sums of Fourier series, the spectral values are weighted with the *rectangular window* $r_\Omega(\omega) = \begin{cases} 1 \text{ for } |\omega| \leqslant \Omega \\ 0 \text{ for } |\omega| > \Omega \end{cases}$. Similar as in the case of Fourier series, the Gibbs phenomenon can again be observed at jump discontinuities of $f$ when approximating by $f_\Omega$ (cf. p. 23). It is sufficient to consider only real-valued functions $f$ for this section.

**Theorem 10.2 (The Gibbs Phenomenon in Fourier Transform)** *If $f$ is continuous in $[a, t_0[$ and $]t_0, b]$, $f(t_0+) - f(t_0-) > 0$ and $f$ otherwise satisfies the conditions of the inversion formula, then*

$$\lim_{\Omega\to\infty}\max_{a<t<t_0} (f(t)-f_\Omega(t)) = \lim_{\Omega\to\infty}\max_{t_0<t<b} (f_\Omega(t)-f(t)) \approx 0.09\cdot(f(t+)-f(t-)).$$

***Proof*** To explain this, we assume that $f$ has only a single jump discontinuity at $t_0 = 0$. Then $f$ is of the form $f(t) = f_c(t) + (f(0+) - f(0-))s(t)$ with a continuous function $f_c$ and the jump function $(f(0+) - f(0-))s(t)$. In terms of Cauchy principal value integrals (cf. p. 271), it follows that

$$f_\Omega(t) = \int\limits_{-\infty}^{+\infty} f_c(s)\frac{\sin(\Omega(t-s))}{\pi(t-s)}\,\mathrm{d}s + (f(0+) - f(0-))\int\limits_0^\infty \frac{\sin(\Omega(t-s))}{\pi(t-s)}\,\mathrm{d}s,$$

where the first integral converges to $f_c(t)$ everywhere for $\Omega \to \infty$. With the substitution $\Omega(t-s) = x$, the second integral results in (cf. p. 26)

$$\int\limits_0^\infty \frac{\sin(\Omega(t-s))}{\pi(t-s)}\,\mathrm{d}s = \int\limits_{-\infty}^{\Omega t} \frac{\sin(x)}{\pi x}\,\mathrm{d}x = \int\limits_{-\infty}^0 \frac{\sin(x)}{\pi x}\,\mathrm{d}x + \int\limits_0^{\Omega t} \frac{\sin(x)}{\pi x}\,\mathrm{d}x$$

$$= \frac{1}{2} + \int\limits_0^{\Omega t} \frac{\sin(x)}{\pi x}\,\mathrm{d}x = \frac{1}{2} + \frac{1}{\pi}\,\mathrm{Si}(\Omega t).$$

This integral has, for any $\Omega > 0$, on the positive half-axis at $t = \dfrac{\pi}{\Omega}$ the maximum $\dfrac{1}{2} + \dfrac{1}{\pi}\,\mathrm{Si}(\pi) \approx 1.09$, corresponding to the minimum of approximately $-1.09$ at $t = -\dfrac{\pi}{\Omega}$. Thus, an overshoot of about 9% of the jump height of $f$ is also observed near the jump discontinuity for arbitrarily large $\Omega$.                                □

**Smoothing by Weight Functions**

If one uses, as in Fejér's theorem, a *triangular window* as a weight function in the spectral domain, the approximation function is smoothed and the Gibbs phenomenon is eliminated (cf. p. 31 and p. 136). As with Fejér's theorem for Fourier series, for continuous $f$ and $\Omega \to \infty$, one even obtains uniform convergence of the smoothed approximation functions $\widetilde{f}_\Omega$ to $f$, if the functions $\widetilde{f}_\Omega$ are defined by

$$\widetilde{f}_\Omega(t) = \frac{1}{2\pi}\int\limits_{-\Omega}^{+\Omega} \widehat{f}(\omega)\left(1 - \frac{|\omega|}{\Omega}\right)e^{j\omega t}\,\mathrm{d}\omega = \frac{1}{2\pi}\int\limits_{-\Omega}^{+\Omega}\int\limits_{-\infty}^{+\infty} f(s)\left(1 - \frac{|\omega|}{\Omega}\right)e^{j\omega(t-s)}\,\mathrm{d}s\,\mathrm{d}\omega.$$

With the interchange of the order of integration and the Fourier transform of the triangular window calculated in Example 2, we get (Exercise)

$$\widetilde{f}_\Omega(t) = \int\limits_{-\infty}^{+\infty} f(s)\frac{2\sin^2(\Omega(t-s)/2)}{\pi\Omega(t-s)^2}\,\mathrm{d}s.$$

Since the convolution kernel is positive, the approximation in the vicinity of a jump discontinuity of $f$ is monotonic. This shows the disappearance of the Gibbs phenomenon. Due to the lower weight of the higher frequency components, the approximation $\widetilde{f}_\Omega$ is less oscillatory compared to $f_\Omega$, and the convergence to $f$ for $\Omega \to \infty$ is improved.

The convolution kernel corresponding to the triangular window $(1 - |\omega|/\Omega)r_\Omega(\omega)$ is again called the *Fejér kernel*. Similar results can also be achieved with other window functions, such as the Gaussian curve.

Results on Fourier transforms with weight functions can be found, for example, in Champeney (1989) or Chandrasekharan (1989). They play an important role when one wants to achieve good pointwise approximations of $f$ within given tolerance ranges in signal processing.

## 10.4 Calculations with Fourier Transforms

*For this section, we make the following general assumption*: $f$ and $g$ are integrable, piecewise continuously differentiable, real- or complex-valued functions with Fourier transforms $\widehat{f}$ and $\widehat{g}$. $f$ and $\widehat{f}$ and $g$ and $\widehat{g}$ are called *correspondence pairs* or *Fourier pairs*, and these correspondences are denoted by

$$f(t) \;\circ\!\!-\!\!\bullet\; \widehat{f}(\omega) \quad \text{and} \quad g(t) \;\circ\!\!-\!\!\bullet\; \widehat{g}(\omega).$$

As with Fourier series and discrete Fourier transform—see again Sects. 4.1–4.6 and p. 97—a series of properties for calculations with Fourier transforms results, which we document in the correspondence notation. We denote the parameter $t$ as time and the parameter $\omega$ as angular frequency.

| | | | |
|---|---|---|---|
| **Linearity** | $(\alpha f + \beta g)(t)$ | $\circ\!\!-\!\!\bullet\;(\alpha \widehat{f} + \beta \widehat{g})(\omega)$ | |
| **Symmetries** | $\widehat{f}(t)$ | $\circ\!\!-\!\!\bullet\;2\pi f(-\omega)$ | (for integrable $\widehat{f}$ ) |
| | $\overline{f}(t)$ | $\circ\!\!-\!\!\bullet\;\overline{\widehat{f}(-\omega)}$ | ($\overline{f}$ complex conjugate to $f$) |
| **Scaling** | $f(\alpha t)$ | $\circ\!\!-\!\!\bullet\;\dfrac{1}{|\alpha|}\widehat{f}\!\left(\dfrac{\omega}{\alpha}\right)$ | ($\alpha \neq 0$) |
| **Time Shift** | $f(t - t_0)$ | $\circ\!\!-\!\!\bullet\;e^{-j\omega t_0}\,\widehat{f}(\omega)$ | |
| **Frequency Shift** | $e^{j\omega_0 t}f(t)$ | $\circ\!\!-\!\!\bullet\;\widehat{f}(\omega - \omega_0)$ | |

These relationships are readily obtained from the definition of the Fourier transform and from the substitution rule for integrals (Exercise).

**Differentiation in Time Domain** *If, in addition to the general condition of this section, the functions $f$ are continuous and $f'$ is integrable, then for the Fourier transform of $f' = \mathrm{d}f/\mathrm{d}t$*

$$\widehat{\frac{\mathrm{d}f}{\mathrm{d}t}}(\omega) = j\omega\widehat{f}(\omega).$$

From the integrability of $f'$, it follows that $\displaystyle\lim_{t\to+\infty} f(t) = f(0) + \int_0^\infty f'(t)\,\mathrm{d}t$ exists and is zero. Accordingly, $\displaystyle\lim_{t\to-\infty} f(t) = 0$. With integration by parts, one then obtains

$$\widehat{\frac{\mathrm{d}f}{\mathrm{d}t}}(\omega) = -\int_{-\infty}^{+\infty}(-j\omega)f(t)\mathrm{e}^{-j\omega t}\,\mathrm{d}t = j\omega\widehat{f}(\omega).$$

**Convolution in Time Domain**  *For the Fourier transform $\widehat{f * g}$ of the convolution $f * g$, it holds*

$$\widehat{f * g} = \widehat{f}\cdot\widehat{g}.$$

This is seen by exchanging the order of integration:

$$\widehat{f * g}(\omega) = \int_{-\infty}^{+\infty}\left(\int_{-\infty}^{+\infty} f(s)g(t-s)\,\mathrm{d}s\right)\mathrm{e}^{-j\omega t}\,\mathrm{d}t$$

$$= \int_{-\infty}^{+\infty}\left(\int_{-\infty}^{+\infty} g(t-s)\mathrm{e}^{-j\omega(t-s)}\,\mathrm{d}t\right)f(s)\mathrm{e}^{-j\omega s}\,\mathrm{d}s$$

$$= \int_{-\infty}^{+\infty}\widehat{g}(\omega)f(s)\mathrm{e}^{-j\omega s}\,\mathrm{d}s = \widehat{f}(\omega)\widehat{g}(\omega).$$

**Remark**  The definition of the Fourier transform is not uniform in the literature. Also common are the following definitions of the Fourier transform of a function $f$:

$$\int_{-\infty}^{+\infty} f(t)\mathrm{e}^{-2\pi j\omega t}\,\mathrm{d}t\,,\qquad \frac{1}{\sqrt{2\pi}}\int_{-\infty}^{+\infty} f(t)\mathrm{e}^{-2\pi j\omega t}\,\mathrm{d}t\,,\qquad \int_{-\infty}^{+\infty} f(t)\mathrm{e}^{+j\omega t}\,\mathrm{d}t.$$

Such differences should be noted in various literature sources. Accordingly, the given correspondences between "time functions" and "spectral functions" should be converted with a factor and the scaling relationship.

**Examples**

1. From Example 3 of p. 276, it follows for $f(t) = \mathrm{e}^{-a|t|} = \mathrm{e}^{-at}s(t) + \mathrm{e}^{at}s(-t)$, $a > 0$, $s(t)$ the unit step function (Fig. 10.4),

$$\widehat{f}(\omega) = \frac{1}{a+j\omega} + \frac{1}{a-j\omega} = \frac{2a}{a^2+\omega^2}.$$

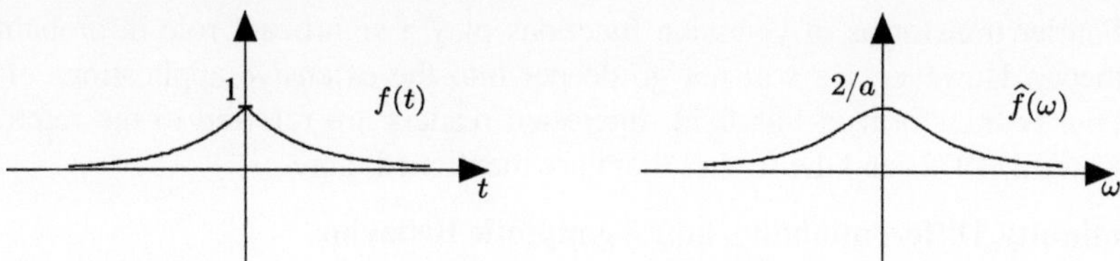

**Fig. 10.4**  The function $f(t) = \mathrm{e}^{-a|t|}$ and its Fourier transform

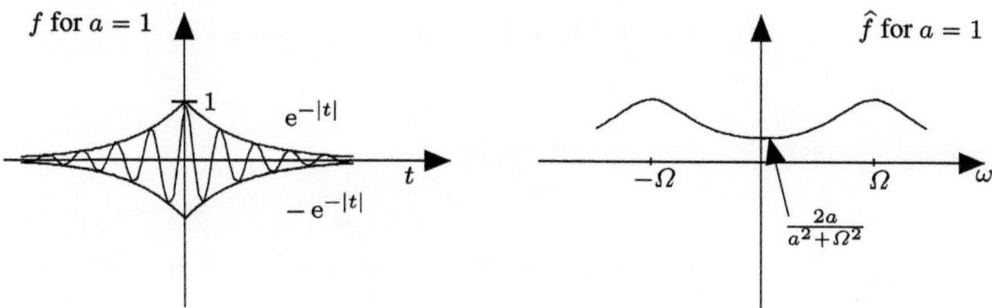

**Fig. 10.5**  An amplitude modulated cosine and its Fourier transform

2. The function $f(t) = \mathrm{e}^{-a|t|}\cos(\Omega t) = \dfrac{1}{2}\mathrm{e}^{-a|t|}(\mathrm{e}^{j\Omega t} + \mathrm{e}^{-j\Omega t})$, $a > 0$, is a cosine oscillation amplitude modulated by $\mathrm{e}^{-a|t|}$. Its spectral function is frequency shifted compared to that of $\mathrm{e}^{-a|t|}$ (Fig. 10.5).

$$\widehat{f}(\omega) = \frac{a}{a^2 + (\omega - \Omega)^2} + \frac{a}{a^2 + (\omega + \Omega)^2}.$$

3. For the Gaussian curve $G_\sigma^m(t) = \dfrac{1}{\sigma\sqrt{2\pi}}\mathrm{e}^{-(t-m)^2/(2\sigma^2)}$, $m \in \mathbb{R}$, $\sigma > 0$, one calculates the Fourier transform according to Example 4 on p. 277 using scaling and shifting rules:

$$\widehat{G}_\sigma^m(\omega) = \mathrm{e}^{-j\omega m - \sigma^2\omega^2/2}.$$

The convolution $G_\sigma^{m_1} * G_\tau^{m_2}$ of two Gaussian functions then has the spectral function

$$\widehat{G}_\sigma^{m_1}(\omega)\widehat{G}_\tau^{m_2}(\omega) = \mathrm{e}^{-j\omega(m_1+m_2)-(\sigma^2+\tau^2)\omega^2/2} = \widehat{G}_{\sqrt{\sigma^2+\tau^2}}^{m_1+m_2}(\omega).$$

The convolution $G_\sigma^{m_1} * G_\tau^{m_2}$ uniquely determined by the spectral function via the inverse Fourier formula is therefore again a Gaussian curve (cf. Exercise A12 in Chap. 8):

$$G_\sigma^{m_1} * G_\tau^{m_2} = G_{\sqrt{\sigma^2+\tau^2}}^{m_1+m_2}.$$

Fourier transforms of Gaussian functions play a significant role in probability theory. However, we will not go deeper into the extensive applications of the Fourier transform in this field. Interested readers are referred to the reference Kawata (1972) and the further literature mentioned there.

**Continuity, Differentiability, and Asymptotic Behavior**

1. From the definition of the spectral function $\widehat{f}$ of $f$, it follows

$$|\widehat{f}(\omega)| \leqslant \int\limits_{-\infty}^{+\infty} |f(t)|\, \mathrm{d}t < \infty \qquad\qquad (\omega \in \mathbb{R}),$$

and by interchanging limit and integration,

$$\lim_{\omega \to \omega_0} \widehat{f}(\omega) = \lim_{\omega \to \omega_0} \int\limits_{-\infty}^{+\infty} f(t)\mathrm{e}^{-j\omega t}\, \mathrm{d}t = \int\limits_{-\infty}^{+\infty} f(t)\mathrm{e}^{-j\omega_0 t}\, \mathrm{d}t = \widehat{f}(\omega_0).$$

*Riemann-Lebesgue Lemma.* From the Riemann-Lebesgue Lemma on page 50, we have

$$\lim_{|\omega| \to \infty} |\widehat{f}(\omega)| = 0.$$

*Thus, the Fourier transform $\widehat{f}$ is a continuous and bounded function, and it vanishes for $|\omega| \to \infty$.*

2. As with Fourier series, there is a relationship between the differentiability properties of a function $f$ and the asymptotic behavior of its spectral function $\widehat{f}$ on the one hand and between the asymptotic behavior of $f$ and the differentiability properties of $\widehat{f}$ on the other hand:

(a) *If in addition to the general assumption of the section $f$ is $k$-times continuously differentiable and $f, f', \ldots, f^{(k)}$ are integrable, then for $\omega \in \mathbb{R}$*

$$|\omega|^k |\widehat{f}(\omega)| \leqslant \int\limits_{-\infty}^{+\infty} |f^{(k)}(t)|\, \mathrm{d}t.$$

(b) *If $g(t) = t^k f(t)$ is integrable, then $\widehat{f}$ is $k$-times continuously differentiable, and for $\omega \in \mathbb{R}$*

$$\widehat{g}(\omega) = j^k \widehat{f}^{(k)}(\omega).$$

Roughly speaking: The smoother $f$ is, the faster $\widehat{f}$ decays; the faster $f$ decays, the smoother $\widehat{f}$ is.

2(a) follows from the differentiation rule, and 2(b) follows inductively by interchanging differentiation and integration from proving the assertion for the case $k = 1$:

$$j\frac{d\widehat{f}}{d\omega}(\omega) = j\int_{-\infty}^{+\infty} f(t)\frac{d}{d\omega}e^{-j\omega t}\,dt = \int_{-\infty}^{+\infty} tf(t)e^{-j\omega t}\,dt = \widehat{g}(\omega).$$

**Plancherel Equation and Multiplication Theorem**

**Theorem 10.3 (Plancherel Equation and Multiplication Theorem)** *If $f$ and $g$ are square-integrable, then:*

1. *Plancherel equation in the two variants*

$$\int_{-\infty}^{+\infty} |f(t)|^2 dt = \frac{1}{2\pi}\int_{-\infty}^{+\infty} |\widehat{f}(\omega)|^2\,d\omega \tag{10.7}$$

$$\int_{-\infty}^{+\infty} f(t)\overline{g(t)}dt = \frac{1}{2\pi}\int_{-\infty}^{+\infty} \widehat{f}(\omega)\overline{\widehat{g}(\omega)}\,d\omega. \tag{10.8}$$

2. *The Multiplication Theorem*

$$\widehat{f\cdot g} = \frac{1}{2\pi}\widehat{f}*\widehat{g}. \tag{10.9}$$

The left-hand side of the Plancherel equation (10.7) is referred to as the normalized signal energy (see page 54). Functions $f$ with $\int_{-\infty}^{+\infty} |f(t)|^2\,dt < \infty$ are also called *energy signals*. The signal energy can also be calculated from the spectral function $\widehat{f}$ according to (10.7). In particular, the spectral function of energy signals is again a square-integrable function. The multiplication theorem plays an important role in amplitude modulation in communication systems and is therefore often referred to as the *modulation theorem*.

***Proof*** (1)    Define $h(s) = \int_{-\infty}^{+\infty} f(t)\overline{f(t-s)}\,dt$. Under our assumptions, the convolution $h$ is integrable, bounded, and continuous (see Appendix B). By the convolution theorem and the symmetry relationship from page 279, $h$ has the Fourier transform

$$\widehat{h}(\omega) = \widehat{f}(\omega)\overline{\widehat{f}(\omega)} = |\widehat{f}(\omega)|^2.$$

We have $h(0) = \displaystyle\int_{-\infty}^{+\infty} |f(t)|^2 \, dt$, and using $\delta_n(t) = \dfrac{n}{\pi(1+n^2t^2)}$ (cf. p. 182), we get

$$h(0) = \lim_{n\to\infty} \int_{-\infty}^{+\infty} h(t)\delta_n(t) \, dt.$$

Since $\delta_n(t)$ is the Fourier transform of $\dfrac{1}{2\pi} e^{-|\omega|/n}$ (see page 280), we get

$$h(0) = \frac{1}{2\pi} \lim_{n\to\infty} \int_{-\infty}^{+\infty} h(t)\left( \int_{-\infty}^{+\infty} e^{-|\omega|/n} e^{-j\omega t} \, d\omega \right) dt$$

$$= \frac{1}{2\pi} \lim_{n\to\infty} \int_{-\infty}^{+\infty} e^{-|\omega|/n} \left( \int_{-\infty}^{+\infty} h(t) e^{-j\omega t} \, dt \right) d\omega$$

$$= \frac{1}{2\pi} \lim_{n\to\infty} \int_{-\infty}^{+\infty} e^{-|\omega|/n} \widehat{h}(\omega) \, d\omega = \frac{1}{2\pi} \int_{-\infty}^{+\infty} |\widehat{f}(\omega)|^2 \, d\omega,$$

thus proving (10.7). The interchange of limit and integral in the last step is possible because the functions $e^{-|\omega|/n}\widehat{h}(\omega) \geqslant 0$ form an increasing sequence of functions (monotone convergence theorem, see Appendix B, p. 493).

To show Eq. (10.8), we use the abbreviation $a(f, g) = \displaystyle\int_{-\infty}^{+\infty} f(t)\overline{g(t)} \, dt$. The *polarization identity* (left to the reader as an exercise) is valid:

$$a(f, g) = \frac{1}{4}\left( a(f+g, f+g) - a(f-g, f-g) \right)$$

$$+ \frac{j}{4}\left( a(f+jg, f+jg) - a(f-jg, f-jg) \right).$$

Using (10.7), all arguments $f+g,\ f-g,\ f+jg,\ f-jg$ can be replaced by their respective Fourier transforms multiplied by the factor $\dfrac{1}{\sqrt{2\pi}}$. Due to the linearity of the Fourier transform, (10.8) then follows: $a(f, g) = \dfrac{1}{2\pi} a(\widehat{f}, \widehat{g})$, i.e.,

$$\int_{-\infty}^{+\infty} f(t)\overline{g(t)} \, dt = \frac{1}{2\pi} \int_{-\infty}^{+\infty} \widehat{f}(\omega)\overline{\widehat{g}(\omega)} \, d\omega.$$

*Proof of* (2)  For $h(t) = \overline{g(t)}e^{j\omega t}$, $\widehat{h}(s) = \widehat{\overline{g}}(s - \omega) = \overline{\widehat{g}(\omega - s)}$ by the symmetry and frequency shifting properties on page 279. From the Plancherel equation (10.8), the multiplication theorem follows:

$$\int\limits_{-\infty}^{+\infty} f(t)g(t)e^{-j\omega t}\, \mathrm{d}t = \int\limits_{-\infty}^{+\infty} f(t)\overline{h(t)}\, \mathrm{d}t$$

$$= \frac{1}{2\pi} \int\limits_{-\infty}^{+\infty} \widehat{f}(s)\overline{\widehat{h}(s)}\, \mathrm{d}s$$

$$= \frac{1}{2\pi} \int\limits_{-\infty}^{+\infty} \widehat{f}(s)\widehat{g}(\omega - s)\, \mathrm{d}s = \frac{1}{2\pi} \widehat{f} * \widehat{g}(\omega).$$

$\square$

**Remark**  The Plancherel equation, the multiplication theorem, and the convolution theorem also hold for arbitrary square Lebesgue-integrable functions and play a central role in many applications of Fourier analysis. For further details, see also Sect. 10.7 and the application examples in Chaps. 11 to 14.

**Examples**

1. Applying the Plancherel equation (10.7) to the rectangular function

$$f(t) = \begin{cases} 1 & \text{for} \quad |t| \leqslant 1 \\ 0 & \text{for} \quad |t| > 1, \end{cases}$$

we immediately compute the integral $\displaystyle\int\limits_{-\infty}^{+\infty} \frac{\sin^2(\omega)}{\omega^2}\, \mathrm{d}\omega = \frac{\pi}{2} \int\limits_{-\infty}^{+\infty} |f(t)|^2\, \mathrm{d}t = \pi.$

2. For the Gaussian function $f(t) = e^{-t^2/2}$, it follows from page 277 $\widehat{f}(\omega) = \widehat{f}(0)f(\omega)$. The value of $\widehat{f}(0)$ is also shown by the Plancherel equation:

$$\int\limits_{-\infty}^{+\infty} |f(t)|^2\, \mathrm{d}t = \frac{\widehat{f}(0)^2}{2\pi} \int\limits_{-\infty}^{+\infty} |f(\omega)|^2\, \mathrm{d}\omega, \quad \text{so} \quad \widehat{f}(0) = \int\limits_{-\infty}^{+\infty} e^{-t^2/2}\, \mathrm{d}t = \sqrt{2\pi}.$$

**Summary**  With the Fourier inversion formula, we have now obtained a representation as a superposition of harmonic functions for integrable, piecewise continuously differentiable functions and recognized important properties of their Fourier transforms, such as similarity, translation, smoothness, and decay properties, Plancherel equation, convolution, and multiplication theorem. However, disadvantages of the

Fourier transform on the function class considered so far are also apparent. Many important functions for applications, such as the unit step function, polynomials, and periodic functions like $\sin(t)$, are not integrable over $\mathbb{R}$ and thus do not have Fourier transforms in the sense considered so far. These remaining limitations will be overcome in the next section, where we introduce the Fourier transform on a sufficiently large class of distributions for many practical questions. In this class, all the abovementioned functions will be contained as regular distributions, and the Fourier transform will be bijective and thus invertible. The transformation rules obtained will remain valid in the distributional sense.

## 10.5   The Fourier Transform of Tempered Distributions

The approach to find a class of functions $\mathcal{S}$ on which the Fourier transform is an invertible mapping is provided by the statements about differentiability and decay properties: For those infinitely differentiable functions that, together with all their derivatives, decay faster than any power of $1/|t|$ as $|t| \to \infty$, the Fourier transforms also have these properties, and the Fourier inversion formula shows that the Fourier transform is bijective on this class of functions.

### *The Fourier Transform of Rapidly Decreasing Functions*

**Definition** A function $\varphi \colon \mathbb{R} \to \mathbb{C}$ is called a rapidly decreasing function, if $\varphi$ is infinitely differentiable and

$$\lim_{|t| \to \infty} t^m \varphi^{(k)}(t) = 0$$

holds for arbitrary natural numbers $k, m \geqslant 0$. The set of all rapidly decreasing functions is denoted by $\mathcal{S}$ and is called Schwartz space in honor of L. Schwartz.

Every test function $\varphi \in \mathcal{D}$ (cf. Chap. 8) belongs to $\mathcal{S}$, so $\mathcal{D} \subset \mathcal{S}$. Other typical examples are functions of the form $P(t)\mathrm{e}^{-t^2}$ with a polynomial $P$. Not belonging to $\mathcal{S}$ are, for example, $1/(1+|t|^2)$ or $\mathrm{e}^{-|t|}$, because the first function decays too slowly and the second is not differentiable. The following properties of $\mathcal{S}$ are established:

1. $\mathcal{S}$ is a function vector space, i.e., linear combinations of functions in $\mathcal{S}$ are again in $\mathcal{S}$.
2. Products of functions in $\mathcal{S}$ are again in $\mathcal{S}$, and products $P\varphi$ between functions $\varphi \in \mathcal{S}$ and polynomials $P$ are in $\mathcal{S}$.
3. Arbitrary derivatives of functions in $\mathcal{S}$ are again in $\mathcal{S}$. Every $\varphi \in \mathcal{S}$ is integrable, because with a (depending on $\varphi$) suitable constant $M$, due to the rapid decay of $\varphi$, it holds that

$$\int\limits_{-\infty}^{+\infty} |\varphi(t)|\, dt \leqslant M \int\limits_{-\infty}^{+\infty} (1+|t|)^{-2}\, dt < \infty.$$

4. For all $\varphi \in S$, the translations $\varphi(t - t_0)$, $t_0 \in \mathbb{R}$, and the products $\varphi(t)e^{\pm j\omega t}$, $\omega \in \mathbb{R}$, belong again to $S$.
5. For every $\varphi \in S$, the Fourier transform $\widehat{\varphi}$ also belongs to $S$.

All formulas for the Fourier transform shown in 10.1 and 10.4 naturally also apply to functions in $S$. The fact that with $\varphi \in S$ also $\widehat{\varphi}$ belongs to $S$, the symmetry property from page 279, and the Fourier inversion formula together result in the following main theorem about the Fourier transform on $S$.

**Theorem 10.4 (The Fourier Transform on $S$)** *The Fourier transform is a linear, bijective mapping $\mathcal{F}\colon S \to S$ with the inverse mapping $\mathcal{F}^{-1}$, given for $\mathcal{F}\varphi = \widehat{\varphi}$ by*

$$\mathcal{F}^{-1}(\mathcal{F}\varphi)(t) = \varphi(t) = \frac{1}{2\pi} \int\limits_{-\infty}^{+\infty} \widehat{\varphi}(\omega)e^{j\omega t}\, d\omega.$$

**Remark** In the inversion formula for functions in $S$, the previously used Cauchy principal value integral can also be understood as an improper Riemann integral or as a Lebesgue integral over $\mathbb{R}$, since on $S$ all three integral concepts lead to the same result (cf. Appendix B, p. 496, theorem of dominated convergence for $\Omega \to \infty$ applied to $1_{[-\Omega,\Omega]}\widehat{\varphi}$).

**Consequences**
1. The symmetry property shows $\widehat{\widehat{\varphi}}(t) = \mathcal{F}(\mathcal{F}\varphi)(t) = 2\pi\,\varphi(-t)$.
2. Since with $f$ and $g$ in $S$ also $\widehat{f}\cdot\widehat{g}$ lies in $S$, the bijectivity of the Fourier transform on $S$ shows that with any two functions $f$ and $g$ in $S$, their convolution $f * g$ is again a rapidly decreasing function in $S$:

$$f * g = \mathcal{F}^{-1}\mathcal{F}(f * g) = \mathcal{F}^{-1}(\mathcal{F}f \cdot \mathcal{F}g) \in S.$$

3. For an integrable function $f$ and $\varphi$ in $S$, more generally also for integrable $\varphi$, one obtains by interchanging the order of integration

$$\int\limits_{-\infty}^{+\infty} \widehat{f}(s)\varphi(s)\, ds = \int\limits_{-\infty}^{+\infty} \varphi(s)\left(\int\limits_{-\infty}^{+\infty} f(t)e^{-jst}\, dt\right) ds$$

$$= \int\limits_{-\infty}^{+\infty} f(t)\left(\int\limits_{-\infty}^{+\infty} \varphi(s)e^{-jst}\, ds\right) dt = \int\limits_{-\infty}^{+\infty} f(t)\widehat{\varphi}(t)\, dt.$$

Thus, the integrand $\widehat{f}\varphi$ yields the same integral as the integrand $f\widehat{\varphi}$. This immediately recalls the definition of the generalized derivative $\dot{f}$ for distributions $f$ (cf. p. 171). There we defined for a distribution $f$ and a test function $\varphi$

$$\langle \dot{f}, \varphi \rangle = \int\limits_{-\infty}^{+\infty} \dot{f}(t)\varphi(t)\, dt = - \int\limits_{-\infty}^{+\infty} f(t)\varphi'(t)\, dt = -\langle f, \varphi' \rangle;$$

thus we analogously replaced the "integrand" $\dot{f}\varphi$ except for the factor $-1$ by $f\varphi'$. This observation conveys the idea to introduce the Fourier transform for distributions according to the same recipe. Instead of the test function set $\mathcal{D}$ and the distributions in $\mathcal{D}'$, one now uses the new, larger test function space $\mathcal{S}$, defines distributions on $\mathcal{S}$ as linear continuous functionals, and introduces their Fourier transform. Regarding the continuity on $\mathcal{S}$, one uses the following convergence definition according to L. Schwartz:

**Definition** A sequence of functions $\varphi_n$ in $\mathcal{S}$ converges in $\mathcal{S}$ to $\varphi \in \mathcal{S}$, if for arbitrary natural numbers $m, k \geqslant 0$ all functions $t^m \varphi_n^{(k)}(t)$ converge uniformly to $t^m \varphi^{(k)}(t)$, i.e.,

$$\sup_{t \in \mathbb{R}} |t^m \varphi_n^{(k)}(t) - t^m \varphi^{(k)}(t)| \to 0 \quad \text{for} \quad n \to \infty.$$

We then write $\varphi = \mathcal{S}\text{-}\lim\limits_{n \to \infty} \varphi_n$.

A sequence of functions $\varphi_n$ thus converges in $\mathcal{S}$ to the zero function if the $\varphi_n$ and all their derivatives converge uniformly to zero and decay faster than any power of $1/|t|$ as $|t| \to \infty$. With $\varphi_n$, also products $P\varphi_n$ with polynomials $P$ converge to zero in $\mathcal{S}$. With this concept of convergence for sequences of functions in $\mathcal{S}$, it follows continuity of the Fourier transform on $\mathcal{S}$.

## *Continuity of the Fourier Transform on $\mathcal{S}$*

**Theorem 10.5** *If a sequence of functions $\varphi_n$ converges in $\mathcal{S}$ to $\varphi$, then for the Fourier transforms $\mathcal{S}\text{-}\lim\limits_{n \to \infty} \widehat{\varphi}_n = \widehat{\varphi}$.*

***Proof*** To prove this, it suffices to consider the case $\mathcal{S}\text{-}\lim\limits_{n \to \infty} \varphi_n = 0$. First, we note that for arbitrary $\psi \in \mathcal{S}$, the following estimate holds:

$$\sup_{\omega \in \mathbb{R}} |\widehat{\psi}(\omega)| \leqslant \sup_{t \in \mathbb{R}} |\pi(1 + t^2)\psi(t)|.$$

This estimate results from $|\widehat{\psi}(\omega)| \leqslant \displaystyle\int\limits_{-\infty}^{+\infty} |\psi(t)|\,\mathrm{d}t$ by integrating the inequality

$|\psi(t)| \leqslant (\pi(1+t^2))^{-1} \sup\limits_{t\in\mathbb{R}} |\pi(1+t^2)\psi(t)|$, noting that

$$\int\limits_{-\infty}^{+\infty} \frac{1}{\pi(1+t^2)}\,\mathrm{d}t = 1$$

is true. Moreover, it is clear (cf. p. 279) that for $m, k \in \mathbb{N}_0$

$$|\omega^m \widehat{\varphi}_n^{(k)}(\omega)| = \Big|\int\limits_{-\infty}^{+\infty} (t^k \varphi_n(t))^{(m)} \mathrm{e}^{-j\omega t}\,\mathrm{d}t\Big| = |\widehat{\psi}_n(\omega)|$$

with $\psi_n(t) = (t^k \varphi_n(t))^{(m)} \in \mathcal{S}$ is true. Because with $\mathcal{S}\text{-}\lim\limits_{n\to\infty}\varphi_n = 0$ also $\mathcal{S}\text{-}\lim\limits_{n\to\infty}\psi_n = 0$ holds, it follows from the shown estimate $\sup\limits_{\omega\in\mathbb{R}} |\widehat{\psi}_n(\omega)| \to 0$ and $\sup\limits_{\omega\in\mathbb{R}} |\omega^m \widehat{\varphi}_n^{(k)}(\omega)| \to 0$ for $n \to \infty$. This means $\mathcal{S}\text{-}\lim\limits_{n\to\infty}\widehat{\varphi}_n = 0$ and shows the continuity of the Fourier transform on $\mathcal{S}$. $\qquad\qquad\square$

## *Tempered Distributions*

**Definition** Every continuous linear mapping $T\colon \mathcal{S} \to \mathbb{C}$ is called a tempered distribution. The set of all tempered distributions is denoted by $\mathcal{S}'$.

Instead of $T(\varphi)$, we use the notation $\langle T, \varphi \rangle$ for the value of a distribution $T \in \mathcal{S}'$ on a test function $\varphi \in \mathcal{S}$ as in Chap. 8. The continuity of a linear functional $T \in \mathcal{S}'$ on $\mathcal{S}$ means that $\lim\limits_{n\to\infty} \langle T, \varphi_n \rangle = \langle T, \varphi \rangle$ if $\varphi = \mathcal{S}\text{-}\lim\limits_{n\to\infty}\varphi_n$. From the definition, it follows:

1. $\mathcal{S}'$ is a vector space.
2. The restrictions of distributions in $\mathcal{S}'$ to $\mathcal{D} \subset \mathcal{S}$ yield distributions in $\mathcal{D}'$. In this sense, $\mathcal{S}' \subset \mathcal{D}'$. For $T \in \mathcal{D}'$ with compact support, $T(\varphi)$, $\varphi \in \mathcal{S}$, is defined and continuous on $\mathcal{S}$. Distributions with compact support are thus understood as elements of $\mathcal{S}'$. Many examples of distributions in Chap. 8 are also examples of distributions belonging to $\mathcal{S}'$. Derivatives $\dot{T}$ and products $fT$ with infinitely often differentiable functions $f$ are defined for $T \in \mathcal{S}'$ as before (see p. 173). For distributions $T \in \mathcal{S}'$, the generalized derivatives $T^{(k)}$, $k \in \mathbb{N}$, also belong to $\mathcal{S}'$.
3. It can be shown that for $T \in \mathcal{S}'$, the product with an infinitely often differentiable function $f$ is tempered if and only if $f$ and all derivatives $f^{(k)}$, $k \in \mathbb{N}$, have at

most polynomial growth, i.e., if for every $k \in \mathbb{N}_0$ there exists a natural number $N$ such that $\lim\limits_{|t| \to \infty} |t|^{-N} |f^{(k)}(t)| = 0$ holds (Schwartz 1957).

We prove only that the product $fT \in \mathcal{S}'$ for such functions $f$ and $T \in \mathcal{S}'$:

For $\varphi \in \mathcal{S}$ we have by the product rule that every derivative $D^p(f\varphi)$ tends to zero for $|t| \to \infty$ faster than every power of $1/|t|$. Thus, $f\varphi \in \mathcal{S}$ and the product $fT$ is well defined. To see the continuity of $fT$, we observe $\| P(t) D^q \varphi_k \|_\infty \to 0$ with $k \to \infty$ for every sequence $\varphi_k \to 0$ in $\mathcal{S}$, every polynomial $P$, and every derivative $D^q$. Since derivatives of $f$ can be majorized by polynomials, we have also $\| t^q D^p(f\varphi_k) \|_\infty \to 0$ for any $p, q \in \mathbb{N}_0$, i.e., $f\varphi_k \to 0$ in $\mathcal{S}$, which proves the continuity of $fT$.

**Definition** The vector space of all infinitely often differentiable functions $f$ with $fT \in \mathcal{S}'$ for $T \in \mathcal{S}'$ is called the space $\mathcal{O}_M$ of *multipliers in $\mathcal{S}'$*. Its elements are called *slowly increasing* or *polynomially bounded*.

Before we give examples of tempered distributions, we introduce the Fourier transform on $\mathcal{S}'$ according to the previously announced procedure.

## *The Fourier Transform on $\mathcal{S}'$*

**Definition** The Fourier transform $\widehat{T}$ of a distribution $T \in \mathcal{S}'$ is defined for $\varphi \in \mathcal{S}$ by $\langle \widehat{T}, \varphi \rangle = \langle T, \widehat{\varphi} \rangle$.

The definition is consistent with the definition of the Fourier transform for integrable functions $f$. If we consider $f$ as a regular distribution $T_f$, then $\widehat{T}_f = T_{\widehat{f}}$ holds:

For $\varphi \in \mathcal{S}$, it follows by changing the order of integration

$$\langle \widehat{T}_f, \varphi \rangle = \langle f, \widehat{\varphi} \rangle = \int\limits_{-\infty}^{+\infty}\int\limits_{-\infty}^{+\infty} f(\omega)\varphi(t) e^{-j\omega t}\, dt\, d\omega = \langle \widehat{f}, \varphi \rangle = \langle T_{\widehat{f}}, \varphi \rangle.$$

The Fourier transform of distributions that do not belong to $\mathcal{S}'$ is not defined here. For more information, see, for example, Gel'fand et al. (1964).

## *Inverse Fourier Transform on $\mathcal{S}'$*

Using the same convergence concept for distributions as on p. 182, we write $T = \mathcal{S}'\text{-}\lim\limits_{n \to \infty} T_n$, if $\lim\limits_{n \to \infty} \langle T_n, \varphi \rangle = \langle T, \varphi \rangle$ for all $\varphi \in \mathcal{S}$. Thus, the Fourier transform $\mathcal{F}$ on $\mathcal{S}'$ is linear and invertible, and $\mathcal{S}'\text{-}\lim\limits_{n \to \infty}\mathcal{F}(T_n) = \mathcal{F}(T)$ holds for $T = \mathcal{S}'\text{-}\lim\limits_{n \to \infty} T_n$. The inverse transform $\mathcal{F}^{-1}$ is given for $T \in \mathcal{S}'$ and $\varphi \in \mathcal{S}$ by

$$\langle \mathcal{F}^{-1}T, \varphi \rangle = \langle T, \mathcal{F}^{-1}\varphi \rangle.$$

These statements hold for $T \in \mathcal{S}'$, $T = \mathcal{S}'\text{-}\lim\limits_{n \to \infty} T_n$, $\varphi \in \mathcal{S}$ as

$$\langle \mathcal{F}\mathcal{F}^{-1}T, \varphi \rangle = \langle T, \mathcal{F}^{-1}\mathcal{F}\varphi \rangle = \langle T, \varphi \rangle,$$
$$\langle \mathcal{F}^{-1}\mathcal{F}T, \varphi \rangle = \langle T, \mathcal{F}\mathcal{F}^{-1}\varphi \rangle = \langle T, \varphi \rangle,$$
$$\langle \widehat{T_n}, \varphi \rangle = \langle T_n, \widehat{\varphi} \rangle \xrightarrow[n \to \infty]{} \langle T, \widehat{\varphi} \rangle = \langle \widehat{T}, \varphi \rangle.$$

The last property is a *continuity property of the Fourier transform on $\mathcal{S}'$*. For mathematicians with knowledge of functional analysis, it should be noted that this is the continuity of $\mathcal{F}$ when $\mathcal{S}'$ is equipped with the so-called weak topology, denoted as $\sigma(\mathcal{S}', \mathcal{S})$-topology. For more details, see, for example, Rudin (1991).

The inversion formulas $\mathcal{F}^{-1}(\mathcal{F}T) = T$ and $\mathcal{F}(\mathcal{F}^{-1}T) = T$ are equations in $\mathcal{S}'$ and usually not pointwise relationships like the inversion formulas for functions because distributions $T$ generally do not have values at individual points.


### *Calculating with Fourier Transforms in $\mathcal{S}'$*


*The transformation rules of p. 181 apply in the sense of coordinate transformations for distributions also in $\mathcal{S}'$.*

We show this statement exemplarily for translations and derivatives:

1. One obtains the correspondence

$$T(t - t_0) \;\multimap\; e^{-j\omega t_0}\, \widehat{T}(\omega)$$

with $A(t) = t - t_0$ and the notation $T_A$ for the translation of $T$, with $\varphi \in \mathcal{S}$ and $e_{t_0}(x) = e^{-jx t_0}$ from the correspondence for a translation of $\widehat{\varphi}$ (see p. 180):

$$\langle \widehat{T_A}, \varphi \rangle = \langle T, \widehat{\varphi} \circ A^{-1} \rangle = \langle T, \widehat{e_{t_0}\varphi} \rangle = \langle e_{t_0}\widehat{T}, \varphi \rangle.$$

2. For generalized derivatives one calculates as follows:

$$T^{(k)}(t) \;\multimap\; (j\omega)^k \widehat{T}(\omega)$$
$$(-jt)^k T(t) \;\multimap\; \widehat{T}^{(k)}(\omega).$$

For $\varphi \in \mathcal{S}$ and the polynomial $P(x) = (jx)^k$, the following holds:

$$\langle \widehat{T^{(k)}}, \varphi \rangle = \langle T, (-1)^k \widehat{\varphi}^{(k)} \rangle = \langle T, \widehat{P\varphi} \rangle = \langle P\widehat{T}, \varphi \rangle,$$
$$\langle (-1)^k \widehat{PT}, \varphi \rangle = \langle (-1)^k PT, \widehat{\varphi} \rangle = (-1)^k \langle T, \widehat{\varphi^{(k)}} \rangle = \langle \widehat{T}^{(k)}, \varphi \rangle.$$

From the symmetry property $\mathcal{F}^2 T(\omega) = 2\pi T(-\omega)$ (see p. 279), one recognizes that the Fourier transform $\mathcal{F}$ satisfies the relation $\mathcal{F}^4(T) = 4\pi^2 T$. Therefore, the inverse transformation $\mathcal{F}^{-1} = (4\pi^2)^{-1}\mathcal{F}^3$ and $\mathcal{F}$ has only four possible eigenvalues $\pm\sqrt{2\pi}, \pm j\sqrt{2\pi}$. Examples of eigenfunctions of the Fourier transform can be found on p. 296.

We now provide some typical examples of distributions in $\mathcal{S}'$ and their Fourier transforms.

### *Examples of Tempered Distributions and Their Fourier Transforms*

1. *The $\delta$-impulse.* The $\delta$-impulse has only the zero point as its support and belongs to $\mathcal{S}'$. From the equation

$$\langle \delta, \widehat{\varphi} \rangle = \widehat{\varphi}(0) = \int\limits_{-\infty}^{+\infty} \varphi(t)\,dt = \langle 1, \varphi \rangle,$$

it follows that its Fourier transform is $\widehat{\delta} = 1$; thus

$$\delta(t) \circ\!\!-\!\!\bullet\ \widehat{\delta}(\omega) = 1.$$

In particular, the constant function $f = 1$ belongs to $\mathcal{S}'$. Conversely, for the Fourier transform of $f = 1$ and $\varphi \in \mathcal{S}$,

$$\langle \widehat{f}, \widehat{\varphi} \rangle = \langle f, \widehat{\widehat{\varphi}} \rangle = 2\pi \int\limits_{-\infty}^{+\infty} \varphi(-t)\,dt = 2\pi \int\limits_{-\infty}^{+\infty} \varphi(t)\,dt = 2\pi\widehat{\varphi}(0) = \langle 2\pi\delta, \widehat{\varphi} \rangle.$$

The resulting Fourier transform (see also p. 184)

$$\widehat{f} = \widehat{1} = 2\pi\delta$$

is often noted in the form $\displaystyle\int\limits_{-\infty}^{+\infty} e^{-j\omega t}\,dt = 2\pi\delta(\omega)$. Note that $e^{-j\omega t}$ is not integrable, and the left side is not defined in the conventional sense and can only be understood as *a symbol for the distribution* $2\pi\delta(\omega)$.

The Fourier transform of the impulse $\delta(t - t_0)$ is given by the shift rule as $e^{-j\omega t_0}$.

2. *The unit step function.* The function $s(t) = \begin{cases} 1 & \text{for } t \geq 0 \\ 0 & \text{for } t < 0 \end{cases}$ is a regular distribution in $\mathcal{S}'$ and has the generalized derivative $\dot{s} = \delta$. From the differentiation rule for Fourier transforms, it follows

$$\widehat{\dot{s}}(\omega) = j\omega\widehat{s}(\omega) = \widehat{\delta}(\omega) = 1.$$

Considering that $j\omega k\delta(\omega) = 0$, we obtain (see p. 174)

$$\widehat{s}(\omega) = \frac{1}{j\omega} + k\delta(\omega).$$

$1/\omega$ denotes the singular distribution $\mathrm{vp}(1/\omega)$, defined by the principal value (p. 167), and $k \in \mathbb{C}$ is a constant. To determine this constant $k$, we apply $\widehat{s}$ to the test function $\varphi(\omega) = \mathrm{e}^{-\omega^2/2} \in \mathcal{S}$. Then, on the one hand,

$$\langle \widehat{s}, \varphi \rangle = \langle s, \widehat{\varphi} \rangle = \langle s(t), \sqrt{2\pi}\,\mathrm{e}^{-t^2/2} \rangle = \sqrt{2\pi} \int\limits_{0}^{\infty} \mathrm{e}^{-t^2/2}\,\mathrm{d}t = \pi,$$

and on the other hand

$$\langle \widehat{s}, \varphi \rangle = -j\langle \frac{1}{\omega}, \mathrm{e}^{-\omega^2/2} \rangle + k\langle \delta(\omega), \mathrm{e}^{-\omega^2/2} \rangle = k,$$

since the first term on the right side is zero, because $1/\omega$ is an odd function and $\mathrm{e}^{-\omega^2/2}$ is an even function. Thus we get $k = \pi$; therefore

$$s(t) \mathrel{\circ\!\!-\!\!\bullet} \widehat{s}(\omega) = \frac{1}{j\omega} + \pi\delta(\omega).$$

For the *sign function* $\mathrm{sgn}(t) = s(t) - s(-t)$, we get

$$\mathrm{sgn}(t) \mathrel{\circ\!\!-\!\!\bullet} \widehat{\mathrm{sgn}}(\omega) = \frac{2}{j\omega}.$$

3. *Slowly growing functions.* Every locally integrable function $f$ that grows slowly, i.e., $|t|^{-N}|f(t)| \to 0$ for $|t| \to \infty$ with an appropriate choice of $N \in \mathbb{N}$, belongs to $\mathcal{S}'$. Choose $C > 0$ and $N$ so that for $t$ outside a suitable bounded neighborhood $U$ of zero the estimate

$$|f(t)| \leq C(1 + |t|^2)^N$$

holds, and consider a sequence of functions $\varphi_n$ in $\mathcal{S}$ with $\mathcal{S}\text{-}\lim\limits_{n\to\infty}\varphi_n = 0$; then it follows

$$\varepsilon_n = \sup_{t\in\mathbb{R}}(1 + |t|^2)^{N+1}|\varphi_n(t)| \to 0 \qquad \text{for } n \to \infty$$

and thus

$$|\langle f, \varphi_n\rangle| \leqslant \sup_{t\in U} |\varphi_n(t)| \int_U |f(t)|\,dt + \int_{\mathbb{R}\setminus U} C(1 + |t|^2)^{N+1}(1 + |t|^2)^{-1}|\varphi_n(t)|\,dt$$

$$\leqslant \sup_{t\in U} |\varphi_n(t)| \int_U |f(t)|\,dt + C\pi\varepsilon_n \to 0.$$

This shows the continuity of the regular distribution corresponding to $f$ on $\mathcal{S}$. In particular, all polynomials and also the functions $e^{j\omega_0 t}$, $\sin(\omega_0 t)$, and $\cos(\omega_0 t)$ belong to the space $\mathcal{S}'$.

Periodic distributions $T$ are, according to the theorem on p. 211, always representable as generalized Fourier series of the form

$$T(t) = \sum_{k=-\infty}^{+\infty} c_k e^{jk\omega_0 t}$$

with polynomially bounded coefficients $c_k$. They are therefore tempered as the sum of a constant and the generalized derivative $f^{(n)}$ of a continuous periodic function $f$ ($n$ suitable, see Sect. 9.1).

Analogously, the slowly growing regular distribution $f(t) = \ln(|t|)$ and its generalized derivatives and thus also the pseudofunctions $\mathrm{pf}(t^{-m})$ for $m \in \mathbb{N}$ belong to $\mathcal{S}'$ and all rational functions understood as pseudofunctions.

Similarly, all integrable and all measurable bounded functions belong to $\mathcal{S}'$. Also all $p$-integrable functions $f$ (i.e., $|f|^p$ is integrable) are for $p > 1$ tempered distributions. This follows with $1/p + 1/q = 1$ from the *Hölder inequality*

$$|\langle f, \varphi_n\rangle| \leqslant \left(\int_{-\infty}^{+\infty} |f(t)|^p dt\right)^{1/p} \left(\int_{-\infty}^{+\infty} |\varphi_n(t)|^q dt\right)^{1/q}$$

since the right-hand side converges for $\varphi_n \to 0$ in $\mathcal{S}$ to zero as well.

**Examples of Distributions That Are Not Tempered**

The functions $e^t$, $e^{-t}$, $e^{t^2}$ are examples of distributions that belong to $\mathcal{D}'$, but due to their large growth as $|t| \to \infty$, they do not belong to $\mathcal{S}'$. They do not possess Fourier transforms in $\mathcal{S}'$.

**Remark** It can be shown that distributions $T \in \mathcal{D}'$ coincide locally, i.e., on open bounded sets, with generalized derivatives of continuous functions of suitably high order. Distributions with support inside a compact set $K$ can be represented as generalized derivatives of continuous functions with support in $K$, of suitably high order. Tempered distributions are generalized derivatives of certain slowly growing continuous functions of a certain order. The proofs of these remarkable theorems on the structure of distributions go beyond the scope of our introduction. They can be found in the fundamental monograph *"Théorie Des Distributions"* by Laurent Schwartz (1957) or in Vladimirov (2002).

As further examples, we calculate the Fourier transforms for $f(t) = \ln(|t|)$, $g(t) = |t|^{-1/2}$, the Fourier transforms of the Hermite functions, of polynomials, generalized Fourier series, and rational functions.

4. *The Fourier Transform of* $f(t) = \ln(|t|)$. From $\dot{f}(t) = \mathrm{vp}(1/t)$, $\widehat{\dot{f}}(\omega) = j\omega\widehat{f}$, and the Fourier transform of $\dot{f}(t) = \mathrm{vp}(1/t)$

$$\mathrm{vp}\left(\frac{1}{t}\right) \circ\!\!-\!\!\bullet \ -j\pi\mathrm{sgn}(\omega)$$

it follows that

$$\widehat{f}(\omega) = -\pi\,\mathrm{pf}(|\omega|^{-1}) + k\delta(\omega).$$

To determine the constant $k$, we calculate the integrals $I_1 = \langle \widehat{f}(\omega), \mathrm{e}^{-\omega^2/2}\rangle$ and $I_2 = \langle \mathrm{pf}(|\omega|^{-1}), \mathrm{e}^{-\omega^2/2}\rangle$. Here, we use properties of the Euler Gamma function $\Gamma(x)$.

From $\Gamma(x) = \int\limits_0^\infty \mathrm{e}^{-u}u^{x-1}\mathrm{d}u$, we obtain, with the substitution $u = t^2/2$ and differentiation under the integral for $x = 1/2$, the value of $\Gamma'(1/2)$:

$$\Gamma'\left(\frac{1}{2}\right) = 2\sqrt{2}\int\limits_0^\infty \mathrm{e}^{-t^2/2}\ln(t)\mathrm{d}t - \sqrt{\pi}\ln(2).$$

With the Euler-Mascheroni constant $\gamma \approx 0.5772$ and $\Gamma'(1/2) = (-\gamma - 2\ln(2))\sqrt{\pi}$, we have

$$I_1 = \langle \ln(|t|), \sqrt{2\pi}\,\mathrm{e}^{-t^2/2}\rangle = 2\sqrt{2\pi}\int\limits_0^\infty \mathrm{e}^{-t^2/2}\ln(t)\mathrm{d}t$$

$$= \sqrt{\pi}\Gamma'\left(\frac{1}{2}\right) + \pi\ln(2) = -\pi(\gamma + \ln(2)).$$

For $I_2$, with the substitution $\omega = t^{1/2}$ and integration by parts (see p. 168)

$$I_2 = 2 \int_0^1 \frac{e^{-\omega^2/2} - 1}{\omega} d\omega + 2 \int_1^\infty \frac{e^{-\omega^2/2}}{\omega} d\omega$$

$$= \frac{1}{2} \int_0^\infty \ln(t) e^{-t/2} dt - \frac{1}{2} \int_1^\infty \ln(t) e^{-t/2} dt + \int_1^\infty \frac{e^{-t/2}}{t} dt.$$

Since the last two terms add up to zero, with $t = 2u$

$$I_2 = \int_0^\infty \ln(2u) e^{-u} du = \ln(2) + \Gamma'(1) = \ln(2) - \gamma.$$

Thus, the constant $k = -2\pi\gamma$, and we have the result

$$\ln(|t|) \; \circ\!\!-\!\!\bullet \; -\pi\,\mathrm{pf}(|\omega|^{-1}) - 2\pi\gamma\delta(\omega).$$

5. *The Fourier Transform of $f(t) = |t|^{-1/2}$.* The even function $f(t) = |t|^{-1/2}$ is regular and belongs to $\mathcal{S}'$. It holds that

$$\widehat{f}(\omega) = 2 \int_0^\infty t^{-1/2} \cos(|\omega|t) dt = 2|\omega|^{-1/2} \int_0^\infty u^{-1/2} \cos(u) du.$$

With the known value $\sqrt{\pi/2}$ of the *Fresnel integral* $\int_0^\infty u^{-1/2} \cos(u) du$, we obtain the correspondence

$$|t|^{-1/2} \; \circ\!\!-\!\!\bullet \; \sqrt{2\pi}|\omega|^{-1/2}.$$

$|t|^{-1/2}$ is thus a generalized eigenfunction of the Fourier transform with eigenvalue $\sqrt{2\pi}$.

6. *The Fourier Transforms of Hermite Functions.* The Hermite polynomials $H_n$ are defined for $n \geqslant 0$ by

$$H_n(t) = (-1)^n e^{t^2} \frac{d^n}{dt^n} e^{-t^2}.$$

One obtains from

$$e^{-t^2} H_n(t) = (-1)^n \frac{d^n}{dt^n} e^{-t^2} = -\frac{d}{dt}\left(e^{-t^2} H_{n-1}(t)\right)$$

$$= e^{-t^2}\left(2t H_{n-1}(t) - H'_{n-1}(t)\right)$$

the relationship

$$H_n(t) = 2t\, H_{n-1}(t) - H'_{n-1}(t).$$

The *Hermite functions* $h_n(t) = e^{-t^2/2} H_n(t)$ then satisfy the relation

$$h'_n(t) = -t e^{-t^2/2} H_n(t) + e^{-t^2/2} H'_n(t).$$

From this, with the above equation for $H_n$ and $H'_n$

$$h_{n+1}(t) = t\, h_n(t) - h'_n(t).$$

We show that the functions $h_n$ are eigenfunctions of the Fourier transform with eigenvalues $(-j)^n \sqrt{2\pi}$. For $n = 0$, i.e., for $h_0(t) = e^{-t^2/2}$, this claim is already shown. The symmetry property $h_n(t) = (-1)^n h_n(-t)$ holds.
With induction and the notation $f_n(t) = jt h_n(t)$, one obtains by Fourier transforming the last equation for $h_{n+1}$

$$\widehat{h}_{n+1}(\omega) = -j\,\widehat{f}_n(\omega) - j\omega\widehat{h}_n(\omega)$$

$$= -j\left( j^n \sqrt{2\pi}(-1)^{n+1} h'_n(\omega) + (-j)^n \sqrt{2\pi}\,\omega h_n(\omega) \right)$$

$$= (-j)^{n+1}\sqrt{2\pi}\left(-h'_n(\omega) + \omega h_n(\omega)\right) = (-j)^{n+1}\sqrt{2\pi}\, h_{n+1}(\omega).$$

Thus, the claim is shown: $\widehat{h}_n = (-j)^n \sqrt{2\pi}\, h_n$ for all $n \in \mathbb{N} \cup \{0\}$.

It should be noted that the normalized Hermite functions $(2^n n! \sqrt{\pi})^{-1/2} h_n$ form a complete orthonormal system of eigenfunctions of the Fourier transform in $L^2(\mathbb{R})$. Suitably scaled, they are also eigenfunctions of the harmonic oscillator. For details on special functions, see Folland (1992) or Triebel (1986).

7. *Polynomials.* For each monomial $P(t) = t^m$ and $\varphi \in \mathcal{S}$, it holds that

$$\langle \widehat{P}, \varphi \rangle = \langle 1, \omega^m \widehat{\varphi}(\omega) \rangle = \langle 1, (-j)^m \widehat{\varphi^{(m)}}(\omega) \rangle = \langle j^m \widehat{1}^{(m)}, \varphi \rangle = \langle 2\pi j^m \delta^{(m)}, \varphi \rangle.$$

We thus obtain

$$Q(t) = \sum_{k=0}^{N} a_k t^k \;\multimap\; \widehat{Q}(\omega) = 2\pi \sum_{k=0}^{N} j^k a_k \delta^{(k)}(\omega).$$

8. *Trigonometric Functions and Generalized Fourier Series.* From Example 1 and from $\cos(\omega_0 t) = (e^{j\omega_0 t} + e^{-j\omega_0 t})/2$ and $\sin(\omega_0 t) = (e^{j\omega_0 t} - e^{-j\omega_0 t})/(2j)$, the following correspondences are obtained:

$$e^{j\omega_0 t} \; \circ\!\!-\!\!\bullet \; 2\pi\delta(\omega - \omega_0)$$

$$\cos(\omega_0 t) \; \circ\!\!-\!\!\bullet \; \pi[\delta(\omega + \omega_0) + \delta(\omega - \omega_0)]$$

$$\sin(\omega_0 t) \; \circ\!\!-\!\!\bullet \; j\pi[\delta(\omega + \omega_0) - \delta(\omega - \omega_0)]$$

$$\sum_{k=-\infty}^{+\infty} c_k\, e^{jk\omega_0 t} \; \circ\!\!-\!\!\bullet \; 2\pi \sum_{k=-\infty}^{+\infty} c_k\delta(\omega - k\omega_0).$$

The last correspondence holds for polynomially bounded coefficients $c_k$. This shows that periodic functions and distributions have a discrete spectrum (cf. p. 32). The Fourier transform is a sequence of equidistant impulses, whose strengths, except for the factor $2\pi$, are precisely the Fourier coefficients $c_k$. Observing the results of Sect. 9.1, we notice that periodic sequences of impulses—often called impulse train—again have impulse trains as Fourier transforms:

$$\sum_{k=-\infty}^{+\infty} \delta(t - kp) = \frac{1}{p} \sum_{k=-\infty}^{+\infty} e^{j2k\pi t/p} \; \circ\!\!-\!\!\bullet \; \frac{2\pi}{p} \sum_{k=-\infty}^{+\infty} \delta(\omega - 2k\pi/p)$$

$$\sum_{k=-\infty}^{+\infty} c_k\delta(t - kp) \; \circ\!\!-\!\!\bullet \; \sum_{k=-\infty}^{+\infty} c_k\, e^{-jkp\omega}.$$

From Example 2 and the rule for frequency shifts, with the unit step function $s(t)$, we obtain

$$e^{j\omega_0 t}\, s(t) \; \circ\!\!-\!\!\bullet \; \pi\delta(\omega - \omega_0) + \frac{1}{j(\omega - \omega_0)}$$

$$\cos(\omega_0 t)s(t) \; \circ\!\!-\!\!\bullet \; \frac{\pi}{2}[\delta(\omega - \omega_0) + \delta(\omega + \omega_0)] + \frac{j\omega}{\omega_0^2 - \omega^2}$$

$$\sin(\omega_0 t)s(t) \; \circ\!\!-\!\!\bullet \; \frac{\pi}{2j}[\delta(\omega - \omega_0) - \delta(\omega + \omega_0)] + \frac{\omega_0}{\omega_0^2 - \omega^2}.$$

The occurring rational functions are to be understood here and also in the following example, if they have poles for real $\omega$, as *pseudofunctions* (cf. p. 168).

9. *Rational Functions.* Functions of particular importance in linear system theory are those whose Fourier transforms have the form $Q(j\omega)/P(j\omega)$ with polynomials $P$ and $Q$ (cf. Sect. 11.2). To determine them, one needs to perform the inverse transformation of the typical partial fractions of $Q(j\omega)/P(j\omega)$ (cf. Appendix A, p. 488 for the partial fraction decomposition of rational functions).

For $b \in \mathbb{R}$, $r \in \mathbb{R} \setminus \{0\}$, and $k \in \mathbb{N}$, the following correspondences hold ( $\mathrm{sgn}(t)$ denotes the sign function and $s(t)$ the unit step function):

$$\frac{1}{2} e^{jbt} \frac{t^{k-1}}{(k-1)!} \operatorname{sgn}(t) \circ\!\!-\!\!\bullet \frac{1}{(j\omega - jb)^k}$$

$$-\operatorname{sgn}(r)\frac{t^{k-1}}{(k-1)!} e^{(r+jb)t} s(-\operatorname{sgn}(r)t) \circ\!\!-\!\!\bullet \frac{1}{(j\omega - (r+jb))^k}.$$

*Proof.* With $\widehat{\operatorname{sgn}}(\omega) = \dfrac{2}{j\omega}$ it follows from the equation

$$\frac{1}{(j\omega)^k} = \frac{j^{k-1}}{(k-1)!} \frac{\mathrm{d}^{k-1}}{\mathrm{d}\omega^{k-1}} \left(\frac{1}{j\omega}\right)$$

and the rules for frequency shifts and derivatives (cf. p. 291), immediately the first correspondence. With for $a \in \mathbb{R} \setminus \{0\}$ (cf. p. 276) and the rule for differentiation, we obtain

$$g(t) = \frac{t^{k-1}}{(k-1)!} e^{-|a|t} s(t) \circ\!\!-\!\!\bullet \frac{1}{(j\omega + |a|)^k} = \frac{\operatorname{sgn}(a)^k}{(j \operatorname{sgn}(a)\omega + a)^k},$$

$$g(\operatorname{sgn}(a)t) \circ\!\!-\!\!\bullet \frac{\operatorname{sgn}(a)^k}{(j\omega + a)^k} = \hat{g}\left(\frac{\omega}{\operatorname{sgn}(a)}\right).$$

With $r = -a \neq 0$ and again the rule for frequency shifts, the second aforementioned correspondence is obtained.

We note that the inverse Fourier transform of a rational function $Q(j\omega)/P(j\omega)$ is a causal distribution (i.e., it has its support in the half-axis $[0, \infty[$) if and only if all poles of $Q/P$ have real parts $r < 0$. $Q(j\omega)/P(j\omega)$ belongs to the space $\mathcal{O}_M$ of multipliers in $\mathcal{S}'$ if and only if no zeros of $P$ lie on the imaginary axis (cf. p. 290).

10. *Impulse Sequence of the Discrete Sign Function.* A discrete analog to the sign function is the impulse sequence $T(t) = \displaystyle\sum_{k\in\mathbb{Z}\setminus\{0\}} \operatorname{sgn}(k)\delta(t-k)$. It has as Fourier transform the generalized Fourier series

$$\widehat{T}(\omega) = \sum_{k\in\mathbb{Z}\setminus\{0\}} \operatorname{sgn}(k)e^{-jk\omega} = -2j\sum_{k=1}^{\infty} \sin(k\omega).$$

This is a regularization of the function $- j\cot(\omega/2)$. In a closed interval $I$ around the origin, $I \subset ]-2\pi, 2\pi[$, we can see that $\widehat{T}$ is the regularization by the principal value of $- j\cot\left(\dfrac{\omega}{2}\right)$. By Exercise A6 in Chap. 7, it holds

$$f(\omega) = -2j \ln\left|2\sin\left(\frac{\omega}{2}\right)\right| = 2j\sum_{k=1}^{\infty} \frac{\cos(k\omega)}{k}$$

for $\omega \neq 2\pi m$, $m \in \mathbb{Z}$. Therefore, we have

$\widehat{T}(\omega) = f'(\omega) = -j \cot\left(\dfrac{\omega}{2}\right)$ for $\omega \neq 2\pi m$, $m \in \mathbb{Z}$. Thus, in $I$ it holds

$$\widehat{T}(\omega) = -j \operatorname{vp}\left(\cot\left(\frac{\omega}{2}\right)\right),$$

because for $\varphi$, $\operatorname{supp}(\varphi) \subset I$, and the even function $p(\omega) = \omega \cot\left(\dfrac{\omega}{2}\right)$, we have

$$\left\langle \frac{p(\omega)}{\omega}, \varphi(\omega)\right\rangle = \int_0^\infty \frac{p(\omega)\varphi(\omega) - p(-\omega)\varphi(-\omega)}{\omega}\,d\omega$$

$$= \int_0^\infty (\varphi(\omega) - \varphi(-\omega))\cot\left(\frac{\omega}{2}\right)\,d\omega.$$

More on regularizations can be found in Gel'fand et al. (1964). In contrary to $\sum_{k\in\mathbb{Z}} e^{jk\omega} = 2\pi \sum_{k\in\mathbb{Z}} \delta(\omega - 2\pi k)$, this $\widehat{T}$ is not a measure.

## 10.6   Fourier Transform of Convolutions

*Under suitable conditions on the distributions $T$ and $G$ in $\mathcal{S}'$, the convolution $T * G$ is again a tempered distribution, and for its Fourier transform, the following holds:*

$$\widehat{T * G} = \widehat{T} \cdot \widehat{G}.$$

This equation forms an important basis for calculating convolutions and for many applications of the Fourier transform. However, because convolutions and products for two arbitrary distributions $T$ and $G$ cannot generally be defined and the Fourier transform is only introduced on $\mathcal{S}'$, additional conditions are required for the validity of the convolution equation. The following theorem provides such conditions, which are necessary for our later application examples.

**Theorem 10.6** *Sufficient for the validity of the equation $\widehat{T * G} = \widehat{T} \cdot \widehat{G}$ with distributions $T$ and $G$ in $\mathcal{S}'$ is any of the following conditions:*

1. *$T$ and $G$ are integrable functions.*
2. *$T$ and $G$ are square-integrable functions.*
3. *One of the two distributions $T$ or $G$ has a Fourier transform that belongs to the space $\mathcal{O}_M$ of multipliers in $\mathcal{S}'$.*

*The same conditions on $\widehat{T}$ and $\widehat{G}$ in place of $T$ and $G$ are sufficient for the validity of the multiplication theorem*

$$\widehat{T \cdot G} = \frac{1}{2\pi}\widehat{T} * \widehat{G}.$$

*In particular, all the indicated convolutions are possible and belong to $\mathcal{S}'$.*

**Explication** In general, for the validity of the convolution theorem, first the convolution $T * G$ must be possible and belong to $\mathcal{S}'$, and second, the product $\widehat{T}\widehat{G}$ must be defined and also belong to $\mathcal{S}'$. If the first condition holds, it follows from Fubini's theorem (Appendix B) that the convolution $T * G$ is also an integrable function. The convolution theorem then follows directly by interchanging the order of integration (compare Exercise A14 in Chap. 9). If the second condition holds, then $T * G$ is a continuous bounded function and $\widehat{T * G} = \widehat{T}\widehat{G}$ is an integrable function. For the Fourier transform of square-integrable functions, see the following section. In the third condition with $\widehat{T} \in \mathcal{O}_M$, we have that this is a multiplier in $\mathcal{S}'$. The distributions $T$, whose Fourier transforms $\widehat{T} \in \mathcal{O}_M$, are so-called *rapidly decreasing distributions*. The space of rapidly decreasing distributions is denoted as $\mathcal{O}'_C$, and it holds $\mathcal{F}(\mathcal{O}_M) = \mathcal{O}'_C$. Typical cases for distributions $T \in \mathcal{O}'_C$, which occur in our application examples, are causal fundamental solutions of asymptotically stable linear differential equations with constant coefficients (see Sect. 9.2 and Chap. 11). The convolution $T * G$ exists then for all distributions $G \in \mathcal{S}'$ and belongs to $\mathcal{S}'$. This condition for $\widehat{T * G} = \widehat{T}\widehat{G}$ includes the cases with $T \in \mathcal{S}$ or with $T \in \mathcal{E}'$ having a compact support because then $\widehat{T} \in \mathcal{O}_M$. The following inclusions hold for the spaces considered:

$$\begin{array}{ccccccc} \mathcal{D} & \subset & \mathcal{S} & \subset & \mathcal{O}_M & \subset & \mathcal{E} \\ \cap & & \cap & & \cap & & \cap \\ \mathcal{E}' & \subset & \mathcal{O}'_C & \subset & \mathcal{S}' & \subset & \mathcal{D}' \end{array}$$

($\mathcal{E}$ is the space of infinitely often differentiable functions, equipped with the topology such that $\mathcal{E}'$ is its dual space. We do not use it further on.)

Under the mentioned conditions, the convolution $T * G$, as in Sect. 8.7, p. 190, can be defined with test functions $\varphi \in \mathcal{D}$ by $(T * G)(\varphi) = T(\check{G} * \varphi)$. This equation can then be extended to a definition of the convolution of distributions in $\mathcal{S}'$.

All statements 1–3 also apply to translations and generalized derivatives of the involved distributions $T$ and $G$, if $T$ and $G$ meet one of the stated conditions.

The proof of statement 3 of the theorem requires advanced knowledge of the structure of distributions, which goes beyond the scope of this book (see also the remark after Example 3, p. 295). Interested readers are referred to Schwartz (1957) or Vladimirov (2002). There, the convolution theorem is also proven for further classes of distributions, which we do not address here. Further details on the Fourier transform of convolutions and on products of distributions can be found in Hirata and Ogata (1958), Champeney (1989), and Oberguggenberger (1992), some results of integration theory on the convolution of functions in Appendix B.

## *Examples*

1. *The Unit Step Function.* The convolution $s * s(t) = ts(t)$ for the unit step function $s(t)$ exists in $\mathcal{S}'$ and has the Fourier transform $\dot{js}$, i.e.,

$$\widehat{s * s}(\omega) = -\frac{1}{\omega^2} + j\pi\dot{\delta}(\omega).$$

However, the square of the distribution $\widehat{s}(\omega) = \dfrac{1}{j\omega} + \pi\delta(\omega)$ cannot be formed with the developed calculus.

The term $-1/\omega^2$ of $\widehat{s * s}$ is to be understood as the second generalized derivative of $\ln(|\omega|)$ and denotes the singular distribution $\mathrm{pf}(-1/\omega^2)$ (see p. 168).

2. *Integral Functions and Smoothing.* For functions $f$ from $\mathcal{S}$, the integral function

$$F(t) = \int\limits_{-\infty}^{t} f(s)\, \mathrm{d}s$$

belongs to $\mathcal{S}'$. It can be written as a convolution with the unit step function $s(t)$:

$$F(t) = \int\limits_{-\infty}^{+\infty} f(u)s(t - u),\, \mathrm{d}u = (f * s)(t).$$

Both $\widehat{f * s}$ and $\widehat{f} \cdot \widehat{s}$ belong to $\mathcal{S}'$, and the convolution theorem holds and yields

$$\widehat{F}(\omega) = \frac{\widehat{f}(\omega)}{j\omega} + \pi\, \widehat{f}(0)\delta(\omega).$$

Similarly, with the Fourier transform $\widehat{r_T}(\omega) = 2T\dfrac{\sin(\omega T)}{\omega T}$ of the rectangular function $r_T(t)$ (see p. 275), the smoothing

$$G(t) = \frac{1}{2T} \int\limits_{t-T}^{t+T} f(u)\, \mathrm{d}u = \frac{1}{2T} f * r_T(t)$$

of an integrable function $f$ has the following Fourier transform:

$$\widehat{G}(\omega) = \widehat{f}(\omega)\frac{\sin(\omega T)}{\omega T}.$$

The convolution theorem is applicable because $r_T$ has a bounded support.

3. *Convolution of the Principal Value.* An interesting example is the convolution $f * f$ for the principal value $f(t) = \mathrm{vp}(1/t)$. The convolution exists because $f$ can be represented as the sum of a distribution with bounded support and a square-integrable function: To do this, choose a test function $\varphi \in \mathcal{D}$ that is constantly

one in a neighborhood of zero; then $f\varphi + f(1 - \varphi)$ is such a representation of $f$. Using the convolution theorem of the Fourier transform, it follows that $\widehat{f * f}(\omega) = (-j\pi\,\mathrm{sgn}(\omega))^2 = -\pi^2$; thus $(f * f)(t) = -\pi^2\delta(t)$. Note that $\mathrm{supp}(f) = \mathbb{R}$, but $\mathrm{supp}(f * f) = \{0\}$.

4. *Fourier Transform of Time-Limited and BandLimited Signals.* If $T$ is a time-limited signal, i.e., a distribution in $\mathcal{S}'$ with bounded support, then the Fourier transform $\widehat{T}$ is a multiplier in $\mathcal{S}'$ and thus infinitely often differentiable. Using complex function theory, it can be shown that $\widehat{T}$ can be represented everywhere by its Taylor series (see, e.g., Rudin 1991). From the identity theorem for power series, it follows that the Fourier transform $\widehat{T}$ of a time-limited signal $T \neq 0$ cannot vanish completely on any interval. Similarly, it can be seen that a bandlimited signal $T$, i.e., a signal whose generalized spectral function $\widehat{T}$ has bounded support, is always an infinitely differentiable function that does not vanish on any interval for $T \neq 0$.

5. *Convolutions of Impulse Trains and Products of Fourier Series.* Impulse trains and their convolutions are of fundamental importance in discrete linear filters. We consider four variants that are relevant for applications (see later Sect. 11.6).

(a) *Convolution of a Rapidly Decreasing Impulse Train with a Tempered Impulse Train.*

$$\text{Let } T(t) = \sum_{k=-\infty}^{+\infty} c_k\delta_k \text{ and } G(t) = \sum_{k=-\infty}^{+\infty} d_k\delta_k \text{ be given.}$$

Here, $\delta_k(t) = \delta(t - ka)$ denotes an impulse at $ka$ with a fixed $a > 0$. We assume that the *coefficients $c_k$ are rapidly decreasing*, i.e., $|k|^m c_k \to 0$ for every $m \in \mathbb{N}$ and $|k| \to \infty$. Then $T$ has the Fourier transform

$$\widehat{T}(\omega) = \sum_{k=-\infty}^{+\infty} c_k e^{-jk\omega a}.$$

$\widehat{T}$ is an infinitely differentiable $2\pi/a$-periodic function (see p. 51 and p. 298) and thus a multiplier in $\mathcal{S}'$. The distribution $T$ is an example of a rapidly decreasing distribution. We now further assume that the sequence of *coefficients $d_k$ is polynomially bounded* for $|k| \to \infty$. Then $G \in \mathcal{S}'$ and

$$\widehat{G}(\omega) = \sum_{k=-\infty}^{+\infty} d_k e^{-jk\omega a}.$$

According to condition no. 3 on p. 300, the convolution theorem applies: $\widehat{T * G}(\omega) = \widehat{T}(\omega) \cdot \widehat{G}(\omega)$. The coefficients $h_k$ of the impulse train resulting from the convolution

$$T * G(t) = \sum_{k=-\infty}^{+\infty} h_k\delta_k$$

and the generalized Fourier series $\widehat{T}(\omega) \cdot \widehat{G}(\omega) = \sum_{k=-\infty}^{+\infty} h_k e^{-jk\omega a}$ are given by the *discrete convolution of the coefficients $c_k$ and $d_k$* (see also p. 199):

$$h_k = \sum_{n=-\infty}^{+\infty} c_n d_{k-n}.$$

To prove this, we have to show that the generalized Fourier series $\widehat{T} \cdot \widehat{G}$ has the given coefficients $h_k$.

For this, we consider the partial sums $T_N(t) = \sum_{n=-N}^{+N} c_n \delta_n$ and the convolutions $T_N * G$. For $\varphi \in \mathcal{S}$,

$$\langle T_N * G, \varphi \rangle = \sum_{n=-N}^{+N} \sum_{k=-\infty}^{+\infty} c_n d_k \varphi((n+k)a) = \sum_{m=-\infty}^{+\infty} h_m(N) \varphi(ma)$$

with $h_m(N) = \sum_{n=-N}^{N} c_n d_{m-n}$, since the above series converges absolutely and can be rearranged with the index transformation $n + k = m$. Thus,

$$\widehat{T}_N(\omega) = \sum_{n=-N}^{N} c_n e^{-jn\omega a} \quad \text{and} \quad \widehat{T}_N \cdot \widehat{G}(\omega) = \sum_{k=-\infty}^{+\infty} h_k(N) e^{-jk\omega a}.$$

It holds that $\widehat{T}_N \widehat{G} \to \widehat{T}\widehat{G}$ in $\mathcal{S}'$ and $h_k(N) \to h_k$ as $N \to \infty$: First, the series $\sum_{n=-\infty}^{+\infty} c_n d_{k-n}$ converges absolutely for all $k$ due to the growth conditions of the coefficients $c_n$ and $d_n$, i.e., all $h_k$ are well defined. For each $\varphi \in \mathcal{S}$, second, $\langle (\widehat{T} - \widehat{T}_N)\widehat{G}, \varphi \rangle = \langle \widehat{G}, (\widehat{T} - \widehat{T}_N)\varphi \rangle$. We have that $(\widehat{T} - \widehat{T}_N)\varphi$ in $\mathcal{S}$ converges to zero as $N \to \infty$. This follows from the boundedness of the functions $\omega^p \varphi^{(q)}(\omega)$ $(p, q \in \mathbb{N})$ and the fact that arbitrary derivatives of $\widehat{T} - \widehat{T}_N$ converge uniformly to zero (see p. 135). Thus, as $N \to \infty$, it follows that $\widehat{T}_N \widehat{G} \to \widehat{T}\widehat{G}$ in $\mathcal{S}'$. If $h_{-k}$ now denote the Fourier coefficients of $\widehat{T}\widehat{G}$, then for $N \to \infty$

$$(\widehat{T} - \widehat{T}_N)\widehat{G}(\omega) = \sum_{k=-\infty}^{+\infty} (h_k - h_k(N)) e^{-jk\omega a} \to 0 \text{ in } \mathcal{S}'.$$

Then the inverse Fourier transforms also converge

$$\sum_{k=-\infty}^{+\infty} (h_k - h_k(N)) \delta_k \to 0 \text{ in } \mathcal{S}' \text{ for } N \to \infty.$$

For $k \in \mathbb{Z}$, let $\varphi_k$ be a test function with support in $[(k - 1/2)a, (k + 1/2)a]$, which is one in a neighborhood of $ka$. Then $(h_k - h_k(N))\varphi_k(ka) \to 0$ as $N$ tends to $\infty$, finally for each $k \in \mathbb{Z}$, hence

$$h_k = \lim_{N \to \infty} h_k(N) = \sum_{n=-\infty}^{+\infty} c_n d_{k-n}.$$

(b) *Convolution of an Impulse Sequence with Summable Coefficients with an Impulse Sequence that Has Bounded Coefficients.*

Another variant, which will be interesting for discrete filters in the application examples of Sect. 11.6 in the next chapter, is the convolution of two impulse sequences

$$T(t) = \sum_{k=-\infty}^{+\infty} c_k \delta_k \quad \text{and} \quad G(t) = \sum_{k=-\infty}^{+\infty} d_k \delta_k,$$

of which we assume that $T$ has absolutely summable coefficients and $G$ has bounded coefficients, i.e.,

$$\sum_{k=-\infty}^{+\infty} |c_k| = C < \infty \quad \text{and} \quad |d_n| < M$$

for suitable constants $C$ and $M$ and all $n \in \mathbb{Z}$. This example shows that under suitable conditions, the convolution theorem $\widehat{T * G} = \widehat{T} \cdot \widehat{G}$ can be used to define the product on the right-hand side, even if neither factor $\widehat{T}$ nor $\widehat{G}$ is a multiplier in $\mathcal{S}'$.

Under the mentioned conditions, for $h_k = \sum_{n=-\infty}^{+\infty} c_n d_{k-n}$ and $\varphi \in \mathcal{S}$

because $\sum_{k=-\infty}^{+\infty} |\varphi(ka)| \leqslant A < \infty$ with a suitable constant $A > 0$

$$\sum_{k=-\infty}^{+\infty} |h_k||\varphi(ka)| \leqslant \sum_{k=-\infty}^{+\infty} \left( \sum_{n=-\infty}^{+\infty} |c_n||d_{k-n}| \right) |\varphi(ka)| \leqslant ACM < \infty.$$

The series $\sum_{k=-\infty}^{+\infty} \sum_{n=-\infty}^{+\infty} c_n d_{k-n} \varphi(ka)$ is therefore absolutely convergent, and any rearrangement converges to the same limit. Since the coefficients $h_k$ are bounded, $|h_k| \leqslant MC$, the convolution defined by

$$T * G(t) = \sum_{k=-\infty}^{+\infty} h_k \delta_k = \mathcal{F}^{-1}\left( \sum_{k=-\infty}^{+\infty} h_k e^{-jk\omega a} \right)$$

is a tempered distribution. One can now, as in the previous example, define the product of the series $\widehat{T}$ and $\widehat{G}$ by $\widehat{T}(\omega) \cdot \widehat{G}(\omega) = \sum_{k=-\infty}^{+\infty} h_k \mathrm{e}^{-jk\omega a}$ and thus also in this case obtain the convolution relationship

$$\widehat{T * G}(\omega) = \sum_{k=-\infty}^{+\infty} h_k \mathrm{e}^{-jk\omega a} = \widehat{T}(\omega) \cdot \widehat{G}(\omega).$$

*The following should be noted about the product defined in this way:*
It can be shown that the vector space of continuous functions $f$ on $[0, 2\pi/a]$ with absolutely summable Fourier coefficients $c_k$ is a complete normed space $\mathcal{A}$ if one introduces the norm $\|f\|_{\mathcal{A}} = \sum_{k \in \mathbb{Z}} |c_k|$. The Fourier series $\widehat{T}$ belongs to $\mathcal{A}$. This space is an algebra, i.e., with $f$ and $g$ also $f \cdot g$ belongs to $\mathcal{A}$ (see p. 78), and it holds that $\|fg\|_{\mathcal{A}} \leqslant \|f\|_{\mathcal{A}} \|g\|_{\mathcal{A}}$. One can consider $\widehat{G}$ with the Fourier coefficients $d_k$ then as a continuous linear functional on $\mathcal{A}$ by $\widehat{G}(f) = \sum_{k \in \mathbb{Z}} d_k c_k$. In the literature, $\widehat{G}$ is referred to as a *pseudo-measure* (see Edwards (1982) and the references therein). Because $\mathcal{A}$ is an algebra, one can define the product $f\widehat{G}$ for $f \in \mathcal{A}$ by $f\widehat{G}(p) = \widehat{G}(fp)$. It is also a continuous linear functional on $\mathcal{A}$ due to the norm inequality above. For $f = \widehat{T}$, $f\widehat{G}$ agrees with the product of $\widehat{T}$ and $\widehat{G}$ introduced above. The convolution theorem, also called Fourier exchange theorem, and a way to introduce products of two distributions in the exchange theorem were studied in Hirata and Ogata (1958).

(c) *Convolution of Impulse Sequences with Absolutely Summable Coefficients*
For $T$ and $G$ as in the preceding example, assume now that both distributions have absolutely summable coefficients. Since the coefficients are then bounded, it follows with the same notations as before the validity of

$$\widehat{T * G}(\omega) = \sum_{k=-\infty}^{+\infty} h_k \mathrm{e}^{-jk\omega a} = \widehat{T}(\omega) \cdot \widehat{G}(\omega).$$

Additionally, the proof of Wiener's $1/f$-Theorem (cf. p. 78) shows that the coefficients $h_k$ are also absolutely summable and that the product $\widehat{T}\widehat{G}$ belongs to the normalized algebra $\mathcal{A}$ introduced there. If, for example, $\widehat{G}$ has no zero, then the quotient $\widehat{T}/\widehat{G}$ also lies in $\mathcal{A}$ and is the Fourier transform of a convolution of two impulse sequences with absolutely summable coefficients.

(d) *Convolution of Impulse Sequences with Square Summable Coefficients and the Multiplication of Fourier Series in $L^2([0, 2\pi/\omega_0])$*

(i) First, we investigate the product of Fourier series $f(t) = \sum_{k=-\infty}^{+\infty} c_k \mathrm{e}^{jk\omega_0 t}$

and $g(t) = \sum_{k=-\infty}^{+\infty} d_k \mathrm{e}^{jk\omega_0 t}$ with square summable coefficients. The product $fg$ is integrable over $[0, 2\pi/\omega_0]$, since $|fg| \leqslant |f|^2 + |g|^2$ and

$f$ and $g$ are square-integrable on $[0, 2\pi/\omega_0]$. If $f_N(t) = \sum\limits_{k=-N}^{+N} c_k e^{jk\omega_0 t}$

and $g_N(t) = \sum\limits_{k=-N}^{+N} d_k e^{jk\omega_0 t}$, then as on p. 78 the convergence of $f_N g_N$

to $fg$ follows in the norm of $L^1([0, 2\pi/\omega_0])$ and the Fourier coefficients

of $fg$ are $h_k = \sum\limits_{n=-\infty}^{+\infty} c_n d_{k-n}$. This series is absolutely convergent.

Since $fg$ is integrable over $[0, 2\pi/\omega_0]$, it follows that $\lim\limits_{|k|\to\infty} h_k = 0$

(Riemann-Lebesgue Lemma, p. 282). With $f$ and $g$, the product $fg$ is
also a $2\pi/\omega_0$-periodic distribution in $\mathcal{S}'$.

(ii)  For two impulse sequences $T(t) = \sum\limits_{k=-\infty}^{+\infty} c_k \delta_k$ and $G(t) = \sum\limits_{k=-\infty}^{+\infty} d_k \delta_k$

with square summable coefficients $c_k$ and $d_k$, the convolution is defined

as in the previous examples by $T * G(t) = \sum\limits_{k=-\infty}^{+\infty} h_k \delta_k$ with $h_k$ as above,

and from i) we obtain the validity of the convolution theorem $\widehat{T * G} = \widehat{T} \cdot \widehat{G}$.


## 10.7   Fourier Transform of Square-Integrable Functions

Signals of finite energy play an important role in many applications, mathematically
speaking square-integrable functions. Every such function can also be regarded as a
regular distribution in $\mathcal{S}'$. The vector space $L^2(\mathbb{R})$ of all square-integrable functions
is contained in $\mathcal{S}'$. The Fourier transform of tempered distributions thus also gives
the Fourier transform of square-integrable functions and many of their properties.
Two square integrable functions $f$ and $g$ are considered equal if they represent the
same distribution in $\mathcal{S}'$, i.e., if for all $\varphi \in \mathcal{S}$ it holds that

$$\langle f, \varphi \rangle = \int_{-\infty}^{+\infty} f(t)\varphi(t)\, dt = \int_{-\infty}^{+\infty} g(t)\varphi(t)\, dt = \langle g, \varphi \rangle.$$

The same statement holds with $\mathcal{D}'$ in place of $\mathcal{S}'$, i.e., if the last equation holds for all
$\varphi \in \mathcal{D}$. It can be proven that this is exactly the case if $f(t) = g(t)$ holds for almost
all $t \in \mathbb{R}$, i.e., if $f(t) \neq g(t)$ holds at most on a Lebesgue null set (cf. Appendix B).
On $L^2(\mathbb{R})$, an *inner product* and a *norm* are defined by

$$\langle f | g \rangle = \int_{-\infty}^{+\infty} f(t)\overline{g(t)}\, dt \quad \text{and} \quad \|f\|_2 = \langle f | f \rangle^{1/2}$$

(cf. also p. 62 and later 14.1). Two square-integrable functions $f$ and $g$ represent the same element in $L^2(\mathbb{R})$ if and only if the norm $\|f - g\|_2 = 0$. Without proofs, we state the following important statements, for the verification of which one mainly uses theorems of integration theory (see, for instance, Triebel 1992 or Weidmann 1980):

1. *The vector space $L^2(\mathbb{R})$ is complete with respect to the norm defined above, i.e., every sequence of functions $f_n$ in $L^2(\mathbb{R})$, $n \in \mathbb{N}$, with $\|f_n - f_m\|_2 \to 0$ for $n, m \to \infty$ converges to an element $f \in L^2(\mathbb{R})$.*
2. *The Cauchy-Schwarz inequality holds: For $f, g \in L^2(\mathbb{R})$*

$$|\langle f | g \rangle| \leqslant \|f\|_2 \, \|g\|_2.$$

   *The inner product is continuous in both variables. In the Cauchy-Schwarz inequality equality holds if and only if $f$ and $g$ are linearly dependent, i.e., if $f = \alpha g$ for some $\alpha \in \mathbb{C}$.*
3. *For every $f \in L^2(\mathbb{R})$ there is a sequence of rapidly decreasing functions $f_n \in \mathcal{S}$ that converges to $f$ in $L^2(\mathbb{R})$, i.e., $\lim\limits_{n \to \infty} \|f_n - f\|_2 = 0$.*

With these properties, it is shown that the Fourier transform maps the space $L^2(\mathbb{R})$ onto itself and that the Plancherel equation (p. 283) can be extended to all of $L^2(\mathbb{R})$.

**Theorem 10.7 (The Fourier Transform on $L^2(\mathbb{R})$)** *For the Fourier transform on $L^2(\mathbb{R})$, the following assertions hold:*

1. *If $f \in L^2(\mathbb{R})$ and $f = \lim\limits_{n \to \infty} f_n$ in $L^2(\mathbb{R})$, $f_n \in \mathcal{S}$ for $n \in \mathbb{N}$, then the Fourier transforms $\widehat{f_n}$ converge in $L^2(\mathbb{R})$ to $\widehat{f}$. In particular, $\widehat{f} \in L^2(\mathbb{R})$.*
2. *For any two functions $f$ and $g$ in $L^2(\mathbb{R})$, the Plancherel equation holds*

$$\langle f \,|\, g \rangle = \frac{1}{2\pi} \, \langle \widehat{f} \,|\, \widehat{g} \rangle.$$

   *It is an orthogonality relation: The functions $f$ and $g$ are orthogonal if and only if their Fourier transforms are orthogonal.*
3. *The Fourier transform is continuous, bijective, and continuously invertible on $L^2(\mathbb{R})$.*

**Proof**

1. Because of the Plancherel equation in $\mathcal{S}$ (p. 283) for $f_n$, $f_m$ from $\mathcal{S}$, it holds that

$$\|\widehat{f_n} - \widehat{f_m}\|_2^2 = 2\pi \, \|f_n - f_m\|_2^2.$$

For $f_n \in \mathcal{S}$ with $\lim\limits_{n \to \infty} f_n = f$ in $L^2(\mathbb{R})$, the functions $\widehat{f_n}$ form a Cauchy sequence in $L^2(\mathbb{R})$, which, due to the completeness of $L^2(\mathbb{R})$, converges to a square-integrable function $g$. On the other hand, this sequence converges in $\mathcal{S}'$ to

$\widehat{f}$. Since both convergence in $L^2(\mathbb{R})$ and convergence in $\mathcal{S}'$ imply convergence in $\mathcal{D}'$, $g$ and the Fourier transform $\widehat{f}$ coincide.

2. For $f$ and $g$ from $L^2(\mathbb{R})$, let $f = \lim_{n\to\infty} f_n$ and $g = \lim_{m\to\infty} g_m$ in $L^2(\mathbb{R})$ with appropriate sequences of rapidly decreasing functions $f_n$ and $g_m$. The Plancherel equation on $L^2(\mathbb{R})$ follows from 1) by the continuity of the inner product:

$$2\pi \langle f \mid g \rangle = 2\pi \lim_{n,m\to\infty} \langle f_n \mid g_m \rangle = \lim_{n,m\to\infty} \langle \widehat{f_n} \mid \widehat{g_m} \rangle = \langle \widehat{f} \mid \widehat{g} \rangle.$$

In particular, the Fourier transform $\mathcal{F}$ on $L^2(\mathbb{R})$ is injective and continuous.

3. If $f = \lim_{n\to\infty} f_n$, $f_n \in \mathcal{S}$, then $f = \mathcal{F}(\lim_{n\to\infty} \mathcal{F}^{-1}(f_n))$. Therefore, the Fourier transform is also surjective on $L^2(\mathbb{R})$.

$\square$

The convolution formula and the multiplication theorem hold in the sense of distributions (cf. p. 301 and Appendix B, p. 501). For $f$ and $g \in L^2(\mathbb{R})$, $f * g$ and $\widehat{f} * \widehat{g}$ are bounded continuous functions, which do not necessarily need to be integrable or square-integrable. However, their Fourier transforms are defined when considering $f * g$ or $\widehat{f} * \widehat{g}$ as distributions in $\mathcal{S}'$. The products $fg$ and $\widehat{f}\widehat{g}$ are integrable functions.

**Theorem 10.8** *For all functions $f$ and $g$ in $L^2(\mathbb{R})$, the following convolution relationships hold:*

$$\widehat{f * g} = \widehat{f} \cdot \widehat{g},$$

$$\widehat{f \cdot g} = \frac{1}{2\pi} \widehat{f} * \widehat{g}.$$

***Proof*** For $f_n$ and $g_n$ in $\mathcal{S}$ with $\lim_{n\to\infty} f_n = f$ and $\lim_{n\to\infty} g_n = g$ in $L^2(\mathbb{R})$, we have $\widehat{f_n * g_n} = \widehat{f_n}\widehat{g_n}$ in any case. Using the Cauchy-Schwarz inequality and the Plancherel equation for the $L^1$-norm (cf. p. 501) of $\widehat{fg} - \widehat{f_n}\widehat{g_n}$,

$$\|\widehat{fg} - \widehat{f_n}\widehat{g_n}\|_1 \leqslant \|\widehat{f} - \widehat{f_n}\|_2 \|\widehat{g}\|_2 + \|\widehat{g} - \widehat{g_n}\|_2 \|\widehat{f_n}\|_2$$

$$= 2\pi (\|f - f_n\|_2 \|g\|_2 + \|g - g_n\|_2 \|f_n\|_2).$$

Thus, $\|\widehat{fg} - \widehat{f_n}\widehat{g_n}\|_1$ converges to zero as $n \to \infty$. Since for integrable functions $h$, $\|\widehat{h}\|_\infty \leqslant \|h\|_1$, it follows that

$$\mathcal{F}^{-1}(\widehat{f_n}\,\widehat{g_n}) \longrightarrow \mathcal{F}^{-1}(\widehat{f}\,\widehat{g}) \text{ uniformly as } n \to \infty.$$

Moreover, using Young's inequality (cf. p. 501),

$$\|f * g - f_n * g_n\|_\infty \leqslant \|f - f_n\|_2 \|g\|_2 + \|g - g_n\|_2 \|f_n\|_2,$$

we get

$$\mathcal{F}^{-1}(\widehat{f_n}\,\widehat{g_n}) = f_n * g_n \to f * g \quad \text{uniformly as } n \to \infty.$$

Therefore, the first convolution relationship $\widehat{f * g} = \widehat{f} \cdot \widehat{g}$ follows. The multiplication theorem is proven analogously. $\qquad\square$

**Remark** It can be shown that for $f \in L^2(\mathbb{R})$, the functions

$$\widehat{f_N}(\omega) = \int\limits_{-N}^{+N} f(t)\mathrm{e}^{-j\omega t}\,\mathrm{d}t$$

converge to $\widehat{f}$ in $L^2(\mathbb{R})$, i.e., $\lim\limits_{N\to\infty}\|\widehat{f_N} - \widehat{f}\|_2 = 0$. Analogously, the functions

$$f_N(t) = \frac{1}{2\pi}\int\limits_{-N}^{+N} \widehat{f}(\omega)\mathrm{e}^{j\omega t}\,\mathrm{d}\omega \text{ converge to } f \text{ in } L^2(\mathbb{R}).$$

**Examples**

1. For $\widehat{f}(\omega) = \omega^{-1/4}1_{]0,1]}(\omega)$, the product $\widehat{f}^2$ is integrable but does not belong to $L^2(\mathbb{R})$. Therefore, the convolution $f * f = \mathcal{F}^{-1}(\widehat{f}^2)$ also does not belong to $L^2(\mathbb{R})$ ($1_{]0,1]}$ is the indicator function of $]0, 1]$).
2. The function $f(t) = \sin(at)/t, a > 0$, is not absolutely integrable but is square-integrable. It has the Fourier transform

$$\widehat{f}(\omega) = \begin{cases} \pi & \text{for } |\omega| \leqslant a \\ 0 & \text{otherwise.} \end{cases}$$

This follows from Example 1 on p. 275 using the symmetry rule from p. 279.

Before studying applications in the next chapter, we briefly discuss how the Fourier transform can also be introduced for functions of several variables and how the essential transformation rules extend to this case.

## 10.8   The Fourier Transform for Functions of Several Variables

**Definition** For integrable, complex-valued functions in $p$ variables, the Fourier transform $\widehat{f}$ is defined by

$$\widehat{f}(\boldsymbol{\omega}) = \widehat{f}(\omega_1, \ldots, \omega_p) = \int\limits_{\mathbb{R}^p} f(\mathbf{x})\mathrm{e}^{-j\boldsymbol{\omega}\cdot\mathbf{x}}\,\mathrm{d}\lambda^p(\mathbf{x}).$$

Here, $\boldsymbol{\omega} \cdot \mathbf{x}$ denotes the usual dot product of the vectors $\boldsymbol{\omega} = (\omega_1, \ldots, \omega_p)$ and $\mathbf{x} = (x_1, \ldots, x_p)$ in $\mathbb{R}^p$, $\mathrm{d}\lambda^p(\mathbf{x}) = \mathrm{d}x_1 \mathrm{d}x_2 \ldots \mathrm{d}x_p$ is the $p$-dimensional volume element in Cartesian coordinates, and subsequently $|\mathbf{x}|$ denotes the Euclidean norm of a vector $\mathbf{x}$.

**Example** The function $f(x_1, x_2) = \mathrm{e}^{-(x_1^2 + x_2^2)/2}$ of the two variables $x_1$ and $x_2$ has the Fourier transform

$$
\widehat{f}(\omega_1, \omega_2) = \int\limits_{-\infty}^{+\infty} \int\limits_{-\infty}^{+\infty} \mathrm{e}^{-(x_1^2 + x_2^2)/2} \mathrm{e}^{-j(\omega_1 x_1 + \omega_2 x_2)} \, \mathrm{d}x_1 \mathrm{d}x_2
$$

$$
= \int\limits_{-\infty}^{+\infty} \mathrm{e}^{-j\omega_1 x_1 - x_1^2/2} \, \mathrm{d}x_1 \int\limits_{-\infty}^{+\infty} \mathrm{e}^{-j\omega_2 x_2 - x_2^2/2} \, \mathrm{d}x_2
$$

$$
= 2\pi \mathrm{e}^{-\omega_1^2/2} \mathrm{e}^{-\omega_2^2/2} = 2\pi \mathrm{e}^{-(\omega_1^2 + \omega_2^2)/2}.
$$

Using the notations introduced on p. 187 with the help of multi-indices, the Schwartz space $\mathcal{S}(\mathbb{R}^p)$ of rapidly decreasing functions is defined analogously to Sect. 10.5:

**Definition** $\varphi \colon \mathbb{R}^p \to \mathbb{C}$ belongs to $\mathcal{S}(\mathbb{R}^p)$, if $\varphi$ is infinitely differentiable and if for arbitrary multi-indices $m = (m_1, \ldots, m_p)$ and $k = (k_1, \ldots, k_p)$ in $\mathbb{N}_0^p$ the following holds:

$$
\|\varphi\|_{m,k} = \sup_{\mathbf{x} \in \mathbb{R}^p} \left| \mathbf{x}^m \partial^k \varphi(\mathbf{x}) \right| < \infty.
$$

In other words, the product of an arbitrary polynomial in $p$ variables and any derivative of $\varphi$ remains a bounded function, and every derivative of $\varphi$ decreases for $|\mathbf{x}| \to \infty$ faster than $1/|\mathbf{x}|^N$ for any $N \in \mathbb{N}$.

A sequence of functions $\varphi_n \in \mathcal{S}(\mathbb{R}^p)$ converges to zero, if for arbitrary $m$ and $k \in \mathbb{N}_0^p$ it holds that $\lim\limits_{n \to \infty} \|\varphi_n\|_{m,k} = 0$. With this concept of convergence on $\mathcal{S}(\mathbb{R}^p)$, the set $\mathcal{S}'(\mathbb{R}^p)$ of tempered distributions is defined.

**Definition** A tempered distribution $T \in \mathcal{S}'(\mathbb{R}^p)$ is a continuous, linear mapping $T$ from $\mathcal{S}(\mathbb{R}^p)$ to $\mathbb{C}$.

The statements of Sect. 10.5 about $\mathcal{S} = \mathcal{S}(\mathbb{R})$ and $\mathcal{S}' = \mathcal{S}'(\mathbb{R})$ apply correspondingly also to $\mathcal{S}(\mathbb{R}^p)$ and $\mathcal{S}'(\mathbb{R}^p)$. In particular, the properties of the Fourier transform discussed can be extended to the case of multiple variables. We summarize them in the following table (p. 317). The proofs of these properties are obtained in an analogous manner to the case of a single variable. The Fourier transform is continuous and bijective on $\mathcal{S}(\mathbb{R}^p)$ as in the case $p = 1$. As in Sect. 10.7, the Fourier transform is introduced on the vector space $L^2(\mathbb{R}^p)$ of square-integrable functions. The statements formulated there also hold in $L^2(\mathbb{R}^p)$.

In particular, the Plancherel equation holds again

$$
\|f\| = \left( \int_{\mathbb{R}^p} |f(\mathbf{x})|^2 \, d\lambda^p(\mathbf{x}) \right)^{1/2} = (2\pi)^{-p/2} \left( \int_{\mathbb{R}^p} |\widehat{f}(\mathbf{x})|^2 \, d\lambda^p(\mathbf{x}) \right)^{1/2}
$$

$$
= (2\pi)^{-p/2} \|\widehat{f}\|.
$$

**Theorem 10.9** *The mapping* $(2\pi)^{-p/2}\mathcal{F}$ *with the Fourier transform* $\mathcal{F}$ *is norm-preserving and bijective on the vector space* $L^2(\mathbb{R}^p)$.

**Theorem 10.10 (The Fourier Inversion Formula on $\mathcal{S}(\mathbb{R}^p)$)** *For* $\varphi \in \mathcal{S}(\mathbb{R}^p)$ *and* $\mathbf{x} \in \mathbb{R}^p$, *the following holds:*

$$
\varphi(\mathbf{x}) = \frac{1}{(2\pi)^p} \int_{\mathbb{R}^p} \widehat{\varphi}(\boldsymbol{\omega}) e^{j\boldsymbol{\omega}\cdot\mathbf{x}} \, d\lambda^p(\boldsymbol{\omega}).
$$

***Proof*** To prove the inversion formula, we set $h(\mathbf{s}) = e^{-|\mathbf{s}|^2/2}$. Using the substitution rule for integrals and the similarity relation of the Fourier transform, it follows that

$$
\int_{\mathbb{R}^p} \widehat{\varphi}(\mathbf{s}) h\left(\frac{\mathbf{s}}{n}\right) d\lambda^p(\mathbf{s}) = \int_{\mathbb{R}^p} \varphi(\mathbf{s}) n^p \widehat{h}(n\mathbf{s}) \, d\lambda^p(\mathbf{s}) = \int_{\mathbb{R}^p} \varphi\left(\frac{\mathbf{s}}{n}\right) \widehat{h}(\mathbf{s}) \, d\lambda^p(\mathbf{s});
$$

thus, for $n \to \infty$, taking the limit under the integral (according to the dominated convergence theorem, p. 496) and with $\int_{\mathbb{R}^p} \widehat{h}(\mathbf{x}) \, d\lambda^p(\mathbf{x}) = (2\pi)^p$, we obtain

$$
h(\mathbf{0}) \int_{\mathbb{R}^p} \widehat{\varphi}(\mathbf{s}) \, d\lambda^p(\mathbf{s}) = \int_{\mathbb{R}^p} \widehat{\varphi}(\mathbf{s}) \, d\lambda^p(\mathbf{s}) = \varphi(\mathbf{0}) \int_{\mathbb{R}^p} \widehat{h}(\mathbf{s}) \, d\lambda^p(\mathbf{s}) = (2\pi)^p \varphi(\mathbf{0}).
$$

This is the inversion formula at the point $\mathbf{x} = 0$. The general case follows from the translation rule with $\psi(\boldsymbol{\omega}) = \varphi(\boldsymbol{\omega} + \mathbf{x})$

$$
\varphi(\mathbf{x}) = \psi(\mathbf{0}) = \frac{1}{(2\pi)^p} \int_{\mathbb{R}^p} \widehat{\psi}(\boldsymbol{\omega}) \, d\lambda^p(\boldsymbol{\omega}) = \frac{1}{(2\pi)^p} \int_{\mathbb{R}^p} \widehat{\varphi}(\boldsymbol{\omega}) e^{j\boldsymbol{\omega}\cdot\mathbf{x}} \, d\lambda^p(\boldsymbol{\omega}).
$$

$\square$

### *The Jordan Inversion Formula*

The proof shows that the inversion formula also holds for continuous integrable functions $\varphi$ with an integrable Fourier transform $\widehat{\varphi}$. This variant of the inversion formula goes back to C. Jordan (1838–1922) and does not require differentiability conditions compared to the inversion theorem on p. 271.

**Theorem 10.11 (Jordan's Theorem)** *For any integrable continuous function $f$ on $\mathbb{R}^p$ with an integrable Fourier transform $\widehat{f}$ and all $\mathbf{x} \in \mathbb{R}^p$, the following holds:*

$$f(\mathbf{x}) = \frac{1}{(2\pi)^p} \int\limits_{\mathbb{R}^p} \widehat{f}(\boldsymbol{\omega}) e^{j\boldsymbol{\omega}\cdot\mathbf{x}} \, d\lambda^p(\boldsymbol{\omega}).$$

## *The Fourier Transform for Tempered Distributions on $\mathbb{R}^p$*

**Definition** Following the same procedure as in Sect. 10.5, we define the Fourier transform $\widehat{T}$ of a distribution $T \in \mathcal{S}'(\mathbb{R}^p)$ by

$$\langle \widehat{T}, \varphi \rangle = \langle T, \widehat{\varphi} \rangle.$$

As in Sect. 10.5, properties 1–6 of the following table on p. 317 also apply to tempered distributions $T$ instead of $\varphi$. The convolution relationships from point 7 of the table apply under analogous conditions as those mentioned in the case of one variable on p. 300. Point 8 of the table becomes irrelevant for distributions. The Plancherel equation from point 9 b) corresponds to the distributional form for $T \in \mathcal{S}'(\mathbb{R}^p)$ and $\varphi \in \mathcal{S}(\mathbb{R}^p)$

$$\langle T, \overline{\varphi} \rangle = \frac{1}{(2\pi)^p} \langle \widehat{T}, \overline{\widehat{\varphi}} \rangle.$$

**Examples**

1. *The Rectangular Aperture.* For $a > 0$, $b > 0$ the function

$$f(\mathbf{x}) = \begin{cases} 1 & \text{for } \mathbf{x} \in [-a, a] \times [-b, b] \\ 0 & \text{otherwise} \end{cases}$$

has the Fourier transform

$$\widehat{f}(\boldsymbol{\omega}) = \int\limits_{-a}^{a} e^{-j\omega_1 x_1} \, dx_1 \int\limits_{-b}^{b} e^{-j\omega_2 x_2} \, dx_2 = 4ab \frac{\sin(a\omega_1)}{a\omega_1} \cdot \frac{\sin(b\omega_2)}{b\omega_2}.$$

Fourier transforms of functions of two variables play an important role in *optical diffraction theory*. For example, if coherent light with the amplitude distribution $f$ is diffracted by an aperture in the $(x_1, x_2)$ plane, then in the case of Fraunhofer diffraction, the intensity distribution of the diffracted light on a screen is proportional to $|\widehat{f}|^2$. For the above amplitude distribution $f$ with the rectangular aperture $[-1, 1] \times [-2, 2]$, one obtains $|\widehat{f}|$ (not squared) as shown in

**Fig. 10.6** Surface plot of $|\widehat{f}|$

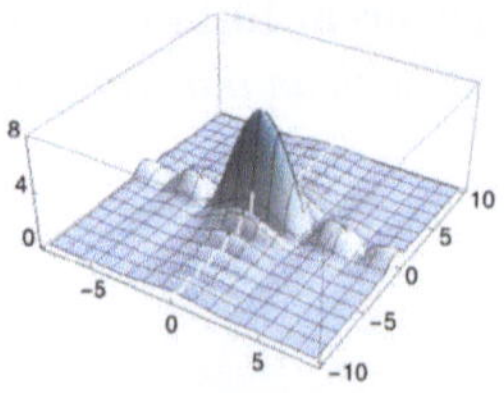

**Fig. 10.7** Visible Fraunhofer
diffraction pattern

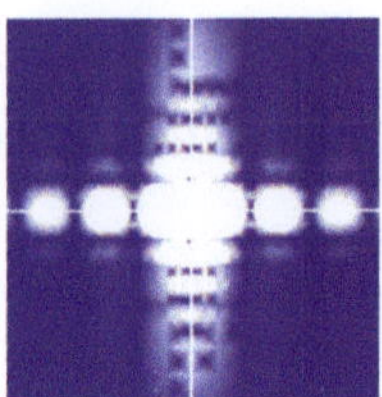

Fig. 10.6 for $|\omega_1| \leqslant 8$, $|\omega_2| \leqslant 10$ and a diffraction pattern by its square as shown
in Fig. 10.7. In this sense, one can *"see the Fourier transform."* We do not go
deeper into the extensive applications in coherent optics here but refer interested
readers to the textbooks Papoulis (1968) or Walker (1988). Related applications
in antenna theory can be found in Bracewell (1999).

2. *The Circular Aperture.* The function $f(x, y) = 1$ for $x^2 + y^2 \leqslant R$, $f(\mathbf{x}) = 0$
   otherwise, is an example of a radial function $f(\mathbf{x}) = f(|\mathbf{x}|)$ ( $\mathbf{x} = (x, y)$ ). With
   polar coordinates $x = r\cos(\phi)$, $y = r\sin(\phi)$, $\omega_1 = \varrho\cos(\psi)$, $\omega_2 = \varrho\sin(\psi)$,
   and $f_0(r) = f(r\cos\phi, r\sin\phi)$, it holds for radial functions $f$ of two variables
   generally that

$$\widehat{f}(\boldsymbol{\omega}) = \int_{-\infty}^{+\infty}\int_{-\infty}^{+\infty} f(x, y)\mathrm{e}^{-j\mathbf{x}\cdot\boldsymbol{\omega}}\,\mathrm{d}x\,\mathrm{d}y = \int_{0}^{2\pi}\int_{0}^{\infty} f_0(r)\mathrm{e}^{-jr\varrho\cos(\phi-\psi)}r\,\mathrm{d}r\,\mathrm{d}\phi.$$

Substituting $\theta = \phi - \psi - \pi/2$ shows that the integral is independent of $\psi$, i.e.,
$\widehat{f}$ is also radial.

$$\widehat{f}(\boldsymbol{\omega}) = \int_{0}^{\infty}\int_{0}^{2\pi} f_0(r)\mathrm{e}^{jr\varrho\sin(\theta)}r\,\mathrm{d}\theta\,\mathrm{d}r = 2\pi\int_{0}^{\infty} f_0(r)J_0(r\varrho)r\,\mathrm{d}r = \widetilde{f}_0(\varrho).$$

Here, $J_0(r\varrho) = \dfrac{1}{2\pi}\displaystyle\int_{0}^{2\pi} \mathrm{e}^{jr\varrho\sin(\theta)}\,\mathrm{d}\theta$ is the *Bessel function of order zero.* The
function $\widetilde{f}_0(\varrho)$ is the *Hankel transform of* $f_0(r)$ *of order zero.*

The two illustrations Figs. 10.8 and 10.9 show for $R = 1$ a cross-section of
$|\widetilde{f}_0|^2$ and a plot of the function $|\widetilde{f}_0|$, whose square corresponds to the according
diffraction pattern.

**Fig. 10.8** Cross-section of $|\widetilde{f_0}|^2$

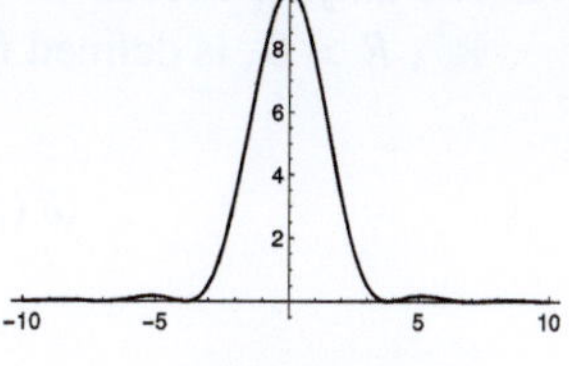

**Fig. 10.9** 3D plot of $|\widetilde{f_0}|$

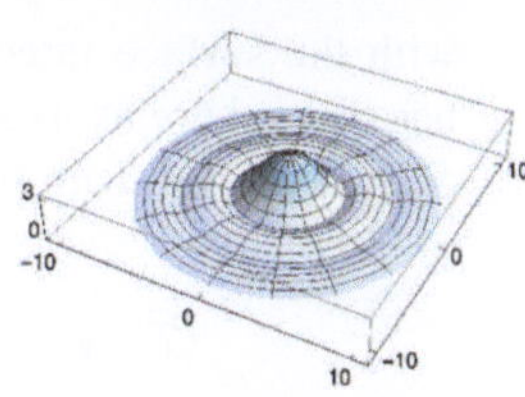

**Fig. 10.10** 3D plot of $|\widehat{f}(\boldsymbol{\omega})|^2$ as in Example 1 with $b \gg a$

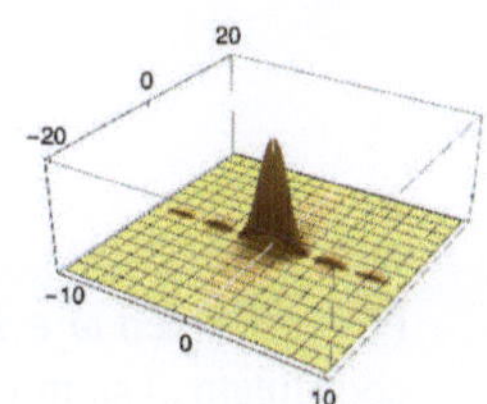

In the given example with $f_0(r) = \begin{cases} 1 & \text{for } 0 \leqslant r \leqslant R \\ 0 & \text{otherwise,} \end{cases}$ we have

$$\widehat{f}(|\boldsymbol{\omega}|) = \widetilde{f_0}(\varrho) = 2\pi \int_0^R J_0(r\varrho)r\,\mathrm{d}r.$$

Specifically, $\widehat{f}(\mathbf{0}) = \pi R^2$ is the area of the disk with radius $R$.

It is easy to see that rotationally invariant functions $f$ on $\mathbb{R}^p$, i.e., $f(A\mathbf{x}) = f(\mathbf{x})$ for matrices $A = (A^*)^{-1}$ with $\det A = 1$ ($A^*$ the transposed matrix), also have a rotationally invariant Fourier transform: $\widehat{f \circ A}(\boldsymbol{\omega}) = \widehat{f}(A\boldsymbol{\omega})$. For detailed treatments of the *Bessel functions* and other *special functions* and for transformations related to the Fourier transform, such as the *Hankel* or the *Mellin transform*, refer to Bracewell (1999), Folland (1992), or Vladimirov (2002).

3. *The Vertical Slit Aperture.*

Consider an infinite slit $g(x, y) = 1_{[-a,a]}(x)1(y)$ of half width $a$. Its Fourier transform is $\widehat{g}(\omega_1, \omega_2) = \widehat{1_{[-a,a]}}(\omega_1)\widehat{1}(\omega_2) = \dfrac{4\pi}{\omega_1} \sin(a\omega_1)\delta(\omega_2)$. We cannot build the square of it. With a slit modeled as a rectangular aperture $f(\mathbf{x})$ as in Example 1 with $a \ll b$, one obtains $|\widehat{f}(\boldsymbol{\omega})|^2$ as shown above in Fig. 10.10 for $a = 1, b = 1000$, and $|\omega_1| \leqslant 10, |\omega_2| \leqslant 20$:

4. *The Surface Measure.* The distribution $T(\mathbf{x}) = \delta(|\mathbf{x}| - R)$, $\mathbf{x} = (x_1, x_2, x_3)$ in $\mathbb{R}^3$, $R > 0$, is defined for $\varphi \in \mathcal{S}(\mathbb{R}^3)$ by

$$\langle \delta(|\mathbf{x}| - R), \varphi(\mathbf{x})\rangle = \int\limits_{|\mathbf{x}|=R} \varphi(\mathbf{x})\, do(\mathbf{x})$$

with the surface integral over the sphere $|\mathbf{x}| = R$. $T$ belongs to $\mathcal{S}'(\mathbb{R}^3)$ and has this sphere as its support. We calculate the Fourier transform of $T$. For $\varphi \in \mathcal{S}(\mathbb{R}^3)$ it holds with the volume element $d\lambda^3(\boldsymbol{\omega}) = d\omega_1 d\omega_2 d\omega_3$

$$\langle \widehat{T}, \varphi\rangle = \langle T, \widehat{\varphi}\rangle = \int\limits_{|\mathbf{x}|=R} \int\limits_{\mathbb{R}^3} \varphi(\boldsymbol{\omega}) e^{-j\boldsymbol{\omega}\cdot\mathbf{x}}\, d\lambda^3(\boldsymbol{\omega}) do(\mathbf{x})$$

$$= \int\limits_{\mathbb{R}^3} \varphi(\boldsymbol{\omega}) \int\limits_{|\mathbf{x}|=R} e^{-j\boldsymbol{\omega}\cdot\mathbf{x}}\, do(\mathbf{x}) d\lambda^3(\boldsymbol{\omega}).$$

For each fixed $\boldsymbol{\omega} \in \mathbb{R}^3$ we now choose $\boldsymbol{\omega}/|\boldsymbol{\omega}|$ as the "north pole" of the spherical coordinate system $(r, \theta, \phi)$ for the calculation of the surface integral.
With $\boldsymbol{\omega} \cdot \mathbf{x} = |\boldsymbol{\omega}||\mathbf{x}|\cos(\theta)$, the angle $\theta$ between $\mathbf{x}$ and $\boldsymbol{\omega}$, and $|\mathbf{x}| = R$, it follows that

$$\langle \widehat{T}, \varphi\rangle = R^2 \int\limits_{\mathbb{R}^3} \varphi(\boldsymbol{\omega}) \int\limits_0^{\pi} \int\limits_0^{2\pi} e^{-jR|\boldsymbol{\omega}|\cos(\theta)} \sin(\theta)\, d\phi d\theta d\lambda^3(\boldsymbol{\omega})$$

$$= 2\pi R^2 \cdot \frac{1}{jR} \int\limits_{\mathbb{R}^3} \frac{\varphi(\boldsymbol{\omega})}{|\boldsymbol{\omega}|} \int\limits_0^{\pi} \frac{\partial}{\partial\theta}\left(e^{-jR|\boldsymbol{\omega}|\cos(\theta)}\right) d\theta d\lambda^3(\boldsymbol{\omega})$$

$$= \frac{4\pi R}{2j} \int\limits_{\mathbb{R}^3} \frac{\varphi(\boldsymbol{\omega})}{|\boldsymbol{\omega}|}\left(e^{jR|\boldsymbol{\omega}|} - e^{-jR|\boldsymbol{\omega}|}\right) d\lambda^3(\boldsymbol{\omega})$$

$$= 4\pi R \int\limits_{\mathbb{R}^3} \varphi(\boldsymbol{\omega}) \frac{\sin(R|\boldsymbol{\omega}|)}{|\boldsymbol{\omega}|}\, d\lambda^3(\boldsymbol{\omega}).$$

Thus, the Fourier transform of $\delta(|\mathbf{x}| - R)$ is calculated, and we obtain the correspondence

$$\delta(|\mathbf{x}| - R) \;\circ\!\!-\!\!\bullet\; 4\pi R \frac{\sin(R|\boldsymbol{\omega}|)}{|\boldsymbol{\omega}|}.$$

This example will be the key to solving initial value problems of the three-dimensional wave equation in Sect. 12.8.

## *Summary*

The Fourier transform can be extended from the class of functions studied in the first sections of this chapter to tempered distributions. This applies in the case of a one-dimensional or $p$-dimensional underlying parameter space. In the examples, we have recognized that a large class of generalized functions, which occur in applications, has tempered Fourier transforms and that the calculation rules familiar from the classical case can also be translated in an appropriate manner (Table 10.1). The spectral concepts and transformation properties provided offer a powerful tool for solving many practical problems. We will explore the scope of Fourier methods associated with distribution theory in a selection of typical applications in the next chapters.

**Table 10.1**  Properties of the Fourier transform on $\mathcal{S}\,(\mathbb{R}^p)$

| $\varphi(\mathbf{x}) = \varphi(x_1, \ldots, x_p) \in \mathcal{S}\,(\mathbb{R}^p)$ | $\widehat{\varphi}(\boldsymbol{\omega}) = \widehat{\varphi}(\omega_1, \ldots, \omega_p)$ |
|---|---|
| 1. Linearity | |
| $\alpha\varphi_1(\mathbf{x_1}) + \beta\varphi_2(\mathbf{x_2})$ | $\alpha\widehat{\varphi}_1(\boldsymbol{\omega}_1) + \beta\widehat{\varphi}_2(\boldsymbol{\omega}_2)$ |
| 2. Symmetry | |
| $\widehat{\varphi}(\mathbf{x})$ | $(2\pi)^p \varphi(-\boldsymbol{\omega})$ |
| $\overline{\varphi}(\mathbf{x})$ | $\overline{\widehat{\varphi}(-\boldsymbol{\omega})}$ |
| 3. Similarity, scaling | |
| $\varphi(\alpha\mathbf{x}) \qquad (\alpha \neq 0)$ | $\dfrac{1}{\lvert \alpha \rvert^p}\, \widehat{\varphi}\left(\dfrac{\boldsymbol{\omega}}{\alpha}\right)$ |
| 4. Translations | |
| $\varphi(\mathbf{x} - \mathbf{x}_0)$ | $e^{-j\boldsymbol{\omega}\cdot\mathbf{x}_0}\widehat{\varphi}(\boldsymbol{\omega})$ |
| $e^{j\boldsymbol{\omega}_0\cdot\mathbf{x}}\varphi(\mathbf{x})$ | $\widehat{\varphi}(\boldsymbol{\omega} - \boldsymbol{\omega}_0)$ |
| 5. Differentiation ($k$ multi-index, cf. p. 317) | |
| $\partial^k \varphi(\mathbf{x}),\ \ k \in \mathbb{N}_0^p$ | $(j\boldsymbol{\omega})^k\widehat{\varphi}(\boldsymbol{\omega})$ |
| 6. Multiplication with a polynomial | |
| $\mathbf{x}^k\varphi(\mathbf{x}),\ \ k \in \mathbb{N}_0^p$ | $j^{\lvert k\rvert}\partial^k\,\widehat{\varphi}(\boldsymbol{\omega})$ |
| 7. Convolution and modulation | |
| $(\varphi_1 * \varphi_2)(\mathbf{x})$ | $\widehat{\varphi}_1(\boldsymbol{\omega})\widehat{\varphi}_2(\boldsymbol{\omega})$ |
| $\varphi_1(\mathbf{x})\varphi_2(\mathbf{x})$ | $\dfrac{1}{(2\pi)^p}\,(\widehat{\varphi}_1 * \widehat{\varphi}_2)(\boldsymbol{\omega})$ |
| 8. Continuity, Riemann-Lebesgue Lemma | |
| For $\varphi(\mathbf{x}) \in \mathcal{S}\,(\mathbb{R}^p)$ | $\widehat{\varphi}(\boldsymbol{\omega})$ is continuous and bounded and $\displaystyle\lim_{\lvert\boldsymbol{\omega}\rvert\to\infty} \widehat{\varphi}(\boldsymbol{\omega}) = 0$ holds |
| 9. Plancherel equation | |
| (a) $\quad \displaystyle\int_{\mathbb{R}^p} \lvert\,\varphi(\mathbf{x})\,\rvert^2\, d\lambda^p(\mathbf{x})$ | $= \dfrac{1}{(2\pi)^p}\displaystyle\int_{\mathbb{R}^p} \lvert\widehat{\varphi}(\boldsymbol{\omega})\rvert^2\, d\lambda^p(\boldsymbol{\omega})$ |
| (b) $\quad \displaystyle\int_{\mathbb{R}^p} \varphi_1(\mathbf{x})\overline{\varphi_2(\mathbf{x})}\, d\lambda^p(\mathbf{x})$ | $= \dfrac{1}{(2\pi)^p}\displaystyle\int_{\mathbb{R}^p} \widehat{\varphi}_1(\boldsymbol{\omega})\overline{\widehat{\varphi}_2(\boldsymbol{\omega})}\, d\lambda^p(\boldsymbol{\omega})$ |

For the Fourier transform of arbitrary distributions in $\mathcal{D}'$, please refer to the textbooks by Schwartz (1957), Gel'fand et al. (1964), Horváth (1966), and Zemanian (2010).

## 10.9   Exercises

**(A1)** Compute the Fourier transforms of

$$f_1(t) = \mathrm{sgn}(t),$$

$$f_2(t) = t^2 \,\mathrm{sgn}(t) \qquad (\mathrm{sgn}(t) \text{ is the sign function}),$$

$$f_3(t) = t^2 e^{-t^2},$$

$$f_4(t) = e^{-at^2 + bt + c} \qquad (a > 0),$$

$$f_5(t) = \frac{t}{1 - jt}.$$

**(A2)** Which function $f$ corresponds to the following spectral function?

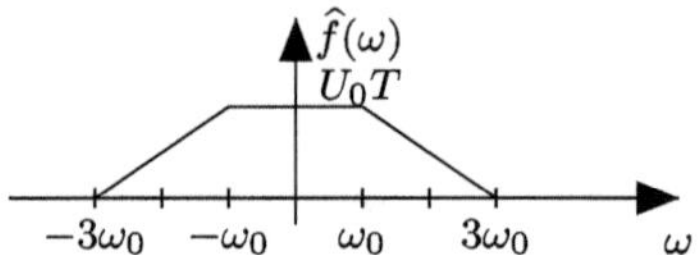


Hint: $\widehat{f}$ is the sum of two triangular frequency windows.

**(A3)** For $a > 0$, let

$$F_a(x) = \frac{1}{\pi} \frac{a}{a^2 + x^2} = \frac{1}{a} F_1\left(\frac{x}{a}\right).$$

Compute for $a > 0$, $b > 0$ the Fourier transforms $\widehat{F_a * F_b}$ and $\widehat{F_{a+b}}$.

**(A4)** Use the Plancherel equation to compute the integral $\displaystyle\int_0^\infty \frac{\sin(ax)\sin(bx)}{x^2}\, dx$.

**(A5)** The Plancherel equation holds for all square-integrable functions $f$. For which bandlimited function $g$, i.e., $\widehat{g}(\omega) = 0$ for $|\omega| > \Omega$, is the mean square error $\varepsilon = \int_{-\infty}^{+\infty} |f(t) - g(t)|^2 \, dt$ minimal when approximating $f$ by $g$?

**(A6)** Compute the Fourier transform of the distribution $S(t) = \dfrac{1}{t} * r_T(t)$.
($r_T$ is the rectangular function $r_T(t) = 1$ for $|t| < T$, $r_T(t) = 0$ for $|t| \geqslant T$).

**(A7)** (a) Find examples of functions and distributions $f$ and $g$, such that their convolution exists, but the convolution theorem for the Fourier transform does not hold.

(b) Find examples of functions $f$ such that for the integral function $f * s$ the formula $\widehat{f * s}(\omega) = \dfrac{\widehat{f}(\omega)}{j\omega} + \pi\,\widehat{f}(0)\delta(\omega)$ does not hold (where $s(t)$ is the unit step function).

**(A8)** Compute the Fourier transform of $f(t) = \cos(t)\,(s(t+1) - s(t-1))$, and justify that the multiplication theorem from p. 300 holds.

**(A9)** (a) Compute the causal solution $h$ of $au^{(3)} + b\dot{u} = \delta$ for $a, b > 0$.

(b) Compute $\mathcal{F}^{-1}(\widehat{f})$ for $\widehat{f}(\omega) = \dfrac{1}{(j\omega)^3 a + j\omega b}$.

(c) Compute $\mathcal{F}^{-1}(\widehat{f})$ for

$$\widehat{f}(\omega) = \frac{1}{(j\omega)^3 a + j\omega b} + \frac{\pi}{b}\delta(\omega) - \frac{\pi}{2b}(\delta(\omega - \omega_0) + \delta(\omega + \omega_0))$$

with $\omega_0 = \sqrt{b/a}$ for positive $a, b$.

**(A10)** (a) Compute $\mathcal{F}^{-1}(\widehat{f})$ for $\widehat{f}(\omega) = \dfrac{5\omega + 9 - 10j}{\omega^2 - 4j\omega - 13}$, and (b) $\mathcal{F}^{-1}(\widehat{h}_{RC,\alpha})$ for

$$\widehat{h}_{RC,\alpha}(\omega) = \begin{cases} 1 & \text{for } |\omega| \leqslant b - a \\ & \qquad (0 < a < b, \alpha = a/b) \\ 0 & \text{for } b + a \leqslant |\omega| \\ \cos^2\left(\pi\dfrac{|\omega| - (b - a)}{4a}\right) & \text{for } b - a \leqslant |\omega| \leqslant b + a. \end{cases}$$

Sketch $\widehat{h}_{RC,\alpha}$ and $h_{RC,\alpha}$, and discuss the meanings of the parameters $a$, $b$, and $\alpha$ (see also *Raised Cosine Pulse*, p. 390 and p. 398). Hint: $\widehat{h}_{RC,\alpha}$ is real and even (see p. 273). Use addition theorems for the $\cos^2$ function.

**(A11)** Compute the Fourier transforms of

$$f_1(x, y) = xye^{-(x^2+y^2)} \quad \text{and} \quad f_2(x, y) = \begin{cases} 1 & \text{for } (x-1)^2 + (y-1)^2 \leqslant 1 \\ 0 & \text{otherwise.} \end{cases}$$

**(A12)** Reflect on the fact that the generalized derivative $\dot{f}$ of $f(t) = \sin\left(e^{t^2}\right)$ is indeed a tempered distribution, but that generally for $\varphi \in \mathcal{S}$, the equation

$$\langle \dot{f}, \varphi \rangle = \int_{-\infty}^{+\infty} 2te^{t^2} \cos\left(e^{t^2}\right) \varphi(t)\, dt \text{ does not hold.}$$

**(A13)**  For $a, b > 0$ compute the Fourier transforms of $f(t) = \dfrac{e^{-a|t|} - e^{-b|t|}}{t}$ and

$g(t) = \dfrac{1 - e^{-b|t|}}{t}$. Hint: For $\widehat{g}$ use $\lim\limits_{a\to 0+} e^{-a|t|}$ and $\lim\limits_{a\to 0+} \arctan(\omega/a)$.

**(A14)**$^\star$  Compute the Fourier transforms of the following distributions:

$$f(t) = |t|, \quad g(t) = t^n \mathrm{sgn}(t), \quad h(t) = ts(t),$$

$$p(t) = \mathrm{pf}(t^{-n}) \text{ for } n \in \mathbb{N}, \quad q(t) = \mathrm{pf}(|t|^{\lambda-1}) \ (0 < \lambda < 1).$$

Hint: Use in the last example $e^{-rt} t^{\lambda-1} s(t) \longrightarrow t^{\lambda-1} s(t)$ in $\mathcal{S}'$ for $r \to 0+$

and $\displaystyle\int_0^\infty t^{\lambda-1} e^{-(r+j\omega)t}\, dt = (r+j\omega)^{-\lambda} \Gamma(\lambda)$ for $r > 0$ with the well-known

Gamma function.

**(A15)**$^\star$  Compute the Fourier transform of $f(x) = \dfrac{1}{\cosh(ax)}$ for $a > 0$.

For which $a$ is $f$ an eigenfunction of the Fourier transform and to what eigenvalue?

(Hint: Use the residue theorem and integrals over rectangle paths with base $[-R, R]$ and upper edge $[-R + j\pi/a, R + j\pi/a]$ with $R \to \infty$.)

**(A16)**$^\star$  (a)  *Fourier Transform Under Linear Coordinate Transformation.*

Let $A$ be a real regular $(p \times p)$ matrix and $B$ be the transposed matrix of $A^{-1}$. Show that for $T \in \mathcal{S}'(\mathbb{R}^p)$ holds (see notation on p. 187)

$$\widehat{T_A} = |\det(B)|\, (\widehat{T})_B.$$

(b)  Let $P$ be the parallelogram bounded in the plane by the lines $y = \pm 1$ and $y = x \pm 1$. Compute the Fourier transform of the function $f_p$, which is one on $P$ and zero elsewhere.

**(A17)**  Plot $f(\omega, n) = -2 \sum\limits_{k=1}^{n} \sin(k\omega)$ for large $n$ (say $n = 30$) and $-\cot(\omega/2)$ to illustrate the regularization as on p. 299. Zoom in around zero, and calculate some integrals around zero with $f(\omega, n)\varphi(\omega)$, $\varphi \in \mathcal{D}$ such that it has $\mathrm{supp}(\varphi) \subset\ ]-2\pi, 2\pi[$ including the singularity of $\cot(\omega/2)$. Compare the approximation $\langle f(\omega, n), \varphi \rangle$ with $\langle -\mathrm{vp}(\cot(\omega/2)), \varphi \rangle$. Take, for example, $\varphi(\omega) = e^{-1/(1-\omega^2)} 1_{[-1,1]}(\omega)$, any derivative of it or product with a smooth function.

# Chapter 11
# Basics of Linear Filters

**Abstract** This chapter treats in detail the application of the Fourier transform in linear filtering as an important part in Electrical Engineering and other fields. The first part is devoted to the continuous case, e.g., analog linear circuits. A fundamental theorem on the convolution representation of causal translation-invariant linear systems (LTI systems) is the starting point. All systems are considered as operators on a signal space of distributions. It is shown how different types of linear filters can be mathematically designed. The examples are lowpass, bandpass, allpass, and bandstop filters with their mathematical representation and with their realization by standard circuits. Conditions for stability of linear filters by their frequency response are deduced. The second part on discrete linear filters starts again with a theorem that causal systems have a convolution representation by their impulse response. Counterexamples are given, if causality is missing and the input signal space is the space of bounded discrete signals. The z-transform is introduced for a treatment of discrete LTI systems with their transfer functions. Invertibility and design of causal linear phase FIR filters is studied. Causal IIR filters are calculated with the bilinear transform. All topics are completed with examples and exercises.

## 11.1  Signals

The aim of this chapter is to introduce some fundamental concepts and results of linear system theory. Important mathematical tools in this context are the calculus of distributions and Fourier analysis, whose significance for this field will be presented below. The subject of system theory is the study of signals and systems that transmit signals. There are myriads of such systems in devices of our everyday life.

Transmission systems play an important role in physics, communication, and control engineering. These are physical systems that process time- or space-dependent input signals and output them in a modified form as output signals. A distinction is made between analog, i.e., continuous, and discrete signal processing. Examples of transmission systems include electrical circuits, telephone systems that transmit conversations, a guitar string that is set into vibration, an object that is



R. Brigola, *Fourier Analysis and Distributions*, Texts in Applied Mathematics 79,
https://doi.org/10.1007/978-3-031-81311-5_11

irradiated with light and partially absorbs and partially reflects it, audio and video transmission systems, and many more.

The permissible and possible input and output signals for a given system are generally restricted to certain classes. Therefore, in the first section, we introduce suitable signal spaces that allow both analog and discrete linear systems to be treated mathematically uniformly, and in the second section, we describe the linear systems that will be studied subsequently.

The mathematical model for a transmission system is a transformation $L$ between a set $\mathcal{Z}$ of permissible input signals and a set $\mathcal{A}$ of possible output signals. In deterministic models—which we will restrict ourselves to in the following—the sets $\mathcal{Z}$ and $\mathcal{A}$ are vector spaces of functions or distributions $f$.

The variable of a signal is often time. When considering continuous parameters, one speaks of *analog signals* $f$. If the parameter is discrete, then $f$ is called a *discrete signal*. The values of the signals, in the discrete mathematical model subsequently the coefficients of an impulse sequence, are assumed to be real or complex numbers. If the coefficient range of discrete signals is also discretized— as in technical sampling systems of digital communication—one speaks of *digital signals*. We will not discuss the effects of quantization, i.e., the approximation of real or complex values by discrete values, but refer to the extensive literature on digital signal processing for this. Furthermore, we will restrict ourselves to the case of one-dimensional parameter sets from $\mathbb{R}$ or from the set $a\mathbb{Z}$ with a fixed chosen $a > 0$. The number $1/a$ then represents the sampling frequency of a discrete system.

For continuous parameters, we use the following signal spaces for $\mathcal{Z}$ or $\mathcal{A}$:

1. $\mathcal{D}$ and $\mathcal{D}'$, the spaces of test functions and distributions (see Chap. 8)
2. $\mathcal{D}'_r$, the subvector space of distributions in $\mathcal{D}'$ with support in $[r, \infty[$, and $\mathcal{D}'_R$, the space of causal distributions:

$$\mathcal{D}'_R = \bigcup_{r \in \mathbb{R}} \mathcal{D}'_r$$

3. $\mathcal{D}'_p$, the space of $p$-periodic distributions
4. $\mathcal{S}$ and $\mathcal{S}'$, the spaces of rapidly decreasing test functions and tempered distributions, which we introduced with the Fourier transform in Chap. 10
5. $\mathcal{S}'_r$, the space of tempered distributions in $\mathcal{S}'$ with support in $[r, \infty[$, $r \in \mathbb{R}$, and $\mathcal{S}'_R$, the space of causal tempered distributions:

$$\mathcal{S}'_R = \bigcup_{r \in \mathbb{R}} \mathcal{S}'_r$$

6. $\mathcal{E}'$, the space of distributions with bounded support
7. $\mathcal{O}'_C$, the space of rapidly decreasing distributions. A distribution $T$ belongs to $\mathcal{O}'_C$ if and only if its Fourier transform $\widehat{T}$ is a multiplier in $\mathcal{S}'$.
(See p. 290, p. 299, p. 301, and the example on p. 303.)

8. $L^p(\mathbb{R})$ $(1 \leq p < \infty)$ and $L^\infty(\mathbb{R})$, the spaces of $p$-integrable and that of measurable essentially bounded functions on $\mathbb{R}$ (see Appendix B, p. 491). They can be considered as subspaces of $\mathcal{S}'$.

Except for $\mathcal{D}'_r$ and $\mathcal{S}'_r$, all these spaces have the property that for each element $f$, every time-shifted element $f_\tau$ (see p. 186) with a translation $\tau(t) = t + t_0$ $(t, t_0 \in \mathbb{R})$ is again contained in the respective same signal space. For $\mathcal{D}'_r$ and $\mathcal{S}'_r$, this holds only for translations to the right.

For statements about the continuity of operators on these signal spaces, one needs assumptions about convergence, mathematically speaking about the topology in these spaces. For the spaces $\mathcal{D}'$, $\mathcal{E}'$, and $\mathcal{S}'$ and their subspaces, we use the already introduced convergence concepts from p. 182, p. 202, and p. 290. For the convergence of a sequence $T_n \to T$ as $n \to \infty$ in $\mathcal{D}'_R$ (respectively, $\mathcal{S}'_R$), we require that $\langle T_n, \varphi \rangle \to \langle T, \varphi \rangle$ for all $\varphi$ in $\mathcal{D}$ (respectively, $\mathcal{S}$) and additionally that all $T_n$ and $T$ belong to $\mathcal{D}'_r$ (respectively, $\mathcal{S}'_r$) for a suitable $r \in \mathbb{R}$. More precisely, $\mathcal{D}'_R$ and $\mathcal{S}'_R$ should have the so-called topology of the inductive limit of the spaces $\mathcal{D}'_r$ or $\mathcal{S}'_r$. The significance of this topology will become clear later in an example on p. 327. Convergence in the spaces $L^p$ and $L^\infty$ is the usual norm convergence in these spaces with the norms given in Appendix B, p. 500, which we denote by $\|.\|_p$ and $\|.\|_\infty$.

When dealing with discrete parameters, we use the following signal spaces for $\mathcal{Z}$ or $\mathcal{A}$:

1. $\mathcal{X} = \{ x = \sum\limits_{n=-\infty}^{+\infty} x_n \delta_n \mid x_n \in \mathbb{C} \}$, the space of discrete signals.

   *Here, $\delta_n$ denotes the Dirac functional $\delta(t - na)$.* The step size $a > 0$ is arbitrary and assumed to be fixed from now on. All signal spaces specified below are to be understood as subspaces of $\mathcal{X}$ with the same step size $a$. The space $\mathcal{X}$ is endowed with the topology induced by $\mathcal{D}'$. A sequence $(x_N)_{N \in \mathbb{N}}$ of discrete signals $x_N = \sum\limits_{n=-\infty}^{+\infty} x_{N,n} \delta_n$ converges in $\mathcal{X}$ to $x = \sum\limits_{n=-\infty}^{+\infty} x_n \delta_n$ if and only if $\lim_{N \to \infty} x_{N,n} = x_n$ for all $n \in \mathbb{Z}$, meaning it converges pointwise to $x$.

2. $\mathcal{X} \cap \mathcal{D}'_k$, the space of discrete signals in $\mathcal{D}'$ with support in $[ka, \infty[$ for $k \in \mathbb{Z}$, $a$ as above, and $\mathcal{X} \cap \mathcal{D}'_R$, the space of causal discrete signals with the topology induced by $\mathcal{D}'_R$

3. $\mathcal{X} \cap \mathcal{S}'$, $\mathcal{X} \cap \mathcal{S}'_k$, and $\mathcal{X} \cap \mathcal{S}'_R$, the corresponding spaces of tempered discrete signals in $\mathcal{S}'$ with their induced convergence concepts

4. $\mathcal{X} \cap \mathcal{E}'$, the space of discrete signals with finite support, with $\mathcal{E}'$ convergence

5. $\mathcal{X} \cap \mathcal{O}'_C$, the space of discrete signals with rapidly decaying coefficients

6. $l_d^p = \{x \in \mathcal{X} \mid \sum_{n=-\infty}^{+\infty} |x_n|^p < \infty \}$, for $1 \le p < \infty$, the space of discrete signals $x$ with coefficients $x_n$ that are $p$-summable. The norm of $x$ is $\|x\|_p = (\sum_{n=-\infty}^{+\infty} |x_n|^p)^{1/p}$.

7. $l_d^\infty = \{x \in \mathcal{X} \mid \sup_{n \in \mathbb{Z}} |x_n| < \infty \}$, the space of discrete signals $x$ with bounded coefficients. The norm of $x$ with coefficients $x_n$ is $\|x\|_\infty = \sup\{|x_n| : n \in \mathbb{Z}\}$.

Except for $\mathcal{X} \cap \mathcal{D}_k'$ and $\mathcal{X} \cap \mathcal{S}_k'$, each of these signal spaces includes every translation $x_\tau$ of any element $x$. The translations $\tau$ here are mappings on $a\mathbb{Z}$ of the form $\tau(na) = (n - m)a$, $m \in \mathbb{Z}$. *A translation $x_\tau$ of $x \in \mathcal{X}$ by $ma$ to the right with $m \in \mathbb{N}$ is simply the convolution $x_\tau = x * \delta_m$ of $x$ with $\delta_m$.*

$\mathcal{X} \cap \mathcal{D}_k'$ and $\mathcal{X} \cap \mathcal{S}_k'$ also contain all right translations of their elements. Convergence of a sequence $(x_N)$ to $x$ in $\mathcal{X} \cap \mathcal{D}_R'$ (respectively, $\mathcal{X} \cap \mathcal{S}_R'$) means, in addition to pointwise convergence, that all $x_N$ and $x$ lie in a space $\mathcal{X} \cap \mathcal{D}_k'$ (respectively, $\mathcal{X} \cap \mathcal{S}_k'$) for some suitable $k \in \mathbb{Z}$. The spaces $l_d^p$ ($1 \le p \le \infty$) can be equipped with norms from the respective spaces $l^p(\mathbb{Z})$ (cf. Appendix B, p. 501). The mapping $x = \sum_{k=-\infty}^{+\infty} x_k \delta_k \to (x_k)_{k \in \mathbb{Z}}$ is then an isometry between $l_d^p$ and $l^p(\mathbb{Z})$, and the embeddings of $l_d^p$ spaces into $\mathcal{X}$ are continuous. Examples of signals from different signal spaces will be seen in the following sections.

## 11.2  Translation-Invariant Linear Systems

In the following definition, we assume that the parameter set $I$ satisfies $I = \mathbb{R}$ or $I = a\mathbb{Z}$. The definition then applies equally to continuous and discrete linear systems. The signal spaces $\mathcal{Z}$ and $\mathcal{A}$ are spaces from the lists in the previous section according to this distinction of the parameter set.

**Definition**

1. A system $L \colon \mathcal{Z} \to \mathcal{A}$ is called linear if $L$ is a linear operator, i.e., if for $f_1, f_2 \in \mathcal{Z}$ and arbitrary constants $c_1$ and $c_2$ it holds

$$L(c_1 f_1 + c_2 f_2) = c_1 L f_1 + c_2 L f_2.$$

2. It is called translation-invariant if $L$ can be interchanged with translations $\tau$ of the parameter set $I$ with the translation operator $T_\tau$ on $\mathcal{D}'$, defined by $T_\tau f = f_\tau$, i.e., if for all $f \in \mathcal{Z}$ and $Lf = g$, it holds

$$L(T_\tau f) = T_\tau(Lf) = g_\tau.$$

A shifted input signal then results in a correspondingly shifted output signal.

3. The system is called causal or realizable if an output signal can only be observed when an input signal is present, i.e., if for all $t_0 \in I$ and all $f \in \mathcal{Z}$ with support in $[t_0, \infty[$, it holds that the support of $Lf$ is also in $[t_0, \infty[$.

**Remark** If the (continuous or discrete) parameter of the signals represents time, then one also speaks of *time-invariant* instead of translation-invariant linear systems, also denoted as LTI Systems.

**Examples**

1. Examples of analog time-invariant linear systems include a delay line, described by $Lf(t) = f(t - t_0)$, $t_0 > 0$, or a differentiator, described by $Lf(t) = c\dot{f}(t)$, $c \in \mathbb{R}$. Both systems are causal, and the signal classes $\mathcal{Z}$ and $\mathcal{A}$ can be chosen as the spaces of distributions $\mathcal{D}'$ or $\mathcal{S}'$. Other examples can be found in Sect. 9.3, where causal time-invariant linear systems are given by linear differential equations with constant coefficients. The associated operators $L$ in all these examples are given by convolution with the causal impulse response. Permissible input signals are all distributions that can be convolved with the impulse response, for example, all causal input signals $f \in \mathcal{D}'_R$. For $\mathcal{A}$, one can then choose $\mathcal{A} = \mathcal{D}'_R$, in the case of stable systems (cf. later p. 334) also $\mathcal{Z} = \mathcal{A} = \mathcal{S}'_R$.

2. Let an asymptotically stable, linear differential equation $P(D)u = f$ be given. Here, $P$ is a polynomial with real constant coefficients, whose zeros have negative real parts, and $D = \mathrm{d}/\mathrm{d}t$. $\mathcal{D}'_p$ denotes the space of $p$-periodic generalized Fourier series.

   The system $L : \mathcal{D}'_p \to \mathcal{D}'_p$, where each $p$-periodic generalized Fourier series $f(t) = \sum\limits_{k=-\infty}^{\infty} c_k\, e^{j2\pi kt/p}$ is assigned the periodic solution $L(f)(t) =$

   $\sum\limits_{k=-\infty}^{\infty} h_k c_k\, e^{j2\pi kt/p}$ with $h_k = 1/P(2\pi kj/p)$, is an analog time-invariant linear system (cf. p. 65 and p. 214).

3. (a) The operator $L : \mathcal{X} \to \mathcal{X}$ on the space $\mathcal{X}$ of discrete signals, $Lx = y$ with $y_n = x_{n-k}$ $(k \in \mathbb{Z})$, describes a discrete translation-invariant linear system. Similarly, averaging $L : \mathcal{X} \to \mathcal{X}$ on the space $\mathcal{X}$ of discrete signals, for example, $Lx = y = x * \left( \dfrac{1}{2}\delta_0 + \dfrac{1}{2}\delta_1 \right)$ with $y_n = \dfrac{1}{2}x_n + \dfrac{1}{2}x_{n-1}$.

   (b) Analogously, all discrete recursive systems $L : \mathcal{X} \to \mathcal{X}$, $Lx = y$ with

   $$y_n = \sum_{k=0}^{N} a_k x_{n-k} + \sum_{l=1}^{M} b_l y_{n-l}, \quad (N \geq 0,\ M \geq 1,\ a_k,\ b_l \in \mathbb{R})$$

   are linear, translation-invariant, and causal (cf. also p. 114), if the initial rest state is assumed, i.e., $y = 0$ for $x = 0$.

## 11.3   Analog Linear Filters, Continuity, and Causality

All analog systems mentioned in the first example above have in common that the operator $L$ can be represented as a *convolution with the impulse response* $h = L\delta$. For $f \in \mathcal{Z}$ it holds

$$Lf = L(f * \delta) = f * L\delta = f * h. \tag{11.1}$$

The operators $L$ in these examples with the signal spaces $\mathcal{D}'_R$ or $\mathcal{S}'_R$ for $\mathcal{Z}$ and $\mathcal{A}$ are interchangeable with the convolution (cf. p. 196) and are also *continuous*: For $n \to \infty$ it holds

$$f_n \to f \text{ in } \mathcal{Z} \Longrightarrow Lf_n = f_n * h \to f * h = Lf \text{ in } \mathcal{A}. \tag{11.2}$$

Time-invariant linear systems with these two properties are called *linear filters*.

So, for a linear system $L : \mathcal{Z} \to \mathcal{A}$ to be a linear filter, it requires first the convolution representation (11.1) and second the continuity property (11.2) for suitably given signal spaces $\mathcal{Z}$ and $\mathcal{A}$.

If one assumes that in practice causal systems and signals with "finite past" play a role, i.e., signals whose support is bounded below, then the global importance of linear filters is shown in the following theorem, in which we summarize results from Schwartz (1957), Albrecht and Neumann (1979), and Neumann (1980). In this theorem, the above mathematical continuity requirement on $L$ is replaced by the *causality condition* for a linear system, which is physically justified in numerous applications. The space $\mathcal{Z}$ of input signals is the vector space $\mathcal{D}'_R$ of all distributions on $\mathbb{R}$ whose support is bounded below or the space $\mathcal{E}'$ of distributions with bounded support. The image space of $L$ is the vector space $\mathcal{D}'$ of all distributions. A linear system of the form $Lf = f * h$ is causal if and only if the support of $h$ is in the interval $[0, \infty[$.

### *Automatic Continuity of Causal Time-Invariant Linear Systems*

**Theorem 11.1 (Theorem of Albrecht-Neumann)** *On the spaces* $\mathcal{Z} = \mathcal{E}'$ *or* $\mathcal{Z} = \mathcal{D}'_R$, *every causal time-invariant linear system* $L: \mathcal{Z} \to \mathcal{D}'$ *is automatically continuous and is represented by the convolution with the impulse response* $h = L\delta$, *i.e., for every* $f \in \mathcal{Z}$ *it holds* $Lf = f * h$.

The proof of this fundamental result requires functional analytic methods and cannot be carried out here. In the aforementioned works by E. Albrecht and M. Neumann, one can find further variants of this theorem for other signal spaces as well. The theorem shows that many physically relevant, realizable time-invariant linear systems are linear filters and always possess the required continuity property (11.2). Their transfer behavior can thus be empirically determined by measuring the impulse response. Convolution representations are also obtained for linear dissipative multiparameter systems. Readers interested in the topic may refer to the works of König (1959), Day (1961), Weiss (1991), the textbooks by Zemanian (1995), Vladimirov (2002), Partington (2004), and Pohl and Boche (2010), and the references found therein for further study.

For continuity, the convergence concepts in $\mathcal{E}'$ and $\mathcal{D}'_R$ introduced on p. 202 are crucial, in other words, the topology of these spaces. In the following example, we provide a causal system, either on $\mathcal{Z} = \mathcal{D}'_R$ or on $\mathcal{Z} = \mathcal{E}'$, defined by a convolution operator, which is not continuous if the chosen space $\mathcal{Z}$ is equipped with $\mathcal{D}'$-convergence, i.e., with the coarser topology induced by $\mathcal{D}'$.

**Example** Consider $L : \mathcal{Z} \to \mathcal{D}'$, defined by $Lf = f * h$ with $h = \displaystyle\sum_{n=0}^{\infty} \delta_n$ and $\mathcal{Z} = \mathcal{D}'_R$ or $\mathcal{Z} = \mathcal{E}'$. The *accumulator* $L$ is a causal convolution operator.

For $n \in \mathbb{N}$ let

$$f_n = -\delta_{-n} + \delta_0.$$

Then it holds

$$\mathcal{D}'\text{-}\lim_{n \to \infty} f_n = \delta_0,$$

but (compare also the example on p. 200)

$$Lf_n = - \sum_{k=-n}^{-1} \delta_k \quad \text{and thus} \quad \mathcal{D}'\text{-}\lim_{n \to \infty} Lf_n = - \sum_{k=-\infty}^{-1} \delta_k \neq L\delta_0 = h.$$

On the other hand, the sequence $f_n$ is not convergent in the topologies defined on p. 202 in $\mathcal{D}'_R$ or in $\mathcal{E}'$, and $L$ is continuous on these spaces with these topologies according to the preceding theorem.

For specifically given operators $L$, which are usually directly defined as convolution operators in applications of system theory, it can often also be shown directly that the necessary continuity property is present, even if the permissible signal class $\mathcal{Z}$ does not coincide with $\mathcal{D}'_R$ or $\mathcal{E}'$. These systems are then also linear filters, even if they do not satisfy the causality condition. Linear filters, i.e., continuous convolution systems on suitable signal spaces, are studied below.

### The Frequency Response of Analog Linear Filters

We now consider convolution systems $Lf = f * h$, for which $h$ is a tempered distribution (cf. p. 289). If, for example, the system is described by an ordinary

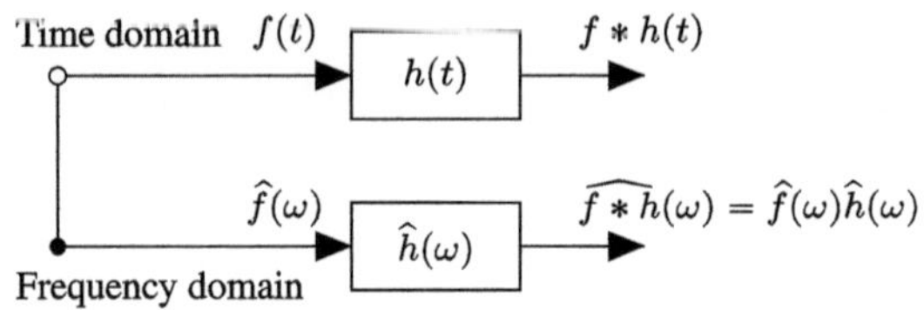



**Fig. 11.1** Schematic linear filter

differential equation as in Sect. 9.2, then its causal fundamental solution is tempered if and only if it contains no exponentially growing components (cf. p. 64 and p. 217), i.e., if no zero of the characteristic polynomial has a positive real part. For the input signals $f$, we assume that the convolutions $f * h$ exist and that the convolution theorem $\widehat{f * h} = \hat{f}\,\hat{h}$ holds for the Fourier transforms. This is the case, for instance, if the signals $f$ are "time-limited" distributions, i.e., if they have bounded supports, if $f$ and $h$ are square-integrable functions (cf. p. 300), or if $f$ belongs to $\mathcal{S}'$ and $\hat{h}$ is a multiplier in $\mathcal{S}'$ (cf. p. 290). From the continuity of the Fourier transformation on the associated signal spaces and the convolution theorem, it then follows directly the continuity of the considered convolution operators $Lf = f * h$.

**Definition** For a linear filter $Lf = f * h$ that satisfies the above requirements, the Fourier transform $\hat{h}$ is referred to as the frequency response of the filter.

If the $\delta$-impulse belongs to the space $\mathcal{Z}$ of input signals, then the frequency response is the Fourier transform of the impulse response $L\delta = h$. The transmission behavior of the system can then be schematically described as in Fig. 11.1:

With the inverse Fourier transform $\mathcal{F}^{-1}\left(\hat{f}\,\hat{h}\right)$ of $\widehat{f * h}$, the system response in the time domain is given by the equation

$$f * h = \mathcal{F}^{-1}\left(\hat{f}\,\hat{h}\right). \tag{11.3}$$

A descriptive interpretation of this formula, which is to be read as an equation between tempered distributions, is obtained if we assume more specific conditions on $h$ and $f$. If, for example, $f$ is a piecewise continuously differentiable, time-limited signal and $|h|^2$ is also an integrable function with $h$, then the output signal $f * h$ is continuous and integrable and has an integrable Fourier transform (cf. p. 301). The Jordan inversion theorem from p. 312 and the convolution theorem then give the pointwise representation of the system response $f * h$ to the input quantity $f$:

$$f * h(t) = \frac{1}{2\pi} \int_{-\infty}^{+\infty} \widehat{f * h}(\omega)\, e^{j\omega t}\, d\omega = \frac{1}{2\pi} \int_{-\infty}^{+\infty} \hat{f}(\omega)\hat{h}(\omega)\, e^{j\omega t}\, d\omega. \tag{11.4}$$

The input signal $f(t) = \dfrac{1}{2\pi} \int_{-\infty}^{+\infty} \hat{f}(\omega)\, e^{j\omega t}\, d\omega$ is a superposition of harmonic oscillations whose amplitudes and phases are expressed as a function of frequency by the spectral function $\widehat{f}(\omega)$. The amplitudes $|\widehat{f}(\omega)|$ of the oscillations involved are amplified or attenuated by the factor $|\widehat{h}(\omega)|$ during transmission, and the phases $\arg\left(\widehat{f}(\omega)\right)$ are additively changed by the phases $\arg\left(\widehat{h}(\omega)\right)$. *The frequency response $\widehat{h}$ contains information about amplitude and phase changes during the transmission of $f$. It thus provides crucial information for the analysis and design of linear transmission systems, whose properties can thus be specified.*

It can be seen, due to $\widehat{\delta} = 1$ and $f * \delta = f$, that the representation (11.4) also holds if $h$ contains additional $\delta$ components. Examples that meet the special conditions mentioned above can be found in Sects. 5.2 and 9.3. For more general situations, for example, if $h$ also contains derivatives $\dot{\delta}$ of $\delta$, the pointwise representation (11.4) requires additional differentiability properties of $f$. The growth of $\widehat{h}(\omega)$ can then be compensated by sufficiently rapid decay of $|\widehat{f}(\omega)|$ for $|\omega| \to \infty$ in the case of correspondingly smooth input signals $f$, so that the product $\widehat{f}\,\widehat{h}$ results in an integrable function and the representation (11.4) remains valid. For time-limited distributions $f$ and tempered distributions $h$ without further assumptions, the somewhat more abstract distribution equation (11.3) is always valid—and that is the advantage of the distribution method. In particular, we recognize that no new frequencies are generated when filtering a signal $f$: If $\widehat{f}(\omega) = 0$, then also $\widehat{f\,h}(\omega) = 0$.

**Definition** For a filter with a piecewise differentiable frequency response

$$\widehat{h}(\omega) = |\widehat{h}(\omega)|\,e^{j\,\arg(\widehat{h}(\omega))},$$

we refer to the function $A(\omega) = |\widehat{h}(\omega)|$ as amplitude response. The function $\Phi(\omega) = \arg(\widehat{h}(\omega))$ is called the phase response. The almost everywhere defined function $D(\omega) = -\mathrm{d}\Phi(\omega)/\mathrm{d}\omega$ is called phase delay or group delay.

For a frequency-dependent group delay, signals with a large bandwidth are transmitted through the filter with correspondingly strong phase changes. For example, group delay changes of about 1–3 ms in signals in audio systems are already perceptibly above the threshold of audibility and require phase equalization in high-fidelity components. To illustrate regular frequency responses—for example, with the help of a computer algebra system—one can graphically represent the amplitude response, the phase response, or the phase delay separately. For reasons of symmetry, it is sufficient to represent over the half-axis $\omega \geq 0$ (cf. again p. 273). In electrical engineering, the frequency response is often represented as a locus curve in the complex plane dependent on the parameter $\omega$. Signals $f$ in this field of application are often time-dependent voltage waveforms.

**Examples**

1. *Differentiator.*   The frequency response $\widehat{h}$ of the ideal differentiator, considered as a filter on $\mathcal{S}'$ with the causal impulse response $a\dot{\delta}(t)$, is given by the differentiation rule of the Fourier transform as $\widehat{h}(\omega) = j\omega a$. This shows that the differentiator is very sensitive to high-frequency noise in the input signal.

They are amplified by the factor $\omega a$. For a time-limited, twice continuously differentiable input signal $f$, it follows with Jordan's inversion formula from p. 312

$$f * h(t) = a\dot{f}(t) = \frac{1}{2\pi} \int\limits_{-\infty}^{+\infty} j\omega a\,\widehat{f}(\omega)\,\mathrm{e}^{j\omega t}\ \mathrm{d}\omega = \frac{a}{2\pi} \int\limits_{-\infty}^{+\infty} \widehat{\dot{f}}(\omega)\,\mathrm{e}^{j\omega t}\ \mathrm{d}\omega .$$

We have already encountered an approximate realization as an electrical circuit on p. 156. If one wants a system that acts as a differentiator for signal components with low frequencies and strongly attenuates high-frequency signal components, then one can choose as frequency response

$$\widehat{h}_b(\omega) = \frac{j\omega a}{(1 + jb\omega)^2}$$

with $b > 0$. Amplitude and phase changes in the transmission correspond to those of the ideal differentiator up to the angular frequency $\omega = 1/b$. Above this angular frequency, the amplitude drops to zero, and the phase shifts from $\pi/2$ to $-\pi/2$. The impulse response of this approximate differentiator is (exercise)

$$h_b(t) = \frac{a(1 - t/b)}{b^2}\ \mathrm{e}^{-t/b}\,s(t)$$

with the unit step function $s(t)$. The system is causal.

2. *Integrator.*    An ideal integrator is a linear system $L$, which is defined with a real constant $K$ by $Lf(t) = K \int\limits_{-\infty}^{t} f(x)\mathrm{d}x$ (Fig. 11.2).

Its response to the unit step function $s(t)$ is $a(t) = Kts(t)$, and its impulse response is $\dot{a}(t) = Ks(t)$. Its frequency response is $\widehat{h}(\omega) = K/(j\omega) + K\pi\delta(\omega)$.

Here, $1/\omega$ denotes the principal value $\mathrm{vp}(\omega^{-1})$ (cf. p. 167). If one chooses the space $\mathcal{S}$ or the set of integrable functions with bounded support as the input signal class $\mathcal{Z}$, then the convolution theorem holds for the Fourier transform. Integrators are causal linear systems. They can be approximately realized in a manner similar to differentiators (cf. p. 156) using suitably configured operational amplifiers in electrical circuits.

3. *Ideal Lowpass Filter.*    If the amplitude response $|\widehat{h}(\omega)|$ of a filter becomes small as soon as $|\omega|$ exceeds a cutoff frequency $\omega_c > 0$, then it is called a lowpass filter. The angular frequency $\omega_c$ is referred to as the cutoff frequency, the interval $]-\omega_c, \omega_c[$ as the passband, and its complement as the stopband. Low-pass filters play an important role in communications technology. We will see in Sect. 12.1

**Fig. 11.2** Schematic integrator

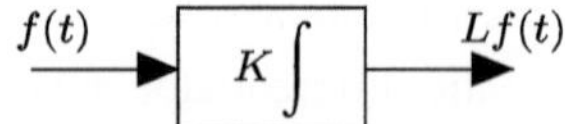

that they can be used for the reconstruction of a continuous signal from suitable samples. The *ideal lowpass filter,* also called the *Küpfmüller lowpass,* operates distortion-free in the passband, i.e., its response $Lf$ to an input signal $f$ has the same form as $f$, but possibly time-delayed. It is defined by

$$Lf(t) = A_0 f(t - t_0) \qquad (A_0 > 0,\ t_0 > 0)$$

for all $f$ with $\mathrm{supp}(\widehat{f}) \subset\ ]-\omega_c, \omega_c[$. Components of a signal $f$ with frequencies $\omega$, $|\omega| \geq \omega_c$, are cut off. The frequency response is therefore

$$\widehat{h}(\omega) = A_0\, e^{-j\omega t_0}\, r_{\omega_c}(\omega),$$

where $r_{\omega_c}(\omega) = 1$ for $|\omega| < \omega_c$, $r_{\omega_c}(\omega) = 0$ for $\omega \geqslant \omega_c$.

If one chooses the input signal space $\mathcal{Z}$ as the space of square-integrable functions $L^2(\mathbb{R})$ or the space $\mathcal{E}'$ of time-limited distributions, then the convolution theorem holds for the Fourier transform. For a time-limited input signal $f$, $\widehat{f}$ is an infinitely differentiable function (see p. 303), and application of $f * h$ to a test function $\varphi \in \mathcal{S}$ shows

$$\langle f * h, \varphi \rangle = \langle \widehat{f}\,\widehat{h}, \mathcal{F}^{-1}\varphi \rangle = \frac{A_0}{2\pi} \int_{-\infty}^{+\infty} \int_{-\omega_c}^{+\omega_c} \widehat{f}(\omega)\, e^{j\omega(t-t_0)}\, d\omega\ \varphi(t)\, dt.$$

Thus,

$$f * h(t + t_0) = \frac{A_0}{2\pi} \int_{-\omega_c}^{+\omega_c} \widehat{f}(\omega)\, e^{j\omega t}\, d\omega.$$

From this representation of the system response in the time domain, one can see, in addition to the time delay of $t_0$ and the band limitation in the frequency domain, that the output signal $f * h$ is an infinitely differentiable function. No matter how irregular $f$ is, band limitation produces a smoothed signal, which, however, is no longer time-limited (Illustration Fig. 11.3).

The ideal lowpass filter is *not causal* and therefore cannot be realized as a transmission system by any electrical circuit. The filter, considered in $\mathcal{E}'$, has the impulse response

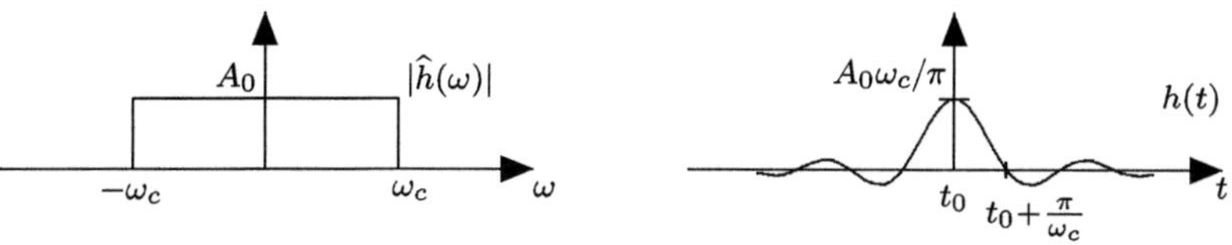


**Fig. 11.3** Frequency and impulse response of an ideal lowpass filter

$$h(t) = \frac{A_0 \sin\big(\omega_c(t - t_0)\big)}{\pi(t - t_0)}.$$

It would be present for any previous time interval for an exciting pulse at time $t = 0$. Despite the noncausality, the Küpfmüller lowpass is very useful in many theoretical studies and also in the design of practical transmission systems due to its simple form. In practice, filters that are realizable and stable and that approximately have the same frequency response are sought. *Stability* means that a bounded input signal also results in a bounded output signal.

If one wants to construct linear filters with a causal impulse response by specifying their frequency response, it should be noted that its real and imaginary parts, in other words amplitude distortion and phase distortion, cannot be chosen independently (see p. 274). There are criteria that indicate for which frequency responses there are realizable, i.e., causal filters. Mention should be made of the theorems of R. Paley and N. Wiener and the Hilbert transform. References on this topic include Paley and Wiener (1934), Dym and McKean (1985), Papoulis (1987), Stein and Weiss (1971), or Pohl and Boche (2010). We will not go further into this topic mathematically, but instead show in the following example how to construct *realizable, stable approximation filters with a rational frequency response*.

## *Butterworth Lowpass Filter*

Realizable, stable lowpass filters for signals in $L^\infty(\mathbb{R})$ or more generally from $\mathcal{S}'$ are obtained with the principal design method as frequency response $\widehat{h}$ a rational function

$$\widehat{h}(\omega) = \frac{K}{P(j\omega)}$$

to be set, where $K$ is a real constant and $P$ is a polynomial of degree $n$ with real positive coefficients, whose zeros have negative real parts:

$$P(j\omega) = (j\omega)^n + a_{n-1}(j\omega)^{n-1} + \ldots + a_1(j\omega) + a_0.$$

The constant $K$, the order $n$ of the filter, and the coefficients $a_0, \ldots, a_{n-1}$ are to be determined such that the filter meets the desired requirements for gain, passband, and stopband in each case. For example, $|\widehat{h}(\omega)|$ should lie in the shaded area, i.e., meet the following tolerance scheme. Since $|\widehat{h}|$ is even, it suffices to consider the half-axis $\omega \geq 0$ (cf. Fig. 11.4).

The realizability arises from the fact that a filter with a rational frequency response $\widehat{h}(\omega) = K/P(j\omega)$ can be described by a differential equation of the form

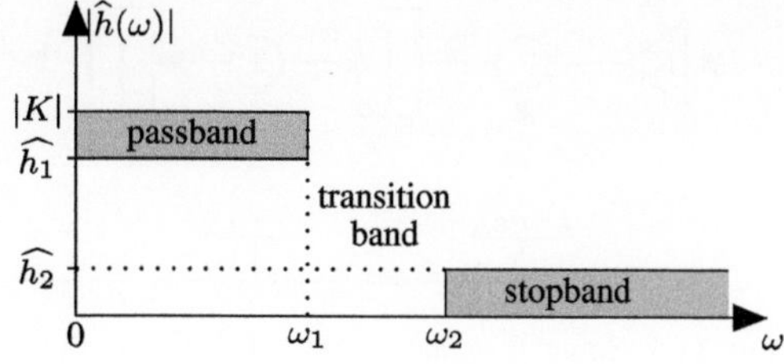



**Fig. 11.4** Tolerance scheme of a Butterworth lowpass filter

$$u^{(n)} + a_{n-1}u^{(n-1)} + \ldots + a_1\dot{u} + a_0u = Kf.$$

This equation has a unique causal fundamental solution and unique causal solutions $u$ for right-hand sides $f$ with $\mathrm{supp}\,(f) \subset [0, \infty[$ (see Sect. 9.2). The corresponding causal system is given by $Lf = f * h = u$.

Substituting the causal fundamental solution $h$ and taking the Fourier transform of the differential equation result in

$$\widehat{h^{(n)}} + a_{n-1}\widehat{h^{(n-1)}} + \ldots + a_0\widehat{h} = ((j\omega)^n + a_{n-1}(j\omega)^{n-1} + \ldots + a_0)\widehat{h} = K\widehat{\delta} = K.$$

The causal fundamental solution $h$ of the differential equation has the Fourier transform $\widehat{h}(\omega) = K/P(j\omega)$ exactly if all zeros of $P$ have negative real parts (see also p. 298). Since then the regular distribution $\widehat{h}$ belongs to the space $\mathcal{O}_M$ of multipliers in $\mathcal{S}'$, the convolution theorem for the Fourier transform holds (see p. 290 and p. 300).

For a right-hand side $Kf$ with $\mathrm{supp}\,(f) \subset [0, \infty[$, we now substitute

$$\dot{v}_0 = u^{(n)} + a_{n-1}u^{(n-1)} + \ldots + a_1\dot{u} = Kf - a_0u$$

and consider $v_0$ as the causal response of the system $\dot{v}_0 = g$ to the input quantity $g = Kf - a_0u$. By integration follows

$$v_0 = u^{(n-1)} + a_{n-1}u^{(n-2)} + \ldots + a_2\dot{u} + a_1u.$$

Since $\mathrm{supp}\,(v_0) \subset [0, \infty[$, no integration constant occurs. With further substitutions we obtain the following first-order differential equation system:

$$\dot{v}_0 = Kf - a_0u\,, \quad \dot{v}_k = v_{k-1} - a_ku \text{ for } 1 \leqslant k \leqslant n-1, \quad u = v_{n-1}.$$

*This system with the "state variables" $v_0, \ldots, v_{n-1}$ can be replicated as an electronic circuit with proportional elements, adders, and integrators* (see also p. 231). References to implementations in circuit technology can be found at the end of the section.

The circuit is described by the following signal flow diagram in Fig. 11.5:

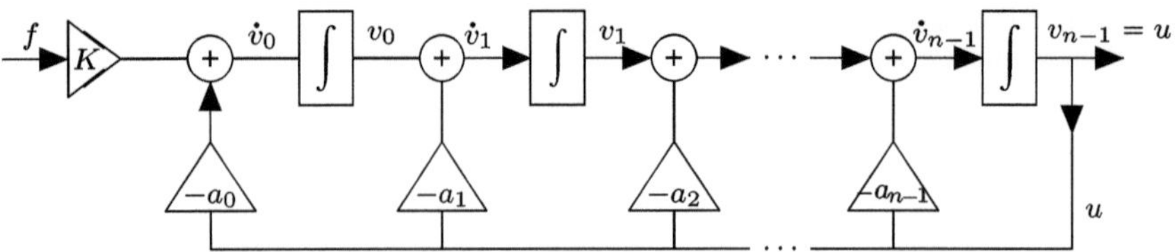

**Fig. 11.5** Signal flow and realizability by a circuit

The *stability* of the system can be characterized by the fact that its impulse response $h$ is *an integrable function*. For a signal $f$ whose amplitudes are bounded by $|f(t)| \leqslant M$ ($t \in \mathbb{R}$), the filter response $f * h$ is then also bounded with $|f * h(t)| \leqslant M \int_{-\infty}^{+\infty} |h(t)| \, dt$, i.e., the filter is stable. Integrability of $h$ means that the zeros of the characteristic polynomial $P$ of the differential equation must have negative real parts; otherwise components of the form $\alpha t^m e^{\beta t} s(t)$ with $\Re(\beta) \geqslant 0$, $m \in \mathbb{N}_0$, $\alpha \in \mathbb{R}$, would be present in the impulse response. These components would not be integrable. This property of the zeros of $P$, already assumed in the approach, thus ensures the desired stability. Because the polynomial $P$ has real coefficients according to the approach,

$$|\widehat{h}(\omega)|^2 = \frac{K^2}{P(j\omega)P(-j\omega)},$$

and the location of the zeros of the polynomial $Q(z) = P(z)P(-z)$ shows symmetry to both the real and the imaginary axis in the complex plane. Conversely, given zeros $z_1, \ldots, z_{2n}$ (all $z_k \neq 0$) that exhibit this double symmetry, they form a polynomial $Q$ that can be factored as $Q(z) = P(z)P(-z)$, where $P$ has positive real coefficients and all zeros of $P$ have negative real parts. Accordingly, $\widehat{h}(\omega) = K/P(j\omega)$ is the frequency response of a realizable stable filter.

For the *Butterworth lowpass* the function $|\widehat{h}|^2$ should run as horizontally as possible below the cutoff frequency $\omega_c > 0$ for $\omega \geq 0$. Since $\omega/\omega_c < 1$ there, this requirement is best fulfilled if $|\widehat{h}|^2$ depends only on the highest power of $\omega/\omega_c$. For $\omega/\omega_c < 1$ the lower powers of $\omega/\omega_c$ provide large contributions to the denominator of $|\widehat{h}|^2$, causing a drop in gain. With the substitution of the parameter $j\omega$ by the complex variable $z$, the Butterworth lowpass filter, named after Butterworth (1930), is given by

$$Q(z) = P(z)P(-z) = 1 + \left(\frac{z}{j\omega_c}\right)^{2n}.$$

The order $n$ and $\omega_c$ must be determined from the filter requirements. To determine the coefficients of $P$, we consider the zeros of $Q$. For a given order $n$ of the filter and given cutoff frequency $\omega_c$, they are given by

$$z_k = j\omega_c \, e^{j(2k+1)\pi/(2n)} \qquad (k = 0, \ldots, 2n - 1).$$

The zeros with negative real part are the $z_k$ with $k = 0, \ldots, n - 1$. Since $P$ is assumed to have positive coefficients (the same signs are necessary for stability), it follows from the assumption $P(0) = Q(0) = 1$.

With $(-z_0)(-z_1) \cdots (-z_{n-1}) = \omega_c^n$ and $Q(j\omega) = 1 + (\omega/\omega_c)^{2n}$, it follows that

$$\widehat{h}(\omega) = \frac{K\omega_c^n}{(j\omega - z_0)(j\omega - z_1) \cdots (j\omega - z_{n-1})} \quad \text{and} \quad |\widehat{h}(\omega)| = \frac{|K|}{\sqrt{1 + (\omega/\omega_c)^{2n}}}.$$

To give an example with specific filter requirements, we demand the following values: DC gain $K = 1$, passband edge $\omega_1/(2\pi) = 3\,\text{kHz}$, minimum passband gain $\widehat{h}_1 = 0.9$, stopband edge $\omega_2/(2\pi) = 10\,\text{kHz}$, and maximum stopband gain $\widehat{h}_2 = 0.1$.

The required order $n$ of the filter results from $|\widehat{h}(\omega)|^2 = K^2/(1 + (\omega/\omega_c)^{2n})$ as $(\omega_1/\omega_c)^{2n} \leqslant K^2/\widehat{h}_1^2 - 1 = \alpha_1$ and $(\omega_2/\omega_c)^{2n} \geqslant K^2/\widehat{h}_2^2 - 1 = \alpha_2$: For $n$ it must hold that $n \geqslant (\alpha_1/\alpha_2)/(2\ln(\omega_1/\omega_2))$. From $(\omega_1/\omega_c)^{2n} \leqslant \alpha_1$, $(\omega_2/\omega_c)^{2n} \geqslant \alpha_2$, it follows that $\omega_c$ must lie in the interval $[\omega_1 \, e^{-\ln(a_1)/(2n)}, \ \omega_2 \, e^{-\ln(a_2)/(2n)}]$. It is common to choose the geometric mean of the interval bounds for $\omega_c$ to ensure that the filter requirements are met even with slight variations in filter parameters, e.g., in the capacitances of the implementing circuit. With the given data of the example, it follows that $n = 3$, $\omega_c/(2\pi) = 4.2\,\text{kHz}$, $z_0, z_1, z_2$ are the zeros of $Q$ with negative real parts as given on p. 334, and thus

$$\widehat{h}(\omega) = \frac{1}{(1 + j\omega/\omega_c)(1 + j\omega/\omega_c - (\omega/\omega_c)^2)}$$

meets the filter requirements. The corresponding impulse response can be found on p. 232. A graphical representation of $|\widehat{h}|$ and the *group delay* (see Fig. 11.6)

$$D(\omega) = -\frac{d}{d\omega} \arg(\widehat{h}(\omega)) = -\sum_{k=0}^{n-1} \Re(z_k)/(\Re(z_k)^2 + (\omega - \Im(z_k))^2)$$

show that the filter is suitable for nearly distortion-free transmission in the range up to about $\omega_c/2$, i.e., up to a frequency of approximately $2.1\,\text{kHz}$. In this range, the filter matches very well with an ideal lowpass. *Delay* $-\arg(\widehat{h}(\omega))/\omega$ and *group delay* of the filter are always less than $0.11\,\text{ms}$, the attenuation at $\omega_c$ according to the design with $K = 1$ is about $3\,\text{dB}$, $-20\log_{10}(|\widehat{h}(\omega_c)|) = 10\log_{10}(2)$, with Volt input and reference amplitude $1\,\text{V}$.

Butterworth filters (as analog filters or in discrete form, see p. 377) are widely used in audio high-fidelity systems or in communication technology in WLAN receivers and many more. The third-order *normalized transfer function* ($\omega_c = 1$) is $H(s) = 1/((1 + s)(1 + s + s^2))$, $s \in \mathbb{C}$, with associated frequency response $H(j\omega)$.

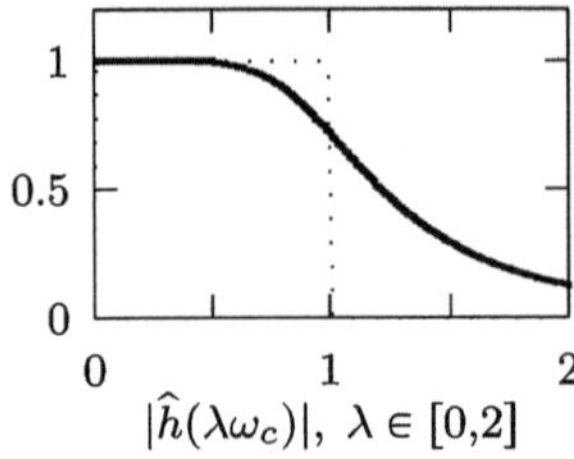
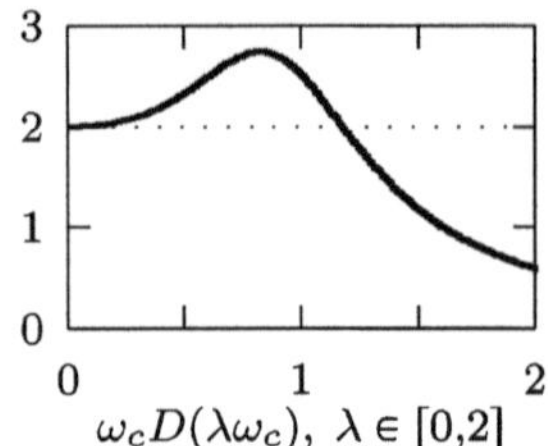



**Fig. 11.6** Attenuation $|\widehat{h}|$ and $\omega_c$ times group delay

In an analogous manner, *Chebyshev lowpass filters* are also constructed. For their frequency response, one requires

$$|\widehat{h}(\omega)| = \frac{K}{\sqrt{1 + \varepsilon^2 T_n^2(\omega/\omega_c)}}.$$

Here, $T_n$, $n \in \mathbb{N}$, are the Chebyshev polynomials from p. 107. The constants $K$, $\varepsilon$, and the filter order $n$ are to be determined according to the attenuation specifications for the filter. Chebyshev lowpass filters are more rippled in the passband compared to Butterworth filters due to the zeros of the Chebyshev polynomials, but they attenuate more strongly in the stopband for the same order (see pp. 113–114). However, they have greater and more frequency-dependent delays. All details for designing such filters can be found in Exercise A26 of Chap. 5 on p. 127.

## 11.4  Analog Filters with Rational Frequency Responses

A large class of typical linear transmission elements can be described analogously to the Butterworth filter by differential equations of the form

$$u^{(m)} + \alpha_{m-1} u^{(m-1)} + \cdots + \alpha_0 u = \beta_n f^{(n)} + \beta_{n-1} f^{(n-1)} + \ldots \beta_0 f$$

with real coefficients $\alpha_0, \alpha_1, \ldots, \alpha_{m-1}, \alpha_m = 1$, and $\beta_0, \ldots, \beta_{n-1}, \beta_n \neq 0$. With the polynomials $P(z) = \sum_{k=0}^{m} \alpha_k z^k$ and $Q(z) = \sum_{k=0}^{n} \beta_k z^k$ and the differential operator $D = \dfrac{\mathrm{d}}{\mathrm{d}t}$, the differential equation is briefly noted as

$$P(D)u = Q(D)f.$$

Choosing a signal class $\mathcal{Z}$ with input signals $f$, which can be convolved with the causal solution $h$ of $P(D)h = Q(D)\delta$, such as $\mathcal{Z} = \mathcal{S}'_R$, then $Lf = f * h$ defines a causal time-invariant system, which is referred to as an analog filter. Here, we consider $Q(j\omega)/P(j\omega)$ as a pseudofunction in $\mathcal{S}'$ (see p. 298).

Such filters are implemented in countless applications in electrical engineering and are now widespread as inexpensive components in mass products. In the following, frequency response, stability, and realizability will be discussed, and some standard examples will be presented.

### *Common Linear Factors of the Polynomials P and Q*

Assume that $P(z) = \prod_{k=1}^{m}(z - z_k)$ and $Q(z) = \beta_n \prod_{k=1}^{n}(z - b_k)$. If $P$ and $Q$ have a common linear factor $(z - z_1) = (z - b_1)$, then let $\widetilde{P}$ and $\widetilde{Q}$ be given by $\widetilde{P}(z) = \prod_{k=2}^{m}(z - z_k)$ and $\widetilde{Q}(z) = \beta_n \prod_{k=2}^{n}(z - b_k)$. It is evident that *the causal solutions of* $P(D)u = Q(D)f$ *and* $\widetilde{P}(D)u = \widetilde{Q}(D)f$ *are identical.* Common linear factors of $P$ and $Q$ thus have no influence on the solution of the described system when initial values are zero and therefore also do not affect the frequency response and stability properties of the system discussed subsequently. We therefore assume in the following that $P$ and $Q$ have no common linear factors.

### *Frequency Response and Transfer Function of the Causal System*

The described causal system has a frequency response $\widehat{h}(\omega)$ if and only if its impulse response $h(t)$ is a tempered distribution, i.e., if all zeros of the characteristic polynomial $P$ have real parts $r \le 0$. For $h \in \mathcal{S}'$, i.e., without exponentially growing parts, it holds that

$$\widehat{h}(\omega) = \mathcal{F}(Q(D)g)(\omega) = \sum_{k=0}^{n} \beta_k (j\omega)^k \widehat{g}(\omega)$$

with the causal fundamental solution $g$ of the differential equation $P(D)g = \delta$. For $\widehat{h} \in \mathcal{S}'$, the Fourier transform of the differential equation $P(D)h = Q(D)\delta$ shows that

$$P(j\omega)\widehat{h}(\omega) = Q(j\omega).$$

If $P$ has zeros with real parts $r = 0$, then $1/P(j\omega)$ is not a multiplier in $\mathcal{S}'$ (cf. 298), the convolution theorem for the Fourier transform then generally no longer applies, and $Q(j\omega)/P(j\omega)$ is then *not the frequency response* of the described *causal* system. Although $\mathcal{F}^{-1}(1/P(j\omega))$ is indeed a fundamental solution of $P(D)g = \delta$, it is not causal, and this fundamental solution can no longer be convolved with arbitrary causal right-hand sides of the differential equation. In other words, the equation $P(j\omega)\widehat{h}(\omega) = Q(j\omega)$ cannot be solved *in* $\mathcal{F}(\mathcal{S}'_R)$ for $\widehat{h}(\omega)$ by multiplying with $1/P(j\omega)$. If the zeros of $P$ have real parts $r < 0$, then $1/P(j\omega)$ is a multiplier in $\mathcal{S}'$, the impulse response $\mathcal{F}^{-1}(Q(j\omega)/P(j\omega))$ is a rapidly decreasing causal distribution, the convolution theorem applies for arbitrary tempered disturbances $f$, and the frequency response of the system is

$$\widehat{h}(\omega) = \frac{Q(j\omega)}{P(j\omega)}.$$

**Definition**  The function $H(z) = Q(z)/P(z)$, closely associated with the frequency response, is called the *transfer function* of the system.

**Examples**

1. The differential equation $\dot{u} = Kf$ describes an integrator (cf. p. 330) with transfer function $H(z) = K/z$. Its frequency response, however, is not $K/(j\omega)$, but $\widehat{h}(\omega) = K/(j\omega) + K\pi\delta(\omega)$. The equation $au^{(3)} + b\dot{u} = f$ describes, with $a, b > 0$, $\omega_0 = \sqrt{b/a}$, a causal filter on $\mathcal{S}'_R$ with impulse response $h(t) = s(t)(1 - \cos(\omega_0 t))/b$. The characteristic polynomial has the roots $\lambda_1 = 0$, $\lambda_{2,3} = \pm j\omega_0$. The transfer function is $H(z) = 1/P(z) = 1/(az^3 + bz)$, but the frequency response is (this calculation is left as a good exercise for the reader; cf. pp. 293 and 298)

$$\widehat{h}(\omega) = \frac{1}{(j\omega)^3 a + j\omega b} + \frac{\pi}{b}\delta(\omega) - \frac{\pi}{2b}[\delta(\omega - \omega_0) + \delta(\omega + \omega_0)] \neq \frac{1}{P(j\omega)}.$$

   The convolution theorem does not hold if, for example, $f(t) = U_0 s(t)$.

2. The causal filter for the differential equation $\dot{u} - au = af$ can be realized by an adder, an integrator, and a proportional element (cf. p. 333). It has the transfer function $H(z) = a/(z - a)$ and for positive $a \in \mathbb{R}$ the impulse response

$$h(t) = a\,e^{at}\,s(t).$$

   It is *not tempered*, the system is *unstable* (see below), and it has *no frequency response* that by definition should belong to $\mathcal{S}'$.

   The filter with the frequency response $\dfrac{a}{j\omega - a}$ has for $a > 0$ the impulse response

$$\widetilde{h}(t) = -a\,e^{at}\,s(-t)$$

and is *not causal.*

These examples show that the equation $\widehat{h}(\omega) = H(j\omega)$ for the frequency response of a system of the form $P(D)u = Q(D)f$ is only correct if all poles of $H$ have negative real parts.

## *Stability of the Causal System*

A causal time-invariant linear system of the form $P(D)u = Q(D)f$ is called *stable* if a bounded input signal results in a bounded output signal (BIBO-stable, "bounded input bounded output").

For polynomials $P$ and $Q$ without common linear factors, the system is certainly *unstable* if $P$ has zeros with positive real parts. It is also *unstable* if $P$ has zeros on the imaginary axis. For a zero at the origin, consider, for example, $f(t) = s(t)$ as input signal. For zeros of the form $\pm jb$, $b > 0$, consider, for example, $f(t) = \cos(bt)s(t)$ as input signal: The convolution with the part $(c_1 \cos(bt) + c_2 \sin(bt))s(t)$ of the impulse response corresponding to the zero pair $\pm jb$ is unbounded for $t \to \infty$ (with suitable $c_1, c_2$).

We now assume that all zeros of $P$ have negative real parts. Then the preceding considerations show that the frequency response is given by

$$\widehat{h}(\omega) = \frac{Q(j\omega)}{P(j\omega)}.$$

If $\deg Q = n > m = \deg P$, then the impulse response $h$ contains additive terms of the form $c_k \delta^{(k)}(t)$ with $k \geq 1$ and certain constants $c_k \neq 0$ (cf. partial fraction decomposition and Fourier transforms of derivatives on p. 291). The system is in this case *unstable*, because then, for example, the excitation $f(t) = \sin((\omega_0 t)^2)s(t)$ results in an unbounded system response. In summary we obtain the following:

**Theorem 11.2** *For polynomials $P$ and $Q$ without common zeros, the causal time-invariant linear system described by $P(D)u = Q(D)f$ is stable if and only if $\deg Q \leq \deg P$ holds and all poles of $Q/P$ have negative real parts. This is exactly the case when the impulse response $h$ has the form $h(t) = p(t) + c\delta(t)$ with a suitable constant $c$ and an integrable function $p(t)$. The frequency response of a stable system of this form is $Q(j\omega)/P(j\omega)$. The term $c\delta(t)$ in the impulse response appears when $\deg Q = \deg P$. The integrable function $p(t)$ is under these conditions even a rapidly decreasing distribution, and the convolution theorem of the Fourier transform applies to all tempered input signals $f$.*

## *Realization of the Causal System*

We continue to consider the differential equation from p. 336 with the polynomials $P$ and $Q$ given there. Let again $Q(z) = \beta_n \prod\limits_{k=1}^{n} (z - b_k)$ be the factorization of

$Q(z) = \sum\limits_{k=0}^{n} \beta_k z^k$. For $n \geq \deg P = m$, let $Q_1(z) = \beta_n \prod\limits_{k=1}^{m-1} (z - b_k) = \sum\limits_{k=0}^{m-1} c_k z^k$ and $Q_2 = \dfrac{Q}{Q_1}$. For $n < m$ let $Q_1(z) = Q(z)$, $c_k = \beta_k$ for $0 \leq k \leq n$, $c_k = 0$ for $k > n$ and $Q_2(z) = 1$.

The right side of $P(D)u = Q(D)f$ can thus be represented by

$$Q(D)f = Q_2(D)[Q_1(D)f].$$

With the causal fundamental solution $g$ of the system, one obtains for an input signal $f$ with $\mathrm{supp}\,(f) \subset [0, \infty[$ the output signal

$$u = g * Q(D)f = g * [Q_2(D)Q_1(D)f] = Q_2(D)[g * Q_1(D)f].$$

This input-output relation can be schematically represented by a serial connection of two systems:

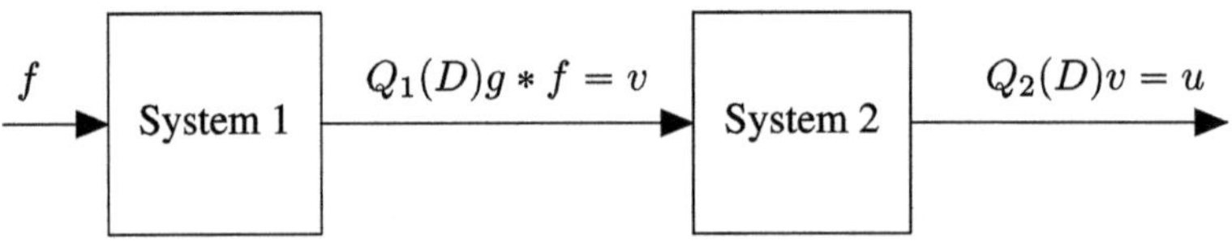


System 2 can be electrically simulated by proportional elements, adders, and differentiating elements. We consider System 1 and assume without loss of generality that $n = m - 1$. With the coefficients $c_k$ of $Q_1$, it is given by

$$v^{(m)} + \alpha_{m-1} v^{(m-1)} + \cdots + \alpha_0 v = c_{m-1} f^{(m-1)} + c_{m-2} f^{(m-2)} + \cdots + c_0 f.$$

As on p. 333 it is transformed into a first-order differential equation system: Let $\dot{x}_0 = c_0 f - \alpha_0 v$. Then

$$v^{(m)} + \alpha_{m-1} v^{(m-1)} + \cdots + \alpha_1 \dot{v} - c_{m-1} f^{(m-1)} - \cdots - c_1 \dot{f} = \dot{x}_0.$$

With the causality of the system, integration yields

$$v^{(m-1)} + \alpha_{m-1} v^{(m-2)} + \cdots + \alpha_1 v - c_{m-1} f^{(m-2)} - \cdots - c_2 \dot{f} - c_1 f = x_0.$$

With $\dot{x}_1 = x_0 + c_1 f - \alpha_1 v$ and $x_{m-1} = v$, repeated integration and continuation of the process results in the following *state description by a first-order system*:

$$\dot{\mathbf{x}}(t) = \mathbf{A}\mathbf{x}(t) + \mathbf{C}f(t)$$

with $\mathbf{x}(t) = (x_0(t), x_1(t), \ldots, x_{m-1}(t))^T$, $\quad \mathbf{C} = (c_0, c_1, \ldots, c_{m-1})^T$ and

$$\mathbf{A} = \begin{pmatrix} 0 & \cdots\cdots\cdots\cdots\cdots & -\alpha_0 \\ 1 & 0 \cdots & \cdots & -\alpha_1 \\ & 1 \cdots & & \vdots \\ & \ddots\cdots & & \vdots \\ & \cdots\cdots & & \vdots \\ 0 & \quad 1 \ 0 & -\alpha_{m-2} \\ & \quad 1 & -\alpha_{m-1} \end{pmatrix}.$$

In control theory, this state description is called the *observer canonical form* of the system. The corresponding block diagram has already been introduced in Sect. 9.3 on p. 231.

**Examples**

1. *Allpass Filter.* An allpass filter $L$ is a filter with a constant amplitude response. When considering the filter in the signal space $L^2(\mathbb{R})$, it follows from the Plancherel equation that the energy $\int_{-\infty}^{\infty} |Lf(t)|^2 dt$ of an output signal $Lf$ is proportional to the energy of an input signal $f$ of the filter.

   Allpass filters can be used for *signal delay* and *phase equalization* of transformed signals. For example, one can derive from the frequency response of a lowpass filter of the form

$$\widehat{h}_{\mathrm{LP}}(\omega) = \frac{K}{P(j\omega)} = \frac{K}{|P(j\omega)|\, \mathrm{e}^{-j\Phi(\omega)}}$$

   the frequency response $\widehat{h}_{\mathrm{AP}}$ of an allpass filter by replacing the numerator $K$ with the complex conjugate of the denominator. This results in

$$\widehat{h}_{\mathrm{AP}}(\omega) = \mathrm{e}^{2j\Phi(\omega)}.$$

   The amplitude response is then constantly 1, and the phase delay is doubled compared to the lowpass filter (Allpass Delay Equalizer).

2. *Highpass Filter.* Highpass filters have their stopband $0 \le \omega < \omega_c$ below a cutoff angular frequency $\omega_c$ and their passband above it. One way to obtain a highpass frequency response is the *Lowpass to Highpass Transformation*, i.e., replacing the normalized parameter $s = j\omega/\omega_c$ in the frequency response of a lowpass filter with the cutoff frequency $\omega_c/(2\pi)$ by the parameter $1/s$. For

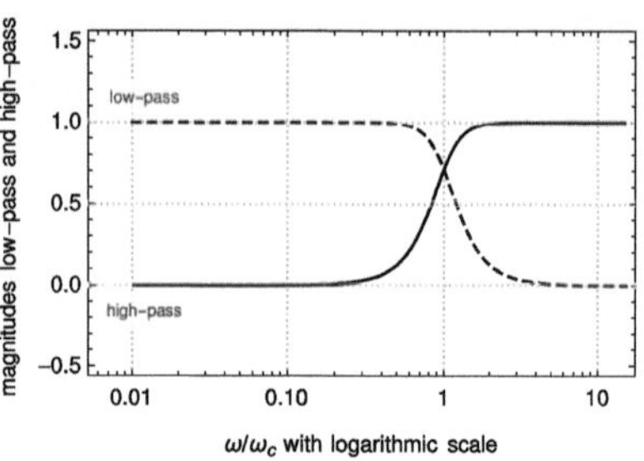



**Fig. 11.7** Lowpass-highpass transform

example, from the frequency response $\widehat{h}_{\mathrm{LP}}$ of the Butterworth lowpass filter from p. 335, one derives the rational frequency response $\widehat{h}_{\mathrm{HP}}$ of a highpass filter of the same order with the same cutoff frequency:

$$\widehat{h}_{\mathrm{HP}}(\omega) = \frac{1}{(1 - j\omega_c/\omega)(1 - j\omega_c/\omega - (\omega_c/\omega)^2)} = \frac{(j\omega)^3}{\omega_c^3}\,\widehat{h}_{\mathrm{LP}}(\omega).$$

In general, if the transfer function (cf. p. 338) of a stable lowpass filter has poles $z_k$, then $\omega_c^2/z_k$ are the corresponding poles of the highpass filter's transfer function (see the image below right). The highpass filter is therefore also stable, and its frequency response has the same magnitude as the lowpass filter at $\omega_c$, in this example $1/\sqrt{2}$, i.e., about 3 dB attenuation. For the amplitude responses, the following holds $|\widehat{h}_{\mathrm{HP}}(\alpha\,\omega_c)| = |\widehat{h}_{\mathrm{LP}}(-\omega_c/\alpha)| = |\widehat{h}_{\mathrm{LP}}(\omega_c/\alpha)|$ for $\alpha > 0$. Lowpass and highpass filters have the same order. In the case of Butterworth filters, lowpass and derived highpass filters have the same group delays, because for Butterworth filters, the poles $z_k$ of the lowpass transfer function have the magnitude $\omega_c$. Verifying these statements is given to readers here as a small calculation exercise.

From $\log(\alpha\,\omega_c) = \log(\omega_c) + \log(\alpha)$, $\log(\omega_c/\alpha) = \log(\omega_c) - \log(\alpha)$, it can be seen that the transformation mirrors the amplitude response of the lowpass filter for $\omega > 0$ at the cutoff frequency on a *logarithmic frequency axis* (Fig. 11.7).

3. *Bandpass Filter.*   Bandpass filters can be obtained by serially connecting highpass and lowpass filters or by using the *Lowpass to Bandpass Transformation*. In this transformation, the normalized parameter $j\omega_n = j\omega/\omega_c$ in the frequency response of a lowpass filter is replaced using the Joukowsky transform (see p. 127) with

$$\frac{1}{B}\left(j\omega_n + \frac{1}{j\omega_n}\right).$$

In filter design, besides the cutoff frequency $\omega_c$, the normalized bandwidth $0 < B < 1$ is also freely selectable. The quantity $Q = 1/B$ is referred to as the quality factor of the bandpass filter. A first-order lowpass filter with the frequency response

**Fig. 11.8** How the filter transformations above and below map a lowpass pole $P$

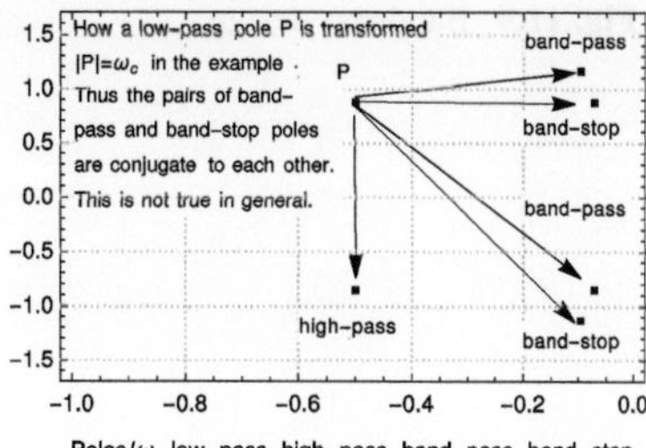



$$\widehat{h}_{\mathrm{LP}}(\omega) = \frac{K}{(1 + j\omega_n)}$$

corresponds to, for example, a second-order bandpass filter with the frequency response

$$\widehat{h}_{\mathrm{BP}}(\omega) = \frac{j\omega_n B K}{1 + j\omega_n B - \omega_n^2}.$$

In this transformation, it generally applies: The order of the lowpass filter is doubled and $\widehat{h}_{\mathrm{BP}}(\omega_c) = \widehat{h}_{\mathrm{LP}}(0)$ holds. As with the lowpass filter, the bandpass filter is also stable. This follows from the mapping properties of the *Joukowsky transform* (see p. 127 and Exercise A5 at the end of the chapter). Figure 11.8 shows as a typical example the transformation of the pole $z_0 = j\omega_c\,e^{j\pi/6}$ of the transfer function of the Butterworth lowpass filter from p. 335 into a pair (generally not conjugate) of complex poles of the transfer function of the generated bandpass filter with $B = 1/3$. The other lowpass pole $z_2 = \overline{z}_0$ then leads to the conjugate complex poles for this pair in the bandpass filter.

The passband limits $\omega_1$ and $\omega_2$, $\omega_2 > \omega_1 > 0$, with $|\widehat{h}_{\mathrm{BP}}(\omega_1)| = |\widehat{h}_{\mathrm{BP}}(\omega_2)| = |\widehat{h}_{\mathrm{LP}}(\omega_c)|$, are $\omega_1 = \omega_c(-B + \sqrt{4 + B^2})/2$, $\omega_2 = \omega_c(B + \sqrt{4 + B^2})/2$. Therefore, $(\omega_2 - \omega_1)/\omega_c = B$ is the normalized bandwidth and $\omega_1 \cdot \omega_2 = \omega_c^2$ holds. For $\alpha > 0$ it holds that $|\widehat{h}_{\mathrm{BP}}(\alpha\,\omega_c)| = |\widehat{h}_{\mathrm{BP}}(\omega_c/\alpha)|$, i.e., on a logarithmic frequency axis and normalized frequency $\omega_n = \omega/\omega_c$, the amplitude response for $0 < \omega_n < 1$ is mirrored to the side $\omega_n > 1$ and $\omega_{1,n}$, $\omega_{2,n}$ are symmetrically located around the center frequency $\omega_n = 1$. See Fig. 11.9, starting from the Butterworth lowpass filter on p. 335 with normalized transfer function $H_{LP}$, so that the bandpass transfer function is $H_{BP}(s) = H_{LP}((s + 1/s)/B)$, the magnitude $|H_{BP}(j\lambda)|$ for $0.4 \leq \lambda \leq 2.5$ is plotted. Readers are asked to verify all statements in Exercise A5 at the end of the chapter.

4. *Bandstop Filter.*   In the *Lowpass to Bandstop Transformation*, the parameter $j\omega_n = j\omega/\omega_c$ in a lowpass frequency response is replaced by $B/(j\omega_n + 1/j\omega_n)$. The order $n$ of the lowpass filter is thereby doubled. The quantity $B \in\ ]0, 1[$ is the normalized bandwidth of the stopband. The bandstop filter (syn. notch filter) is then also stable with the lowpass filter and has an $n$-fold zero in the amplitude response at $\omega_c$. The poles corresponding to the lowpass pole $z_0 = j\omega_c\,e^{j\pi/6}$ as in

**Fig. 11.9** Bandpass design

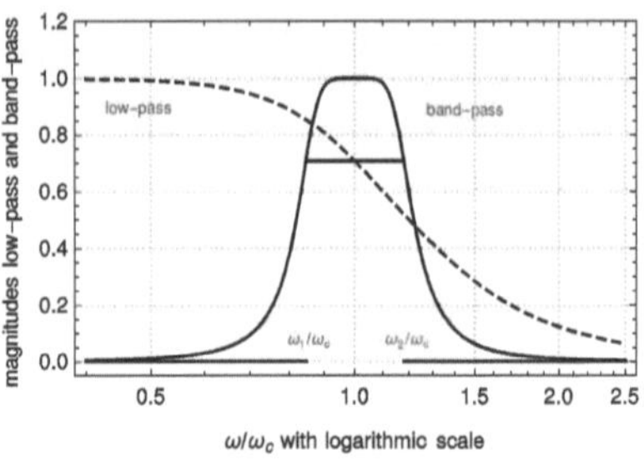

**Fig. 11.10** Bandstop design

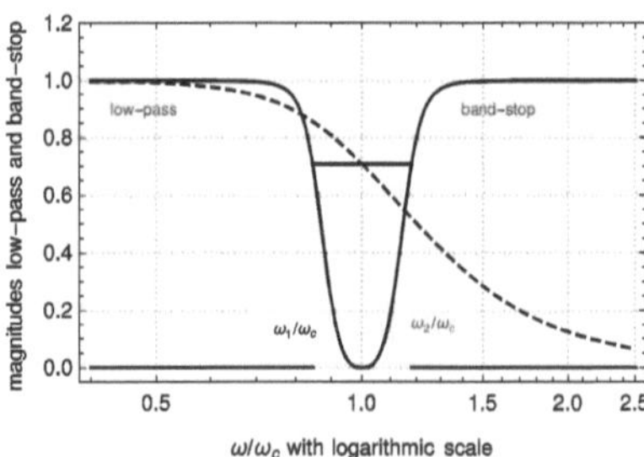

the previous example with $B = 1/3$ are also shown in Fig. 11.8 on the preceding page. Readers are asked to verify the statements about this transformation in Exercise A5. In Fig. 11.10—with analogous data as in the previous example—the amplitude response $|\widehat{h}_{BS}|$ of the generated sixth-order bandstop filter is shown. The band limits $\omega_{1,n}$, $\omega_{2,n}$ are the same as in the previous bandpass example.

To implementation possibilities of different filter types using operational amplifiers, resistors, and capacitors—in particular, universally adjustable filters that can be set as lowpass, highpass, bandpass, or bandstop filters depending on the purpose—we refer to the instructive source Tietze and Schenk (2008).

## 11.5   Periodic Signals and Stationary Filter Response

We now consider a stable, causal time-invariant linear system $Lf = h * f$ on $\mathcal{S}'$, which is given by a differential equation of the form $P(D)u = f$ with a polynomial $P$ and by the causal fundamental solution $h$ of the equation. Transient processes decay with increasing time. We calculate the filter response to a signal $f(t)s(t)$ turned on at time $t = 0$ with $f(t) = U_0\, e^{j\omega_0 t}$ and the unit step function $s(t)$:

$$(fs * h)(t) = s(t) \int_{-\infty}^{t} U_0\, e^{j\omega_0(t-x)}\, h(x)\mathrm{d}x$$

$$= U_0 \, e^{j\omega_0 t} \, s(t) \left[ \int\limits_{-\infty}^{+\infty} h(x) \, e^{-j\omega_0 x} \, dx - \int\limits_{t}^{+\infty} h(x) \, e^{-j\omega_0 x} \, dx \right]$$

$$= U_0 \widehat{h}(\omega_0) \, e^{j\omega_0 t} \, s(t) + r(t).$$

The function

$$r(t) = -U_0 \, e^{j\omega_0 t} \int\limits_{t}^{+\infty} h(x) \, e^{-j\omega_0 x} \, dx$$

vanishes due to the integrability of $h$ for $t \to \infty$. It is the transient filter response, i.e., it represents the decaying transient process.

**Theorem 11.3** *For $t > 0$, $t \to \infty$, the stationary filter response is $U_0 \widehat{h}(\omega_0) \, e^{j\omega_0 t}$. It is the uniquely determined periodic solution of $P(D)u = U_0 \, e^{j\omega_0 t}$.*

In many applications, it is common to study the filter response to periodic input signals. If one chooses a periodic function $f(t) = U_0 \, e^{j\omega_0 t}$ as a mathematical model of an excitation, it is implicitly assumed that the signal $f$ has been present for an infinite amount of time at all times, so the system is always in a steady-state condition. It then holds that

$$(f * h)(t) = U_0 \widehat{h}(\omega_0) \, e^{j\omega_0 t} \, .$$

Under our stability assumption, $\widehat{h}(\omega) = 1/P(j\omega)$ is a multiplier in $\mathcal{S}'$ (see p. 290 and p. 300), and more generally, it follows for a $T$-periodic distribution

$$f(t) = \sum_{k=-\infty}^{+\infty} c_k \, e^{j2\pi k t / T}$$

with the convolution theorem of the Fourier transform

$$h_k = \widehat{h}(2\pi k/T) \quad \text{and} \quad \widehat{f\,h}(\omega) = 2\pi \sum_{k=-\infty}^{+\infty} c_k h_k \delta(\omega - 2\pi k/T)$$

the periodic filter response (see p. 294–298)

$$(f * h)(t) = \mathcal{F}^{-1}(\widehat{f\,h})(t) = \sum_{k=-\infty}^{+\infty} c_k h_k \, e^{j2\pi k t / T} \, .$$

With the impulse response $h$ of a stable causal system of the form $P(D)u = f$, we thus obtain the periodic solution for periodic $f$ also through periodic convolution:

**Theorem 11.4** *For a $T$-periodic distribution $f$, the convolution $f * h$ is precisely the filter response, which also results from the $T$-periodic convolution of the excitation $f$ with the $T$-periodic transfer function $h_T$,*

$$h_T(t) = \sum_{k=-\infty}^{+\infty} h_k \, e^{j2\pi kt/T} \, .$$

*The coefficients $h_k$ are the sample values $\widehat{h}(2\pi k/T)$ of the frequency response $\widehat{h}$ and $f * h$ is again a $T$-periodic distribution.*

Refer once more also to Sects. 5.2 and 9.1 for this.

**Remark** We note that the functions $e^{j\omega t}$ are *eigenfunctions* of the convolution operator $L$, which mathematically describes the filter. The associated eigenvalues are the factors $\widehat{h}(\omega)$:

$$L\, e^{j\omega_0 t} = e^{j\omega_0 t} * h(t) = \mathcal{F}^{-1}(2\pi \widehat{h}(\omega)\delta(\omega - \omega_0)) = \widehat{h}(\omega_0)\, e^{j\omega_0 t} \, .$$

The effect of $L$—usually complicated in applications— is reduced by the Fourier transform to the simple algebraic multiplication operation: $\widehat{Lf} = \widehat{h}\widehat{f}$. This significantly simplifies the analysis of $L$. For linear differential operators on $\mathcal{S}'$ of the form

$$L = \sum_{k=0}^{n} c_k \frac{d^k}{dt^k},$$

it holds similarly that $L\, e^{j\omega t} = \sum_{k=0}^{n} c_k (j\omega)^k\, e^{j\omega t}$ and in general

$$\widehat{Lf}(\omega) = \sum_{k=0}^{n} c_k (j\omega)^k \, \widehat{f}(\omega).$$

This fact is the reason for the usefulness of the Fourier transform in solving differential equations. We will return to this point in Sects. 12.8 and 14.1.

The preceding considerations show how the frequency response $\widehat{h}$ of stable time-invariant linear systems can be approximately determined using measurement techniques: Sine waves of different frequencies are used as test signals on the transmission system, the steady-state condition is waited for, and finally the values of the frequency response $\widehat{h}$ for the test frequencies are obtained from the measured magnitude and phase changes during transmission. From these discrete values, an approximation for $\widehat{h}$ is constructed through interpolation.

## *Periodization in the Time Domain and Sampling in the Frequency Domain*

If a locally integrable function $f(t)$ is cut out by a time window, yielding a time-limited course $f_0(t) = \begin{cases} f(t) \text{ for } -T/2 \leqslant t < T/2 \\ 0 \quad \text{otherwise} \end{cases}$ and $f_0$ is continued to a $T$-periodic function $f_T$, then $f_T$ can be represented as a convolution with a pulse train $\delta_T(t) = \sum\limits_{k=-\infty}^{+\infty} \delta(t - kT)$:

$$f_T(t) = \sum_{k=-\infty}^{+\infty} f_0(t + kT) = (f_0 * \delta_T)(t).$$

The convolution theorem can be applied, and the generalized spectral function $\widehat{f_T}$ of $f_T$ is obtained by (cf. p. 298)

$$\widehat{f_T}(\omega) = \widehat{f_0}(\omega) \cdot \widehat{\delta_T}(\omega) = 2\pi \sum_{k=-\infty}^{+\infty} \frac{1}{T} \widehat{f_0}\left(\frac{2\pi k}{T}\right) \delta\left(\omega - \frac{2\pi k}{T}\right).$$

The coefficients $\dfrac{1}{T} \widehat{f_0}\left(\dfrac{2\pi k}{T}\right) = \dfrac{1}{T} \int\limits_{-T/2}^{+T/2} f_0(t)\, e^{-j2\pi kt/T}\, dt$ are precisely the Fourier coefficients of the Fourier series of $f_T$. They result from the sample values of the continuous spectral function $\widehat{f_0}$ of the time-limited section $f_0$. One says: *Periodization in the time domain corresponds to sampling in the frequency domain. Analogously, sampling in the time domain corresponds to periodization in the frequency domain.* Refer to Fig. 11.11 for illustration and later Sect. 12.2.

## *Numerical Approximations for Fourier Transforms*

The relation between the Fourier coefficients of the time-limited signal section $f_0$ of $f$ and sampled values of the Fourier transform of $f_0$ allows approximations for $\widehat{f}$ to be calculated using the discrete Fourier transform. The coefficients $\widehat{c}_k$ of the discrete Fourier transform calculated from sampled values of $f_0$ (cf. Sect. 6) are used to approximate sampled values of $\widehat{f_0}$. Interpolation of the points $(2\pi k/T, \widehat{c}_k T)$ results in an approximation for $\widehat{f_0}$ and thus for $\widehat{f}$ (see the remark on p. 310 and (9.2) on p. 270). The quality of this approximation depends on the number of sampled values and how well $f$ is approximated by $f_0$ (see Exercise A6 on p. 380, where an initial approximation is improved by so-called *zero padding* in the discrete Fourier transform).

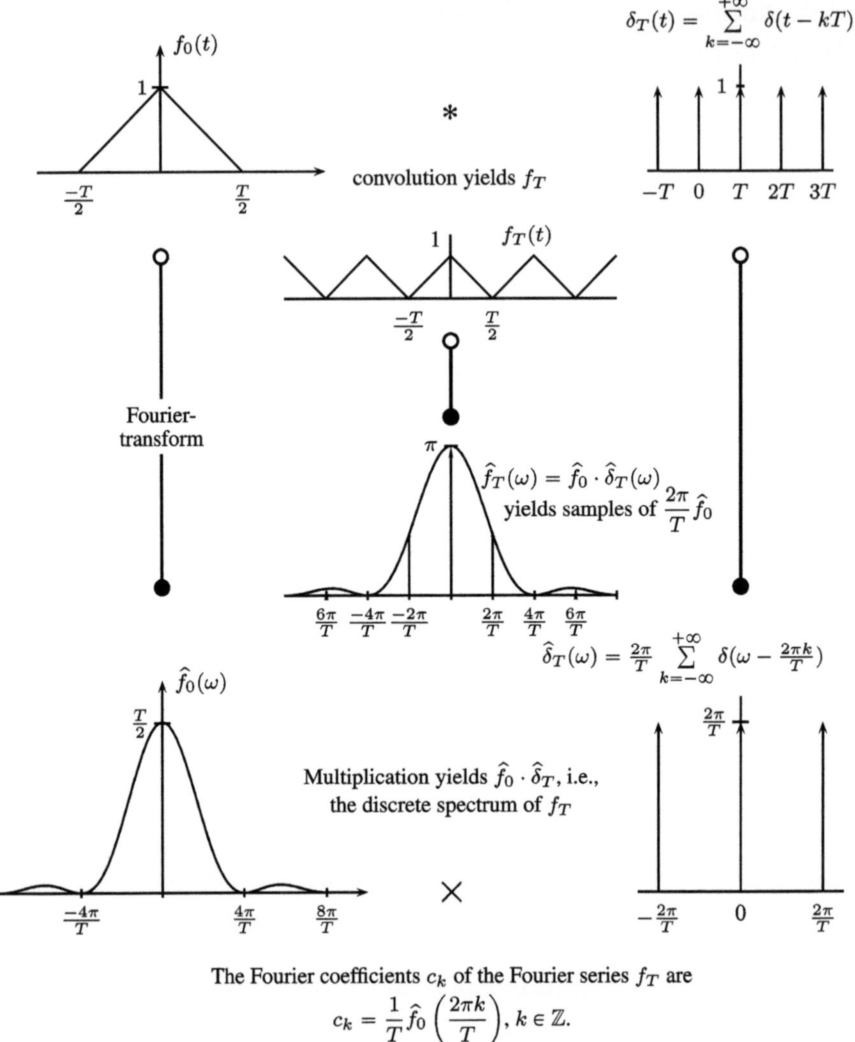



The Fourier coefficients $c_k$ of the Fourier series $f_T$ are

$$c_k = \frac{1}{T}\widehat{f}_0\left(\frac{2\pi k}{T}\right),\, k \in \mathbb{Z}.$$

**Fig. 11.11** Periodization in time and sampling in frequency domain

## *The Poisson Summation Formula*

Using the previously used notations, the application of the inverse Fourier transform
to $\widehat{f}_T(\omega)$ yields the Fourier series representation of the periodic function $f_T$, and we
obtain the following *Poisson Summation Formula*:

**Theorem 11.5**  *It holds* $f_T(t) = \displaystyle\sum_{k=-\infty}^{+\infty} f_0(t+kT) = \frac{1}{T}\sum_{k=-\infty}^{+\infty} \widehat{f_0}\!\left(\frac{2\pi k}{T}\right) e^{j2\pi kt/T}.$

The Poisson summation formula also applies to many functions $f_0$ that do not have a bounded support. For example, one easily sees that it also holds if $f_0$ is replaced by a rapidly decreasing function $\varphi \in \mathcal{S}$. Due to the rapid decrease of $\varphi$ as $t \to \infty$, the series $\displaystyle\sum_{k=-\infty}^{+\infty} \varphi(t+kT)$ is absolutely and uniformly convergent and represents the associated periodic function pointwise as well. One of the most well-known applications is obtained with $\varphi(t) = e^{-\alpha t^2}$, $\alpha > 0$, and $T = 1$. The Poisson summation formula then yields the following *functional equation for the theta function*:

$$\sum_{k=-\infty}^{+\infty} e^{-\alpha(t+k)^2} = \sqrt{\frac{\pi}{\alpha}}\sum_{k=-\infty}^{+\infty} e^{j2\pi kt}\, e^{-\pi^2 k^2/\alpha} = \sqrt{\frac{\pi}{\alpha}}\left(1+2\sum_{k=1}^{+\infty} e^{-\dfrac{\pi^2 k^2}{\alpha}}\cos(2\pi kt)\right).$$

The Poisson summation formula, which can also be shown in a corresponding form for functions of several variables, has a variety of different applications, for example, in solving heat conduction problems, in the development of sampling and quadrature formulas, or in the description of crystal structures. Interested readers are referred, for example, to Strichartz (1994).

## *Application Examples*

There are countless applications of linear filters in today's technologies because a wide range of signals are linearly transformed for use. Here, we consider only three examples.

1. *Lowpass filters* generally have a smoothing effect. Thus, they can be used to a certain extent for denoising of signals. As an example, in Fig. 11.12 there is an illustration with the function $f(t) = A\cos(\omega_0 t/20)$, $\omega_0 = 1$ rad/s, A=1 V, with additive uniform noise in $[-1, 1]$. This signal is lowpass filtered with a cutoff angular frequency of 0.1 rad/s in Fig. 11.13. Analogously, lowpass filters can be used in image processing to unmask images, smooth them, etc. In digital communication, lowpass filters are used to transform a discrete into a continuous signal as we will see with the Shannon sampling theorem in Sect. 12.1.
2. *Matched filters* are implemented in myriads of systems like mobile phones, WLAN, and many more, where transmitted signals are noisy at the receiver side and their arrival must be detected. We show that matched filters have an optimal signal-to-noise ratio. A matched filter has the same shape as the transmitted signal.

**Fig. 11.12** Noisy
signal

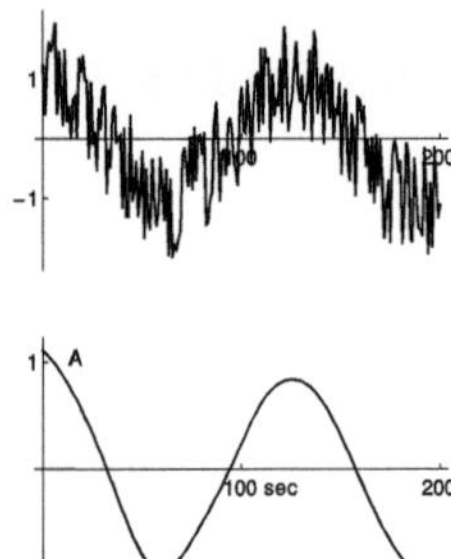

**Fig. 11.13** This signal
lowpass filtered

Let a received signal be $r(t) = s(t) + n(t)$ with $s \in L^2(\mathbb{R})$ an energy signal and $n(t)$ a Gaussian white noise with (one-sided) power spectral density $N_0$. Equalization can possibly be performed before the receive filter so that only white noise remains in the signal. Then, the filter output of the receiver is $d(t) = d_s(t) + d_n(t)$ with the signal part $d_s$ and noise part $d_n$. The receiver is a linear filter with impulse response $h$. Its frequency response $\widehat{h}$ shall be optimized so that the instantaneous signal-to-noise ratio $\mathrm{SNR}(T_D) = \dfrac{d_s^2(T_D)}{\sigma_d^2}$ is maximal. $T_D$ is a suitably chosen detection time that estimates a signal delay and $\sigma_d^2$ the variance of the noise part $d_n$.

Since $d_s(T_D) = (s * h)(T_D)$ and $\sigma_d^2 = \dfrac{N_0}{2} \cdot \dfrac{1}{2\pi} \int\limits_{-\infty}^{+\infty} |\widehat{h}(\omega)|^2 \mathrm{d}\omega$, independent of $T_D$, the task is to maximize $\mathrm{SNR}(T_D) = \dfrac{\left| \dfrac{1}{2\pi} \int\limits_{-\infty}^{+\infty} \widehat{s}(\omega)\widehat{h}(\omega)\, e^{j\omega T_D}\, \mathrm{d}\omega \right|^2}{\dfrac{N_0}{4\pi} \int_{-\infty}^{+\infty} |\widehat{h}(\omega)|^2 \mathrm{d}\omega}.$

The Cauchy-Schwarz inequality implies $\mathrm{SNR}(T_D) \leq \dfrac{1}{N_0 \pi} \int\limits_{-\infty}^{+\infty} |\widehat{s}(\omega)|^2 \mathrm{d}\omega.$

Thus, *the matched filter $\widehat{h}_{MF}$ with the same shape as the transmit signal $s$, i.e., $\widehat{h}_{MF}(\omega) = K\overline{\widehat{s}}(\omega)\, e^{-j\omega T_D}$, $h_{MF}(t) = K s(T_D - t)$ with a constant $K$ has maximal* SNR. It is left to the reader to calculate the SNR of some filters for concrete signals. In communication systems, the information is transmitted in sequences of pulses $s(t)$, whose shapes, e.g., raised cosine windows, are known at the receiver (cf. p. 390) so that matched filters can be used in such systems.

3. *Filter banks* are families of bandpass filters, which are used sequentially or in parallel in signal processing for various purposes, from signal detection, signal transmission, pattern recognition to radar applications, etc., in practice mostly as discrete filters for discretized signals. We will see an example of their use for discrete processing of a continuous wavelet transform with Matlab in Sect. 14.2, p. 484.

## 11.6  Discrete Linear Filters and z-Transform

Analogous to Sect. 11.3, in the following we investigate fundamental questions regarding the representability of discrete linear systems through convolutions, stability, and realization possibilities. I hope to encourage readers to engage with the ideas and methods of this subfield of discrete mathematics by studying further literature. This subfield significantly influences many areas of our technical and application-oriented environment. Implementations of discrete filters exist in processors and many programs for discrete data processing in technology and business.

We consider translation-invariant linear systems $L : \mathcal{Z} \to \mathcal{A}$, where the signal spaces $\mathcal{Z}$ and $\mathcal{A}$ are subspaces of the space $\mathcal{X}$ of discrete signals introduced on p. 323, which need to be specified in each case. As with analog filters, the question arises about representability through convolution with the impulse response

$$Lx = L(x * \delta) = x * L\delta$$

and the question about the continuity of $L$ with respect to suitable topologies on $\mathcal{Z}$ and $\mathcal{A}$. A continuous system that can be described by a convolution operator $L$ is referred to, as before, as a *linear filter*. A first representation theorem readily follows from the properties of convolutions already discussed in Sect. 8.7.

**Theorem 11.6** *Let $\mathcal{Z}$ be the space $\mathcal{X}$ of all discrete signals or one of the spaces from list items 1–6 on page 323. Then every continuous linear translation-invariant operator $L : \mathcal{Z} \to \mathcal{X}$, which has a finite impulse response $h = L\delta \in \mathcal{X} \cap \mathcal{E}'$, is a linear filter*

$$Lx = x * h.$$

***Proof*** Let $h = \displaystyle\sum_{n=-M}^{+M} h_n \delta_n$ be the impulse response $L\delta$. For a signal $x = \displaystyle\sum_{k=-\infty}^{+\infty} x_k \delta_k$, let $x_N$ be the partial sum $x_N = \displaystyle\sum_{k=-N}^{+N} x_k \delta_k$. From the linearity and translation invariance of the operator $L$, it follows

$$Lx_N = \sum_{k=-N}^{+N} x_k \left( \sum_{n=-M}^{+M} h_n \delta_{n+k} \right) = \sum_{m=-(N+M)}^{N+M} \left( \sum_{k=-N}^{+N} x_k h_{m-k} \right) \delta_m.$$

In the last equation, we regrouped the sum with the index transformation $n + k = m$. Here, $h_n = 0$ for $|n| > M$. $Lx_N$ is thus the convolution $x_N * h$ (one may refer again to the example in item 6 on p. 199). For $N \to \infty$, it follows from the continuity of $L$ with $x_N \to x$ in $\mathcal{Z}$ that $Lx = \mathcal{D}'\text{-}\lim_{N\to\infty}(x_N * h) = x * h$. The last equation results from the convergence of the convolutions $x_N * h \to x * h$ for $h \in \mathcal{X} \cap \mathcal{E}'$ (see item 7(b) on p. 197 and Young's inequality for $l^p$ spaces in Appendix B).  $\qquad\square$

**Remark** In the case $\mathcal{Z} = l_d^\infty$ (item 7 of the list on p. 324), the above proof fails because the partial sums $x_N = \sum_{k=-N}^{+N} x_k \delta_k$ of $x \in l_d^\infty$ generally do not converge to $x$ in the norm of $l_d^\infty$, as can be seen from the example $x = \sum_{k=-\infty}^{+\infty} \delta_k$. For this space, there is also a counterexample to the statement of the above theorem on p. 355.

As with analog filters, discrete *causal* translation-invariant linear systems on the space $\mathcal{X} \cap \mathcal{D}'_R$ are automatically continuous and always linear filters. For practical applications in signal processing with causal translation-invariant linear operators and causal signals, the desired convolution representation always results. The subsequent two theorems are due to Albrecht (2011). Since the proofs in the discrete case are significantly simpler than with analog systems, we can also provide them within the scope of our introduction to the field.

## *Automatic Continuity of Causal Linear Discrete Systems*

**Theorem 11.7 (Automatic Continuity of Causal Linear Discrete Systems)** *Every causal linear operator $L : \mathcal{X} \cap \mathcal{D}'_R \to \mathcal{X}$ is automatically continuous, and there is a uniquely determined infinite lower triangular matrix $A_L = (a_{m,n})_{m,n\in\mathbb{Z}}$ (i.e., $a_{m,n} = 0$ for $m < n$), such that for all $x = \sum_{n=-\infty}^{+\infty} x_n \delta_n$ in $\mathcal{X} \cap \mathcal{D}'_R$ and $y = Lx$,*

$$y = \sum_{m=-\infty}^{+\infty} y_m \delta_m \text{ the relation } y_m = \sum_{n=-\infty}^{+\infty} a_{m,n} x_n \text{ holds. Due to the causality of } L,$$

*this series is a finite sum.*

***Proof*** A linear mapping $L : \mathcal{X} \cap \mathcal{D}'_R \to \mathcal{X}$ is, by definition of convergence in $\mathcal{X} \cap \mathcal{D}'_R$, continuous if and only if its restrictions $L|_{\mathcal{Z}_k}$ on all spaces $\mathcal{Z}_k = \mathcal{X} \cap \mathcal{D}'_k$ are continuous. The elements of $\mathcal{Z}_k$ have their supports in $[ka, \infty[$, $a > 0$ as chosen fixed on p. 323. We consider for $m \in \mathbb{Z}$ and $x = \sum_{n=-\infty}^{+\infty} x_n \delta_n$ the linear operator $P_m : \mathcal{X} \to \mathbb{C}$, defined as the projection $P_m x = x_m$ onto the component with index $m$. By definition of convergence in $\mathcal{X}$, it follows that $L$ is continuous if and only if the compositions $P_m \circ L|_{\mathcal{Z}_k} : \mathcal{Z}_k \to \mathbb{C}$ are continuous for all $k$ and $m$ from $\mathbb{Z}$.

To prove this property for a causal linear operator $L : \mathcal{X} \cap \mathcal{D}'_R \to \mathcal{X}$, we introduce for $m, k \in \mathbb{Z}$ with $m \geq k$ the notation

$$Q_{k,m} : \mathcal{Z}_k \to \{x \in \mathcal{Z}_k : x_n = 0 \text{ for all } n > m\}$$

for the canonical projection. For $x \in \mathcal{Z}_k$ and $m \geq k$, it is therefore $Q_{k,m}(x) = \sum_{n=k}^{m} x_n \delta_n$. This projection is continuous on $\mathcal{Z}_k$ and has a finite-dimensional range.

Now, let $L : \mathcal{X} \cap \mathcal{D}'_R \to \mathcal{X}$ be a linear causal operator. For $m < k$, the mappings $P_m \circ L|_{\mathcal{Z}_k}$ are zero due to the causality of $L$ and are therefore continuous. For $m \geq k$, due to the causality of $L$ for $x \in \mathcal{Z}_k$, $P_m(L(E_k - Q_{k,m})x)) = 0$ holds. Here, $E_k$ denotes the identity on $\mathcal{Z}_k$. Therefore, $P_m \circ L|_{\mathcal{Z}_k} = P_m \circ L \circ Q_{k,m}$ on $\mathcal{Z}_k$. Since the image space of $Q_{k,m}$ is finite-dimensional, the continuity of $P_m \circ L|_{\mathcal{Z}_k}$ follows in this case as well. In other words, $L$ is continuous because every output value $y_m = P_m(Lx)$ for $x \in \mathcal{Z}_k$ is determined only by values $x_n$ up to the "time" $ma$, i.e., by the signal segment $Q_{k,m}x$.

Next, define $a_{m,n} = P_m(L\delta_n)$ for $m, n \in \mathbb{Z}$. Then, with the continuity of $L$, it follows for all $x = \sum_{n=-\infty}^{+\infty} x_n \delta_n$ in $\mathcal{X} \cap \mathcal{D}'_R$ and $y = Lx$

$$y_m = P_m(Lx) = \sum_{n=-\infty}^{+\infty} x_n P_m(L\delta_n) = \sum_{n=-\infty}^{+\infty} a_{m,n} x_n,$$

and due to the causality of $L$, all $a_{m,n} = 0$ for $n > m$. Additionally, since every $x$ from $\mathcal{X} \cap \mathcal{D}'_R$ has a bounded-below support, the series is finite.                 $\square$

This immediately results in the following theorem for translation-invariant systems:

**Theorem 11.8** *Each translation-invariant causal linear operator* $L : \mathcal{X} \cap \mathcal{D}'_R \to \mathcal{X}$ *is continuous, and with* $h = L\delta_0 = \sum_{n=0}^{+\infty} h_n \delta_n$ *the convolution representation holds*

$$Lx = h * x = \sum_{n=-\infty}^{+\infty} \left( \sum_{k=-\infty}^{+\infty} x_k h_{n-k} \right) \delta_n.$$

***Proof*** The translation invariance results for the matrix representation of $L$ shown in the previous theorem in a matrix $A_L = (a_{m,n})_{m,n \in \mathbb{Z}}$, such that

$$a_{m+1,n+1} = P_{m+1}(L\delta_{n+1}) = P_m(L\delta_n) = a_{m,n}$$

holds for all $m, n \in \mathbb{Z}$. For $h = L\delta_0 = \sum_{n=0}^{+\infty} h_n \delta_n$, it follows from $h_m = P_m(L\delta_0) = a_{m,0}$ that $h_{m-n} = a_{m-n,0} = a_{m,n}$ and thus the convolution representation for

$$y = L\left(\sum_{n=-\infty}^{+\infty} x_n \delta_n\right) = \sum_{m=-\infty}^{+\infty} y_m \delta_m \text{ with the finite sum } y_m = \sum_{n=-\infty}^{+\infty} a_{m,n} x_n =$$

$$\sum_{n=-\infty}^{+\infty} h_{m-n} x_n. \qquad\qquad\qquad\qquad\qquad\qquad\qquad\qquad\qquad \square$$

**Remark** It can be shown that operators $L$ as in the last two theorems are also continuous if the image space is replaced by $\mathcal{X} \cap \mathcal{D}'_R$ with its finer topology. See Albrecht and Neumann (1979), Remark 1.5.

Again, the concept of convergence introduced on $\mathcal{X} \cap \mathcal{D}'_R$ is crucial for continuity (see p. 323). We reconsider the example from p. 327.

**Example** The causal discrete linear system $Lx = x * h$ with $h = \sum_{n=0}^{\infty} \delta_n$ on $\mathcal{X} \cap \mathcal{D}'_R$ is called an *accumulator*. For $x = \sum_{n=-N}^{+\infty} x_n \delta_n$ and $y = \sum_{k=-N}^{\infty} y_k \delta_k$ with $y = Lx$, it is $y_k = x_k + y_{k-1}$. As already shown on p. 327, for $f_n = -\delta_{-n} + \delta_0$, $\mathcal{D}'\text{-}\lim_{n\to\infty} f_n = \delta_0$ holds, but $\mathcal{D}'\text{-}\lim_{n\to\infty} Lf_n \neq L\delta_0$. However, in $\mathcal{X} \cap \mathcal{D}'_R$, the sequence of $f_n$ in the topology introduced there is not convergent, and $L$ is continuous with respect to the $\mathcal{X} \cap \mathcal{D}'_R$ topology according to the last theorem. Thus, in typical examples from engineering disciplines, it is crucial to consider not only the mere operator definition but also the signal spaces and their topology.

As we have already seen with analog filters, translation-invariant linear systems $L : \mathcal{Z} \to \mathcal{A}$ for applications are often defined, respectively, designed with a desired impulse response $h$ or a desired frequency response $\widehat{h}$ *directly as convolution operators* $Lx = x * h$. The same applies to discrete systems. In this situation, the following theorem shows the continuity of a large class of such systems. We summarize some cases relevant for applications.

**Summary** A linear convolution operator $L : \mathcal{Z} \to \mathcal{A}$, $Lx = x * h$ is continuous, i.e., a linear filter in the following cases:

1. $\mathcal{Z} = \mathcal{X}, h \in \mathcal{X} \cap \mathcal{E}', \mathcal{A} = \mathcal{X}$
2. $\mathcal{Z} = \mathcal{X} \cap \mathcal{E}', h \in \mathcal{X}, \mathcal{A} = \mathcal{X}$
3. $\mathcal{Z} = \mathcal{X} \cap \mathcal{D}'_R, h \in \mathcal{X} \cap \mathcal{D}'_R, \mathcal{A} = \mathcal{X} \cap \mathcal{D}'_R$
4. $\mathcal{Z} = \mathcal{X} \cap \mathcal{S}', h \in \mathcal{X} \cap \mathcal{O}'_C, \mathcal{A} = \mathcal{X} \cap \mathcal{S}'$
5. $\mathcal{Z} = l_d^\infty, h \in l_d^1, \mathcal{A} = l_d^\infty$
6. $\mathcal{Z} = l_d^1, h \in l_d^\infty, \mathcal{A} = l_d^\infty$
7. $\mathcal{Z} = l_d^2, h \in l_d^2, \mathcal{A} = l_d^\infty$

In all cases, for the coefficients $y_m$ of the convolution $y = x * h = \sum\limits_{m=-\infty}^{+\infty} y_m \delta_m$ with

$x = \sum\limits_{n=-\infty}^{+\infty} x_n \delta_n$ and $h = \sum\limits_{k=-\infty}^{+\infty} h_k \delta_k$, the representation $y_m = \sum\limits_{k=-\infty}^{+\infty} h_{m-k} x_k$ holds.

***Proof*** Cases 1 and 2 follow from No. 7 on p. 197. Case 3 was shown above. In cases 4–7 according to our examples under No. 5, p. 303, the convolution equation $\widehat{x * h} = \widehat{x}\widehat{h}$ of the Fourier transform holds, and the continuity of $L$ follows from the continuity of the Fourier transform on $\mathcal{S}'$, respectively, from Young's inequality for the $l^p$ spaces (Appendix B, p. 501). From this inequality, further combination possibilities result with $l^1(\mathbb{Z}) \subset l^2(\mathbb{Z}) \subset \cdots \subset l^\infty(\mathbb{Z})$, which we will not discuss further. In all cases, $y = Lx$ has the mentioned form, as also stated in the examples on p. 199, p. 303, and p. 501. Compare also the illustration on p. 200.  $\square$

## *Continuous, Causal, and Stable Translation-Invariant Linear Systems That Cannot Be Represented as Convolutions*

All the systems considered so far have been assumed to be convolution operators. It should be noted, however, that there are indeed continuous translation-invariant linear systems that cannot be represented as convolution systems and thus are not characterized by their impulse response. We will consider an example of such a system on $l_d^\infty$, which is continuous, causal, and stable in the sense of the following section. This example dates back to Stefan Banach (1932) and his construction of Banach limits and was provided by Sandberg (2001). We follow a proof by Albrecht (2011). In doing so, we use some arguments from functional analysis, particularly the *Hahn-Banach extension theorem* and a *fixed-point theorem of A. Markov and S. Kakutani* (see Rudin, 1991, Theorems 3.5 and 5.23).

**Example** Consider the closed subspace $M \subset l_d^\infty$ of all $x = \sum\limits_{n=-\infty}^{+\infty} x_n \delta_n$, such that $\lim_{n \to -\infty} x_n = 0$ holds. $M$ is invariant under translations $T_m$. For $m \in \mathbb{Z}$, let $T_m$ be the translation by $ma$, that is, $x = \sum\limits_{k=-\infty}^{\infty} x_k \delta_k \to T_m x = \sum\limits_{k=-\infty}^{\infty} x_{m+k} \delta_k$. According to the *Hahn-Banach extension theorem*, there exists a continuous linear functional $P \neq 0$ on $l_d^\infty$, such that $P(M) = \{0\}$ and $\|P\| = 1 = P(1)$. Here, 1 denotes the impulse sequence, all of whose coefficients are 1. Each such functional $P$ is an extension of the limit functional $p(x) = \lim_{n \to -\infty} x_n$, defined for such $x \in l_d^\infty$, whose coefficients have a limit as $n \to -\infty$. *Among all these, there is a translation-invariant functional $P_0$*, i.e., $P_0(T_m x) = P_0(x)$ for $m \in \mathbb{Z}$, $x \in l_d^\infty$. To see this, one needs an additional argument, which will be provided next. The operator then defined by $L(x) = \sum_{k=-\infty}^{+\infty} P_0(x)\delta_k$ is linear, translation-invariant, trivially causal, and continuous as a mapping $L : l_d^\infty \to l_d^\infty$ and thus also stable in the sense of the following paragraph. In particular, $L\delta_0 = 0$. The operator $L$ can therefore not be represented as a convolution.

Readers with knowledge of functional analysis can understand the required additional argument as follows: Consider the convex bounded set

$$K := \{P \in l_d^{\infty\prime} : P(M) = \{0\}, \ \|P\| = 1 = P(1)\} \neq \emptyset.$$

Here, $l_d^{\infty\prime}$ is the *dual space of* $l_d^\infty$, i.e., the Banach space of all continuous linear functionals on $l_d^\infty$. The weak-* topology on $l_d^{\infty\prime}$ is the coarsest topology, such that all $x \in l_d^\infty$ are continuous as functionals on $l_d^{\infty\prime}$. The unit ball $b_1'$ of $l_d^{\infty\prime}$ is weak-* compact, and the norm is weak-* lower semicontinuous. A weak-* convergent net $(P_\lambda)_{\lambda \in \Lambda}$ in $K$ thus converges to a functional in $K$. Hence, $K \subset b_1'$ is weak-* closed and therefore also weak-* compact.

The adjoint mappings $T_m' : l_d^{\infty\prime} \to l_d^{\infty\prime}$ of the translations $T_m$, defined by $T_m' P(x) = P(T_m x)$, are weak-* continuous, map $K$ affine-linearly into itself, and form a commuting family. According to the aforementioned *fixed-point theorem of A. Markov and S. Kakutani*, there exists a common fixed point $P_0 \neq 0$ in $K$ for the family of all $T_m'$, $m \in \mathbb{Z}$. This functional $P_0$ is then translation-invariant (cf. Day, 1961).

An entirely analogous example also shows that there are continuous linear causal translation-invariant operators on $L^\infty(\mathbb{R})$ that are not convolution operators. The same holds in the case of several variables on $L^\infty(\mathbb{R}^n)$ with a suitably adjusted definition of the causality condition. Interested readers are referred to the works of Rudin (1972) on " Invariant Means in $L^\infty$", Albrecht and Neumann (1979), and the references cited therein. Another example, also mentioned by Sandberg (2001) and understandable with the same arguments as above, can be found in Exercise A14 at the end of the chapter.

In the following, we consider linear operators $L : \mathcal{Z} \to \mathcal{A}$ between subspaces $\mathcal{Z}$ and $\mathcal{A}$ of $\mathcal{X}$, which are assumed to be convolution operators, i.e., discrete linear filters.

## *Stability and Realizability of Discrete Linear Filters*

**Definition** *A discrete linear filter* $L : \mathcal{Z} \to \mathcal{A}$ *is called stable if there exists a constant* $C > 0$ *such that for every input signal* $x \in \mathcal{Z} \cap l_d^\infty:$ $\|Lx\|_\infty \leq C\|x\|_\infty.$

For stable filters, bounded input signals $x \in l_d^\infty$, i.e., those with bounded coefficients $x_n$, also result in bounded output signals $Lx$, and moreover, the maximum values of the coefficients of $Lx$ are never greater in magnitude than the absolute maximum value of the $x_n$ $(n \in \mathbb{Z})$ multiplied by $C$. The following theorem characterizes stable and realizable discrete filters by the properties of their impulse response. Stability means that small disturbances in $x$ only have small effects on the filter response $Lx$.

**Theorem 11.9** *Let* $L : \mathcal{Z} \to \mathcal{A}$ *be a discrete linear filter that belongs to one of the cases 1–7 of the above summary. Its impulse response $h$ has the coefficients $h_k$, $k \in \mathbb{Z}$. Then the following statements hold:*

1. *$L$ is stable if and only if* $\displaystyle\sum_{k=-\infty}^{+\infty} |h_k| < \infty.$
2. *$L$ is causal if and only if $h_k = 0$ for all $k < 0$.*

***Proof*** For 1. If $h \in l_d^1$, then for $x \in l_d^\infty$ and $y = Lx$ with coefficients $y_k$, it holds that $\displaystyle |y_k| \le \sum_{n=-\infty}^{+\infty} |h_n||x_{k-n}| \le \sup_{n \in \mathbb{Z}} |x_n| \sum_{n=-\infty}^{+\infty} |h_n|$; thus $\|y\|_\infty \le \|h\|_1 \|x\|_\infty$, i.e., $L$ is stable.

Conversely, let $L$ be stable. For cases 1, 4, and 5, $h \in l_d^1$, and nothing needs to be shown. For other cases, we define a sequence of signals $x_N$, $N \in \mathbb{N}$, whose $k$th coefficients $x_{N,k}$ are given using the complex conjugate filter coefficients:

$$x_{N,k} = \begin{cases} \overline{h}_{N-k}/|h_{N-k}| & \text{for } 0 \le k \le 2N \text{ and } h_{N-k} \ne 0 \\ 0 & \text{otherwise.} \end{cases}$$

Since for all $N$ the signals $x_N$ are finite and always $\|x_N\|_\infty \le 1$, they belong to $\mathcal{Z} \cap l_d^\infty$ for the considered spaces $\mathcal{Z}$, and it holds that $\|Lx_N\|_\infty \le C$ with a positive constant $C$. For every $N \in \mathbb{N}$ and $y_N = Lx_N$, it follows that

$$y_{N,N} = \sum_{k=-\infty}^{+\infty} h_k x_{N,N-k} = \sum_{k=-N}^{+N} |h_k| \le C$$

and thus the absolute summability of the filter coefficients $h_k$.

For 2. If the filter is realizable, then due to $h = L\delta$ by definition $h_k = 0$ for $k < 0$. Conversely, it follows from $\mathrm{supp}(h) \subset a\mathbb{N} \cup \{0\}$ and $\mathrm{supp}(x * h) \subset \mathrm{supp}(x) + \mathrm{supp}(h)$ the causality of a filter with such an impulse response $h$. $\qquad\square$

**Remark** In many textbooks on linear system theory, stability is defined in a weaker sense than above. Often it is only required that a bounded input signal should result in a bounded output signal. It should be noted, therefore, that a filter that only meets this weaker condition does not generally have an impulse response with absolutely summable coefficients. Consider, for example, the accumulator with impulse response $h = \displaystyle\sum_{k=0}^{+\infty} \delta_k$ on $X \cap \mathcal{E}'$, or note that the convolutions $x * h$ in cases

6 and 7 always lie in $l_d^\infty$ and there are elements $h$ in $l_d^2 \subset l_d^\infty$ that do not belong to $l_d^1$. According to our definition, stability is equivalent to the continuity of $L$, if both the subspace $\mathcal{Z} \cap l_d^\infty$ of input signals and the space $\mathcal{A} \cap l_d^\infty$ of output signals are endowed with the $l_d^\infty$ norm.

## *Frequency Response and Transfer Function of Discrete Linear Filters and z-Transforms of Discrete Signals*

Before we look at specific examples of filters, we will discuss the $z$-transform, which is a common mathematical tool when working with discrete filters.

**Definition**

1. If the impulse response $h = \displaystyle\sum_{k=-\infty}^{+\infty} h_k \delta_k$ of a linear filter $L : \mathcal{Z} \to \mathcal{A}$ belongs to $\mathcal{X} \cap \mathcal{S}'$, then the Fourier transform $\widehat{h}$ of $h$ is called the frequency response of the filter.
2. For a discrete signal $x \in \mathcal{X} \cap \mathcal{S}'$ with coefficients $x_k$, $\widehat{x}$ is called the spectrum of $x$ and $X(z) = \displaystyle\sum_{k=-\infty}^{+\infty} x_k z^{-k}$ its $z$-transform.
3. If the $z$-transform $H(z)$ of a discrete linear filter with impulse response $h$ converges for certain $z \in \mathbb{C}$, then $H$ is called the transfer function of the filter.

If a discrete filter satisfies the convolution relationship $\widehat{x * h} = \widehat{x}\widehat{h}$, then the filter can be described through properties of the frequency response or designed according to given requirements for the frequency response, just as in the analog case. We will return to this in examples later. In the discrete case, the frequency response $\widehat{h}$ and spectra $\widehat{x}$ are periodic distributions.

In many applications of discrete filters, it is common to work with the transfer function instead of the frequency response. $H(z)$ is a Laurent series. Laurent series are fundamental to function theory. Assuming that the coefficients $h_k$ are exponentially bounded, i.e., $|h_k| \leq c_1 r^k$ and $|h_{-k}| \leq c_2 \varrho^k$ for all $k \geq k_0$ with suitable $c_1, c_2, k_0$, then this Laurent series converges in the annulus $A = \{z \in \mathbb{C} \,|\, r < |z| < R = 1/\varrho\}$ and diverges outside of $A$ (comparison criterion with the geometric series). For $\varrho = 0$, set $R = +\infty$. For $r \geq R$, $A$ is the empty set. The series converges absolutely and uniformly in any closed annulus $r_1 \leq |z| \leq R_2$ in $A$. On the boundary $|z| = r$ and $|z| = R$, no general statement about convergence can be made.

For causal filters, $H(z)$ is defined, if the power series $\displaystyle\sum_{k=0}^{\infty} h_k x^k$ has a radius of convergence $\varrho > 0$. Then the region of convergence of $H$ is given by $|z| > r$, where $r = 1/\varrho \neq 0$ or $r = 0$ for $\varrho = +\infty$. It then holds that $H(z) \to h_0$ for $|z| \to \infty$.

A causal impulse response $h$ of a discrete filter, which belongs to the space $\mathcal{X} \cap \mathcal{S}'$, has polynomially bounded coefficients, as $\widehat{h}$ is a generalized Fourier series (see Sect. 9.1). Its $z$-transform $H$ then converges in any case for $|z| > 1$.

If the circle $|z| = 1$ belongs to the convergence annulus of a $z$-transform $X$, then the $2\pi/a$-periodic spectrum $\widehat{x}$ of the corresponding signal $x \in l_d^1$ is continuous and given by the values of $X$ on the unit circle ($a$ is the step size chosen on p. 323). The coefficients $x_k$ of $\widehat{x}$ are precisely the coefficients of the Laurent series $X$. We then have

$$\widehat{x}(\omega) = \sum_{k=-\infty}^{+\infty} x_k \, e^{-jk\omega a} = X(e^{j\omega a}) \quad \text{and} \quad x_k = \frac{a}{2\pi} \int_0^{2\pi/a} X(e^{j\omega a}) \, e^{jk\omega a} \, d\omega.$$

*For all signals $x$, $h$, etc., used in the following, it is generally assumed from now on that they have exponentially bounded coefficients as described above, i.e., that their $z$-transforms converge in an annulus $r < |z| < R$, if $r \neq R$.*

For explicitly given signals, the region of convergence of their $z$-transforms can often be calculated using the ratio test or the root test known from the basic analysis.

**Examples**

1. For $R > 0$ and $r > 0$, let $x = \displaystyle\sum_{k=-\infty}^{+\infty} x_k \delta_k$ be given by $x_k = \begin{cases} R^k & \text{for } k < 0 \\ r^k & \text{for } k \geq 0. \end{cases}$

   The $z$-transform of $x$ is then

$$X(z) = \sum_{k=-\infty}^{-1} R^k z^{-k} + \sum_{k=0}^{+\infty} r^k z^{-k} = \frac{z}{R-z} + \frac{z}{z-r},$$

   where both geometric series converge when both $|z/R| < 1$ and $|r/z| < 1$. The $z$-transform is thus defined in the annulus $r < |z| < R$ and represents a holomorphic function there if $r < R$.

   Observe: For $x = \displaystyle\sum_{k=-\infty}^{+\infty} \delta_k$, the $z$-transform is not defined.

2. The discrete analog to the unit step function is $u$ with coefficients $u_k = 0$ for $k < 0$ and $u_k = 1$ for $k \geq 0$. Then the $z$-transform is

$$U(z) = \sum_{k=0}^{+\infty} z^{-k} = \frac{1}{1-z^{-1}}.$$

   The series converges for $|z| > 1$, thus in the exterior of the unit disk.

## Basic Properties of the z-Transform

We restrict ourselves to those elementary properties of the $z$-transform that we use in our application examples. A further treatment of its properties and numerous applications, as well as closely related other transforms, can be found in the books by Jury (1973), Oppenheim (1978) or Oppenheim and Schafer (2013).

1. *Linearity*

   The $z$-transform is linear because convergent series can be added term by term and multiplied by scalars. For the summation of two $z$-transforms, it must be assumed that a common convergence annulus exists.

2. *z-Transform of Translations*

   If a signal $x$ has $z$-transform $X(z)$, then the definition of $X(z)$ shows that the translation $x * \delta_k$ by $ka$ has $z$-transform $z^{-k} X(z)$.

   Example: Let $x$ be the $2N$-periodic rectangular signal starting at $k = 0$ with $x_k = 1$ for $k = 0, \ldots, N-1$, $x_k = 0$ for $k = N, \ldots, 2N - 1$, and $x_{k+2N} = x_k$ $(k \geq 0)$, so $x = \sum_{k=0}^{N-1} \delta_k * \sum_{k=0}^{\infty} \delta_{2Nk}$. Then the $z$-transform is

   $$X(z) = \frac{z^{N+1}}{(z - 1)(z^N + 1)}.$$

3. *Differentiation of z-Transforms*

   Laurent series may be differentiated term by term in their annulus of convergence. If a discrete signal $x = \sum_{k=-\infty}^{+\infty} x_k \delta_k$ has $z$-transform $X(z)$, then $-zX'(z)$ is $z$-transform of $\sum_{k=-\infty}^{+\infty} k x_k \delta_k$.

4. *z-Transform of Complex Conjugate Signals*

   If a signal $x$ has $z$-transform $X(z)$, then the complex conjugate signal $\overline{x}$ has $z$-transform $\overline{X(\overline{z})}$.

5. *z-Transform of Convolutions*

   Given a discrete linear filter $L : \mathcal{Z} \to \mathcal{A}$ and $h = L\delta$ in one of the considered seven cases from p. 354. If $X(z)$ converges in the annulus $A_1$ and $H(z)$ in the annulus $A_2$ and $A = A_1 \cap A_2$ is not empty, then for $y = x * h$ and $z \in A$ it holds that

   $$Y(z) = H(z) \cdot X(z).$$

***Proof*** For $x$ and $h$ with coefficients $x_k$ and $h_k$ and $z \in A$, we have that

$$\sum_{n,k\in\mathbb{Z}} |h_k||x_{n-k}||z|^{-n} \le \left(\sum_{k=-\infty}^{+\infty} |h_k||z|^{-k}\right)\left(\sum_{n=-\infty}^{+\infty} |x_n||z|^{-n}\right) < +\infty.$$

Since the function $(n,k) \to h_k x_{n-k} z^{-k}$ is summable over $\mathbb{Z}^2$, it can be summed in any order according to Fubini's theorem (Appendix B, p. 496). Thus, for $y = x * h$, one gets for all $z \in A$

$$Y(z) = \sum_{n=-\infty}^{+\infty}\left(\sum_{k=-\infty}^{+\infty} h_k x_{n-k}\right) z^{-n} = \sum_{k=-\infty}^{+\infty} h_k z^{-k} \sum_{n=-\infty}^{+\infty} x_{n-k} z^{-(n-k)}.$$

6. *Inversion of the z-Transform*

According to the identity theorem for Laurent series, the $z$-transform is invertible. The coefficients $x_n$ of a signal $x$ given the $z$-transform $X$ can be calculated in different ways:

(a) *By Laurent series expansion of $X$.*
(b) *By a contour integral using the residue theorem.*
    The function $f : \mathbb{C} \setminus \{0\} \to \mathbb{C}$ with $f(z) = z^k$, $k \in \mathbb{Z}$, is continuous in a neighborhood of the circle $\gamma(t) = r\,e^{jt}$ with $r > 0$ and $0 \le t < 2\pi$. The contour integral in the mathematically positive sense is

$$\oint_\gamma z^k\,dz = \int_0^{2\pi} r^k\,e^{jkt}\,jr\,e^{jt}\,dt$$

$$= jr^{k+1}\int_0^{2\pi} e^{j(k+1)t}\,dt = \begin{cases} 0 & \text{for } k \ne -1, \\ 2\pi j & \text{for } k = -1. \end{cases}$$

In particular, it is independent of the radius $r$. A $z$-transform $X$ is absolutely and uniformly convergent along a positively traversed circle $\gamma$ in its annulus of convergence. Interchanging integration and summation gives for the coefficients of the signal $x$ representations as contour integrals

$$\frac{1}{2\pi j}\oint_\gamma X(z)\cdot z^{n-1}\,dz = \frac{1}{2\pi j}\sum_{k=-\infty}^{+\infty} x_k \oint_\gamma \cdot z^{n-1-k}\,dz = x_n.$$

The evaluation of the contour integrals can be done using the residue theorem from complex analysis (see Appendix A). In practice, in linear filters mainly rational transfer functions are used. The inverse transform of their partial fractions can then be done using geometric series (see subsequent Example 4 and Example 1 on p. 368) or using tables available in formularies. Computer algebra systems like Maple, Mathematica, or the engineering fields widely used Matlab can calculate $z$-transforms and their inverses as well. Thus, applications of the $z$-transform are also available to readers who have not yet experience with the theorems of complex analysis used in the following examples.

7. *$z$-Transform of the Impulse Response of Stable Causal Discrete Filters*
   A discrete filter with transfer function $H$ is stable if and only if the unit circle is included in the annulus of convergence of the series. For causal filters, this is the case if and only if all singularities of $H$ lie inside the unit disk. This follows immediately from the theorem on page 357 characterizing stable causal filters because transfer functions $H$ of causal filters with exponentially bounded coefficients have a convergence region of the form $|z| > r \geq 0$ (cf. p. 358) and belong to a stable filter if and only if $r < 1$. $\qquad\qquad\square$


## First Application Examples


With the transformation rules, one easily finds new pairs of correspondences between signals and their $z$-transforms. The $z$-transform of convolutions allows, in causal linear filters described by difference equations, to immediately specify the transfer function. The first example shows this application to convolutions.

1. *$z$-Transform for Difference Equations.* The coefficients $y_n$ of the system response of a causal filter $y = x * h$ on $\mathcal{X}$ with impulse response $h = \delta_0 + \delta_1$ satisfy the difference equation $y_n = x_n + x_{n-1}$ for all $n \in \mathbb{Z}$. The transfer function of the filter is $H(z) = 1 + z^{-1} = (z+1)/z$ with the region of convergence $|z| > 0$.

   The difference equation $y_n + y_{n-1} = x_n + 2x_{n-1} + x_{n-2}$ is also satisfied for the coefficients of $x$ and $y$. *This relationship, valid for all coefficients of $x$ and $y$, corresponds to the following convolution equation for the signals in total:*

$$y * (\delta_0 + \delta_1) = x * (\delta_0 + 2\delta_1 + \delta_2).$$

*$z$-Transform of both sides of the equation leads to*

$$Y(z)(1 + z^{-1}) = X(z)(1 + 2z^{-1} + z^{-2}),$$

and thus to the same transfer function $H$ for $|z| > 0$

$$H(z) = \frac{1 + 2z^{-1} + z^{-2}}{1 + z^{-1}} = \frac{z + 1}{z}.$$

The equation $Y(z) = H(z)X(z)$ holds for $z$ with $|z| > 0$ in the region of convergence of $X$.

2. *Application of the Differentiation Rule.* The discrete signal $u = \sum\limits_{k=0}^{+\infty} \delta_k$ has the

z-transform $U(z) = \dfrac{z}{z-1}$ with $|z| > 1$. Then $X(z) = -zU'(z) = \dfrac{z}{(z-1)^2}$

with $|z| > 1$ is the z-transform of $x = \sum\limits_{k=0}^{+\infty} k\delta_k$.

3. *Application of Complex Conjugation.* Given a real discrete signal $s$, which arises from sampling a cos function with a sampling interval $a > 0$,

$$s = \sum_{k=0}^{+\infty} \cos(\omega_0 ka + \varphi)\delta_k.$$

The signal $s$ is then the real part of the complex signal $s_c = \sum\limits_{k=0}^{+\infty} e^{j(\omega_0 ka + \varphi)}\delta_k$.

The z-transform of $s_c$ is

$$S_c(z) = e^{j\varphi} \sum_{k=0}^{+\infty} \left(\frac{e^{j\omega_0 a}}{z}\right)^k = \frac{z\,e^{j\varphi}}{z - e^{j\omega_0 a}}.$$

Then the z-transform $S$ of $s = (s_c + \overline{s_c})/2$ is

$$S(z) = \frac{1}{2}\left(\frac{z\,e^{j\varphi}}{z - e^{j\omega_0 a}} + \frac{z\,e^{-j\varphi}}{z - e^{-j\omega_0 a}}\right) = \frac{z^2 \cos(\varphi) - z\cos(\omega_0 a - \varphi)}{z^2 - 2z\cos(\omega_0 a) + 1}.$$

4. *Laurent Series Expansion.* Given $X(z) = \dfrac{z^{-3}}{z+4}$, there are the expansions into geometric series

$$X(z) = z^{-4} \sum_{k=0}^{\infty} (-4z^{-1})^k \quad \text{for } |z| > 4 \text{ or}$$

$$X(z) = \frac{z^{-3}}{4} \sum_{k=0}^{\infty} \left(-\frac{z}{4}\right)^k \quad \text{for } 0 < |z| < 4.$$

In the first case, the signal $x$ with z-transform $X$ in the region $|z| > 4$ is given by the coefficients

$$x_k = \begin{cases} (-4)^{k-4} & \text{for } k \geq 4, \\ 0 & \text{otherwise.} \end{cases}$$

In the second case, the signal $x$ with $z$-transform $X$ in the annulus $0 < |z| < 4$ has the coefficients

$$x_k = \begin{cases} \dfrac{1}{4}\left(-\dfrac{1}{4}\right)^{3-k} & \text{for } k \le 3, \\ 0 & \text{otherwise.} \end{cases}$$

If a signal $x \in \mathcal{X}$ has a $z$-transform, then the $z$-transform is uniquely determined. However, the example shows that different signals can indeed have the same algebraic expression as $z$-transform. Therefore, to determine the inverse, the relevant region of convergence of the Laurent series must be given. If, for the abovementioned rational function $X$, it is additionally known that it is the $z$-transform of a causal signal, then only the Laurent series of the first case is possible.

5. *Inversion with the Contour Integral and the Residue Theorem.*
   As an example, we consider the $z$-transform $X(z) = z(z - z_1)^{-1}$, $z_1 \ne 0$, and $|z| > |z_1|$ as the region of convergence. Denoting by $f_n(z) = X(z)z^{n-1}$ the integrand of the contour integral above in 7(b), it follows from the residue theorem (see Appendix A, p. 486):
   (a) For $n \ge 0$ with the pole at $z_1$,

$$x_n = \frac{1}{2\pi j} \oint_{|z|=2|z_1|} f_n(z)\,dz = \mathrm{Res}(f_n, z_1) = \lim_{z \to z_1} (z - z_1) f_n(z) = z_1^n.$$

(b) For $n < 0$, $f_n$ has another pole at $z = 0$. The residue for $z = 0$ is obtained through the Laurent series expansion around the origin

$$f_n(z) = \frac{z^n}{z - z_1} = -\frac{z^n}{z_1}\frac{1}{1 - z/z_1} = -\sum_{k=0}^{+\infty} \frac{z^{n+k}}{z_1^{k+1}}.$$

The residue is the coefficient of $z^{-1}$, thus $-z_1^n$. With a curve $C$ in the residue theorem (p. 486), which for $n < 0$ encloses both poles of $f_n$, the residues for $z = 0$ and $z = z_1$ compensate each other in the sum.

*The inverse $z$-transform of $X$ is therefore* $x = \displaystyle\sum_{n=0}^{+\infty} z_1^n \delta_n.$

In the same way, one obtains with the residue theorem for $X(z) = z(z - z_1)^{-2}$ and $n \ge 0$ with $f_n$ as above

$$x_n = \lim_{z \to z_1} \frac{d}{dz}[(z - z_1)^2 f_n(z)] = n z_1^{n-1}$$

and analogous formulas for poles of higher order.

## *Causal Filters with Rational Transfer Function and Difference Equations*

In view of applications, we now consider discrete filters $y = h * x$ for causal signals $x \in l_d^\infty$, where the relationship between the coefficients of $y$ and $x$ is given by a linear difference equation with constant coefficients:

$$\sum_{k=0}^{M} b_k y_{n-k} = \sum_{m=0}^{N} a_m x_{n-m} \text{ with } b_0 = 1.$$

In order for the output signal $y$ to be uniquely determined, we also assume that the filter is causal. This excludes nontrivial solutions of the homogeneous equation. The $z$-transform of the associated convolution equation for $x$ and $y$ can then be directly read off the difference equation (see Example 1 on p. 362):

$$\left( \sum_{k=0}^{M} b_k z^{-k} \right) Y(z) = \left( \sum_{m=0}^{N} a_m z^{-m} \right) X(z).$$

By assumption, $X(z)$ converges for $|z| > 1$ and $H(z)$ for $z$ with a sufficiently large magnitude (see p. 359). From this, it follows, according to the convolution theorem (p. 360), the rational transfer function $H$ of the filter

$$H(z) = \frac{Q(z)}{P(z)} = \frac{\displaystyle\sum_{m=0}^{N} a_m z^{-m}}{\displaystyle\sum_{k=0}^{M} b_k z^{-k}}.$$

The assumed causality of the filter implies for the input impulse $x = \delta_0$ that for indices $k < 0$ all filter coefficients $h_k = 0$ must be zero. Specifically, the initial values $h_{-M} = \cdots = h_{-1} = 0$ are set for the recursive solution of the difference equation, so that all $h_k$ for $k \geq 0$ and thus the uniquely determined impulse response $h$ follow. We obtain for the causal filter with the rational transfer function $H(z) = Q(z)/P(z)$ the equation

$$\left( \sum_{n=0}^{+\infty} h_n z^{-n} \right) \left( \sum_{k=0}^{M} b_k z^{-k} \right) = \sum_{m=0}^{N} a_m z^{-m}.$$

Calculation of the coefficients in the series product on the left side, comparison of coefficients, and solving for the sought coefficients $h_n$ yield the Following:

**Recursion Equations for the Coefficients of the Impulse Response of the Filter**

$$h_0 = a_0 \ \text{ and } \ h_n = a_n - \sum_{k=1}^{n} b_k h_{n-k} \ \text{ for } n = 1, 2, \ldots \quad (b_0 = 1),$$

where $a_n = 0$ for $n > N$ and $b_k = 0$ for $k > M$ are set.

**Realization of Filters with Rational Transfer Function**

The following illustration shows a typical circuit network in electrical engineering (in the image domain of the $z$-transform), with multipliers $\otimes$, adders $\oplus$, and delay elements for one time step $a$, denoted by $z^{-1}$. The block diagram shows a possible realization—as a circuit or by means of software—of the previously discussed causal filter, where we can assume without loss of generality $M = N$. By rearranging the difference equation, the block diagram in Fig. 11.14 can be easily understood:

$$y_n = a_0 x_n + \sum_{m=1}^{N} (a_m x_{n-m} - b_m y_{n-m}).$$

There are also other possible realizations of the same filter (characterized by different parentheses in the difference representation or variations in the representation of $H$). For this, refer to the previously cited literature on circuit design such as, for example, Tietze and Schenk (2008).

**Remark** Other step sizes $a > 0$ for the considered discrete signals $x$ result in correspondingly different periods $2\pi/a$ in the spectra $\widehat{x}$ and different bandwidths of the considered filters. Multiplication of a $z$-transformed $X$ by $z^{-k}$ corresponds for $k > 0$ with a step size $a > 0$ to a delay by $ka$. The block diagram also shows that it is possible to realize very different transfer functions with the same circuit or with the same software by variable selection of the coefficients $a_k$ and $b_k$—a fact that opens

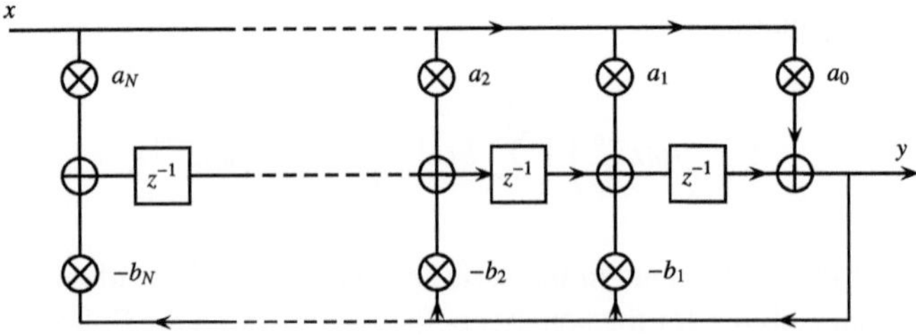


**Fig. 11.14** Block diagram of a discrete filter, realizable through software

up extensive technical possibilities. The same applies to analog frequency responses with the block diagram from p. 231. These simple realization possibilities explain the prominent role that filters with rational transfer functions play in engineering.

**Example (Solution of a Difference Equation for Fibonacci Numbers)** The Fibonacci numbers $f_n$, appearing at the beginning of the thirteenth century in the book Liber Abaci by Leonardo da Pisa, also known as Leonardo Fibonacci, have numerous applications. Interested readers can easily research such applications, such as in the runtime calculation of algorithms like the Euclidean algorithm for calculating greatest common divisors and many more. We consider the causal filter for the difference equation

$$f_n - f_{n-1} - f_{n-2} = x_{n-1}.$$

The coefficients of the impulse response of the filter for $n \geq 0$, noted here as $f_n$, form the Fibonacci sequence

$$(f_0, \, f_1, \, f_2, \ldots) = (0, 1, 1, 2, 3, 5, 8, \ldots).$$

With the $z$-transform, a closed representation of the Fibonacci numbers is easily found. As a transfer function $H$, whose coefficients are the Fibonacci numbers $f_n$, we get

$$H(z) = \frac{z^{-1}}{1 - z^{-1} - z^{-2}} = \frac{z}{(z - z_1)(z - z_2)},$$

with

$$z_1 = \frac{1 + \sqrt{5}}{2} \quad \text{and} \quad z_2 = \frac{1 - \sqrt{5}}{2}.$$

Partial fraction decomposition of $H(z) = \dfrac{1}{\sqrt{5}} \left( \dfrac{z_1}{z - z_1} - \dfrac{z_2}{z - z_2} \right)$ and then development of the partial fractions into Laurent series in the region $|z| > |z_1|$ as in previous examples (see Example 1 on p. 359) yield for $n \geq 0$

$$f_n = \frac{1}{\sqrt{5}}(z_1^n - z_2^n).$$

The quotients $f_{n+1}/f_n$ converge for $n \to \infty$ to $g = (1 + \sqrt{5})/2$. For two intervals with lengths $L$ and $S < L$, such that $S = \sqrt{L(L - S)}$, $S$ is the *Golden Ratio*, and it holds that $S = L/g$. It holds that $g^{n-2} \leq f_n$ for $n \geq 2$ (exercise).

**Causality and Stability of Filters with Rational Transfer Functions**

A rational transfer function $H$ with poles $z_1, \ldots, z_M$ is the $z$-transform of a filter
with causal impulse response $h$ if and only if one sets the region of convergence $A$
of $H$ to be $A = \{z \in \mathbb{C} : |z| > r\}$ with $r = \max\{|z_1|, \ldots, |z_M|\}$ and if the degree
of the numerator polynomial of $H$ does not exceed the degree of the denominator
polynomial. (Note: In examples, do not write $H$ as before with powers of $z^{-1}$, but
as a polynomial fraction with powers $z^k$, $k \geq 0$.)

Justification: It is immediately apparent that $h$ cannot be causal if the degree
of the numerator of $H$ is greater than the degree of the denominator, and thus
$H$ has a polynomial component. Conversely, the Laurent series expansion of the
partial fractions for rational $H$ in the region $A$ shows the causality of the resulting
impulse response $h$ if the degree of the numerator does not exceed the degree of the
denominator. A corresponding expansion in an annular region with $|z| < r$ (i.e., with
a pole in the exterior region) leads to a Laurent series with an analytic power series
component and thus to a noncausal associated impulse response $h$ (cf. Example 4,
p. 363 and Exercise 12 at the end of the chapter).

A causal filter with rational transfer function $H$ has only poles as singularities
and is stable if and only if they lie within the unit circle (No. 7, p. 362).

## *Stable Inverse Filters and Stable Signal Reconstruction*

We continue with the consideration of causal filters $y = h * x$ with a rational transfer
function $H$ and a corresponding difference equation as on p. 365. $H$ then has at most
as many zeros as poles. The input signals $x$ are again causal signals from $l_d^\infty$.

The *inverse filter* arises when the roles of input and output signals are reversed.
First, examples are considered to show what needs to be taken into account.

**Examples**

1. (*Causal Stable Inverse Filters*)  Given is the causal stable filter $y = h * x$ with a
   finite impulse response $h = \delta_0 - \delta_1/4 - \delta_2/8$ and transfer function

   $$H(z) = 1 - \frac{1}{4}z^{-1} - \frac{1}{8}z^{-2} = \frac{(z - 1/2)(z + 1/4)}{z^2}$$

   for $|z| > 0$ and a corresponding difference equation

   $$y_n = x_n - \frac{1}{4}x_{n-1} - \frac{1}{8}x_{n-2}.$$

   *The number of zeros and poles of $H$ matches, and all zeros and poles lie within
   the unit circle.* Therefore, there exists a causal stable inverse filter: Solving the
   difference equation for $x_n$ gives $x_n = y_n + x_{n-1}/4 + x_{n-2}/8$ and thus the inverse

filter with the transfer function

$$H_{\text{inv}}(z) = \frac{1}{1 - z^{-1}/4 - z^{-2}/8} = \frac{z^2}{(z - 1/2)(z + 1/4)} = \frac{1}{H(z)}.$$

$H_{\text{inv}}$ has the following partial fraction decomposition and Laurent series representation for $|z| > 1/2$ (presented here as a small calculation exercise for readers, see p. 489 and p. 363):

$$H_{\text{inv}}(z) = 1 + \frac{1/3}{z - 1/2} - \frac{1/12}{z + 1/4} = 1 + \frac{1}{12} \sum_{k=0}^{\infty} \left( \frac{4}{2^k} + (-1)^{k+1} \frac{1}{4^k} \right) z^{-(k+1)}.$$

*The inverse filter with impulse response $h_{\text{inv}}$, the inverse z-transform of $H_{\text{inv}}(z)$ ($|z| > 1/2$), is therefore causal and also stable* (see p. 368). However, it has an infinitely long impulse response and is thus a so-called IIR filter (*Infinite Impulse Response Filter*). It holds that $h * h_{\text{inv}} = h_{\text{inv}} * h = \delta_0$, and for causal signals $x$ from $l_d^{\infty}$ the convolutions $h_{\text{inv}} * (h * x) = (h_{\text{inv}} * h) * x = x$ are associative (see p. 196).

The frequency response $\widehat{h}_{\text{inv}}$ is like $\widehat{h}$ due to the exponentially fast decay of its Fourier coefficients infinitely often differentiable, as is generally the case with stable filters with rational transfer functions (see p. 51, the $1/f$-Theorem of N. Wiener in Sect. 5.6 and the Laurent series of the partial fractions of $H$ and $H_{\text{inv}}$).

If for $y = h * x,\ \displaystyle\sum_{k=0}^{\infty} y_k \delta_k = \sum_{k=0}^{\infty} h_k \delta_k * \sum_{k=0}^{\infty} x_k \delta_k$ the input signals $x$ have a fixed length $N + 1$, i.e., $x_k = 0$ for $k > N$, and one sets $\underline{x} = (x_0, x_1, \ldots, x_N)^T$, then $\underline{x}$ can already be reconstructed with $\underline{y} = (y_0, y_1, \ldots, y_N)^T$ and the first $N + 1$ coefficients of $h_{\text{inv}}$, which subsequently also appear in the matrix inverse to $\underline{\underline{H}}$:

$$\underline{y} = \underline{\underline{H}}\,\underline{x} \quad \text{and} \quad \underline{x} = \underline{\underline{H}}^{-1}\,\underline{y},$$

$$\underline{\underline{H}} = \begin{pmatrix} h_0 & 0 & 0 & \ldots & 0 \\ h_1 & h_0 & 0 & \ldots & 0 \\ h_2 & h_1 & h_0 & \ldots & 0 \\ \vdots & & & \ddots & 0 \\ h_N & h_{N-1} & \ldots & & h_0 \end{pmatrix}, \quad \underline{\underline{H}}^{-1} = \begin{pmatrix} h_{\text{inv},0} & 0 & 0 & \ldots & 0 \\ h_{\text{inv},1} & h_{\text{inv},0} & 0 & \ldots & 0 \\ h_{\text{inv},2} & h_{\text{inv},1} & h_{\text{inv},0} & \ldots & 0 \\ \vdots & & & \ddots & 0 \\ h_{\text{inv},N} & h_{\text{inv},N-1} & \ldots & & h_{\text{inv},0} \end{pmatrix}.$$

2. (*Causal Unstable Inverse Filters*) Inversion of the stable causal filter $y = h * x$ with $y_n = x_n - x_{n-1}$ yields by solving for $x_n$ the inverse filter with the difference equation $x_n = y_n + x_{n-1}$ for the input signal $y$ and output signal $x$.

The inverse filter, the accumulator from p. 354, is *causal, but unstable*, since the zero $z = 1$ of $H$ becomes a pole of $H_{\text{inv}} = 1/H$ on the unit circle.

With erroneous values $y_n$, all errors in $y_k$ for $k \leq n$ are then added up in $x_n$ during the inversion of $y = h * x$. This can result in an outcome that has little in common with the original signal $x$.

3. (*Noncausal Stable Inverse Filters*) $H(z) = z^{-1}$, $|z| > 0$, gives with $h = \delta_1$ a causal stable filter, $H_{\text{inv}}(z) = z$, $z \in \mathbb{C}$, with $h_{\text{inv}} = \delta_{-1}$ a *stable noncausal* filter. Here the degree of the numerator of $H$ is less than the degree of the denominator.

   The transfer function $H(z) = 1 - 2z^{-1} = (z - 2)/z$ ($|z| > 0$) of the causal filter with impulse response $h = \delta_0 - 2\delta_1$ has the zero $z = 2$. The Laurent series of $1/H$ in the region $|z| < 2$ as the $z$-transform of the inversion with the impulse response $h_{\text{inv}} = -\sum_{k=1}^{\infty} 2^{-k}\delta_{-k}$ corresponds to a *stable noncausal* filter (see p. 368).

4. (*Noncausal Unstable Inverse Filters*) Choosing in Example 1 for the impulse response $h_{\text{inv}}$ the coefficients of a Laurent series expansion of $1/H$ in a region $A$ with $|z| < |z_0|$ for a zero $z_0$ of $H$ and all $z \in A$, then $h_{\text{inv}}$ is *noncausal*. Since the unit circle is not in $A$, the corresponding filter is *unstable*.

**Conclusions**

(1)  If the degree of the numerator polynomial of $H$ is less than the degree of the denominator polynomial, then the inverse filter is noncausal. If one chooses for the Laurent series expansion of $1/H$ an annulus of convergence with a zero of $H$ in the outer region, then the corresponding inverse filter becomes noncausal. The stability of the inverse filter is retained according to the $1/f$-Theorem of N. Wiener (see p. 80) if the unit circle is in such a region (proof: series expansion of the partial fractions of $1/H$).

(2)  If one has as the region of convergence for the Laurent series expansion of $1/H$ the region $A = \{z \in \mathbb{C} : |z| > r\}$ with $r = \max\{|z_1|, \ldots, |z_N|\}$ with as many zeros $z_1, \ldots, z_N$ of $H$ as $H$ has poles, then the corresponding impulse response $h_{\text{inv}}$ is causal. If the unit circle is not included in this region, the inverse filter is unstable. However, if it belongs to this region, then the inverse filter is also stable. One can then reconstruct the input signal in a stable way from the values of $y = h * x$ with knowledge of $h$, i.e., bounded disturbances of $y$ lead to only bounded errors in the reconstruction of $x$. The combination of causality and stability of a filter with a rational transfer function $H$ is preserved upon inversion only if all zeros and poles of $H$ lie within the unit disk and $H$ has as many zeros as poles. Filters with this property are so-called *minimum phase filters* (see Exercise A13 at the end of the chapter).

## *Amplitude Response, Phase Response, and Group Delay*

For a discrete filter with a piecewise differentiable frequency response $\widehat{h}(\omega)$, $|\widehat{h}(\omega)|$ is the *amplitude response*, $\Phi(\omega) = \arg(\widehat{h}(\omega))$ is the *phase response*, and the function $D(\omega) = -\,\mathrm{d}\Phi(\omega)/\,\mathrm{d}\omega$ defined except for finitely many points is its *group delay* (cf. p. 329).

A rational transfer function $H \neq 0$ of a filter as on p. 365 is often represented in factorized form. As there, let the coefficient of the denominator polynomial be $b_0 = 1$. Further, let $r$ be the smallest index such that $a_r$ among the coefficients of the numerator $Q$ is not zero. For $N > r$ and $M \geq 1$, $H$ has the form with the zeros $c_k$, the poles $d_k$, and possibly the origin as a $|M - N|$-fold pole or zero:

$$H(z) = a_r z^{M-N} \frac{\displaystyle\prod_{k=1}^{N-r} (z - c_k)}{\displaystyle\prod_{k=1}^{M} (z - d_k)}.$$

Experienced engineers can often quickly recognize characteristic properties of the filter by the location of the zeros $c_k$ and the poles $d_k$. The impulse response can be obtained by partial fraction decomposition of $H$ for given zeros and poles, as already shown in examples. If the filter is stable with the frequency response $\widehat{h}(\omega) = H(\mathrm{e}^{j\omega a})$, then taking the logarithm of the amplitude response gives

$$20\log_{10} |\widehat{h}(\omega)| = 20\log_{10} |a_r| + \sum_{k=1}^{N-r} 20\log_{10} |\,\mathrm{e}^{j\omega a} - c_k| - \sum_{k=1}^{M} 20\log_{10} |\,\mathrm{e}^{j\omega a} - d_k|.$$

Here, the logarithmic unit dB (decibel) is used, and it shows the following:

The amplitude attenuation in dB for the angular frequency $\omega$ is composed additively from the constant $|a_r|$ and the lengths of the *zero vectors* $n_k = \mathrm{e}^{j\omega a} - c_k$ minus the lengths of the *pole vectors* $p_k = \mathrm{e}^{j\omega a} - d_k$, *all measured in dB.* An amplitude attenuation of 6 dB corresponds to an approximate halving of the amplitude. Similarly, the phase response behaves additively. For $0 \leq \omega < 2\pi/a$, we have

$$\Phi(\omega) = \Phi\left(a_r\, \mathrm{e}^{j\omega a(M-N)}\right) + \sum_{k=1}^{N-r} \Phi\left(\mathrm{e}^{j\omega a} - c_k\right) - \sum_{k=1}^{M} \Phi\left(\mathrm{e}^{j\omega a} - d_k\right).$$

Corresponding relationships for the above excluded cases $N = r$ or $M = 0$ are seen analogously. For $N = r$, the numerator of $H$ becomes $a_r z^{M-r}$, and for $M = 0$ in non-recursive filters, the denominator of $H$ becomes 1. The representations of the amplitude attenuation in dB and the phase response are then to be adjusted accordingly.

For practical applications, filters with linear phase are of great importance. Compare, for example, the remark on p. 329 about phase distortions of only a few milliseconds in filters for audio applications, or consider transmission systems that use phase modulation techniques. For filters $y = h * x$ with linear phase, the group delay defined except for possible jumps in the phase function is the constant frequency-independent phase delay of the filter.

## *Filter Examples and Filter Design*

The previous explanations should now of course be supplemented by concrete filter examples. A main task of filter design in practice is to approximate ideal, often non-realizable filters within given tolerance ranges with realizable stable filters. We limit ourselves in the following to one significant example each for the two essential filter types of *FIR filters* (Finite Impulse Response Filters) and *IIR filters* (Infinite Impulse Response Filters). If the readers' interest is aroused to deepen the fundamental concepts presented with the rich further literature on discrete signal processing and its applications, they will find excellent sources in the books by Oppenheim and Schafer (2013), Jury (1973), and Tietze and Schenk (2008).

### Causal FIR Filters with Real Coefficients and Linear Phase

FIR filters are filters with finite impulse response. Such filters are always stable, as their frequency responses are trigonometric polynomials. We first show that causal FIR filters with real coefficients and constant group delay can be constructed by imposing certain symmetry conditions on the filter coefficients. Consider a filter of length $N \geq 1$ with a transfer function $H$ of the form

$$H(z) = \sum_{n=0}^{N-1} h_n z^{-n}$$

*with real coefficients $h_n$ and the symmetry $h_n = h_{N-1-n}$. It follows that*

$$2H(z) = \sum_{n=0}^{N-1} h_n (z^{n+1-N} + z^{-n}) = z^{(1-N)/2} \sum_{n=0}^{N-1} h_n \left( z^{n-(N-1)/2} + z^{-n+(N-1)/2} \right).$$

With $z = e^{j\omega a}$ (*a* our time step width) we obtain the corresponding frequency response $\widehat{h}$ with the *constant group delay* $(N-1)a/2$:

$$\widehat{h}(\omega) = H\left(e^{j\omega a}\right) = e^{j(1-N)\omega a/2} \sum_{n=0}^{N-1} h_n \cos\left(\left(\frac{N-1}{2} - n\right)\omega a\right).$$

**Remark** FIR filters with the coefficient symmetry used above are called FIR filters of type I when $N$ is odd and FIR filters of type II when $N$ is even. Similarly, filters of the so-called types III and IV can be constructed if the above symmetry condition is replaced by the condition $h_n = -h_{N-1-n}$. FIR filters are also referred to as non-recursive filters, while IIR filters are called recursive filters.

Since the number of coefficients of the filter corresponds to the number of multipliers in a circuit implementation or programming (see p. 366), a more efficient representation in terms of the number of multiplications, utilizing the given symmetry, is useful. As an exercise, calculate that for an even filter length $N \geq 2$ the transfer function of our FIR filter is given by

$$H(z) = \sum_{n=0}^{(N-2)/2} h_n(z^{-n} + z^{-N+n+1})$$

and for odd $N \geq 3$ by

$$H(z) = h_{(N-1)/2}\, z^{-(N-1)/2} + \sum_{n=0}^{(N-3)/2} h_n(z^{-n} + z^{-N+n+1}).$$

**Design of FIR Filters by Approximation with a Window Function**

Since the frequency responses of discrete filters are periodic, it is natural to approximate them by weighted partial sums of their Fourier series expansions. For input signals of finite duration, i.e., in $\mathcal{X} \cap \mathcal{E}'$, or signals from $l_d^2$, we consider the $\Omega$-periodic frequency response of an ideal lowpass filter with cutoff angular frequency $0 < \omega_c < \Omega/2 = \pi/a$

$$\widehat{g}(\omega) = \begin{cases} 1 & \text{for } |\omega| \leq \omega_c \\ 0 & \text{for } \omega_c < |\omega| < \Omega/2. \end{cases}$$

With $g_n = \dfrac{1}{\Omega} \displaystyle\int_{-\omega_c}^{+\omega_c} e^{jn\omega a}\, d\omega$ the transfer function $G$ would be given by

$$G(z) = \sum_{n=-\infty}^{+\infty} z^{-n} g_n = \frac{2\omega_c}{\Omega} \sum_{n=-\infty}^{+\infty} z^{-n}\, \frac{\sin(n\omega_c a)}{n\omega_c a}.$$

This transfer function has infinitely many coefficients and is also noncausal. An approximation with a partial sum, i.e., an approximation by multiplying $G$ with a rectangular window, results in the Gibbs phenomenon in the frequency response (see p. 25 and p. 136). Analogous to the elimination of the Gibbs phenomenon by Fejér means, i.e., by weighting with a triangular window (see p. 31), other window functions can be used. There are many different window functions that are used in practice. All windows share the common feature that the weights $w_n$ used decrease to zero toward the higher frequencies, thus counteracting the Gibbs phenomenon. Compare also Sect. 12.6, in which we examine window effects in the discrete Fourier transform in more detail. A detailed discussion of window functions can be found in Slepian (1983) or Harris (1978).

With a symmetric discrete window of the form $w = \sum_{n=1-N}^{N-1} w_n \delta_n$ and $w_{-n} = w_{+n}$, the modified transfer function $\widetilde{G}$ is formed from $G$

$$\widetilde{G}(z) = \sum_{n=1-N}^{N-1} w_n z^{-n} g_n.$$

From this noncausal transfer function $\widetilde{G}$, a causal transfer function $H$ is finally obtained, with which the lowpass filter is approximated. By delay, here multiplication by $z^{1-N}$, and using the symmetry of the values $g_n$ and $w_n$, the causal approximating transfer function $H$ is obtained with

$$|H(\mathrm{e}^{j\omega a})| = |\widetilde{G}(\mathrm{e}^{j\omega a})|$$

$$H(z) = z^{1-N}\widetilde{G}(z) = \sum_{n=0}^{2(N-1)} z^{-n}\, w_{N-n-1}\, g_{N-n-1}.$$

**Theorem 11.10 (FIR Filter with Constant Group Delay)** *With $H$, we have obtained a transfer function of a causal filter of length $(2N - 1)$, which has real coefficients and the constant group delay $(N - 1)a$ ($a$ was the time step used in our discrete signals).*

The following representation shows the amplitude response of this filter in dB for two different windows. We choose $N = 20$ as the number $N$ of filter coefficients, $\omega_c = \Omega/4$, and two cos-windows of the form

$$w_n = \begin{cases} \alpha + \beta \cos\left(\dfrac{n\pi}{N-1}\right) + \gamma \cos\left(\dfrac{2n\pi}{N-1}\right) & \text{for } 1 - N \leq n \leq N - 1 \\[2ex] 0 & \text{otherwise.} \end{cases}$$

The first example uses $\alpha = 0.54$, $\beta = 0.46$, $\gamma = 0$, the commonly used so-called *Hamming window* (Fig. 11.15), the second with $\alpha = 0.42$, $\beta = 0.5$, $\gamma = 0.08$, and

**Fig. 11.15** Lowpass,
Hamming window

**Fig. 11.16** Lowpass,
Blackman window

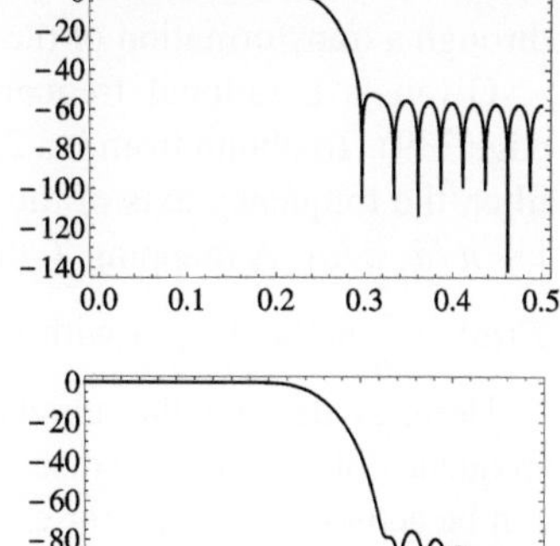

the so-called *Blackman window* with slightly higher attenuation in the stopband at
the cost of a less steep roll-off in the transition band compared to the design with
the Hamming window (Fig. 11.16). Due to periodicity, the frequency response $\widehat{h}$ of
a discrete filter is generally only usable up to half the sampling frequency $1/(2a)$.
Plotted is the *amplitude response in* dB as a function of $s\Omega$, $0 \le s \le 0.5$, and the
cutoff angular frequency $\omega_c$ is therefore at $s = 1/4$ on the abscissa.

Alternative design methods for FIR filters are based on a tolerance scheme as
shown on p. 332 and calculate the filter coefficients according to given optimality
criteria.

Such criteria can be the minimization of the maximum approximation error or,
for example, frequency-weighted error criteria. A frequently used method of this
type is the Parks-McClellan algorithm. It is discussed in detail in the book by
Oppenheim and Schafer (2013), to which we refer here.

Today, typical filter lengths $N$ of FIR filters are in the range of $N = 58$ in multi-
band graphic equalizers to $N = 160$ and more when used in CD players. Particularly
important applications of FIR filters are adaptive filtering on the receiving side to
compensate for distortions in transmission channels and multi-rate signal processing
in systems with different sampling rates. In this context, FIR filters are used for
decimation and interpolation. Again, we refer to the specialized literature on digital
signal processing already cited above.


### Design of IIR Filters Using the Bilinear Transformation


Rational transfer functions, whose denominator polynomial is not constant, have an
infinitely long impulse response due to the feedbacks seen in the block diagram on
p. 366. They are therefore called *IIR filters* (Infinite Impulse Response). As with FIR
filters, there are different design methods for IIR filters depending on the purpose.
As an example, we explain the method of bilinear transformation, by which a stable

discrete filter is constructed from the frequency response of a stable analog filter through a transformation of the frequency axis.

Given is a rational frequency response $R(j\omega)$ of a stable analog filter (cf. page 339). To obtain from it a $2\pi/a$-periodic frequency response of a stable discrete filter, the frequency axis of the analog filter is bijectively mapped onto the interval $]-\pi/a, \pi/a[$. A mapping $T$ that accomplishes this is

$$T(\omega) = \frac{2}{a} \arctan\left(\frac{\omega a}{2V}\right) \text{ with inverse mapping } T^{-1}(\Omega) = \frac{2}{a} V \tan\left(\frac{\Omega a}{2}\right) = \omega.$$

Here, $\Omega$ denotes the angular frequency in the discrete case, $1/a$ the sampling frequency of the intended discrete system, and $V$ a factor with which a pre-distortion can be achieved, so that, for example, a desired cutoff frequency $\omega_c/(2\pi)$ is a fixed point under the mapping $T$. Thus, we define the desired frequency response $\widehat{h}(\Omega)$ of the sought discrete filter by $\widehat{h}(\Omega) = R(jT^{-1}(\Omega))$ for $\Omega \in \left]-\frac{\pi}{a}, \frac{\pi}{a}\right[$. That this frequency response $\widehat{h}$ is rational in $z = e^{j\Omega a}$ can be seen as follows:

The *Möbius transform* $B : \overline{\mathbb{C}} \to \overline{\mathbb{C}}$ of the compactified complex plane by adding the point $\infty$, defined by

$$B(z) = \frac{2V}{a} \frac{1-z^{-1}}{1+z^{-1}} = s \text{ for } z \in \mathbb{C}, \quad B(-1) = \infty, \quad B(\infty) = \frac{2V}{a},$$

is bijective with the inverse mapping $B^{-1}(s) = (2V/a + s)/(2V/a - s) = z$ for $s \in \mathbb{C}$. This mapping $B$ is called *bilinear transformation*.

One easily verifies the following properties of the bilinear transformation $B$, which ensure the stability of the discrete filter designed with it: For the real part of $s = B(z)$, the equivalences

$$\Re(s) = \Re(B(z)) < 0 \iff |z| < 1$$

$$\Re(s) = \Re(B(z)) = 0 \iff |z| = 1$$

$$\Re(s) = \Re(B(z)) > 0 \iff |z| > 1$$

hold. For $z = e^{j\Omega a}$, it follows from $j\tan(x) = (1 - e^{-2jx})/(1 + e^{-2jx})$ that $\widehat{h}$ is rational in $z$: $\widehat{h}(\Omega) = R(jT^{-1}(\Omega)) = R\left(\frac{2V}{a}\frac{1-z^{-1}}{1+z^{-1}}\right) = R(B(z))$.

**Theorem 11.11 (Bilinear Transform)** $H(z) = R(B(z))$ *is the rational transfer function of the discrete linear filter with the $2\pi/a$-periodic frequency response* $\widehat{h}(\Omega) = H\left(e^{j\Omega a}\right)$. *If $R(j\omega)$ is the frequency response of a stable analog filter as assumed, then the discrete filter with the transfer function $H$ is also stable.*

**Example** We demonstrate the approach using the example of the third-order Butterworth filter, whose frequency response we calculated on p. 335. Its cutoff frequency was $\omega_c/(2\pi) = 4.2$ kHz. To keep this cutoff frequency invariant for the discrete lowpass filter generated as above, we choose as prewarping $V = \omega_c a/2 \cot(\omega_c a/2)$. Thus, it follows that $T(\omega_c) = \omega_c = T^{-1}(\omega_c)$ and $\widehat{h}(\omega_c) = R(j\omega_c)$. The frequency

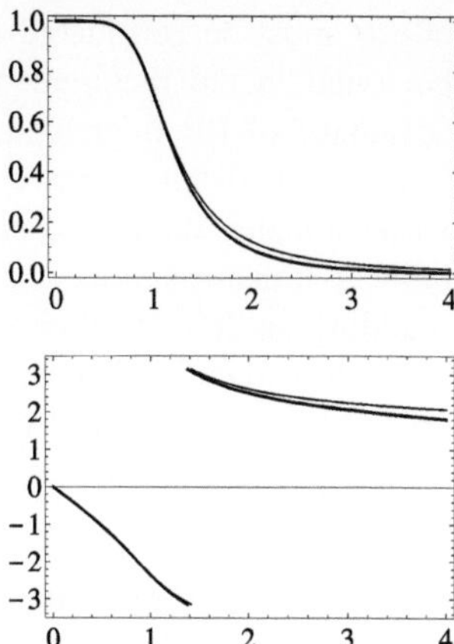

**Fig. 11.17** Amplitude response, analog and discrete Butterworth lowpass compared

**Fig. 11.18** Phase response, analog and discrete Butterworth lowpass compared

response $R(j\omega)$ of our example was

$$R(j\omega) = \frac{1}{(1 + j\omega/\omega_c)(1 + j\omega/\omega_c + (j\omega/\omega_c)^2)}.$$

Replacing $j\omega$ with $B(z)$ yields with $c = \omega_c a$ the transfer function $H(z)$ of the discrete analog, which can be realized according to the block diagram on p. 366:

$$H(z) = \left(1 + \frac{2V}{c}\frac{1 - z^{-1}}{1 + z^{-1}}\right)^{-1}\left(1 + \frac{2V}{c}\frac{1 - z^{-1}}{1 + z^{-1}} + \frac{4V^2}{c^2}\frac{(1 - z^{-1})^2}{(1 + z^{-1})^2}\right)^{-1}.$$

The following illustrations—each as a function of $s\omega_c$, $0 \le s \le 4$—show in Fig. 11.17 the amplitude response of the analog Butterworth filter and the $(2\pi)/a$-periodic amplitude response of the corresponding discrete filter generated as above and in Fig. 11.18 the phase responses of the two filters. A sampling frequency of 44.1 kHz was chosen as $1/a$. The thicker plotted curves are those of the discrete filter. The cutoff angular frequency $\omega_c$ is at $s = 1$ on the abscissa.

For comparison, some key figures are as follows: At the passband edge 3 kHz, the amplitude of the analog filter's frequency response is 0.9395, and that of the discrete filter is 0.9443. The corresponding values at the stopband edge 10 kHz are 0.0739 for the analog filter and 0.0455 for the discrete filter. At half the cutoff frequency 2.1 kHz, the value is 0.9923 for the analog filter and 0.9932 for the discrete filter. At the cutoff frequency 4.2 kHz, both frequency responses have the same amplitude value of 0.7071 due to the chosen prewarping.

In an analogous way, other filter types than lowpass filters can also be obtained using the bilinear transformation. As alternative design methods, approaches are to be mentioned in which a corresponding discrete filter is constructed from the sample values of the impulse response or the frequency response of an analog filter to be emulated. In doing so, as with the discrete Fourier transform and the sampling theorem of Shannon discussed in the following section, aliasing effects must be considered at too low sampling frequencies. These methods can be found in the literature on digital signal processing already cited above. An advantage of IIR filters compared to FIR filters is that a much lower filter order is required to approximate a desired amplitude response than with FIR filters. Disadvantages include nonlinear distortions of the frequency axis as with the bilinear transformation, a hardly achievable linear phase response, feedback of rounding and quantization errors affecting stability, and other issues that require great care when implementing IIR filters. If there is interest, one should therefore refer to the extensive literature on the subject.

## *Notes on Applications of Noncausal Discrete Filters*

Since our main focus so far has been on causal linear filters, it should not go unmentioned at the end of this chapter that noncausal filters can also be used in many application areas. If one disregards real-time signal processing and instead thinks of processing complete datasets, such as audio data from music tracks on a CD or image data, it becomes immediately clear that noncausal filters can also be used for processing. Most readers will already be familiar with a whole range of different filters from image editing programs—such as for edge sharpening or smoothing, etc. Reference to literature on special fields is again made here.

Noncausal filters can also be used for time series in "near real-time" processing with sufficient data buffering. Applications of noncausal smoothing filters in radio telemetry, as in Hurd (1997) and in other sources, are such as the Deep Space Network Galileo Telemetry System on the NASA Technical Reports Server.

**Summary** Fourier analysis and distribution theory have enabled the *unified* and effective representation and treatment of analog and discrete translation-invariant linear systems $L : \mathcal{Z} \to \mathcal{A}$ using the same mathematical tools. *Such systems differ in their mathematical model essentially only by the choice of the signal spaces $\mathcal{Z}$ and $\mathcal{A}$.* Fundamental system properties, under appropriate signal spaces and continuity conditions, follow from representations as linear filters and their characterization by the impulse response and frequency response. In examples, we have seen how linear filters can be constructed according to given criteria. In the mathematical model, functions or distributions are processed by the operator $L$, and in the special case of discrete systems, sequences of impulses are processed. In practical applications, the idealized models are realized approximately, with analog filters shown by suitable circuits as exemplified by the Butterworth lowpass filter. For discrete systems, the coefficients of the impulse sequences in the mathematical model are very often proportional to quantized samples of analog signals in practice. Signal processing by linear filters can then be performed in processors by processing only the coefficients. Coefficient sequences resulting from output signals can be fed back into analog systems through digital-to-analog conversion. Examples include signal processing in images or music and in applications such as WLAN, DSL, DVB-T, and many others. A schematic representation in a signal flow diagram is seen in Sect. 12.2 on sampling and interpolations with linear filters. All discussed filters are available today as components at low cost or as software in countless applications.

## 11.7  Exercises

For the following exercises, it is useful to use mathematical software.

**(A1)** When connecting electrical devices to the power grid, an interference filter is often used, which is intended to keep high-frequency oscillations away from the user:

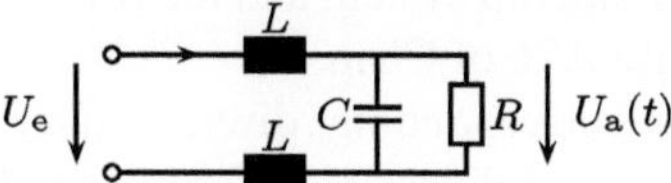


What is the amplitude of $U_a$ at low frequencies ($|\omega| \to 0$) and at high frequencies ($|\omega| \to \infty$) for an input quantity $U_e(t) = U_0 \sin(\omega t)$?

**(A2)** Calculate a Butterworth lowpass filter (cf. p. 332) for DC gain $K = 1$, passband edge $\omega_1/(2\pi) = 3\,\text{kHz}$, and stopband edge $\omega_2/(2\pi) = 5\,\text{kHz}$, with minimum passband gain $\hat{h}_1 = 0.9$ and maximum stopband gain $\hat{h}_2 = 0.1$.

**(A3)** The active circuit depicted with an operational amplifier will become a second-order lowpass filter (Sallen-Key Biquad Filter type) with appropriate choice of capacitors $C1$ and $C2$ and resistors $R1$ and $R2$.

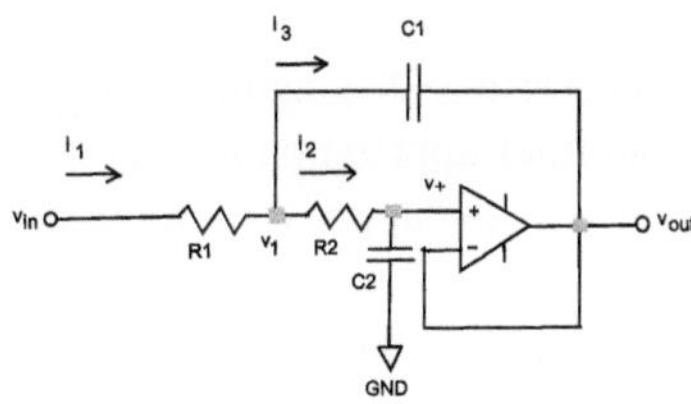


(a) Derive the frequency response of the circuit.

(b) Compute the resistors $R1$ and $R2$ with $C1 = 8\,\text{nF}$, $C2 = 4\,\text{nF}$ and cutoff frequency $10\,\text{kHz}$ so that the circuit yields a Butterworth second-order lowpass filter with $\widehat{h}_{\text{LP}}(\omega) = \dfrac{1}{1 + j\sqrt{2}\,\omega/\omega_c - \omega^2/\omega_c^2}$.

(c) If you interchange the resistors and capacitors in the circuit (with analogous numbering), you obtain a second-order highpass filter. Compute its transfer function, amplitude, and phase response with $C1 = C2 = 100\,\text{nF}$ and $\omega_c/(2\pi) = 1000\,\text{Hz}$, so that the filter has frequency response

$$\widehat{h}_{\mathrm{HP}}(\omega) = \frac{(j\omega)^2}{\omega_c^2}\widehat{h}_{\mathrm{LP}}(\omega)$$

with the Butterworth response $\widehat{h}_{\mathrm{LP}}$ from (b) (lowpass to highpass transformation, p. 342).

(**A4**) Verify the statements in Example 2 on p. 342.

(**A5**) (a) Verify the statements in Examples 3 and 4 on p. 343.

(b) Compute the amplitude and phase responses of the bandpass and bandstop filters in the graphs on p. 343 using the given data.

(c) Compute the poles with $B = 1/3$ in the right image on p. 342, which were mentioned in Examples 2–4 on pages 342–343 for the transfer functions of highpass, bandpass, and bandstop filters. Use a computer algebra system and the inverse of the Joukowsky mapping as in Exercise A26 of Chap. 5.

(d) Create a corresponding discrete filter from the bandpass filter in (b) using the bilinear transformation. Plot the amplitude and phase responses of the filter. Compare analog and discrete filters.

(**A6**) Consider the triangular function $f(t) = (1 - |t|)(u(t + 1) - u(t - 1))$, where $u(t)$ is the unit step function.

(a) Compute a DFT (cf. Sect. 6) with eight equidistant samples of $f$ and step size $1/4$ in the interval $[-1, 1]$. Graphically represent a first approximation for the (real, even) Fourier transform of $f$ using the computed DFT coefficients by a polygonal line (cf. p. 275 and p. 347). Pay attention to the correct assignment of frequencies and the duration $T$ of sampling.

(b) Then, extend the vector of samples of $f$ by appending 2040 zeros (zero padding). Perform a DFT again with this extended vector, and use it to represent another approximation for $\widehat{f}$ as above. Compare the results, and explain why the approximation has improved with zero padding.

(**A7**) A realizable discrete filter has the transfer function $H(z) = \dfrac{z^2 + 1}{z^2 - 1}$. Provide the impulse response.

(**A8**) Let $X(z)$ be the $z$-transform of the signal $x = \displaystyle\sum_{k=-\infty}^{+\infty} x_k \delta_k$. Compute relationships of the $z$-transforms of

$$x_- = \sum_{k=-\infty}^{+\infty} x_{-k}\delta_k, \quad x_\alpha = \sum_{k=-\infty}^{+\infty} \alpha^k x_k \delta_k \quad (\alpha \neq 0) \text{ and } v = \sum_{k=-\infty}^{+\infty} k x_k \delta_k.$$

**(A9)** An analog frequency response of a bandstop filter (notch filter), which blocks the angular frequency $\omega_c$, is $R(j\omega) = \dfrac{A_0(1 + (j\omega/\omega_c)^2)}{1 + j\omega/(\omega_c Q) + (j\omega/\omega_c)^2}$.

- (a) Present the amplitude and phase response of the filter for $A_0 = 1$ and for two different quality factors $Q_1 = 0.5$ and $Q_2 = 10$ graphically.
- (b) *Discrete Bandstop Filters.* Develop a program that implements a discrete bandstop filter using the bilinear transformation with quality factor $Q$, sampling frequency $1/a$, and stop frequency $\omega_c/(2\pi)$.
- (c) Acquire an audio recording of a football match from the 2010 World Cup in South Africa with strong vuvuzela noise. Create a cascade filter of appropriate quality to block the fundamental frequency of the vuvuzela around 233 Hz (note A sharp) and its three subsequent harmonics (octave, fifth, and octave), and limit the output signal through lowpass filtering to the voice bandwidth. Apply your "anti-vuvuzela filter" to the recording, and test the result.

**(A10)** *Discrete Chebyshev Lowpass Filters.* From the analog filter of Task 26 in Chap. 5, generate a discrete Chebyshev filter using the bilinear transformation with the same cutoff frequency and attenuation specifications.

**(A11)** Calculate the impulse responses and frequency responses of the filters with the transfer functions $H(z) = \dfrac{60 - 8z^{-1} - 4z^{-2}}{60 + 15z^{-2}}$ and $H_{\text{inv}} = 1/H(z)$.

**(A12)** *Discrete Allpass Filters.* In this task, all filters are assumed to be discrete causal and stable filters with a rational transfer function $H$. It then holds that $|\widehat{h}(\omega)|^2 = C(e^{j\omega a}) = H(e^{j\omega a})H^*(e^{-j\omega a})$ with $H^*(z) = \overline{H(\overline{z})}$.

- (a) Assume the transfer function is $H(z) = a_0 z^{-n} \dfrac{\prod\limits_{k=1}^{N}(1 - c_k z^{-1})}{\prod\limits_{k=1}^{M}(1 - d_k z^{-1})}$, which is assumed to be completely simplified with $c_k \neq 0,\, d_k \neq 0,\, M, N \in \mathbb{N}$, $n \in \mathbb{N}_0$, and $a_0 \in \mathbb{C}$. Show that $|\widehat{h}(\omega)|^2 = 1$ holds precisely when $H$ takes the form

$$H(z) = e^{j\varphi}\, z^{-n} \prod_{k=1}^{M} \frac{z^{-1} - \overline{a_k}}{1 - a_k z^{-1}}$$

  with $a_k \in \mathbb{C},\, |a_k| < 1$, and $\varphi \in [0, 2\pi[$.
- (b) Show that an allpass filter as in (a) has a positive group delay.

**(A13)*** *Discrete Minimum Phase Filters.* Discrete causal stable filters with a rational transfer function, having as many zeros as poles and with their zeros and poles inside the unit circle, are called *minimum phase filters* (cf. Oppenheim & Schafer, 2013). They have causal stable inverses.

(a) Show that discrete causal stable filters with a rational transfer function $H$ can be represented by $H = H_{\min} H_{\text{all}}$ or $H = H_{\min} H_{\text{uc}} H_{\text{all}}$ with the transfer function $H_{\min}$ of a minimal-phase filter, the transfer function $H_{\text{uc}}$ of a filter whose zeros lie on the unit circle, and the transfer function $H_{\text{all}}$ of an allpass filter. Demonstrate the decomposition with the help of some pole-zero diagrams.

(b) Conclude that a minimal-phase filter has the smallest group delay among all filters with the same amplitude response.

**(A14)⋆** *Ideal DC Blocking Operator.* Consider the operator $Lx = \displaystyle\sum_{m=-\infty}^{+\infty} y_m \delta_m$ with $y_m = x_m - m_{(C,1)}(x)$, defined for $x \in l_d^\infty$ where the limit

$$m_{(C,1)}(x) = \lim_{N \to \infty} \frac{1}{2N+1} \sum_{k=-N}^{+N} x_k$$

of the Cesáro mean exists. Extend $L$ to a continuous linear translation-invariant operator on the entire space $l_d^\infty$.

Consider the set $M = \{x \in l_d^\infty : m_{(C,1)}(x) = 0\}$ and the set $K = \{P \in l_d^{\infty'} : P(M) = 0, \|P\| = 1 = P(1)\}$ as in Example on page p. 355, and show as there that there exists a translation-invariant functional $P_0$ in $K$. Show that $Lx = x - 1 \cdot P_0(x)$ is continuous and translation-invariant, but not a convolution operator. Consider $L\delta_0$ and $Lu$ for $u = \displaystyle\sum_{k=0}^{\infty} \delta_k$.

# Chapter 12
# Further Applications of the Fourier Transform

**Abstract** Further applications of Fourier analysis are examined. Shannon's sampling theorem is proven and discussed. The spectral properties of sampling applications are considered and a basic digital transmission system is shown. The transmission of signals with a linear multi-carrier system, such as WLAN or mobile data transmission, is treated as a current everyday application. The method is orthogonal frequency division multiplexing (OFDM), which uses the FFT and linear filters. Further sections examine the Heisenberg uncertainty principle and its consequences for the time-bandwidth product of signals. Closely related to this is the windowed Fourier transform (STFT) as a tool for time-frequency analysis. Inversion formulas for the STFT with continuous and discrete parameters are proven. The use of time windows in the DFT to reduce alias effects is discussed. In further sections, initial value problems for the homogeneous and inhomogeneous wave and heat equations in two and three dimensions are solved. The Fourier transform of distributions is used to solve these equations. The Huygens' principle for waves is explained. For the heat equation, an inhomogeneous boundary value 3D problem is solved approximately as a further application of the FEM method and the solution is displayed graphically.

## 12.1  Shannon's Sampling Theorem

The theoretical starting point for signal transmission methods, where discrete approximations of $f(t)$ are transmitted instead of a continuous analog signal $f(t)$, is Shannon's sampling theorem (1949). It states that a signal $f(t)$ can be reconstructed from its samples under suitable conditions. A recommended read on the history of the theorem and its developments, with a wealth of relevant references, is the article *"Sampling—50 Years After Shannon"* by Unser (2000).

© The Author(s), under exclusive license to Springer Nature Switzerland AG 2025
R. Brigola, *Fourier Analysis and Distributions*, Texts in Applied Mathematics 79,
https://doi.org/10.1007/978-3-031-81311-5_12

## *Shannon's Sampling Theorem for Bandlimited Functions*

**Theorem 12.1 (Sampling Theorem)** *If $f$ is an integrable function that is bandlimited by $\omega_c > 0$, i.e., $\widehat{f}(\omega) = 0$ for $|\omega| > \omega_c$, and if $tf(t)$ is also integrable with $f(t)$, then for all $t \in \mathbb{R}$ with $t_a = \pi/\omega_c$, the following sampling formula holds:*

$$f(t) = \sum_{k=-\infty}^{+\infty} f\left(\frac{k\pi}{\omega_c}\right) \frac{\sin(\omega_c t - k\pi)}{\omega_c t - k\pi} = \sum_{k=-\infty}^{+\infty} f(kt_a) \frac{\sin(\omega_c(t - kt_a))}{\omega_c(t - kt_a)},$$

*the series being absolutely and uniformly convergent.*

**Proof** From the assumptions, it follows that the spectral function $\widehat{f}$ is continuously differentiable (cf. p. 282). Hence, it is represented pointwise by its Fourier series in $[-\omega_c, \omega_c]$, and this series is absolutely and uniformly convergent (cf. p. 28):

$$\widehat{f}(\omega) = \sum_{k=-\infty}^{+\infty} c_k \, e^{-jk\omega t_a} \qquad (t_a = \pi/\omega_c, \ |\omega| \leqslant \omega_c)$$

$$c_k = \frac{1}{2\omega_c} \int_{-\omega_c}^{+\omega_c} \widehat{f}(\omega) \, e^{jk\omega t_a} \, d\omega = \frac{\pi}{\omega_c} f(kt_a).$$

Term-by-term integration of the series is possible because the series converges uniformly. Since bandlimited functions are infinitely differentiable (cf. p. 282), the sampling theorem follows from the Fourier inversion formula:

$$f(t) = \frac{1}{2\pi} \int_{-\omega_c}^{+\omega_c} \widehat{f}(\omega) \, e^{j\omega t} \, d\omega = \frac{1}{2\pi} \int_{-\omega_c}^{+\omega_c} \sum_{k=-\infty}^{+\infty} c_k \, e^{-jk\omega t_a} \, e^{j\omega t} \, d\omega$$

$$= \sum_{k=-\infty}^{+\infty} \frac{1}{2\omega_c} f(kt_a) \int_{-\omega_c}^{+\omega_c} e^{j\omega(t - kt_a)} \, d\omega = \sum_{k=-\infty}^{+\infty} f\left(\frac{k\pi}{\omega_c}\right) \frac{\sin(\omega_c t - k\pi)}{\omega_c t - k\pi}.$$

$\square$

**Remark** As a reference for variants and generalizations of the theorem, see Jerri (1977) or Butzer et al. (1988). For example, with theorems of Paley and Wiener (1934) it can be shown that the sampling series converges absolutely and uniformly for bandlimited square-integrable functions.

The sampling theorem provides a formula that allows for the interpolation of the values of $f$ at times $t \neq k\pi/\omega_c$, given all discrete values $f(k\pi/\omega_c)$, $k \in \mathbb{Z}$. The sampling frequency must be at least twice the cutoff frequency $\omega_c/(2\pi)$. By increasing the sampling frequency $\omega_c/\pi$, the formula applies to signals of correspondingly higher frequency bandwidth. With a lower sampling frequency than $\omega_c/\pi$ and given bandwidth $\omega_c$ of $f$, aliasing effects occur with the sampling series (see p. 387).

For direct practical implementation in signal transmission, the formula is not suitable because it is not causal. To reconstruct $f(t_0)$ at time $t_0$, one would also need all values $f(k\pi/\omega_c)$, $k\pi/\omega_c > t_0$. However, a function $f$ whose spectrum $\widehat{f}$ vanishes outside an interval $[-\omega_c, \omega_c]$ is not time-limited (cf. p. 303), meaning the sampling formula requires nonzero values of $f$ from the entire future $t > t_0$. Nevertheless, the sampling theorem is a starting point for practical approximation methods for reconstructing $f$ from sample values. In these methods, a realizable filter for interpolation is used, replacing the impulse response of the ideal lowpass filter used below.

To illustrate, consider finitely many samples $f(k\pi/\omega_c)$, $-M \le k \le N$. From the impulse sequence $\dfrac{\pi}{\omega_c} \displaystyle\sum_{k=-M}^{+N} f\left(k\dfrac{\pi}{\omega_c}\right) \delta\left(t - k\dfrac{\pi}{\omega_c}\right)$ as the input signal for an ideal lowpass filter with the frequency response $\widehat{h}(\omega) = A_0\, e^{-j\omega t_0}$ for $|\omega| \le \omega_c$, $\widehat{h}(\omega) = 0$ for $|\omega| > \omega_c$, then at the output of the lowpass filter, we get (see Fig. 12.1)

$$\frac{\pi}{\omega_c} \sum_{k=-M}^{+N} f\left(k\frac{\pi}{\omega_c}\right) \delta\left(t - \frac{k\pi}{\omega_c}\right) * \frac{A_0 \sin\left(\omega_c(t - t_0)\right)}{\pi(t - t_0)}$$

$$= A_0 \sum_{k=-M}^{+N} f\left(k\frac{\pi}{\omega_c}\right) \frac{\sin\left(\omega_c(t - t_0) - k\pi\right)}{\omega_c(t - t_0) - k\pi}.$$

Except for a factor and the time delay of $t_0$, the right side is an approximation of $f$ that converges to $A_0 f(t - t_0)$ as $N, M \to \infty$.

**Remarks**

(1) The functions $e_k(t) = \sqrt{\dfrac{\omega_c}{\pi}}\, \dfrac{\sin(\omega_c t - k\pi)}{\omega_c t - k\pi}$ ($k \in \mathbb{Z}$) form a complete orthonormal system in the space of $L^2$ functions bandlimited by $\omega_c$. The sampling formula is thus precisely the development of $f$ with respect to these basis functions. The proof of the sampling theorem shows that $\widehat{e_k}(\omega) =$

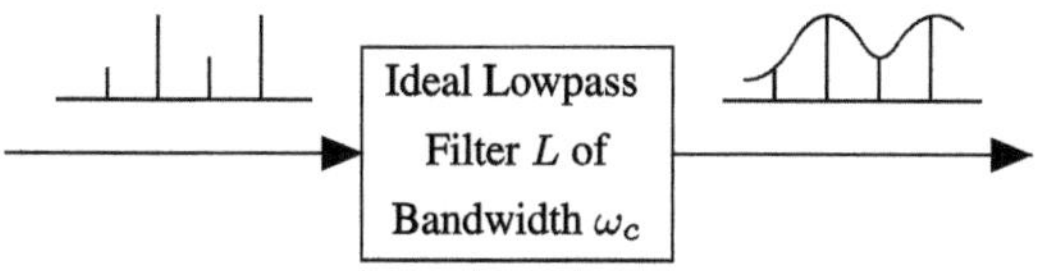

$$\frac{\pi}{\omega_c} \sum_{k=-M}^{+N} f\left(k\frac{\pi}{\omega_c}\right) \delta\left(t - k\frac{\pi}{\omega_c}\right) \xrightarrow{\ L\ } A_0 \sum_{k=-M}^{+N} f\left(k\frac{\pi}{\omega_c}\right) \frac{\sin\left(\omega_c(t - t_0) - k\pi\right)}{\omega_c(t - t_0) - k\pi}$$

**Fig. 12.1** Schematic digital-to-analog conversion

$\sqrt{\pi/\omega_c}\,e^{-j\omega k\pi/\omega_c}$ holds for $|\omega| \leq \omega_c$, $\widehat{e_k}(\omega) = 0$ otherwise. Completeness and orthogonality of the functions $e_k(t)$ therefore follow from Plancherel's equation (p. 308) and the fact that the functions $\widehat{e_k}(\omega)$ form a complete orthogonal system in $L^2([-\omega_c, \omega_c])$.

(2) The sampling series converges very slowly because the interpolation function $\sin(t)/t$ decays slowly for $|t| \rightarrow \infty$. With oversampling, i.e., replacing the sampling points $k\pi/\omega_c$ by $k\pi/(\alpha\omega_c)$ with $\alpha > 1$, one can obtain an interpolation function that decays like $1/t^2$ for $|t| \rightarrow \infty$. To see this precisely, solve the corresponding exercise A1 in the exercise part 12.9.

(3) With additional assumptions about $f$, such as information about the energy distribution of $f$ or its decay behavior, error estimates for the truncation error of the above approximation can be shown. Similarly, there are error estimates for the case when the sampling points are not exactly maintained and instead of $f(k\pi/\omega_c)$ the values $f(k\pi/\omega_c + \varepsilon_k)$ are sampled (the so-called jitter errors). In addition, in practical transmission systems, the sampled values are not transmitted continuously, but the value range is discretized and only a finite number of rounded values are transmitted. The resulting signal distortion, called quantization noise, corresponds in the time domain to the addition of an impulse train with the consequence of a broadband noise spectrum. There are also studies on the rounding errors resulting from this. For readers interested in error analysis, it is recommended to start with works such as Jerri (1977).

(4) Sampling methods with irregularly distributed sampling points (irregular sampling) play a role, for example, in radar technology. For this, see the works of H.-G. Feichtinger and K. Gröchenig (detailed references can be found in Unser (2000)).

*Generalizations*

There are numerous generalizations of the presented sampling theorem. These include, in particular, sampling theorems for time-limited, generally non-bandlimited functions with statements about the approximation quality of the considered sampling series. In general, representations of the form

$$f(t) = \sum_{k\in\mathbb{Z}} f\left(k\frac{\pi}{\Omega}\right)\varphi(\Omega t - k\pi) \quad \text{or} \quad f(t) = \lim_{\Omega\to\infty} \sum_{k\in\mathbb{Z}} f\left(k\frac{\pi}{\Omega}\right)\varphi(\Omega t - k\pi)$$

are sought for functions $f$ of certain function classes and bandwidths $\Omega$, and the approximation properties of the sampling series are derived from the assumptions about the function $f$ and the properties of the kernels $\varphi$. Such properties can be time or band limitation, decay behavior, etc. This topic will not be further addressed here, but rather reference is made to further literature such as Butzer and Stens (1992) or Unser (2000) and the references cited therein. Further aspects can also be found in the following Sect. 12.5 on time-frequency analysis (see p. 415) and Sect. 14.2 on wavelets (see p. 468).

## 12.2   Sampling as the Basis of Digital Transmission Technology

### *Sampling, Critical Sampling, Over-, and Undersampling*

Equidistant sampling of a bandlimited—and therefore infinitely differentiable—slowly increasing function $f \in \mathcal{O}_M$, with a sampling frequency $1/t_a$, can be described as the multiplication of $f$ with a sequence of impulses at the times $kt_a$. With the note on p. 163 and impulse strengths $t_a f(kt_a)$ the resulting discrete signal $f_d$ and $f$ as well as $\widehat{f_d}$ and $\widehat{f}$ each have the same physical units (see also p. 388 for the reconstruction of $f$ from sampled values or literature on the functionality of D/A converters for voltage signals in volts). For $f_d$—understood as a distribution with time parameter $t$—the following applies

$$f_d(t) = f(t) \cdot \sum_{k=-\infty}^{+\infty} t_a \delta(t - kt_a) = t_a \sum_{k=-\infty}^{+\infty} f(kt_a)\delta(t - kt_a).$$

Therefore, from the theorems on the convergence of convolutions (p. 197) and on Fourier transforms of impulse trains (p. 298) and of products (p. 300), the following fundamental relationship between the spectrum $\widehat{f}$ of $f$ and the periodic spectrum of the discrete signal $f_d$ follows (cf. 347):

**Theorem 12.2** *The spectrum of the discrete signal $f_d$, which is generated by sampling a bandlimited function $f \in \mathcal{O}_M$ with sampling frequency $1/t_a$, is given by*

$$\widehat{f_d}(\omega) = \widehat{f} * \sum_{k=-\infty}^{+\infty} \delta(\omega - 2\pi k/t_a) = \sum_{k=-\infty}^{+\infty} \widehat{f}(\omega - 2\pi k/t_a).$$

For $k \neq 0$, the spectra $\widehat{f}(\omega - 2\pi k/t_a)$ are *replicas* of $\widehat{f}$. In the case of *critical sampling* with the sampling rate $1/t_a = \omega_c/\pi$, referred to as the *Nyquist frequency*, these replicated spectra immediately adjoin each other. A reconstruction of $f$ from the sample values using a realizable lowpass filter is generally not possible, as this requires a transition region from the passband to the stopband (cf. p. 335). This transition region only arises at sampling rates $1/t_a > \omega_c/\pi$, i.e., through *oversampling* (Fig. 12.2). In the case of *undersampling* with rates $1/t_a < \omega_c/\pi$, overlaps of the replicated spectra occur in the spectrum of $f_d$. A reconstruction of $f$ from the corresponding sample values is then not possible, as aliasing effects occur in the signal spectrum, especially at higher frequencies (Fig. 12.3). The following schematic diagram shows the first graphic as a magnitude spectrum of $f_d$ with oversampling $t_a < \pi/\omega_c$, the second as a magnitude spectrum of $f_d$ with undersampling $t_a > \pi/\omega_c$.

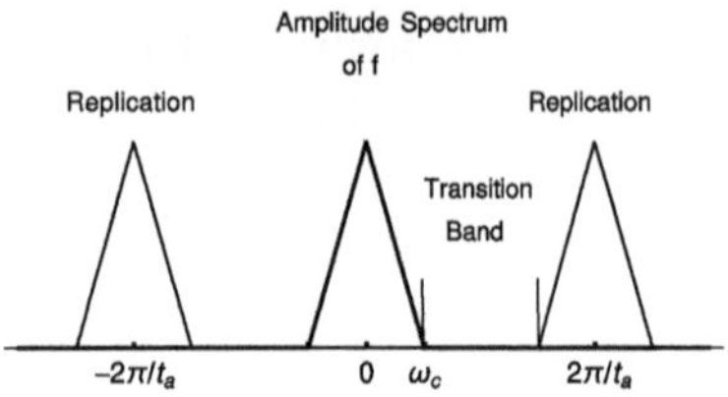

**Fig. 12.2** Sampling without aliasing

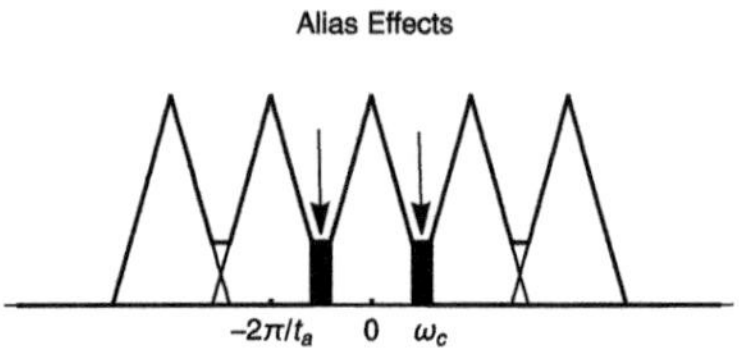

**Fig. 12.3** Sampling with aliasing



## *The Scheme of Digital Transmission in Practice*

In practical implementations, the sampling theorem suggests the recipe that lowpass filtering of the impulse sequence obtained from the sample values yields an approximation of the continuous signal. An impression is given by the following diagram. Sampling is technically done through sample and holds circuits (*S&H*). The values of the resulting step function are proportional to the quantized sample values of the signal. The impulse sequence for *reconstruction* from the quantized sample values is approximated by a sequence of rectangular pulses. An impulse $\delta(t - kt_a)$ at sampling frequency $FS = 1/t_a$ (*DAC sampling frequency clock FS*) is replaced by the rectangle $R(t - kt_a)/t_a$, $R$ being the indicator function of $[0, t_a[$ ("rectangle area" equal to one), i.e., an impulse $t_a\delta(t - kt_a)$ of strength $t_a$ is replaced by the convolution $t_a\delta(t - kt_a) * R(t)/t_a = R(t - kt_a)$. This creates a step function with the quantized sample values, i.e., in addition to quantization errors, there are distortions compared to the spectrum of the discrete signal model $f_d$, as the spectrum $t_a\,e^{-j\omega kt_a}$ of an impulse $t_a\delta(t - kt_a)$ is multiplied by the spectrum $e^{-j\omega t_a/2}\sin(\omega t_a/2)/(\omega t_a/2)$ of the rectangle $R(t)/t_a$ (cf. p. 275). These distortions can be compensated by a correction filter (*inverse* $\sin(x)/x$ *filter*) with digital filtering before the D/A conversion or afterward with an analog filter (see Fig. 12.4).

Specifically, for example, in digital telephony with ISDN, and similarly in newer methods like Voice over IP, a frequency range up to 3700 Hz is transmitted and filtered with a stopband starting at 4000 Hz. In standard telephone quality, speech signals are sampled at 8 kHz according to the sampling theorem, i.e., at time intervals of 125 μs. Only quantized, rounded values are transmitted, which can be encoded as 8-bit-long digital code words. In Voice over IP, optionally lossy compressions are also used, similar to the MP3 encoding mentioned later, i.e., code words with less than 8 bits per sample value are used to ultimately reduce the required bandwidth during transmission.

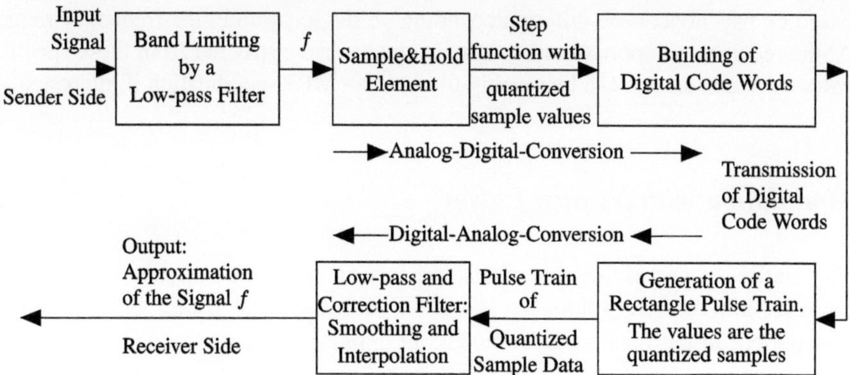

Fig. 12.4  Illustratively a digital transmission system

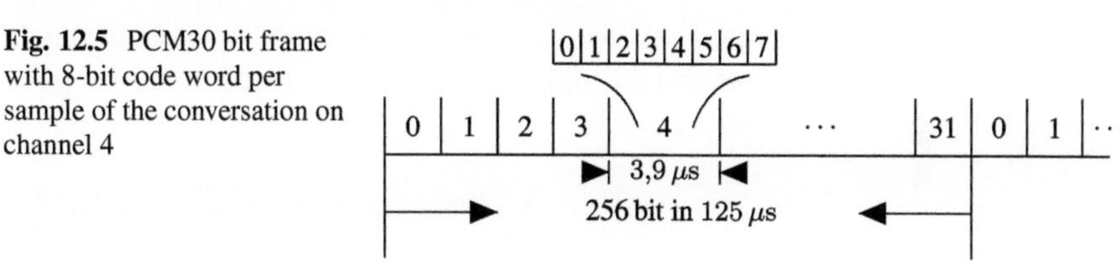

Fig. 12.5  PCM30 bit frame with 8-bit code word per sample of the conversation on channel 4

In the European PCM30 system (PCM stands for Pulse Code Modulation) for landline connections, 32 channels per transmission device are transmitted in a bit frame of $32 \times 8\,\text{bit} = 256\,\text{bit}$ per pulse frame, i.e., every $125\,\mu\text{s}$. One of the 32 channels contains a frame synchronization word, another channel contains signaling information (e.g., dialed phone numbers), and the remaining 30 channels contain the voice signals of 30 different conversations, which can be transmitted over shared line routes through cyclic aggregation (multiplexing; see the following graphic for a PCM30 bit frame).

The simultaneous transmission of multiple signals over a shared line between switching centers is possible through time utilization between the sampling points of a signal. During this time, other signals are sampled and transmitted. In this technique, the bit rate per telephone channel is $8 \times 8\,000\,\text{bit} = 64\,\text{kbit/s}$, resulting in a bit rate for the transmission device for 32 channels of $32 \times 64\,\text{kbit/s} = 2.048\,\text{Mbit/s}$ (see Fig. 12.5). The economic benefit in digital telephony is an increase in switching capacity and the high utilization of expensive lines between the switching centers. This technology is still in use but is increasingly being replaced by the aforementioned Internet telephony, which saves costs for operators through higher bandwidth usage and cheaper equipment in the switching centers.

A well-known additional application example of the sampling theorem is audio files in the so-called WAV format. Here, sampling is done with 44100 values per second, i.e., a bandwidth of about $20\,\text{kHz}$ is achieved. In the MP3 format, the frequency range is divided into several subbands. The FFT values of time sections of the acoustic signal are then quantized and transmitted with a varying number of bits based on psychoacoustic criteria depending on the location in the frequency range. At the receiver, an approximation of the signal is reconstructed from these spectral values. Details on this can be quickly found in an MP3 specification if interested.

## *Modulation with Nyquist Pulses*

The starting point of discrete signal processing in today's digital transmission methods are discrete values $x_k$, $k \in \mathbb{Z}$, in which the useful information is transported. In the mathematical model, an impulse sequence $x_d = t_a \sum_{k \in \mathbb{Z}} x_k \delta_k = t_a \sum_{k \in \mathbb{Z}} x_k \delta(t - kt_a)$ is present, from which a continuous signal $s = x_d * h$ is generated through a linear filter with regular impulse response $h$. We assume that the convolution $x_d * h$ is possible and all sampling values $h(nt_a)$ exist (e.g., $\mathrm{supp}(x_d)$ bounded, $h \in \mathcal{S}'$ continuous).

$$s(t) = (x_d * h)(t) = \sum_{k \in \mathbb{Z}} x_k t_a h(t - kt_a).$$

With $h$, the transmission and reception filters and a linear filter describing the transmission channel are combined, i.e., $s(t)$ is the received signal. It is immediately apparent that the sampling values $s(nt_a) = x_n$ yield exactly the desired useful information if $t_a h(0) = 1$ and $h(nt_a) = 0$ for $n \neq 0$. Filters $h$, also called *pulse shapes* with this property (*zero crossing property*), are called *Nyquist Pulses*.

**Example** If the values $x_k$ are sampled values $x_k = f(kt_a)$ of a function $f$ bandlimited by $\omega_c$ as in the previous Shannon sampling theorem and the function $h(t) = \dfrac{\sin(\omega_c t)}{\pi t}$ with $\omega_c = \pi/t_a$, then according to the proof of the sampling theorem $s(t) = f(t)$. The function $h$ is a Nyquist pulse, as are products of $h$ with functions $g$ that have the value $g(0) = 1$ at zero. The so-called "*raised cosine filter*" $h_{\mathrm{RC},\alpha}$, which in practice is often used and falls off much faster than $h$ for $|t| \to \infty$, is an example of this (see also later on p. 398, where it is given as a pulse shape in the frequency domain):

$$h_{\mathrm{RC},\alpha}(t) = \frac{\sin(\pi t/t_a)}{\pi t} \cdot \frac{\cos(\pi \alpha t/t_a)}{1 - (2\alpha t/t_a)^2}.$$

The parameter $\alpha$ controls the bandwidth extension (*excess bandwidth*) compared to the spectrum of the sinc pulse $h$. The spectrum $\widehat{h}_{\mathrm{RC},\alpha}$ with falling cosine flanks, from which this pulse shape gets its name, is given in Exercise A10 to Chap. 10, p. 319 (there $\pi/t_a = b$, $a = \alpha b$). Applications of various Nyquist pulses can be found in Proakis and Salehi (2013).

Nyquist pulses $h$ allow the reconstruction of the values $x_k$ by sampling $s(t)$ even if $h$ and therefore $s$ are not bandlimited, thus despite aliasing effects in the spectrum of the pulse train $s_d(t) = \sum\limits_{k=-\infty}^{+\infty} t_a s(kt_a)\delta(t - kt_a)$. In applications, $h$ is often a function with bounded support (see later p. 394). If the $x_k$ are referred to as symbols to be transmitted, then transmissions with Nyquist pulses are free from *intersymbol interference* (abbreviated ISI in literature).

**Modulations with Pulse Shapes That Are Not Nyquist Pulses**

If we consider the task of reconstructing the values $x_k$ from the sampled values $s_k = s(kt_a)$, $k \in \mathbb{Z}$, of the received signal $s(t)$ as a discrete linear filter problem, then with the results on inverse discrete filters, we obtain (see 11.6, p. 368 and 8.7, p. 195):

**Theorem 12.3** *If a pulse shape $h$ corresponds to a discrete filter with impulse response $h_d = t_a \sum\limits_{k \in \mathbb{Z}} h(kt_a)\delta_k$ that has an inverse with impulse response $h_{d,inv} = \sum\limits_{k \in \mathbb{Z}} g_k \delta_k$, so that $(x_d * h_d) * h_{d,inv} = x_d * (h_d * h_{d,inv})$ is associative,[1] then $x_d = t_a \sum\limits_{k \in \mathbb{Z}} x_k \delta_k$ is reconstructed by the discrete convolution $x_d = s_d * h_{d,inv}$, i.e.,*

$$x_n = \sum_{k \in \mathbb{Z}} s_k g_{n-k} \text{ for } n \in \mathbb{Z}.$$

Theorems on discrete filters in different signal spaces, on stability, causality, invertibility, and possible design methods for FIR or IIR filters were already presented in Sect. 11.6. Starting from the modulation of discrete information with various pulse shapes as impulse responses of linear filters, a variety of signal processing algorithms have been developed. Some aspects of this follow in Sect. 12.5 on time-frequency analysis and Sect. 14.2 on wavelets. For an in-depth study of various methods of application-specific signal processing, reference is made here only to the extensive literature on the subject, for example, Papoulis (1977) on signal analysis, Proakis and Salehi (2013), Couch (2012) on digital communication systems, Salditt et al. (2017) on imaging methods in biomedicine, or the works mentioned and referenced at the end of Sect. 12.1.

A study of digital signal processing, which today permeates almost every area of life and all fields of science, requires specialized training in dealing with the mathematical methods and ultimately with the technology through which designed algorithms can be implemented.

---

[1] The $z$-transforms of $x_d$, $h_d$, $h_{d,inv}$ must have a common region of convergence.

## 12.3  The Basic Idea of Multi-Carrier Transmission with OFDM

This section describes the basic idea of the OFDM multi-carrier method as a far-reaching technology. OFDM (English) stands for *Orthogonal Frequency Division Multiplexing*. Nearly everyone uses this method almost around the clock today, because OFDM is comprehensively used for the transmission of WLAN, DSL, digital radio (DAB), and TV (DVB), in powerline communication and mobile communication with LTE, LTE+, and 5G standards. An OFDM application in optical transmission systems with bandwidths up to 1 Tb/s is in development (see, for example, Ma et al. 2010). The history of Frequency Division Multiplexing (FDM), back then with analog technology, goes back to the first patents on multitone telegraphy in the years 1875–1876 by Alexander Graham Bell, Elisha Gray, and Thomas Edison. A readable account of the development of OFDM methods can be found in Weinstein (2009).

Today's digital OFDM methods, referred to as DMT (*Discrete Multitone Transmission*) in ADSL and VDSL, go back to works by Chang (1966) and Weinstein and Ebert (1971). They are a combination of applications of the DFT, the sampling theorem with filter technology, together with the use of coding and encryption algorithms. Additionally, methods for estimating the properties of transmission channels are included, based on which transmission errors are to be corrected at the receiver to recover the user information. Physically, the OFDM methods are implemented with highly developed hardware in electrical and communications engineering.

Characteristic of OFDM transmission is that large parts of the required transmission and reception technology consist of discrete signal processing, which can be cost-effectively realized with uniquely developed algorithms on integrated circuits (ICs) compared to analog technology. Only because of this, today one can get a WLAN USB stick or digital media devices including necessary software as mass products for relatively low costs.

In the literature on communications engineering, there are a number of easily searchable textbooks dedicated in detail to the OFDM methods. Therefore, only the essential ideas will be presented here in all necessary brevity, as far as they can be easily understood with the methods of Fourier analysis treated in the present text. They may serve as an incentive for readers to deepen their knowledge with specialized literature if interested.

### *Mathematical Components of an OFDM Transmission System*

1. *From the Coded Bit Stream with QAM to Trigonometric Polynomials*

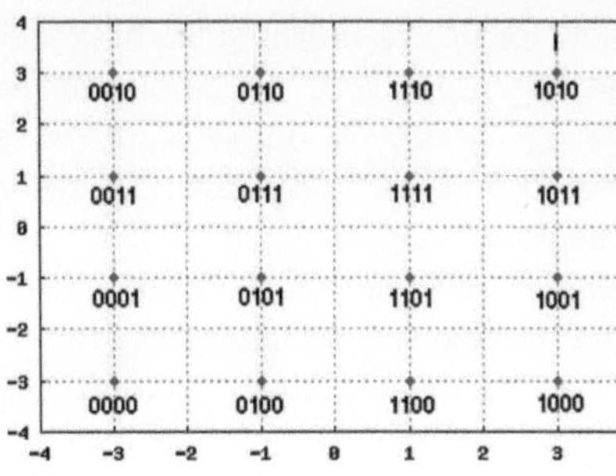


On the left is a mapping of 4 bits each with a Gray code to numbers in the complex plane (abscissa: real part, ordinate: imaginary part). For a further example, we set four 16-QAM mappings:

$$1110 \rightarrow c_1 = A(1 + 3j)$$
$$1011 \rightarrow c_2 = A(3 + j)$$
$$0010 \rightarrow c_3 = A(-3 + 3j)$$
$$0101 \rightarrow c_4 = A(-1 - j)$$

**Fig. 12.6** 16-QAM Gray Coded Symbol Mapping without factor $A$

On the sender side, there is a bit stream, i.e., a 01 sequence of data that is to be transmitted. The data is usually encoded and encrypted (keywords: error-correcting codes, interleaving, possibly WPA2 encryption, etc.).

To explain OFDM at hand of an example, let us assume that transmission is to be done with 16-QAM modulation. 16-QAM stands for *Quadrature Amplitude Modulation* with an alphabet of 16 complex numbers. From the bit stream, blocks of 4 bits each are injectively mapped to a set of 16 complex numbers, also referred to as QAM symbols (see Fig. 12.6).

A 16-QAM modulation with its assignments is shown below. All complex values are multiplied by the scaling factor $A = 1/\sqrt{10}$. This normalizes the power of a 16-QAM modulated uniformly distributed 01 random bit sequence in the transmit signal to one (cf. Couch 2012).

To generate an OFDM symbol $S_i$ with $N$ carriers for a time period from $iT$ to $(i+1)T$, the sequential bit stream is parallelized into $n \leq N$ 4-bit blocks, which are mapped with 16-QAM to $n$ complex numbers $c_{i,k}$ as shown above.

With the $N$ carriers $e^{j2\pi kt/T}$ and the complex amplitudes $c_{i,k}$, a trigonometric polynomial with a bandwidth $B \leq (N-1)/T$ Hz is formed, which is limited in duration to the interval $[iT, (i+1)T]$ by multiplication with a time window $T g_{i,T}(t) = T g_T(t - iT)$, resulting in the *OFDM symbol $S_i$ in the baseband*:

$$S_i(t) = T \sum_{k=0}^{N-1} c_{i,k}\, e^{j2\pi kt/T}\, g_{i,T}(t).$$

$N - n$ carriers, whose frequencies are agreed upon between the transmitter and receiver, remain unoccupied with $c_{i,k} = 0$, or they can be used with predetermined amplitudes as *pilot carriers* or—prefixed to the symbol—as *preambles* at the receiver for channel estimation and synchronization (cf. also Example 4, p. 94). The useful information of $4n$ bits is thus contained in the assigned complex amplitudes $c_{i,k}$ of $S_i$. The function $g_T$ is the impulse response of the transmit filter.

We initially assume a rectangular window for $g_{i,T}$ to illustrate the basic idea of OFDM and consider only a single time step with $i = 0$. To simplify notation, the index $i$ is therefore omitted, and $T g_{i,T} = w_T = 1_{[0,T[}$ is set ($w_T$ thus being the indicator function of the interval $[0, T[$). The trigonometric polynomial component in the OFDM symbol $S$ occupies the frequency band $[0, (N-1)/T]$. However, the product with a time window $w_T$ is no longer bandlimited, i.e., the spectrum of $S$ results in out-of-band interference. In implementations, one would therefore choose other windows $w_T$ whose amplitude spectrum falls off faster than that of a rectangular window.

The functions $e^{j2\pi kt/T} w_T(t)$, $k = 0, \ldots, N-1$, form an orthogonal system in the space $L^2([0, T])$, the frequency band $[0, (N-1)/T]$ is divided by the carrier frequencies with fixed frequency spacing $1/T$, and all QAM values are transmitted together during the symbol duration $T$. Because of these properties, the method is called *Orthogonal Frequency Division Multiplexing*, abbreviated OFDM. Due to the orthogonality of the carriers, transmission interference at one of the carrier frequencies has no effect on the other carriers, i.e., there is no *inter-carrier interference* (ICI)—at least as long as the transmission channel does not cause frequency dispersion due to Doppler effects in moving receivers as in mobile communications, and it is neglected that a signal, which is not bandlimited due to the rectangular window $w_T$, is transmitted over a bandlimited channel. Distortions of a linear time-invariant channel, for example, due to multipath propagation and superpositions of multiple delayed signal sections arriving at the receiver, can be corrected there—with moderate noise— by estimating the channel impulse response.

The following Fig. 12.7 shows the magnitude spectra of $c_k e^{j2\pi kt/T} w_T(t)$ for $k = 1$ and $k = 4$. It can be seen that the spectra are Nyquist pulses in the frequency domain (cf. p. 391), i.e., $|c_k \widehat{w}_T(\omega - 2\pi k/T)| = 0$ at each maximum point $\omega$ of the magnitude spectra $|c_n \widehat{w}_T(\omega - 2\pi n/T)|$ for $n \neq k$ ($0 \leq k, n \leq N-1$). Figure 12.8 shows in advance the shape of a typical WLAN spectrum, with 48 data carriers and here for visibility exaggeratedly large 4 pilot carriers. A transmission with a rectangle time window had the blue spectrum. Transmission with a common raised cosine window, explained a little later, has much less out-of-band emission, as is seen in the red spectrum.

**Fig. 12.7** Magnitude of two carriers

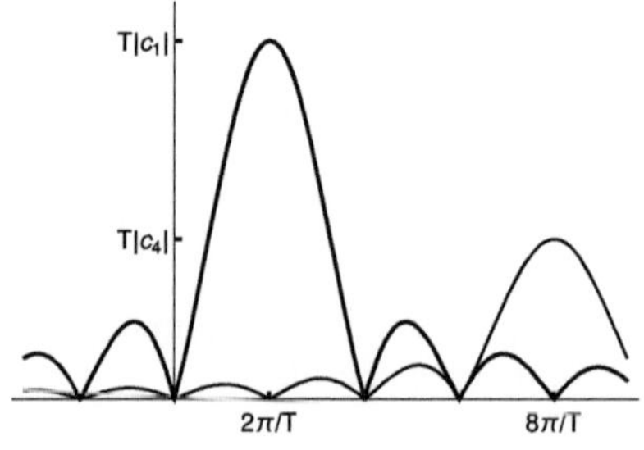

**Fig. 12.8** Illustratively a
WLAN spectrum

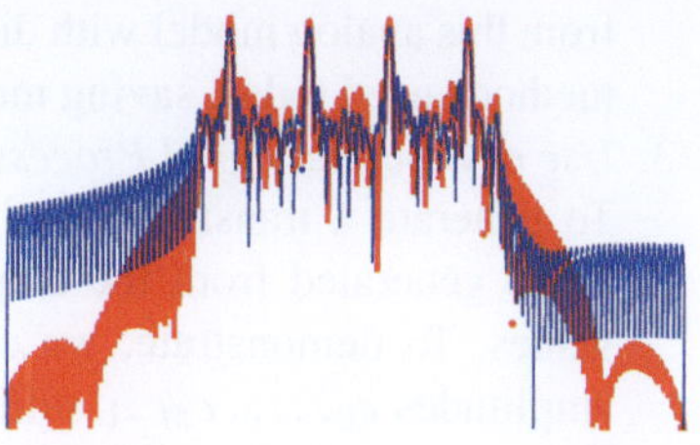

2. *Real-Valued Transmission Signal, Quadrature Modulation, and Demodulation*
   A real transmission signal is obtained from $S$ through quadrature modulation
   (QM) with an intermediate frequency $\omega_c$. The generated real-valued signal $S_{\mathbb{R}}$ is

$$S_{\mathbb{R}}(t) = \Re \left( e^{j\omega_c t} \sum_{k=0}^{N-1} c_k \, e^{j2\pi kt/T} \right) w_T(t).$$

We obtain the following representation, where $I(t)$ is called the *in phase* and
$Q(t)$ the *quadrature component* of $S(t)$:

$$S_{\mathbb{R}}(t) = \cos(\omega_c t) \sum_{k=0}^{N-1} \left( \Re(c_k) \cos(2\pi kt/T) - \Im(c_k) \sin(2\pi kt/T) \right) w_T(t)$$

$$- \sin(\omega_c t) \sum_{k=0}^{N-1} \left( \Re(c_k) \sin(2\pi kt/T) + \Im(c_k) \cos(2\pi kt/T) \right) w_T(t)$$

$$= (I(t) \cos(\omega_c t) - Q(t) \sin(\omega_c t)) w_T(t).$$

Pairwise orthogonality of the carriers and signal bandwidth are preserved in QM.
The signal spectrum is shifted to the intermediate frequency $f_c = \omega_c/(2\pi)$ (for
WLAN, $f_c$ is about 2.4 GHz or 5 GHz). Repeated QM and lowpass filtering
to suppress high-frequency remnants at the receiver return the complex-valued
function $S$ in the baseband, assuming no influences from the transmission
channel distort the signal. Using addition theorems for cosine and sine functions,
we find

$$2\cos(\omega_c t) S_{\mathbb{R}}(t) = I(t) + \text{high-frequency remnant}$$

$$2\sin(\omega_c t) S_{\mathbb{R}}(t) = Q(t) + \text{high-frequency remnant}$$

$$S(t) = I(t) + jQ(t).$$

From samples of $S$, the receiver can then use a DFT to reconstruct the
amplitudes $c_k$ and, with inversion of the 16-QAM mapping, finally reconstruct
the transmitted bit groups. In the following, it will be explained how to transition

from this analog model with discrete signal processing to the digital transmission methods used today, saving much expensive analog technology.

3. *Use of Discrete Signal Processing with an IDFT*

   To generate a transmission signal with discrete signal processing, samples of $S$ are generated from the amplitudes $c_k$ of the OFDM symbol with $M \geq N$ values. To demonstrate, we choose $M = 8$. An IDFT (see Sect. 6.1) of the amplitudes $c_0, \ldots, c_{M-1}$ suffices for this purpose without extensive hardware. For "upsampling" to $M$ values, a zero sequence of length $M - N$ is simply appended to the amplitudes $c_k$. The subsequent IDFT has length $M$ and provides $M$ samples $(y_n)_{0 \leq n \leq M-1}$. For radix-2 IFFT/FFT algorithms used in practice, $M$ is a power of two, such as $M = 2048$ for LTE with $20\,\text{MHz}$ bandwidth, 1201 used carriers per OFDM symbol, $15\,\text{kHz}$ carrier spacing, and $30.72\,\text{MHz}$ sampling frequency.

   With sufficient samples, potentially after further "upsampling" the IDFT values with *CIC filters* (Cascade Integrator Comb Filter), quadrature modulation can also be performed discretely without analog mixers by multiplying the real and imaginary parts of the IDFT list with the values of the modulating cosine and sine functions at the corresponding sample points and subtracting the resulting lists point by point. For that, one can use DDS components (Direct Digital Synthesizer) matching the bandwidth and bit resolution of the subsequent D/A converter. The result is samples of the real transmission signal $S_{\mathbb{R}}$, which are fed to a D/A conversion (see p. 385–387).

**Example** With the data $T = 1/2\,\text{s}$, $c_0 = 0$, $c_1, \ldots c_4$ as on p. 393, $M = 64$, $\omega_c = 64\pi\,\text{rad/s}$, the following graph shows in Fig. 12.9 the QM modulation of the analog signal $S(t)$ (thin line) and the approximation (thick line), which is generated from IDFT values and discrete QM modulation as described. The second curve is drawn with a value offset of $+0.4$ for visible distinction. Figure 12.10 shows the amplitude spectrum of the first curve in dB. For both cases, the *rectangular time window $w = 1_{[0,T[}$* was used, and for interpolation of the samples in the second case, the series from the Shannon sampling theorem with bandwidth $B = 128\pi\,\text{rad/s}$. The moderate roll-off of the amplitude spectrum, which does not meet practical requirements for permissible out-of-band emissions, is apparent.

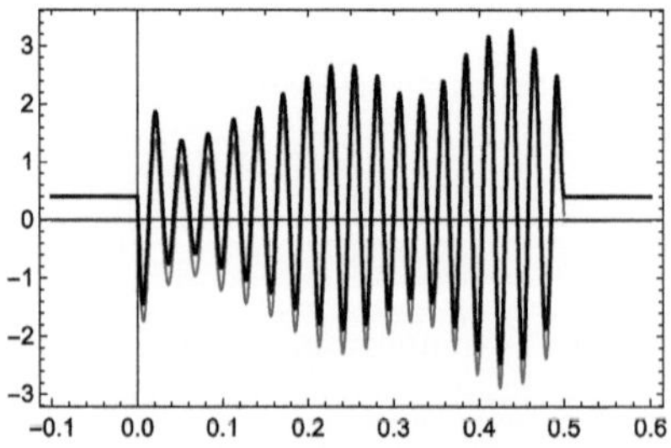

**Fig. 12.9** QM-modulated signals over time $t$, analog signal, and approximation from samples (thick) with offset

**Fig. 12.10** Amplitude
spectrum in dB, plotted over
frequency $\omega_c/(2\pi) = 32\,\text{Hz}$

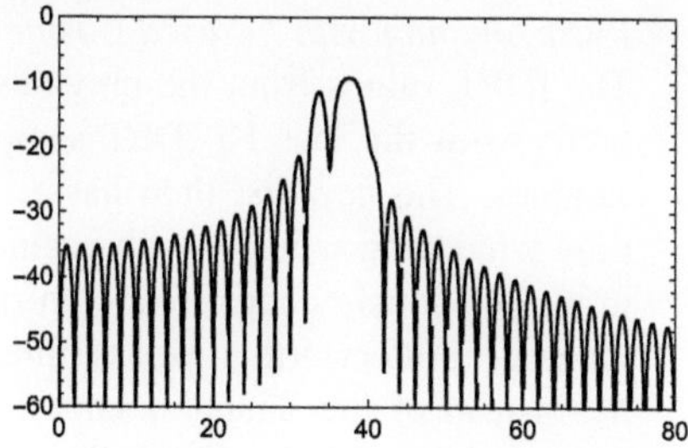

4. *Other Pulse Shapes Reducing Out-of-Band Emissions, Cyclic Prefix, and Postfix*
   In practice, stronger damping in the out-of-band region is required. Specifications
   for WLAN, DAB, etc., define spectral masks that must be adhered to for
   transmission.

   There are various methods to achieve better damping in the out-of-band region.
   One can quickly find a large number of publications on this topic under keywords
   like "OFDM Pulse Shaping." Methods such as smoothed time windows (W-
   OFDM for Windowed OFDM), filter banks for pulse shaping (FBMC, "Filter
   Bank Based Multicarrier Systems"), and other variants of the multi-carrier
   method can be used. Important aspects are the Heisenberg's uncertainty principle
   and the Balian-Low theorem regarding the time-frequency localization of signals
   (see the following sections of this chapter).

   In the following example, it is shown how the signal spectrum in the out-of-
   band region can be attenuated by extending the signal duration $T$ and using a
   time window $w_{T(1+\alpha)}$ of duration $T(1 + \alpha)$ with rounded edges instead of the
   previously used rectangular window. At the same time, a "Cyclic Prefix" (CP)
   and a "Postfix" are introduced by cyclically extending the signal. While $\widehat{w}_T$
   is a Nyquist pulse in the frequency domain, this is not the case for $\widehat{w}_{T(1+\alpha)}$.
   As a result, the transmit signal experiences some inter-carrier interference
   (ICI) because the carriers $e^{j2\pi kt/T}\, w_{T(1+\alpha)}(t)$, $k = 0, \ldots, N - 1$, are no
   longer orthogonal to each other. This method is, on the other hand, easy to
   implement and is also used in real systems (cf. Montreuil et al. (2013), Broadcom
   Recommendations for Tx Symbol Shaping). An important advantage of the
   cyclic prefix is that the convolution with the channel impulse response $h$ of a
   time-invariant transmission channel can be represented as a cyclic convolution
   if this impulse response does not last longer than the prefix. This allows for
   interference suppression using the sample values $\widehat{h}(2\pi k/T)$ of the estimated
   channel frequency response $\widehat{h}$. More on this below in point 7. The time interval
   with the cyclic prefix is called the *Guard Interval* (GI). It often comprises $1/4$ of
   the core symbol duration (e.g., LTE[2]).

---

[2] LTE 4G with 64-QAM, 20 MHz bandwidth: 1201 subcarriers, core symbol $T = 66.67\,\mu\text{s}$, GI
(long) $16.67\,\mu\text{s}$.

*Pulse Shaping with "Raised Cosine Window," Cyclic Prefix, and Postfix*
The IDFT values from the previous example are now cyclically extended by a prefix with the last 16 IDFT samples and a postfix with the first four IDFT samples. The new list then has $L = 84$ elements. Instead of the rectangular time window, a window with cosine edges ("*raised cosine window*") is used as in the previously cited Broadcom recommendations.[3] Additionally, a realizable (analog) Butterworth lowpass filter is used for interpolating the samples of $S_\mathbb{R}$ instead of the Shannon sampling series to model the D/A conversion. For the weighting of the IDFT list with time window values, the time window $w_{T(1+\alpha)}(t) = Tp(t)$ is used with $\alpha = T/16$ and

$$
p(t) = \begin{cases} 1/T & \text{for } 0 \le |t| < \dfrac{T(1-\alpha)}{2} \\[2mm] \dfrac{1}{2T}\left(1+\cos\left(\dfrac{\pi}{\alpha T}\left(|t|-\dfrac{T(1-\alpha)}{2}\right)\right)\right) & \text{for } \dfrac{T(1-\alpha)}{2} \le |t| \le \dfrac{T(1+\alpha)}{2} \\[2mm] 0 & \text{otherwise.} \end{cases}
$$

Weighting the cyclically extended IDFT list with values of $Tp\left(t-\dfrac{T(1+\alpha)}{2}\right)$ at times $t_n = T(1+\alpha)(2n+1)/(2L)$, $n = 0,\dots L-1$, results in the first three prefix values and the last three postfix values entering the cosine edges of the window, thus "rounding" the OFDM pulse shape. Below in Fig. 12.11 is the interpolated signal with lowpass delay and cyclic extensions at the beginning and end, and on the right, the resulting amplitude spectrum is shown in bold compared to the thinly drawn one using the rectangular window. (Try to examine the example yourself with a computer algebra system.) See also the figure on p. 395.

From the Fourier transform $\widehat{p}$ (cf. Exercise A10, p. 319), it can be seen that the amplitude spectrum of the OFDM signal in the out-of-band region is now much more attenuated than with the rectangular window, as can also be clearly seen in the following Fig. 12.12:

**Fig. 12.11** QM-modulated signal with the raised cosine window over time $t$ in seconds, after interpolation with Butterworth lowpass filter, group delay in the passband $\sim$21 ms; signal extended by pre- and postfix

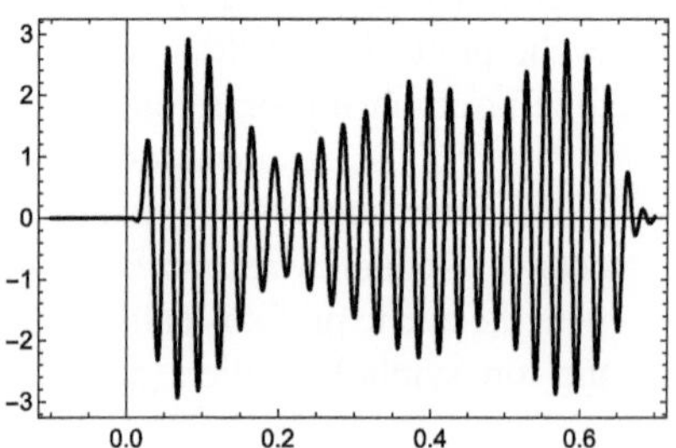

---

[3] Broadcom is a supplier, e.g., for DSL in wired connections (DSLAMs) of various providers.

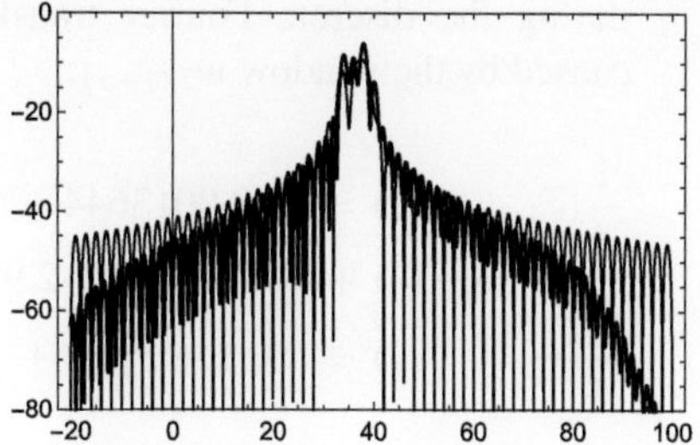

**Fig. 12.12** dB amplitude spectra over frequency $f$ ($f_c = \omega_c/(2\pi) = 32\,\text{Hz}$), bold with raised cosine window, thin with rectangular window. For $f < f_c$ window effect, beyond the stopband edge 96 Hz, additional lowpass effect

$$\widehat{p}(\omega) = \frac{\sin(\omega T/2)}{\omega T/2}\left(\frac{\cos(\alpha\omega T/2)}{1 - (\alpha\omega T/\pi)^2}\right).$$

The Butterworth filter for interpolation was designed with the following data: passband edge 48 Hz, stopband edge 96 Hz, minimum passband gain 0.9, maximum stopband gain 0.005, and DC gain $K = 1$ (Fig. 12.11). The order of the filter is 9, and the group delay approximately 21 ms in the passband (cf. p. 335). The attenuation gain in the range up to the intermediate frequency is due to the Raised Cosine Window, and the even stronger falloff of the spectrum in the right part of the last image is due to the additional attenuation in the stopband of the lowpass filter.

5. *Reconstruction of the Information with a DFT for Identical Reception of $S_{\mathbb{R}}$*
   With identical reception without distortions caused by the transmission channel, the receiver retrieves the in-phase component $I(t)$ and the quadrature component $Q(t)$ of the signal by extracting a time interval of duration $T$ and inverting the quadrature modulation. After an $N$-point DFT of $I(t) + jQ(t)$ over this time interval, it can be determined which complex amplitudes $c_k$ of the OFDM alphabet correspond to the DFT result. One primary advantage of periodicity with prefix and postfix is that moderate synchronization errors (phase offset) can be easily corrected if pilot values among the amplitudes are known.

**Example** Sampling in the previously calculated example[4] with sampling times $t_n = nT/N + \Delta t$, $N = 5$, $n = 0,\ldots,N-1$, and $\Delta t = 0.226\,\text{s}$, i.e., asynchronously starting in the prefix, initially yields the DFT list dft

$$\text{dft} = (-0.00136 + 0.00096j,\ -2.10649 + 2.35922j,\ -2.37870 + 2.08454j,$$
$$+ 3.02492 - 2.97252j,\ -0.37881 + 1.36073j).$$

Using $c_1$ as a pilot value results in the corrected phase (cf. Example 4, p. 94) giving the result $(\widetilde{c}_0,\ldots,\widetilde{c}_4)$, which can now be compared with the values $(c_0,\ldots,c_4)$ on p. 393. The deviations between the two lists are due to rounding

---

[4] Recommendation: Compute the example yourself using a computer algebra system.

during the discrete Fourier transforms and slight inter-carrier interferences caused by the window $w_{T(1+\alpha)}$:

$$\frac{1}{A}(\tilde{c}_0,\ldots,\tilde{c}_4) = (\,-0.0013644 + 0.00095966j,\ 1.00016 + 3.00049j,$$
$$3.0015 + 0.997246j,\ -2.99428 + 3.00339j,\ -1.00272 - 0.994806j\,)$$
$$\frac{1}{A}(c_0,\ldots,c_4) = (\,0,\ 1 + 3j,\ 3 + 1j,\ -3 + 3j,\ -1 - 1j\,).$$

6. *Energy Density and Spectral Power Density of an OFDM Transmission*
   An OFDM transmission, like the segment under consideration, is time-limited and thus an energy signal. Therefore, there is no power density in the usual sense other than zero. The squared magnitude of $\widehat{S_\mathbb{R}}(\omega)$ in the example represents an energy density according to its physical dimension. The spectrum of the approximation for $S_\mathbb{R}$ is, as seen in the example, essentially determined by the variance of the OFDM alphabet and the Fourier transforms of the window function and the interpolating lowpass filter. In theoretical approaches, an OFDM transmission can be modeled as an infinitely lasting cyclostationary stochastic process, and an average spectral power density can be specified. Calculations under various assumptions about the modulation method can be found in Couch (2012) or Gardner et al. (2008). The specifications of the transmission methods (WLAN, DVB-T, etc.) provide *spectral masks for spectral power densities* that must be adhered to—for signals in volts, then with the unit $V^2$/Hz. In practice, this evidence is provided through simulations for spectral estimation. For this purpose, several thousand OFDM symbols are generated with a random bit sequence, and the interpolation of the (discrete) DFT spectra is averaged for spectral estimation. Various averaging methods are in use (averaged periodogram technique, Bartlett's method, Welch's method, etc.). A comprehensive treatment of the topic *spectral estimation* with statements on consistency, unbiasedness, and variance of various estimates can be found, for example, in the textbook on digital signal processing by Kammeyer and Kroschel (2012).

7. *Effects of the Transmission Channel, Preambles, Pilots, and Cyclic Prefix*
   The transmission channel for OFDM, whether wired or wireless, has various effects on the functionality and quality of the transmission. These are summarized under the term "fading." These include echo effects and time delays in multipath propagation, leading to possible cancelations or intersymbol interference (ISI), attenuation for various reasons (transmitter distance, weather), frequency dispersion due to Doppler effects with moving transmitters or receivers. Additionally, noise, nonlinear distortions of the RF amplifiers, especially in mass products designed with cost and energy optimizations, jitter effects in oscillators in components, and much more, which communications engineers must master for a robust functioning overall system during design and implementation. An initial impression of the topic of channel equalization is given by the following considerations:

Assume that the receiver has the sampled values $r(nT/N)$ $(0 \leq n \leq N-1)$ of the received signal segment. The channel is assumed to be a noiseless causal linear filter. Its impulse response $h$ should decay within the duration of a cyclic prefix preceding the OFDM symbol. Through the extension of the symbol with prefix and postfix as in the example and the decay of transient components of the convolution of the signal $S$ with $h$ within the prefix duration, the values $r(nT/N)$ can be regarded as sampled values of the convolution of $h$ with the $T$-periodic Fourier series $S_T$, which arises through the periodicity of the OFDM symbol. Since $\mathrm{supp}(h)$ is bounded, $\widehat{h}$ is a multiplier in $\mathcal{S}'$. With sampled values of the channel frequency response $\widehat{h}$, the periodized received signal $r_T$ satisfies (cf. p. 300 and p. 344)

$$r_T(t) = S_T * h = \sum_{k=0}^{N-1} c_k \widehat{h}(2\pi k/T)\, e^{j2\pi kt/T}\,.$$

It follows that $r(nT/N) = r_T(nT/N) = \displaystyle\sum_{k=0}^{N-1} c_k \widehat{h}(2\pi k/T)\, e^{j2\pi kn/N}$, i.e., $r(nT/N)$ is the $n$th component of an IDFT of $\left(c_k \widehat{h}(2\pi k/T)\right)_{k=0,\dots,N-1}$. A DFT of the values $r(nT/N)$ with result $\widehat{r}_k$ $(0 \leq k \leq N-1)$ in the $k$th component then shows for a desired complex amplitude $c_k$ of the OFDM symbol:

$$c_k = \frac{\widehat{r}_k}{\widehat{h}(2\pi k/T)}\,.$$

Preambles and pilot symbols known to the receiver can be used for synchronization and for estimating the impulse response $h$ or the channel frequency response $\widehat{h}$. However, this seemingly simple channel equalization must be approached with caution. The necessary duration of the prefix and the channel frequency response $\widehat{h}$ must be reliably estimated; the values $\widehat{h}(2\pi k/T)$ must not become zero (cf. Sects. 5.6 and 11.6 on convolution inverses), and the convolution equation $r = S*h$ is a typical ill-posed problem (cf. Excercise A12 in Chap. 9). Very small values of $\widehat{h}(2\pi k/T)$ in the quotient for $c_k$ cause a strong increase in noise components in the received signal that were not considered. Therefore, in real practice, there are various modified equalization algorithms, with known $\widehat{h}$—for example, in wired transmissions—partially with preequalization already at the transmitter, otherwise at the receiver side.

8. *Final Remarks and Notes on Advanced Communication Technology*
There are many other topics that need to be mastered for practical real-time transmission (within a few μs per OFDM symbol, see footnote[11]) with OFDM or modifications of the procedure (OFDMA, COFDM, FBMC, GFDM, etc.). These include, in particular, peak reduction (with many equal amplitudes $c_k$ in the OFDM symbol, *peak-to-average power ratio reduction, PAPR reduction*), channel equalization with multiple frequency-selective channels (Doppler effects

with moving transmitters or receivers with frequency dispersion), and many others. Despite simple principles, it is a long way to robust technology, demanding high skill from engineers and computer scientists. Since this text can by no means be a comprehensive introduction to the art of communication technology, but only aims to describe some fundamental ideas originating in Fourier analysis and essentially provide suggestions for acquiring further knowledge on the subject if interested, the already frequently cited specialized literature on communication technology is recommended once again for everything else mentioned in the text.

## 12.4  Heisenberg's Uncertainty Principle

Already in earlier sections, we had qualitatively observed that the spectral width of a signal is greater the shorter the signal duration is. Conversely, the impulse response of a lowpass filter lasts perceptibly longer in time the smaller the cutoff frequency of the filter is. The same aspect shows up in the scaling property $f(\alpha t) \; \circ\!\!-\!\!\bullet \; \dfrac{1}{|\alpha|} \widehat{f}\left(\dfrac{\omega}{\alpha}\right)$ of the Fourier transform for $\alpha \neq 0$ or in the fact that a time-limited signal $f$ has a Fourier transform $\widehat{f}$ that does not completely vanish in any frequency interval (see p. 303). For illustration, one can look again at the examples in Sects. 10.1 and 11.2, such as rectangular or triangular functions with their Fourier transforms on p. 275 or the correspondences $\delta(t) \; \circ\!\!-\!\!\bullet \; 1$ and $1 \; \circ\!\!-\!\!\bullet \; 2\pi \delta(\omega)$.

To obtain quantitative statements about the observed coupling of compression and expansion in the time-frequency domain, a measure is needed for the duration and bandwidth of signals. Although there is no uniform definition of duration and bandwidth for the immense variety of possible signals, the definition of dispersion is suitable for introducing these concepts for a large class of signals. For the signals $f$ considered below, we assume that the functions $f$ are continuous and piecewise continuously differentiable. Furthermore, $tf(t)$ and $\dot{f}(t)$ should also be square-integrable along with $f(t)$. We interpret the parameter $t$ as a time parameter.

**Definition**

1. The dispersion $\Delta_a^2(f)$ of $f \neq 0$ around the point $a$ is defined by

$$\Delta_a^2(f) = \frac{\displaystyle\int_{-\infty}^{+\infty} (t-a)^2 |f(t)|^2 \, dt}{\displaystyle\int_{-\infty}^{+\infty} |f(t)|^2 \, dt}.$$

2. The effective duration $D_t(f)$ of $f \neq 0$ is defined by

$$D_t(f) = \Delta_a(f) \quad \text{with} \quad a = \frac{\int\limits_{-\infty}^{+\infty} t|f(t)|^2 \, dt}{\int_{-\infty}^{+\infty} |f(t)|^2 \, dt}.$$

3. The effective bandwidth $D_\omega(f)$ of $f \neq 0$ is defined by

$$D_\omega(f) = \Delta_b(\widehat{f}) \quad \text{with} \quad b = \frac{\int\limits_{-\infty}^{+\infty} \omega|\widehat{f}(\omega)|^2 \, d\omega}{\int\limits_{-\infty}^{+\infty} |\widehat{f}(\omega)|^2 \, d\omega}.$$

The dispersion $\Delta_a^2(f)$ is a measure of how well or poorly $f$ is "concentrated around $a$." If $|f(t)|$ is very small outside a small neighborhood of $a$, then the factor $(t-a)^2$ makes the numerator of $\Delta_a^2(f)$ small compared to the denominator, and the dispersion is small. If $|f(t)|$ is large for $(t-a)^2 > 1$, the same factor causes an increase in the numerator compared to the denominator, and the dispersion becomes large. If we interpret the function $|f(t)|^2$ as mass density, then $S = \int\limits_{-\infty}^{+\infty} t|f(t)|^2 \, dt / \int_{-\infty}^{+\infty} |f(t)|^2 \, dt$ is the center of mass and $\Delta_S^2(f)$ is the moment of inertia with respect to the center of mass. If we interpret the function $|f(t)|^2 / \int_{-\infty}^{+\infty} |f(t)|^2 \, dt$ as the density of a probability distribution, then $S$ is the expected value and $\Delta_S^2(f)$ is the variance of the probability distribution.

## *Examples*

1. For $f(t) = (2\pi\sigma^2)^{-1/4} \, e^{-(t-m)^2/(4\sigma^2)}$, $\sigma > 0$, the function $|f(t)|^2$ is the density of the Gaussian distribution known from probability theory with mean $m$ and variance $\sigma^2$. The effective duration of $f$ is therefore according to the previous remark $D_t(f) = \sigma$. Using *Plancherel's theorem* and the *differentiation rule* for the Fourier transform, the effective bandwidth follows:

$$D_\omega^2(f) = \Delta_0^2(\widehat{f}) = \frac{1}{2\pi} \int\limits_{-\infty}^{+\infty} \omega^2|\widehat{f}(\omega)|^2 \, d\omega = \frac{1}{2\pi} \int\limits_{-\infty}^{+\infty} |\widehat{\dot{f}}(\omega)|^2 \, d\omega$$

$$= \int\limits_{-\infty}^{+\infty} |\dot{f}(t)|^2 \, dt = \frac{1}{4\sigma^4} \int\limits_{-\infty}^{+\infty} (t-m)^2|f(t)|^2 \, dt = \frac{1}{4\sigma^2}.$$

The product of effective duration and bandwidth yields $D_t(f)D_\omega(f) = \dfrac{1}{2}$.

2. For the triangular function $f(t) = \begin{cases} A(1 - |t|/T) & \text{for } |t| \leq T \\ 0 & \text{for } |t| > T \end{cases}$, one calculates

$$\int_{-\infty}^{+\infty} |f(t)|^2 \, dt = \frac{2A^2 T}{3} \quad \text{and the effective duration is} \quad D_t(f) = \frac{T}{\sqrt{10}}.$$

According to Plancherel's theorem, $\int_{-\infty}^{+\infty} |\widehat{f}(\omega)|^2 \, d\omega = \frac{4\pi A^2 T}{3}$. The center of mass of $|\widehat{f}|^2$ is zero because $|\widehat{f}|^2$ is an even function. As in the first example, it follows that

$$\int_{-\infty}^{+\infty} \omega^2 |\widehat{f}(\omega)|^2 \, d\omega = \int_{-\infty}^{+\infty} |\widehat{\dot{f}}(\omega)|^2 \, d\omega = 2\pi \int_{-\infty}^{+\infty} |\dot{f}(t)|^2 \, dt = \frac{4\pi A^2}{T}.$$

The effective bandwidth is thus $D_\omega(f) = \sqrt{3}/T$. The time-bandwidth product is

$$D_t(f)D_\omega(f) = \sqrt{3}/\sqrt{10} \approx 0.548,$$

i.e., about 9.6% larger than in the first example with the Gaussian function.

**Illustratively** The calculations in these examples show us that $\omega \widehat{f}(\omega)$ is square-integrable if and only if $\dot{f}(t)$ has this property. *Since the bandwidth is given by an integral over the squared magnitude of the derivative $\dot{f}$ of a time signal $f$, a compression of the signal $f$ must cause an increase in the bandwidth through simultaneously growing slopes.* Therefore, the functions $f$ and $\widehat{f}$ cannot be simultaneously concentrated near individual points. A quantitative description of this fact is provided by Heisenberg's uncertainty principle. It was discovered by W. Heisenberg in 1927 in quantum mechanics. Its significance for signal transmission was investigated by Gabor (1946) . In our context, it reads as follows:

## *Uncertainty Principle for the Time-Bandwidth Product*

**Theorem 12.4 (Time-Bandwidth Product)** *For square-integrable signals $f \neq 0$ and any $a, b \in \mathbb{R}$ the following holds*

$$\Delta_a^2(f)\Delta_b^2(\widehat{f}) \geq \frac{1}{4}.$$

*In particular, for the time-bandwidth product, it always holds that $D_t(f)D_\omega(f) \geq \frac{1}{2}$. Equality $D_t(f)D_\omega(f) = 1/2$ holds if and only if $|f|$ is a Gaussian function, i.e., if $f(t) = c\,e^{jat}\,e^{-(t-m)^2/(4\sigma^2)}$ with any real constants $a$, $m$, $c \neq 0$, $\sigma \neq 0$.*

**Proof** We can assume that with $f$, also $tf(t)$ and $\dot{f}(t)$ are square-integrable, otherwise $\Delta_a^2(f) = \infty$ or $\Delta_b^2(\widehat{f}) = \infty$ would hold, and the inequality would be trivially satisfied. For $a = b = 0$, integration by parts yields

$$\int_\alpha^\beta \overline{tf(t)}\,\dot{f}(t)\,\mathrm{d}t = t|f(t)|^2\Big|_\alpha^\beta - \int_\alpha^\beta \left(|f(t)|^2 + tf(t)\overline{\dot{f}(t)}\right)\,\mathrm{d}t,$$

so

$$\int_\alpha^\beta |f(t)|^2\,\mathrm{d}t = -2\Re\left(\int_\alpha^\beta \overline{tf(t)}\,\dot{f}(t)\,\mathrm{d}t\right) + t|f(t)|^2\Big|_\alpha^\beta.$$

Due to the assumptions about $f$ (cf. p. 402), the limits of the integrals exist for $\alpha \to -\infty$, $\beta \to +\infty$, and it holds that $\lim_{\alpha\to-\infty} \alpha|f(\alpha)|^2 = \lim_{\beta\to+\infty} \beta|f(\beta)|^2 = 0$. Thus, it follows

$$\int_{-\infty}^{+\infty} |f(t)|^2\,\mathrm{d}t = -2\Re\left(\int_{-\infty}^{+\infty} \overline{tf(t)}\,\dot{f}(t)\,\mathrm{d}t\right).$$

Using the Cauchy-Schwarz inequality and Plancherel's theorem we obtain

$$\left(\int_{-\infty}^{+\infty} |f(t)|^2\,\mathrm{d}t\right)^2 \leq 4\left(\int_{-\infty}^{+\infty} t^2|f(t)|^2\,\mathrm{d}t\right)\left(\int_{-\infty}^{+\infty} |\dot{f}(t)|^2\,\mathrm{d}t\right)$$

$$= 4\left(\int_{-\infty}^{+\infty} t^2|f(t)|^2\,\mathrm{d}t\right)\left(\frac{1}{2\pi}\int_{-\infty}^{+\infty} \omega^2|\widehat{f}(\omega)|^2\,\mathrm{d}\omega\right)$$

and hence the uncertainty relation $\Delta_0^2(f)\Delta_0^2(\widehat{f}) \geq \frac{1}{4}$.

The general case for $a \neq 0$ or $b \neq 0$ can be obtained with $g(t) = e^{-jbt}\,f(t+a)$. Since $\Delta_a^2(f) = \Delta_0^2(g)$ and $\Delta_b^2(\widehat{f}) = \Delta_0^2(\widehat{g})$ hold, it follows

$$\Delta_a^2(f)\Delta_b^2(\widehat{f}) = \Delta_0^2(g)\Delta_0^2(\widehat{g}) \geq \frac{1}{4}.$$

The Cauchy-Schwarz inequality above becomes an equality if and only if $tf(t)$ and $\dot{f}(t)$ are linearly dependent, i.e., if the differential equation $ktf(t) = \dot{f}(t)$ holds (cf.

p. 308). The only nontrivial, square-integrable solutions of this differential equation are of the form $f(t) = c\,e^{kt^2/2}$ with $c \neq 0$, $k < 0$. With $k = -1/(2\sigma^2)$, the last statement of the theorem follows.                                    $\square$

**Remark** The smoothness assumptions on $f$ from p. 402 can be omitted as with Plancherel's theorem (cf. p. 285), i.e., the Heisenberg uncertainty relation holds for any square-integrable functions $f$. A proof of this more general statement was given in 1931 by W. Pauli and H. Weyl. It can be found, for example, in the textbook of Dym and McKean (1985).

## *Application Examples*

1. *Resolution in the Time Domain.* In electrical measurement technology, it is known that, for example, with an oscilloscope of 100 MHz effective bandwidth ($D_\omega = 2\pi \cdot 100$ MHz), only a temporal resolution in the order of 1 ns is possible, with $D_\omega$ as above, $D_t \geq 1/(2D_\omega) \approx 0.8 \cdot 10^{-9}$ s. For signals of shorter effective duration, the oscilloscope acts as a lowpass filter, and the signals are no longer exactly reproduced, but smoothed in reproduction and prolonged in duration. Start and stop pulses for measuring time intervals below the duration given by the uncertainty principle then merge in the reconstruction; a time measurement of such short intervals is therefore no longer possible.
2. *Resolution in the Frequency Domain.* The effective duration $D_t$ for which one must sing a tone or play an instrument to assign it a pitch or frequency with the accuracy $D_\omega$ is, according to the uncertainty principle, at least $1/(2D_\omega)$. For instance, if $D_\omega = 2\pi \cdot 1$ Hz, then $D_t$ must be greater than about $8 \cdot 10^{-2}$ s. In very fast passages of a musical piece, slight intonation weaknesses of the virtuosos cannot be noticed. Therefore, amateur musicians are recommended to choose the fastest possible musical pieces for a potential performance.
3. *Ultra-Short-Pulse Laser of High Bandwidth.* The pulse duration of today's mode-locked short-pulse lasers is in the range of a few femtoseconds (1 fs $= 10^{-15}$ s) with typical pulse repetition rates of 80–100 MHz up to 20–30 GHz. The corresponding enormous bandwidths enable time-resolved spectroscopy, for example, in the analysis of chemical reactions. In terahertz time-domain spectroscopy (THz-TDS), a noninvasive broadband method for investigating material properties in the far infrared is available. Application areas include investigations of crystal structures, biomedical diagnostics, or pharmaceutical quality control. Readers interested in laser technology are referred to the textbook Rullière (1998).

## *Heisenberg's Uncertainty Principle in Quantum Mechanics*

Since the Copenhagen Conference in 1927, atomic physics has undergone a probabilistic interpretation with quantum mechanics. In this interpretation, experimental experiences, their theoretical description, and interpretation were brought together, unifying both the wave model and the particle model of matter without contradiction. Historical developments of quantum mechanics and its mathematical foundations can be found in works by P. A. M. Dirac (1958), Messiah (2003), or other relevant literature.

Heisenberg's uncertainty principle holds a central position in the development and interpretation of quantum theory. To formulate it in the language of quantum mechanics, we consider a free electron moving along the $x$-axis. However, its state at a fixed time cannot be specified by a position $x_0 \in \mathbb{R}$ and a momentum $p_0 \in \mathbb{R}$ as in classical mechanics but is described by a complex-valued, square-integrable wave function $\psi(x)$, whose $L^2$ norm is

$$\|\psi\|_2 = \left( \int\limits_{-\infty}^{+\infty} |\psi(x)|^2 \, dx \right)^{1/2} = 1.$$

The functions $x\psi(x)$ and $\dot{\psi}(x)$ are also assumed to be square-integrable with $\psi(x)$. The function $|\psi|^2$ is interpreted as the probability density of the electron's presence. The position of the particle is thus a random variable with the expected value $a = \int\limits_{-\infty}^{+\infty} x|\psi(x)|^2 \, dx$ and the variance $\Delta_a^2(\psi)$. The probability that a position measurement in the state $\psi$ yields a value $x \in [x_1, x_2]$ is $\int_{x_1}^{x_2} |\psi(x)|^2 \, dx$.

The variance $\Delta_a^2(\psi)$ is a measure of the uncertainty of the position, as the larger the variance, the greater the probability of presence in intervals that do not contain the expected value $a$. If the variance is very small, then the position is said to be sharply determined. The probability of presence in very small intervals around $a$ is then large because the density function $|\psi|^2$ is concentrated around $a$ for small variance.

The momentum of the electron is essentially given by the Fourier transform of $\psi$, namely by the function

$$\widetilde{\psi}(p) = (2\pi\hbar)^{-1/2}\widehat{\psi}(p/\hbar).$$

The constant $\hbar$ is the reduced Planck constant. The momentum is also a random variable. The function $|\widetilde{\psi}|^2$ is interpreted as the probability density for the momentum distribution. The expected value $b$ for a momentum measurement is then

$$b = \int\limits_{-\infty}^{+\infty} p\,|\tilde{\psi}(p)|^2\,\mathrm{d}p = \frac{\hbar}{2\pi}\int\limits_{-\infty}^{+\infty} p\,|\widehat{\psi}(p)|^2\,\mathrm{d}p.$$

The variance $\Delta_b^2(\tilde{\psi})$ is a measure of the uncertainty of the particle's momentum. The sharper the momentum is determined, the smaller $\Delta_b^2(\tilde{\psi})$ is. For the product of the variances of $\psi$ and $\widehat{\psi}$, the uncertainty relation $\Delta_a^2(\psi)\Delta_{b/\hbar}^2(\widehat{\psi}) \geq 1/4$ holds according to p. 404. From

$$\Delta_b^2(\tilde{\psi}) = \frac{1}{2\pi\hbar}\int\limits_{-\infty}^{+\infty} (p-b)^2 \left|\widehat{\psi}\left(\frac{p}{\hbar}\right)\right|^2 \mathrm{d}p$$

$$= \frac{\hbar^2}{2\pi}\int\limits_{-\infty}^{+\infty} \left(p-\frac{b}{\hbar}\right)^2 |\widehat{\psi}(p)|^2\,\mathrm{d}p = \hbar^2 \Delta_{b/\hbar}^2(\widehat{\psi})$$

the following uncertainty principle, discovered by W. Heisenberg (1901–1976) in 1927, results. It is one of the fundamental statements of quantum mechanics.

**Heisenberg's Uncertainty Principle** *Position and momentum of an electron in the state $\psi$ are not simultaneously sharply defined, but rather afflicted with an uncertainty. For the wave functions $\psi$ and $\tilde{\psi}$, the following uncertainty relation holds*

$$\Delta_a(\psi)\Delta_b(\tilde{\psi}) \geq \hbar/2.$$

The statement also applies to wave functions in three-dimensional space. One just needs to apply the Fourier transform for functions of several variables. The uncertainty relation is *not* based on limits of measurement accuracy but is a *general property of functions*. For example, one can speak of the frequency of an oscillation "in the pure sense" only if the oscillation process is periodic and thus particularly unlimited in time. A duration is then entirely undefined. Conversely, the shorter the process lasts, the more questionable it is to speak of periodicity and thus of frequency; the concept itself becomes fuzzy, the process must be mathematically described by the corresponding spectral function instead of a pure frequency, and uncertainty relations arise. In electrical engineering, this fact is known, as we have seen, in the case of the time-duration-bandwidth product. Quantum mechanics shows that even position and momentum in the physical description of atomic particles through probability densities are subject to such uncertainties. The same applies to other quantities whose product yields an action. For example, one obtains an analogous uncertainty relation for the product of energy and duration of an atomic event. Applications of the uncertainty principle to questions in physics, such as the explanation of the tunneling effect, can be found in the according literature.

## 12.5   Time-Frequency Analysis, Windowed Fourier Transforms

For many applications in signal processing, the Fourier transform in its original form is not suitable. Because the Fourier integral extends over the entire time axis, a full knowledge of the signal's time course would be necessary to analyze the spectral properties of a signal, including knowledge of all future signal values $t > t_0$ for analysis at a fixed time $t_0$. Furthermore, the asymptotic properties of the Fourier transform show that even temporally narrow disturbances affect the entire spectrum (see p. 282). In its classical form, the Fourier transform also does not allow for simultaneous time-frequency analysis. For example, speech or a piece of music in our everyday experience has a specific "time pattern" and at the same time a specific "frequency pattern." However, the spectral function of a signal does not show at what times and with what respective amplitudes a specific angular frequency $\omega$ occurs in a signal $f$, but rather accumulates contributions of the same angular frequency $\omega$ over the entire time course of $f$ in $\widehat{f}(\omega)$. D. Gabor (1900–1979) already noticed these disadvantages for signal processing purposes, and in 1946 in his work "Theory of Communication," he proposed time-frequency localization through Fourier transforms with window functions.

To obtain information about the "time-frequency pattern" of a signal, one determines not the spectral function $\widehat{f}$ of the entire signal, but the spectral functions for time segments of $f$. Time segments of a signal $f$ are obtained by multiplying $f$ with functions of finite effective duration. Such functions are referred to as window functions or time windows.

### *Windowed Fourier Transforms, Gabor Transform*

All signals $f$ and window functions $w$ are assumed to be piecewise continuously differentiable and square-integrable. For window functions $w$, we assume that $w \neq 0$ and furthermore that with $w(t)$ and $\widehat{w}(\omega)$, both $tw(t)$ and $\omega\widehat{w}(\omega)$ are also square-integrable. The window functions $w$ then have finite effective duration and bandwidth (see Sect. 12.4). In particular, $|t|^{1/2}w(t)$ and $(1 + |t|)w(t)$ are square-integrable, and the Cauchy-Schwarz inequality for the product $(1 + |t|)^{-1}(1 + |t|)w(t)$ shows that $w(t)$ is integrable.

Analogously, $\widehat{w}(\omega)$ is integrable. The functions $w(t)$ and $\widehat{w}(\omega)$ are then also continuous. As in previous sections (see p. 311), we use the notations $\langle f(t)|g(t)\rangle = \int_{-\infty}^{+\infty} f(t)\overline{g(t)}\,dt$ for the inner product of square-integrable functions and $\|f\| = \langle f(t)|f(t)\rangle^{1/2}$ for the norm of $f$ in $L^2(\mathbb{R})$. The quantities

$$t^* = \langle tw(t)|w(t)\rangle/\|w\|^2 \quad \text{or} \quad \omega^* = \langle \omega\widehat{w}(\omega)|\widehat{w}(\omega)\rangle/\|\widehat{w}\|^2$$

**Fig. 12.13** From "Syrinx" of Claude Debussy

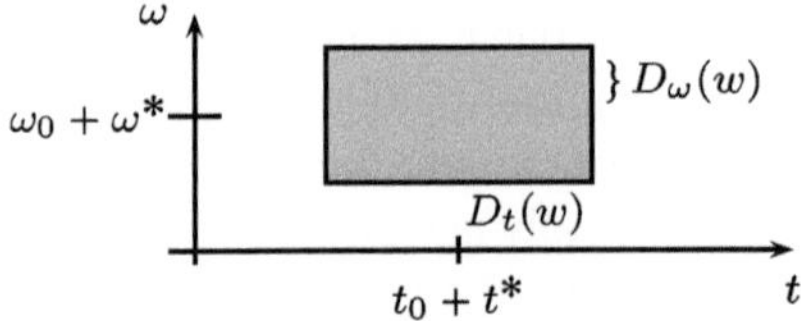


**Fig. 12.14** Time-frequency localization of the "note" $w_{\omega_0,t_0}$ in the time-frequency window $[t_0 + t^* - D_t(w), t_0 + t^* + D_t(w)] \times [\omega_0 + \omega^* - D_\omega(w), \omega_0 + \omega^* + D_\omega(w)]$

are referred to as the *time center* and *frequency center* of a window $w \neq 0$ (see p. 403).

**Definition** The transform $\mathcal{G}_w$, which maps a signal $f$ to the function $\mathcal{G}_w f = \tilde{f}$, defined by $\tilde{f}(\omega, t) = \langle f(s) | w(s - t) \, e^{j\omega s} \rangle = \int\limits_{-\infty}^{+\infty} f(s) \overline{w(s - t)} \, e^{-j\omega s} \, ds$, is called the windowed Fourier transform with the time window $w$. It is also abbreviated as STFT (Short-Time Fourier Transform). The windowed Fourier transform with the Gaussian window $w(t) = g_\alpha(t) = (4\pi\alpha)^{-1/2} \, e^{-t^2/(4\alpha)}$, $\alpha > 0$, is referred to as the Gabor transform $\mathcal{G}_\alpha$.

Instead of periodic harmonic oscillations $e^{j\omega t}$, the windowed Fourier transform uses translations of amplitude-modulated oscillations with the envelope $w$. To give an illustrative interpretation of $\tilde{f}(\omega_0, t_0)$ for fixed $\omega_0$ and $t_0$, consider the first two bars[5] of the wonderful flute piece " Syrinx" of Claude Debussy (1862–1918) (Fig. 12.13):

The "time-frequency pattern" is given in musical notation by the positions and note values of the individual notes, supplemented by dynamic indications such as "forte" or "piano." Similarly, one can consider the function

$$w_{\omega_0,t_0}(s) = w(s - t_0) \, e^{j\omega_0 s}$$

for a time window $w$ with the time center $t^*$ and the frequency center $\omega^*$ as a "note", which is localized in the frequency range around $\omega_0 + \omega^*$ with the effective bandwidth $D_\omega(w)$ and in the time range around $t_0 + t^*$ with the effective duration $D_t(w)$ (Fig. 12.14).

---

[5] The invention of musical notation is—like others—an ingenious human achievement and its expressive possibilities are inexhaustible.

The complex number $\widetilde{f}(\omega_0, t_0) = \langle f(s)|w_{\omega_0,t_0}(s)\rangle$ then indicates (see p. 60), the extent to which the "note" $w_{\omega_0,t_0}$ is present in the signal $f$, i.e., whether approximately at the time $t_0 + t^*$ the angular frequency $\omega_0 + \omega^*$ is represented in the signal, and if so, with what amplitude and phase.

The approximation error is due to the time duration $D_t(w) > 0$ and the bandwidth $D_\omega(w) > 0$ of the window $w \neq 0$, and thus due to the fact that the values of $w_{\omega_0,t_0}$ and $\widehat{w}_{\omega_0,t_0}$ in the corresponding time-frequency window (see figure) with appropriate weight enter into the integral

$$\widetilde{f}(\omega_0, t_0) = \langle f(s)|w_{\omega_0,t_0}(s)\rangle = (2\pi)^{-1}\langle \widehat{f}(\omega)|\widehat{w}_{\omega_0,t_0}(\omega)\rangle.$$

The smaller $D_t(w)$ is, the better $\widetilde{f}(\omega, t_1)$ and $\widetilde{f}(\omega, t_2)$ can be distinguished for adjacent time points $t_1$ and $t_2$, i.e., the more easily the frequencies present in the signal can be assigned to the different times at which they occur. Therefore, the smaller $D_t(w)$ is, the better the time resolution by $\widetilde{f}$. The smaller the bandwidth $D_\omega(w)$ is, the better the corresponding resolution of different frequencies. However, as we saw in the last section, the quality of a simultaneous time-frequency localization is limited by the uncertainty relation $D_t(w)D_\omega(w) \geq 1/2$. The best compromise with regard to the uncertainty relation is therefore the windowed Fourier transform with Gaussian windows proposed by Dennis Gabor (1900–1979), known as the Gabor transform (see p. 404).

**Example** A short-term model for a siren tone or chirp is approximately the function $f(t) = A\sin(g(t))$ with $g(t) = 2\pi t\left(\alpha t + \beta t^2\right)$ for $0 \leq t \leq 10$ s and constants $A$, $\alpha$, $\beta$. The derivative of the argument $g'(t) = 2\pi t(2\alpha + 3\beta t)$ can be considered as the instantaneous angular frequency at time $t$. The magnitude spectrum, approximately calculated with $A = 1$, $\alpha = 4$ [$1/s^2$], $\beta = -4/15$ [$1/s^3$] over $T = 10$ s, shows a multitude of frequencies up to the maximum frequency 20 Hz, but not the parabolic frequency modulation and not the instantaneous frequencies at different times (left image below). The graph of an approximation for $|\widetilde{f}|$, $\widetilde{f}$ the windowed Fourier transform of $f$ with the "Hann window" $w(t) = 0.5 - 0.5\cos(\omega_0 t)$ for $0 \leq t \leq 1$ s, $w(t) = 0$ otherwise ($\omega_0 = 2\pi$ rad/s), on the other hand, clearly shows the rise and fall of the instantaneous frequencies and corresponds to our usual impression of the variable frequency of the siren tone (right image). For the calculation of the approximations of $|\widehat{f}|$ and $|\widetilde{f}|$ with the DFT, also compare p. 347 and the following Sect. 12.6. For Fig. 12.15 a 512-point DFT was used over a total of $T = 10$ s, with the DFT coefficients $|\widehat{c}_k T|$ plotted as an approximation for $|\widehat{f}(2\pi k/T)|$. In the second case, 50 Hann windows of duration 1 s were used at intervals of 0.2 s each. Per time segment, a 128-point DFT was performed, and the resulting (single-sided) DFT magnitude spectra were combined to form the second image in Fig. 12.16. Neither representation shows the constant amplitude $A = 1$. One reason is the strong aliasing effects due to the frequency modulation. The sum of the $|\widehat{c}_k|^2$ of the left image agrees numerically very well with the quadratic mean of $f$ in $[0, T]$ (in both cases, the value is about 0.5). Numerical integration to calculate $|\widetilde{f}|$ for 20 Hz at $t_0 = 5$ s results in approximately 0.24, as shown in the spectrogram

**Fig. 12.15** DFT of the siren signal

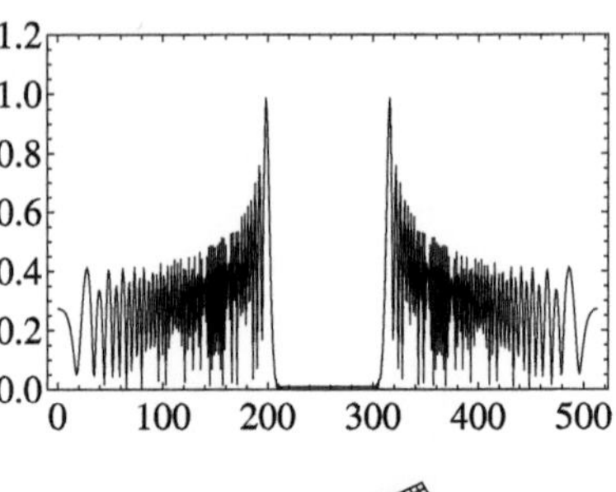

**Fig. 12.16** STFT of the siren signal

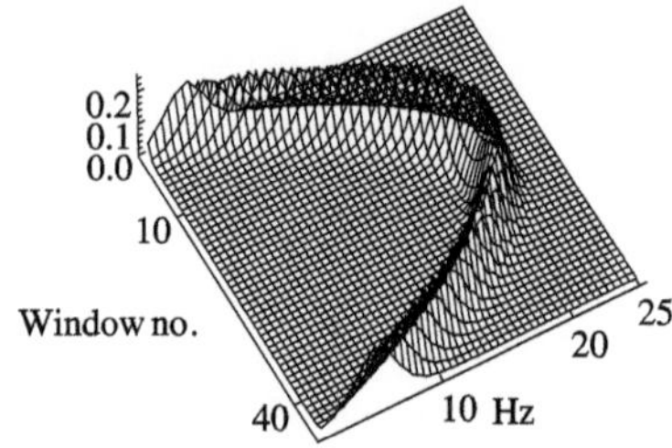


in Fig. 12.16. The signal values (and thus A) can only be approximately recovered from the signal DFT using an interpolation polynomial or the formula for discrete reconstruction from the data on page 415. At end of the book on page 483 you will see another spectrogram of a piece of music like the right image here.

## *Reconstruction of a Signal from Its Windowed Fourier Transform*

For a fixed chosen window $w \neq 0$, let now the transform $\mathcal{G}_w f = \widetilde{f}$ of a signal $f$ be given. An inverse formula for reconstructing the original signal $f$ from the values of $\widetilde{f}$ can be obtained by representing the local part $f_s(t) = \overline{w(t-s)} f(t)$ as a Fourier integral. For fixed $s \in \mathbb{R}$, $\widetilde{f}(\omega, s)$ is the Fourier transform of $f_s(t)$:

$$\widetilde{f}(\omega, s) = \widehat{f_s}(\omega) = \langle f(t) | w(t-s) \, e^{j\omega t} \rangle = \langle f(t)\overline{w(t-s)} | \, e^{j\omega t} \rangle.$$

According to our assumptions about $f$ and $w$ (cf. p. 409), $f_s$ is integrable in $t$ and piecewise continuously differentiable, so that with the Fourier inversion formula for each continuity point $t$ of $f$ (cf. p. 271), it holds

$$\overline{w(t-s)} f(t) = f_s(t) = \frac{1}{2\pi} \int\limits_{-\infty}^{+\infty} \widetilde{f}(\omega, s) \, e^{j\omega t} \, d\omega.$$

Multiplying both sides of this equation with $w(t - s)$, then integrating with respect to $s$, and dividing by $\|w\|^2$ give, due to $\int_{-\infty}^{+\infty} |w(t - s)|^2 \, ds = \|w\|^2$, the desired reconstruction formula.

**Theorem 12.5 (Pointwise Reconstruction Formula)**  *At each continuity point $t$ of a piecewise continuously differentiable, square-integrable function $f$, the value $f(t)$ can be recovered from the windowed Fourier transform of $f$ by*

$$f(t) = \frac{1}{2\pi \|w\|^2} \int_{-\infty}^{+\infty} \int_{-\infty}^{+\infty} \widetilde{f}(\omega, s) w(t - s) \, e^{j\omega t} \, d\omega \, ds.$$

*At discontinuity points $t$ of $f$, the right side gives, as in the Fourier inversion formula, the value $[f(t+) + f(t-)]/2$.*

If $\widetilde{f}(\omega, s) = \langle f(t) | w_{\omega,s}(t) \rangle$ is understood as the projection of the signal onto its time-frequency components, then the reconstruction formula is the "back-projection" by which the signal is recovered from the superposition of its components.

**Remarks**  From $\|w\|^2 \langle f(t) | f(t) \rangle = \int_{-\infty}^{+\infty} \int_{-\infty}^{+\infty} |f_s(t)|^2 \, dt \, ds < \infty$ it follows that the functions $f_s$ are square-integrable with respect to $t$ for almost all $s$ (cf. Appendix B, Fubini's theorem). Applying the Plancherel equation to the inner integral then gives with $w \neq 0$

$$\|f\|^2 = \frac{1}{2\pi \|w\|^2} \int_{-\infty}^{+\infty} \int_{-\infty}^{+\infty} |\widetilde{f}(\omega, s)|^2 \, d\omega \, ds = \frac{1}{2\pi \|w\|^2} \|\widetilde{f}\|^2.$$

This equation corresponds to the Plancherel equation for the Fourier transform and implies that the windowed Fourier transform $\mathcal{G}_w$ can be extended to a continuous injective mapping defined on the whole $L^2(\mathbb{R})$ into $L^2(\mathbb{R}^2)$ (cf. p. 308). The image $V = \mathcal{G}_w(L^2(\mathbb{R}))$ is a closed subspace of $L^2(\mathbb{R}^2)$, and any function $h \in L^2(\mathbb{R}^2)$ can be uniquely decomposed (cf. p. 61 and later 14.1, p. 449) in the form $h = h_V + h_V^\perp$ with $h_V \in V$ and

$$\langle v | h_V^\perp \rangle = \int_{-\infty}^{+\infty} \int_{-\infty}^{+\infty} v(\omega, s) \overline{h_V^\perp(\omega, s)} \, d\omega \, ds = 0.$$

The function $h_V$ is the orthogonal projection of $h$ onto $V$. The adjoint operator $\mathcal{G}_w^*$ to the operator $\mathcal{G}_w$ is defined by the equation $\langle f | \mathcal{G}_w^* g \rangle = \langle \mathcal{G}_w f | g \rangle$. From $\|f\|^2 = (2\pi \|w\|^2)^{-1} \|\widetilde{f}\|^2$ it follows with the polarization equation (p. 284) that $\langle f_1 | f_2 \rangle = (2\pi \|w\|^2)^{-1} \langle f_1 | \mathcal{G}_w^* \mathcal{G}_w f_2 \rangle$ for all $f_1, f_2 \in L^2(\mathbb{R})$. Thus, $f = (2\pi \|w\|^2)^{-1} \mathcal{G}_w^* \mathcal{G}_w f$

holds for all $f \in L^2(\mathbb{R})$. The inverse transform to $\mathcal{G}_w$ is therefore the restriction of $(2\pi\|w\|^2)^{-1}\mathcal{G}_w^*$ to the image $V$ of $\mathcal{G}_w$. For the signals we consider, it is given as an integral transformation by the right side of the reconstruction formula. If $h \in L^2(\mathbb{R}^2)$ has the decomposition $h = h_V + h_V^\perp$ with the part $h_V^\perp$ orthogonal to $V$, then $\langle f | \mathcal{G}_w^* h_V^\perp \rangle = \langle \mathcal{G}_w f | h_V^\perp \rangle = 0$ for all $f \in L^2(\mathbb{R})$, thus $\mathcal{G}_w^* h_V^\perp = 0$. With $f = (2\pi\|w\|^2)^{-1}\mathcal{G}_w^* h_V$, $\mathcal{G}_w f = h_V$, it follows

$$(2\pi\|w\|^2)^{-1}\mathcal{G}_w\mathcal{G}_w^* h = (2\pi\|w\|^2)^{-1}\mathcal{G}_w\mathcal{G}_w^* h_V = h_V.$$

The orthogonal projection of $L^2(\mathbb{R}^2)$ onto $V$ is thus the mapping $(2\pi\|w\|^2)^{-1}\mathcal{G}_w\mathcal{G}_w^*$. For more detailed information about adjoint operators and orthogonal projections, see for example Weidmann (1980).

With these remarks, it can be seen how desired time-frequency properties can be approximated in signal processing.


### Signal Processing with Windowed Fourier Transforms


Given a windowed Fourier transform $\mathcal{G}_w$ for a fixed chosen window $w \neq 0$, since the functions $\mathcal{G}_w f$, $f \in L^2(\mathbb{R})$, are bounded and $L^2(\mathbb{R}^2)$ also contains unbounded functions, not every square-integrable function $h(\omega, t)$ can be the windowed Fourier transform of a function $f \in L^2(\mathbb{R})$:

$V = \mathcal{G}_w(L^2(\mathbb{R})) \neq L^2(\mathbb{R}^2)$. Otherwise, signals with arbitrary time-frequency properties could be constructed—in contradiction to Heisenberg's uncertainty principle. However, one can proceed as follows to obtain signals that approximate the desired time-frequency properties as closely as possible:

For a given signal $f(t)$, the windowed Fourier transform $\tilde{f} = \mathcal{G}_w f$ is computed and $\tilde{f}$ is processed as desired to $h$ from $L^2(\mathbb{R}^2)$, for example, by filtering, shifting values, amplifying, etc. The function $h$ is the model of the desired time-frequency properties. However, in general, there is no signal $g$ such that $h = \mathcal{G}_w g$. The signal $f_h$ in $L^2(\mathbb{R})$, whose time-frequency properties are very close to those of $h$, is $f_h = (2\pi\|w\|^2)^{-1}\mathcal{G}_w^* h$, because according to the preceding remarks, the function $\tilde{f}_h$ as the orthogonal projection of $h$ onto $V$ minimizes the mean square error $\|h - \tilde{f}\|$, $f \in L^2(\mathbb{R})$ (cf. also later Sect. 14.1, p. 449).


### Discrete Windowed Fourier Transform


Of great importance for numerical approximation and thus for digital signal processing is the question of whether a signal can be reconstructed from the sampled values of its windowed Fourier transform. We present a sampling theorem and illustrate some fundamental aspects of discrete time-frequency analysis with windowed Fourier transforms.

Under the same conditions as for the reconstruction formula on p. 413, we assume that the window function $w \neq 0$ vanishes outside an interval $[a, b]$. For a fixed value of $s$, the support of $f_s(t) = \overline{w(t-s)} f(t)$ is contained in $[a+s, b+s]$. Fourier series expansion of $f_s$ in this interval yields for each continuity point $t$ of $f$ in $[a+s, b+s]$

$$f_s(t) = \sum_{k=-\infty}^{+\infty} c_k(s)\, e^{jk\omega_0 t} \quad \text{with} \quad \omega_0 = \frac{2\pi}{b-a},$$

$$c_k(s) = \frac{1}{b-a} \int_{a+s}^{b+s} f(t)\overline{w(t-s)}\, e^{-jk\omega_0 t}\, \mathrm{d}t = \frac{\omega_0}{2\pi} \widetilde{f}(k\omega_0, s).$$

Multiplying $f_s(t)$ by $w(t-s)$ results in the functions $w_{k\omega_0,s}$, which were defined by $w_{k\omega_0,s}(t) = w(t-s)\, e^{jk\omega_0 t}$:

$$|w(t-s)|^2 f(t) = \frac{\omega_0}{2\pi} \sum_{k=-\infty}^{+\infty} \widetilde{f}(k\omega_0, s) w_{k\omega_0,s}(t).$$

Instead of integrating this equation over $s$ and dividing by $\|w\|^2$ as in the reconstruction formula on p. 413, we form a discrete approximation for $\|w\|^2 = \int_{-\infty}^{+\infty} |w(t-s)|^2\, \mathrm{d}s$ by $A_{t_0}(t) = t_0 \sum_{n=-\infty}^{+\infty} |w(t-nt_0)|^2$ and sum over $s_n = nt_0, n \in \mathbb{Z}$:

$$A_{t_0}(t) f(t) = \frac{\omega_0 t_0}{2\pi} \sum_{n=-\infty}^{+\infty} \sum_{k=-\infty}^{+\infty} \widetilde{f}(k\omega_0, nt_0) w_{k\omega_0,nt_0}(t).$$

Due to the limited support of $w$, the series for $A_{t_0}(t)$ has only finitely many nonzero terms. Now we obtain the desired sampling theorem, i.e., a discrete reconstruction formula under the condition $A_{t_0}(t) \neq 0$:

**Theorem 12.6 (Discrete Reconstruction)** *If $A_{t_0}(t) \neq 0$ everywhere, then the signal $f$ is given at each continuity point $t$ by*

$$f(t) = \frac{\omega_0 t_0}{2\pi} \sum_{n=-\infty}^{+\infty} \sum_{k=-\infty}^{+\infty} \widetilde{f}(k\omega_0, nt_0) w_{k\omega_0,nt_0}(t) A_{t_0}(t)^{-1}.$$

The better the time-frequency localization of $w$, the faster the values $|w_{k\omega_0,nt_0}(t)|$ will decrease. In practice, for bandlimited signals $f$, finite partial sums of the right side with discrete approximations of the values of $\tilde{f}$ yield good approximations for $f(t)$.

From the derivation of the formula, the following conditions for a stable reconstruction are determined:

1. For numerically stable reconstruction, it is not enough to require $A_{t_0}(t) > 0$, but $\inf_{t \in \mathbb{R}} A_{t_0}(t) > 0$. Otherwise, small errors in the calculation of the values $\tilde{f}(k\omega_0, nt_0)$ would lead to very large errors in $f(t)$ at points $t$ where the value of $A_{t_0}(t)$ is very close to zero. This is a condition on the sampling rate because $\lim_{t_0 \to 0+} A_{t_0}(t) = \|w\|^2 \neq 0$ holds for all $t$ if the window is assumed to be continuous. Thus, this stability condition can be maintained for sufficiently small $t_0$.
2. A necessary condition for $A_{t_0}(t) > 0$ is $0 < t_0 \leq b - a$, otherwise $A_{t_0}(t) = 0$ for $b < t < a + t_0$. The given discrete reconstruction is therefore certainly *not* possible if $\omega_0 t_0 > 2\pi$.

Such conditions are typical when searching for stable discrete reconstruction formulas. Considering window functions $w$ that are not time-limited, we will analogously demand, as in point 1, that $\sup_{t \in \mathbb{R}} A_{t_0}(t) < \infty$ and that both this upper bound and the lower bound from point 1 converge to $\|w\|^2$ as $t_0 \to 0+$.

The mathematical task in the search for sampling formulas is to find conditions on the window function and the set of sampling points $(k\omega_0, nt_0)$, $k, n \in \mathbb{Z}$, such that the operator $\mathcal{G}_w^{\omega_0, t_0}$, which maps a signal to the sequence $\left( \langle f(t) | w_{k\omega_0, nt_0}(t) \rangle \right)_{k,n \in \mathbb{Z}}$, is injective. To obtain numerically stable formulas, $\mathcal{G}_w^{\omega_0, t_0}$ must additionally be continuous in an appropriate sense and have a continuous inverse mapping. This task leads in modern signal processing to the study of complete orthonormal systems in suitable function spaces. Instead of pointwise convergent sampling series, we then consider series that approximate the analyzed signals in the norm of the function space used. The considered signals can also be functions $f(t, \mathbf{x})$ that depend not only on time $t$ but also on a spatial variable $\mathbf{x}$. Such signals appear, for example, in image processing. Accordingly, systems of functions with multiple variables are used. Readers who find this section a motivation to delve deeper into the subject, given the importance of digital signal processing in audio and video technology but also in many other areas of engineering and natural sciences, are referred to further literature, such as Daubechies (1992), Feichtinger and Strohmer (2003), Gröchenig (2001), or Meyer (1993).

*Finally, some central results of discrete time-frequency analysis with windowed Fourier transforms are cited:*

1. If the product $\omega_0 t_0 > 2\pi$ holds, then for any choice of window $w$, there are always signals $f \in L^2(\mathbb{R})$, $f \neq 0$, that are orthogonal to all functions $w_{k\omega_0, nt_0}$. Therefore, a reconstruction of such signals from their windowed Fourier transforms $\mathcal{G}_w^{\omega_0, t_0} f$ is not possible. Discrete reconstruction formulas are generally subject to the condition $\omega_0 t_0 \leq 2\pi$.
2. If the function system $w_{k\omega_0, nt_0}$, $k, n \in \mathbb{Z}$, forms a complete orthogonal system in $L^2(\mathbb{R})$, then necessarily $\omega_0 t_0 = 2\pi$ must hold.

3. Even for $\omega_0 t_0 = 2\pi$, the functions $w_{k\omega_0, nt_0}$, $k, n \in \mathbb{Z}$, with the Gaussian window proposed by D. Gabor $w(t) = (2\pi)^{-1/4} e^{-t^2/4}$, do not form an orthonormal system in $L^2(\mathbb{R})$. It can be shown that

$$\inf\left\{ \|f\|^{-2} \sum_{k,n \in \mathbb{Z}} |\langle f | w_{k\omega_0, nt_0}\rangle|^2 \; : \; f \in L^2(\mathbb{R}),\, f \neq 0 \right\} = 0$$

holds. Although the functions $w_{k\omega_0, nt_0}$, $k, n \in \mathbb{Z}$, form a *complete system* in $L^2(\mathbb{R})$ (i.e., any $f \in L^2(\mathbb{R})$ can be approximated arbitrarily well by linear combinations of the $w_{k\omega_0, nt_0}$ with respect to the norm of $L^2(\mathbb{R})$), a numerically stable reconstruction of signals $f \in L^2(\mathbb{R})$ from the coefficients $\langle f | w_{k\omega_0, nt_0}\rangle$ is generally not possible.

4. While orthogonality relations for the functions $w_{k\omega_0, nt_0}$ would be desirable, practical requirements for good time-frequency localization of the windows even force $\omega_0 t_0 < 2\pi$, i.e., higher sampling rates are necessary than those that allow for the orthogonality of the system $w_{k\omega_0, nt_0}$. This statement is contained in the **uncertainty principle of R. Balian and F. Low**:

   *If the functions $w_{k\omega_0, nt_0}$ form a complete orthonormal system in $L^2(\mathbb{R})$ for a window $w \in L^2(\mathbb{R})$ with $\omega_0 t_0 = 2\pi$, then it holds that*

$$\int\limits_{-\infty}^{+\infty} t^2 |w(t)|^2 \, \mathrm{d}t = \infty \quad \text{or} \quad \int\limits_{-\infty}^{+\infty} \omega^2 |\widehat{w}(\omega)|^2 \, \mathrm{d}\omega = \infty.$$

5. For $\omega_0 t_0 < 2\pi$, there are windows $w$ and corresponding complete function systems (the so-called *Gabor frames*) $w_{k\omega_0, nt_0}$, $k, n \in \mathbb{Z}$ that enable stable reconstruction with very good time-frequency localization, i.e., with

$$\int\limits_{-\infty}^{+\infty} t^2 |w(t)|^2 \, \mathrm{d}t < \infty \quad \text{and} \quad \int\limits_{-\infty}^{+\infty} \omega^2 |\widehat{w}(\omega)|^2 \, \mathrm{d}\omega < \infty.$$

A derivation and detailed discussion of these results can be found, for example, in the already mentioned book of Daubechies (1992) or in Gröchenig (2001). Aspects of the window functions when using the DFT to approximate windowed Fourier transforms are discussed in the following section.


## 12.6   Time Windows with the Discrete Fourier Transform

In practice, the spectrum of a signal $f$ can often not be calculated exactly. Instead, one usually uses the spectral function of a signal segment $f w_T$ with a time window $w_T \neq 0$ as an approximation. Also, when analyzing unknown signals $f$, the observation duration $T$ is necessarily finite, so that only information about time segments $f w_T$ can be processed. In time-frequency analysis, as in the last section, the interest is also in the spectral functions of such time segments of the signal.

For the approximate calculation of the Fourier transform of $f w_T$, the discrete Fourier transform is often used (see p. 347 and Sect. 6). The spectrum $\widehat{f w_T}$ of the signal segment $f w_T$ is different from the actual spectrum $\widehat{f}$ of $f$. If $w_T$ is a time window with support in $[0, T]$, then according to the modulation theorem from p. 283 for square-integrable or bandlimited signals $f$

$$\widehat{f w_T} = \frac{1}{2\pi} \widehat{f} * \widehat{w_T}. \tag{12.1}$$

The spectral function of $f w_T$ is compared to $\widehat{f}$ "smeared, smoothed, and blurred" due to the convolution of $\widehat{f}$ with $\widehat{w_T}$. The shorter the observation duration $T$, the greater the bandwidth of $w_T$ according to the uncertainty principle, and the worse the frequency localization of $w_T$ and thus of $f w_T$ (see Sects. 12.4 and 12.5). A typical problem is then, for example, the resolution of periodic signal components of closely adjacent frequencies, especially when these signal components have very different amplitudes.

The observation duration $T$ and the shape of the time window $w_T$ also have an impact on the quality of the approximations for $\widehat{f w_T}$, which are obtained with a finite discrete Fourier transform from sampled values of $f w_T$. Therefore, when using the discrete Fourier transform, some fundamental aspects of the interaction between the observation duration $T$, the properties of the weighting function $w_T$, and the sampling rate used for the discrete Fourier transform must be considered.

## *Truncation Effects in the Discrete Fourier Transform*

In the discrete Fourier transform, from finitely many values $y_n = f(n\Delta t)$, $\Delta t > 0$, $n = 0, \ldots, N - 1$, of a signal $f$, the Fourier coefficients

$$\widehat{c_k} = \frac{1}{N} \sum_{n=0}^{N-1} y_n \, e^{-jkn2\pi/N}$$

are calculated for $k = 0, \ldots, N - 1$ (see 6, p. 86). We assume the signal $f$ to be continuous in $[0, T[$ with the limit value $f(T-)$ and piecewise continuously differentiable. The sampled time section of $f$ beyond the sampling period of duration $T_a = (N - 1)\Delta t$ can be arbitrarily extended to a periodic function $f_p$ with the period $p = N\Delta t = T$, for example as in the following figure, where we have added a straight segment between $T_a$ and $T$ so that $f_p$ becomes continuous (see Fig. 12.17).

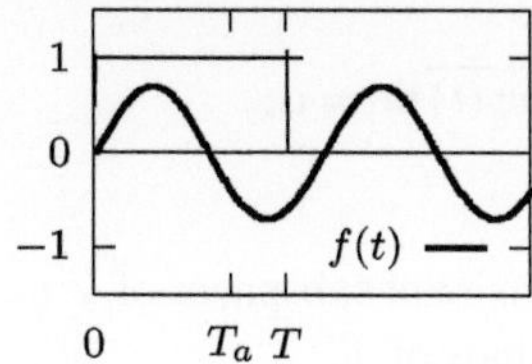


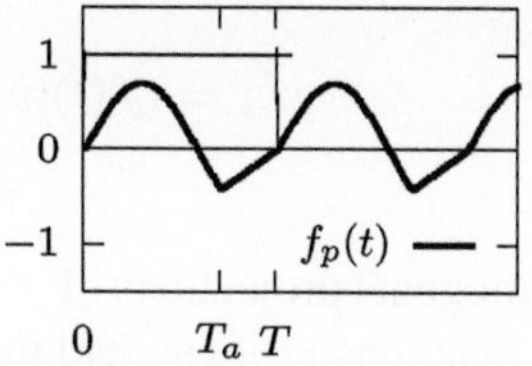


**Fig. 12.17** Last sample time is $T_a < T$, and a possible $T$-periodic extension could be $f_p$

Let us denote by $w_T$ the *rectangle window*

$$w_T(t) = 1 \text{ for } 0 \le t < T, \ w_T(t) = 0 \text{ otherwise,}$$

then the quantities $\widehat{c}_k$ are on one hand approximations for Fourier coefficients $c_k(f_p)$ of $f_p$, and on the other hand, according to p. 347, they also yield approximations for sample values of $\widehat{f w_T}$ and thus approximations for the Fourier coefficients $c_k = c_k(f w_T)$ of $f w_T$. In many applications $\widehat{c}_k T$ also serves as an estimator for $\widehat{f}(2\pi k/T)$ (see also Exercise A4 of Chap. 11).

If, for instance, $N$ is an even number, then one uses the value $\widehat{c}_k$ for the indices $k = 0, \ldots, (N-2)/2$, respectively, as an approximation for the Fourier coefficient $c_k$ of $f w_T$. For $k = (N+2)/2, \ldots, N-1$, $\widehat{c}_k$ serves accordingly as an approximation for $c_{-N+k}$ and $\widehat{c}_{N/2}$ as an approximation for $(c_{-N/2} + c_{N/2})/2$ (see p. 88). The corresponding oscillations to the fundamental circular frequency $\omega_0 = 2\pi/T$

$$v_0(t) = 1, \ v_1(t) = e^{j\omega_0 t}, \ldots, v_{(N-2)/2}(t) = e^{j(N-2)\omega_0 t/2}, \ v_{N/2}(t)$$

$$= \cos(N\omega_0 t/2), v_{(N+2)/2}(t) = e^{-j(N-2)\omega_0 t/2}, \ldots\ldots, v_{N-1}(t) = e^{-j\omega_0 t},$$

generate an $N$-dimensional function vector space $V$ in $L^2([0, T])$ (see p. 12).

For the rectangle window $w_T$, the $T$-periodic extension of $f w_T$ has jump discontinuities at $t = kT$, $k \in \mathbb{Z}$, if $f(0) \ne f(T-)$. According to p. 87, with continuous $f_p$ as above, the aliasing relationships hold

$$\widehat{c}_k = \sum_{m=-\infty}^{+\infty} c_{k+mN}(f w_T) + \frac{1}{2N}(f(0) - f(T-)) = \sum_{m=-\infty}^{+\infty} c_{k+mN}(f_p). \quad (12.2)$$

If the signal $f$ is a mixture of harmonic oscillations with circular frequencies $k\omega_0$, $k = 0, \ldots, N/2$, i.e., if $f(t) = \sum_{k=0}^{N-1} \alpha_k v_k(t)$ is a linear combination of the functions $v_0, \ldots, v_{N-1}$, then $f(0) = f(T-)$, and with the inner product from p. 12, it follows from the aliasing relationship (12.2)

$$\widehat{c_k} = \langle f(t)|v_k(t)\rangle = \frac{1}{T}\int_0^T f(t)\overline{v_k(t)}\,dt = \alpha_k. \tag{12.3}$$

The orthogonal projections of $f$ onto the one-dimensional subspaces of $V$ generated by the functions $v_k$ then yield the exact spectral values of $f$.

It is different if the periodic extension of $f w_T$ has a jump discontinuity at $t = T$ or if the originally observed signal $f$ contains harmonic oscillations whose period duration does not match $T$. In practice, this will often be the case when analyzing unknown signals $f$, which are sampled over an arbitrarily chosen time period. Simple examples of such cases are given by the functions $f_1(t) = \cos(t)$ and $f_2(t) = -\cos(t/2) + \cos(t)/2$. For $T = \pi$, the $T$-periodic extension of $f_1 w_T$ with the rectangle window $w_T$ has a jump discontinuity at $T$, while that of $f_2 w_T$ is continuous, but $f_2$ is not $T$-periodic.

If $f w_T(0) \neq f w_T(T-)$, then every $T$-periodic extension, $T = N\Delta t$, of $f$ beyond the interval $[0, T_a]$, $T_a = (N - 1)\Delta t$, has jump discontinuities or steep flanks in the vicinities of the points $kT$, $k \in \mathbb{Z}$ (see last figure). From considerations on the asymptotics of Fourier coefficients (p. 48), it follows that the magnitudes of the coefficients $c_k$ of a $T$-periodic extension of the signal section for $|k| \to \infty$ decrease only slowly. Consequences are, according to Eq. (12.2), aliasing effects in the coefficients $\widehat{c_k}$ of the discrete Fourier transform.

Even if by chance $f w_T(0) = f w_T(T-)$ as in the example $f_2 w_T$ above, aliasing effects arise as soon as $f$ contains oscillation components with frequencies $\nu \neq k/T$, and also if they lie within the Nyquist interval with the cutoff frequency $N/(2T)$.

*Every signal component with a circular frequency $\omega_1 \neq 2\pi k/T$ has nonzero projections in all subspaces of $L^2([0, T])$, which are generated by the functions $v_k$ for $k = 0, \ldots, N - 1$:*

$$\langle e^{j\omega_1 t}\, w_T(t)|v_k(t)\rangle \neq 0 \text{ for all } k = 0, \ldots, N - 1.$$

**Example** Consider for example the signal $g(t) = A\,e^{j\omega_1 t}$, and then for the $k$th Fourier coefficient $c_k(g w_T)$ of $g w_T$ with the rectangle window $w_T$ for the interval $[0, T[$ according to (12.1) and p. 347 with $\widehat{g}(\omega) = 2\pi A\delta(\omega - \omega_1)$:

$$c_k(g w_T) = \frac{1}{T}\widehat{g w_T}\left(\frac{2\pi k}{T}\right) = \frac{1}{2\pi T}(\widehat{g} * \widehat{w_T})\left(\frac{2\pi k}{T}\right) \tag{12.4}$$

$$= \frac{A}{T}\,e^{-j(2\pi k/T - \omega_1)T/2}\,\widehat{w_T}\left(\frac{2\pi k}{T} - \omega_1\right)$$

$$= (-1)^k A\,e^{j\omega_1 T/2}\,\frac{\sin(\pi k - \omega_1 T/2)}{\pi k - \omega_1 T/2}.$$

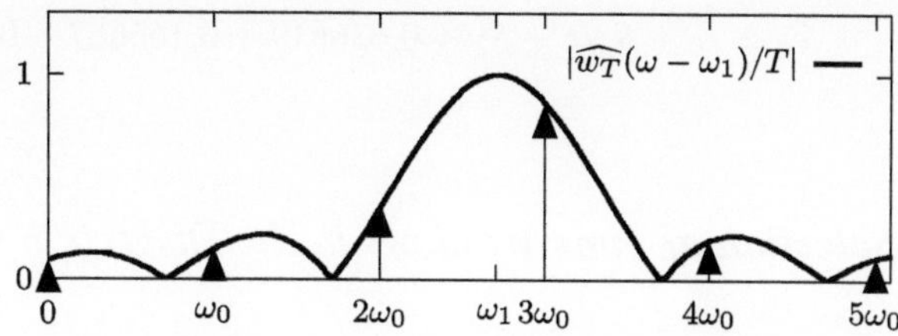



**Fig. 12.18** The arrows show the absolute weights $g_k$ at $k\omega_0$

These coefficients distort, as per (12.2), the amplitudes and phases of the estimates $\widehat{c}_k$ of signal components at frequencies $k/T$, $k \le N/2$, when $\omega_1 \ne 2\pi k/T$. When using the rectangular window according to (12.2) and (12.4), they contribute as alias effects to all DFT coefficients $\widehat{c}_k$. Thus, they are "spread" onto the oscillations at frequencies $k/T$ (see the next figure). This phenomenon is referred to in signal processing as the *spectral leakage effect*. Additionally, for all $\widehat{c}_k$, there is a constant additive component $(g(0) - g(T-))/(2N)$ if the $T$-periodic extension of $gw_T$ at $T$ has a jump discontinuity.

The spectral leakage effect occurs with modified coefficients $c_k(gw_T)$ even when using other window functions $w_T$ instead of the rectangular window, and it results from the uncertainty principle for the time-duration-bandwidth product of the window $w_T$.

Figure 12.18 shows some absolute weights $g_k = |c_k(gw_T)/A|$, through which the amplitude $A$ of $gw_T$ is distributed onto the Fourier coefficients of frequencies adjacent to $\omega_1 \ne 2\pi k/T$, $k \in \mathbb{Z}$ by the periodicity induced by $w_T$.

We briefly consider an example that illustrates the discussed truncation effects due to the rectangular window using specific data for a given signal $f$.

**Example** For the $4\pi$-periodic function $f(t) = \cos(t/2)$, the segment $fw_T$ with the rectangular window $w_T$ of length $T = \pi$ has the spectrum $c_k = -\dfrac{2 + 8kj}{\pi(16k^2 - 1)}$. The $\pi$-periodic extension with $fw_T(0) = fw_T(\pi)$ has the mean value $c_0 = 2/\pi$ on $[0, \pi[$ and jump discontinuities of height $S_1 = 1$ at $t = k\pi$ $(k \in \mathbb{Z})$. A 3-point DFT on $[0, \pi[$ yields the DFT coefficients $\widehat{c}_0 = (3 + \sqrt{3})/6$ and $\widehat{c}_1 = \overline{\widehat{c}_2} = (3 - \sqrt{3})/12 - j(3 - \sqrt{3})/12$.

We specifically examine $\widehat{c}_0$. The series $\displaystyle\sum_{m=1}^{\infty}(c_{3m} + c_{-3m}) = -\frac{4}{\pi}\sum_{m=1}^{\infty}\frac{1}{144m^2 - 1}$

has, according to (11.2), the limit $S_2 = \widehat{c}_0 - c_0 - S_1/6 = -\dfrac{2}{\pi} + \dfrac{2 + \sqrt{3}}{6}$. Using known equations for the digamma function $\Psi = \Gamma'/\Gamma$, $S_2$ can also be obtained as

$$S_2 = \frac{\Psi(11/12) - \Psi(1/12) - 12}{6\pi} = -\frac{2}{\pi} + \frac{\cot(\pi/12)}{6}.$$

In decimal approximation, for $\widehat{c}_0$, the decomposition now yields

$$\widehat{c}_0 = c_0 + S_1/6 + S_2 \approx 0.636619 + 0.166667 - 0.014611 = 0.788675.$$

## *Selection of Time Windows in the Discrete Fourier Transform*

By choosing an appropriate window function $w_T$, one can achieve a reduction of the distortion effects in the spectrum of the discrete Fourier transform and thereby reduce the error in estimating the spectrum of $f w_T$ or of $f$. The frequency localization is better, according to the considerations on the uncertainty principle in Sect. 12.4, the faster $|\widehat{w_T}(\omega)|$ decreases for increasing $|\omega|$:

1. One usually chooses a window function $w_T \neq 0$ that is as smooth as possible with support in $[0, T]$ and $w_T(0) = w_T(T) = 0$. Then the $T$-periodic extension of $f w_T$ for continuous signals $f$ has no jump discontinuities, and the aliasing effects described by formula (12.2) are reduced if the Fourier coefficients of this extension decrease rapidly (cf. 4.5). One then obtains a better estimate with $\widehat{c}_k T$ than with the rectangular window for the value $\widehat{f}(2\pi k/T)$, which is often sought in applications.
2. One chooses the observation duration $T$ to be as long as possible. The smaller $T$ is, the larger the bandwidth of $w_T$, i.e., the worse the frequency localization.
3. One chooses the number $N$ of samples to be as high as possible. More signal frequencies are then resolved exactly (cf. Eq. (12.3)). For fully observed time-limited signals, "zero padding" improves the approximations for $\widehat{f}$.
4. The leakage effect is less significant, the faster the side lobes of $|\widehat{w_T}|$ decrease compared to the main lobe (cf. the preceding image). Therefore, window functions are often chosen where these side lobes of $|\widehat{w_T}|$ decrease rapidly.

In practice, many different weighting functions $w_T$ are used. The use of special window functions and thus the compromise that must always be made due to the uncertainty principle depend on the aim of the respective application. Criteria besides the decay behavior of $\widehat{w_T}$ and the bandwidth of the window include, for example, its energy concentration in a given frequency band or simple calculation and implementation possibilities in software applications. A detailed comparative discussion of commonly used window functions can be found, for example, in Slepian (1983) or in Harris (1978).

**Example**  To conclude, we consider as an illustrative example the signal

$$f(t) = A_1 \cos(2\pi \nu_1 t) + A_2 \cos(2\pi \nu_2 t)$$

with $A_1 = 1$, $A_2 = 0.03$, $\nu_1 = 10.25\,\text{Hz}$, and $\nu_2 = 12\,\text{Hz}$. Figure 12.19 shows the discrete Fourier transform with the rectangular window $w_T$, $T = 2\,\text{s}$, for $N = 128$. The signal frequency $\nu_2$ cannot be detected. With the same $T$ and $N$, the often-used *Hann window $w_T$,*

**Fig. 12.19**  128-point DFT,
$T = 2$s, rectangular window

**Fig. 12.20**  128-point DFT,
$T = 2$s, Hann window

**Fig. 12.21**  1024-point DFT,
$T = 5$s, Hann window

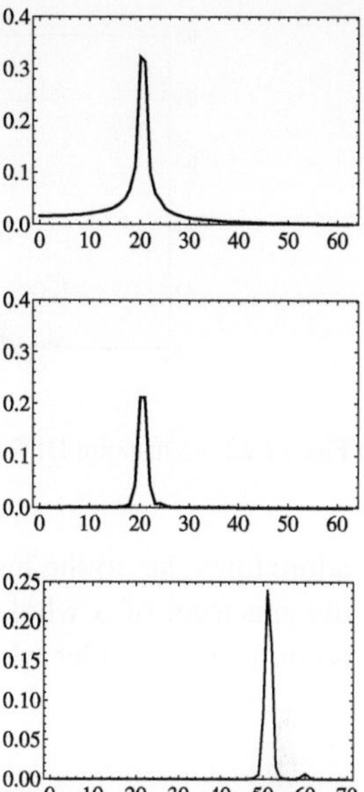

$$w_T(t) = 0.5 - 0.5\cos(2\pi t/T), \;\; 0 \le t \le T$$

is used in the second Fig. 12.20. In the third Fig. 12.21, this window is used again
with $T = 5$ s and $N = 1024$. From the result of this DFT in the third image, the
12 Hz signal frequency can at least be suspected. Displayed is the single-sided DFT
magnitude spectrum.

One notices from the graphs that the height of the "peaks" does not correspond
to the actual (half) amplitude values of the two oscillations. This is a consequence
of the aliasing effect and the attenuations due to the added weighting functions.
Therefore, caution is required, and additional information about the nature of a
problem is needed to reasonably interpret DFT spectra of unknown signals, which
are far more complex in practice than this small example and often affected by
disturbances.

Next is another, still simple example of a DFT spectrum of a real signal,
calculated with the rectangular window. Figure 12.22 shows the 8820-point DFT
of an audio signal of 4 s duration, consisting only of the tones F4, A4, C5, F5 of the
F major chord, played on the piano and enriched with the tones F4, Eb5, F5, played
on the alto saxophone.

The tones have the frequencies in equal temperament: F4=349.23, A4=440,
C5=523.25, Eb5=622.25, and F5=698.46 Hz. With prior knowledge about the
signal, one recognizes the played notes (the second octave requires intonation
adjustment on the alto sax; the author unfortunately intonated about 8 Hz too high
for F5). Likewise, one sees a whole series of resonating overtones (octaves and fifths
upward), but also subharmonic frequencies (F3, C4, Eb4) and a broad spectrum of

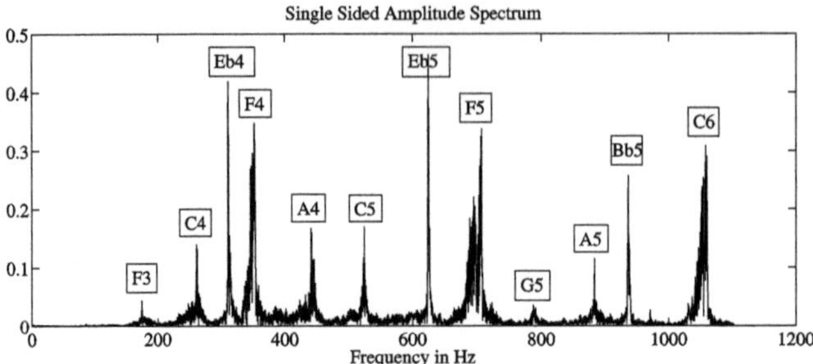


**Fig. 12.22**  8820-point DFT in 4s with piano and saxophone

admixtures due to the instrument characteristics and DFT aliasing effects. Imagine the spectrum of a whole orchestra or a band with drums, guitar, bass, and brass section, and consider what a well-trained ear can distinguish while enjoying music.

## 12.7   Initial Value Problems for Stable LTI Systems

In Sect. 9.2 we discussed causal initial value problems for differential equations of the form $P(D)u = Q(D)f$ with polynomials $P$ and $Q$ for $t \geq 0$ and distributional right-hand sides $f \in \mathcal{D}'_+$. Such problems occur in time-invariant linear transmission systems that have energy storage elements charged at the initial time $t = 0$. The correspondences of the Fourier transform of rational functions on p. 298 show that such problems can also be solved using the Fourier transform if $f$ belongs to the space $\mathcal{S}'_+$, i.e., $f \in \mathcal{S}'$ and $\operatorname{supp}(f) \subset [0, \infty[$, and if further the polynomials $Q$ and $P$ do not have common linear factors and all poles of $Q/P$ have negative real parts.

**Example**  We will once again treat the $RLC$ oscillating circuit from Example 3 on p. 230 as an example. The differential equation

$$\ddot{U}_a + \frac{2}{\sqrt{LC}}\dot{U}_a + \frac{1}{LC}U_a = U_1\dot{\delta} \quad \text{and} \quad U_a(0-) = U_0 \,, \quad \dot{U}_a(0-) = 0$$

described the oscillating circuit at critical damping ($R^2 = 4L/C$) with input voltage $U_e(t) = U_1 s(t)$ and given initial values. The solution is the voltage progression across the inductance. The homogeneous differential equation is asymptotically stable and the right-hand side is tempered. As in Sect. 9.2, we are interested in the solution from the initial time $t = 0$ onward, excluding the past $t < 0$.

The unique solution $T \in \mathcal{S}'$ with $\operatorname{supp}(T) \subset [0, \infty[$ is obtained according to the theorem on p. 223 from the distributional equation

$$\ddot{T} + \frac{2}{\sqrt{LC}}\dot{T} + \frac{1}{LC}T = U_1\dot{\delta} + \frac{2U_0}{\sqrt{LC}}\delta + U_0\dot{\delta}.$$

Since under the given conditions $1/P$ has only poles with negative real parts and hence $1/P(j\omega)$ is a multiplier in $\mathcal{S}'$, Fourier transform of this equation and solving for $\widehat{T}$ give

$$\widehat{T}(\omega) = \frac{1}{P(j\omega)}\left((U_1 + U_0)j\omega + \frac{2U_0}{\sqrt{LC}}\right).$$

The inverse Fourier transform of the partial fractions using the correspondences from p. 298, left as an exercise for the reader, then yields the same solution $T$ with support in $[0, \infty[$ as on p. 231. Here, as there, $s(t)$ denotes the Heaviside step function.

$$T(t) = \left(U_0 + U_1 + \frac{(U_0 - U_1)t}{\sqrt{LC}}\right) e^{-t/\sqrt{LC}} s(t).$$

For the solution method shown, the condition that all poles of $Q/P$ have negative real parts is not sufficient if $Q$ and $P$ have common linear factors with zeros whose real part $r \geq 0$. We consider the following example.

**Example** The causal time-invariant system on $\mathcal{S}'_+$, described by the differential equation

$$P(D)u = \ddot{u} + \dot{u} - 2u = \ddot{f} + 2\dot{f} - 3f = Q(D)f,$$

is stable with vanishing initial values. It has the impulse response

$$h(t) = \delta(t) + e^{-2t} s(t)$$

and the frequency response $\widehat{h}(\omega) = \dfrac{j\omega + 3}{j\omega + 2}$. The general solution of the homogeneous equation $P(D)u = 0$ is $u_H(t) = k_1 e^t + k_2 e^{-2t}$ with $k_1, k_2 \in \mathbb{R}$. The solution $u_H$ is not tempered, and for nonvanishing initial values $c_0, c_1$ the corresponding causal initial value problem (cf. p. 223)

$$\ddot{u} + \dot{u} - 2u = \ddot{f} + 2\dot{f} - 3f + (c_0 + c_1)\delta + c_0\dot{\delta}$$

in $\mathcal{S}'_+$ cannot generally be solved by Fourier transform. The reason is that with vanishing initial values the common linear factors compensate at the zero $z = 1$ of $P$ and $Q$, while with nonvanishing initial values and proceeding as above, a noncausal solution is obtained.

How initial value problems for certain partial differential equations can also be solved using the Fourier transform is shown in the following section.

## 12.8  Initial Value Problems for 3D Wave and Heat Equations

In previous sections, we solved some boundary value problems for the wave and heat equations using Fourier series. As an application of the Fourier transform, we now obtain solutions of initial value problems for wave and heat equations in unbounded space. Because the Fourier transform converts differentiation into a simple algebraic multiplication operation, it transforms the respective partial differential equations into easily solvable ordinary differential equations.

### *The Initial Value Problem for the 3D Homogeneous Wave Equation*

The homogeneous wave equation describes, for example, the propagation of small disturbances in frictionless, compressible fluids or gases in the absence of external forces. In homogeneous unbounded isotropic space, the corresponding initial value problem in Cartesian coordinates is given for $u : \mathbb{R}^3 \times [0, \infty[ \to \mathbb{R}$ by

$$\frac{\partial^2 u}{\partial t^2}(\mathbf{x}, t) = c^2 \Delta_{\mathbf{x}} u(\mathbf{x}, t),$$

$$u(\mathbf{x}, 0) = f(\mathbf{x}), \quad \frac{\partial u}{\partial t}(\mathbf{x}, 0+) = g(\mathbf{x}), \quad u(\mathbf{x}, t) = 0 \text{ for } t < 0. \quad (12.5)$$

Here, $\mathbf{x} \in \mathbb{R}^3$ and $\Delta_{\mathbf{x}}$ is the Laplace operator related to the spatial parameters. If the equation describes, for instance, sound propagation, then $u(\mathbf{x}, t)$ is the pressure deviation at time $t$ from the normal atmospheric pressure at location $\mathbf{x}$. The solution $u$ depends on the initial conditions, which we assume to be in $\mathcal{S}(\mathbb{R}^3)$.

The Fourier transform of the equations with respect to the spatial coordinates gives, by interchanging the Fourier integral with differentiation in $t$,

$$\frac{\partial^2 \widehat{u}}{\partial t^2}(\boldsymbol{\omega}, t) = \frac{\partial^2}{\partial t^2} \int_{\mathbb{R}^3} u(\mathbf{x}, t)\, e^{-j\boldsymbol{\omega}\cdot\mathbf{x}}\, d\lambda^3(\mathbf{x}) = c^2 \int_{\mathbb{R}^3} \Delta_{\mathbf{x}} u(\mathbf{x}, t)\, e^{-j\boldsymbol{\omega}\cdot\mathbf{x}}\, d\lambda^3(\mathbf{x})$$

$$= -c^2 |\boldsymbol{\omega}|^2 \widehat{u}(\boldsymbol{\omega}, t), \quad \widehat{u}(\boldsymbol{\omega}, 0) = \widehat{f}(\boldsymbol{\omega}), \quad \frac{\partial \widehat{u}}{\partial t}(\boldsymbol{\omega}, 0+) = \widehat{g}(\boldsymbol{\omega}). \quad (12.6)$$

For each fixed $\boldsymbol{\omega}$, this is an initial value problem for an ordinary differential equation in $t$. We impose the condition $\widehat{u}(\boldsymbol{\omega}, t) = 0$ for $t < 0$. Then we have with the unit step $s(t)$ the unique solution (cf. Theorem 9.5)

$$\widehat{u}(\boldsymbol{\omega}, t) = \left( \widehat{f}(\boldsymbol{\omega}) \cos(ct|\boldsymbol{\omega}|) + \widehat{g}(\boldsymbol{\omega}) \frac{\sin(ct|\boldsymbol{\omega}|)}{c|\boldsymbol{\omega}|} \right) s(t).$$

Since $\cos(ct|\boldsymbol{\omega}|) = \dfrac{\mathrm{d}}{\mathrm{d}t}\dfrac{\sin(ct|\boldsymbol{\omega}|)}{c|\boldsymbol{\omega}|}$, it is sufficient to determine the inverse Fourier transform of $\dfrac{\sin(ct|\boldsymbol{\omega}|)}{c|\boldsymbol{\omega}|}$. We already did this on p. 316 and obtain for $t > 0$:

$$\frac{1}{4\pi c^2 t}\delta(|\mathbf{x}| - ct) \;\circ\!\!-\!\!\bullet\; \frac{\sin(ct|\boldsymbol{\omega}|)}{c|\boldsymbol{\omega}|}. \tag{12.7}$$

$\delta(|\mathbf{x}| - ct)$ is the singular distribution given by the integral over the spherical surface $|\mathbf{x}| = ct$. By the convolution theorem, we obtain $u$ referred to as a wave. Since the Fourier transform is one to one, this $u$ is a unique solution of (12.5).

**Theorem 12.7** *The initial value problem (12.5) for the wave equation in space has for $\mathbf{x} \in \mathbb{R}^3$ and $t > 0$ the solution*

$$u(\mathbf{x}, t) = \frac{\partial}{\partial t}\frac{1}{4\pi c^2 t}\int\limits_{|\mathbf{y}|=ct} f(\mathbf{x} - \mathbf{y})\,\mathrm{d}o(\mathbf{y}) + \frac{1}{4\pi c^2 t}\int\limits_{|\mathbf{y}|=ct} g(\mathbf{x} - \mathbf{y})\,\mathrm{d}o(\mathbf{y})$$

$$= \frac{\partial}{\partial t}\frac{t}{4\pi}\int\limits_{|\mathbf{n}|=1} f(\mathbf{x} + ct\mathbf{n})\,\mathrm{d}o(\mathbf{n}) + \frac{t}{4\pi}\int\limits_{|\mathbf{n}|=1} g(\mathbf{x} + ct\mathbf{n})\,\mathrm{d}o(\mathbf{n}). \tag{12.8}$$

For the integral transformation see p. 498. The assumptions on the initial conditions can be relaxed. If $f$ is three times and $g$ is twice continuously differentiable, it results in a classical solution $u$ that is twice continuously differentiable. The interchange of differentiations and integrals made in (12.6) is allowed. The solution formula (12.8) shows that initial disturbances spread through space over time and that with initial conditions $f$ and $g$ in $\mathcal{S}(\mathbb{R}^3)$ or those with bounded supports, the solution $u$ decays at least as fast as $1/t$ for increasing times $t$. The solution $u(\mathbf{x}, t)$ at a point $\mathbf{x}$ depends at time $t$ only on the values of the initial conditions on the spherical surface around $\mathbf{x}$ with radius $ct$. We also observe that $\delta(|\mathbf{x}| - ct)/(4\pi c^2 t)$ is a fundamental solution for inhomogeneous problems.

**Propagation of Local Disturbances** *A spatially bounded initial disturbance leads to a time-limited effect in wave propagation in space.*

To explain this, we consider an initial disturbance whose support is a bounded set $U$ with the boundary surface $\partial U$. Then $f(\mathbf{x}) = g(\mathbf{x}) = 0$ outside of $U$. Specifically, imagine a sound that is generated in $U$ at time $t = 0$. Now, let $\mathbf{x}$ be a point outside of $U$ and $d$ and $D$ be the minimum and maximum distances between $\mathbf{x}$ and the points of $U$, respectively. For $t < d/c$, the sphere $S_{ct}(\mathbf{x})$ around $\mathbf{x}$ with radius $ct$ lies outside of $U$, $f$ and $g$ are zero there, and it follows that $u(\mathbf{x}, t) = 0$ for $t < d/c$. For $t = d/c$, $S_{ct}(\mathbf{x})$ touches the set $U$, the wave reaches $\mathbf{x}$: For times $t$ between $d/c$ and $D/c$, $S_{ct}(\mathbf{x})$ and $U$ intersect, at $\mathbf{x}$ effects $u(\mathbf{x}, t) \neq 0$ can occur. For times $t > D/c$, $U$ lies within the sphere $S_{ct}(\mathbf{x})$, and it follows again that $u(\mathbf{x}, t) = 0$, i.e., the disturbance has passed $\mathbf{x}$. Therefore, at $\mathbf{x}$ an effect $u(\mathbf{x}, t) \neq 0$ is noticeable only

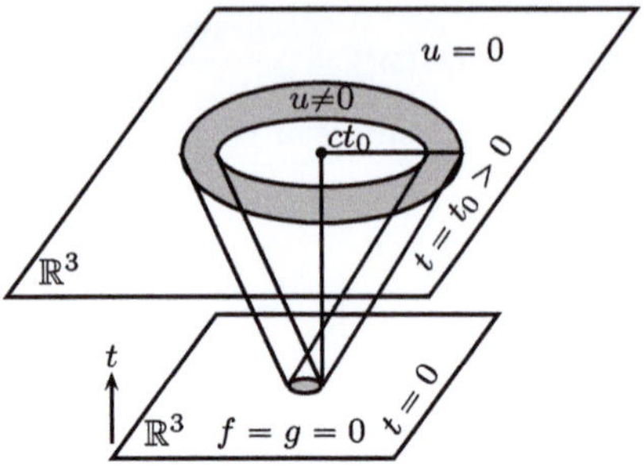

**Fig. 12.23** Illustration of wave propagation in space

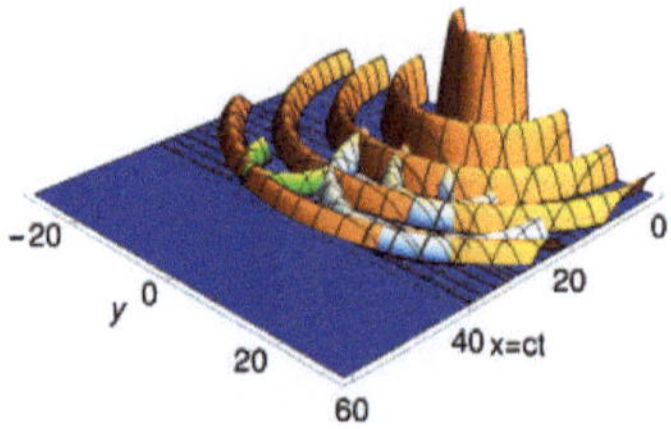

**Fig. 12.24** Illustration of wave propagation and Huygens' principle



in the time interval $d/c \le t \le D/c$. There is a primary and a secondary wavefront. At a given time $t$, the primary wavefront takes the form of a surface that separates those points that have not yet been reached by the wave from the points where the disturbance is acting or has already acted. The points of this surface have a distance $ct$ from the boundary $\partial U$ of $U$ and thus lie on the envelope of all spheres with centers on $\partial U$ and radii $ct$ (*Huygens' principle*). Similarly, the secondary wavefront separates those points that are no longer affected by the disturbance from all others. The constant $c$ is the finite propagation speed of the wavefronts. As illustrations consider Figs. 12.23 and 12.24.

It is therefore possible to transmit signals as sharply bounded waves in three-dimensional space, whose support has a spherical or shell-like shape. This is an extremely significant fact for communication transmission.

The second Fig. 12.24 illustrates a radial wave emitted by a sine source $A\sin(\omega t)s(t)$ at the origin. It shows the spatially decreasing amplitudes (for example air pressure or electric field strength) in the plane $0 < x = ct < 60, |y| \le 30, z = 0$ for values $\ge 0$. The used data are $A = 1, \omega = \pi/2, c = 2$ with their according physical units. Shown is $u(x, y, 0, 20)$ at $t = 20$, when the wave has not yet reached $x > 40$. **Huygens' principle**: The wavefront can be seen as the envelope of the wavefronts by sources of the same type, but starting at different places and times before $t = 20$ with the correspondingly decreased amplitudes (blue, green in the figure for two such waves).

For the solution of the *inhomogeneous wave equation* $\Box u = \dfrac{1}{c^2}\partial_t^2 u - \Delta_{\mathbf{x}} u = f$, see Exercise A7. There we use the Lorentz gauge so that for an electric scalar potential $u$ we have $\Box u = \varrho/\varepsilon_0$ with the D'Alembert operator $\Box$. The result is the so-called *retarded potential*. The distribution $g_1(\mathbf{x}, t) = \delta(|x| - ct)/(4\pi c^2 t)$ in (12.7) is a fundamental solution of Eq. (12.5). Thus, it has to be multiplied by

$c^2$ to obtain a fundamental solution $g$ for the D'Alembert operator $\Box$. With a twice continuously differentiable source $f$, which is zero for $t < 0$, the retarded solution $f * g$ of $\Box u = f$ is also zero for $t < 0$, twice continuously differentiable, and can be written as integral (see also (9.4), p. 235. For the integrals, see p. 497 and Exercise A7)

$$u(\mathbf{x}, t) = f * \frac{\delta(|\mathbf{x}| - ct)}{4\pi t} = c^2 \int_{-\infty}^{t} \frac{(t-s)^2}{4\pi} \int_{|\mathbf{n}|=1} \frac{f(\mathbf{x} + c(t-s)\mathbf{n}, s)}{t-s} \, do\,(\mathbf{n}) \, ds$$

$$= \int_{\mathbb{R}^3} \frac{f\left(\mathbf{y}, t - \dfrac{|\mathbf{x} - \mathbf{y}|}{c}\right)}{4\pi |\mathbf{x} - \mathbf{y}|} \, d\lambda^3(\mathbf{y}) = f * \frac{\delta\left(t - \dfrac{|\mathbf{x}|}{c}\right)}{4\pi |\mathbf{x}|}. \tag{12.9}$$

The equations express mathematically precisely the Huygens principle (Fig. 12.24).

### *The Initial Value Problem for the 2D Homogeneous Wave Equation*

The wave equation in the plane describes problems where the initial conditions $f$ and $g$ depend only on two spatial coordinates. We consider functions $f$ that are three times continuously differentiable and functions $g$ that are twice continuously differentiable, which depend on $\mathbf{x} = (x_1, x_2, x_3)$ only on $x_1$ and $x_2$, interpret the corresponding initial value problem (12.5) in the plane as a spatial problem with the symmetry axis $x_1 = x_2 = 0$, and use its already known solution (12.8):

We calculate the surface integrals in (12.8) by setting $\varphi = f$ or $\varphi = g$ and integrating using spherical coordinates. Then for $\mathbf{x} = (x_1, x_2, x_3)$ and functions $\varphi(\mathbf{x}) = \varphi(x_1, x_2)$ independent of $x_3$ we have

$$\int_{|\mathbf{n}|=1} \varphi(\mathbf{x} + ct\mathbf{n}) \, do\,(\mathbf{n}) = \int_0^{2\pi} \int_0^{\pi} \varphi(x_1 + ct \sin\theta \cos\phi, x_2 + ct \sin\theta \sin\phi) \sin\theta \, d\theta \, d\phi.$$

We integrate over the upper and lower hemisphere surfaces separately, that is, we divide the integration range of the inner integral at $\pi/2$ into two subintervals. Using the substitution $\theta = \arcsin r$, $d\theta = (1 - r^2)^{-1/2} \, dr$, it follows

$$\int_0^{\pi/2} \varphi(x_1 + ct \sin\theta \cos\phi, x_2 + ct \sin\theta \sin\phi) \sin\theta \, d\theta$$

$$= \int_0^1 \frac{\varphi(x_1 + ctr\cos\phi,\, x_2 + ctr\sin\phi)}{\sqrt{1-r^2}}\, r\, dr.$$

For the second subintegral from $\pi/2$ to $\pi$, one obtains the same result. Substituting into the surface integral then gives for $\mathbf{x} = (x_1, x_2, x_3)$

$$\int_{|\mathbf{n}|=1} \varphi(\mathbf{x} + ct\mathbf{n})\, do(\mathbf{n}) = 2 \int_0^{2\pi} \int_0^1 \frac{\varphi(x_1 + ctr\cos\phi,\, x_2 + ctr\sin\phi)}{\sqrt{1-r^2}}\, r\, dr\, d\phi.$$

This is an integral independent of $x_3$ over the unit disk in the $x_1 x_2$-plane. So if $f$ and $g$ for $\mathbf{x} = (x_1, x_2, x_3)$ depend only on the first two coordinates, then the solution of the initial value problem (12.5) is also independent of $x_3$. Because the integrands are independent of the height $x_3$, the surface integrals in (12.8) can be expressed as two identical integrals over the unit disk in the plane $x_3 = 0$. Thus, we obtain the solution of the initial value problem (12.5) for planar problems and those with the symmetry axis $x_1 = x_2 = 0$.

**Theorem 12.8** *The initial value problem (12.5) for the wave equation in the plane has the solution for $\mathbf{x} = (x_1, x_2) \in \mathbb{R}^2$ and $t > 0$*

$$u(\mathbf{x}, t) = \frac{\partial}{\partial t} \frac{t}{2\pi} \int_{|\mathbf{y}|\leq 1} \frac{f(\mathbf{x} + ct\mathbf{y})}{\sqrt{1 - |\mathbf{y}|^2}}\, d\lambda^2(\mathbf{y}) + \frac{t}{2\pi} \int_{|\mathbf{y}|\leq 1} \frac{g(\mathbf{x} + ct\mathbf{y})}{\sqrt{1 - |\mathbf{y}|^2}}\, d\lambda^2(\mathbf{y}).$$

$$(12.10)$$

*For three-dimensional problems with the symmetry axis $x_1 = x_2 = 0$, the solution at time $t > 0$ at a point $\mathbf{z} = (x_1, x_2, x_3) = (\mathbf{x}, x_3)$ with $u(\mathbf{z}, t) = u(\mathbf{x}, t)$ is also given by formula (12.10).*

**Propagation of Local Disturbances** *For the initial value problem of the wave equation in the plane, an initial disturbance bounded in space at any point leads to a timely unlimited effect.*

The solution $u$ at a point $\mathbf{x} = (x_1, x_2)$ in the plane depends at time $t > 0$ on the values of the initial conditions in the entire disk around $\mathbf{x}$ with radius $ct$. Local disturbances in the plane propagate with the speed $c$ and then continuously affect points once reached by the wave. For example, if you place an autumn leaf on a still water surface and throw a stone into the water, the outgoing wave will reach the leaf and continue to spread. The leaf will continue to sway long after it has been passed by the propagation front. This may provide an illustration of the situation, even though water waves are only very roughly described by the two-dimensional wave equation (12.5). The difference to the previously discussed propagation of spatially local disturbances is easily understood when the planar problem is viewed as a three-dimensional problem with the symmetry axis $x_1 = x_2 = 0$. An initial condition with bounded support in the plane corresponds to a disturbance whose support is an infinitely extended cylinder in space. Even at arbitrarily large times, a point $(x_1, x_2)$ in the plane will be reached by disturbances from great heights $x_3$.

## *The Initial Value Problem for the Homogeneous Heat Equation*

The initial value problem for the homogeneous heat equation in homogeneous unbounded isotropic space is given for $u : \mathbb{R}^p \times [0, \infty[ \to \mathbb{R}$ by

$$\frac{\partial u}{\partial t}(\mathbf{x}, t) = k\Delta_{\mathbf{x}} u(\mathbf{x}, t), \quad u(\mathbf{x}, 0) = f(\mathbf{x}). \tag{12.11}$$

Here, $u(\mathbf{x}, t)$ is the absolute temperature at a location $\mathbf{x} \in \mathbb{R}^p$ at time $t \geq 0$. The spatial dimension $p$ is arbitrary. The constant $k > 0$ is the thermal diffusivity. As with the wave equation, we initially assume the initial temperature $f \geq 0$ to be a smooth, rapidly decreasing function and obtain with the Fourier transform of the equations in (12.11) with respect to the spatial coordinates the ordinary differential equation

$$\frac{\partial \widehat{u}}{\partial t}(\omega, t) = -k|\omega|^2 \widehat{u}(\omega, t), \quad \widehat{u}(\omega, 0) = \widehat{f}(\omega).$$

Imposing $\widehat{u} = 0$ for $t < 0$ its unique solution is $\widehat{u}(\omega, t) = \left(\widehat{f}(\omega)\,\mathrm{e}^{-k|\omega|^2 t}\right) s(t)$. With the inverse Fourier transform $K_t(\mathbf{x}) = (4\pi kt)^{-p/2}\,\mathrm{e}^{-|\mathbf{x}|^2/(4kt)}$ of $\mathrm{e}^{-k|\omega|^2 t}$ for $t > 0$, the solution of (12.11) follows by convolution of $f$ with $K_t$.

**Theorem 12.9** *The initial value problem (12.11) for the homogeneous heat equation has for $\mathbf{x} \in \mathbb{R}^p$ and $t > 0$ the solution*

$$u(\mathbf{x}, t) = (4\pi kt)^{-p/2} \int_{\mathbb{R}^p} f(\mathbf{y})\,\mathrm{e}^{-|\mathbf{x}-\mathbf{y}|^2/(4kt)}\,\mathrm{d}\lambda^p(\mathbf{y}). \tag{12.12}$$

Due to the rapid decay of the heat kernel $K_t(\mathbf{x})$, a smooth solution still results for initial conditions $f \in \mathcal{S}'(\mathbb{R}^p)$. It can be proven (cf. for example John (1981)) that for $f \geq 0$, the solution $u$ in (12.12) is the unique nonnegative solution of the heat problem (12.11). If until time $t = 0$ the temperature is zero everywhere and at a location $\mathbf{y}$ at time $t = 0$ the temperature $f(\mathbf{y})$ is produced, then the density function $f(\mathbf{y})K_t(\mathbf{x} - \mathbf{y})$ describes the temperature in $\mathbf{x}$ at time $t$ produced at $\mathbf{y}$. The heat kernel thus shows the equalization of temperature spatially and temporally, the convolution integral (12.12), and the superposition of the influences that act in $\mathbf{x}$ at time $t$ by the initial temperatures of all spatial points $\mathbf{y}$.

Inhomogeneous initial value problems for wave and heat equations in space can also be solved using the Fourier method. For this, one has to determine fundamental solutions of the equations. For the heat equation, this is posed as Exercise A6, for the 3D wave equation as Exercise A7, and for the Schrödinger equation as Exercise A8 at the end of the chapter. For the 3D potential equation, we have already derived a fundamental solution on p. 234, in the 2D case in Exercise A7 of Chap. 9. Fundamental solutions to other problems can be found, for example, in Triebel (1992), Folland (1995), Hörmander (2003), and Ortner and Wagner (2015).

## *Inhomogeneous Boundary Value Problems for the Heat Equation*

In applied numerical mathematics, inhomogeneous heat equations with various initial and boundary conditions for complex 3D regions can approximately be solved with the Finite Element Method as introduced in Sect. 9.5. If the problems are time dependent, one can also use it, when solutions are calculated in progressive discrete time steps. For theory on the (distributional) solutions for such problems, it is referred to the extensive literature about partial differential equations and FEM methods, for example, to Dautray and Lions (1992).

As an example, the temperature distribution in a pump casing is shown, computed with Elmer FEM (see https://research.csc.fi/web/elmer). I have chosen this example because interested readers can easily reproduce it themselves, as the software and the data can be downloaded free of charge from the Elmer homepage. There can be found other examples too. Figure 12.25 shows the used mesh.

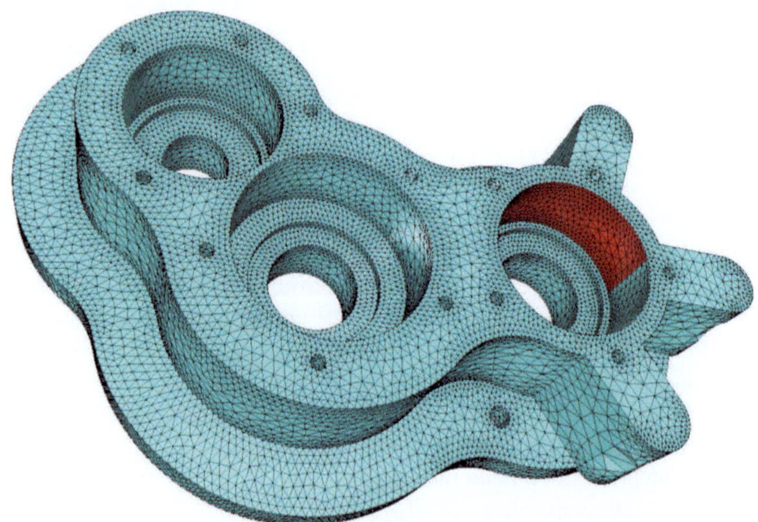

**Fig. 12.25**  3D-FEM model of a pump casing with the used mesh

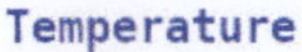

**Fig. 12.26** Temperature in the pump casing generated during operation

Heat is being generated internally in the casing during operation and being cooled at parts of the upper boundary, providing a steady-state temperature. Thus, no initial conditions were needed. Figure 12.26 shows the approximate FEM solution. It is not differentiable and can only be understood as a weak solution in an appropriate Sobolev space (cf. p. 245). In the image, the surface of the model is smoothed. The solver needed 9 s on my old notebook to compute the solution.

The following input data have been used:

1. The model has 181214 volume elements and 58761 edge elements.
2. The material is assumed to be aluminum.
3. During operation the pump is cooled to 293K on parts of the upper surface.
4. The inner heat source is assumed to be constant 0.017 W/kg. The temperature scale is given in degrees Kelvin, i.e., $293K = 19.85°C = 67.73°F$.

## 12.9 Exercises

(**A1**) Assume that a function $f \in L^2(\mathbb{R})$ satisfies the conditions of the sampling theorem of p. 384 with $\hat{f}(\omega) = 0$ for $|\omega| > \omega_c$. Furthermore, let $\alpha > 1$:

(a) Show that for $|\omega| \le \alpha\omega_c$ the following holds:

$$\widehat{f}(\omega) = \frac{\pi}{\alpha\omega_c} \sum_{k=-\infty}^{+\infty} f\left(\frac{k\pi}{\alpha\omega_c}\right) e^{-jk\pi\omega/(\alpha\omega_c)}.$$

(b) Let $\widehat{w_\alpha}$ be a spectral window function whose graph sketched below:

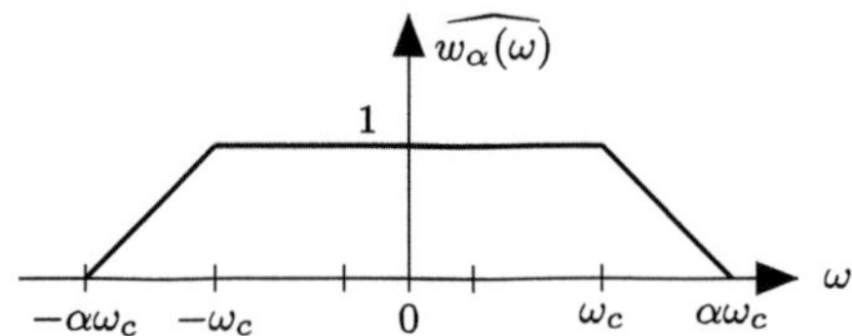


Show that $w_\alpha(t) = \dfrac{\cos(\omega_c t) - \cos(\alpha\omega_c t)}{\pi(\alpha - 1)\omega_c t^2}$ holds.

(c) It holds that $\widehat{f} = \widehat{w_\alpha}\,\widehat{f}$. Using this, show the sampling formula

$$f(t) = \frac{\pi}{\alpha\omega_c} \sum_{k=-\infty}^{+\infty} f\left(\frac{k\pi}{\alpha\omega_c}\right) w_\alpha\left(t - \frac{k\pi}{\alpha\omega_c}\right),$$

i.e., a sampling formula with oversampling, in which the basis functions $w_\alpha(t)$ decay like $1/t^2$ for $|t| \to \infty$ (cf. Remark 2 on p. 386).

(A2) Plot the graph of an approximation for $|\widetilde{f}|$ with the help of a computer algebra system, $\widetilde{f}$ being the windowed Fourier transform of $f(t) = \sin(40\pi t^2)$, $0 \le t \le 10$. Use the Hann window from the example on p. 411.

(A3) What is the effective bandwidth of the Hann window $w_T$ used in Sects. 12.5 and 12.6?

$$w_T(t) = \begin{cases} 0.5 - 0.5\cos(2\pi t/T) & \text{for } 0 \le t \le T \\ 0 & \text{otherwise.} \end{cases}$$

(A4) What is the formula corresponding to Eq. (12.4) from p. 420 for the window function

$$w_T(t) = \begin{cases} 1 - 2|t - T/2|/T & \text{for } 0 \le t \le T \\ 0 & \text{otherwise} \end{cases} \quad ?$$

With what weights does the spectral leakage effect impact a discrete Fourier transform performed with the triangle window?

(A5) Solve the following causal initial value problem for $t \ge 0$ using the Fourier transform as in Sect. 12.7:

$$x^{(3)}(t) + 4x''(t) + 6x'(t) + 4x(t) = \sin(t)s(t) + \delta(t)$$

with the unit step function $s(t)$ under the following initial conditions:

$$x(0-) = 1, \quad x'(0-) = 2 \text{ and } x''(0-) = 1.$$

Determine the right-sided limits $x(0+)$, $x'(0+)$, $x''(0+)$ of the solution.

**(A6)**$^\star$ (a) Determine the fundamental solution $g$ of the heat equation with

$$g(\mathbf{x}, t) = 0 \text{ for } t < 0$$

$$\frac{\partial}{\partial t} g(\mathbf{x}, t) - k\Delta_{\mathbf{x}} g(\mathbf{x}, t) = \delta(\mathbf{x}) \otimes \delta(t).$$

(b) Show that the inhomogeneous heat equation

$$\frac{\partial u}{\partial t}(\mathbf{x}, t) = k\Delta_{\mathbf{x}} u(\mathbf{x}, t) + F(\mathbf{x}, t), \quad u(\mathbf{x}, 0) = 0, \quad F(\mathbf{x}, t) = 0 \text{ if } t < 0$$

for $\mathbf{x} \in \mathbb{R}^3$, $t > 0$, and $F$ such that the convolution integral exists, is solved by

$$u(\mathbf{x}, t) = \int_0^t \int_{\mathbb{R}^3} K_s(\mathbf{y}) F(\mathbf{x} - \mathbf{y}, t - s) \, d\lambda^3(\mathbf{y}) \, ds,$$

$K_s(\mathbf{y}) = (4\pi k s)^{-3/2} e^{-|\mathbf{y}|^2/(4ks)}$ being the kernel of the homogeneous equation. Give a sufficient condition for $F$ such that the above convolution integral exists.

(c) Solve the corresponding problem for $u(\mathbf{x}, 0) = f(\mathbf{x}) \geq 0$, $f \in \mathcal{S}(\mathbb{R}^3)$.

**(A7)**$^\star$ Verify the fundamental solution $g$ with $g(\mathbf{x}, t) = 0$ for $t < 0$ of the inhomogeneous wave equation $\Box u = f$ and the integral transformations in Eq. (12.9). Specify the solution of $\Box u = f$ for a time-varying source $f$ in the origin, e.g., $f(\mathbf{x}, t) = A\sin(\omega t)s(t) \otimes \delta(\mathbf{x})$. Again $s(t)$ denotes the unit step function.

**Remark** The distribution $g$ is called the *retarded fundamental solution*. In electrodynamics an inhomogeneity $F$ for the wave equation can be a time-varying electric charge density or current density. The field $F * g$ is called the *retarded potential*, signifying the fact that the observed field is retarded at an observation position $\mathbf{x}$, i.e., delayed in time by $\Delta t = |\mathbf{x} - \mathbf{x}'|/c$ relative to a source variation at $\mathbf{x}'$ due to the finite speed $c$ of wave propagation. The distribution $\tilde{g}$ defined by $\tilde{g}(\mathbf{x}, t) = g(\mathbf{x}, -t)$ is called the advanced fundamental solution and $F * \tilde{g}$ the *advanced potential* accordingly. For a discussion of advanced potentials, see the literature on electrodynamics, e.g., the Feynman Lectures on Physics, which are available online from Caltech websites.

(**A8**)⋆  The *Schrödinger equation*

$$\frac{\partial \psi}{\partial t}(\mathbf{x}, t) = j\frac{\hbar}{2m}\Delta_{\mathbf{x}}\psi(\mathbf{x}, t)$$

describes in nonrelativistic quantum mechanics the wave function $\psi$ of a free particle of mass $m$ in the absence of external forces. Solve this equation for $t > 0$ with the initial condition $\psi(\mathbf{x}, 0) = \psi_0(\mathbf{x}) \in \mathcal{S}(\mathbb{R}^3)$.

# Chapter 13
# The Malgrange-Ehrenpreis Theorem

**Abstract** In this chapter an elementary proof for the famous Malgrange-Ehrenpreis theorem is given. The proof uses only the product rule for derivatives, the Fourier transform of generalized derivatives, the Taylor formula, and Cramer's rule for solving regular linear systems of equations. The theorem states that every linear partial differential equation with constant coefficients has a fundamental solution. An abstract version and a constructive version of the theorem are proven.

## 13.1  Preliminaries

The aim of this chapter is to prove the famous theorem of Malgrange-Ehrenpreis about the existence of fundamental solutions of linear partial differential equations with constant coefficients. This theorem is, in the author's view, a highlight of Fourier analysis and distribution theory in general and therefore also of the introductory text on this topic presented here to the readers.

For the reading of this section, we first recall the notation for differential operators with multiple variables using multi-indices, which was introduced in Sect. 8.6. A polynomial of degree $\leq m$ in $\xi = (\xi_1, \ldots, \xi_n)$ is noted using multi-indices $k = (k_1, \ldots, k_n) \in \mathbb{N}_0^n$ with $|k| = k_1 + k_2 + \cdots + k_n$ by

$$P(\xi) = P(\xi_1, \ldots, \xi_n) = \sum_{|k| \leq m} a_k \xi_1^{k_1} \xi_2^{k_2} \cdots \xi_n^{k_n} = \sum_{|k| \leq m} a_k \xi^k.$$

If $\mathbb{C}[\xi]$ denotes the set of all polynomials, then $\mathbb{C}[\partial]$ is the set of all linear differential operators $P(\partial) = \sum_{|k| \leq m} a_k \partial^k$ with constant coefficients. Here $m \in \mathbb{N}$ and the coefficients $a_k \in \mathbb{C}$ for $k \in \mathbb{N}_0^n$ are arbitrarily chosen. For an index $k \in \mathbb{N}_0^n$, $\xi^k = \xi_1^{k_1} \cdots \xi_n^{k_n}$ and $\partial^k = \partial_1^{k_1} \cdots \partial_n^{k_n}$ for $\partial = (\partial_1, \ldots, \partial_n)$, where $\partial_i = \partial/\partial x_i$. The distribution $\delta$ is, as before, the Dirac measure at the origin of $\mathbb{R}^n$.



R. Brigola, *Fourier Analysis and Distributions*, Texts in Applied Mathematics 79, https://doi.org/10.1007/978-3-031-81311-5_13

## 13.2  The Malgrange-Ehrenpreis Theorem

In this section, for the sake of clarity, we will omit the previously used vector arrows in the notation of variables from $\mathbb{R}^n$ or $\mathbb{C}^n$ and simply write $x$ instead of $\mathbf{x}$. The fundamental result of Malgrange-Ehrenpreis now reads as follows:

**Theorem 13.1 (Malgrange-Ehrenpreis Theorem, Abstract Version)** *For every nonconstant polynomial $P \in \mathbb{C}[\xi]$, the partial differential equation $P(\partial)T = \delta$ has a solution $T$ in $\mathcal{D}'(\mathbb{R}^n)$.*

This fundamental result was initially proven by Ehrenpreis (1954) and Malgrange (1956). Since then, there have been several other proofs, for example, by Rudin (1991), König (1994), Ortner and Wagner (1994, 1997), and Wagner (2009).

The proof presented here was inspired by the work of Wagner (2009) and was developed by my colleague H. Leinfelder (2012). This proof essentially follows the approach given by Wagner (2009) and is entirely elementary, relying only on the product rule, Fourier transform of generalized derivatives, Taylor's theorem, and Cramer's rule for solving regular linear systems of equations.

Before we proceed to the proof of Theorem 13.1, let us recall some additional notations. The letter $n$ denotes the dimension of the underlying space $\mathbb{R}^n$. As before, $j$ denotes the complex unit with $j^2 = -1$. The product $\xi x$ on $\mathbb{R}^n$ is shorthand for the expression $\xi_1 x_1 + \cdots + \xi_n x_n$. We use the symbol $\cdot$ as a placeholder for variables in $\mathbb{R}^n$, so that $e^{\zeta \cdot}$ and $j \cdot$ denote the functions $(e^{\zeta \cdot})(x) = e^{\zeta x}$ for $\zeta, x \in \mathbb{R}^n$ and $(j \cdot)(\xi) = j\xi$ for $\xi \in \mathbb{R}^n$. A linear differential operator $P(\partial) = \sum_{|k| \leqslant m} a_k \partial^k$ has degree $m$ if its principal part $P_m(\partial) = \sum_{|k|=m} a_k \partial^k$ does not vanish. In the Malgrange-Ehrenpreis theorem, we consider only the interesting case of nonconstant polynomials $P$. Regarding the notation for distributions, we refer to Chaps. 8 and 10. For the Fourier transform on $\mathcal{S}'(\mathbb{R}^n)$, we use the notation $\mathcal{F}$, as in Chap. 10.

For $P \in \mathbb{C}[\xi]$ and $\zeta \in \mathbb{R}^n$, the following *operator identities* hold on $\mathcal{D}'(\mathbb{R}^n)$ and $\mathcal{S}'(\mathbb{R}^n)$:

$$e^{-\zeta \cdot} P(\partial) e^{\zeta \cdot} = P(\partial + \zeta) \tag{13.1}$$

$$\mathcal{F} P(\partial) \mathcal{F}^{-1} = P(j \cdot). \tag{13.2}$$

To understand the first formula, note that by the product rule for derivatives

$$\partial_i (e^{\zeta \cdot} T) = \zeta_i \, e^{\zeta \cdot} T + e^{\zeta \cdot} \partial_i T = e^{\zeta \cdot} (\partial_i + \zeta_i)T.$$

By repeated application, $\partial_i^{k_i}(e^{\zeta \cdot} T) = e^{\zeta \cdot}(\partial_i + \zeta_i)^{k_i} T$; hence

$$\partial^k (e^{\zeta \cdot} T) = e^{\zeta \cdot}(\partial + \zeta)^k T.$$

Multiplying by $a_k$ and summing give for $P(\xi) = \displaystyle\sum_{|k| \leqslant m} a_k \xi^k$

$$P(\partial)\, e^{\zeta \cdot}\, T = e^{\zeta \cdot}\, P(\partial + \zeta)T.$$

Multiplying both sides of this equation by $e^{-\zeta \cdot}$ demonstrates the identity (13.1).

The second formula (13.2) follows immediately from the linearity of the Fourier transform, as stated in relation No. 5 in the table on page 317.

The proof of the Malgrange-Ehrenpreis theorem relies on two auxiliary lemmata, which are also of interest on its own. In the following, we employ three technical tools from analysis and linear algebra. These three statements, denoted by (C1)–(C3), and their proofs are collected in a short appendix at the end of the section.

First, it is noted that for polynomials $P$, $Q \in \mathbb{C}[\xi]$ with every fundamental solution $E$ of $P(\partial)$, i.e., $P(\partial)E = \delta$, there exists a distributional solution $T = Q(\partial)E$ for $P(\partial)T = Q(\partial)\delta$ (cf. p. 135). In a way, a kind of converse to this statement is the following lemma.

**Lemma 13.1** *Let $P$ and $Q$ be polynomials in $\mathbb{C}[\xi]$ of degrees $m \in \mathbb{N}$ and $p \in \mathbb{N}$, respectively, and let $Q_p$ denote the principal part of $Q$. For $\omega \in \mathbb{R}^n$ such that $Q_p(\omega) \neq 0$ and for pairwise distinct $\eta_0, \eta_1, \ldots, \eta_p \in \mathbb{R}^n$ on the line $\mathbb{R}\omega$, suppose that certain $E_{\eta_0}, E_{\eta_1}, \ldots, E_{\eta_p}$ in $\mathcal{D}'(\mathbb{R}^n)$ are solutions of the partial differential equation*

$$P(\partial)E = Q(\partial + \eta)\delta \tag{13.3}$$

*when substituting $\eta = \eta_k$ for $0 \leq k \leq p$ into (13.3).*

*Choosing $\lambda_k$ such that $\eta_k = \lambda_k \omega$, and $a_k = \displaystyle\prod_{q=1, q \neq k}^{p} (\lambda_k - \lambda_q)^{-1}$ $(0 \leqslant k \leqslant p)$,*

*then $E = \dfrac{1}{Q_p(\omega)} \displaystyle\sum_{k=0}^{p} a_k\, E_{\eta_k}$ is a fundamental solution of $P(\partial)$.*

**Proof** We use the second formula from (C1), there with $P = Q$, $\eta = \lambda \omega$, and we obtain the operator formula

$$Q(\partial + \eta) = Q(\partial + \lambda \omega) = \lambda^p Q_p(\omega) + \sum_{q=0}^{p-1} \lambda^q V_q(\partial) \tag{13.4}$$

with $V_q(\partial) = \displaystyle\sum_{|\alpha|=q} \omega^\alpha\, Q^{(\alpha)}(\partial)/\alpha!$. Under the given assumptions, we now substitute $\eta_k = \lambda_k \omega$ for $k = 1, \ldots, p$ into Eq. (13.3). This results, due to the operator equation (13.4) with distributions $T_q = V_q(\partial)\delta$, in the identities

$$P(\partial)E_{\eta_k} = \lambda_k^p\, Q_p(\omega)\delta + \sum_{q=0}^{p-1} \lambda_k^q\, T_q. \tag{13.5}$$

We multiply Eq. (13.5) by the coefficients $a_k$ provided in Lemma 13.1 above and sum over $k = 0, \ldots, p$. This leads to the equation

$$P(\partial)\left(\sum_{k=0}^{p} a_k\, E_{\eta_k}\right) = \left(\sum_{k=0}^{p} a_k \lambda_k^p\right) Q_p(\omega)\,\delta + \sum_{q=0}^{p-1}\left(\sum_{k=0}^{p} a_k\, \lambda_k^q\right) T_q. \quad (13.6)$$

Due to the choice of coefficients $a_k$, according to result (C2) from the appendix (there with $m = p$), we have

$$\sum_{k=0}^{p} a_k\, \lambda_k^p = 1 \quad\text{and}\quad \sum_{k=0}^{p} a_k\, \lambda_k^q = 0 \quad (0 \leqslant q \leqslant p - 1).$$

Thus, from (13.6) it follows that

$$P(\partial)\left(\sum_{k=0}^{p} a_k\, E_{\eta_k}\right) = Q_p(\omega)\delta\,,$$

i.e., $P(\partial)E = \delta$, and thus $E$ is a fundamental solution of $P(\partial)$.     $\square$

According to Lemma 13.1, a proof of the Malgrange-Ehrenpreis theorem is obtainable if the solvability of (13.3) for some nonzero polynomial $Q$ is guaranteed for all $\eta \in \mathbb{R}^n$. Equivalent to this is the solvability of

$$P(\partial)E = Q(\partial - 2\eta)\delta \qquad (13.7)$$

for all $\eta \in \mathbb{R}^n$. To solve (13.7) for any chosen $\eta \in \mathbb{R}^n$, we use for $E$

$$E = e^{\zeta\,\cdot}\, \mathcal{F}^{-1} S \qquad (13.8)$$

with $\zeta \in \mathbb{R}^n$ and an unknown $S \in \mathcal{S}'(\mathbb{R}^n)$. With the relations

$$e^{\zeta\,\cdot}\,\delta = \delta \text{ and } \mathcal{F}\delta = 1,$$

Eq. (13.7) for the sought distribution $S \in \mathcal{S}'(\mathbb{R}^n)$ reads

$$P(\partial)\, e^{\zeta\,\cdot}\, \mathcal{F}^{-1} S = Q(\partial - 2\eta)\, e^{\zeta\,\cdot}\, \mathcal{F}^{-1} 1\,. \qquad (13.9)$$

Multiplying (13.9) on the left first by $e^{-\zeta\,\cdot}$ and applying the Fourier transform $\mathcal{F}$, we get

$$\mathcal{F}\left(e^{-\zeta\,\cdot}\, P(\partial)\, e^{\zeta\,\cdot}\right)\mathcal{F}^{-1} S = \mathcal{F}\left(e^{-\zeta\,\cdot}\, Q(\partial - 2\eta)\, e^{\zeta\,\cdot}\right)\mathcal{F}^{-1} 1\,. \qquad (13.10)$$

Applying formulas (13.1) and then (13.2) from the beginning of this section to Eq. (13.10) yields the relationship

$$P(j\cdot + \zeta)S = Q(j\cdot + \zeta - 2\eta)\,. \qquad (13.11)$$

It is now crucial that $\zeta$ and $Q \in \mathbb{C}[\xi]$ can still be arbitrarily chosen. We set $\zeta = \eta$ and $Q(\xi) = \widetilde{P}(-\xi)$. Here, $\widetilde{P}(\xi)$ denotes the polynomial $\overline{P(\bar\xi)}$, where the complex conjugation applies only to the coefficients of $P$. Equation (13.11) is then satisfied by the regular distribution

$$S = \frac{\widetilde{P}(-j\cdot + \eta)}{P(j\cdot + \eta)} = \frac{\overline{P(j\cdot + \eta)}}{P(j\cdot + \eta)}\,.$$

Note that for $P \neq 0$, the zero set $N(P(j\cdot + \eta))$ is a Lebesgue null set (see (C3)). Since $|S| \leqslant 1$ almost everywhere, $S \in \mathcal{S}'(\mathbb{R}^n)$. Inserted into (13.8), this distribution $S$ thus yields a solution $E$ of (13.7).

Thus, we have shown the following result:

**Lemma 13.2** *Choosing for $P \in \mathbb{C}[\xi]$ the polynomial $Q \in \mathbb{C}[\xi]$ such that $Q(\xi) = \widetilde{P}(-\xi)$, the equation*

$$E = E_\eta = e^{\eta\cdot}\,\mathcal{F}^{-1}\left(\overline{\frac{P(j\cdot + \eta)}{P(j\cdot + \eta)}}\right) \tag{13.12}$$

*is a distributional solution of $P(\partial)E = Q(\partial - 2\eta)\delta$ for each $\eta \in \mathbb{R}^n$.*

**Remark** Equation (13.7) also has a distributional solution $E$ for $Q \in \mathbb{C}[\xi]$ given by $Q(\xi) = \widetilde{P}(-\xi)R(\xi)$ with any $R \in \mathbb{C}[\xi]$.

*Now we obtain the proof of the Malgrange-Ehrenpreis theorem (Theorem 13.1).*

**Proof** For a nonconstant polynomial $P$ in Lemma 13.2, setting $\eta = -\widetilde{\eta}/2$, the partial differential equation $P(\partial)E = Q(\partial + \widetilde{\eta})\delta$ is solvable for all $\widetilde{\eta} \in \mathbb{R}^n$, where $Q(\xi) = \widetilde{P}(-\xi)$. Thus, the assumption (13.3) in Lemma 13.1 is satisfied for $\omega$ with $P_m(\omega) \neq 0$ and appropriately chosen $\eta_k \in \mathbb{R}\omega$. According to Lemma 13.1, $P(\partial)$ has a fundamental solution. $\qquad\square$

With a bit more effort, a constructive, explicit formulation of the Malgrange–Ehrenpreis theorem can be derived from Lemmata 13.1 and 13.2.

**Theorem 13.2 (Malgrange-Ehrenpreis Theorem, Constructive Formulation)** *Let $P \in \mathbb{C}[\xi]$ be a nonconstant polynomial of degree $m$. Let $\omega \in \mathbb{R}^n$ such that $P_m(\omega) \neq 0$, and let $\lambda_0, \lambda_1, \ldots, \lambda_m \in \mathbb{R}$ be pairwise distinct. For $0 \leqslant k \leqslant m$, let*

$$\eta_k = \lambda_k\omega \text{ and } a_k = \prod_{q=0,q\neq k}^{m} (\lambda_k - \lambda_q)^{-1}. \text{ Then}$$

$$E = \frac{1}{P_m(2\omega)}\sum_{k=0}^{m} a_k\,e^{\eta_k\cdot}\,\mathcal{F}^{-1}\left(\overline{\frac{P(j\cdot + \eta_k)}{P(j\cdot + \eta_k)}}\right) \in \mathcal{D}'(\mathbb{R}^n)$$

*is a fundamental solution of $P(\partial)$, i.e., $P(\partial)E = \delta$.*

**Proof** We set $\widetilde{\omega} = -2\omega$, $\widetilde{\eta}_k = \lambda_k\widetilde{\omega}$, and $Q(\xi) = \widetilde{P}(-\xi)$. The distributions $E_{\eta_k}$ determined from (13.12) in Lemma 13.2 then satisfy the equations $P(\partial)E_{\eta_k} = Q(\partial + \widetilde{\eta}_k)\delta$. According to Lemma 13.1, the distribution $E = \dfrac{1}{Q_m(\widetilde{\omega})}\displaystyle\sum_{k=0}^{m} a_k\,E_{\eta_k}$

is a fundamental solution of $P(\partial)$. Note that $Q_m(\widetilde{\omega}) = \widetilde{P}_m(-\widetilde{\omega}) = \widetilde{P}_m(2\omega) = \overline{P_m(2\omega)} \neq 0$.    $\square$

The significance of the Malgrange-Ehrenpreis theorem lies in the solvability of partial differential equations. Here is a typical result.

**Corollary** *If $P \in \mathbb{C}[\xi]$, $E$ is a fundamental solution of $P(\partial)$, and $F$ is a distribution with compact support, then the distribution $T = E * F$ is a solution of $P(\partial)T = F$.*

**Proof** $P(\partial)T = P(\partial)(E * F) = (P(\partial)E) * F = \delta * F = F$.    $\square$

**Concluding Remarks**

(1) Theorem 2 illustrates the great variety of possible fundamental solutions in partial differential equations. However, even though an explicit formula for a fundamental solution $E$ of $P(\partial)$ was given in this theorem, it is only suitable to a limited extend for the specific computation of $E$ due to the difficulty in computing $\mathcal{F}^{-1}(\overline{P(j\cdot + \eta_k)}/P(j\cdot + \eta_k))$. Already in the case of any nonconstant polynomial $P$ in one variable $\xi = t$, it can be guessed that other methods lead more easily (though certainly not trivially) to the goal. While in this case, the right-hand side of formula (13.8) can still be computed, a causal fundamental solution $E$, without any calculation, can be given using the right-sided Laplace transform $\mathcal{L}$, widely used by engineers, as $E = \mathcal{L}^{-1}(1/P)$ or also with the Fourier transform by $E = \mathcal{F}^{-1}(1/P(j\cdot))$, as in this case $1/P(j\cdot)$ always belongs to $\mathcal{S}'$. This fundamental solution is not causal if $P$ has zeros with nonnegative real parts (see p. 337).

For $1/P(j\cdot)$ in $\mathcal{S}'(\mathbb{R}^n)$, a fundamental solution for $P(\partial)$ is $\mathcal{F}^{-1}(1/P(j\cdot))$. Refer to Eqs. (13.11) and (13.8) with $Q = 1$, $\zeta = 0$ for details. If $1/P(j\cdot)$ is integrable, then $\mathcal{F}^{-1}(1/P(j\cdot))$ is the only tempered fundamental solution of $P(\partial)$. It then belongs to $\mathcal{O}'_C(\mathbb{R}^n)$. For details on this and statements about regularity properties of fundamental solutions, refer to Hörmander (2003) and other literature on partial differential equations.

(2) In convolution equations of linear system theory, the question arises as to which convolution kernels $T$ have fundamental solutions $E$, i.e., when $T * E = \delta$ is solvable and thus the question of solvability of convolution equations. Statements about convolution kernels with compact support are also found in the aforementioned work of Wagner (2009) and Hörmander (2003). Fundamental solutions for specific differential operators can be found in Ortner and Wagner (2015).

## Appendix: Technical Resources

**(C1)** *Let $P \in \mathbb{C}[\xi]$ be a polynomial of degree $m \in \mathbb{N}$ and $\lambda \in \mathbb{R}$ and $x, \omega \in \mathbb{R}^n$.*

*Then $P(x + \lambda\omega) = \lambda^m P_m(\omega) + \sum_{k=0}^{m-1} \lambda^k \left( \sum_{|\alpha|=k} \frac{P^{(\alpha)}(x)}{\alpha!} \omega^\alpha \right)$ with $P_m$ being the*

*main part of $P$, $P^{(\alpha)} = \partial^\alpha P \in \mathbb{C}[\xi]$ and $\alpha! = \alpha_1!\alpha_2! \cdots \alpha_n!$. Similarly, in terms of* *operators on $\mathcal{S}'(\mathbb{R}^n)$,*

$$P(\partial + \lambda\omega) = \lambda^m P_m(\omega) + \sum_{k=0}^{m-1} \lambda^k \left( \sum_{|\alpha|=k} \frac{P^{(\alpha)}(\partial)}{\alpha!} \, \omega^\alpha \right).$$

***Proof*** The second formula in (C1) follows from the first formula using the linearity of the Fourier transform, by replacing $x$ with $jx$ and applying formula (13.2) in the form $P(\partial) = \mathcal{F}^{-1} P(j\cdot)\mathcal{F}$. To prove the first formula in (C1), we abbreviate $y = \lambda\omega$. Using the Taylor series, we obtain

$$P(x+y) = \sum_{|\alpha|\leqslant m} \frac{(\partial^\alpha P)(x)}{\alpha!} \, y^\alpha = \sum_{k=0}^{m} \left( \sum_{|\alpha|=k} \frac{(\partial^\alpha P)(x)}{\alpha!} \, \omega^\alpha \right) \lambda^k.$$

Writing $P(x) = \sum_{|\alpha|\leqslant m} a_\alpha x^\alpha$, we see that $(\partial^\alpha P)(x) = a_\alpha \alpha!$ for $|\alpha| = m$; hence

$$\sum_{|\alpha|=m} \frac{(\partial^\alpha P)(x)}{\alpha!} \, \omega^\alpha = \sum_{|\alpha|=m} a_\alpha \, \omega^\alpha = P_m(\omega) \text{ with } P_m \text{ being the main part of } P.$$

For $y = \lambda\omega$, we obtain $P(x + \lambda\omega) = \lambda^m P_m(\omega) + \sum_{k=0}^{m-1} \left( \sum_{|\alpha|=k} \frac{P^{(\alpha)}(x)}{\alpha!} \, \omega^\alpha \right) \lambda^k.$

$\qquad\square$

**(C2)** *If $\lambda_0, \lambda_1, \ldots, \lambda_m \in \mathbb{C}$ are pairwise distinct, then the system of equations*

$$\sum_{k=0}^{m} a_k \lambda_k^q = \delta_{qm} \quad (0 \leqslant q \leqslant m)$$

*for unknowns $a_k$ with $0 \leq k \leq m$ has the unique solution $a_k = \displaystyle\prod_{q=0, q\neq k}^{m} (\lambda_k - \lambda_q)^{-1}.$*
*Here, $\delta_{qm}$ is the Kronecker delta symbol, i.e., $\delta_{qm} = 0$ for $q \neq k$ and $\delta_{mm} = 1$.*

***Proof*** The linear equation system above is associated with a Vandermonde determinant that does not vanish, as $\lambda_0, \lambda_1, \ldots, \lambda_m$ are pairwise distinct. Therefore, the unknowns $a_0, a_1, \ldots, a_m$ can be computed using Cramer's rule. For example, for the interesting case $m > 1$, the unknown $a_0 = \det A / \det B$ is calculated as follows: With

$$A = \begin{pmatrix} 0 & 1 & \ldots & 1 \\ 0 & \lambda_1 & \ldots & \lambda_m \\ \vdots & \vdots & & \vdots \\ 0 & \lambda_1^{m-1} & \ldots & \lambda_m^{m-1} \\ 1 & \lambda_1^m & \ldots & \lambda_m^m \end{pmatrix} \quad \text{and} \quad B = \begin{pmatrix} 1 & 1 & \ldots & 1 \\ \lambda_0 & \lambda_1 & \ldots & \lambda_m \\ \vdots & \vdots & & \vdots \\ \lambda_0^{m-1} & \lambda_1^{m-1} & \ldots & \lambda_m^{m-1} \\ \lambda_0^m & \lambda_1^m & \ldots & \lambda_m^m \end{pmatrix},$$

$$a_0 = (-1)^m \left( \prod_{1 \le k < q \le m} (\lambda_q - \lambda_k) \right) \left( \prod_{0 \le k < q \le m} (\lambda_q - \lambda_k) \right)^{-1} = \prod_{q=1}^{m} (\lambda_0 - \lambda_q)^{-1}.$$

Analogous formulas for $a_k$ $(k = 1, \ldots, m)$ are obtained. $\qquad\square$

**(C3)** *For every $P \in \mathbb{C}[\xi]$, $P \ne 0$, the set of zeros $N(P) = \{\xi \in \mathbb{R}^n \,|\, P(\xi) = 0\}$ is a Lebesgue null set.*

**Proof** We consider only the relevant case $n > 1$ and first show that for every orthogonal matrix $C \in \mathbb{R}^{n \times n}$ with $N(P)$, also $N(P(\cdot C))$ is a Lebesgue null set and vice versa. Due to $|\det(C)| = 1$, this follows directly from the invariance properties of the Lebesgue measure $\lambda$ (see the change of variables theorem on p. 497):

$$\lambda(N(P)) = \lambda(N(P(\cdot C))).$$

Next, we construct an appropriate orthogonal matrix $C \in \mathbb{R}^{n \times n}$ for $P \in \mathbb{C}[\xi]$, such that $\lambda(N(P(\cdot C)))$ can be conveniently computed. With $\zeta = (\zeta_1, \zeta') \in \mathbb{R}^n$ and an initially arbitrary matrix $C \in \mathbb{R}^{n \times n}$, consisting of the rows $c_1, c_2, \ldots, c_n$, we have

$$\zeta\, C = \zeta_1\, c_1 + \zeta_2\, c_2 + \cdots + \zeta_n\, c_n = \zeta_1\, c_1 + \zeta'\, C'.$$

where the matrix $C' \in \mathbb{R}^{(n-1) \times n}$ has the rows $c_2, \ldots, c_n$. We now apply (C1) with $x = \zeta' C'$. $\lambda = \zeta_1$ and $\omega = c_1$, obtaining with the main part $P_m$ of $P$

$$P(\zeta C) = \zeta_1^m\, P_m(c_1) + \sum_{k=0}^{m-1} \zeta_1^k \left( \sum_{|\alpha|=k} \frac{P^{(\alpha)}(\zeta' C')}{\alpha!}\, c_1^\alpha \right).$$

Setting $a = P_m(c_1)$ and $P_k(\zeta') = P_{k,C}(\zeta') = \sum_{|\alpha|=k} P^{(\alpha)}(\zeta' C')/\alpha!$, this can be written in the form $P(\zeta C) = a\, \zeta_1^m + \sum_{k=0}^{m-1} P_k(\zeta')\, \zeta_1^k.$

We now choose $c_1$ so that $a = P_m(c_1) \ne 0$ and $\sum_{k=1}^{n} c_{1,k}^2 = 1$ and add $c_2, \ldots, c_n$ in such a way that $(c_1, c_2, \ldots, c_n)$ forms an orthonormal basis of $\mathbb{R}^n$; thus $C$ represents an orthogonal matrix. The fact that $N(P(\cdot C))$ is a Lebesgue null set can now be seen with the indicator function $1_{N(P(\cdot C))}$ of this set:

$$\lambda(N(P(\cdot C))) = \int_{\mathbb{R}^n} 1_{N(P(\cdot C))}(\zeta)d\zeta = \int_{\mathbb{R}^{n-1}} \left( \int_{\mathbb{R}} 1_{N(P(\cdot C))}(\zeta_1, \zeta')d\zeta_1 \right) d\zeta'$$

$$= \int_{\mathbb{R}^{n-1}} 0\, d\zeta' = 0.$$

Note here that for every $\zeta' \in \mathbb{R}^{n-1}$, the set $\{\zeta_1 \in \mathbb{R} \,|\, P(\zeta C) = 0\}$ has at most $m$ elements and is therefore a null set. $\qquad\square$

# Chapter 14
# Outlook on Further Concepts

**Abstract** Further developments of the ideas in Fourier analysis are shown in this chapter. First, the basic definitions and properties of Hilbert spaces are presented. Examples are given with their inner products, e.g., spaces of square integrable functions and square summable sequences. As orthonormal bases in such spaces, the Haar system, the trigonometric system, the sinc system, Legendre polynomials, Hermite and Laguerre functions, and spherical harmonics are considered. As application of the spherical harmonics and the Laguerre functions, an outline of quantum mechanical results for the nonrelativistic hydrogen atom is given, so that the periodic system of elements and the hybridization of a water molecule can be explained by eigenfunctions of the according Schrödinger operator.

A final section treats continuous and discrete wavelet transforms. A pointwise reconstruction formula for the continuous wavelet transform is proven, and the algorithm of Mallat for multi-resolutions is shown. For explanations the Haar wavelet is used. Further examples are image compressions and image denoising with Daubechies wavelets. Finally, a spectrogram with an STFT and a wavelet scalogram of an audio piece are computed for comparison and graphically shown. A filterbank with a series of bandpass filters is used for the scalogram.

## 14.1 Hilbert Spaces and Special Complete Orthogonal Systems

A fundamental aspect of the applications of Fourier analysis presented in the preceding chapters is the representation of functions or distributions as a superposition of functions of a given function system. The "building blocks" $e^{j2\pi kt/T}$, $k \in \mathbb{Z}$, or $e^{j\omega t}$, $\omega \in \mathbb{R}$, lead to series or integral representations of the form

$$f(t) = \sum_{k=-\infty}^{+\infty} \langle f(t)|e^{j2\pi kt/T}\rangle e^{j2\pi kt/T} \quad \text{or} \quad f(t) = \frac{1}{2\pi} \int_{-\infty}^{+\infty} \widehat{f}(\omega)e^{j\omega t}\, d\omega.$$



R. Brigola, *Fourier Analysis and Distributions*, Texts in Applied Mathematics 79, https://doi.org/10.1007/978-3-031-81311-5_14

For square-integrable functions $f$ from $L^2([0, T])$ or $L^2(\mathbb{R})$, approximations

$$f_N(t) = \sum_{k=-N}^{+N} \langle f(t)|e^{j2\pi kt/T}\rangle e^{j2\pi kt/T} \quad \text{or} \quad f_\Omega(t) = \frac{1}{2\pi} \int_{-\Omega}^{+\Omega} \widehat{f}(\omega)e^{j\omega t}\, d\omega$$

for the represented function $f$ result, which can be understood as orthogonal projections onto subspaces with the inner product in $L^2([0, T])$ or $L^2(\mathbb{R})$ (cf. p. 62 and p. 308). They therefore have the smallest mean square error to $f$ among all approximations in these subspaces. In the first case, $f_N$ is the orthogonal projection onto the subspace of $L^2([0, T])$ generated by the trigonometric polynomials up to degree $N$. In the second case, $f_\Omega$ is the orthogonal projection onto the subspace of $L^2(\mathbb{R})$ consisting of functions bandlimited by $\Omega$ (cf. p. 318). Due to the truncation of high-frequency components, the approximations $f_N$ or $f_\Omega$ can also be understood as smoothings of the respective original function $f$. Technically speaking, they are the result of a lowpass filtering of $f$ (cf. 11.2).

The functions of the trigonometric system appear, for example, as eigenfunctions in differential equations (cf. 1.2 and 7.4) and in time-invariant linear systems $L$ in the steady state (cf. 5.2 and 11.2). An *eigenfunction* of a linear operator $L$ is a function $e \neq 0$ with $Le = \lambda e$. The factor $\lambda$ is the *eigenvalue* of $L$ for the eigenfunction $e$.

## *Schematically*

For $e_k(t) = e^{j2\pi kt/T}$ and a stable, time-invariant linear filter $L$ with rational frequency response $\widehat{h}$, it holds in the steady state (cf. p. 345) for $f \in L^2([0, T])$:

$$f(t) = \sum_{k=-\infty}^{+\infty} c_k e_k(t) \qquad\qquad Lf(t) = \sum_{k=-\infty}^{+\infty} c_k h_k e_k(t)$$

$$\boxed{\quad L \quad}$$

$$c_k = \langle f\,|\,e_k\rangle \qquad\qquad\qquad h_k = \widehat{h}(2\pi k/T)$$

The functions $e_k(t)$ from $L^2([0, T])$ are eigenfunctions of $L$, and the spectral values $h_k$ are the corresponding eigenvalues of the operator $L$ on $L^2([0, T])$. The right side $\sum_{k=-\infty}^{+\infty} h_k \langle f|e_k\rangle e_k$ is called a *spectral representation of* $L$. The coefficients $h_k\langle f|e_k\rangle$ are square summable, and the series converges in $L^2([0, T])$ to the function $Lf$.

**Remark** Consider for rapidly decreasing excitations of the system the operator $L$ on $\mathcal{S}$ (cf. 10.5); then the Fourier integral representation

$$L\varphi(t) = \frac{1}{2\pi} \int\limits_{-\infty}^{+\infty} \widehat{\varphi}(\omega)\widehat{h}(\omega)e^{j\omega t} \, d\omega$$

for $\varphi \in S$ can also be interpreted as expansion of $L\varphi$ in terms of eigenfunctions of $L$. Although the functions $e_\omega(t) = e^{-j\omega t}$ belong neither to $S$ nor to $L^2(\mathbb{R})$, they can be viewed as elements of $S'$ in the following sense as *generalized eigenfunctions*: For all $\varphi \in S$, it holds by the convolution theorem that $\langle L\varphi, e_\omega \rangle = \widehat{L\varphi}(\omega) = \widehat{h}(\omega)\langle \varphi, e_\omega \rangle$. Based on this equation, if one denotes $e_\omega$ as a generalized eigenfunction of $L$ for the eigenvalue $\widehat{h}(\omega)$, then the Fourier inversion formula above can be read as a spectral representation of $L$ with generalized eigenfunctions. Detailed information on generalized eigenfunctions and their applications can be found in Gel'fand et al. (1964).

The spectral representations achieved with Fourier analysis facilitate the understanding of the effect of $L$ (cf. p. 328 and p. 345). The representation of $L$ as a multiplication operator, which is achieved in this way, is analogous to the principal axis transformation of symmetric matrices $A$ known from linear algebra, i.e., to the representation of the linear mapping $A$ as a multiplication operator on the eigenspaces of $A$.

Many technical or physical problems can be mathematically described in linear approximation by equations of the form $Lf = g$, where $L$ is a linear operator on a suitable function space $H$, to which the right side $g$ and the sought function $f$ belong. If $H$ like $L^2([0, T])$ or $L^2(\mathbb{R})$ is a complete vector space with an inner product $\langle h_1 | h_2 \rangle$ for $h_1, h_2 \in H$, and the corresponding norm $\|h\| = \langle h | h \rangle^{1/2}$, $h \in H$, and if $\{e_k | k \in \mathbb{N}_0\}$ is *any orthonormal system* of functions in $H$, i.e., $\langle e_k | e_k \rangle = 1$ and $\langle e_k | e_m \rangle = 0$ for $k \neq m$, then the following statements result in complete analogy to classical Fourier analysis:

1. Every element $g$ from the closed subspace of $H$ generated by the $e_k$ can be expanded in a series with the functions $e_k$, which converges in $H$ to $g$:

$$g = \sum_{k=0}^{\infty} \langle g | e_k \rangle e_k.$$

This series is called the Fourier series of the function $g$ with respect to the orthonormal system $\{e_k | k \in \mathbb{N}_0\}$. This representation of $g$ generalizes the familiar decomposition of vectors in finite-dimensional vector spaces into their components with respect to an orthonormal basis and the Fourier series expansion of functions from $L^2([0, T])$.

2. If a linear operator $L$ on $H$ has the spectral representation

$$Lf = \sum_{k=0}^{\infty} \lambda_k \langle f | e_k \rangle e_k$$

for $f \in H$ with eigenvalues $\lambda_k \neq 0$, then one can solve the equation $Lf = g$ for a right side $g \in H$ with $g = \sum_{k=0}^{\infty} \langle g | e_k \rangle e_k$ by a comparison of coefficients, provided that $\sum_{k=0}^{\infty} \left| \dfrac{\langle g | e_k \rangle}{\lambda_k} \right|^2 < \infty$. A solution $f$ is given by

$$f = \sum_{k=0}^{\infty} \lambda_k^{-1} \langle g | e_k \rangle e_k.$$

The series converges in $H$ to $f$ if and only if its coefficients are square summable (cf. Parseval equation, p. 451).

**Remark** In many specific problems of this type, the eigenvalues $\lambda_1 \geqslant \lambda_2 \geqslant \ldots$, ordered by magnitude, form a sequence converging to zero. The problem $Lf = g$ is then a so-called *ill-posed inverse problem*, because even small data errors, i.e., small deviations in the coefficients $\langle g | e_k \rangle$, are enormously amplified by the divisions by $\lambda_k$ as $\lambda_k \to 0$. Instead of the above "naive" series solution, approximations that are less sensitive to errors are used. Such approximations are obtained through so-called *regularization methods* (cf. also Exercise 13 of Chap. 9). Presentations of regularization methods and solutions of specifically given problems from various fields of application—e.g., computed tomography, image reconstruction, soil exploration in geology, spectroscopy, and much more—can be found in Engl and Groetsch (1987) or Groetsch (1993) and the references therein,

Even these few remarks show that the methods of Fourier analysis can be transferred to problems described by linear operators $L$ with spectral representations in suitable function spaces $H$. The arising eigenfunctions $e_k$ of $L$ do not necessarily have to be trigonometric functions, thus also offering no immediate interpretation through oscillations and frequencies. The "visual intuitiveness" of the familiar frequency concepts from classical Fourier analysis is replaced by the versatility of simple representation and computation possibilities using other "building blocks" $e_k$, which replace the trigonometric functions.

Before we consider examples that demonstrate the usefulness of series expansions in terms of functions $e_k$, we define the concept of the *Hilbert space*. It is fundamental to the theory of linear problems, to which David Hilbert (1862–1943) made groundbreaking contributions.

### Hilbert Spaces

In the following, we consider vector spaces $H$ over the fields of $\mathbb{R}$ of real or $\mathbb{C}$ of complex numbers.

## Definition

1. An inner product on a space $H$ is a mapping $\langle . | . \rangle : H \times H \to \mathbb{C}$, such that for all $x, y, z \in H$ and $\alpha, \beta \in \mathbb{C}$ the following conditions are satisfied:

$$\begin{aligned}
\text{a)} \quad & \langle \alpha x + \beta y | z \rangle = \alpha \langle x | z \rangle + \beta \langle y | z \rangle \\
& \langle x | \alpha y + \beta z \rangle = \overline{\alpha} \langle x | y \rangle + \overline{\beta} \langle x | z \rangle \\
\text{b)} \quad & \langle x | y \rangle = \overline{\langle y | x \rangle} \\
\text{c)} \quad & \langle x | x \rangle > 0 \quad \text{for } x \neq 0
\end{aligned}$$

2. A Hilbert space $H$ is a vector space with an inner product, which is complete with respect to the norm $\|x\| = \langle x | x \rangle^{1/2}$.

The completeness of a Hilbert space $H$ means that every Cauchy sequence of elements $x_n$ in $H$ converges to an element $x \in H$, i.e., $\lim_{n \to \infty} x_n$ exists in $H$ if and only if $\|x_n - x_m\| \to 0$ for $n, m \to \infty$. It is immediately apparent that the finite-dimensional vector spaces $\mathbb{R}^p$ and $\mathbb{C}^p$ with the usual scalar product are examples of Hilbert spaces. Infinite-dimensional examples are function spaces such as $L^2([0, T])$ or $L^2(\mathbb{R})$ (cf. 10.7 and 12.5).

As with the scalar product of vectors in $\mathbb{R}^p$, for elements $x, y \in H$ the *Cauchy-Schwarz inequality* holds

$$|\langle x | y \rangle| \leqslant \|x\| \, \|y\|.$$

Elements $x, y \in H$ are called *orthogonal* if $\langle x | y \rangle = 0$. Every closed subspace $V$ of a Hilbert space $H$ is again a Hilbert space with the inner product restricted to $V$. Its *orthogonal complement* $V^{\perp}$ is the set of all $x \in H$ with $\langle x | v \rangle = 0$ for all $v \in V$. $V^{\perp}$ is again a closed subspace of $H$. Every $x \in H$ has a unique decomposition $x = x_V + x_V^{\perp}$ with $x_V \in V$ and $x_V^{\perp} \in V^{\perp}$. The mapping $P_V : H \to H, x \mapsto x_V$, is called the *orthogonal projection* of $H$ onto $V$.

The central importance of Hilbert spaces lies in the connection of analytical and geometrical concepts such as angle and orthogonality, enabled by the inner product in $H$. As an example, consider the following theorem (cf., for instance, Weidmann (1980)), which is well known in finite-dimensional vector spaces.

**Theorem 14.1** *For every element $x$ of a Hilbert space $H$ and for every closed subspace $V$ of $H$, there is a uniquely determined best approximation $x_V \in V$ for $x$, i.e., $\|x - x_V\| < \|x - y\|$ for all $y \neq x_V$ from $V$. This best approximation for $x$ in $V$ is the orthogonal projection $P_V(x)$ of $x$ onto $V$.*

Applications of the theorem in function spaces can be found in Sects. 9.5 and 12.5. Without proof of the defined properties, some typical examples of Hilbert spaces that occur in applications are given.

### Examples of Some Hilbert Spaces

1. *Square-Integrable Functions.* For a domain $G$ in $\mathbb{R}^p$, the vector space $L^2(G)$ of all square-integrable functions on $G$ is a Hilbert space with the inner product

$$\langle f|g\rangle = \int_G f(\mathbf{x})\overline{g(\mathbf{x})}\,\mathrm{d}\lambda^p(\mathbf{x})\,.$$

2. *Bandlimited Functions.* The vector space $PW_\Omega$ of all functions $f$ bandlimited by $\Omega > 0$ in $L^2(\mathbb{R})$, i.e., all real-valued square-integrable $f$ with $\mathrm{supp}(\widehat{f}) \subset [-\Omega, \Omega]$, is a closed subspace of $L^2(\mathbb{R})$, thus a Hilbert space. It is denoted by $PW_\Omega$ after R. Paley (1907–1933) and N. Wiener (1894–1964), who showed significant results on bandlimited functions (see, e.g., Rudin (1991) or Young (1980)).

3. *Sobolev Spaces.* For a domain $G$ in $\mathbb{R}^2$, an inner product is defined on the space $\mathcal{D}(G)$ of real-valued test functions on $G$ (cf. 8.6) by

$$\langle \varphi_1|\varphi_2\rangle = \int_G (\varphi_1(x, y)\varphi_2(x, y) + \mathrm{grad}\,\varphi_1(x, y) \cdot \mathrm{grad}\,\varphi_2(x, y))\,\mathrm{d}x\,\mathrm{d}y\,.$$

Just as the set $\mathbb{Q}$ of rational numbers can be extended to the complete vector space $\mathbb{R}$, $\mathcal{D}(G)$ can be extended to a vector space, which is complete with respect to the norm associated with the inner product and thus is a Hilbert space. It is denoted by $H_0^1(G)$ and is contained in $L^2(G)$. As abstract as this construction may seem, it is important for applications. $H_0^1(G)$ is the *Sobolev space*, in which in Sect. 9.5 the solution and the approximate solutions with finite elements for the Dirichlet problem $-k\Delta u = f$ on $G$, $u = 0$ on $\partial G$ were found. Its elements are the functions that vanish on the boundary $\partial G$ of $G$ and together with their generalized first derivatives are square-integrable.

More generally, Sobolev spaces can also be introduced for domains in $\mathbb{R}^p$. They are vector spaces of regular distributions, whose derivatives up to a certain order are also regular. As above, derivatives can then be included in the inner product (see also Appendix B, p. 502). Sobolev spaces are of fundamental importance in the study of partial differential equations and in approximation theory. See, for example, Triebel (1986, 1992), Atkinson and Han (2005), and the references mentioned in Sect. 9.5.

4. *Square-Summable Sequences.* The set $l^2(\mathbb{N})$ of all sequences of complex numbers $z_k$, which are square summable, i.e., for which $\sum_{k=1}^{\infty} |z_k|^2 < \infty$ holds, is a Hilbert space with the following inner product:

$$\langle (a_k)_k|(b_k)_k\rangle = \sum_{k=1}^{\infty} a_k\overline{b_k}\,.$$

The vector space operations in $l^2(\mathbb{N})$ are defined componentwise. This Hilbert space was used by W. Heisenberg for the formulation of quantum mechanics (cf., for instance, Messiah (2003)).

## Complete Orthonormal Systems in Hilbert Spaces

**Definition** An orthonormal system of elements $e_k$, $k \in \mathbb{N}_0$, of a Hilbert space $H$ is called complete if for every $x \in H$

$$x = \sum_{k=0}^{\infty} \langle x | e_k \rangle e_k.$$

Thus, with respect to a complete orthonormal system, every $x \in H$ can be represented by its Fourier series with the Fourier coefficients $\langle x | e_k \rangle$. The following theorems can be proven (cf. Triebel 1992 or Weidmann 1980):

**Theorem 14.2** *For an orthonormal system $(e_k)_{k \in \mathbb{N}_0}$ in a Hilbert space $H$, the following are equivalent:*

1. *The system of $e_k$, $k \in \mathbb{N}_0$, is complete.*
2. *For $x \in H$, it holds that $x = 0$ if and only if $\langle x | e_k \rangle = 0$ for all $k \in \mathbb{N}_0$.*
3. *For every $x \in H$, the Parseval equation $\|x\|^2 = \sum_{k=0}^{\infty} |\langle x | e_k \rangle|^2$ holds.*

**Theorem 14.3** *For an orthonormal system $(e_k)_{k \in \mathbb{N}_0}$ in a Hilbert space $H$ and coefficients $c_k$, the series $\sum_{k=0}^{\infty} c_k e_k$ converges in $H$ if and only if the coefficients $c_k$ are square summable, i.e., if*

$$\sum_{k=0}^{\infty} |c_k|^2 < \infty.$$

**Theorem 14.4** *Let $(e_k)_{k \in \mathbb{N}_0}$ be an orthonormal system in a Hilbert space $H$ and $V$ the closed subspace of $H$ generated by it. The orthogonal projection $x_V$ of an element $x \in H$ onto $V$ is then given by*

$$x_V = \sum_{k=0}^{\infty} \langle x | e_k \rangle e_k.$$

Without proofs, we give some examples of how the expansion according to complete orthonormal systems in Hilbert spaces can be applied. Instead of the countable index set $\mathbb{N}_0$ as above, other countable sets can also be chosen as index sets of the orthonormal system.

## *Examples of Specific Complete Orthonormal Systems in Hilbert Spaces*

1. *The Haar system.* The step function $\psi(t) = s(t) - 2s(t - 1/2) + s(t - 1)$ (s(t) the unit step function) is the Haar wavelet, which we will use in the next section. The functions $\psi_{n,k}(t) = 2^{n/2}\psi(2^n t - k)$ with $n, k \in \mathbb{Z}$ build a classical complete orthonormal system in $L^2(\mathbb{R})$. The Haar system has applications, for example, in the numerical solution of linear integral equations.
2. *Trigonometric Functions.* In the Hilbert space $L^2([0, T])$, the functions

$$1, \ \sqrt{2}\cos(2\pi kt/T) \text{ and } \sqrt{2}\sin(2\pi kt/T) \text{ for } k \in \mathbb{N}$$

   with the inner product defined on p. 12 form a complete orthonormal system. For the Hilbert space of complex-valued $L^2$ functions on $[0, T]$, the functions $e^{j2\pi kt/T}$ with $k \in \mathbb{Z}$ form such a system.
3. *Chebyshev Polynomials.* The Chebyshev polynomials $T_n$, for $n \in \mathbb{N}_0$ given by

$$T_n(x) = \cos(n \arccos x),$$

   form a complete orthogonal system in the real Hilbert space $L^2_w([-1, 1])$ with the weight function $w(x) = 1/\sqrt{1 - x^2}$ and the inner product introduced in Sect. 6 on p. 108. There, applications in interpolation, approximation, numerical integration, and the design of low-pass filters were shown.
4. *Hardy Functions.* In the Hilbert space $PW_\Omega$ of $L^2$ functions bandlimited by $\Omega$, equipped with the inner product of $L^2(\mathbb{R})$, the functions

$$e_k(t) = \sqrt{\frac{\Omega}{\pi}} \frac{\sin(\Omega t - k\pi)}{\Omega t - k\pi}$$

   form a complete orthonormal system ($k \in \mathbb{Z}$), named after G. H. Hardy (1877–1947). The series expansion $f = \sum_{k=-\infty}^{+\infty} \langle f | e_k \rangle e_k$ yields the Shannon sampling theorem (cf. 12.1) for functions $f$ from $PW_\Omega$.
5. *Hermite Functions.* The Hermite functions $h_n$, $n \geqslant 0$, on $\mathbb{R}$ are given by (see also p. 296)

$$h_n(x) = (-1)^n (2^n n! \sqrt{\pi})^{-1/2} e^{x^2/2} \frac{d^n}{dx^n} e^{-x^2}.$$

   They form a complete orthonormal system in $L^2(\mathbb{R})$. In quantum mechanics, a one-dimensional harmonic oscillator with a quadratic potential $x^2 \kappa/2$ is described by the Hamilton operator $\mathcal{H}$ with the reduced Planck constant $\hbar = h/(2\pi)$

$$\mathcal{H}f(x) = -\frac{\hbar^2}{2m} f''(x) + \frac{\kappa}{2} x^2 f(x)$$

in $L^2(\mathbb{R})$. It has the functions $h_n(\alpha x)$, $\alpha = (\kappa m)^{1/4}/\hbar^{1/2}$, $n \geqslant 0$, as eigenfunctions. The corresponding eigenvalues are the discrete energy levels $E_n = \hbar\sqrt{\kappa/m}\,(n+1/2)$, which the harmonic oscillator can have in the stationary state. The study of the harmonic oscillator was a starting point for quantum mechanics for the pioneers including E. Schrödinger and is closely linked to Fourier analysis (see p. 407). Still today the harmonic oscillator is a relevant model with profound applicability across physics. Countless systems exhibit nearly harmonic behavior making the model relevant, from the study of atoms and subatomic particles to quantum computing.

6. *Legendre Polynomials and Spherical Harmonics.* For $k \geqslant 0$, the $k$-th Legendre polynomial $P_k$ on $[-1, 1]$ is defined as the function

$$P_k(t) = \frac{1}{2^k k!} \frac{\mathrm{d}^k}{\mathrm{d}t^k}(t^2 - 1)^k.$$

The Legendre polynomials form a complete orthogonal system in the real vector space $L^2([-1, 1])$ with the inner product $\langle f | g \rangle = \int\limits_{-1}^{+1} f(t)g(t)\,\mathrm{d}t$. They allow polynomial approximations with the smallest mean square error for functions in this Hilbert space. Their norm in $L^2([-1, 1])$ is $\|P_k\| = \sqrt{2/(2k + 1)}$.

With the Legendre polynomials, the spherical harmonics $Y_{l,m}$, $l \in \mathbb{N}_0$, $m \in \mathbb{Z}$, $|m| \leqslant l$, are defined. For spherical coordinates $\theta \in [0, \pi]$, $\phi \in [0, 2\pi[$, they are

$$Y_{l,m}(\theta, \phi) = \sqrt{\frac{2l + 1}{4\pi} \frac{(l - |m|)!}{(l + |m|)!}}\, P_{l,|m|}(\cos \theta)\mathrm{e}^{jm\phi}$$

with $P_{l,m}(t) = (-1)^m(1-t^2)^{m/2}\dfrac{\mathrm{d}^m}{\mathrm{d}t^m}P_l(t)$ for $0 \leqslant m \leqslant l$. The spherical harmonics $Y_{l,m}$, $l \in \mathbb{N}_0$, $|m| \leqslant l$, on the unit sphere $S$ of $\mathbb{R}^3$ form a complete orthonormal system in $L^2(S)$. They find applications in solving potential problems with given boundary values on a spherical surface and in series expansions of potentials generated by spatially limited charge distributions. These series are known as *multipole expansions* in physics.

7. *Laguerre Functions.* The $n$-th Laguerre polynomial is

$$L_n(x) = \sum_{k=0}^{n} \binom{n}{k} \frac{(-1)^k}{k!} x^k.$$

We find $L_0(x) = 1$, $L_1(x) = 1 - x$, $L_2(x) = (2 - 4x + x^2)/2 \ldots$

The functions $f_n(x) = \mathrm{e}^{-x/2}L_n(x)$, $x \geqslant 0$, build a complete orthonormal system in the Hilbert space $L^2([0, \infty[)$. The associated Laguerre polynomials $L_n^m$ for $m \in \mathbb{N}_0$ are

$$L_n^m(x) = (-1)^m \frac{\mathrm{d}^m}{\mathrm{d}x^m} L_{n+m}(x) = \sum_{i=0}^{n} \binom{n+m}{n-i} \frac{(-x)^i}{i!}.$$

For example, $L_1^1(x) = 2 - x$, $L_2^1(x) = (6 - 6x + x^2)/2 \ldots$

The Laguerre functions $R_{n,l}$ in $L^2([0, \infty[)$ are then defined for $n \geqslant l + 1$ by

$$R_{n,l}(x) = \mathrm{e}^{-x/2} x^l L_{n-l-1}^{2l+1}(x).$$

They play an important role in quantum mechanics. Quantum physics describes atoms and particles on a microscopic scale, and quantum effects permeate our everyday lives. Quantum effects range from light to atomic clocks in GPS satellites, computers and phones with transistors, semi-conductors, etc. Molecular structures in chemistry can be investigated experimentally using quantum mechanical models. The experimental methods for that are, for example, X-Ray Crystallography, Electron Diffraction, NMR Spectroscopy, et.al. Known molecule structures, their geometric shape, or orbital configurations (see below) form the cornerstone of modern synthetic chemistry with its thousands of compounds and products. Therefore, some fundamental facts are outlined below with the nonrelativistic Schrödinger equation and its Hamilton operator for the hydrogen atom. The Hamilton operators in quantum mechanics are closely connected with the Fourier transform (cf. Sect. 12.4).

**Laguerre Functions and Spherical Harmonics in Quantum Mechanics**
Spherical harmonics and Laguerre functions are fundamental in quantum mechanics. The wave functions for atoms, i.e., the eigenfunctions of the corresponding Hamilton operators, determine—with corresponding eigenvalues and dimensions of the eigenspaces—the probability densities for the localization of electrons by a measurement, e.g., in the *shell model* in chemistry. These densities are also called *Orbitals* and can be approximately illustrated by density plots for regions, where electrons are most likely in. The term " Orbital" is also used to indicate such a region. The Hamilton operator $\mathcal{H}$ for the neutral hydrogen atom (or atoms similar to it, e.g., $He^+$, $Li^{++}$) without outer forces and without spin is the basic model. Atomic hydrogen constitutes about 75% of the baryonic mass of the universe.

The nonrelativistic, time-independent Schrödinger equation in this model is given by

$$\mathcal{H}\psi = \left( -\frac{\hbar^2}{2M} \Delta - \frac{Ze^2}{4\pi\epsilon_0 r} \right) \psi = E\psi,$$

where $\hbar = h/(2\pi)$ is the reduced Planck constant, $M$ the reduced mass of the atom, $Ze$ the nuclear charge, and $Ze^2/(4\pi\epsilon_0 r)$ the according potential (cf. p. 233). The atomic nucleus is assumed to be at rest. The operator $\Delta$ is the Laplace differential operator, and the $\psi$ are functions with spherical variables $(r, \theta, \phi)$. The solutions $\psi$

of the equation are eigenfunctions of $\mathcal{H}$ with according eigenvalues $E$. We retain the term wave function, even if the eigenfunctions in the example are time-independent. The eigenvalues of $\psi$ are the possible energies of the atom in state $\psi$. Without going into details—see, for example, Dirac (1958), Messiah (2003), or Triebel (1992) for that—some summarized facts are as follows:

The equation can be solved in spherical coordinates by separation of the variables, and one obtains for the bounded state of the atom the system of eigenfunctions

$$\psi_{n,l,m}(r, \theta, \phi) = C_{n,l,m} R_{n,l}(2Zr/(na_0))Y_{l,m}(\theta, \phi).$$

Therein, $n$ is the principal quantum number, which describes the shell of an electron, the quantum numbers $l = 0, \ldots, n - 1$ denote subshells with the orbital angular momentum $L = \hbar^2 l(l + 1)$. The magnetic quantum number $m$, $|m| \leqslant l$, describes a specific orbital within a subshell and yields the projection of the orbital angular momentum along a specified axis, if an external magnetic field is applied. The constant $a_0$ is the Bohr radius. All quantum numbers are integers. The constants $C_{n,l,m}$ have to be chosen so that the functions $|\psi_{n,l,m}|^2$ are probability densities. The eigenvalue for an eigenfunction with principal quantum number $n$ is

$$E_n = -\frac{Me^4 Z^2}{8\epsilon_0^2 h^2 n^2} = -R_y \frac{Z^2}{n^2} \qquad (n \in \mathbb{N}).$$

$R_y = Me^4/(8\epsilon_0^2 h^2)$ is the Rydberg energy, and $E_n$ is the energy level of the shell $n$. The according eigenspace has dimension $n^2$ and is spanned by the orthogonal functions $\psi_{n,l,m}$ for $l = 0, \ldots, n - 1$ and $-l \leqslant m \leqslant +l$. The obtained normalized eigenfunctions are products of Laguerre functions for the radial part $R_{n,l}$ and spherical harmonics $Y_{l,m}$ for the angular part. They are explicitly given by

$$\psi_{n,l,m}(r, \theta, \phi) = \left\{ \left( \frac{2Z}{a_0 n} \right)^3 \frac{(n - l - 1)!}{2n(n + l)!} \right\}^{1/2} R_{n,l}\left( \frac{2Zr}{a_0 n} \right) Y_{l,m}(\theta, \phi).$$

We observe that the support of a function $\psi_{n,l,m}$ is unbounded. Thus, if the atom is in such a state, then the probability for the location of an electron in the associated shell for any region in space is positive as a consequence of the probabilistic interpretation of quantum mechanics. We also notice that any other function in the $n^2$-dimensional eigenspace of $\psi_{n,l,m}$ is also an eigenfunction with the same eigenvalue $E_n$. The factor $a_0^{-3/2}$ with the unit length$^{-3/2}$ of $R_{n,l}$ is necessary for a dimensionless probability density when we integrate $|\psi_{n,l,m}|^2$ over space.

In chemistry and spectroscopy, the probability densities with quantum number $l = 0$ are denoted as s-orbitals, for $l = 1$ as p-orbitals, for $l = 2$ as d-orbitals, and $l = 3$ as f-orbitals. The notation $2s^2$, for example, then denotes the s-orbital with two electrons for the quantum numbers $n = 2$ and $l = 0$. The orbital $2p_z$ has quantum numbers $n = 2, l = 1, m = 0$, and $2p_x, 2p_y$ are those for $n = 2, l = 1, m = \pm 1$.

We describe the wave functions for the orbitals $1s$, $2s$, and the three $2p$ orbitals, and see below a density plot of the probability distribution of the $2p_z$ orbital. For $n = 1$ we have $E_1 = -R_y Z^2$ and the wave function with $\rho = Zr/a_0$

$$1s : \qquad \psi_{1,0,0}(r, \theta, \phi) = \frac{1}{\sqrt{\pi}} \left(\frac{Z}{a_0}\right)^{3/2} \mathrm{e}^{-\rho}.$$

For $n = 2$ we have $E_2 = -R_y Z^2/4$ and $n^2 = 4$ wave functions, again with $\rho = Zr/a_0$. In the angular parts of $\psi_{2,1,-1}$ and $\psi_{2,1,1}$, the two real orthogonal functions $\sqrt{2}\sin(\phi)$ and $\sqrt{2}\cos(\phi)$ are chosen instead of the two complex valued functions $\mathrm{e}^{\pm j\phi}$ on $[0, 2\pi]$.

$$2s \;\; : \psi_{2,0,0}(r, \theta, \phi) \;\; = \frac{1}{4\sqrt{2\pi}} \left(\frac{Z}{a_0}\right)^{3/2} (2 - \rho)\mathrm{e}^{-\rho/2}$$

$$2p_z : \psi_{2,1,0}(r, \theta, \phi) \;\; = \frac{1}{4\sqrt{2\pi}} \left(\frac{Z}{a_0}\right)^{3/2} \rho\mathrm{e}^{-\rho/2} \cos(\theta)$$

$$2p_y : \psi_{2,1,-1}(r, \theta, \phi) = \frac{1}{4\sqrt{2\pi}} \left(\frac{Z}{a_0}\right)^{3/2} \rho\mathrm{e}^{-\rho/2} \sin(\theta) \sin(\phi)$$

$$2p_x : \psi_{2,1,1}(r, \theta, \phi) \;\; = \frac{1}{4\sqrt{2\pi}} \left(\frac{Z}{a_0}\right)^{3/2} \rho\mathrm{e}^{-\rho/2} \sin(\theta) \cos(\phi).$$

For hydrogen with $Z = 1$ the energies are approximately $E_1 = -13.6$ eV, and for excited states $n = 2$ and $n = 3$ we have $E_2 = -3.4$ eV, $E_3 = -1.51$ eV, i.e., the energy levels of the orbitals increase with $n$. Illustrations for eigenvalues and orbitals are shown in Figs. 14.1, 14.2 and 14.3.

### Hydrogen Spectrum

When the H-atom absorbs energy, such that the electron changes its position with principle quantum number $n$ to an orbital with a quantum number $k > n$, i.e., with greater energy, it emits electromagnetic waves of the frequencies $\nu_{n,k}$ given in the next formula. As in Fig. 14.1, we can observe the wavelengths of the visible Balmer series in the hydrogen spectrum (J. J. Balmer, 1885).

$$\nu_{n,k} = \frac{1}{h}(E_k - E_n) = R\left(\frac{1}{n^2} - \frac{1}{k^2}\right) \text{ with } 1 \leqslant n < k.$$

$R$ is the Rydberg frequency $R = R_y/h = 3.2898419602500(36) \cdot 10^{15}$ Hz. The measured wavelengths of the Balmer series in vacuum $\lambda = c/\nu$, $c$ the light speed, with $n = 2$ nm are $\lambda_1 = 656.4628$ (red, $k = 3$), $\lambda_2 = 486.2711$ (turquoise, $k = 4$), $\lambda_3 = 434.1687$ (indigo, $k = 5$), $\lambda_4 = 410.2882$ (violet, $k = 6$), $\lambda_5 = 397.1187$ (violet, $k = 7$)... Minor differences in high-precision measurements compared with the above formula are due to the simplifications in the mathematical model. In a constant magnetic field, the spectrum splits and is therefore more diverse with correspondingly more energy levels.

**Fig. 14.1**  Visible spectrum of the Balmer series

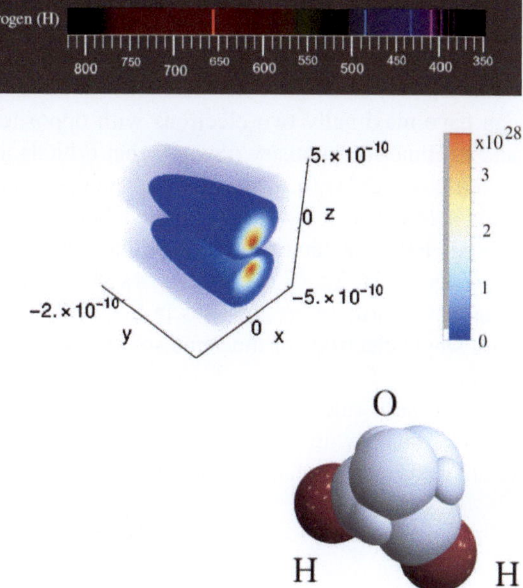


**Fig. 14.2**  Density plot of the $2p_z$ orbital in the half-space $y \leqslant 0$

**Fig. 14.3**  Isosurfaces of tetrahedrally $sp^3$ hybridized $H_2O$ orbitals

For the water molecule $H_2O$ in Fig. 14.3, please see the end of the section. The angles between the orbitals are slightly different from those in a regular tetrahedron.

## *Atoms with Multiple Electrons and the Periodic Table of Elements*

Atoms with multiple electrons have potentials too complicated for an analytic solution of the corresponding eigenvalue problem. Thus, the hydrogen model and its orbitals, as described above, are also used as approximations for atoms of the elements in chemistry with additional rules of W. Pauli (1900–1958) for electrons with spin and the rules of E. Madelung (1881–1972) and F. Hund (1896–1997) for electron configurations. The spin of electrons was introduced in 1925 by W. Pauli in agreement with results as in the Stern-Gerlach experiment in 1922. Pauli considered the spin as an abstract particle property without a concrete physical interpretation. The spin is denoted as a fourth quantum number $m_s$, which has for electrons the two possible values $\pm 1/2$. It is common to denote it also by two different arrows for the two cases, thus by an $\uparrow$ or a $\downarrow$. Further modifications and theoretical extensions came with Dirac's work on relativistic quantum theory (cf. Dirac 1958). We leave all this aside for a good reason (it would require much more advanced mathematics) and briefly see how the periodic table in chemistry can be described to a large extent using the nonrelativistic model so far with additional occupation rules for orbitals with electrons according to W. Pauli, E. Madelung, and F. Hund.

The rules for electron configurations in orbitals say that positions of electrons in the ground state are in the orbitals of the lowest possible energy. The Pauli principle says that two electrons each in an atom must have different sets of the four quantum numbers $(n, l, m, m_s)$. Thus, each suborbital, described by its quantum numbers, can have maximally two electrons with opposite spin. The rules of E. Madelung and F. Hund in chemistry now say that orbitals are filled with electrons in the so-called $n + l$ order ($n, l$ the quantum numbers as above). This means that orbitals with a lower $n + l$ value are filled before those with a higher $n + l$. If two orbitals have the same $n + l$, then that with lower $n$ is occupied. Thus, we have the order $1s, 2s, 2p, 3s, 3p, 4s, 3d, 4p, 5s \ldots$ in the following illustration along the diagonals. Furthermore, orbitals in that order with equal energy are occupied first with single electrons of the same spin before electrons with opposite spin can join them.[1]

With these rules one can find the electron configurations of the atoms in the periodic table, with some exceptions due to electron interactions, shielding, and relativistic effects, which are neglected in this model.

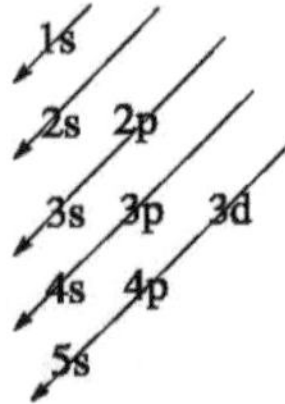


Let us give examples for a few elements with their respective four quantum numbers: Helium [He] has the configuration $(1, 0, 0, \uparrow)$ and $(1, 0, 0, \downarrow)$, i.e., $1s^2$. Lithium [Li] with three electrons has the configuration [He] plus $2s^1$, i.e., [He] plus $(2, 0, 0, \uparrow)$. Oxygen with eight electrons has the configuration $1s^2, 2s^2$, and $2p^4$, which means $2p_z^2, 2p_x^1, 2p_y^1$. Exceptions, for example, are Copper [Cu], Silver [Ag], Gold [Au], and some others.

Chemical reactions like ionization and bonds take place between outer, partially unoccupied orbitals of the atoms involved. Therefore, elements (for low quantum numbers $n$) with the same number of electrons in their outer shell are chemically similar and form a group in a column of the periodic table. Each row in the table ends with an atom having a fully occupied outer shell. In this group with number 18 in the table are the noble gases from He to Og, which are inert from He to Ar. From Kr to Og the elements are able to form compounds under certain conditions.


### *The $H_2O$ Water Molecule*


There exist two main theories of structure and bonding in molecules, the first being Valence Bond Theory, which began in 1927 with the work of W. Heitler

---

[1] It is worth watching the videos on chemistry in the MIT OpenCourseWare project.

and F. London, and the second the Molecule Orbital Theory (MOT), for which R. S. Mulliken was awarded the 1966 Nobel Prize in Chemistry. Both theories are used widely today with molecular modeling computer programs, though for usually different purposes. References, as an incentive to learn more on molecules and their chemical reactions, could be Huheey et al. (1993) for structures in inorganic chemistry, Shaik and Hiberty (2010) for Valence Bond Theory, and Fleming (2009) for Molecular Orbital Theory.

The last point of this section is a short explanation of the above image right of a hybridized oxygen atom in valence bond with two hydrogen atoms to our life spending water molecule $H_2O$.

The concept of hybridized atoms was developed about 1932 by L. Pauling (1901–1994) to describe observed molecule formations like $CH_4$ or $H_2O$ in terms of quantum mechanics. The key lies in the fact that wave functions, i.e., eigenfunctions of $\mathcal{H}$, are replaced in the same eigenspace by suitable linear combinations of them, and thus the corresponding energy is retained. For the water molecule shown above, this means that the four wave functions $2s^2$: $\psi_s = \psi_{2,0,0,\uparrow\downarrow}$, $2p_z^2$: $\psi_{p_z} = \psi_{2,1,0,\uparrow\downarrow}$, $2p_x^1$: $\psi_{p_x} = \psi_{2,1,1,\uparrow}$, $2p_y^1$: $\psi_{p_y} = \psi_{2,1,-1,\uparrow}$ of oxygen are replaced by four linear combinations of them in the same eigenspace for quantum number $n = 2$, which in turn represent an orthonormal basis of the eigenspace. According to Valence Bond Theory, the $H_2O$ molecule consists of the three 1s orbitals of oxygen and hydrogen, and linear combinations of eigenfunctions in four so-called $sp^3$ hybrid orbitals, built of the $2s$ and $2p$ orbitals of oxygen. A hybridization of $H_2O$ is confirmed experimentally using various methods. A new eigenspace basis with four $sp^3$ orthonormal hybrid orbitals is given, for example, by the following mixing:

$$\psi_1 = \frac{1}{2}\left(\psi_s + \psi_{p_x} + \psi_{p_y} + \psi_{p_z}\right), \quad \psi_2 = \frac{1}{2}\left(\psi_s - \psi_{p_x} - \psi_{p_y} + \psi_{p_z}\right),$$

$$\psi_3 = \frac{1}{2}\left(\psi_s + \psi_{p_x} - \psi_{p_y} - \psi_{p_z}\right), \quad \psi_4 = \frac{1}{2}\left(\psi_s - \psi_{p_x} + \psi_{p_y} - \psi_{p_z}\right).$$

This mathematical hybridization results in orbitals having the shape of a regular tetrahedron as shown in the right image above. However, the experimentally confirmed angles between the orbitals are slightly different from those of a regular tetrahedron, usually explained by electron repulsions in the molecule between the bonding and the nonbonding orbitals ($104.5°$ between the two bonding orbitals instead of $109.47°$). This is one of the reasons that other mixings of the $s$ and $p$ orbitals than above have been studied to predict the shape and angles of the $H_2O$ and other molecules. In Molecule Orbital Theory electrons are not assigned to individual orbitals in chemical bonds but can be delocalized and moving under the influence of the atomic nuclei in the whole molecule. Without going further into details beyond the scope of this text and outside of the author's experience, it is referred to the abovementioned books on chemistry covering the present state of the theories.

As a fact, especially organic molecules with bonds of $C$, $H$, and $O$ ($C$ for Carbon) have many hybridized atoms, and knowing them allows chemists—with a whole host of other methods—to predict their geometric shape and to plan new molecules.

**Final Remarks**

 (i) The presented theory is not relativistically invariant. With Dirac's theory one obtains a split hydrogen spectrum, which depends on Sommerfeld's fine structure constant $\alpha$. Physicists then speak of the fine structure of the H-spectrum. For that, readers are referred to the literature on relativistic quantum mechanics and quantum field theory.

 (ii) All linear combinations in an eigenspace are of course also possible states of the atom with the same energy levels. The true states of atoms or molecules and their orbitals must therefore be determined and confirmed by experiments in physics.

(iii) All assertions in quantum mechanics must be interpreted in a probabilistic sense, and we remember that existing orbitals have unbounded support. When an orbital of our mathematical model is not occupied by an electron, then of course the probability to measure it is zero and thus that orbital simply does not exist in physical reality. The outlined model does not make any predictions about the shape of molecules. Many additional studies are therefore necessary, for example, on the influence of electron interactions, the influence of nuclei, the influence of external fields, and relativistic effects, to name but a few. And there are still scientific discussions about the scope and limitations of quantum physics. All this is left to interested readers for further work.

To summarize, this section has given us an idea of the scope of the Hilbert space concept and its applications in very different areas. Orthonormal systems of *special functions* enable solutions of approximation problems, classical differential and integral equations of mathematical physics, signal processing, and a first understanding of quantum mechanics.


## 14.2   Wavelets

A more recent development in applied mathematics is wavelet theory. Wavelet theory is again formulated in the framework of Hilbert spaces and closely linked to Fourier analysis. A wavelet is a special function that, under suitable conditions, allows to construct a complete orthonormal system in a Hilbert space $H$ like $L^2(\mathbb{R})$. Elements $f$ of $H$ can then be expanded into series with this orthonormal system as described in the previous section. Wavelets and associated series expansions have been used since A. Haar (1885–1933) in mathematics, physics, and engineering for various applications, with a unified theory emerging since the late 1970s. An overview of the historical development of the theory is given by Meyer (1993). Detailed presentations can be found, for example, in Daubechies (1992), Chui (1992), Grafakos (2008), Grafakos (2010), Holschneider (1999), Mallat (2009), Meyer (1995), or Blatter (2003).

Wavelets have become particularly well known due to their success in signal compression. These successes are especially based on fast recursive algorithms for computing wavelet decompositions. The objectives of this section are to present some fundamental concepts of wavelet theory in the vector space $L^2(\mathbb{R})$ and to develop the algorithms for fast wavelet transform. Finally, we will show an example of image data compression with wavelets and image reconstruction from the compressed data, an example of image denoising, and a comparison of a spectrogram with an STFT to a corresponding wavelet scalogram of a short piece of music. For this purpose, we initially refer to time-frequency analysis treated in Sect. 12.5.

### *Time-Frequency Analysis with the Windowed Fourier Transform*

The classical Fourier transform uses the periodic functions $e^{j\omega t}$, $\omega \in \mathbb{R}$, for signal analysis. All these functions are derived by scaling from the function $e^{jt}$. The scale parameter is the angular frequency $\omega$. The spectral function $\widehat{f}$ of a signal $f$ shows summarily the components of the oscillations $e^{j\omega t}$ in the signal but does not provide any information about the time-frequency pattern of the analyzed signal $f$ (cf. p. 409). A method for time-frequency analysis is the windowed Fourier transform. For a chosen time window $w \neq 0$, the corresponding windowed Fourier transform

$$\mathcal{G}_w f = \tilde{f} \text{ with } \tilde{f}(\omega, t) = \langle f(s) | w(s-t) e^{j\omega s} \rangle = \int\limits_{-\infty}^{+\infty} f(s) \overline{w(s-t)} e^{-j\omega s} \, ds$$

is a continuously invertible mapping from $L^2(\mathbb{R})$ to $L^2(\mathbb{R}^2)$ (cf. p. 413). In the inversion formula ($\|w\|$ is the $L^2$-norm of $w$)

$$f(t) = \frac{1}{2\pi \|w\|^2} \int\limits_{-\infty}^{+\infty} \int\limits_{-\infty}^{+\infty} \tilde{f}(\omega, s) w(t-s) e^{j\omega t} \, d\omega \, ds$$

from p. 413, the signal $f$ is represented as a superposition of the functions $w_{\omega,s}(t) = w(t-s)e^{j\omega t}$. The "building blocks" $w_{\omega,s}(t)$ of the windowed Fourier transform are again derived from scalings $e^{j\omega t}$ of the "mother function" $w(t)$, so that the functions $w_{\omega,s}$ belong to $L^2(\mathbb{R})$. By translating the window with the parameter $s$, the entire time axis is covered.

Under suitable conditions on the window function $w$ and the sampling points $(k\omega_0, nt_0)$ in the time-frequency domain ($k, n \in \mathbb{Z}$), complete systems of functions $w_{k\omega_0,nt_0}$ can be found in $L^2(\mathbb{R})$, which allow for discrete reconstruction formulas, i.e., series representations for signals $f \in L^2(\mathbb{R})$ with the "building blocks" $w_{k\omega_0,nt_0}$ (cf. p. 415). However, the drawbacks of the windowed Fourier transform are shown in the statements on p. 416. One disadvantage is also the always equal duration and bandwidth of all functions $w_{\omega,s}$ (cf. p. 410). The uncertainties in the time and frequency resolution determined by the window $w$ are always the same for all functions $w_{\omega,s}$ across the entire time-frequency domain. However, since the frequency is proportional to the number of oscillations per unit of time, a precise analysis of short-term high-frequency signal components requires a sharper time resolution, i.e., a shorter duration of the window than for signal components with large wavelengths, which require a wide time window (cf. also 12.8). Therefore, the windowed Fourier transform is poorly suited for studying signals that contain both very high and very low frequencies.

## *Time-Scale Analysis with the Wavelet Transform*

In wavelet theory, instead of using amplitude-modulated oscillations $w_{\omega,s}$ to analyze signals, one uses scalings and translations of a single, with an admissibility condition freely selectable "window function" $\psi$. The function $\psi$ is called a wavelet.

**Definition** A function $\psi \in L^2(\mathbb{R})$ that satisfies the admissibility condition[2]

$$0 < C_\psi = \int_{-\infty}^{+\infty} \frac{|\widehat{\psi}(\omega)|^2}{|\omega|}\, d\omega < \infty$$

is called a wavelet. The wavelet transform $\mathcal{W}_\psi f$ of a function $f \in L^2(\mathbb{R})$ is defined with $\mathbb{R}^* = \mathbb{R} \setminus \{0\}$, $(a, b) \in \mathbb{R}^* \times \mathbb{R}$, and $\psi_{a,b}(t) = |a|^{-1/2} \psi\left(\dfrac{t - b}{a}\right)$ by

$$\mathcal{W}_\psi f(a, b) = \langle f(t) | \psi_{a,b}(t) \rangle = \int_{-\infty}^{+\infty} f(t) |a|^{-1/2} \overline{\psi\left(\dfrac{t - b}{a}\right)}\, dt .$$

**Some Fundamental Properties of the Wavelet Transform**

1. With the admissibility condition for a wavelet $\psi$, it can be shown that the wavelet transform $\mathcal{W}_\psi$ has a continuous inverse mapping. Integrable wavelets $\psi$ fulfill $\widehat{\psi}(0) = \int_{-\infty}^{+\infty} \psi(t)\, dt = 0$, because the admissibility condition implies that the

---

[2] The value of the constant $C_\psi$ depends on the used convention for the Fourier transform.

integrand $|\widehat{\psi}(\omega)|^2/\omega$ must not have a singularity at $\omega = 0$. This can only be true, if $\widehat{\psi}(0) = 0$. With that condition we also recognize that the functions $\psi_{a,b}$ act as bandpass filters. If $\psi \neq 0$ has a bounded support, then the graph of $\psi$ looks like a "small wave" with values above and below the real axis due to this condition, explaining the term "wavelet." It can be shown that the set of functions $\psi_{a,b}$ is dense in $L^2(\mathbb{R})$. From $\|\psi_{a,b}\| = \|\psi\|$ it follows that $\lim\limits_{(a,b)\to(a_0,b_0)} \|\psi_{a,b} - \psi_{a_0,b_0}\| = 0$. Thereby, with the Cauchy-Schwarz inequality one obtains that $W_\psi f$ is continuous as a function in $(a, b)$ (Exercise).

2. If a wavelet $\psi$ has a time duration $D_t(\psi)$ with the time center $t^*$ (see p. 410) and a bandwidth $D_\omega(\psi)$ with the frequency center $\omega^* \neq 0$, then the "analysis building blocks" $\psi_{a,b}$ have the time duration $D_t(\psi_{a,b}) = aD_t(\psi)$ with the time center $b + at^*$ and the bandwidth $D_\omega(\psi_{a,b}) = a^{-1}D_\omega(\psi)$ with the frequency center $a^{-1}\omega^*$. Thus, the scale parameter $a$ changes the time-frequency localization. The time duration and bandwidth of $\psi_{a,b}$ are measures of the uncertainty in time and frequency resolution when analyzing signals with the family $\psi_{a,b}$ (see 12.5).

To illustrate, if one imagines a wavelet $\psi$ as a time window that vanishes outside a bounded interval, then the parameter $b$ shifts the wavelet so that $W_\psi f(a, b) = \langle f | \psi_{a,b} \rangle$ contains local information about $f$ around the time point $b + at^*$. The parameter $a$ controls the width of the window. Short-term high-frequency components of $f$, for example, with the angular frequency $\omega_0$, are localized with high temporal resolution by small parameters $a = \omega^*/\omega_0$, since the time duration $aD_t(\psi_{a,b})$ of the window $\psi_{a,b}$ becomes small with $a$. Low-frequency signal components can be correspondingly localized with large parameter values $a$, wide time windows, and high-frequency resolution. This "*zoom property,*" i.e., adjusting the window width in different frequency ranges, is a decisive advantage of the wavelet transform over the windowed Fourier transform when analyzing the time-frequency pattern of signals.

3. From the Plancherel equation follows the "*frequency representation*" for the wavelet transform of a signal $f$ in $L^2(\mathbb{R})$

$$W_\psi f(a, b) = \frac{|a|^{1/2}}{2\pi} \langle \widehat{f}(\omega) | \widehat{\psi}(a\omega)\mathrm{e}^{-j\omega b} \rangle = \frac{|a|^{1/2}}{2\pi} \int\limits_{-\infty}^{+\infty} \widehat{f}(\omega)\overline{\widehat{\psi}(a\omega)}\mathrm{e}^{j\omega b}\,\mathrm{d}\omega \,.$$

If $\widehat{\psi}$ is concentrated around $\omega^* \neq 0$, then $\widehat{\psi}(a\omega)$ is concentrated around $\omega^*/a$. For a fixed $a$, the wavelet transform $W_\psi f(a, b)$ as a function of $b$ is primarily determined by the frequencies of $f$ around $\omega^*/a$. Associating different frequencies with details of different sizes, the significance of the scale parameter $a$ becomes clear: For a fixed $a$, the wavelet transform $W_\psi f(a, b)$ contains information about details of "size" $\omega^*/a$ that the signal $f$ contains in a time neighborhood of $b + at^*$. Since the detail resolution is determined by the scale parameter $a$, the signal analysis with the wavelet transform is called *time-scale analysis. The wavelet transform $W_\psi f$ corresponds for fixed $a$ to a filtering of $f$ with the frequency response $|a|^{1/2}(2\pi)^{-1}\overline{\widehat{\psi}(a\omega)}$* (see 11.2). The larger $a$ is, the more $f$ is smoothed; the smaller $a$ becomes, the more details of $f$ become visible through the "optics" of this filter. Thus, in applications with software for signal processing, continuous wavelet transforms are realized by *filter banks*, i.e., arrays of bandpass filters based on wavelet frequency responses as described above, and a signal is processed with the filters of the array. The smaller $a > 0$ is, the greater the bandwidth of these filters. By software, the transform is accomplished with signal samples and discrete bandpass filters.

## *The Haar Wavelet*

The *Haar wavelet* is the function

$$\psi(t) = \begin{cases} 1 & \text{for} & 0 \leqslant t < 1/2 \\ -1 & \text{for} & 1/2 \leqslant t < 1 \\ 0 & \text{otherwise.} \end{cases}$$

The function system $\psi_{n,k}$, named after A. Haar (1885–1933), $n$ and $k$ from $\mathbb{Z}$,

$$\psi_{n,k}(t) = 2^{n/2}\psi(2^n t - k),$$

is a classical complete orthonormal system in the real vector space $L^2(\mathbb{R})$, generated by scalings and translations of $\psi$. The proof of this statement can be found, for example, in Daubechies (1992) or can be provided by the readers themselves. The Fourier transform of $\psi$ is

$$\widehat{\psi}(\omega) = \frac{\sin(\omega/4)^2}{\omega/4}e^{-j(\omega-\pi)/2}.$$

The constant $C_\psi$ (see p. 462) is $C_\psi = 2\ln(2)$. Figures 14.4, 14.5, and 14.6 show the Haar wavelet $\psi$, the function $\psi_{2,12}(t) = 2\psi(4(t-3))$, the function $f$

$$f(t) = \begin{cases} t & \text{for} & 0 \leqslant t < 1 \\ 2 - t & \text{for} & 1 \leqslant t < 2 \\ 1 & \text{for} & 3 \leqslant t < 4 \\ 0 & \text{otherwise,} \end{cases}$$

and the wavelet transform of $f$ for $0 < a < 2$ and $0 < b < 4$ (right). The wavelet transform has larger magnitudes here due to the coarse structure of $f$ for higher values of $a$ and vanishes for $a \to 0$, as $f$ does not exhibit "short-term fine structures." For $a = 2$ we observe a smoothed version of $f$. More precise interpretations of wavelet transforms, for example, concerning relationships between local properties of functions $f$ and growth properties of their wavelet

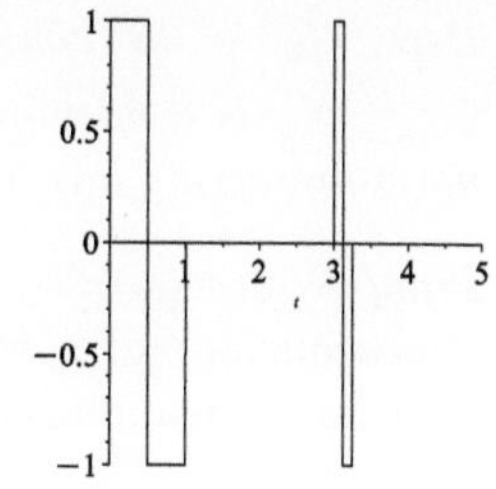

**Fig. 14.4**  Haar wavelets

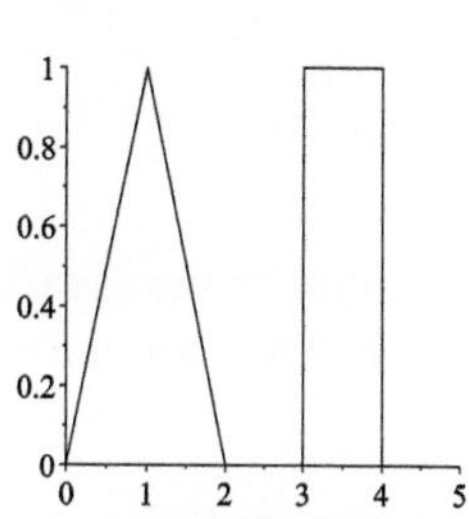

**Fig. 14.5**  A function $f$

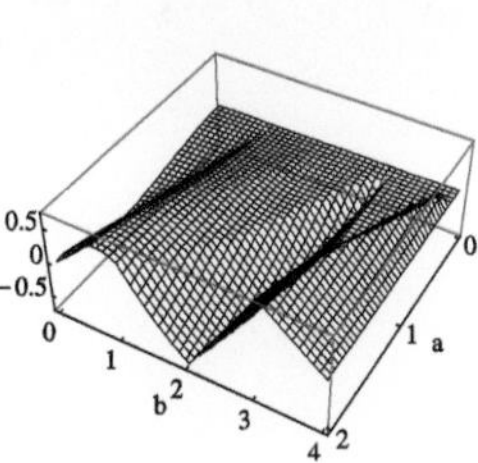

**Fig. 14.6**  Wavelet transform
of $f$

transforms, can be found in the aforementioned additional literature. Readers are
encouraged to experience and interpret wavelet transforms of own examples with
mathematics software.

## Pointwise Inversion Formula for the Wavelet Transform

In a similar way to the windowed Fourier transform, an inversion formula can be
proven for the wavelet transform. In the following, we denote $\mathbb{R}^* = \mathbb{R} \setminus \{0\}$ and
$\mathbb{R}_-^2 = \mathbb{R}^* \times \mathbb{R}$. The vector space $W = L^2(\mathbb{R}_-^2, |a|^{-2}\,\mathrm{d}a\,\mathrm{d}b))$ is a Hilbert space with
the inner product

$$\langle u | v \rangle_W = \int_{\mathbb{R}_-^2} u(a, b)\overline{v(a, b)}|a|^{-2}\,\mathrm{d}a\,\mathrm{d}b\,.$$

**Theorem 14.5**  *For $f$ and $g$ in $L^2(R)$ and an admissible wavelet $\psi$, it holds the
orthogonality relation*

$$\langle \mathcal{W}_\psi f | \mathcal{W}_\psi g \rangle_W = C_\psi \langle f | g \rangle.$$

*Here, $C_\psi$ is the constant in the admissibility condition for $\psi$. In particular, $C_\psi^{-1/2} \mathcal{W}_\psi$ is a norm-preserving linear mapping from $L^2(\mathbb{R})$ to W, which has a norm-preserving inverse mapping.*

**Proof** With $F(\omega) = \widehat{f}(\omega)\overline{\widehat{\psi}(a\omega)}$, $G(\omega) = \widehat{g}(\omega)\overline{\widehat{\psi}(a\omega)}$, $a \neq 0$, and the "frequency representation" of a wavelet transformed signal (see p. 463), we have with the Plancherel equation and the Fourier transform with a scaled variable

$$\frac{2\pi|a|}{a^2} \int\limits_{-\infty}^{+\infty} \mathcal{W}_\psi f(a,b)\overline{\mathcal{W}_\psi g(a,b)}\, \mathrm{d}b = \frac{1}{2\pi} \int\limits_{\mathbb{R}} \widehat{\widehat{G}}(b)\overline{\widehat{\widehat{F}}(b)}\, \mathrm{d}b = \int\limits_{\mathbb{R}} \overline{G}(\omega)F(\omega)\, \mathrm{d}\omega\,.$$

Again with the Plancherel equation and interchanging the order of integrations by the Fubini-Tonelli theorem (cf. Appendix B), we obtain (with substitution $a\omega = y$)

$$\langle\, \mathcal{W}_\psi f | \mathcal{W}_\psi g \rangle_W = \int\limits_{\mathbb{R}^*} \left\{ \int\limits_{\mathbb{R}} \mathcal{W}_\psi f(a,b)\overline{\mathcal{W}_\psi g(a,b)}\, \mathrm{d}b \right\} |a|^{-2}\, \mathrm{d}a$$

$$= \frac{1}{2\pi} \int\limits_{\mathbb{R}} \left\{ \widehat{f}(\omega)\overline{\widehat{g}(\omega)} \int\limits_{\mathbb{R}^*} \frac{|\widehat{\psi}(a\omega)|^2}{|a|}\, \mathrm{d}a \right\} \mathrm{d}\omega = C_\psi \frac{1}{2\pi} \int\limits_{\mathbb{R}} \widehat{f}(\omega)\overline{\widehat{g}(\omega)}\, \mathrm{d}\omega = C_\psi \langle f | g \rangle.$$

$$\square$$

Here, $C_\psi$ is the constant in the admissibility condition for $\psi$. In particular, $C_\psi^{-1/2} \mathcal{W}_\psi$ is a norm-preserving linear mapping from $L^2(\mathbb{R})$ to W, which has a norm-preserving inverse mapping. Because $C_\psi^{-1/2} \|\mathcal{W}_\psi f\|_W = \|f\|$, $C_\psi^{-1/2}|\mathcal{W}_\psi f|^2$ can be interpreted as the *energy density* of the signal $f$ in the $(a, b)$ plane. $\|f\|$ is the norm of $f$ in $L^2(\mathbb{R})$.

Now, the functions $h_\alpha(t) = \dfrac{1}{2\sqrt{\pi\alpha}}e^{-t^2/(4\alpha)}$ build an approximate identity in $L^1(\mathbb{R})$ (cf. Exercise A14 in Chap. 9) and $f * h_\alpha(t) \to f(t)$ for continuous $f \in L^1(\mathbb{R})$ and $\alpha \to 0+$. Thus, with $h_{\alpha,t}(s) = h_\alpha(s - t)$ for integrable continuous wavelets $\psi$ and integrable continuous signals $f$,

$$C_\psi^{-1} \langle \mathcal{W}_\psi f | \mathcal{W}_\psi h_{\alpha,t} \rangle_W = \langle f | h_{\alpha,t} \rangle \xrightarrow[\alpha \to 0+]{} f(t),$$

and on the other hand, $\overline{\mathcal{W}_\psi h_{\alpha,t}}(a,b) \to |a|^{-1/2}\psi\left(\dfrac{t-b}{a}\right)$ holds for $\alpha \to 0+$. In summary, we have proven the following theorem about the inversion of the wavelet transform. Other inversion theorems can be found in Daubechies (1992).

**Theorem 14.6 (Pointwise Inversion Formula)** *For an admissible wavelet $\psi$, the wavelet transform $\mathcal{W}_\psi$ is a continuous, continuously invertible mapping from $L^2(\mathbb{R})$ to*

$$W = L^2(\mathbb{R}^2_-, a^{-2}\, da\, db) \text{ with } \|\mathcal{W}_\psi f\|_W = C_\psi^{1/2}\|f\|.$$

*For continuous signals $f \in L^1(\mathbb{R}) \cap L^2(\mathbb{R})$ and continuous $\psi \in L^1(\mathbb{R}) \cap L^2(\mathbb{R})$, it holds at every $t$*

$$f(t) = C_\psi^{-1} \int\limits_{-\infty}^{+\infty} \int\limits_{-\infty}^{+\infty} \mathcal{W}_\psi f(a,b)|a|^{-1/2}\psi\left(\frac{t-b}{a}\right)\frac{da\, db}{a^2}.$$

Analogous to the inversion formulas of the classical Fourier transform and the windowed Fourier transform, here $f$ is represented with the wavelet transform as a superposition of the "building blocks" $\psi_{a,b}$ with the "amplitudes" $\langle f|\psi_{a,b}\rangle = \mathcal{W}_\psi f(a,b)$. Mathematical applications of the continuous wavelet transform include, for example, the study of local and global regularity properties of functions (see, e.g., Holschneider 1999). Over the years, many different wavelet families have been developed with different properties (disappearing moments, compact support, smoothness properties, etc.) for various purposes. Practical applications of the continuous wavelet transform range from music software (e.g., pitch-shifting and time-stretching) to real-time signal analysis in the military sector.

## Discrete Wavelet Transform and Multiscale Analysis

In numerical applications, values $\mathcal{W}_\psi f(a,b)$ of a wavelet transformed signal $f$ can usually only be calculated for parameter values $(a,b)$ from a countable, discrete set $S$. The question then arises under what conditions on the wavelet $\psi$ and on the set $S$ of sample points a stable reconstruction of the signal $f$ from the values $\mathcal{W}_\psi f(a,b) = \langle f|\psi_{a,b}\rangle$, $(a,b) \in S$, of its wavelet transform is possible.

Stability means that small perturbations in the coefficients $\langle f|\psi_{a,b}\rangle$ also result in only small deviations from $f$ in the reconstruction. This is exactly the case when the mapping $f \mapsto (\langle f|\psi_{a,b}\rangle)_{(a,b)\in S}$ from $L^2(\mathbb{R})$ to $l^2(S)$ (see p. 450) is invertible and continuous in both directions, i.e., if there are constants $A > 0$ and $B > 0$ such that for all $f \in L^2(\mathbb{R})$ the following inequality holds:

$$A\|f\|^2 \leqslant \sum_{(a,b)\in S} |\langle f|\psi_{a,b}\rangle|^2 \leqslant B\|f\|^2.$$

Families $(\psi_{a,b})_{(a,b)\in S}$, which span $L^2(\mathbb{R})$ and fulfill these inequalities are so-called *frames* and build not necessarily an orthogonal system. They can be overcomplete but offer many applications, for example, in irregular sampling. If interested, readers can find theory and applications of frames in Strohmer (2000) or O. Christensen (2000). If the family $\psi_{a,b}$ with $(a,b)$ in $S$ is a complete orthonormal system in $L^2(\mathbb{R})$, stability is always present due to the Plancherel equation. Here, we restrict our considerations in the following to such orthonormal systems and leave aside more general systems like Riesz bases or frames. Series expansions based on complete orthonormal systems thus allow for stable discrete reconstruction formulas.

A first example of a discrete reconstruction results with the Haar wavelet $\psi$ (see p. 464). The functions

$$\psi_{n,k}(t) = 2^{n/2}\psi(2^n t - k) \qquad (n,\ k \in \mathbb{Z})$$

form a complete orthonormal system in $L^2(\mathbb{R})$. *Every signal $f \in L^2(\mathbb{R})$ therefore has the discrete wavelet decomposition*

$$f = \sum_{n=-\infty}^{+\infty} \sum_{k=-\infty}^{+\infty} \langle f | \psi_{n,k} \rangle \psi_{n,k}. \tag{14.1}$$

Series representations of the same form are obtained with other wavelets $\psi$, if suitably scaled translations $\psi_{n,k}$ of $\psi$ with $n,\ k \in \mathbb{Z}$ again form a complete orthonormal system in $L^2(\mathbb{R})$.

For the direct application of the series expansion, the integrals $\langle f | \psi_{n,k} \rangle$ would have to be calculated. However, in the years 1986–1989, S. Mallat and Y. Meyer developed a new method that allows for discrete wavelet expansions to be carried out completely recursively. This method, ideal for calculations, is the *multiscale analysis*. The algorithms of the fast wavelet transform that arise from the multiscale analysis, and their applications in signal processing, for example, in data compression, have made wavelet analysis a significant mathematical tool in technical disciplines within a short period of time.

We now describe the multiscale analysis, also known as multiresolution analysis, with the Haar wavelet and then present results that show that the exemplified algorithms also apply to other suitably constructed wavelets.

## *Multiscale Analysis with the Haar Wavelet*

Understanding multiscale analysis and the resulting fast wavelet algorithms requires some preliminary considerations about the structure of the vector spaces generated by the wavelets and about the structure of the wavelet itself. For now, all considerations refer to the Haar wavelet.

### The Vector Spaces Generated by the Haar Wavelet

The properties of the subspaces in $L^2(\mathbb{R})$ generated by the wavelets $\psi_{n,k}$ are crucial for discrete wavelet analysis. For each fixed $n \in \mathbb{Z}$, the functions $\psi_{n,k}$, $k \in \mathbb{Z}$, of the Haar system form a complete orthonormal system of a closed subspace $W_n$ of $L^2(\mathbb{R})$. Any two subspaces $W_n$ and $W_m$ are orthogonal to each other for $n \neq m$, and all vector spaces $W_n$, $n \in \mathbb{Z}$, together span $L^2(\mathbb{R})$ (cf. (14.1)). It is said that $L^2(\mathbb{R})$ is the *direct sum* of the $W_n$:

$$L^2(\mathbb{R}) = \bigoplus_{n=-\infty}^{+\infty} W_n,$$

i.e., every $f \in L^2(\mathbb{R})$ has exactly one representation $f = \sum_{n=-\infty}^{+\infty} f_n$ with $f_n \in W_n$.

According to (14.1) it holds that $f_n = \sum_{k=-\infty}^{+\infty} \langle f | \psi_{n,k} \rangle \psi_{n,k}$. For each $n \in \mathbb{Z}$, we now form the vector spaces

$$V_n = \bigoplus_{l=-\infty}^{n-1} W_l = \cdots \oplus W_{n-3} \oplus W_{n-2} \oplus W_{n-1}.$$

The elements $v_n \in V_n$ then have the form

$$v_n = \sum_{l=-\infty}^{n-1} \sum_{k=-\infty}^{+\infty} \langle v_n | \psi_{l,k} \rangle \psi_{l,k}.$$

From the shape of the wavelet $\psi$, it follows that the elements of $V_n$ are *limits of simple functions* which are constant on the intervals $[2^{-n}k, 2^{-n}(k+1)[$, $k \in \mathbb{Z}$. In other words, their "smallest details" have the width $2^{-n}$.

In the following, let $\varphi = 1_{[0,1[}$ denote the characteristic function of the unit interval and $\varphi_{n,k}$, $n$ and $k$ in $\mathbb{Z}$, their normalized scaled shifts

$$\varphi_{n,k}(t) = 2^{n/2} \varphi(2^n t - k).$$

Because each step function $\psi_{l,k} \in V_n$ can be read as the sum of its individual steps, $v_n$ can also be represented by the following series:

$$v_n = \sum_{k=-\infty}^{+\infty} \langle v_n | \varphi_{n,k} \rangle \varphi_{n,k}. \tag{14.2}$$

*In other words*, the functions $\varphi_{n,k}$, $k \in \mathbb{Z}$, form a complete orthonormal system in $V_n$. The orthogonality condition $\langle \varphi_{n,k} | \varphi_{n,k'} \rangle = 0$ for $k \neq k'$ is satisfied because the supports of $\varphi_{n,k}$ and $\varphi_{n,k'}$ are disjoint.

Any signal $f \in L^2(\mathbb{R})$ can now, due to $L^2(\mathbb{R}) = \bigoplus_{n=-\infty}^{+\infty} W_n$ and $V_n = \bigoplus_{l=-\infty}^{n-1} W_l$, be decomposed with any arbitrary $M \in \mathbb{Z}$ into its orthogonal projection $P_M f$ onto

$V_M$ and a uniquely determined function $Q_M f$ in the orthogonal complement of $V_M$ in $L^2(\mathbb{R})$:

$$f = P_M f + Q_M f,$$

$$f = \sum_{k=-\infty}^{+\infty} \langle f | \varphi_{M,k} \rangle \varphi_{M,k} + \sum_{l=M}^{+\infty} \sum_{k=-\infty}^{+\infty} \langle f | \psi_{l,k} \rangle \psi_{l,k}. \tag{14.3}$$

The first series in this expansion represents the projection $P_M f$ (cf. p. 451) and is a coarse approximation of $f$ by a step function with minimum step width $2^{-M}$. The series $\sum_{k=-\infty}^{+\infty} \langle f | \psi_{l,k} \rangle \psi_{l,k}$ are functions in $W_l$ for $l = M, M+1, \ldots$ and add finer details to $P_M f$, each with a step width $2^{-l-1}$. The projection $P_M f$ shows the averaged course of $f$ over the intervals $[2^{-M}k, 2^{-M}(k+1)[$, $k \in \mathbb{Z}$. For $M \to -\infty$, these intervals become wider and the projections $P_M f$ become less detailed. Conversely, for $M \to +\infty$, increasingly finer details of $f$ are retained in $P_M f$.

One recognizes the excellent time localization of the wavelet representation (14.3), because both the supports of the functions $\varphi_{M,k}$ for sufficiently large $M$ and the supports of the Haar wavelets $\psi_{l,k}$, $l \geq M$ and $k \in \mathbb{Z}$, are only small intervals. If $f$ vanishes on an interval $I$, then the coefficients of those functions $\varphi_{M,k}$ and $\psi_{l,k}$ that have their supports in $I$ are zero. In practice, this means that short-term disturbances of a signal $f$ affect only a few coefficients in (14.3), while such disturbances typically alter the entire spectrum in classical Fourier analysis (cf. 4.5 and 10.4). We summarize:

*The vector spaces $V_n$ have the following properties:*

**(E1)** $V_n \subset V_{n+1}$ *for all* $n \in \mathbb{Z}$, $\{0\} \swarrow \ldots V_{-2} \subset V_{-1} \subset V_0 \subset V_1 \subset \cdots \nearrow L^2(\mathbb{R})$.

**(E2)** *The union of all $V_n$ is dense in $L^2(\mathbb{R})$, i.e., every $f \in L^2(\mathbb{R})$ can be approximated arbitrarily well by its orthogonal projections $P_n f$ onto $V_n$, $n \in \mathbb{Z}$, in the norm of $L^2(\mathbb{R})$:*

$$\lim_{n \to \infty} \| f - P_n f \| = 0.$$

**(E3)** *The intersection of all $V_n$ is the null space:* $\bigcap_{n=-\infty}^{+\infty} V_n = \{0\}$.

**(E4)** *For all $n \in \mathbb{Z}$, it holds that $V_{n+1} = V_n \oplus W_n$.*

**(E5)** *For all $n \in \mathbb{Z}$, it holds $f(t) \in V_n$ if and only if $f(2t) \in V_{n+1}$.*

**(E6)** *For each $n \in \mathbb{Z}$, the functions*

$$\varphi_{n,k}(t) = 2^{n/2} \varphi(2^n t - k) \qquad (k \in \mathbb{Z})$$

*form a complete orthonormal system of $V_n$.*

In contrast to the pairwise orthogonal spaces $W_n$, the spaces $V_n$ are nested as described in (E1). (E2) is a result of integration theory (cf. Appendix B). (E3) means that with decreasing $n \in \mathbb{Z}$, the energy of the signal projections $P_n f$ onto $V_n$ decays: $\lim_{n \to -\infty} \| P_n f \| = 0$ for $f \in L^2(\mathbb{R})$.

**Multiscale Analysis over the Scale of Vector Spaces $V_n$**

We now consider a signal $f$, which for a suitable $M \in \mathbb{Z}$ is a function in one of the spaces $V_M$. The function $f$ has the coefficients $f_k^{(M)}$:

$$f = \sum_{k=-\infty}^{+\infty} f_k^{(M)} \varphi_{M,k}.$$

From a practical perspective, let $f$ be a sufficiently good approximation for an observed signal $S$, whose mean values $m_k = 2^{M/2} f_k^{(M)}$ over the intervals
$[2^{-M}k, 2^{-M}(k+1)[$ for $k \in \mathbb{Z}$ are known. Any signal $S \in L^2(\mathbb{R})$ can be approximated arbitrarily well by such a simple function with a sufficiently large $M$ (cf. (E2) above). Signal processing then extends to the coefficients $f_k^{(M)}$ of the approximation function $f$.

The minimum step width of $f$ is therefore $2^{-M}$. According to (E4) in the preceding considerations, $f$ can be uniquely decomposed into

$$f = v_{M-1} + w_{M-1}$$

with $v_{M-1} \in V_{M-1}$ and $w_{M-1} \in W_{M-1}$. The step function $v_{M-1}$ has the minimum step width $2^{-M+1}$ and thus generally a coarser, less detailed course than $f$. The finer variations of $f$ that are lost in the orthogonal projection of $f$ onto $v_{M-1}$ are captured as "difference information" in $w_{M-1} = f - v_{M-1}$. The minimum step width of $w_{M-1}$ matches that of $f$. Compare this with the subsequent example.

By further decomposing $v_{M-1}$ in the form $v_{M-1} = v_{M-2} + w_{M-2}$ with $v_{M-2} \in V_{M-2}$ and $w_{M-2} \in W_{M-2}$, that is,

$$f = v_{M-2} + w_{M-2} + w_{M-1},$$

one obtains information about the course of $f$ from the function $v_{M-2}$, which varies even less compared to $v_{M-1}$, and from the functions $w_{M-1}$ and $w_{M-2}$, which complementarily contain the finer details of $f$ and $v_{M-1}$. This procedure can be continued and, as we will see, using property (E5) results in an effective algorithm for representing the signal $f$ through information about its coarse structure using a projection $v_{M-N} \in V_{M-N}$ and its fine structure using the functions $w_{M-1}, \ldots, w_{M-N}$, $N > 0$:

$$f = v_{M-N} + w_{M-1} + w_{M-2} + \cdots + w_{M-N}. \tag{14.4}$$

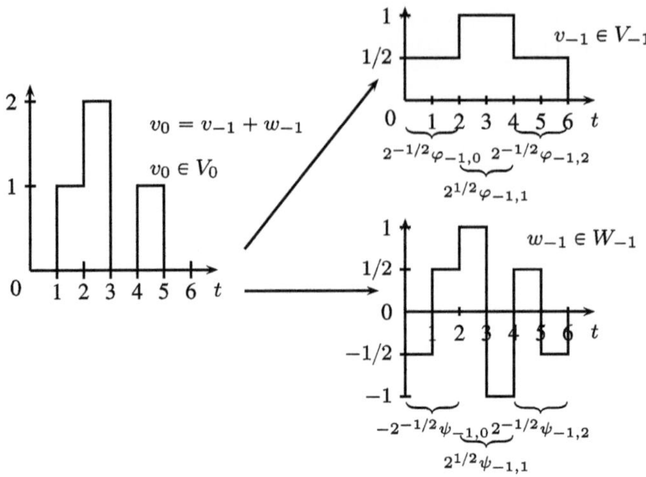

**Fig. 14.7** Example of the decomposition of $v_0$ into $v_{-1}$ and $w_{-1}$

The right side decomposes $f$ into components of varying levels of detail. The levels of detail of $v_{M-N}, w_{M-1}, \ldots, w_{M-N}$ are determined by the different values of the scale parameter $2^{-p}$ in the wavelets $\psi_{p,k}$ with $p = M - N, \ldots, M - 1, k \in \mathbb{Z}$. Therefore, the decomposition is called a *multiscale analysis of the signal $f$*. The fundamental task in the numerical processing of the decomposed signal is then the effective calculation of the coefficients in the series expansions of the functions $v_{M-N}, w_{M-1}, \ldots, w_{M-N}$. Opportunities for data compression in signal processing arise when many of these coefficients are so small that they can be replaced by zero without noticeable quality loss in the reconstruction.

**Example** The function $v_0 \in V_0$ in Fig. 14.7 shows details of step width 1, and the function $v_{-1}$ shows only coarser details of step width 2. Both functions can be described using the step functions $\varphi_{0,k}$ and $\varphi_{-1,k}$ by

$$v_0 = \sum_{k=0}^{5} \langle v_0 | \varphi_{0,k} \rangle \varphi_{0,k} \quad \text{and} \quad v_{-1} = \sum_{k=0}^{2} \langle v_0 | \varphi_{-1,k} \rangle \varphi_{-1,k}.$$

$v_{-1}$ *is an averaged version of $v_0$ over the intervals* $[0, 2[, [2, 4[,$ *and* $[4, 6[$. The function $w_{-1}$ contains the complementary information about details of $v_0$ with step width 1 and is represented by the Haar wavelet through

$$w_{-1} = \sum_{k=0}^{2} \langle v_0 | \psi_{-1,k} \rangle \psi_{-1,k}.$$

The information about $v_0$ is contained in the six coefficients $\langle v_0 | \varphi_{0,k} \rangle$, $k = 0, \ldots 5$, or alternatively in the combined six coefficients of $v_{-1}$ and $w_{-1}$.

**Scaling Function and Scaling Equation of Multiscale Analysis**

The key of multiscale analysis and resulting fast algorithms lies in the observation that not only the spaces $W_n$, $n \in \mathbb{Z}$, can be generated by a single function $\psi$, namely with the resulting basis functions $\psi_{n,k}$, but also the scale of the spaces $V_n$ is generated by a single function $\varphi$. Furthermore in our example with the Haar wavelet, the wavelet $\psi$ is determined by this function $\varphi$. The function $\varphi$ is the characteristic function $\varphi = 1_{[0,1[}$ of the unit interval. Its scaled translations $\varphi_{n,k}$, $k \in \mathbb{Z}$, span the vector spaces $V_n$ according to (E6). $\varphi$ is called the *scaling function* or *father function* for the wavelet $\psi$. The wavelet $\psi$ is called the *mother wavelet* of the Haar multiscale analysis.

Since both $\varphi$ and the wavelet $\psi$ belong to $V_1$ and $V_1$ is generated by the functions $\varphi_{1,k}(t) = 2^{1/2}\varphi(2t - k)$, the following equations hold:

$$\varphi(t) = \sqrt{2} \sum_{k=-\infty}^{+\infty} h_k \varphi(2t - k) \tag{14.5}$$

and for the wavelet $\psi$

$$\psi(t) = \sqrt{2} \sum_{k=-\infty}^{+\infty} g_k \varphi(2t - k) \tag{14.6}$$

with suitable coefficients $h_k$ and $g_k$, $k \in \mathbb{Z}$.

Equation (14.5) is called the *scaling equation of multiscale analysis* for the scaling function $\varphi$. According to Eq. (14.6), the mother wavelet $\psi$ is determined by the function $\varphi$. In our example with the Haar wavelet, $h_0 = h_1 = 2^{-1/2}$, $h_k = 0$ otherwise. With the coefficients $g_k$, the relationship is

$$g_k = (-1)^k \overline{h_{1-k}} \quad \text{for } k \in \mathbb{Z}. \tag{14.7}$$

With a view to other examples, we note in (14.5) and (14.6) series instead of finite sums and complex conjugates $\overline{h_{1-k}}$ in (14.7), because later on complex-valued functions and wavelets will also be generally allowed. Equations (14.5)–(14.7) are basic for fast algorithms with which the coefficients in multiscale decompositions of signals $f$ can be calculated. They are also the starting point for multiscale analyses with other scaling functions $\varphi$ and wavelets $\psi$, which can be constructed from $\varphi$ according to (14.6) and (14.7) (Theorem of S. Mallat, p. 476). With the results of Sect. 11.6 on discrete filters, we can see that the coefficients of $\phi$ define a lowpass filter and those of $\psi$ a highpass filter.

With Eqs. (14.5)–(14.7), we now obtain the algorithms of fast wavelet transform to compute the coefficients in the wavelet decomposition (14.4) of our step function $f \in V_M$ and also to reconstruct $f$ from these coefficients.

**Fast Wavelet Transform with the Haar Wavelet**

Given are the coefficients $f_k^{(M)}$ of a signal $f = \sum\limits_{k=-\infty}^{+\infty} f_k^{(M)} \varphi_{M,k}$ from $V_M$. To compute the coefficients in the wavelet decomposition (14.4) of p. 471

$$f = v_{M-N} + w_{M-1} + w_{M-2} + \cdots + w_{M-N},$$

we denote the coefficients of a projection $P_{M-l} f = v_{M-l}$, $1 \leqslant l \leqslant N$, with $f_k^{(M-l)}$, those of the functions $w_{M-l}$ with $d_k^{(M-l)}$. The coefficients $h_k$ and $g_k$ are given by (14.5)–(14.7).

The desired coefficients $f_k^{(M-N)}$ and $d_k^{(M-l)}$ can now be computed completely recursively without integrations. The algorithms of the fast wavelet transform according to Mallat (1989) are as follows:

**Mallat's Decomposition Algorithm**
*For $l = 1, \ldots, N$ and $k \in \mathbb{Z}$ we have*

$$f_k^{(M-l)} = \sum_{m=-\infty}^{+\infty} f_m^{(M-l+1)} \overline{h_{m-2k}}, \qquad (14.8)$$

$$d_k^{(M-l)} = \sum_{m=-\infty}^{+\infty} f_m^{(M-l+1)} \overline{g_{m-2k}} \qquad (14.9)$$

$$f_k^{(M)} \longrightarrow f_k^{(M-1)} \longrightarrow f_k^{(M-2)} \longrightarrow \cdots \longrightarrow f_k^{(M-N)}$$
$$\qquad\quad d_k^{(M-1)} \qquad\quad d_k^{(M-2)} \qquad \cdots \qquad d_k^{(M-N)}$$

Analogously, one obtains an algorithm for the reconstruction of the coefficients $f_k^{(M)}$ of $f$ from the coefficients $f_k^{(M-N)}$ and $d_k^{(M-1)}, d_k^{(M-2)}, \ldots, d_k^{(M-N)}$.

*Mallat's Reconstruction Algorithm*

*For $l = N, N-1, \ldots, 1$, and $k \in \mathbb{Z}$, we have*

$$f_k^{(M-l+1)} = \sum_{m=-\infty}^{+\infty} \left[ f_m^{(M-l)} h_{k-2m} + d_m^{(M-l)} g_{k-2m} \right]. \qquad (14.10)$$

$$\begin{array}{ccccccccc}
f_k^{(M-N)} & \longrightarrow & f_k^{(M-N+1)} & \longrightarrow & f_k^{(M-N+2)} & \cdots & \cdots & \longrightarrow & f_k^{(M)} \\
d_k^{(M-N)} & \nearrow & d_k^{(M-N+1)} & \nearrow & d_k^{(M-N+2)} & & d_k^{(M-1)} & \nearrow &
\end{array}$$

*Sketch of the Proof:*  The proof of the algorithms (14.8) and (14.9) is done by induction. We show only the base case for (14.8). All other steps of the proof for (14.8) and (14.9) can be done analogously.

For $l = 1$ and $k \in \mathbb{Z}$, due to $h_k = \sqrt{2} \int\limits_{-\infty}^{+\infty} \varphi(t)\overline{\varphi(2t - k)}\, dt$

$$f_k^{(M-1)} = \langle f | \varphi_{M-1,k} \rangle = \left\langle \sum_{m=-\infty}^{+\infty} f_m^{(M)} \varphi_{M,m} \Big| \varphi_{M-1,k} \right\rangle$$

$$= \sum_{m=-\infty}^{+\infty} f_m^{(M)} \int\limits_{-\infty}^{+\infty} 2^{M/2}\varphi(2^M t - m) 2^{(M-1)/2}\overline{\varphi(2^{M-1}t - k)}\, dt$$

$$= \sum_{m=-\infty}^{+\infty} f_m^{(M)} \int\limits_{-\infty}^{+\infty} \sqrt{2}\varphi(2x - (m - 2k))\overline{\varphi(x)}\, dx$$

$$= \sum_{m=-\infty}^{+\infty} f_m^{(M)} \overline{h_{m-2k}}.$$

To prove (14.10), we consider the case $l = N$ and

$$v_{M-N+1} = \sum_{k=-\infty}^{+\infty} f_k^{(M-N+1)} \varphi_{M-N+1,k}.$$

From the scaling equations corresponding to (14.5) and (14.6) for $\varphi(2^{M-N}t - m)$ and $\psi(2^{M-N}t - m)$, it follows with a short calculation

$$v_{M-N+1} = v_{M-N} + w_{M-N}$$

$$= \sum_{m=-\infty}^{+\infty} 2^{(M-N)/2} \left[ f_m^{(M-N)}\varphi(2^{M-N}t - m) + d_m^{(M-N)}\psi(2^{M-N}t - m) \right]$$

$$= \sum_{k=-\infty}^{+\infty} \sum_{m=-\infty}^{+\infty} \left[ f_m^{(M-N)} h_{k-2m} + d_m^{(M-N)} g_{k-2m} \right] \varphi_{M-N+1,k}.$$

Comparison of coefficients gives (14.10) for $l = N$, and a similar induction gives the reconstruction algorithm.

## *Multiscale Analysis with Other Wavelets*

The wavelet analysis with the Haar wavelet has the disadvantage that the Haar functions are discontinuous. Therefore, every finite partial sum of the developments (14.1) or (14.3) is discontinuous, no matter how smooth the approximated observed signal may be. Another disadvantage is the poor frequency localization of the Haar wavelet. Its Fourier transform decays only as $1/|\omega|$ for $|\omega| \to \infty$. The breakthrough of wavelet theory in applications occurred when it was shown that multiscale analysis is also possible with other suitably constructed wavelets. We summarize some important results. In the following, $L^2(\mathbb{R})$ is the vector space of complex-valued, square-integrable functions on $\mathbb{R}$.

**Definition** A function $\varphi$ is a scaling function of a multiscale analysis if it generates a sequence $(V_n)_{n\in\mathbb{Z}}$ of closed subspaces in $L^2(\mathbb{R})$ with properties *(E1) to (E6)* from page 470. *The sequence of subspaces $V_n$ is called the multiscale analysis corresponding to $\varphi$.*

If $W_n$ denotes the orthogonal complement of $V_n$ in $V_{n+1}$, then the following key theorem of Mallat (1989) holds. The proof can also be found in Daubechies (1992).[3]

**Theorem 14.7 (Mallat's Theorem)** *For every multiscale analysis $(V_n)_{n\in\mathbb{Z}}$ with scaling function*

$$\varphi(t) = \sqrt{2} \sum_{k=-\infty}^{+\infty} h_k\varphi(2t - k),$$

*there exists a wavelet $\psi$ such that $\psi(t) = \sqrt{2} \sum_{k=-\infty}^{+\infty} g_k\varphi(2t - k)$ with $g_k = (-1)^k\overline{h_{1-k}}$. For each $n \in \mathbb{Z}$, the functions $\varphi_{n,k}(t) = 2^{n/2}\varphi(2^n t - k)$, $k \in \mathbb{Z}$, form a complete orthonormal system in $V_n$, and the functions $\psi_{n,k}(t) = 2^{n/2}\psi(2^n t - k)$, $k \in \mathbb{Z}$, form a complete orthonormal system in $W_n$. For each $n \in \mathbb{Z}$, the set $\{\varphi_{n,k}, \psi_{m,k} \mid k \in \mathbb{Z}, m \geqslant n\}$ is a complete orthonormal system in $L^2(\mathbb{R})$.*

With the same proofs as outlined in the explained example, the algorithms (14.8) to (14.10) of the fast wavelet transform also generally follow for any multiscale analysis with scaling function $\varphi$ from Eqs. (14.5) to (14.7).

If the coefficient sequence $h_k$, $k \in \mathbb{Z}$, of the scaling function $\varphi$ has finite length, then (14.8) to (14.10) are *non-recursive filters*, i.e., FIR filters (Finite Impulse Response Filters, see p. 372). Otherwise, (14.8) to (14.10) are IIR filters (Infinite Impulse Response Filters). In electrical engineering, the algorithms are also referred to as "*Subband Filtering Schemes*" (see Daubechies (1992)). The difficulties of multiscale analysis lie in the concrete construction of $\varphi$, $\psi$, and the filter sequence $h_k$, $k \in \mathbb{Z}$. In addition to the previously cited literature, we refer to the overview article "How To Make Wavelets" by Strichartz (1993).

---

[3] Observe: The indexing of the spaces $V_n$ and $W_n$ is reversed there.

### *Daubechies Wavelets*

There are several well-known continuous scaling functions that lead to a multiscale analysis. In 1988, I. Daubechies succeeded in constructing a family of multiscale analyses with continuous scaling functions and wavelets $\psi_m$, $m > 0$, that have compact support and coefficient sequences $(h_k)_{k \in \mathbb{Z}}$ of finite length. The support properties allow for good time-frequency localization. The remarkable result, shown in Daubechies (1992), is the following theorem:

**Theorem 14.8 (Theorem of I. Daubechies)**  *For every positive integer $k$ there exist functions $\psi \in C^k(\mathbb{R})$ with compact support so that for $n$, $k$ in $\mathbb{Z}$ the functions*

$$\psi_{n,k}(t) = 2^{n/2}\psi(2^n t - k)$$

*constitute a complete orthonormal system in $L^2(\mathbb{R})$.*

The approximation, regularity, and localization properties of the corresponding wavelet decompositions depend on the increasing filter length with $m$ of the respective Daubechies wavelets $\psi_m$. The wavelets $\psi_m$ cannot be given explicitly but are defined algorithmically. The corresponding coefficients $h_k$ can be tabulated. Such tables can be found in Daubechies (1992). Figure 14.8 shows the Daubechies wavelet with filter length 4, in Daubechies (1992), p. 197, with the notation $_2\Psi$.

For signals of length $n$ (i.e., in (14.8) $n = \max\{k | f_k^{(M)} \neq 0\} - \min\{k | f_k^{(M)} \neq 0\}$), the fast wavelet transform with finite filter length and a decomposition depth $N \ll n$ has a computational cost that grows linearly with the signal length and is therefore more efficient than the cost for the fast Fourier transform (see p. 119).

There are further developments of the multiscale analysis, where instead of a wavelet basis $\{\psi_{n,k} | n, k \in \mathbb{Z}\}$ in $L^2(\mathbb{R})$ two so-called biorthogonal bases $\{\psi_{n,k} | n, k \in \mathbb{Z}\}$ and $\{\widetilde{\psi_{n,k}} | n, k \in \mathbb{Z}\}$ are used. The example at the end of the section, which is intended to demonstrate the power of wavelets in data compression, was processed with such biorthogonal wavelet bases. A detailed presentation of the theory of "biorthogonal decompositions" can be found in Cohen et al. (1992).

**Signal Compression**

One of the most well-known application areas for discrete wavelet transformation is signal compression. A signal $f \in L^2(\mathbb{R})$ is represented by finitely many coefficients of an approximation $v_M$ in the space $V_M$ of a multiscale analysis. If $f$ is smooth and $v_M = v_{M-N} + w_{M-1} + \cdots + w_{M-N}$ is the wavelet decomposition of

**Fig. 14.8** Daubechies
wavelet $_2\Psi$

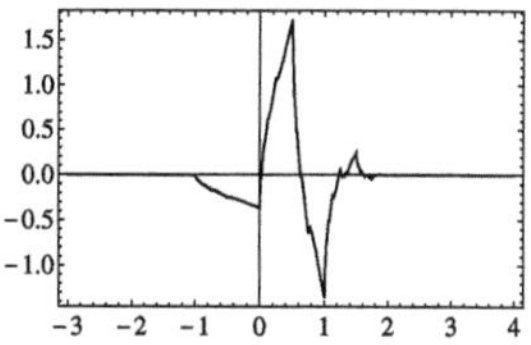

its approximation $v_M$ $(N > 0)$, then the coefficients of the wavelet expansions $w_{M-n}$ decrease for increasing $n$. The coefficients of $v_{M-N}, w_{M-1}, \ldots, w_{M-N}$ are quantized, i.e., their value range is divided into intervals, and all coefficients that lie in the same interval are rounded to the same value. Sufficiently small values, often all coefficients from a certain "detail level" $w_{M-N+l}, 0 \leqslant l \leqslant N - 1$, can then often be replaced by zero without significant loss of detail in the reconstruction of $v_M$. The data compression results from such quantizations and an effective encoding of the quantized data. The JPEG2000 standard for image data compression mentioned at the end of Chap. 5 and the file format DjVu use such wavelet methods.

**Image Data Processing and Two-Dimensional Multiscale Analysis**

In image data processing, a multiscale analysis of signals $f \in L^2(\mathbb{R}^2)$, which depend on two spatial variables, is needed. Complete orthonormal systems in $L^2(\mathbb{R}^2)$ can be constructed with the help of *tensor wavelets*. We summarize the essential results. More details can be found in Daubechies (1992) or Meyer (1995).

1. From a multiscale analysis $(V_n)_{n \in \mathbb{Z}}$ with scaling function $\varphi$ and mother wavelet $\psi$, the tensor products

$$\Psi^{(h)}(x, y) = \varphi(x)\psi(y), \quad \Psi^{(v)}(x, y) = \psi(x)\varphi(y), \quad \Psi^{(d)}(x, y) = \psi(x)\psi(y)$$

   create a *complete orthonormal system* $\Psi^{(q)}_{n,k_1,k_2}$ of $L^2(\mathbb{R}^2)$. The functions $\Psi^{(q)}_{n,k_1,k_2}$ are defined for $n, k_1, k_2 \in \mathbb{Z}$, $q \in \{h, v, d\}$ by

$$\Psi^{(q)}_{n,k_1,k_2}(x, y) = 2^n \Psi^{(q)}(2^n x - k_1, 2^n y - k_2).$$

2. For each $n \in \mathbb{Z}$, the linear combinations of the tensor products (see p. 191) $f \otimes g(x, y) = f(x)g(y)$, $f$ and $g$ from $V_n$, generate the closed subspace $\mathbb{V}_n = V_n \widehat{\otimes} V_n$ of $L^2(\mathbb{R}^2)$. *The sequence* $(\mathbb{V}_n)_{n \in \mathbb{Z}}$ *is a multiscale analysis for* $L^2(\mathbb{R}^2)$: The properties (E1)–(E4) from p. 470 apply analogously, if $\mathbb{W}_n$ is, as there, the orthogonal complement of $\mathbb{V}_n$ in $\mathbb{V}_{n+1}$. As in (E5), a function $f(x, y) \in \mathbb{V}_n$ is in $\mathbb{V}_n$ if and only if $f(2x, 2y) \in \mathbb{V}_{n+1}$. Analogous to (E6), the tensor products

$$\Phi_{n,k_1,k_2}(x, y) = 2^n \varphi(2^n x - k_1)\varphi(2^n y - k_2)$$

   form a *complete orthonormal system of* $\mathbb{V}_n$ for $(k_1, k_2) \in \mathbb{Z}^2$.

3. Accordingly, $\mathbb{W}_n$ is generated by the orthonormal system of the functions $\Psi^{(q)}_{n,k_1,k_2}$ (with $(k_1, k_2) \in \mathbb{Z}^2$, $q \in \{h, v, d\}$), and $L^2(\mathbb{R}^2)$ is the direct sum of the spaces $\mathbb{W}_n$, $n \in \mathbb{Z}$. Due to the construction of the spaces $\mathbb{V}_n$ from tensor products, for all $n \in \mathbb{Z}$ it holds that

$$\mathbb{W}_n = (V_n \widehat{\otimes} W_n) \oplus (W_n \widehat{\otimes} V_n) \oplus (W_n \widehat{\otimes} W_n).$$

The basis functions $\Psi^{(q)}_{n,k_1,k_2}$ with $q = h$ generate the component $V_n \widehat{\otimes} W_n$, those with $q = v$ generate $W_n \widehat{\otimes} V_n$, and those with the index $q = d$ finally generate the subspace $W_n \widehat{\otimes} W_n$ ($h$ stands for "horizontal," $v$ for "vertical," and $d$ for "diagonal").

4. Mallat's algorithms can be extended to the multidimensional case.

In practice, image data (gray values of pixels) are stored in a matrix $C^{(0)}$. The coefficients $c^{(0)}_{k_1,k_2}$ of $C^{(0)}$ correspond to the coefficients $\langle f^{(0)}|\Phi_{0,k_1,k_2}\rangle$ of a signal $f^{(0)} = \sum_{k_1,k_2} \langle f^{(0)}|\Phi_{0,k_1,k_2}\rangle \Phi_{0,k_1,k_2}$ from the space $\mathbb{V}_0$ of a multiscale analysis of $L^2(\mathbb{R}^2)$. In the first step of a multiscale decomposition

$$f^{(0)} = f^{(-1)} + w^{(-1,h)} + w^{(-1,v)} + w^{(-1,d)}$$

with $f^{(-1)} \in \mathbb{V}_{-1}$, $w^{(-1,h)} \in V_{-1}\widehat{\otimes}W_{-1}$, $w^{(-1,v)} \in W_{-1}\widehat{\otimes}V_{-1}$, $w^{(-1,d)} \in W_{-1}\widehat{\otimes}W_{-1}$; $f^{(-1)}$ is then a coarser version of $f^{(0)}$, while the additional finer details of $f^{(0)}$ are captured in horizontal direction ($x$-direction) in $w^{(-1,h)}$, in vertical direction in $w^{(-1,v)}$, and those in diagonal direction in $w^{(-1,d)}$. If the initial matrix $C^{(0)}$ is a $2^N \times 2^N$ matrix, then the corresponding coefficient matrices

$$C^{(-1)} = \left(\langle f^{(0)}|\Phi_{-1,k_1',k_2'}\rangle\right), \quad D^{(-1,h)} = \left(\langle f^{(0)}|\Psi^{(h)}_{-1,k_1',k_2'}\rangle\right),$$

similarly $D^{(-1,v)}$ and $D^{(-1,d)}$, have the size $2^{N-1} \times 2^{N-1}$. As in the one-dimensional case, the multiscale analysis can be continued with a decomposition of $f^{(-1)}$.

Schematically illustrated for a one-level decomposition of an image of J. B. Fourier (see p. 8) in Fig. 14.9 with the Haar wavelet, one obtains:

In practice, the coefficients of the functions $f^{(-m)}$, $w^{(-m,q)}$, $m \geqslant 1$, $q \in \{h, v, d\}$, often decrease rapidly with increasing index $m$. Suitable quantization and coding techniques then lead to excellent results in image data compression. In the example above, the darker the pixels are, the smaller the corresponding coefficients, where black corresponds to zero.

Figures 14.10, 14.11, and 14.12 present an example of data compression. The image data matrix of the first image has a size of $2048 \times 2048$ (about 4.2 MB). The second image shows the result of data compression using the previous JPEG method. The original data was compressed to approximately 14.5 KB. The third image shows the reconstruction from wavelet-compressed data. The data was analyzed with the Cohen-Daubechies-Feauveau 7/9 tap filter, quantized, and compressed so that the data after quantization and coding also occupied about 14.5 KB. The coding algorithms used are from Tian and Wells (1996). Analogous wavelet algorithms are used in the JPEG2000 standard.

**Example for Denoising**

In Figs. 14.13 and 14.14 you see the two images from page 101 that were used there for a test with a watermark in the frequency domain. Gaussian white noise was added to the left JPEG compressed image so that we obtained the right image.

$$C^{(0)} \rightarrow \begin{array}{|c|c|} \hline C^{(-1)} & D^{(-1,h)} \\ \hline D^{(-1,v)} & D^{(-1,d)} \\ \hline \end{array} \rightarrow \begin{array}{|c|c|} \hline \begin{array}{|c|c|} C^{(-2)} & D^{(-2,h)} \\ D^{(-2,v)} & D^{(-2,d)} \end{array} & D^{(-1,h)} \\ \hline D^{(-1,v)} & D^{(-1,d)} \\ \hline \end{array}$$

**Fig. 14.9** Wavelet compressed J. B. Fourier

**Fig. 14.10** Claudia size 4.2 MB

For this denoising example in Fig. 14.15, Matlab's wavelet toolbox with the biorthogonal spline wavelet bior4.4 was used. The applied method FDR (False Discovery Rate) works with a threshold rule based on controlling the expected ratio

**Fig. 14.11**  JPEG
compressed size 14.5 KB

**Fig. 14.12**  Wavelet
compressed 14.5 KB

**Fig. 14.13**  Clara original

**Fig. 14.14**  Noisy Clara

of false positive detections to all positive detections. We find that noise reduction is difficult as soon as broadband noise has entered the signal. This applies particularly to noisy audio examples, since our ears are much more sensitive than our eyes.

**Fig. 14.15** Clara, with
wavelet denoised

## Further Areas of Application

There are many other areas of application for wavelet analysis. ECG analysis,
pattern recognition and edge detection, and applications in the regularization of
ill-posed inverse problems, for example, in computed tomography, finite element
methods using wavelet approaches, and applications in the study of wave propagation, such as radar or sonar waves, are just a few examples. Depending on the
respective scientific field, countless approaches have been developed with wavelets
that were previously treated with conventional Fourier analysis. References can be
found in the literature cited in this section and a whole range of references for
specific purposes by searching the internet.

## At the End

For convenience, there are two appendices. Appendix A contains the residue
theorem referred to in Chap. 11 with relation to the $z$-transform, estimates for
the location of zeros of polynomials, and the recipe to obtain the partial fraction
decomposition of rational functions. Appendix B covers some basics of integration
theory and of convolutions. Appendix C contains solutions to the exercises.

Finally, I return once again to the example of musical notation given on p. 410
and below as an ingenious time-frequency description. The two bars last 11
seconds. In Fig. 14.16, you see a spectrogram from a windowed Fourier transform,
analogously computed with Mathematica as the figure on p. 412. It shows quite well
the line of notes in the piece from B flat (932 Hz) to D flat (554 Hz) and the overtones
in its presentation. Alias effects are also typical from sampling. Better estimates of
the pitches can be seen with an FFT, which shows that the artist tunes a few Hz
higher, which is common in German orchestras.

The first two bars of the wonderful flute piece "Syrinx", Claude Debussy (1862–1918)

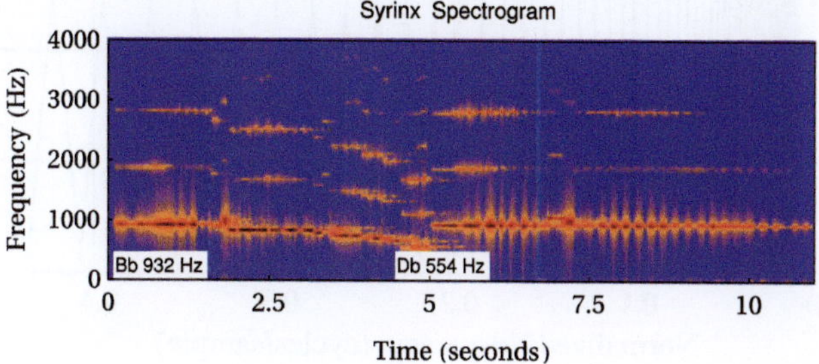


**Fig. 14.16**  Spectrogram made with an STFT

**Fig. 14.17**  Wavelet scalogram made with a filter bank

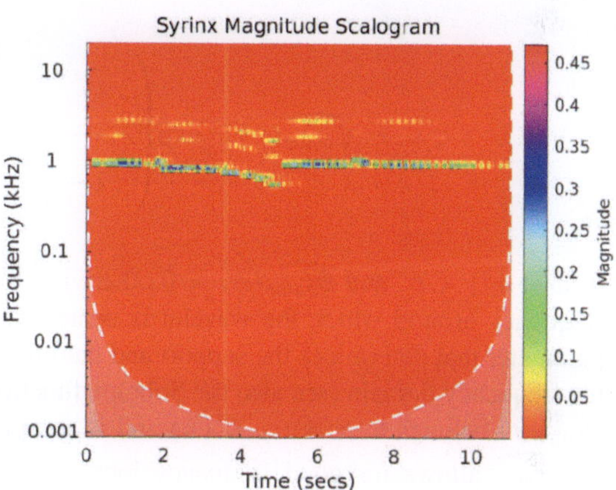


For comparison, at last a corresponding wavelet scalogram from the audio recording with the brilliant flutist E. Pahud.[4] The line of notes is very clear. One observes the excellent intonation of the artist, when the play changes shortly from mezzoforte to forte (green to blue) and from mezzoforte to pianissimo at the end (green to yellow).

Instead of alias effects as in the spectrogram, in Fig. 14.17 we see small rectangles determined by the wavelet's time-frequency localizations due to the uncertainty relation.

The scalogram was computed with an implemented routine for a continuous wavelet transform in Matlab's wavelet toolbox using the bump wavelet $\psi_{\text{bump}}$. Its Fourier transform is (exp the exponential function)

---

[4] Easy to find at YouTube by a search for "Pahud Syrinx".

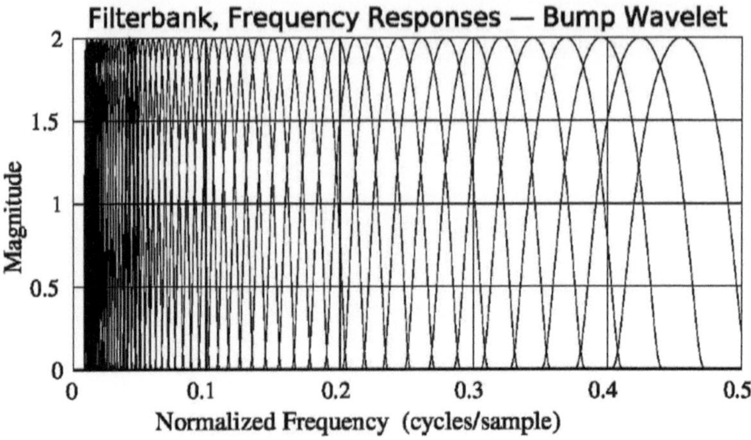



**Fig. 14.18** The filter bank used for the above scalogram

$$\widehat{\psi_{\mathrm{bump}}}(\omega) = \exp\left(-\frac{1}{1 - \left(\dfrac{\omega - \omega^*}{\sigma}\right)^2}\right),$$

if $|\omega - \omega^*| < \sigma$, and $\widehat{\psi_{\mathrm{bump}}}(\omega) = 0$ otherwise. Therein, $\omega^*$ is the central angular frequency around which the wavelet is centered (cf. p. 463), and $\sigma$ is the width parameter that determines the support and therefore the bandwidth of the wavelet in the frequency domain (see also the Sobolev function, used on p. 157). The function $\psi_{\mathrm{bump}}(t)$ belongs to $\mathcal{S}(\mathbb{R}) \subset L^1(\mathbb{R})$ and is complex-valued. The compact support of $\widehat{\psi_{\mathrm{bump}}}$ allows for good frequency localization. The shown frequency is then $\omega^*/(2\pi a)$, $a$ the scaling parameter.

To compute the scalogram, Matlab uses a filterbank with that wavelet, here an array of 57 bandpass filters, normalized corresponding to the sampling rate and duration of the music piece and maximum magnitude 2 (see p. 463, no. 3, p. 464, and p. 342). Figure 14.18 illustrates this filterbank with the bump wavelet.

To summarize, we can end up understanding why wavelet analysis has found its way into many different fields of application, from medicine, biology, physics, chemistry, and engineering to finance and the military. By searching the Internet, you can quickly find many examples for signal analysis with wavelets, particularly in frequency bands that lie outside our perception range.

I hope you could experience how the ideas of Fourier analysis from the first classical problems have developed to modern computational methods and a myriad of applications in our everyday life. I thank you for your attention, if you have read my text up to this point, and hope you have enjoyed reading it and are ready to deepen your knowledge with further reading.

# Appendix A
# The Residue Theorem and the Fundamental Theorem of Algebra

## The Residue Theorem

A function $f$, which is analytic in a region $G \setminus \{z_0\}$ of the complex plane, has for $z$, $0 < |z - z_0| < r$, the Laurent series expansion

$$f(z) = \sum_{k=1}^{\infty} \frac{c_{-k}}{(z - z_0)^k} + \sum_{k=0}^{\infty} c_k (z - z_0)^k.$$

It then follows from the *Cauchy integral formula* that the coefficients $c_k$ for $k \in \mathbb{Z}$ are given by

$$c_k = \frac{1}{2\pi j} \oint_C \frac{f(z)}{(z - z_0)^{k+1}} \mathrm{d}z,$$

where $C$ is a closed, piecewise continuously differentiable curve in $G \setminus \{z_0\}$, which is oriented counterclockwise around $z_0$.

**Definition** The part $H(f, z_0) = \sum_{k=1}^{\infty} \frac{c_{-k}}{(z-z_0)^k}$ in the Laurent series of $f$ in the annulus $0 < |z - z_0| < r$ is called the principal part of $f$ at the expansion point $z_0$. The coefficient $c_{-1}$ of $\frac{1}{z-z_0}$ in the Laurent series of $f$ is called the residue of $f$ at $z_0$, denoted by

$$\mathrm{Res}(f, z_0) = c_{-1} = \frac{1}{2\pi j} \oint_C f(z) \mathrm{d}z,$$

with a closed curve $C$ as described above.

A generalization of this relationship between the contour integral of $f$ and the residue at a singularity of $f$ is the residue theorem.

© The Author(s), under exclusive license to Springer Nature Switzerland AG 2025
R. Brigola, *Fourier Analysis and Distributions*, Texts in Applied Mathematics 79,
https://doi.org/10.1007/978-3-031-81311-5

**Theorem** *Let a function $f$ be analytic in a region $G \subset \mathbb{C}$ except for isolated singularities. Let $C$ be a closed, piecewise continuously differentiable, and positively oriented curve, which encircles finitely many singularities $z_1, z_2, \ldots, z_n$ once, without passing through any singularity itself. Then the following holds:*

$$\oint_C f(z)\mathrm{d}z = 2\pi j \sum_{k=1}^{n} \mathrm{Res}(f, z_k).$$

The residue theorem has many applications in the computation of real integrals, in integrals that count zeros and poles, and also in the stability analysis of feedback systems. We referred to this theorem in Chap. 11 when computing $z$-transformations.

**Example for Computing Residues** The residue at an $m$-fold pole $z_0$ of a function $f$ is $\mathrm{Res}(f, z_0) = \frac{1}{(m-1)!} \lim_{z \to z_0} \frac{\mathrm{d}^{m-1}}{\mathrm{d}z^{m-1}} (z - z_0)^m f(z)$, because from the Laurent series of $f$ in an annulus $0 < |z - z_0| < r$ arises

$$(z - z_0)^m f(z) = c_{-m} + \cdots + c_{-1}(z - z_0)^{m-1} + c_0(z - z_0)^m + \ldots.$$

In this Taylor series, the coefficient $c_{-1}$ appears as the factor of $(z - z_0)^{m-1}$ and is thus uniquely determined by the given formula.

Specifically, for $f(z) = \frac{z}{(z-1)^2(z+1)}$ with the double pole at $z = 1$, the residue is $\mathrm{Res}(f, 1) = \lim_{z \to 1} \left( \frac{z}{z+1} \right)' = \frac{1}{4}$.

## Analytic Functions and the Fundamental Theorem of Algebra

The fundamental theorem of algebra plays a crucial role wherever polynomials are used. It states that every nonconstant polynomial has at least one complex root. There are many different proofs of this theorem, varying in approach and with different prerequisites for the reader regarding the underlying areas of mathematics. A recommendable book on this topic is the source B. Fine, G. Rosenberger (1997). First, some theorems from complex analysis about analytic functions $f : G \to \mathbb{C}$, i.e., complex differentiable functions on a simply connected region $G$ in $\mathbb{C}$, will be summarized, whose proofs can be quickly found in the already cited literature.

**Properties of Analytic Functions**
Assume that $G$ is a simply connected region in $\mathbb{C}$, $C$ is a piecewise smooth, simple closed curve in $G$ with positive orientation, and $f : G \to \mathbb{C}$ is analytic.

1. The Cauchy integral theorem holds: $\oint_C f(z)\mathrm{d}z = 0$.
2. The Cauchy integral formulas hold: $f^{(n)}(z) = \frac{n!}{2\pi j} \oint_C \frac{f(\zeta)}{(\zeta - z)^{n+1}} \mathrm{d}\zeta$

for all $z$ from the interior of the region enclosed by $C$ and $n \in \mathbb{N}_0$.

3. For $a \in G$, the Taylor series expansion $f(z) = \sum_{n=0}^{\infty} \frac{f^{(n)}(a)}{n!}(z-a)^n$ holds.

Using the Cauchy integral formula for $f'$, Liouville's theorem and thus a common proof of the fundamental theorem of algebra quickly follow.

**Theorem** *If $f$ is analytic and bounded on $\mathbb{C}$, then $f$ is constant.*

***Proof*** Let $|f| \leqslant M$. For $z \in \mathbb{C}$ and any simply traversed circle $C$ around $z$ with radius $r$ and positive orientation, it follows from the Cauchy integral formula

$$|f'(z)| = \left| \frac{1}{2\pi j} \oint_C \frac{f(\zeta)}{(\zeta - z)^2} d\zeta \right| \leqslant \frac{1}{2\pi} \int_0^{2\pi} \frac{|f(z + r\,\mathrm{e}^{jt})|}{r} dt \leqslant \frac{M}{r}.$$

With $r \to \infty$, it follows that $f'(z) = 0$ for all $z \in \mathbb{C}$ and therewith the assertion.   $\square$

**Theorem** *Every nonconstant polynomial $P$ has at least one root in $\mathbb{C}$.*

***Proof*** Let $P(z) \neq 0$ for all $z \in \mathbb{C}$. Then $f = 1/P$ is analytic in $\mathbb{C}$ and $\lim_{|z|\to\infty} |f(z)| = 0$. Thus, $f$ is bounded and constant according to Liouville's theorem. A nonconstant polynomial must therefore have a root.   $\square$

By polynomial division and factorization into linear factors, it follows easily that a polynomial of degree $n \geqslant 1$ has exactly $n$ roots in $\mathbb{C}$, which need not be distinct.


## *On Bounds for Roots of Polynomials*


Estimates for the location of roots are often of interest. For a polynomial $P(z) = z^n + Q(z) = z^n + \sum_{k=0}^{n-1} a_k z^k$ of degree $n \geqslant 1$, $r = 1 + |a_{n-1}| + \ldots + |a_0|$, and $|z| \geqslant r$, it follows from the triangle inequality with $|Q(z)| \leqslant |z|^{n-1}(|z| - 1)$ that $|P(z)| \geqslant |z|^{n-1} \geqslant 1$. This inequality shows that all roots of $P$ lie within the open disk around zero with radius $r$. Further estimates for root bounds can be found in M. Dehmer (2006) or J. Stoer, R. Bulirsch (1992). Two estimates from these sources are:

1. *All roots of a polynomial $P(z) = \sum_{k=0}^{n} a_k z^k$ of degree $n \geqslant 1$ with $|a_n| \geqslant |a_k|$ for $k = 0, \ldots, n-1$ lie in the closed disk around zero with radius $r = 2$.*
2. *Let $P(z) = \sum_{k=0}^{n} a_k z^k$ be a polynomial with $a_0 a_n \neq 0$, $n \geqslant 1$. Further, let*

$$\alpha = \max_{1 \leqslant k \leqslant n} \left| \frac{a_{n-k}}{a_n} \right|^{1/k} \quad \text{and} \quad \beta = \max_{1 \leqslant k \leqslant n} \left| \frac{a_{n-k}}{a_n} \right|.$$

*Then for each root $s$ of $P$, the following estimate holds:*

$$|s| < \min\{2\alpha, 1 + \beta\}.$$

Finally, *Gerschgorin's theorem* about the location of the roots of the characteristic polynomial of a square matrix, i.e., about the location of the eigenvalues, should be mentioned here. Gerschgorin's theorem can be found, for example, in the book of J. Stoer, R. Bulirsch (1992) on Numerical Analysis. The theorem states as follows:

**Theorem** *The union of all disks*

$$K_i = \left\{ z \in \mathbb{C} : |z - a_{i,i}| \leqslant \sum_{k=1, k \neq i}^{n} |a_{i,k}| \right\}$$

*contains all eigenvalues of an* $(n \times n)$ *matrix* $A = (a_{i,k})_{1 \leqslant i, k \leqslant n}$.

## Partial Fraction Decomposition of Rational Functions

**Theorem** *Let* $Q/P$ *be a proper rational function, i.e.,* $\deg Q < \deg P$, *and let* $P(z) = c(z - z_1)^{n_1} \cdots (z - z_k)^{n_k}$ *be a factorization of* $P$ *with its zeros* $z_1, \ldots, z_k$ *and their respective multiplicities* $n_1, \ldots, n_k$. *Then the function* $Q/P$ *has a partial fraction decomposition, i.e., a sum representation of the form*

$$\frac{Q(z)}{P(z)} = \frac{a_{11}}{z - z_1} + \frac{a_{12}}{(z - z_1)^2} + \ldots + \frac{a_{1n_1}}{(z - z_1)^{n_1}} +$$

$$+ \frac{a_{21}}{z - z_2} + \frac{a_{22}}{(z - z_2)^2} + \ldots + \frac{a_{2n_2}}{(z - z_2)^{n_2}} +$$

$$\vdots$$

$$+ \frac{a_{k1}}{z - z_k} + \frac{a_{k2}}{(z - z_k)^2} + \ldots + \frac{a_{kn_k}}{(z - z_k)^{n_k}}$$

*with coefficients* $a_{pq} \in \mathbb{C}$, $1 \leqslant p \leqslant k$, $1 \leqslant q \leqslant n_p$. *The representation is unique up to permutation of the zeros of* $P$.

***Proof*** One can prove the theorem by induction on the degree of $P$. If $\deg P = 1$, then $\deg Q = 0$, and the statement is already correct.

Now assume the assertion holds for all proper rational functions for which the degree of the denominator is at most $n - 1$, and let $\deg P = n$, $n > 1$. Then $P(z) = (z - z_1)^{n_1} P_1(z)$ with $P_1(z_1) \neq 0$, $P_1(z) = c(z - z_2)^{n_2} \cdots (z - z_k)^{n_k}$. For $A = Q(z_1)/P_1(z_1)$, we obtain

$$\frac{Q(z)}{P(z)} - \frac{A}{(z - z_1)^{n_1}} = \frac{Q(z) - A P_1(z)}{(z - z_1)^{n_1} P_1(z)}.$$

If $Q = AP_1$, the assertion follows with $Q(z)/P(z) = A/(z - z_1)^{n_1}$ and $a_{pq} = 0$ for $(p, q) \neq (1, n_1)$.

For $Q \neq AP_1$, the numerator on the right side is a polynomial with the zero $z_1$ and therefore of the form $(z - z_1)Q_1(z)$ with a polynomial $Q_1 \neq 0$. Thus, we have

$$\frac{Q(z)}{P(z)} = \frac{A}{(z - z_1)^{n_1}} + \frac{Q_1(z)}{(z - z_1)^{n_1-1}P_1(z)}$$

and $\deg Q_1 < n - 1 = \deg((z - z_1)^{n_1-1}P_1(z))$. Using the induction hypothesis, the assertion about the sum representation of $Q/P$ follows.

To prove the uniqueness of the decomposition, assume that $Q/P$ has another representation with coefficients $b_{pq}$ instead of $a_{pq} \in \mathbb{C}$. Multiplying both representations by $(z - z_1)^{n_1}$ and taking the limit as $z \to z_1$ yield $b_{1n_1} = a_{1n_1}$. Subtracting $a_{1n_1}/(z - z_1)^{n_1} = b_{1n_1}/(z - z_1)^{n_1}$ from $Q(z)/P(z)$ and multiplying the difference by $(z - z_1)^{n_1-1}$, we obtain $a_{1(n_1-1)} = b_{1(n_1-1)}$. Continuing this procedure in an analogous manner yields the equality of all coefficients $a_{pq} = b_{pq}$.                    $\square$

## *Calculation of a Partial Fraction Decomposition*

There are various ways to calculate the partial fraction decomposition for a given rational function. These include the substitution method or the comparison of coefficients between both sides of the above summation representation. It involves solving a linear system of equations for the sought coefficients $a_{pq}$. Another approach that leads to a closed formula for the desired coefficients is the computation using the *Taylor formula* known from the basic analysis. This is explained step by step below and demonstrated at hand of a simple example.

1. Using polynomial division, decompose a given rational function $r$ into $r = g + f$ with a polynomial $g$ and its proper rational part $f = Q/P$. Then calculate the zeros $z_1, \ldots, z_k$ of $P$ and their multiplicities $n_1, \ldots, n_k$.

2. For a zero $z_p$, the coefficients $a_{pq}$, $1 \leqslant q \leqslant n_p$, in the partial fraction decomposition of $f$ are the coefficients of the principal part $H(f, z_p)$ of the Laurent series of $f$ at the expansion point $z_p$. They are obtained from the Taylor expansion of the function $f_1(z) = (z - z_p)^{n_p} f(z)$, which is analytic in a neighborhood of $z_p$, i.e.,
   $f_1(z) = \alpha_{p0} + \alpha_{p1}(z - z_p) + \ldots + \alpha_{p(n_p-1)}(z - z_p)^{n_p-1} + \ldots$ with $\alpha_{pm} = D^m f_1(z_p)/m!$ ($D^m f_1$ is the derivative of order $m$ of $f_1$, $m \in \mathbb{N}_0$). From this, we get

$$H(f, z_p) = \frac{\alpha_{p0}}{(z - z_p)^{n_p}} + \frac{\alpha_{p1}}{(z - z_p)^{n_p-1}} + \ldots + \frac{\alpha_{p(n_p-1)}}{(z - z_p)}.$$

3. **Coefficient Formula.** From 2., we obtain the coefficients $a_{pq}$ of the partial fraction decomposition of $f$ in the form of the preceding theorem for $1 \leqslant p \leqslant k$, $1 \leqslant q \leqslant n_p$:

$$a_{pq} = \alpha_{p(n_p-q)} = \frac{1}{(n_p-q)!} \lim_{z \to z_p} D^{n_p-q}\left( (z-z_p)^{n_p} \frac{Q(z)}{P(z)} \right).$$

**Example** For $\frac{Q(z)}{P(z)} = \frac{1}{(z-z_1)(z-z_2)^2}$ with $z_1 = 0$, $z_2 = j$, the coefficients $a_{pq}$ of the partial fraction decomposition are obtained from the formula, namely $a_{11} = \frac{1}{j^2} = -1$, $a_{21} = -\frac{1}{j^2} = 1$, and $a_{22} = \frac{1}{j} = -j$. Therefore, $\frac{Q(z)}{P(z)} = -\frac{1}{z} + \frac{1}{z-j} - \frac{j}{(z-j)^2}$.

In Chap. 10, p. 298, and in Chap. 11, we referred to partial fraction decompositions to determine the inverse Fourier transforms of rational functions. If the denominator polynomial of a rational function $f = Q/P$ has a degree $P > 4$, then its zeros $z_p$ are calculated approximately, for example, using the *Bairstow method*. The resulting partial fraction decomposition is then an approximation for $Q/P$.

Finally, it should be noted that the considerations for obtaining the principal parts $H(f, z_p)$ of $f$ at the expansion points $z_p$ ($1 \leqslant p \leqslant k$) provide another proof of the partial fraction decomposition, if one takes into account that $F(z) = f(z) - H(f, z_1) - \ldots - H(f, z_k)$ can be extended to an analytic function over the entire $\mathbb{C}$ that vanishes at infinity. $F$ is thus bounded and by Liouville's theorem constantly zero, i.e., $f(z) = H(f, z_1) + \ldots + H(f, z_k)$.

# Appendix B
# Tools from Integration Theory

When working with physical quantities, various measures assigned to these quantities are of fundamental importance. Consider quantities like continuous or point-like distributions of masses or charges, electrical voltages or currents, and measures like average speeds, moments of inertia of masses, effective values of alternating voltages, etc. The mathematical tool for such concepts linked with *mean value formations* is provided by *measure and integration theory*. Therefore, we provide in the following some basics from Lebesgue's integration theory. We will limit ourselves to a compilation of the required concepts and most important theorems. Interested readers can find compact introductions to integration theory in the textbooks by J. Weidmann (1980) or R. Wheeden, A. Zygmund (1977).


## Measures, Null Sets, and Integrals

An *interval* of $\mathbb{R}^n$ is the Cartesian product of $n$ intervals from $\mathbb{R}$. If all of these are bounded, then their product is bounded. The *Lebesgue measure* $\lambda^n(J)$ of an interval $J = I_1 \times I_2 \times \cdots \times I_n$ in $\mathbb{R}^n$ is $\lambda^n(J) = \prod_{k=1}^{n} |I_k|$, where $|I_k| = b_k - a_k$ is the "length" of an interval $I_k = [a_k, b_k]$, and it does not matter whether the $I_k$ are open, half-open, or closed. Single-point intervals $[a, a]$ and the empty set have a "length" of zero. For example, an interval in $\mathbb{R}^3$ is a cuboid, and its Lebesgue measure, if all coordinate axes carry length units, is its volume. Generally, the unit of measure is determined by the respective units of the coordinates.

Let $\mathcal{J}$ be the set of all bounded intervals in $\mathbb{R}^n$; then the function $\lambda^n : \mathcal{J} \to \mathbb{R}_0^+$ obviously has the following properties:

1. $\lambda^n$ *is monotone, i.e.,* $\lambda^n(J_1) \leqslant \lambda^n(J_2)$ *for intervals* $J_1, J_2 \in \mathcal{J}$ *with* $J_1 \subset J_2$.
2. $\lambda^n$ *is additive, i.e.,* $\lambda^n(J_1 \cup J_2) = \lambda^n(J_1) + \lambda^n(J_2)$ *for intervals* $J_1, J_2, J_1 \cup J_2$ *in* $\mathcal{J}$ *with* $J_1 \cap J_2 = \varnothing$.



R. Brigola, *Fourier Analysis and Distributions*, Texts in Applied Mathematics 79,
https://doi.org/10.1007/978-3-031-81311-5

3. *$\lambda^n$ is regular, i.e., for each interval $J_1 \in \mathcal{J}$ and each $\varepsilon > 0$, there is an open interval $J_2 \in \mathcal{J}$ such that $J_1 \subset J_2$ and $\lambda^n(J_1) \leqslant \lambda^n(J_2) < \lambda^n(J_1) + \varepsilon$.*

*Besides the Lebesgue measure, there are many other interval functions $m : \mathcal{J} \to \mathbb{R}_0^+$ that are monotone, additive, and regular. Every such function $m$ is called a measure.*

## *Examples*

1. For a set $M \subset \mathbb{R}^n$ without accumulation points, assign a real number $g(\mathbf{x}) \geqslant 0$ to each point $\mathbf{x} \in M$. The interval function $m$ on $\mathcal{J}$, $m(J) = \sum_{\mathbf{x} \in M \cap J} g(\mathbf{x})$, is a measure. If we consider $g(\mathbf{x})$ as the mass of the point $\mathbf{x} \in M$, while all points in $\mathbb{R}^n \setminus M$ are of mass zero, then $m(J)$ measures the total mass contained in $J$ that is discretely distributed. This $m$ is called a *discrete measure* or a *discrete distribution*.
2. If $m_1$ is a measure in $\mathbb{R}^p$ and $m_2$ is a measure in $\mathbb{R}^q$, then the product $m_1 \otimes m_2(J_1 \times J_2) = m_1(J_1)m_2(J_2)$ with bounded intervals $J_1$ in $\mathbb{R}^p$ and $J_2$ in $\mathbb{R}^q$ is a measure on $\mathbb{R}^{p+q}$. The measure $m_1 \otimes m_2$ is called the *product measure* of $m_1$ and $m_2$. It is immediately evident that the Lebesgue measure $\lambda^{p+q}$ in $\mathbb{R}^{p+q}$ is the product measure of $\lambda^p$ and $\lambda^q$.
3. If $F : \mathbb{R} \to \mathbb{R}$ is any monotonically increasing, right-continuous function, then by

$$
m(J) = \begin{cases}
F(b) - F(a) & \text{for } J = \,]a, b] \\
F(b) - F(a-) & \text{for } J = [a, b] \\
F(b-) - F(a) & \text{for } J = \,]a, b[ \\
F(b-) - F(a-) & \text{for } J = [a, b[
\end{cases} \qquad (a \leqslant b)
$$

a measure on $\mathbb{R}$ is defined. $F$ is called the *distribution function* corresponding to $m$. For instance, $F(x) = x$ is the distribution function corresponding to the Lebesgue measure $\lambda$ in $\mathbb{R}$.

For a measure $m$ in $\mathbb{R}^n$, a set $N$ is an *$m$-null set*, if for every $\varepsilon > 0$, there exists a sequence of intervals $J_i \subset \mathbb{R}^n$, $i \in \mathbb{N}$, such that $N \subset \bigcup_{i=1}^{\infty} J_i$ and $\sum_{i=1}^{\infty} m(J_i) < \varepsilon$. Every subset of an $m$-null set is an $m$-null set, countable unions and intersections of $m$-null sets are again $m$-null sets. For example, finite sets are $\lambda^n$-null sets, the set $\mathbb{Q}$ of rational numbers in $\mathbb{R}$, and thus also $\mathbb{Q}^n$ in $\mathbb{R}^n$ are null sets for the Lebesgue measure in $\mathbb{R}$ or in $\mathbb{R}^n$. To prove this, one uses the countability of the rational numbers $q_1, q_2, \ldots$ and takes the $i$-th number as the center of an interval of length $3^{-i} \cdot \varepsilon/2$. Then $\lambda(\mathbb{Q}) < \varepsilon$. A degenerate interval $[a, b] \times [c, c]$ is a $\lambda^2$-null set; in other words, a line segment has Lebesgue measure zero in $\mathbb{R}^2$. Hyperplanes in $\mathbb{R}^n$, i.e., sets of points $\mathbf{x} = (x_1, \ldots, x_n)$ in $\mathbb{R}^n$ that satisfy an equation of the form $a_0 + a_1 x_1 + \cdots + a_n x_n = 0$, $a_0, \ldots, a_n \in \mathbb{R}$, are $\lambda^n$-null sets. Also, the sphere of a ball in $\mathbb{R}^3$ is a $\lambda^3$-null set. For the discrete measure $m$ in Example 1 above, $\mathbb{R}^n \setminus M$ is a null set.

Two functions $f$ and $g$ on $\mathbb{R}^n$ are *m-almost everywhere equal*, if their values differ at most on an $m$-null set. It is also said that $f(\mathbf{x}) = g(\mathbf{x})$ *for m-almost all* $\mathbf{x}$. A sequence of functions $f_k$ *converges m-almost everywhere* to a function $f$ if there exists an $m$-null set $N$ such that $\lim_{k\to\infty} f_k(\mathbf{x}) = f(\mathbf{x})$ for all $\mathbf{x} \in \mathbb{R}^n \setminus N$.

A function $f : \mathbb{R}^n \to \mathbb{C}$ is a *step function* if there are finitely many, pairwise disjoint, bounded intervals $J_1, \ldots, J_l$ in $\mathbb{R}^n$ such that $f$ is constant on $J_k$, $1 \leqslant k \leqslant l$, and zero outside the $J_k$. For a step function $f$ with intervals of constancy $J_1, \ldots, J_l$, $f(J_k) = c_k$, $f = 0$ outside the $J_k$, $1 \leqslant k \leqslant l$, the *integral with measure m* is defined by

$$\int f\,dm = \sum_{k=1}^{l} c_k m(J_k).$$

*For step functions, the integral with the Lebesgue measure is then equal to the Riemann integral and has the same known properties.*

A function $f : \mathbb{R}^n \to \mathbb{C}$ is called *m-measurable*, if there exists a sequence of step functions that converge $m$-almost everywhere to $f$. All continuous functions and all functions with finitely or countably many discontinuities are measurable with respect to all measures $m$ that we have given in the examples on p. 492. Sums and products of $m$-measurable functions $f$ and compositions $g \circ f$ with continuous functions $g$ are $m$-measurable. Also, limits $f$ of $m$-almost everywhere convergent sequences of $m$-measurable functions are again $m$-measurable. These examples show that with measures as in our examples and with functions that occur in technical fields of application, measurability is not a serious problem. The measurability of the functions considered in the following is a requirement to be able to justify integrals with respect to a diverse class of measures mathematically soundly. The following fundamental theorem of B. Levi (1875-1961) holds:

**Theorem of Monotone Convergence**   *If $(f_k)_{k\in\mathbb{N}}$ is a sequence of real-valued step functions on $\mathbb{R}^n$, such that for each $k \in \mathbb{N}$, $f_k(\mathbf{x}) \leqslant f_{k+1}(\mathbf{x})$ holds m-almost everywhere, and all integrals $\int f_k dm$ are bounded by a common, suitable limit $K \in \mathbb{R}$, then there exists an m-measurable function $f : \mathbb{R}^n \to \mathbb{R}$, such that m-almost everywhere $\lim_{k\to\infty} f_k(\mathbf{x}) = f(\mathbf{x})$ holds.*

With the monotonicity of the sequence $(f_k)_{k\in\mathbb{N}}$ and the uniform boundedness of the integrals, it can be shown that the sequence $(f_k)_{k\in\mathbb{N}}$ diverges at most on an $m$-null set. Furthermore, from the assumptions of the theorem, it follows immediately that the limit $\lim_{k\to\infty} \int f_k dm$ exists. The *m-integral* of the function $f$ in the previous theorem is then defined by

$$\int f\,dm = \lim_{k\to\infty} \int f_k dm.$$

Thus, the integral with respect to a measure $m$ is introduced for all real-valued functions $f$ for which there exists, as in the above theorem, a nondecreasing sequence of step functions converging to $f$ $m$-almost everywhere, with a bounded sequence of integrals. It can be shown that the integral of $f$, as defined above, is independent of the choice of the approximating sequence of step functions. Denote by $\mathcal{T}$ the set of all such functions $f$, then all functions $g$ in the vector space generated by $\mathcal{T}$ are of the form $g = f_1 - f_2$ with $f_1$, $f_2$ in $\mathcal{T}$, and their integral is defined by

$$\int g\,\mathrm{d}m = \int f_1\,\mathrm{d}m - \int f_2\,\mathrm{d}m.$$

The integral of $g$ is again independent of the choice of functions $f_1$ and $f_2$ in $\mathcal{T}$ used to represent $g$.

The *m-integrable functions* in general are those functions $f : \mathbb{R}^n \to \mathbb{C}$ for which both the real part $\Re(f)$ and the imaginary part $\Im(f)$ lie in the real vector space generated by the set $\mathcal{T}$, in other words, those $f : \mathbb{R}^n \to \mathbb{C}$, for which with the imaginary unit $j$ it holds

$$f = \Re(f) + j\Im(f) = (f_1 - f_2) + j(f_3 - f_4) \quad \text{with } f_1,\ f_2,\ f_3,\ f_4 \text{ in } \mathcal{T}.$$

The *m-integral of $f$* is then defined in an obvious way by

$$\int f\,\mathrm{d}m = \int \Re(f)\,\mathrm{d}m + j\int \Im(f)\,\mathrm{d}m.$$

Thus, the concept of the integral has been extended to complex-valued functions. The $m$-integrable functions form a vector space over $\mathbb{C}$. If one identifies two functions, which differ only on a null set, the space $L^1(\mathbb{R}^n, m)$ of respective equivalence classes is obtained. It is common to call the equivalence class of a function $f$ again as function $f$ in $L^1(\mathbb{R}^n, m)$, keeping in mind that it is only determined up to a null set, i.e., $m$-almost everywhere. For the Lebesgue measure $m = \lambda^n$, instead of $L^1(\mathbb{R}^n, \lambda^n)$, the notation $L^1(\mathbb{R}^n)$ is used.

A subset $E$ of $\mathbb{R}^n$ is called an *m-measurable set* if the indicator function $1_E$ is an $m$-measurable function. An $m$-measurable bounded set $E$ has the measure $m(E) = \int 1_E\,\mathrm{d}m$. Every $m$-null set is $m$-measurable. All bounded intervals are $m$-measurable sets. Complements, unions, and intersections of countably many $m$-measurable sets are again $m$-measurable. For discrete measures $m$, every subset $E$ of $\mathbb{R}^n$ is even $m$-measurable. For our examples of measures $m$ and for sets like $\mathbb{R}^n$, intervals, planes, spheres, and their complements, countable unions, and intersections, $m$-measurability is therefore always guaranteed. The integral of $f \in L^1(\mathbb{R}^n, m)$ over an $m$-measurable subset $E$ is

$$\int_E f\,\mathrm{d}m = \int f \cdot 1_E\,\mathrm{d}m.$$

If $f$ is only defined on a subset $E$ of $\mathbb{R}^n$, then $f$ is set to zero outside of $E$ and the integral of $f$ is defined as the integral of $f \cdot 1_E$, if this exists. For instance, for the discrete measure $m$ from Example 1 and an arbitrary subset $E$ of $\mathbb{R}^n$, a function $f : \mathbb{R}^n \to \mathbb{C}$ is $m$-integrable over $E$ with $\int_E f\,\mathrm{d}m = \sum_{\mathbf{x}\in E\cap M} f(\mathbf{x})g(\mathbf{x})$ if and only if the series on the right-hand side converges absolutely. For integrals with the Lebesgue measure, the following notations are common:

$$\int f \mathrm{d}\lambda^n = \int f(\mathbf{x}) \mathrm{d}\lambda^n(\mathbf{x}) = \int f(\mathbf{x}) \mathrm{d}\mathbf{x} = \underbrace{\int \int \cdots \int}_{n-\text{times}} f(x_1, \ldots, x_n) \mathrm{d}(x_1, \ldots, x_n).$$

Similarly, for other measures $\int f \mathrm{d}m = \int f(\mathbf{x}) \mathrm{d}m(\mathbf{x})$, if one wants to highlight the integration variable for clarity. For measures $m$ with distribution function $F$, one also writes $\int f \mathrm{d}F$ instead of $\int f \mathrm{d}m$. Important for practical computations is the following fact:

*All (properly) Riemann-integrable functions are also Lebesgue-integrable. The rules learned for handling Riemann integrals and methods for their calculation apply unchanged to these functions for Lebesgue integrals. Riemann and Lebesgue integrals agree for such functions.*

A first advantage of the Lebesgue integral over the Riemann integral is shown by the following well-known extreme example: The Dirichlet function

$$f(x) = \begin{cases} 0 & \text{for } x \in [0, 1] \setminus \mathbb{Q} \\ 1 & \text{for } x \in [0, 1] \cap \mathbb{Q} \end{cases}$$

is, as is well known, not Riemann integrable. However, it is Lebesgue integrable with

$$\int_{[0,1]} f \mathrm{d}\lambda = \int_0^1 f(x)\, \mathrm{d}x = \int f(x) 1_{[0,1]}(x)\, \mathrm{d}x = 0,$$

because $f$ differs from the zero function only on the $\lambda$ null set $[0, 1] \cap \mathbb{Q}$.

Including null sets already in the definition of integrable functions and their integrals in the sense that the behavior of the functions on null sets becomes completely irrelevant, so that they ultimately only need to be defined almost everywhere, has thus expanded the class of integrable functions. This also enables, for example, an integral calculation for functions with infinitely many discontinuities, which cannot be achieved with the Riemann integral.

We summarize some fundamental theorems from integration theory. Here, $m$ is an arbitrary measure. Proofs of these theorems can be found in the literature cited at the beginning of the chapter.

## *Fundamental Theorems of Integration Theory*

1. (a) *An $m$-measurable function $f$ is $m$-integrable if and only if the function $|f|$ is also $m$-integrable.*
   (b) *For $m$-integrable functions $f$ and $g$ and complex numbers $\alpha$ and $\beta$, we have*

$$\int (\alpha f + \beta g) \mathrm{d}m = \alpha \int f \mathrm{d}m + \beta \int g \mathrm{d}m.$$

(c) *For m-integrable, real-valued functions $f$ and $g$ with $f \leqslant g$ m-almost everywhere, we have*

$$\int f\,\mathrm{dm} \leqslant \int g\,\mathrm{dm}.$$

*In particular, $\left| \int f\,\mathrm{dm} \right| \leqslant \int |f|\,\mathrm{dm}$.*

(d) *For an m-measurable function $f$, we have $f = 0$ m-almost everywhere if and only if $f$ is m-integrable and $\int |f|\,\mathrm{dm} = 0$.*

2. *An improperly Riemann-integrable function $f$ is Lebesgue-integrable if $|f|$ is also improperly Riemann-integrable. Its Lebesgue integral is then equal to the improper Riemann integral.*

3. **Lebesgue Dominated Convergence Theorem.** *For a sequence of m-integrable functions $f_k$ and a function $f$, suppose $\lim_{k\to\infty} f_k(\mathbf{x}) = f(\mathbf{x})$ m-almost everywhere. Furthermore, assume that there exists an m-integrable function $g$ such that $|f_k| \leqslant g$ for every $k \in \mathbb{N}$. Then $f$ is also m-integrable, and taking the limit may be interchanged with the integration:*

$$\int f\,\mathrm{dm} = \lim_{k\to\infty} \int f_k\,\mathrm{dm}.$$

4. **Interchanging the Order of Integration.** The possibility to reduce multiple integrals to the calculation of iterated integrals with only one variable each and the interchangeability of the order of integration is ensured by the following theorem of G. Fubini (1879-1943) and L. Tonelli (1885-1946). Here, the vector space $\mathbb{R}^n$ is regarded as the Cartesian product $\mathbb{R}^n = \mathbb{R}^p \times \mathbb{R}^q$, $p + q = n$. A point $\mathbf{z}$ in $\mathbb{R}^n$ is written in the form $\mathbf{z} = (\mathbf{x}, \mathbf{y})$ with $\mathbf{x} \in \mathbb{R}^p$ and $\mathbf{y} \in \mathbb{R}^q$. On $\mathbb{R}^n$, the product $m_1 \otimes m_2$ of two measures $m_1$ on $\mathbb{R}^p$ and $m_2$ on $\mathbb{R}^q$ is given.

**Fubini-Tonelli Theorem**

(a) *Assume that an $m_1 \otimes m_2$-measurable function $f : \mathbb{R}^n \to \mathbb{C}$ has the following properties:*

   (i) *For $m_1$-almost all $\mathbf{x} \in \mathbb{R}^p$, $f_{\mathbf{x}}(\mathbf{y}) = f(\mathbf{x}, \mathbf{y})$ is $m_2$-integrable on $\mathbb{R}^q$.*

   (ii) *Define $F(\mathbf{x}) = \displaystyle\int_{\mathbb{R}^q} |f(\mathbf{x}, \mathbf{y})|\,\mathrm{dm}_2(\mathbf{y})$, if $f_{\mathbf{x}} \in L^1(\mathbb{R}^q, m_2)$,*

      *$F(\mathbf{x}) = 0$ otherwise. Let the function $F$ be $m_1$-integrable on $\mathbb{R}^p$.*

   *Then $f$ is integrable on $\mathbb{R}^n = \mathbb{R}^p \times \mathbb{R}^q$ with the product measure $m_1 \otimes m_2$.*

(b) *If $f$ is integrable on $\mathbb{R}^n$ with the product measure $m_1 \otimes m_2$, then for $m_1$-almost all $\mathbf{x} \in \mathbb{R}^p$, the function $f_{\mathbf{x}}$ is $m_2$-integrable on $\mathbb{R}^q$. Similarly, for $m_2$-almost all $\mathbf{y} \in \mathbb{R}^q$, the function $f_{\mathbf{y}}(\mathbf{x}) = f(\mathbf{x}, \mathbf{y})$ is $m_1$-integrable on $\mathbb{R}^p$. It holds:*

$$\int\limits_{\mathbb{R}^n} f(\mathbf{x}, \mathbf{y})\mathrm{d}m_1(\mathbf{x}) \otimes m_2(\mathbf{y}) = \int\limits_{\mathbb{R}^p} \left( \int_{\mathbb{R}^q} f(\mathbf{x}, \mathbf{y})\mathrm{d}m_2(\mathbf{y}) \right) \mathrm{d}m_1(\mathbf{x})$$

$$= \int\limits_{\mathbb{R}^q} \left( \int_{\mathbb{R}^p} f(\mathbf{x}, \mathbf{y})\mathrm{d}m_1(\mathbf{x}) \right) \mathrm{d}m_2(\mathbf{y}).$$

5. **Transformation Theorem for Lebesgue Integrals.** *Let $U$ and $V$ be non-empty, open subsets of $\mathbb{R}^n$. Assume $A : U \to V$ is a continuously differentiable, bijective mapping with continuously differentiable inverse $A^{-1} : V \to U$ and $f : V \to \mathbb{C}$ is $\lambda^n$-measurable. The function $f$ is $\lambda^n$-integrable over the set $V$ if and only if $(f \circ A)|\det \partial A|$ is $\lambda^n$-integrable over $U$. Here, $\det \partial A$ is the determinant of the Jacobian matrix $\left( \frac{\partial A_k}{\partial x_i} \right)_{\substack{1 \leqslant k \leqslant n \\ 1 \leqslant i \leqslant n}}$, where $A_k : U \to \mathbb{R}$ are the components of $A = (A_1, \ldots, A_n)$. Then we have*

$$\int\limits_{A(U)=V} f(\mathbf{x})\mathrm{d}\lambda^n(\mathbf{x}) = \int_U (f \circ A)(\mathbf{x})|\det \partial A(\mathbf{x})|\mathrm{d}\lambda^n(\mathbf{x}).$$

*For the case $n = 1$, this is the well-known substitution formula for integrals.*

## *Examples*

1. For intervals $J$ in $\mathbb{R}$, let $m(J) = \begin{cases} \alpha & \text{for } x_0 \in J \\ 0 & \text{for } x_0 \notin J \end{cases}$ be the measure for the point mass $\alpha > 0$ at $x_0 \in \mathbb{R}$ and $\lambda$ the Lebesgue measure on $\mathbb{R}$. For a given function $f : \mathbb{R}^2 \to \mathbb{C}$, let $f_{x_0}(y) = f(x, y)$ be Lebesgue-integrable. Then according to the Fubini-Tonelli Theorem, $f$ is $(m \otimes \lambda)$-integrable with

$$\int f(x, y)\mathrm{d}m(x) \otimes \lambda(y) = \iint f(x, y)\mathrm{d}m(x)\mathrm{d}\lambda(y) = \alpha \int f(x_0, y)\mathrm{d}y.$$

Thus, the integral with this product measure is the Lebesgue integral of $f$ along the line $x = x_0$ in $\mathbb{R}^2$, multiplied by the weight $\alpha$ of $x_0$.

2. *Integration in Spherical Coordinates*: The coordinate transformation $A$ from spherical coordinates to Cartesian coordinates

$$A : U = \{(r, \theta, \phi) \mid r > 0, \ 0 < \theta < \pi, \ 0 < \phi < 2\pi\} \to \mathbb{R}^3$$

$$A(r, \theta, \phi) = (x, y, z) = (r \sin \theta \cos \phi, r \sin \theta \sin \phi, r \cos \theta)$$

is injective and continuously differentiable with $\det \partial A = r^2 \sin(\theta) > 0$ for $\theta \in \left]0, \pi\right[$. The set $\mathbb{R}^3 \setminus A(U) = \{(x, y, z) \in \mathbb{R}^3 \mid y = 0\}$ is a $\lambda^3$-null set. Using the transformation theorem and the Fubini-Tonelli theorem, it follows for $f$ in $L^1(\mathbb{R}^3)$

$$\int_{\mathbb{R}^3} f \, d\lambda^3 = \int_{A(U)} f \, d\lambda^3 = \int_0^\infty \int_0^{2\pi} \int_0^\pi f(A(r, \theta, \phi)) r^2 \sin(\theta) \, d\theta \, d\phi \, dr.$$

## Integration over a Spherical Surface

The *surface integral* on a spherical surface in $\mathbb{R}^3$, familiar from Riemannian integral calculus, can be extended to the Lebesgue integral as follows:

For a subset $E$ of the spherical surface $S_R$ around the origin with radius $R$, set $E^* = \{r\mathbf{x} \mid 0 < r < R, \mathbf{x} \in E\}$. If the spherical sector $E^*$ is $\lambda^3$-measurable, the *surface measure of* $E$ is defined by

$$o(E) = \frac{3}{R} \lambda^3(E^*).$$

Thus, a measure is defined on all subsets $E$ of $S_R$ for which $E^*$ is $\lambda^3$-measurable. These sets $E$ are the $o$-measurable sets on the sphere $S_R$. It follows that

$$\int 1_E \, do = \frac{3}{R} \int 1_{E^*} \, d\lambda^3.$$

Using spherical coordinates for a set $E$ of the form

$$E = \{A(R, \theta, \phi) \mid 0 < \theta_1 \leqslant \theta \leqslant \theta_2 < \pi, \ 0 < \phi_1 \leqslant \phi \leqslant \phi_2 < 2\pi\},$$

$A(r, \theta, \phi)$ as above,

$$\int 1_E \, do = \frac{3}{R} \int_{\phi_1}^{\phi_2} \int_{\theta_1}^{\theta_2} \int_0^R r^2 \sin(\theta) \, dr \, d\theta \, d\phi = \int_{\phi_1}^{\phi_2} \int_{\theta_1}^{\theta_2} R^2 \sin(\theta) \, d\theta \, d\phi.$$

By repeating the process of introducing integrable functions through approximation by step functions on $o$-measurable sets, one obtains the *vector space* $L^1(S_R, o)$ *of $o$-integrable functions $f$ on the sphere* $S_R$, and integration in spherical coordinates shows

$$\int f \, do = \int_0^{2\pi} \int_0^\pi f(R \sin\theta \cos\phi, R \sin\theta \sin\phi, R \cos\theta) R^2 \sin(\theta) \, d\theta \, d\phi.$$

The surface element $do$ in spherical coordinates is $R^2 \sin(\theta)d\theta d\phi$. All bounded $o$-measurable functions on the sphere $S_R$, in particular all continuous functions on $S_R$, are $o$-integrable. Any Lebesgue-measurable function $f : \mathbb{R}^3 \to \mathbb{C}$ has an $o$-measurable restriction to $S_R$, which is $o$-integrable if and only if the iterated integral of its absolute value with respect to $d\theta$ and $d\phi$ on the right side of the last equation exists.

*The theorems of integration theory given in the last section also hold if one of the measures involved is the surface measure $o$.* The special significance of the Lebesgue surface measure $o$ compared to other measures on the sphere lies in its characterization by the validity of the *divergence theorem of Gauss*. We formulate this theorem for the sphere $K_R$ in $\mathbb{R}^3$ with radius $R$. It also holds for more general domains and their boundary measures. For this, compare the literature already mentioned at the beginning of the Appendix.

**Divergence Theorem of Gauss and Green's Formulas**    *If $\mathbf{F} : K_R \to \mathbb{R}^3$ is a vector field that is continuous on the closed sphere $K_R$, continuously differentiable inside the sphere, and $o$-integrable on the sphere $S_R$, and if $\mathbf{F}$ has a $\lambda^3$-integrable divergence* $\operatorname{div}\mathbf{F} = \frac{\partial F_1}{\partial x} + \frac{\partial F_2}{\partial y} + \frac{\partial F_3}{\partial z}$, $\mathbf{F} = (F_1, F_2, F_3)$, *then with the unit outer normal $\mathbf{n}$ of the sphere, it holds that*

$$\int\limits_{K_R} \operatorname{div}\mathbf{F}d\lambda^3 = \int\limits_{S_R} \mathbf{F} \cdot \mathbf{n}do.$$

$\mathbf{F} \cdot \mathbf{n}$ denotes the pointwise scalar product on the sphere between $\mathbf{F}$ and $\mathbf{n}$ in $\mathbb{R}^3$. The unit outer normal on the sphere $S_R$ is given in spherical coordinates by

$$\mathbf{n}(\theta, \phi) = (\sin\theta \cos\phi, \sin\theta \sin\phi, \cos\theta).$$

Set $\mathbf{F} = g\operatorname{grad}\varphi$ with twice continuously differentiable scalar fields $g$ and $\varphi$ in $\mathbb{R}^3$, and note $\operatorname{div}(g\operatorname{grad}\varphi) = g\Delta\varphi + \operatorname{grad}g \cdot \operatorname{grad}\varphi$, $\Delta$ the Laplace operator, so from Gauss's theorem we obtain the **first Green's formula** (here again for the sphere $K_R$):

$$\int\limits_{K_R} (g\Delta\varphi + \operatorname{grad}g \cdot \operatorname{grad}\varphi)d\lambda^3 = \int\limits_{S_R} g\operatorname{grad}\varphi \cdot \mathbf{n}do.$$

Interchanging the roles of $g$ and $\varphi$ and subtracting the two resulting equations yield the **second Green's formula**:

$$\int\limits_{K_R} (g\Delta\varphi - \varphi\Delta g)d\lambda^3 = \int\limits_{S_R} (g\operatorname{grad}\varphi - \varphi\operatorname{grad}g) \cdot \mathbf{n}do.$$

These theorems from vector analysis play a crucial role in solving potential problems in space (cf. Sect. 9.4).

## Measures with Densities

*If $m$ is a measure in $\mathbb{R}^n$ and $f$ is an $m$-measurable, real-valued function such that $f \geqslant 0$ $m$-almost everywhere, then on the $m$-measurable sets $E$ by*

$$\varrho(E) = \int_E f \, \mathrm{d}m$$

*a measure is defined. The function $f$ is called the density of $\varrho$ with respect to $m$. One writes briefly $\varrho = f \cdot m$ or $\mathrm{d}\varrho = f \, \mathrm{d}m$.*

**Example** If $m = \lambda^3$ is the Lebesgue measure in $\mathbb{R}^3$ and a $\lambda^3$-integrable function $f \geqslant 0$ describes a spatial, continuous mass density, then for Lebesgue-measurable sets $E \subset \mathbb{R}^3$ the integral $\int_E \mathrm{d}\varrho = \int_E f \mathrm{d}\lambda^3 = \int_E f(\mathbf{x})\mathrm{d}\mathbf{x}$ is the total mass contained in $E$. Analog models with continuous electric charge distributions according to a Lebesgue-integrable charge density $f = f^+ - f^-$, which becomes negative for negative charges, provide after decomposition of $f$ into positive part $f^+$ and negative part $f^-$ the total charge $\int_E f \mathrm{d}\lambda^3 = \int_E f^+ \mathrm{d}\lambda^3 - \int_E f^- \mathrm{d}\lambda^3$ of $E$, where oppositely charged parts of equal magnitude compensate to zero. The measure $m^+$ with the density $f^+$ is called the positive part and the measure $m^-$ with the density $f^-$ the negative part of the measure $m$. Since the measure $m$ can also take negative values, it is called a *signed measure*.

## L$^p$-Spaces and Convolutions

For $1 \leqslant p < \infty$ and a domain $\Omega \subset \mathbb{R}^n$, $L^p(\Omega)$ denotes the vector space of all Lebesgue-measurable complex-valued functions $f$ on $\Omega$ with

$$\|f\|_p = \left( \int_\Omega |f|^p \mathrm{d}\lambda^n \right)^{1/p} < \infty.$$

The functions $f \in L^p(\Omega)$ are called $p$-integrable with the $p$-norm $\|f\|_p$.
Functions that differ only on a Lebesgue null set are identified. The elements of $L^p(\Omega)$ are therefore, strictly speaking, the equivalence classes of all such identified functions.
For the case $p = \infty$, $L^\infty(\Omega)$ is the vector space of all essentially bounded functions. A Lebesgue-measurable function $f$ on $\Omega$ is essentially bounded if

$$\|f\|_\infty = \inf\left\{ \alpha \in \mathbb{R} \mid \lambda^n(\{\mathbf{x} \in \Omega : |f(\mathbf{x})| \geqslant \alpha\}) = 0 \right\} < \infty.$$

The number $\|f\|_\infty$ is called the essential supremum of $f$, because then $|f(\mathbf{x})| \leqslant \|f\|_\infty$ $\lambda^n$-almost everywhere. With the same identification as above, $\|f\|_\infty$ is a norm, which is referred to as the supremum norm. The $L^p$ spaces are complete normed spaces for all p with $1 \leqslant p \leqslant \infty$.

**Theorem** *The step functions are dense in $L^p(\Omega)$ for all p, $1 \leqslant p < \infty$, and for $p = \infty$, if $\Omega$ has finite measure.*

The following inequalities hold:

**Hölder Inequality** *For $1 \leqslant p \leqslant \infty$ and $\frac{1}{p} + \frac{1}{q} = 1$ (with the convention $\frac{1}{\infty} = 0$), let $f \in L^p(\Omega)$ and $g \in L^q(\Omega)$. Then $fg \in L^1(\Omega)$, and the following holds:*

$$\|fg\|_1 \leqslant \|f\|_p \|g\|_q.$$

**Minkowski Inequality** *For $1 \leqslant p \leqslant \infty$, $f, g \in L^p(\Omega)$, the following holds:*

$$\|f + g\|_p \leqslant \|f\|_p + \|g\|_p.$$

For bounded $\Omega$, $\lambda^n(\Omega) < \infty$, and $1 \leqslant p \leqslant q \leqslant \infty$, we have $L^q(\Omega) \subset L^p(\Omega)$. For unbounded $\Omega$, in general neither $L^q(\Omega) \subset L^p(\Omega)$ nor the reverse inclusion holds.

**Convolutions** For convolutions $f * g$ of functions $f \in L^p(\mathbb{R}^n)$ and $g \in L^q(\mathbb{R}^n)$,

$$f * g(\mathbf{x}) = \int_{\mathbb{R}^n} f(\mathbf{y})g(\mathbf{x} - \mathbf{y})d\lambda^n(\mathbf{y}),$$

the following theorem holds (see, e.g., R. Wheeden, A. Zygmund 1977).

We agree on the following convention:

$$\frac{1}{p} = 0 \text{ for } p = \infty \text{ and } r = \infty \text{ for } \frac{1}{r} = \frac{1}{p} + \frac{1}{q} - 1 = 0.$$

**Theorem** *Let $1 \leqslant p, q \leqslant \infty$, $\frac{1}{p} + \frac{1}{q} \geqslant 1$, and $\frac{1}{r} = \frac{1}{p} + \frac{1}{q} - 1$. For $f \in L^p(\mathbb{R}^n)$ and $g \in L^q(\mathbb{R}^n)$, the convolution $f * g$ belongs to $L^r(\mathbb{R}^n)$, and the Young inequality holds*

$$\|f * g\|_r \leqslant \|f\|_p \|g\|_q.$$

*For $\frac{1}{p} + \frac{1}{q} - 1 = 0$, $f * g$ is continuous and bounded.*


## *Convolutions in Sequence Spaces*

For $1 \leqslant p < \infty$, $l^p(\mathbb{Z})$ is the vector space of all sequences $x = (x_n)_{n \in \mathbb{Z}}$ of complex numbers for which

$$\|x\|_p = \left( \sum_{n=-\infty}^{+\infty} |x_n|^p \right)^{1/p} < \infty.$$

The space $l^\infty(\mathbb{Z})$ is the vector space of all bounded sequences $x = (x_n)_{n\in\mathbb{Z}}$. It is endowed with the norm $\|x\|_\infty = \sup\{|x_n| : n \in \mathbb{Z}\}$. All spaces $l^p(\mathbb{Z})$, each endowed with the norm $\|.\|_p$ for $1 \leqslant p \leqslant \infty$, are complete normed spaces. The Hölder and Minkowski inequalities hold analogously, and the following inclusions are satisfied:

$$l^1(\mathbb{Z}) \subset l^2(\mathbb{Z}) \subset \ldots \subset l^\infty(\mathbb{Z}).$$

**Theorem** *Let* $1 \leqslant p, q \leqslant \infty$, $\frac{1}{p} + \frac{1}{q} \geqslant 1$, *and* $\frac{1}{r} = \frac{1}{p} + \frac{1}{q} - 1$. *For* $x = (x_n)_{n\in\mathbb{Z}} \in l^p(\mathbb{Z})$ *and* $y = (y_n)_{n\in\mathbb{Z}} \in l^q(\mathbb{Z})$, *the discrete convolution*

$$(x * y) = (f_n)_{n\in\mathbb{Z}}, \quad f_n = \sum_{k=-\infty}^{+\infty} x_{n-k} y_k,$$

*exists. This convolution* $x * y$ *belongs to* $l^r(\mathbb{Z})$, *and Young's inequality holds*

$$\|x * y\|_r \leqslant \|x\|_p \|y\|_q.$$


# The Sobolev Space $H_0^1(\Omega)$ and the Poincaré-Friedrichs Inequality

In Sect. 9.5, the Sobolev space $V = H_0^1(\Omega)$, $\Omega \subset \mathbb{R}^2$, a bounded *Lipschitz domain*, was referenced in formulating the Dirichlet boundary value problem for a loaded membrane. Both terms are defined below. Proofs of the following statements can be found, for example, in H. Triebel (1992) or K. Atkinson, W. Han (2005).

**Definition** A bounded domain $\Omega \subset \mathbb{R}^n$ $(n \geqslant 2)$ is called a Lipschitz domain or a domain with Lipschitz boundary $\partial\Omega$, if $\partial\Omega$ can be covered by finitely many open sets $U_1, \ldots, U_m$ such that $\partial\Omega \cap U_k$ for $k = 1, \ldots, m$ is the graph of a Lipschitz continuous function, and for each $k$, the domain $\Omega \cap U_k$ lies on one side of this graph.

Many domains that occur in application problems have a Lipschitz boundary, such as convex polygons, star-shaped non-convex polygons, and many more. For such domains, the solution of a Dirichlet boundary value problem presented in Sect. 9 in the vector space $V = H_0^1(\Omega)$ used there and defined below is possible.

Typical examples of domains that do not have a Lipschitz boundary are circular disks or spheres with a crack.

**Definition** For a domain $\Omega \subset \mathbb{R}^n$, $H^1(\Omega)$ is the space of all real-valued functions $f \in L^2(\Omega)$ whose generalized first-order derivatives also belong to $L^2(\Omega)$, equipped with the norm

$$\|v\|_{H^1} = \left( \sum_{|k| \leqslant 1} \|\partial^k v\|_2^2 \right)^{1/2} \quad \text{and the inner product} \quad \langle u, v \rangle_{H^1} = \sum_{|k| \leqslant 1} \langle \partial^k u, \partial^k v \rangle_{L^2}.$$

Here, $k$ is a multi-index from $\mathbb{N}_0^n$.

The completeness of $H^1(\Omega)$ follows from that of $L^2(\Omega)$: If $(u_n)_{n \in \mathbb{N}}$ is a Cauchy sequence in $H^1(\Omega)$, then for each $k$, $|k| \leqslant 1$, the sequence $(\partial^k u_n)_{n \in \mathbb{N}}$ is a Cauchy sequence in $L^2(\Omega)$ and thus has a limit. The Cauchy-Schwarz inequality yields that these limits converge to the generalized derivatives of the same function. Thus, $H^1(\Omega)$ with this norm is a Hilbert space. For Lipschitz domains $\Omega$, the following theorem about the so-called *trace mapping* $\tau$ holds:

**Theorem** *For a Lipschitz domain $\Omega$, there exists a continuous linear operator*

$$\tau : H^1(\Omega) \to L^2(\partial\Omega),$$

*such that for all $v \in C(\Omega \cup \partial\Omega) \cap H^1(\Omega)$ and $x \in \partial\Omega$ holds:* $(\tau v)(x) = v(x)$.

The existence of a trace allows talking about boundary values of an $H^1$ function and thus defining the space of $H^1$ functions that vanish on the boundary of $\Omega$. A function that belongs to $H^1$ in the interior of a Lipschitz domain has a trace that belongs to $L^2$ on the boundary. We simply write $v|_{\partial\Omega} = \tau v$.

**Definition** For a Lipschitz domain $\Omega \subset \mathbb{R}^n$, the space $H_0^1(\Omega)$, equipped with the norm of $H^1(\Omega)$, is defined by $H_0^1(\Omega) = \{v \in H^1(\Omega) : \tau v = v|_{\partial\Omega} = 0\}$.

*It can be shown that $H_0^1(\Omega)$ is the completion of the space $\mathcal{D}(\Omega)$ of test functions on $\Omega$ with the norm of $H^1(\Omega)$. Thus, $H_0^1(\Omega)$ is also a Hilbert space.*

Finally, it was crucial in solving the variational problem in Sect. 9.5 that the bilinear form $a(u, v)$ used there is positive definite on $H_0^1(\Omega)$. This is ensured by the Poincaré-Friedrichs inequality.

**Poincaré-Friedrichs Inequality.** *Let $\Omega$ be a Lipschitz domain. Then there exists a constant $c$, dependent only on $\Omega$, such that for all $v \in H_0^1(\Omega)$ the Poincaré-Friedrichs inequality*

$$\|v\|_{L^2} \leqslant c\,|v|_{H^1}$$

*holds. Here, $|v|_{H^1} = \left( \sum_{|k|=1} \|\partial^k v\|_2^2 \right)^{1/2}$. In particular, $\|.\|_{H^1}$ and the norm $|.|_{H^1}$ are equivalent on $H_0^1(\Omega)$, because for all $u \in H_0^1(\Omega)$ we have $|u|_{H^1} \leqslant \|u\|_{H^1} \leqslant (c^2 + 1)^{1/2} |u|_{H^1}$.*

The norm $|v|_{H^1}$ agrees up to a factor with the energy norm $a(v, v)^{1/2}$ for elements $v \in H_0^1(\Omega)$ used in Sect. 9.5.

***Proof*** We use that $H_0^1(\Omega)$ is the completion of $\mathcal{D}(\Omega)$ in $H^1(\Omega)$. Let $Q = [-d, d]^n$ be a closed cube such that $\Omega \subset Q$, and choose at first an arbitrary $u \in \mathcal{D}(\Omega)$. We can extend $u$ by zero to $Q$. Now, define $\widetilde{x} \in \mathbb{R}^{n-1}$ by $x = (x_1, \widetilde{x})$ for $x \in \Omega$, and set $\widehat{x} = (-d, \widetilde{x}) \in Q$. Since $u(\widehat{x}) = 0$, we obtain by the fundamental theorem of calculus and the Cauchy-Schwarz inequality

$$u(x) = u(\widehat{x}) + \int_{-d}^{x_1} \frac{\partial u}{\partial y}(y, \widetilde{x}) dy = \int_{-d}^{x_1} 1 \cdot \frac{\partial u}{\partial y}(y, \widetilde{x}) dy$$

$$\leq \left( \int_{-d}^{x_1} 1 dy \right)^{1/2} \left( \int_{-d}^{x_1} \left( \frac{\partial u}{\partial y}(y, \widetilde{x}) \right)^2 dy \right)^{1/2} \leq \sqrt{2d} \left( \int_{-d}^{d} \left( \frac{\partial u}{\partial y}(y, \widetilde{x}) \right)^2 dy \right)^{1/2}.$$

Integrating yields

$$\|u\|_{L^2}^2 = \int_{\Omega} u(x)^2 dx \leq 2d \int_{\Omega} \int_{-d}^{d} \left( \frac{\partial u}{\partial y}(y, \widetilde{x}) \right)^2 dydx$$

$$= 2d \int_{Q} \int_{-d}^{d} \left( \frac{\partial u}{\partial y}(y, \widetilde{x}) \right)^2 dydx = 2d \int_{[-d,d]^{n-1}} \int_{-d}^{d} \int_{-d}^{d} \left( \frac{\partial u}{\partial y}(y, \widetilde{x}) \right)^2 dydx_1 d\widetilde{x}$$

$$= (2d)^2 \int_{[-d,d]^{n-1}} \int_{-d}^{d} \left( \frac{\partial u}{\partial y}(y, \widetilde{x}) \right)^2 dyd\widetilde{x} = (2d)^2 \left\| \frac{\partial u}{\partial x_1} \right\|_{L^2}^2 \leq (2d)^2 |u|_{H^1}^2.$$

Thus, the theorem is proven for $u \in \mathcal{D}(\Omega)$. For $u \in H_0^1$, choose $(u_n)_{n \in \mathbb{N}}$ in $\mathcal{D}(\Omega)$ converging to $u$ in the $H^1$-norm. For $\varepsilon > 0$ and $\|u_k - u\|_{H^1} \leq \varepsilon$, we have

$$\|u\|_{L^2} \leq \|u_k\|_{L^2} + \|u - u_k\|_{L^2} \leq 2d|u_k|_{H^1} + \varepsilon$$

$$\leq 2d|u|_{H^1} + 2d|u_k - u|_{H^1} + \varepsilon \leq 2d|u|_{H^1} + (2d + 1)\varepsilon.$$

Therefore, the Poincaré-Friedrichs inequality is proven, since $\varepsilon$ was arbitrary. By definition, $\|u\|_{H^1}^2 = \|u\|_{L^2}^2 + \sum_{|k|=1} \|\partial^k u\|_{L^2}^2 = \|u\|_{L^2}^2 + |u|_{H^1}^2$ for all $u \in H_0^1(\Omega)$, which shows the asserted equivalence. $\qquad\qquad\square$

Readers interested in the proofs not included in the Appendix are once again recommended the textbooks on functional analysis and partial differential equations mentioned on p. 244 and elsewhere.

# Appendix C
# Solutions to the Exercises

## Exercises of Chap. 3

**(A1)** For parts (a) and (b), the Fourier series $S_f$ of $f$ has the expansion

$$S_f(x) = A\left[\tfrac{3}{4} + \underbrace{\left(-\frac{2}{\pi^2}\cos(x) - \frac{1}{\pi}\sin(x)\right)}_{0.377\,\sin(x-2.575)} + \underbrace{\left(-\frac{1}{2\pi}\sin(2x)\right)}_{0.159\,\sin(2x+3.142)}\right.$$

$$+ \underbrace{\left(-\frac{2}{9\pi^2}\cos(3x) - \frac{1}{3\pi}\sin(3x)\right)}_{0.108\,\sin(3x-2.932)} + \underbrace{\left(-\frac{1}{4\pi}\sin(4x)\right)}_{0.08\,\sin(4x+3.142)}$$

$$\left. + \, 0.064\sin(5x - 3.015) + \ldots \right].$$

The Fourier series converges to $\frac{A}{2}$ at $x = 0$.

**(A2)** The Fourier series $S_u(t)$ of $u(t)$ is given by

$$S_u(t) = \hat{u}\left[\frac{1}{\pi} + \frac{1}{2}\sin(\omega_0 t) - \frac{2}{\pi}\sum_{n=2,4,6,\ldots}\frac{1}{(n^2 - 1)}\cos(n\omega_0 t)\right].$$

**(A3)** (a) The Fourier coefficients are $c_k = (a_k - jb_k)/2$ with

$$a_k = \tfrac{2}{T}\int\limits_{-T/2}^{0}\left|\sin\left(\tfrac{\omega}{2}t\right)\right|\cos(k\omega t)\,dt = -\frac{2}{\pi(4k^2-1)},$$

$$b_k = \tfrac{2}{T}\int\limits_{-T/2}^{0} -\sin\left(\tfrac{\omega t}{2}\right)\sin(k\omega t)\,dt = \frac{4(-1)^k k}{\pi(4k^2-1)}.$$

© The Author(s), under exclusive license to Springer Nature Switzerland AG 2025
R. Brigola, *Fourier Analysis and Distributions*, Texts in Applied Mathematics 79,
https://doi.org/10.1007/978-3-031-81311-5

(b) At $T/2$, it holds for $n \in \mathbb{N}$: $S_f\left(\frac{T}{2}\right) = \frac{2}{\pi}\left(\frac{1}{2} + \sum\limits_{k=1}^{n} \frac{(-1)^{k+1}}{4k^2-1} + R_n\right)$.

Leibniz criterion: $\left|\frac{2R_n}{\pi}\right| < \frac{2}{\pi(4(n+1)^2-1)} < 0.5 \cdot 10^{-3}$ for $n \geqslant 17$.

(A4) For $\varepsilon > 0$, $h > 0$, and $N \geqslant \frac{1}{\varepsilon \sin(\frac{h}{2})} - \frac{1}{2}$, it holds according to the estimate on

page 25 in Sect. 3.1: $\left|S(t) - \sum_{k=1}^{N} \frac{\sin(kt)}{k}\right| \leqslant \frac{1}{(N+\frac{1}{2})\sin(\frac{h}{2})} \leqslant \varepsilon$.

(A5) (a) $\Re(f)$ is even, and $\Im(f)$ is odd. Therefore, the $k$-th Fourier coefficient $c_k$

is $c_k = \frac{1}{\pi} \int\limits_{0}^{\pi} \cos(x\sin(t) - kt)\mathrm{d}t = J_k(x)$. Here, $J_k$ is the Bessel function of

order $k$.

(b) It holds $J_{-k} = (-1)^k J_k$. Thus,

$$\cos(x\sin(t)) = J_0(x) + 2\sum_{k=1}^{\infty} J_{2k}(x)\cos(2kt),$$

$$\sin(x\sin(t)) = 2\sum_{k=1}^{\infty} J_{2k+1}(x)\sin(2(k+1)t).$$

(A6) The function $f$ is the real part and $g$ the imaginary part of $F\left(\mathrm{e}^{jt}\right)$

$$F\left(\mathrm{e}^{jt}\right) = \frac{a}{a - \mathrm{e}^{jt}} = a\frac{(a - \cos(t)) + j\sin(t)}{a^2 - 2a\cos(t) + 1}.$$

(A7) (a) $a_n = \frac{2}{\pi} \int\limits_{0}^{\pi} \mathrm{e}^{at}\cos(nt)\mathrm{d}t = \frac{2}{\pi} \cdot \frac{a}{a^2+n^2}[(-1)^n\mathrm{e}^{a\pi} - 1]$.

(b) $b_n = \frac{2}{\pi} \int\limits_{0}^{\pi} \mathrm{e}^{at}\sin(nt)\mathrm{d}t = \frac{2}{\pi} \cdot \frac{n}{a^2+n^2}[1 - (-1)^n\mathrm{e}^{a\pi}]$.

## Exercises of Chap. 4

(A1) The Fourier series of $f$ for $t \neq k\pi$ is the derivative of

$$F(t) = |\sin t| = -\frac{2}{\pi}\sum_{k=-\infty}^{\infty} \frac{\mathrm{e}^{j2kt}}{4k^2 - 1} = \frac{2}{\pi} - \frac{4}{\pi}\sum_{k=1}^{\infty} \frac{\cos(2kt)}{(2k-1)(2k+1)}.$$

$F$ is continuous, and $f$ is piecewise continuously differentiable. Hence, $f$ has the Fourier series

$$S_f(t) = \frac{4}{\pi}\sum_{k=1}^{\infty} \frac{2k}{(2k-1)(2k+1)}\sin(2kt).$$

**(A2)** (a) $h(t) = \frac{4}{\pi}\left(\frac{\sin(t)}{1} + \frac{\sin(3t)}{3} + \dots\right) = \sum_{\substack{k=-\infty \\ k\neq 0}}^{+\infty} \frac{j}{\pi k}\left((-1)^k - 1\right) e^{jkt}$.

(b) $g(t) = -3h\left(\alpha(t - \frac{1}{2})\right)$ with $\alpha = \pi/2$.

$$\text{Result:} \quad g(t) = (-3) \sum_{\substack{k=-\infty \\ k\neq 0}}^{+\infty} \frac{j}{\pi k}\left((-1)^k - 1\right) e^{-jk\pi/4} \quad \cdot \quad e^{jkt\pi/2}$$

$$\underset{\substack{\text{Amplitude} \\ \text{factor}}}{\uparrow} \qquad\qquad\qquad \underset{\substack{\text{Phase} \\ \text{shifts}}}{\uparrow} \qquad \underset{\substack{\text{Frequency} \\ \text{change}}}{\uparrow}$$

**(A3)** (a) $f(t) = \frac{2}{\pi} - \frac{4}{\pi}\left(\frac{\cos(2t)}{3} + \frac{\cos(4t)}{3\cdot 5}\right)$.

(b) $(2|c_k|)_{k\in\mathbb{Z}} = (\dots, 0, \frac{4}{15\pi}, \frac{4}{3\pi}, \frac{4}{\pi}, \frac{4}{3\pi}, \frac{4}{15\pi}, 0, \dots)$.

(c) The amplitude-modulated function $\cos(6t)f(t) = \cos(3\omega_0 t)f(t)$ has the amplitude spectrum $2|d_k| = |c_{k-3} + c_{k+3}|$.

(d) $g(t) = -2\left(\frac{\cos(2t)}{15\pi} + \frac{\cos(4t)}{3\pi} - \frac{\cos(6t)}{\pi} + \frac{\cos(8t)}{3\pi} + \frac{\cos(10t)}{15\pi}\right)$.

(e) More generally, for $f(t) = \frac{a_0}{2} + \sum_{k=1}^{\infty} a_k \cos(k\omega_0 t)$, it holds

$$\cos(N\omega_0 t)f(t) = \frac{a_N}{2} + \sum_{k=1}^{\infty} \frac{a_{|k-N|} + a_{k+N}}{2} \cos(k\omega_0 t).$$

**(A4)** For $x \neq k\pi, k \in \mathbb{Z}$, we have

$$f(x) = \frac{4}{\pi}\left(\sin(x) + \frac{\sin(3x)}{3} + \frac{\sin(5x)}{5} + \dots\right).$$

According to p. 47, with $c_0 = 0$ and $F_0 = \frac{1}{2\pi}\int_0^{2\pi}\int_0^{t} f(x)\,\mathrm{d}x\,\mathrm{d}t = \frac{\pi}{2}$, we

obtain

$$\int_0^{t} f(x)\,\mathrm{d}x = \frac{\pi}{2} - \frac{4}{\pi}\left(\cos(x) + \frac{\cos(3x)}{3^2} + \frac{\cos(5x)}{5^2} + \dots\right).$$

**(A5)** Substitute in $\sum_{m=-N}^{+N} \sum_{n=-N}^{N} c_m d_n e^{j(m+n)t}$ with $k = m + n$ (compare also with the well-known Cauchy product in power series, whose coefficients also arise from a discrete convolution).

**(A6)** Look for examples in formula collections as L. Råde, B. Westergren (2004).

**(A7)** The Fourier series of the square wave is $F(t) = \frac{4}{\pi}\sum_{k=1}^{\infty} \frac{\sin((2k-1)t)}{2k-1}$.

Term-wise differentiation leads to the series $\frac{4}{\pi}\sum_{k=1}^{\infty} \cos((2k-1)t)$.
For $t = (2m+1)\pi/2, m \in \mathbb{Z}$, and all $k \in \mathbb{N}$, we have

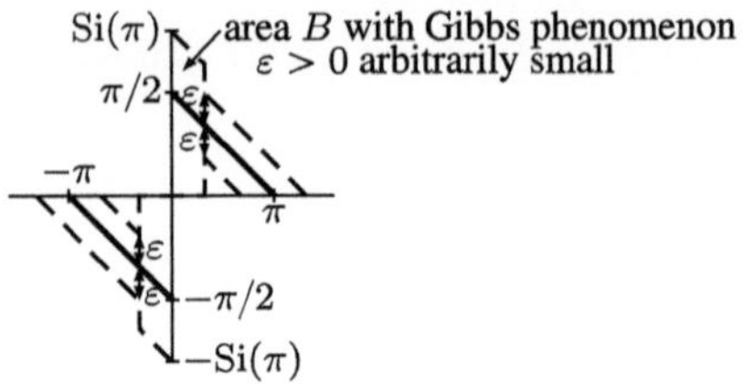


**Fig. C.1** Illustration with regard to Gibbs phenomenon

$\cos((2k-1)(2m+1)\pi/2) = 0$. At these points, the series has a limit of zero. Otherwise, the series diverges:

The assumption $\cos((2k-1)t_0) \to 0$ for some $t_0 \neq (2m+1)\pi/2$, as $k \to \infty$, leads to $\cos(2kt_0)\cos(t_0) = \frac{1}{2}[\cos((2k+1)t_0) + \cos((2k-1)t_0)] \to 0$. Thus, $\cos(2kt_0) \to 0$ as $k \to \infty$; similarly, one would have $\sin(2kt_0) \to 0$ or $\sin(t_0) = 0$. This results in the contradiction $t_0 = m\pi$ for a suitable $m \in \mathbb{Z}$, and thus $|\cos((2k-1)t_0)| = 1$ for $k \in \mathbb{N}$, or $\cos^2(2kt_0)+\sin^2(2kt_0) \to 0$ as $k \to \infty$.

(**A8**) Verify the series expansions by integrations.

(**A9**) Example 1, p. 41: $\sum_{n=1}^{\infty} \frac{1}{4n^2-1} = \frac{1}{2}$.

Example 1, p. 52: $\sum_{n=1}^{\infty}(-1)^{n+1}\frac{1}{n^2} = \frac{\pi^2}{12}$.

With the Fourier series for $f(t) = t(\pi - |t|)$, $t \in (-\pi, \pi)$,

$$\sum_{n=1}^{\infty}(-1)^{n+1}\frac{1}{(2n-1)^3} = \frac{\pi^3}{32}.$$

(**A10**) The proof of the statement—a conjecture by L. Fejér—dates back to E. Landau (1933). Let $S_N(t) = \sum_{k=1}^{N} \frac{\sin(kt)}{k}$. It holds that $S_1(t) > 0$ in $(0, \pi)$. Assuming now that $S_{N-1}(t) > 0$ in $(0, \pi)$ as an induction hypothesis. From the assumption $S_N(t) \leq 0$ for some $t \in (0, \pi)$, it follows that there exists a $t_0$ in $(0, \pi)$ such that $S_N(t_0) \leq 0$ and $S_N$ has a local minimum at $t_0$, and hence $S_N'(t_0) = 0$.

$$S_N'(t) = \frac{\sin((N+1/2)t) - \sin(t/2)}{2\sin(t/2)} \quad \text{for } t \neq 2\pi n, \, n \in \mathbb{Z},$$

thus with $0 < t_0/2 < \pi/2$, $\sin((N+1/2)t_0) = \sin(t_0/2) > 0$, and with phase shift by $\pi/2$ also $|\cos((N+1/2)t_0)| = \cos(t_0/2) > 0$. With $\sin(Nt_0) = \sin((N+1/2)t_0)\cos(t_0/2) - \cos((N+1/2)t_0)\sin(t_0/2)$, thus $\sin(Nt_0) \geq 0$, and from this the contradiction $S_{N-1}(t_0) \leq S_N(t_0) \leq 0$. The functions $S_N$ are odd; hence $S_N(t) < 0$ in $(-\pi, 0)$ for all $N \in \mathbb{N}$. Therefore, the sawtooth can be approximated within a tolerance $\varepsilon > 0$ in area $B$ of the form in Fig. C.1.

(**A11**) (a) This follows with $z = e^{jt}$ by calculating the imaginary part of $\sum_{k=1}^{n} z^k$.

(b) Term-wise integration as on p. 47 gives

$$\int_0^t f(x)\,dx = \frac{a_0}{2}t + \sum_{k=1}^{\infty} \frac{b_k}{k} + \sum_{k=1}^{\infty} \frac{-b_k\cos(kt) + a_k\sin(kt)}{k} < \infty.$$

Note: The integrability of $f$ on $[0, 2\pi]$ would be sufficient, according to a more general version of Dirichlet's theorem, for the representation of the integral function by its Fourier series. It is noteworthy that, unlike the sine coefficients of a classical Fourier series, $\sum_{k=1}^{\infty} a_k/k < \infty$ does not necessarily hold for cosine coefficients. Example: The Fourier series $\sum_{k=2}^{\infty} \cos(kt)/\ln(k)$ (A. Zygmund (2003), therein V.1).

(c) With $a_n = s_n - s_{n-1}, s_0 = 0$, and $\sum_{k=1}^{n} s_{k-1}b_k = \sum_{k=1}^{n} s_k b_{k+1} - s_n b_{n+1}$.

(d) With $r_n = \sum_{k=1}^{n} a_k b_k$ and (c), for $n > m$, it follows that $|r_n - r_m| \leqslant 2Mb_{m+1}$, and from this, using the Cauchy convergence criterion, the statement follows. In particular, if we replace the numbers $a_k$, $b_k$ with functions $a_k(t)$, $b_k(t)$ on a closed interval $I$, such that in $I$ the partial sums $\sum_{k=1}^{n} a_k(t)$ remain uniformly bounded and the $b_k(t)$ converge monotonically and uniformly to zero from above, then $\sum_{k=1}^{\infty} a_k(t)b_k(t)$ converges uniformly on $I$.

(e) The first part from (a) and (d) with $a_k(t) = \sin(kt)$, $b_k(t) = 1/\ln(k)$.

On the other hand, $\sum_{k=2}^{\infty} \frac{1}{k\ln(k)}$ is divergent: $\frac{1}{k\ln(k)} > \int_k^{k+1} \frac{dx}{x\ln(x)}$ for $k \geq 2$

and $\lim_{b\to\infty} \ln(\ln(x))\big|_2^b = \infty$. Hence, from (b) it follows: The series does not represent an integrable function on $[0, 2\pi]$. For more details on when trigonometric series are Fourier series in the classical sense and why Fourier series since Riemann have also prompted thinking about the concept of integration, see A. Zygmund (2003) or the book by C. S. Rees, S. M. Shah, C. V. Stanojević (1981). In Sect. 9.1, it is shown that a distinction between convergent trigonometric series on the one hand and Fourier series on the other hand is no longer necessary when we simply interpret all convergent trigonometric series as periodic distributions.

## Exercises of Chap. 5

(**A1**) (a) $D = 1$,   (b) $D \approx 0.435$.

(**A2**) $(f * g)_{2\pi}(t) = \sum_{k\neq 0} j^{k+1} \frac{a}{\pi^2 k^3} \left((-1)^k - 1\right)^2 e^{jkt}$ is continuously differentiable.

(**A3**) The $T$-periodic transfer functions $g_T$ and $U_a$ are given with $\omega_0 = 2\pi/T$

$$g_T(t) = \sum_{k=-\infty}^{+\infty} \frac{R}{R - k^2\omega_0^2 RLC + jk\omega_0 L} e^{jk\omega_0 t},$$

$$U_a(t) = -\frac{2}{\pi} \sum_{k=-\infty}^{+\infty} \frac{U_0 R}{(4k^2 - 1)(R - k^2\omega_0^2 RCL + jk\omega_0 L)} e^{j2k\omega_0 t}.$$

**(A4)** For $n = \deg P \geq 3$, it follows from the triangle inequality $|P(z)| \geq |z|^{n-1}$ for $|z| > \sum_{k=0}^{n} |a_k| > 1$ ($a_n = 1$); hence $|P(jk)|^{-1} \leq |k|^{-2}$ for sufficiently large $|k|$. For $n=2$, sufficiently large $|k|$: $|P(jk)|^{-1} \leq |k^2 - a_0|^{-1} \leq |k|^{-3/2}$.

**(A5)** Calculate the Fourier coefficients of the function $e^{-a_0 t}$ considered on $]0, 2\pi[$. Deduce from this the statements of the exercise and the one-sided limits given on p. 65 $\pi(\coth(a_0\pi) \pm 1)$ as $t \to 0$ and $t \to 2\pi$.

**(A6)** $u(r, \phi) = \frac{U_0}{2} + \frac{4U_0}{\pi^2} \sum_{k=1}^{\infty} \left(\frac{r}{R}\right)^k \frac{\cos((2k-1)\phi)}{(2k-1)^2}.$

**(A7)** With the coefficients $a_n$ and $b_n$ from page 73 and the addition theorems

$$\cos\left(\frac{cn\pi}{l} t\right) \sin\left(\frac{n\pi}{l} x\right) = \frac{1}{2}\left[\sin\left(\frac{n\pi}{l}(x + ct)\right) + \sin\left(\frac{n\pi}{l}(x - ct)\right)\right],$$

$$\cos\left(\frac{n\pi}{l}(x - ct)\right) - \cos\left(\frac{n\pi}{l}(x + ct)\right) = 2\sin\left(\frac{n\pi}{l} x\right) \sin\left(\frac{cn\pi}{l} t\right),$$

and $a_n = A_n \sin(\varphi_n)$, $b_n = A_n \cos(\varphi_n)$, and the given representations follow.

**(A8)** The approach $u(x, t) = v(x)w(t)$ yields, with $\lambda_n = n\pi\sqrt{k}/l$, the Fourier sine coefficients $b_n$ of $f$, and superposition, $u(x, t) = \sum_{n=1}^{\infty} b_n e^{-\lambda_n^2 t} \sin\left(\frac{n\pi}{l} x\right)$.

**(A9)** With the approach $v(x, t) = \sum_{k=1}^{\infty} v_k(t) \sin(k\pi x/l)$, through coefficient comparison, one obtains the equation $v_k''(t) + \omega_k^2 v_k(t) = F_k(t)$, $\omega_k = ck\pi/l$. From this follows the solution $v_k(t) = \frac{1}{\omega_k} \int_0^t F_k(\tau) \sin(\omega_k(t - \tau))\mathrm{d}\tau$

(variation of constants or later with fundamental solution, p. 217). Calculate the solution explicitly with a computer algebra system for $f = g = 0$, $F(x, t) = A \sin(\omega t)$ with $A > 0$, $\omega > 0$.

**(A10)** $u(x, t) = \sum_{n=1}^{\infty} e^{-\kappa t}(a_n \cos(\lambda_n t) + b_n \sin(\lambda_n t)) \sin\left(\frac{n\pi x}{l}\right),$

$$\lambda_n^2 = \frac{n^2\pi^2 c^2}{l^2} - \kappa^2.$$

Here, $a_n = f_n$ are the Fourier coefficients of the initial displacement $f$, $b_n = (g_n + \kappa f_n)/\lambda_n$ with the Fourier coefficients $g_n$ of $g$. The oscillation decays exponentially with time. For the inhomogeneous problem, see, for example, G. P. Tolstov (1976).

**(A11)** With the $k$-th Bessel function $J_k$ of the first kind, one obtains

$$b_k = \frac{2}{k\pi} \int_0^{\pi} \cos(k\varphi - k\varepsilon \sin(\varphi))\mathrm{d}\varphi = \frac{2}{k}J_k(k\varepsilon).$$

## Exercises of Chap. 6

(**A1**) For $n > N$, $y_n = \sum_{k=0}^{N} a_k \cos((n-k)\omega T) = \Re\left(|\widehat{h}(\omega)| e^{j(n\omega T + \arg(\widehat{h}(\omega)))}\right)$.

(**A2**) $P$ interpolates at the nodes $t_k$ ($k = 0, \ldots, 2m-1$), $W_m(t_n - t_k) = 2m$ for $n = k$, $W_m(t_n - t_k) = 0$ for $n \neq k$, and thus it determines the uniquely defined trigonometric interpolation polynomial $P_2$ for $f$ in the space $V_m$ (see page 99).

(**A3**) With $N = 15$ and $\widehat{c}_0 = (f(0) - f(\pi-))/(2N)$ and with the Fourier coefficients $c_k$ of $f$, $\widehat{c}_1 = \sum_{m=-\infty}^{+\infty} c_{1+mN} + (f(0) - f(\pi-))/(2N)$. The corresponding trigonometric interpolation polynomial interpolates the value 1 at zero and therefore cannot be odd. A DFT after modifying $f$ with the value $f(0) = 0$ results in an odd interpolation function for the then-odd $2\pi$-periodic extendable function, which, however, unlike in Sect. 6 assumed, is not continuous in $[0, \pi[$.

(**A4**) $P_2(t) - f(t) = \sum_{|k| \leqslant N/2}''(\widehat{c}_k - c_k)e^{jkt} - \sum_{|k| \geqslant N/2}'' c_k e^{jkt}$. The statement then follows with the triangle inequality from $\widehat{c}_k - c_k = \sum_{l \in \mathbb{Z}, l \neq 0} c_{k+lN}$. Error estimations according to other criteria can be found, for example, in W. L. Briggs, Van Emden Henson (1995) and other textbooks on numerical mathematics.

(**A5**) At least 320 sampling values are needed in 2 s, thus a sampling frequency $\geqslant 160\,\text{Hz}$.

(**A6**) (a) $\widehat{c}_{24} \neq 0$ and $\widehat{c}_{104} \neq 0$. (b) 28 Hz alias: $\widehat{c}_{28} \neq 0$, $\widehat{c}_{100} \neq 0$.

(**A7**) The DFT values can be generated by any linear combination of oscillations with frequencies 45, 211, 301, 467, or 557 Hz. To avoid such ambiguities, anti-alias filters are used in practice.

(**A8**) With $m = 2000$, $N = 400$, $T = 0.2 \cdot 10^{-3}$s, subsampling through a DFT of the time duration $T$ with $N$ sampling values detects the oscillation $f$ with the DFT coefficients $\widehat{c}_{30} = -j/2$ and $\widehat{c}_{370} = j/2$ (cf. p. 90).

(**A9**) Set $z = 2\tilde{c}_1/A$ with the obtained DFT coefficient $\tilde{c}_1$ for the pilot carrier. The spectrum $(c_k)_{-4 \leqslant k \leqslant 4}$ of $f$ with $N = 8$ is obtained from the DFT coefficients $\tilde{c}_k$ by $c_0 = \tilde{c}_0$, $c_k = \tilde{c}_k z^{-k}$, and $c_{-N/2+k} = \tilde{c}_{N/2+k} z^{N/2-k}$ for $1 \leqslant k < N/2$, and $c_{N/2} = c_{-N/2} = \tilde{c}_{N/2}/(z^{N/2} + z^{-N/2})$.

(**A10**) Apply the IDFT to the convolution of the DFT coefficients of the $x_n$ and $y_n$, and compute analogously as on p. 96. Note potentially different normalizations by other authors and in software.

(**A11**) Use a computer algebra system to solve. You will find that the Clenshaw-Curtis quadrature is generally better than the trapezoidal rule.

(**A12**) Create a program to solve the exercise.

(**A13**) $\langle T_n, T_m \rangle_w$ results in zero for $n \neq m$, $\pi/2$ for $m = n \geqslant 1$, and $\pi$ for $m = n = 0$. It holds $T_k(x) = \Re(z^k)$ for $z = e^{j\varphi}$, $\varphi = \arccos(x)$. From the periodicity of the complex exponential function, it follows that at the points $x_n = \cos(n\pi/m)$, $0 \leqslant n \leqslant m$, all Chebyshev polynomials $T_n$, $T_{2m-n}$, $T_{2m+n}$, $T_{4m-n}$, $T_{4m+n}$, $T_{6m-n} \ldots$ have the same values. The function $f$ thus has zeros at the points $x_n$.

**(A14)** With $L : [-1, 1] \to [-3, 7]$, $L(t) = 5t + 2$, $g$ has the same interpolation polynomial $P$ as in Excercise 22 with 13 Chebyshev abscissas $x_n$ as nodes. The rescaled polynomial $P \circ L^{-1}$ interpolates $f$ at the nodes $L(x_n)$.

**(A15)** (a) Both sides coincide for real variables and are entire functions on $\mathbb{R}$. Therefore, there is a uniquely determined analytical continuation on $\mathbb{C}$. This is then given by (a) (identity theorem for power series).

(b) It always holds $w(z) = z \pm \sqrt{z^2 - 1}$, due to $\sqrt{z^2 - 1} = \pm z\sqrt{1 - 1/z^2}$, thus $w(z) = z \pm z\sqrt{1 - 1/z^2}$. For $|z| > 1$, $1 - 1/z^2$ lies in the circle around one with radius $r = 1$, with the principal value of the root.

Thus, $-\pi/4 < \arg(\sqrt{1 - 1/z^2}) < \pi/4$, hence $\Re(\sqrt{1 - 1/z^2}) > 0$. Therefore, for $w$ as indicated for the case $|z| > 1$, it holds

$$|w(z)| = |z|\,|1 + \sqrt{1 - 1/z^2}| > |z| > 1.$$

For $|z| \leqslant 1$, $z \notin [-1, 1]$, $\sqrt{z^2 - 1} = \pm j\sqrt{1 - z^2}$. In the upper half-plane $\Im(z) > 0$, the positive sign holds: For $j\varepsilon$, $\varepsilon > 0$, it follows that

$$|w(j\varepsilon)| = \varepsilon + \sqrt{1 + \varepsilon^2} > 1.$$

If there were a $z_0$ in the upper half-plane with $|w(z_0)| \leqslant 1$, then a $z_1$ would also lie on the line segment from $j\varepsilon$ to $z_0$ with $|w(z_1)| = 1$ due to the continuity of $|w(z)|$. This would be a contradiction, since then $z_1 \in [-1, 1]$ would hold. Thus, $|w(z)| > 1$ in the upper half-plane. The given formula for the lower half-plane is seen analogously with

$$w(-j\varepsilon) = -j\varepsilon - j\sqrt{1 + \varepsilon^2}$$

by considering the magnitude $|w(-j\varepsilon)| = \varepsilon + \sqrt{1 + \varepsilon^2} > 1$.

To see $T_n(z) = (w^n + w^{-n})/2$, set $z = \cos(x + jy)$, and calculate

$$(w^n + w^{-n})/2.$$

For (c) the poles of $Q$ lie on an ellipse with foci $\pm j\omega_c$ symmetrically to the real and imaginary axes. The solutions of the two equations

$$\cos(n(x + jy)) = \cos(nx)\cosh(ny) - j\sin(nx)\sinh(ny) = \pm j/\varepsilon$$

show all the poles of $Q$. The sought $n$ poles $z_k/(j\omega_c) = \cos(x_k + jy_k)$ with negative real parts are obtained with the given $x_k \in\, ]0, \pi[$ and then always $\sin(x_k) > 0$, if you constantly set $y_k = y = -\mathrm{arsinh}(1/\varepsilon)/n < 0$ from the equation solutions. (d) Program the solution to the exercise.

**(A16)** The values obtained by the DFT for $i_C$ are DC gain $\approx 2.44\,\text{mA}$, RMS value $\approx 2.99\,\text{mA}$, and distortion factor $\approx \sqrt{\dfrac{\sum_{k=2}^{14}|\widehat{c}_k|^2}{\sum_{k=1}^{15}|\widehat{c}_k|^2}} = 0.1816$.

## Exercises of Chap. 7

**(A1)** $\sum_{k=0}^{n-1}\sin((2k+1)\pi t) = \dfrac{2j}{-4}\left(\dfrac{\left(e^{jn\pi t}\right)^2 - 2 + \left(e^{-jn\pi t}\right)^2}{e^{j\pi t}-e^{-j\pi t}}\right) = \dfrac{\sin^2(n\pi t)}{\sin(\pi t)}$.

**(A2)** $\text{Si}(\pi) \approx \sum_{k=0}^{4}(-1)^k \dfrac{\pi^{2k+1}}{(2k+1)!(2k+1)} \approx 1.852$.

**(A3)** $n$-th partial sum of the sawtooth: $S_n'(0) = n$, slope of the Fejér mean $n/2$.

**(A4)** Modify the proof of Fejér's theorem on p. 133 with $\dfrac{1}{T}\displaystyle\int_{-T/2}^{T/2} K_n(t)\mathrm{d}t = 1$.

A number of useful summation kernels can be found, for example, in the textbook of J. S. Walker (1988).

**(A5)** $f(t) = \dfrac{2}{\pi}(S(t) - S(t-\pi))$, $S$ the sawtooth function. This results in the statement with Exercise A9 from Chap. 4.

**(A6)** (a) Review of the proof of Dirichlet's theorem on p. 130 shows that the integrability condition and the existence of the one-sided limits of $f$ and the right- and left-sided derivatives at $t_0$ are sufficient to obtain the convergence of the Fourier series to $(f(t_0+) + f(t_0-))/2$.

   (b) It suffices to show integrability on $[0, \pi/3]$.

$$\int_{\varepsilon}^{\pi/3} f(t)\mathrm{d}t = -\varepsilon\ln\left(2\sin\left(\frac{\varepsilon}{2}\right)\right) - \int_{\varepsilon}^{\pi/3}\frac{t\cos(t/2)}{2\sin(t/2)}\mathrm{d}t.$$

   The first term on the right-hand side vanishes for $\varepsilon \to 0$, and the integrand remains bounded in the second term.

   (c) $f$ is $2\pi$-periodic, even. For the Fourier cosine coefficients, it follows by integration $a_0 = 0$ and for $a_n$, $n \geqslant 1$, with $S(0+) = \pi/2$ and

$$\sin(nt)\cos(t/2) = 1/2 \cdot (\sin((n+1/2)t) + \sin((n-1/2)t)$$

   with integration by parts that $a_n = -1/n$. Since $f$ is differentiable for all $t \neq 2k\pi$, $k \in \mathbb{Z}$, (c) follows from (a).

   (d) With $x = t - \pi$ follows $\ln|2\cos(x/2)| = \ln|2\sin(t/2)|$ and hence the result.

**(A7)** The solution is the Ritz-Galerkin solution $u_4$ on p. 252, Sect. 9.5.

**(A8)** With the continuity of the translation in the $L^2$-norm, the assertion follows directly from $|(f * g)_{2\pi}| \leqslant \|f\|_2\,\|g\|_2$. Elementary proofs of these facts can be found in A. Zygmund (2003), there at I-9.4 and II-1.11.

## Exercises of Chap. 8

**(A1)** $T$ is the Dirac distribution $\delta$. $G$, $H$, and $R$ are continuous on $\mathcal{D}$, but not linear. $S$ is not defined for all $\varphi$ from $\mathcal{D}$, e.g., $\varphi(t) = e^t h(t)$ with a test function $h = 1$ in a neighborhood of zero. $U$ is a distribution.

**(A2)** $t\dot{\delta}(t) = -\delta(t)$, $t^2\dot{\delta}(t) = 0$, $t\ddot{\delta}(t) = -2\dot{\delta}(t)$, $t^2\ddot{\delta}(t) = 2\delta(t)$.

**(A3)** Apply $\mathrm{pf}\,(t^{-2})$, $\mathrm{pf}\,(t_+^{-2})$, and $\delta^{(m)}$ to a test function $t\varphi(t)$ (see p. 168).

**(A4)** With a test function $\alpha \geqslant 0$, $\alpha = 1$ in a neighborhood of zero

$$\mathrm{vp}\,(t^{-1}) = \alpha(t)\,\mathrm{vp}\,(t^{-1}) + (1 - \alpha(t))\frac{1}{t}.$$

**(A5)** The statement follows from the definition of the Cauchy principal value with the mean value theorem and Lebesgue's dominated convergence theorem (Appendix B, p. 496).

**(A6)** $\langle (t_+^\lambda)' , \varphi \rangle = -\lim_{\varepsilon \to 0} \int_\varepsilon^\infty t^\lambda \varphi'(t)\mathrm{d}t$. The statement follows through integration by parts with $\varphi'(t)\mathrm{d}t = \mathrm{d}u$, $u(t) = \varphi(t) - \varphi(0)$, $v(t) = t^\lambda$, and application of the mean value theorem.

**(A7)** $\ddot{f}(t) = -\sin(t)\left[s(t) - s\left(t - \frac{\pi}{2}\right)\right] + \delta(t) + \alpha\dot{\delta}(t) + 2s\left(t - \frac{\pi}{2}\right)$.
Illustrate the result with a sketch.

**(A8)** For $\psi_h(t) = (\varphi(t + h) - \varphi(t))/h$, $h \neq 0$, it holds according to the mean value theorem:

$$|\psi_h(t) - \varphi'(t)| = |\varphi'(t + \lambda h) - \varphi'(t)| \leqslant \sup_{t \in \mathbb{R}} |\varphi''(t)|\,|h|,$$

with a suitable $\lambda \in\, ]0, 1[$. That is, for $h \to 0$ the functions $\psi_h$ converge uniformly to $\varphi'$. Similarly, one shows for derivatives of $\psi_h$ of any order $n$ the uniform convergence to the derivative $\varphi^{(n+1)}$ of $\varphi$ for $h \to 0$. That is, $\psi_h \to \varphi'$ in $\mathcal{D}$ for $h \to 0$. For $\varphi \in \mathcal{D}$, it follows from the continuity of distributions $T$ on $\mathcal{D}$

$$\left\langle \frac{T(t+\Delta t) - T(t)}{\Delta t}, \varphi(t) \right\rangle = \left\langle T(t), \frac{\varphi(t - \Delta t) - \varphi(t)}{\Delta t} \right\rangle \xrightarrow[\Delta t \to 0]{} \langle T(t), -\varphi'(t)\rangle.$$

**(A9)** (a) Substitution $x = nt$ yields

$$\int_{-\infty}^{+\infty} \frac{\sin(nt)}{\pi t}\,\mathrm{d}t = \frac{2}{\pi}\int_0^{+\infty} \frac{\sin(x)}{x}\,\mathrm{d}x = \frac{2}{\pi}\int_0^{+\infty}\int_0^{+\infty} \sin(x)e^{-xy}\,\mathrm{d}y\,\mathrm{d}x.$$

Reverse the order of integration, and use integration by parts. Alternatively, use example 3 of p. 183.

(b) Proceed analogously to the proof of Dirichlet's theorem.

(**A10**) From $T(t) = -\pi \sum_{k=-\infty}^{\infty} \dot{\delta}(t - 2k\pi)$, it follows $\langle T(t), \varphi(t - \frac{1}{2}) \rangle = \pi \frac{16}{9} e^{-4/3}$.

(**A11**) (a)  $\dot{\delta}(t - a) * s(t - b) = \delta(t - (a + b))$.

      (b)  $\dot{\delta}(t - a) * \dot{\delta}(t - b) = \ddot{\delta}(t - (a + b))$.

      (c)  $s(t - a) * f(t) = \int\limits_{-\infty}^{t-a} f(u)\mathrm{d}u$.

      (d)  $(t - (a + b))s(t - (a + b))$.

      (e)  $s(t) * [\ln(t + 1)s(t + 1)] = s(t + 1)[\ln(t + 1) - 1](t + 1)$.

(**A12**) According to the hint

$$G_\sigma^{m_1} * G_\tau^{m_2}(x) = \frac{1}{2\pi\sigma\tau} e^{-v^2/2} \int\limits_{-\infty}^{+\infty} e^{-u^2/2}\mathrm{d}y$$

with

$$u = \frac{\sqrt{\sigma^2 + \tau^2}}{\sigma\tau} \left( y - \frac{\sigma^2 m_2 + \tau^2(x - m_1)}{\sqrt{\sigma^2 + \tau^2}} \right).$$

With $\mathrm{d}y = \frac{\sigma\tau}{\sqrt{\sigma^2+\tau^2}}\mathrm{d}u$, it follows

$$G_\sigma^{m_1} * G_\tau^{m_2}(x) = \frac{1}{2\pi\sqrt{\sigma^2 + \tau^2}} e^{-v^2/2} \int\limits_{-\infty}^{+\infty} e^{-u^2/2}\mathrm{d}u.$$

(**A13**) For $\mathbf{x}$ in the complement of $\mathrm{supp}(f) + \mathrm{supp}(g)$, it holds for any $\mathbf{y} \in \mathbb{R}^n$ due to $\mathbf{x} = \mathbf{y} + (\mathbf{x} - \mathbf{y})$ always $\mathbf{y} \notin \mathrm{supp}(f)$ or $(\mathbf{x} - \mathbf{y}) \notin \mathrm{supp}(g)$, thus $f(\mathbf{y})g(\mathbf{x} - \mathbf{y}) = 0$, and thus $(f * g)(\mathbf{x}) = 0$.

(**A14**) The proof is straightforward using the convergence properties in $\mathcal{D}$. Show that $\psi(x) = \langle T(y), \varphi(x + y) \rangle$ is a continuous function of $x$ and has a continuous derivative with

$$\frac{\psi(x_n) - \psi(x_0)}{x_n - x_0} \longrightarrow \langle T(y), \varphi'(x_0 + y) \rangle$$

for $x_n \to x_0$. Proceeding inductively, one obtains

$$(T * \varphi)^{(k)} = T * \varphi^k = (-1)^k T^{(k)} * \varphi.$$

The result holds also for the case of several variables.

(**A15**) For $x \notin \mathrm{supp}(T)$, there is a neighborhood $U$ of $x$, such that for all $\varphi \in \mathcal{D}$ with $\mathrm{supp}(\varphi) \subset U$ it holds: $\langle T, \varphi \rangle = 0$. If $x \in \mathrm{supp}(\dot{T})$, then there would be a $\varphi \in \mathcal{D}$ with $\mathrm{supp}(\varphi) \subset U$ and $\langle \dot{T}, -\varphi \rangle = \langle T, \varphi' \rangle \neq 0$.

Since $\mathrm{supp}(\varphi') \subset \mathrm{supp}(\varphi)$, this would be a contradiction to $x \notin \mathrm{supp}(T)$. Therefore, $\mathbb{R} \setminus \mathrm{supp}(T) \subset \mathbb{R} \setminus \mathrm{supp}(\dot{T})$.

**(A16)** $(1 * \dot{\delta}) * s = 0 \neq 1 = 1 * (\dot{\delta} * s)$ with the unit step function $s$.

**(A17)** Choose $a > 0$ with $\mathrm{supp}(G) \cap\, ]-a, a[\, = \emptyset$, $\psi \in \mathcal{D}$ with $\mathrm{supp}(\psi) \subset [-a, a]$, $0 \leqslant \psi \leqslant 1$, and $\psi = 1$ in $[-a/2, a/2]$. With $g(t) = (1 - \psi(t))/t$, then $T_0 = gG$ is a particular solution and $T_0 + k\delta$ $(k \in \mathbb{C})$ the general solution of $tT(t) = G(t)$.

**(A18)** For example, $T = \delta^m$ with $m \geqslant n$.

**(A19)** The solutions are $s(t)$, $e^{\alpha t}s(t)$, and $\dot{\delta}$.

**(A20)** It holds $\dot{S} = s' * T = \delta * T = T$. For $U \in \mathcal{D}'_+$ with $\dot{U} = T$, it is $(S - U)' = 0$, and hence $S - U = c$ constant. From the support condition follows $c = 0$; hence $U = S$.

**(A21)** Since one can transform an equation of $n$-th order (also in the sense of distributions) into a system $\mathbf{y}' = A(t)\mathbf{y}$ of the first order, it is sufficient to consider such systems. For any solution $\mathbf{y} \in \mathcal{D}'^n$, a fundamental matrix $F(t)$ of the system with $F(0) = E$ (identity matrix), and $\mathbf{u} = F^{-1}\mathbf{y}$, it follows by differentiation with $\mathbf{y}' - A\mathbf{y} = \mathbf{0}$ that $F\mathbf{u}' = 0$, and hence $\mathbf{u} = \mathbf{c}$ constant. Therefore, $y$ is a linear combination of the columns of $F$, which consist of infinitely differentiable functions.


## Exercises of Chap. 9


**(A1)** $\ddot{f}(t) = \frac{4A}{T}\delta(t + T) - \frac{2A}{T}\delta(t + T/2) - \frac{2A}{T}\delta(t - T/2)$ for $-T \leqslant t < T$. Representation of the corresponding periodic impulse sequence by a generalized Fourier series and twice (generalized) term-by-term differentiation of the Fourier series of $f(t)$ yields with $\omega_0 = \pi/T$ and coefficient comparison

$$f(t) = \frac{3A}{4} + \frac{4A}{\pi^2}\left[\cos(\omega_0 t) - \frac{\cos(2\omega_0 t)}{2} + \frac{\cos(3\omega_0 t)}{9} \mp \cdots\right].$$

**(A2)** *Impulse response.* One solves the homogeneous initial value problem (see p. 217) $LC\ddot{U}(t) + RC\dot{U}(t) + U(t) = 0$, $U(0) = 0$, $\dot{U}(0) = \frac{1}{LC}$. With the roots of the characteristic polynomial $P$ and the notations

$$\omega_d = \frac{1}{\sqrt{LC}} > 0, \quad \alpha = \frac{R}{2L} > 0, \quad \omega_1 = |\alpha^2 - \omega_d^2|^{1/2},$$

three cases arise for the impulse response $h(t)$. The function $s(t)$ denotes the unit step function.

(a) $\omega_d = \alpha$ (double real root of $P$): $h(t) = \omega_d^2 t e^{-\alpha t}s(t)$.

(b) $\omega_d < \alpha$ (two real roots of $P$): $h(t) = \frac{\omega_d^2}{\omega_1}\sinh(\omega_1 t)e^{-\alpha t}s(t)$.

(c) $\omega_d > \alpha$ (complex roots of $P$):   $h(t) = \frac{\omega_d^2}{\omega_1} \sin(\omega_1 t) e^{-\alpha t} s(t)$.

*Step response.*   $a(t) = (h * s)(t)$.

(a) $a(t) = \frac{\omega_d^2}{\alpha^2}\left(1 - e^{-\alpha t} - \alpha t e^{-\alpha t}\right) s(t)$ $\qquad\qquad\qquad$ for $\omega_d = \alpha$.

(b) $a(t) = \left(1 - (\cosh(\omega_1 t) + \frac{\alpha}{\omega_1} \sinh(\omega_1 t)) e^{-\alpha t}\right) s(t)$ $\qquad$ for $\omega_d < \alpha$.

(c) $a(t) = \left(1 - (\cos(\omega_1 t) + \frac{\alpha}{\omega_1} \sin(\omega_1 t)) e^{-\alpha t}\right) s(t)$ $\qquad$ for $\omega_d > \alpha$.

*System response to a rectangular signal*
For $U_e(t) = U_0 s(t) - U_0 s(t - T)$ we have $U_a(t) = U_0(a(t) - a(t - T))$.
For $t \to \infty$, $U_a(t)$ decays $\lim_{t\to\infty} U_a(t) = 0$.
*Periodic solution for a sine excitation*
For the input signal $U_e(t) = U_0 \sin(\omega t)$, one calculates independently of the
values $R$, $C$, and $L$ the solution with the frequency response

$$\widehat{h}(\omega) = \frac{1}{1 + j\omega RC - \omega^2 LC}.$$

The solution $U_a(t) =$

$$U_0 \omega_d^2 \left(\frac{\omega_d^2 - \omega^2}{(\omega_d^2 - \omega^2)^2 + 4\alpha^2\omega^2} \sin(\omega t) - \frac{2\alpha\omega}{(\omega_d^2 - \omega^2)^2 + 4\alpha^2\omega^2} \cos(\omega t)\right).$$

Regarding (c), if you transform a first-order system so that the system
matrix $A$ in the terminology of control engineering is the so-called observer
canonical form as given on p. 232, then you find the impulse response in the
component $a_{31} s(t)$ of $e^{At} s(t)$.

(A3) (a) $y''' + 4y'' + 6y' + 4y = f$ with inhomogeneity $f$.
$\quad$ (b) $y_H(t) = c_1 e^{-2t} + c_2 e^{-t} \sin(t) + c_3 e^{-t} \cos(t)$.
$\quad$ (c) $h(t) = \frac{1}{2}\left(e^{-2t} + e^{-t} \sin(t) - e^{-t} \cos(t)\right) s(t)$, $s(t)$ the unit step function.
$\quad$ (d) $\dot{X} = AX + F + \mathbf{x}_0 \delta$ with initial values $\mathbf{x}_0$, the matrix

$$A = \begin{pmatrix} 0 & 0 & -4 \\ 1 & 0 & -6 \\ 0 & 1 & -4 \end{pmatrix} \quad \text{and} \quad F = \begin{pmatrix} f \\ 0 \\ 0 \end{pmatrix}.$$

Then, as on p. 231, the third component of $X$ is the solution $y$ from (a). A
fundamental matrix is $G(t) = e^{At}$. $G_{31}(t)s(t)$ again shows the impulse
response $h$ from (c). For $\mathbf{x}_0 = \mathbf{0}$ and $\mathrm{supp}(f) \subset [0, \infty[$, the causal
solution of the inhomogeneous problem is the convolution $e^{At} s(t) * F$;
for continuous $F$ this is the formula of variation of constants.

(A4) For $|z| > l$, $u(0, 0, z) = -\gamma\rho_0 \int_{-l}^{+l} \frac{1}{\sqrt{(z-w)^2}} dw = -\gamma\rho_0 \, \mathrm{sgn}(z) \ln\left(\frac{z+l}{z-l}\right)$.

The equipotential surfaces are *rotational ellipsoids*. With the substitutions

$$l_1 = \sqrt{x^2 + y^2 + (z - l)^2}, \quad l_2 = \sqrt{x^2 + y^2 + (z + l)^2}$$

and *elliptical coordinates* $v = \frac{1}{2}(l_1 + l_2), \quad w = \frac{1}{2}(l_1 - l_2)$ follows $l_1 = v + w$, $l_2 = v - w$, $lz = -vw$. For the argument of the potential function (see p. 236, Sect. 9.4) follows

$$\frac{l_1 + l - z}{l_2 - l - z} = \frac{(v + l)l + w(l + v)}{(v - l)l + w(v - l)} = \frac{v + l}{v - l}.$$

Thus, the equipotential surfaces are given by the equations $v = \text{const}$. They are rotational ellipsoids, since $l_1 + l_2 = 2v$ is the sum of the distances of a point $(x, y, z) \notin S$ to the two focal points $(0, 0, l)$ and $(0, 0, -l)$. A power series expansion of the potential shows that $u$ for very large distances $v \gg l$ resembles the potential of a point mass at the origin.

(**A5**)  For the potential $u$ at the origin, according to Coulomb's formula (p. 237)

$$u(0) = \frac{1}{4\pi\varepsilon_0} \int_0^{2\pi} \int_0^{\pi/2} \frac{\sigma_0 \cos(\theta)}{R} R^2 \sin(\theta)\,d\theta\,d\phi = \frac{R\sigma_0}{4\varepsilon_0} \approx 169.5\,\text{kV}.$$

(**A6**)  For the sphere $K$ around the origin with radius $r$ and the internal Dirichlet problem

$$\Delta u = 0 \text{ in } K, \quad u = f \text{ on } \partial K,$$

Green's function is given by $G(\mathbf{x}, \mathbf{y}) = \frac{1}{4\pi}\left(\frac{1}{|\mathbf{x}-\mathbf{y}|} - \frac{r}{|\mathbf{y}|\,|\mathbf{x}-\frac{r^2}{|\mathbf{y}|^2}\mathbf{y}|}\right).$

One can also write $G(\mathbf{x}, \mathbf{y})$ in the following form:

$$G(\mathbf{x}, \mathbf{y}) = \frac{1}{4\pi}\left((|\mathbf{x}|^2 - 2\mathbf{x} \cdot \mathbf{y} + \mathbf{y} \cdot \mathbf{y})^{-1/2} - r(|\mathbf{x}|^2\mathbf{y} \cdot \mathbf{y} - 2r^2\mathbf{x} \cdot \mathbf{y} + r^4)^{-1/2}\right).$$

Partial differentiation with respect to $y_1$, $y_2$, and $y_3$, respectively, yields the gradient $\text{grad}_\mathbf{y}\, G(\mathbf{x}, \mathbf{y})$ for $\mathbf{x} \in K \setminus \partial K$, $\mathbf{y} = (y_1, y_2, y_3) \in \partial K$, $|\mathbf{y}| = r$:

$$\text{grad}_\mathbf{y}\, G(\mathbf{x}, \mathbf{y}) = \frac{(|\mathbf{x}|^2 - r^2)\mathbf{y}}{4\pi r^2|\mathbf{x} - \mathbf{y}|^3}.$$

For $|\mathbf{y}| = r$, $\mathbf{n} = \mathbf{n}(\mathbf{y}) = \mathbf{y}/r$, $\mathbf{x} \in K \setminus \partial K$, the normal derivative follows as

$$\frac{dG}{d\mathbf{n}}(\mathbf{x}, \mathbf{y}) = \text{grad}_\mathbf{y}\, G(\mathbf{x}, \mathbf{y}) \cdot \mathbf{n}(\mathbf{y}) = \frac{|\mathbf{x}|^2 - r^2}{4\pi r|\mathbf{x} - \mathbf{y}|^3}.$$

Using formula (8.5), p. 239, one obtains a representation of the potential $u(\mathbf{x})$ for $\mathbf{x}$ in $K \setminus \partial K$ based on the given boundary values:

$$u(\mathbf{x}) = \frac{r^2 - |\mathbf{x}|^2}{4\pi r} \int\limits_{|\mathbf{y}|=r} \frac{f(\mathbf{y})}{|\mathbf{x}-\mathbf{y}|^3} \mathrm{d}o(\mathbf{y}).$$

This is the *Poisson integral formula* for the sphere. For $\mathbf{x} = \mathbf{0}$ and $R < r$, with the boundary values $u(\mathbf{y})$ for $|\mathbf{y}| = R$, the mean value formula for potential functions follows

$$u(\mathbf{0}) = \frac{1}{4\pi R^2} \int\limits_{|\mathbf{y}|=R} u(\mathbf{y})\mathrm{d}o(\mathbf{y}),$$

more generally $u(\mathbf{x}) = \frac{1}{4\pi R^2} \int\limits_{|\mathbf{y}-\mathbf{x}|=R} u(\mathbf{y})\mathrm{d}o(\mathbf{y})$.

This can be used, for example, to prove the maximum principle for potential functions in $\mathbb{R}^3$ (see p. 70). Interested readers are referred for further information to the literature on potential theory and partial differential equations.

(**A7**) (a)  To show is $\langle \Delta g, h \rangle = \langle g, \Delta h \rangle = h(\mathbf{0})$, for $h \in \mathcal{D}(\mathbb{R}^2)$.

With the Laplace operator in polar coordinates, one calculates

$$\lim_{\varepsilon \to 0+} \int\limits_0^{2\pi} \int\limits_\varepsilon^\infty \ln(r) \left[ \frac{\partial^2}{\partial r^2} + \frac{1}{r}\frac{\partial}{\partial r} + \frac{1}{r^2}\frac{\partial^2}{\partial \phi^2} \right] h(r,\phi)\, r\, \mathrm{d}r\, \mathrm{d}\phi.$$

(i)  With integration by parts it follows

$$\int\limits_\varepsilon^\infty r\ln(r) \frac{\partial^2}{\partial r^2} h(r,\phi)\mathrm{d}r = -\varepsilon\ln(\varepsilon)\frac{\partial h}{\partial r}(\varepsilon,\phi) + (\ln(\varepsilon)+1)h(\varepsilon,\phi)$$

$$+ \int\limits_\varepsilon^\infty \frac{1}{r}h(r,\phi)\mathrm{d}r, \qquad \text{and}$$

$$\int\limits_\varepsilon^\infty \ln(r)\frac{\partial}{\partial r}h(r,\phi)\mathrm{d}r = -\ln(\varepsilon)h(\varepsilon,\phi) - \int\limits_\varepsilon^\infty \frac{1}{r}h(r,\phi)\mathrm{d}r.$$

(ii)  Since $h(r,\phi)$ is $2\pi$-periodic in the variable $\phi$, it follows

$$\int_0^{2\pi} \frac{1}{r}\ln(r)\frac{\partial^2}{\partial\phi^2}h(r,\phi)\mathrm{d}\phi = \frac{1}{r}\ln(r)\frac{\partial}{\partial\phi}h(r,\phi)\Big|_{\phi=0}^{\phi=2\pi} = 0.$$

In summary, one obtains

$$\int_0^{2\pi}\int_\varepsilon^\infty \ln(r)\left[\frac{\partial^2}{\partial r^2} + \frac{1}{r}\frac{\partial}{\partial r} + \frac{1}{r^2}\frac{\partial^2}{\partial\phi^2}\right]h(r,\phi)\,r\mathrm{d}r\mathrm{d}\phi =$$

$$\int_0^{2\pi} h(\varepsilon,\phi)\mathrm{d}\phi - \int_0^{2\pi}\varepsilon\ln(\varepsilon)\frac{\partial}{\partial r}h(\varepsilon,\phi)\mathrm{d}\phi.$$

The first integral converges for $\varepsilon \to 0+$ to $2\pi h(\mathbf{0})$ and the second integral to zero, since $\frac{\partial h}{\partial r}$ is bounded and $\lim_{\varepsilon\to 0+}\varepsilon\ln(\varepsilon) = 0$.

(b) Green's function for the circular disk $K$ around zero with radius $R$ is

$$G(\mathbf{x},\mathbf{y}) = -\frac{1}{2\pi}\left(\ln(|\mathbf{x}-\mathbf{y}|) - \ln\left(\frac{|\mathbf{y}|}{R}\Big|\mathbf{x} - \frac{R^2}{|\mathbf{y}|^2}\mathbf{y}\Big|\right)\right).$$

(c) Analogously to the previous exercise, one finds for $\mathbf{x} \in K$, $|\mathbf{y}| = R$, $\mathbf{n} = \mathbf{y}/R$ the normal derivative

$$\frac{\mathrm{d}G}{\mathrm{d}\mathbf{n}}(\mathbf{x},\mathbf{y}) = \mathrm{grad}_\mathbf{y}\, G(\mathbf{x},\mathbf{y})\cdot\mathbf{n} = \frac{|\mathbf{x}|^2 - R^2}{2\pi R|\mathbf{x}-\mathbf{y}|^2}.$$

Inserting into formula (9.6) on p. 239, where according to Green's formula in the plane the surface integral over the boundary of the sphere is replaced by the line integral over the circle, gives for a point in $K$ with polar coordinates $(r,\phi)$ again the Poisson integral formula with the boundary potential $U$:

$$u(r,\phi) = \frac{R^2 - r^2}{2\pi}\int_0^{2\pi}\frac{U(\psi)}{R^2 + r^2 - 2Rr\cos(\phi-\psi)}\mathrm{d}\psi.$$

(A8) (a) Using the method of image charges, one finds Green's function for the half-space $H = \{(x_1,x_2,x_3) \in \mathbb{R}^3 : x_1 > 0\}$:

$$G(\mathbf{x},\mathbf{y}) = \frac{1}{4\pi|\mathbf{x}-\mathbf{y}|} - \frac{1}{4\pi|\mathbf{x}-\mathbf{y}_s|}.$$

Here, $\mathbf{y}_s = (-y_1, y_2, y_3)$ is the mirror point of $\mathbf{y} = (y_1, y_2, y_3)$.

(b) Its normal derivative with $\mathbf{n} = (-1, 0, 0)$ and $|\mathbf{x} - \mathbf{y}| = |\mathbf{x} - \mathbf{y}_s|$ for $\mathbf{x} = (x_1, x_2, x_3)$, $\mathbf{y} = (0, y_2, y_3)$ is

$$\operatorname{grad}_\mathbf{y} G(\mathbf{x}, \mathbf{y}) \cdot \mathbf{n} = -\frac{\partial G}{\partial y_1}(\mathbf{x}, \mathbf{y}) = -\frac{x_1}{2\pi |\mathbf{x} - \mathbf{y}|^3}.$$

(c) For the solution $u$ of $\Delta u = -\frac{\varrho}{\varepsilon_0}$ in $H$, $u = f$ on the plane $y_1 = 0$, at a point $\mathbf{x} \in H$, formula (9.6), where the boundary measure of the sphere is replaced by the Lebesgue measure in the plane $y_1 = 0$, gives the approach

$$u(\mathbf{x}) = \int\limits_{-\infty}^{+\infty} \int\limits_{-\infty}^{+\infty} f(y_2, y_3) \left(\frac{\partial G}{\partial y_1}(\mathbf{x}, \mathbf{y})\right)_{y_1=0} dy_2 dy_3 + \int\limits_{H} \frac{\varrho(\mathbf{y})}{\varepsilon_0} G(\mathbf{x}, \mathbf{y}) d\lambda^3(\mathbf{y}).$$

(d) With the given data $\varrho(\mathbf{y}) = q\delta(\mathbf{y} - \mathbf{x}_0)$, $\mathbf{x}_0 = (2, 0, 0)$, and $u = 0$ on the plane $y_1 = 0$, for $\mathbf{x} = (x_1, x_2, x_3) \in H$ it follows

$$u(\mathbf{x}) = \frac{q}{4\pi \varepsilon_0} \left(((x_1 - 2)^2 + x_2^2 + x_3^2)^{-1/2} - ((x_1 + 2)^2 + x_2^2 + x_3^2)^{-1/2}\right).$$

(A9) For example, solve the linear system of equations given on p. 257 with Maple, Mathematica, Maxima, or Matlab. The approximate value for $u(L/2, L/2)$ is 0.1827 m. For graphical representation as on p. 258, connect the approximate values of the solution with polygonal lines.

(A10) Solve analogously the problem with an L-shaped membrane as on p. 244. Make use of a triangulation near the edges as fine as possible with your computer. The FEM solution below was computed for $f = 0.5\,\mathrm{N/m^2}$ and $k = 2\,\mathrm{N/m}$.

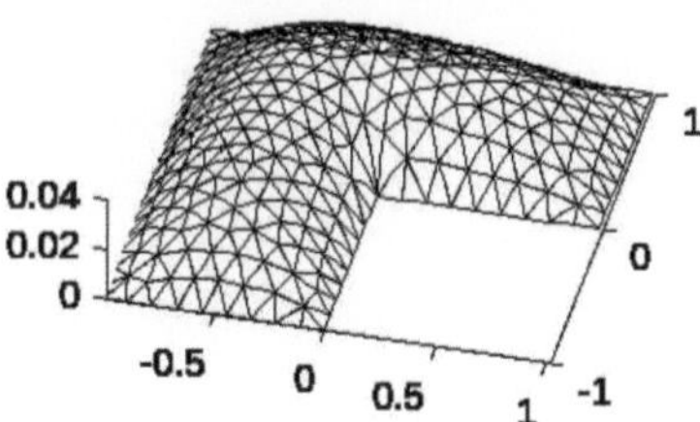

(A11) As in the previous exercise, write a small program for a computer algebra or numerical system. For an interior node $x_k$ of the interval subdivision, the corresponding basis function is

$$v_k(x) = \begin{cases} \frac{Hp}{L}(x - x_{k-1}) & \text{for } x_{k-1} \leq x \leq x_k, \\ -\frac{Hp}{L}(x - x_{k+1}) & \text{for } x_k \leq x \leq x_{k+1}. \end{cases}$$

Using $a(u, v)$ and $l(v)$ from p. 248, set up the system of equations, and calculate the required approximate solutions:

$$T(L/6) \approx 274.271\,°\text{K}, \quad T(L/3) \approx 276.414\,°\text{K}, \quad T(2L/3) \approx 279.628\,°\text{K}.$$

*Note*: The analytical solution of the problem is piecewise linear, continuous, and, in the present case, concave. It is

$$T(x) = \begin{cases} (T_1 - T_0)\dfrac{3k_2}{L(2k_1+k_2)}\, x + T_0 & \text{for } 0 \leqslant x \leqslant L/3, \\[2mm] (T_1 - T_0)\dfrac{3k_1}{L(2k_1+k_2)}\,(x - L) + T_1 & \text{for } L/3 \leqslant x \leqslant L. \end{cases}$$

**(A12)** You can find the Fourier series of $f_1(x)$ using the similarity theorem from formula collections. For the sequence of amplitudes $A_k$, $k \in \mathbb{N}$, it follows

$$(A_1, A_2, A_3, A_4, A_5, \ldots) = \frac{4al}{\pi^2}\left(1, 0, \frac{1}{9}, 0, \frac{1}{25}, 0, \ldots\right).$$

The overtones quickly become very weak compared to the fundamental tone, the octave is absent, and the tone sounds pure and soft.

For the $2l$-periodic, odd extension of $f_2(x)$, the impulse method from p. 214 immediately yields $f_2(x) = \sum_{k=1}^{\infty}(-1)^{k+1}\frac{2n^2h}{(n-1)k^2\pi^2}\sin\left(\frac{k\pi}{n}\right)\sin\left(\frac{k\pi}{l}x\right)$. For $n = 2$, $h = al/2$ as a special case $f_2(x) = f_1(x)$ is included. For the amplitudes $A_k$, it follows, e.g., with $n = 100$ approximately

$$(A_1, A_2, A_3, A_4, \ldots) \approx \frac{6.3428h}{\pi^2}\,(1\,,\, 0.5\,,\, 0.3312\,,\, 0.2484\,,\, \ldots)\,.$$

Comparing these amplitude values with each other shows that for increasing $k$ they approximately decrease as $1/k$, the sound is rather hard and shrill. This corresponds to the experience one has when plucking a guitar near the bridge.

For increasing $n \in \mathbb{N}$, the function sequence $(f_{2,n})_{n\in\mathbb{N}}$ is generated with $f_2$, which converges pointwise to the sawtooth $-\frac{2h}{\pi}S\left(\frac{\pi}{l}(x - l)\right)$ (cf. p. 44).

Accordingly, the amplitudes $A_{k,n} = \left|\frac{2n^2h}{(n-1)k^2\pi^2}\sin\left(\frac{k\pi}{n}\right)\right|$ converge for $n \to \infty$ to the amplitudes $\tilde{A}_k = \frac{2h}{k\pi}$ of this sawtooth. The solutions of the wave equation for the displacements $f_1(x)$ and $f_2(x)$ are not differentiable, thus only to be understood as distribution solutions.

**(A13)** For (a) and (b), the impulse response is $h(t) = ts(t)$, $s(t)$ the unit step. Naive discretization of the convolution equation in the form

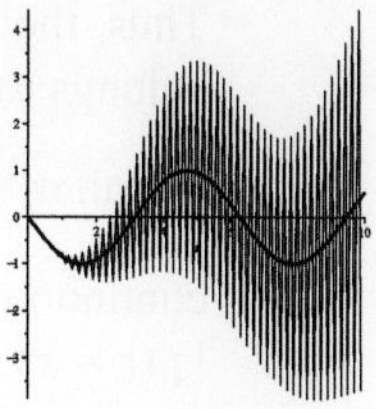

**Fig. C.2**  Naive result

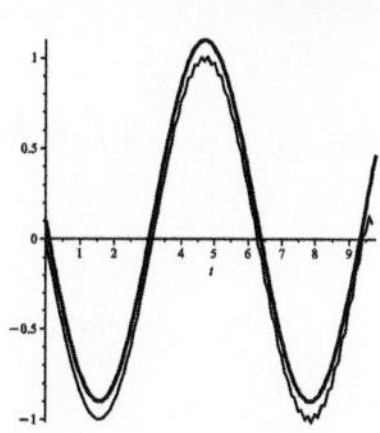

**Fig. C.3**  Regularized
solution

$$x(t_k) = \Delta \sum_{i=1}^{n} h(t_k - s_i) f(s_i)$$

with $\Delta = T/n$, $1 \leqslant i, k \leqslant n$, and equidistant nodes leads to a numerical
solution as shown in Fig. C.2 (exact solution thick, numerical solution
oscillating strongly with growing amplitude). Tikhonov regularization as
given with the same system matrix $A$ leads to the solution in Fig. C.3, where
the exact solution is shown thick with an offset of $+0.1$ so that the two
curves can be clearly distinguished. The determinant of $A$ in the example
is approximately $10^{-200}$.

For (c) the impulse response is $h(t) = \frac{1}{(RC)^2} t e^{-t/(RC)} s(t)$. Distort the
solution $x(t)$ of the convolution equation for $\omega_0 = 2\,\text{rad/s}$, $RC = 1\,\text{s}$, and
$U_0 = 1\,\text{V}$

$$(h * f)(t) = \left( -4\cos(\omega_0 t) - 3\sin(\omega_0 t) + e^{-t/(RC)}(4 + 10\,\omega_0 t) \right) \frac{U_0 s(t)}{25},$$

and thereby calculate numerical approximate solutions for the inverse
problem. You get qualitatively similar results to those in (a) and (b). Use a
computer algebra program like Mathematica or Maple for the computations.

(A14)  $F(t, s) = f(t)g(s)$ is measurable. Further, $T(t, s) = (t - s, s)$ is a linear
transformation, so $H(t, s) = (F \circ T)(t, s) = F(t - s, s) = f(t - s)g(s)$ is
measurable. Therefore $\int\limits_{-\infty}^{+\infty} |H(t, s)|\,dt\,ds =$

$$\int\limits_{-\infty}^{+\infty} \left( \int_{-\infty}^{+\infty} |f(t - s)|\,dt \right) |g(s)|\,ds = \int_{-\infty}^{+\infty} \|f\|_1 |g(s)|\,ds = \|f\|_1 \|g\|_1.$$

Thus, the theorem of Fubini-Tonelli (cf. 496) implies that $f * g$ exists and belongs to $L^1(\mathbb{R})$.

For all $\alpha > 0$, it holds $\int\limits_{-\infty}^{+\infty} h_\alpha(t)dt = 1$ and $\lim_{|t|\to\infty} h_\alpha(t) = 0$. For $t \in \mathbb{R}$, continuous $f \in L^1(\mathbb{R})$, and $\varepsilon > 0$, there exists a $\delta > 0$ such that $|f(t-s) - f(t)| < \varepsilon$ for all $|s| < \delta$. From this it follows

$$|f * h_\alpha(t) - f(t)| = \left| \int\limits_{-\infty}^{+\infty} (f(t-s) - f(t))h_\alpha(s)ds \right|$$

$$\leq \varepsilon \cdot 1 + \|f\|_1 h_\alpha(\delta) + |f(t)| \int\limits_{|s| \geqslant \delta/\sqrt{\alpha}} h_1(s)ds.$$

The last two summands tend to zero for $\alpha \to 0+$, and the assertion follows, since $\varepsilon$ is arbitrary.


## Exercises of Chap. 10

(A1) $\widehat{f_1}(\omega) = \frac{2}{j\omega}$, $\widehat{f_2}(\omega) = \frac{4j}{\omega^3}$, $\widehat{f_3}(\omega) = \sqrt{\pi} \left( \frac{1}{2} - \frac{\omega^2}{4} \right) e^{-\omega^2/4}$,

$\widehat{f_4}(\omega) = \sqrt{\frac{\pi}{a}} e^{-(\omega^2+2j\omega b)/(4a)} e^{c+b^2/(4a)}$, $\widehat{f_5}(\omega) = 2\pi j \left( -e^{-\omega} s(\omega) + \delta(\omega) \right)$.

(A2) $f(t) = \frac{U_0 T}{\pi \omega_0 t^2} \sin(\omega_0 t) \sin(2\omega_0 t)$.

(A3) $\widehat{F_a * F_b} = \widehat{F}_{a+b}$.

(A4) $\int\limits_0^\infty \frac{\sin(ax)\sin(bx)}{x^2} dx = \frac{\pi}{2} \min(a, b)$.

(A5) $g(t) = \frac{1}{2\pi} \int\limits_{-\Omega}^{+\Omega} \widehat{f}(\omega) e^{j\omega t} d\omega$ is the orthogonal projection of $f$ onto the subspace $PW_\Omega$ of $L^2(\mathbb{R})$, which contains functions bandlimited by $\Omega$.

(A6) $\widehat{S}(\omega) = -2\pi j \frac{\sin(\omega T)}{\omega} \operatorname{sgn}(\omega)$.

(A7) Suppose $f(t) = s(t)$ the unit step function for (a) and (b), $g(t) = \cos(\omega t)s(t)$.

(A8) $\widehat{f}(\omega) = \begin{cases} \frac{2(\omega \cos(1)\sin(\omega) - \sin(1)\cos(\omega))}{\omega^2 - 1} & \text{for } |\omega| \neq 1, \\ 1 + \sin(2)/2 & \text{for } |\omega| = 1. \end{cases}$

The multiplication theorem holds, because $\widehat{\cos}$ has compact support and the Fourier transform of $s(t+1) - s(t-1)$ is a slowly increasing infinitely differentiable function.

(A9) For (a) and (c), $f = h$. See also Example 1 on p. 338.

(b) Using $\widehat{f}(\omega) = \frac{1}{j\omega b} - \frac{1}{2b(j\omega - j\sqrt{b/a})} - \frac{1}{2b(j\omega + j\sqrt{b/a})}$ follows with correspondences on page 298

**Fig. C.4** Illustration for
$-\operatorname{vp}(\cot(\omega/2))$

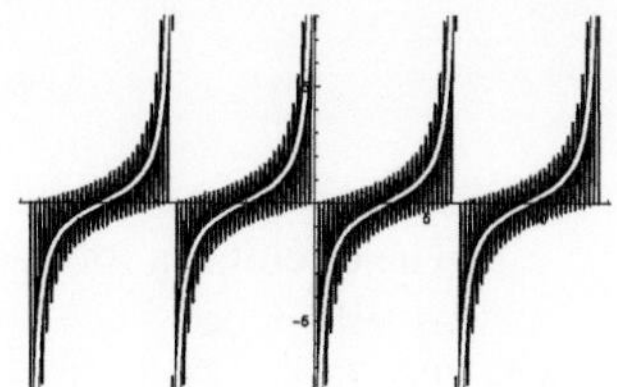

$$f(t) = \frac{1}{2b}\left(1 - \cos\left(\sqrt{\frac{b}{a}}t\right)\right)\operatorname{sgn}(t) = \frac{1}{b}\sin\left(\frac{\sqrt{ab}}{2a}t\right)^2\operatorname{sgn}(t).$$

**(A10)** (a) $\mathcal{F}^{-1}(\widehat{f})(t) = j\left(4e^{j3t} + e^{-j3t}\right)e^{-2t}s(t).$

(b) Using $\cos(x)^2 = (1 + \cos(2x))/2$ and $\cos(x + \pi/2) = \sin(-x)$, one obtains through integration by parts from p. 274 ($\widehat{h}_{\mathrm{RC},\alpha}$ is real-valued and even) $h_{\mathrm{RC},\alpha}(t) = \frac{\sin(bt)}{\pi t}\frac{\cos(at)}{1-(2at/\pi)^2}$, later with $b = \pi/t_a$, $a = \alpha b$ the pulse shape on page 390, with $b = T/2$, $a = \alpha b$ and the rule $\mathcal{F}(\widehat{f})(t) = 2\pi f(-\omega)$ the correspondence between $p$ and $\widehat{p}$ on page 398. Verify despite different forms: $\widehat{h}_{\mathrm{RC},\alpha}(t)/(2b) = \widehat{h}_{\mathrm{RC},\alpha}(t)/T = p(t).$

**(A11)** $\widehat{f}_1(\omega_1, \omega_2) = -\frac{\pi}{4}\omega_1\omega_2 e^{-(\omega_1^2 + \omega_2^2)/4}$. To obtain $\widehat{f}_2$, multiply the Fourier transform of the circular aperture on p. 314 by $e^{-j(\omega_1+\omega_2)}$.

**(A12)** $f$ is continuous and bounded; therefore $\dot{f}$ also belongs to $\mathcal{S}'$. The derivative computed with the chain rule is the generalized derivative of $f$ as a distribution in $\mathcal{D}'$. However, this is not tempered. The generalized derivative of $f$ as a continuous linear functional on $\mathcal{S}$ is an extension of this functional from $\mathcal{D}$ to $\mathcal{S}$.

**(A13)** $\widehat{f}(\omega) = -2j\left(\arctan(\omega/a) - \arctan(\omega/b)\right)$. With the hint, $\widehat{g}(\omega) = -j\pi\operatorname{sgn}(\omega) + 2j\arctan(\omega/b).$

**(A14)** $\widehat{f}(\omega) = -\frac{2}{\omega^2}$, $\widehat{g}(\omega) = \frac{2n!}{(j\omega)^{n+1}}$, $\widehat{h}(\omega) = -\frac{1}{\omega^2} + j\pi\dot{\delta}(\omega),$

$\widehat{p}(\omega) = \frac{(-j)^n\pi\omega^{n-1}}{(n-1)!}\operatorname{sgn}(\omega)$, $\widehat{q}(\omega) = 2\operatorname{pf}(|\omega|^{-\lambda})\cos(\lambda\pi/2)\,\Gamma(\lambda).$

**(A15)** The Fourier transform is $\widehat{f}(\omega) = \frac{\pi}{a\cosh(\pi\omega/(2a))}$. Thus, $f$ is an eigenfunction of the Fourier transform for $a = \sqrt{\pi/2}$ with eigenvalue $\sqrt{2\pi}$.

**(A16)** (a) Use integral transformation (see Appendix B, p. 497). First compute $\mathcal{F}(\varphi \circ A^T)$ for the transposed matrix $A^T$ and $\varphi \in \mathcal{S}(\mathbb{R}^p)$, and then $\mathcal{F}(T_A)$.
(b) Map with a matrix $A$ the parallelogram onto the square $-1 \leqslant x, y \leqslant 1$. Using (a), it follows $\widehat{f}_P(\omega_1, \omega_2) = \frac{4}{\omega_1(\omega_1+\omega_2)}\sin(\omega_1)\sin(\omega_1 + \omega_2).$

**(A17)** (Fig. C.4)

## Exercises of Chap. 11

**(A1)** The resistance $R$ represents the load.

$$|\widehat{h}(\omega)| = \frac{1}{\sqrt{(1 - 2LC\omega^2)^2 + 4\omega^2 L^2/R^2}}.$$

The circuit is a lowpass filter; it holds that $\widehat{U}_a \to \widehat{U}_e$ as $\omega \to 0$ and $\widehat{U}_a \to 0$ as $|\omega| \to \infty$.

(**A2**) The filter order is $n = 6$, and the cutoff frequency $\omega_c/(2\pi)$ is 3.397 kHz.

(**A3**) (a) Using the complex impedances $Z1 = 1/(j\omega C1)$ and $Z2 = 1/(j\omega C2)$ and the standard notation in alternating current calculations with upper-case letters,

$I_1 = I_2 + I_3$ and $\frac{V_{\text{in}}-V_1}{R1} = \frac{V_1}{R2+Z2} + \frac{V_1-V_{\text{out}}}{Z1}$. The voltage $v+$ at the operational amplifier is found similar to a voltage divider using $v_1$: $V+ = V_1\, Z2/(R2 + Z2)$. The "op-amp" is negative feedback, meaning in the steady state the op-amp input voltage is zero, and thus $v_{\text{out}} = v+$. From the last equation follows $V_1 = V_{\text{out}}(R2 + Z2)/Z2$. From this and the second equation $V_{\text{out}} = V_{\text{in}}Z1Z2/(Z1Z2+Z1(R1+R2)+R1R2)$. Finally,

$$\widehat{h}(\omega) = \frac{V_{\text{out}}}{V_{\text{in}}} = \frac{1}{1 + j\omega C2(R1 + R2) - \omega^2 R1R2C1C2}.$$

(b) Coefficient comparison of the solution from (a) with the Butterworth frequency response gives $R1 = R2 = 2813.49\ \Omega$.

(c) One obtains a highpass filter with transfer function

$$H_{HP}(s) = \frac{s^2}{s^2 + s(\frac{1}{R2C1} + \frac{1}{R2C2}) + \frac{1}{R1R2C1C2}}.$$

Coefficient comparison with the Butterworth polynomial analogous to (b) gives $R2 = 2R1$, $R1 = 1125.4\ \Omega$. Plot the amplitude and phase response and the group delay of the filter.

(**A4**) Verify the statements regarding the lowpass to highpass transformation, starting from a lowpass frequency response $\widehat{h}_{\text{LP}}(\omega) = \frac{\prod_{k=1}^{n}(-z_k)}{\prod_{k=1}^{n}(j\omega-z_k)}$ of order $n$ with DC gain $K = 1$ and poles $z_k$ of its transfer function.

(**A5**) (a), (b) Obtain the statements by substituting into the transfer functions, respectively, the frequency responses, starting from a lowpass filter as in the preceding exercise. For example, in the bandpass filter $w_{1,n}$ and $w_{2,n}$ are the positive solutions of the equations $1/B(js + 1/(js)) = -j$ and $1/B(js + 1/(js)) = j$ and for the bandstop filter analogously.

(c) Starting from the lowpass pole $z_0 = j\omega_c e^{j\pi/6}$, calculate with the inverse of the Joukowsky mapping the corresponding poles $z_{\text{BP},1}$ $z_{\text{BP},2}$ for the bandpass filter:

$z_{\text{BP},1}/\omega_c = -0.095278 + j1.15133$, $z_{\text{BP},2}/\omega_c = -0.0713887 - j0.862654$. Since it concerns a Butterworth filter, the bandstop poles are complex

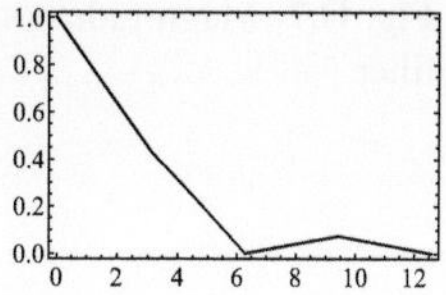

**Fig. C.5**  DFT without zero padding

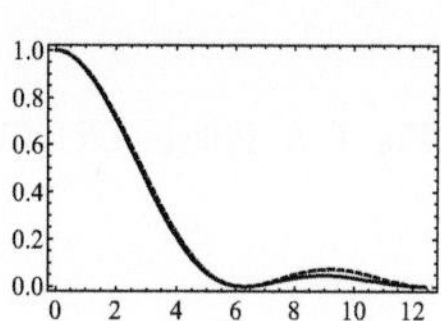

**Fig. C.6**  DFT with zero padding

conjugate to the bandpass poles. The highpass pole is complex conjugate to the lowpass pole.

(d) Use a computer algebra system and the bilinear transformation of p. 376. Compare the discrete with the analog filter.

**(A6)** A DFT approximation of the Fourier transform of $f$ with eight samples is poor (Fig. C.5). In Fig. C.6 you see the Fourier transform represented by the thick line $\widehat{f}$ over the angular frequency, dashed the approximation with 2040 zeros appended to the samples from $\mathrm{supp}(f)$. Deviations of the two approximations from the spectral function $\widehat{f}$ are unavoidable and due to the aliasing effects of the DFT, in the second case, reduced by adding many samples outside the support of $f$.

**(A7)** $h = \delta + 2\sum_{n=1}^{\infty} \delta_{2n}$.

**(A8)** The $z$-transform of $x_-$: $X\left(\frac{1}{z}\right)$, of $x_\alpha$: $X\left(\frac{z}{\alpha}\right)$, and of $v$: $-zX'(z)$.

**(A9)** (a) Plot using a computer algebra system. (b) The transfer function of the discrete notch filter obtained through bilinear transformation is

$$H(z) = \frac{a_0 + a_1 z^{-1} + a_2 z^{-2}}{b_0 + b_1 z^{-1} + b_2 z^{-2}},$$

where $a_0 = \frac{A_0(1+L^2)}{N}$, $N = 1 + \frac{L}{Q} + L^2$, $L = \cot(\omega_c a/2)$, $a_1 = \frac{2A_0(1-L^2)}{N}$, $a_2 = a_0$, $b_0 = 1$, $b_1 = \frac{2(1-L^2)}{N}$, $b_2 = \frac{1-\frac{L}{Q}+L^2}{N}$.

For $A_0 = 1$, $a = 1/44100\mathrm{s}$, $\omega_c = 2\pi \cdot 466\,\mathrm{Hz}$, and $Q = 10$, as shown in Fig. C.7, the amplitude response as a function of frequency is obtained and in Fig. C.8 the phase response of a discrete notch filter that blocks 466 Hz. As part (c) of the exercise, cascade four such filters.

**(A10)** The filter of order $n = 5$ has the following amplitude response and phase response (Figs. C.9 and C.10):

Attenuation at the passband edge is 0.199 dB and at the stopband edge is 49.181 dB.

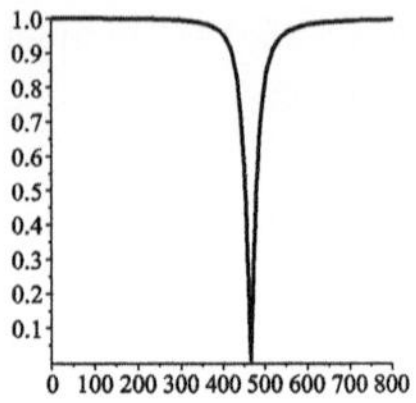

**Fig. C.7**  Magnitude notch
filter

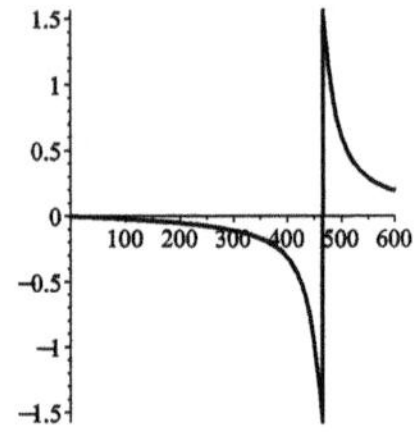

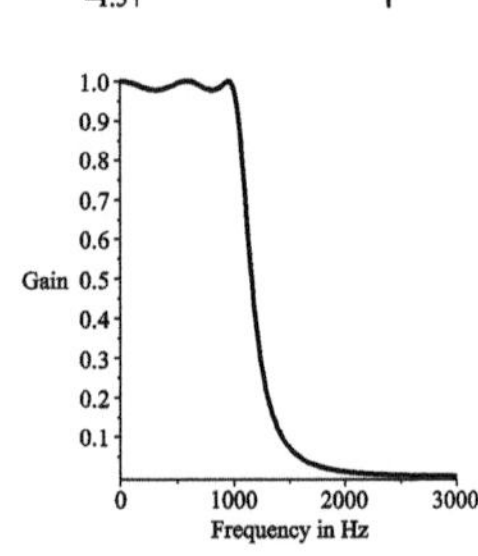

**Fig. C.8**  Phase notch filter

**Fig. C.9**  Gain Chebyshev
filter

**Fig. C.10**  Phase Chebyshev
filter

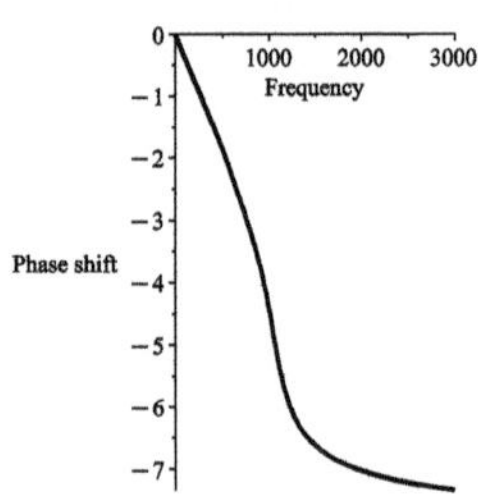

(A11)  The impulse responses of the two causal stable filters with transfer functions
$H$ and $H_{\text{inv}} = 1/H$ are obtained through partial fraction decompositions and
Laurent series expansions of the partial fractions.

$H$ has zeros $z_{0,1} = 1/3$ and $z_{0,2} = -1/5$ and poles $z_{\infty,1} = j/2$ and $z_{\infty,2} = -j/2$. From

$$H(z) = 1 - \frac{4 - 19j}{60(z - j/2)} - \frac{4 + 19j}{60(z + j/2)},$$

the causal impulse response is obtained through Laurent series expansion for
$|z| > 1/2$:

$$h = \delta_0 - \frac{4-19j}{60} \sum_{k=0}^{\infty} \left(\frac{j}{2}\right)^k \delta_{k+1} - \frac{4+19j}{60} \sum_{k=0}^{\infty} \left(\frac{-j}{2}\right)^k \delta_{k+1}.$$

$h$ has real coefficients and can also be rewritten with a bit of calculation into the form:

$$h = \delta_0 + \sum_{k=0}^{\infty} h_{k+1}\delta_{k+1}, \quad \text{where } h_{k+1} = \begin{cases} \dfrac{(-1)^{1+k/2}}{15\cdot 2^{k-1}} & \text{for } k \text{ even,} \\[2em] \dfrac{19\cdot(-1)^{(k+1)/2}}{15\cdot 2^{k+1}} & \text{for } k \text{ odd.} \end{cases}$$

The first filter coefficients—which can also be obtained using recursion equations on p. 366—are

$$(h_0, h_1, h_2, h_3, h_4, h_5, \ldots) = \left(1, -\frac{2}{15}, -\frac{19}{60}, \frac{1}{30}, \frac{19}{240}, -\frac{1}{120}, \ldots\right).$$

Similarly, we have $H_{\text{inv}} = 1 + \frac{65}{96(z-1/3)} - \frac{87}{160(z+1/5)}$ and thus the corresponding causal impulse response

$$h_{\text{inv}} = \delta_0 + \sum_{k=0}^{\infty} \left(\frac{65}{96}\frac{1}{3^k} + (-1)^{k+1}\frac{87}{160}\frac{1}{5^k}\right)\delta_{k+1}.$$

(**A12**) (a) From $|\widehat{h}| = 1$ follows $C(z) = 1$ for all $|z| = 1$. Multiplication with the denominator in that equation yields two equal polynomials, $M = N$, $d_k = 1/\overline{c_k}$, and thus the assertion.

(b) Compute the group delay of a factor $(z^{-1} - \overline{a_k})/(1 - a_k z^{-1})$, and use the assumed causality and stability ($n \geqslant 0$, $|a_k| < 1$).

(**A13**) (a) Consider the following pole-zero plot. The squares denote poles, and the circles denote zeros of the transfer function $H(z)$. Translate this illustration Fig. C.11 into a mathematical proof of the desired statement.

(b) Follows from part (a) of the exercise and from part (b) of the previous exercise A12.

Note: In some literature zeros with magnitude 1 are still allowed for minimum phase filters.

(**A14**) Proceed analogously to the proof of the example on p. 355.


## Exercises of Chap. 12


(**A1**) Compute analogously to the proof of the sampling theorem.

(**A2**) The windowed DFT shows a frequency that increases linearly with time.

(**A3**) For the Hann window, $D_t(w_T)D_\omega(w_T) = \frac{1}{6}(4\pi^2 - 30)^{1/2} \approx 0.513$.

**Fig. C.11**  Diagrams of the
zeros and poles for the filters

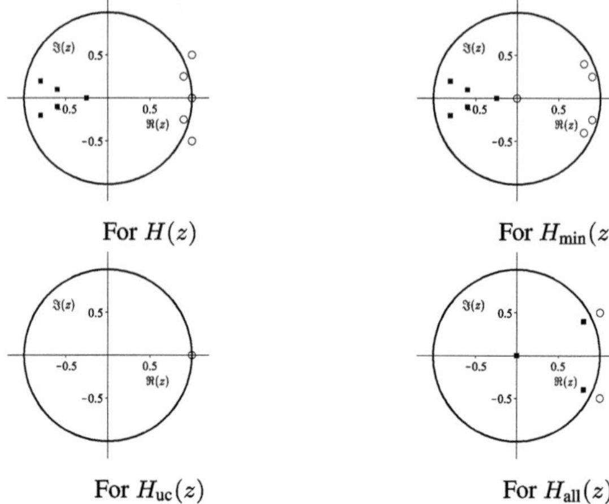

For $H(z)$

For $H_{\min}(z)$

For $H_{\mathrm{uc}}(z)$

For $H_{\mathrm{all}}(z)$

Test with examples the frequency localization of different window functions
with different signals.

(A4)  The weights $g_k$ relevant for the spectral leakage effect for the triangular
window $w_T$ for a signal of the form $f(t) = A e^{j\omega_1 t}$ are given by

$$
g_k = \left| \frac{w_T(2\pi k/T - \omega_1)}{T} \right| = \frac{1}{2} \left( \frac{\sin(k\pi/2 - \omega_1 T/4)}{k\pi/2 - \omega_1 T/4} \right)^2 .
$$

*Additionally:*  Compare the corresponding weights of the rectangular win-
dow (size of local maxima of $\widehat{w}_T$, decay behavior, etc.). Calculate $\|\widehat{w}_T\|$.

(A5)  The zeros of the characteristic polynomial have negative real parts. The
Fourier transform of the equation $x^{(3)} + 4\ddot{x} + 6\dot{x} + 4x = \sin(t)s(t) + 16\delta + 6\dot{\delta} + \ddot{\delta}$ leads by resolving for $\widehat{x}$, partial fraction decomposition, and inverse
transformation to the solution $x(t)$ with support in $[0, \infty[$

$$
x(t) = s(t) \left( \frac{41}{10} e^{-2t} - \frac{29}{10} e^{-t} \cos(t) + \frac{73}{10} e^{-t} \sin(t) - \frac{1}{5} \cos(t) \right).
$$

(A6) (a)  The fundamental solution $g$ corresponds to the temperature distribution
in $\mathbb{R}^3$ at time $t$, resulting when, under vanishing initial conditions at
$t = 0$, the temperature is increased by one unit at the origin. The
Fourier transform with respect to spatial coordinates yields, for fixed $\boldsymbol{\omega}$,
a differential equation in $t$:

$$
\frac{\partial}{\partial t} \widehat{g}(\boldsymbol{\omega}, t) + k|\boldsymbol{\omega}|^2 \widehat{g}(\boldsymbol{\omega}, t) = \delta(t), \quad \widehat{g}(\boldsymbol{\omega}, t) = 0 \text{ for } t < 0.
$$

By Theorem 9.5, p. 217, it has a unique causal fundamental solution. The
inverse Fourier transform (with respect to the coordinates of $\boldsymbol{\omega}$) of the

solution gives the desired fundamental solution for the heat equation, here in the 3D case with $n = 3$,

$$g(\mathbf{x}, t) = (4\pi kt)^{-n/2} e^{-|\mathbf{x}|^2/(4kt)} s(t).$$

(b) The corresponding solution of the inhomogeneous heat equation for right hand sides $F$, for which the convolution integral exists, is given by

$$u(\mathbf{x}, t) = (F * g)(\mathbf{x}, t) = \int_{\mathbb{R}} \int_{\mathbb{R}^3} F(\mathbf{y}, s) g(\mathbf{x} - \mathbf{y}, t - s) \, d\lambda^3(\mathbf{y}) \, ds.$$

If $F(\mathbf{x}, t) = 0$ for $t < 0$, then the time integral extends only over $[0, t]$. A sufficient condition for the validity of the formula is that for every interval $[0, t]$ and $\epsilon > 0$, there exists a constant $C_{t,\epsilon}$ such that for all $\mathbf{x}$: $|F(\mathbf{x}, t)| \le C_{t,\epsilon} e^{\epsilon|\mathbf{x}|^2}$.

(c) The corresponding initial value problem with $u(\mathbf{x}, 0) = f(\mathbf{x})$ has the solution $u = (F * g) + w$, where $w$ is the solution of the homogeneous problem with initial condition $f$ as in formula (12.12) on p. 431.

Test the results with concrete data analytically or numerically with the help of a computer algebra system, and generate graphical outputs.

(A7) Using again the Fourier transform related to the spatial variables for the equation $\frac{\partial^2 u}{\partial t^2} = c^2 \Delta_{\mathbf{x}} u$, $c > 0$, one obtains the fundamental solution $g_1(\mathbf{x}, t)$ with $g_1(\mathbf{x}, t) = 0$ for $t \le 0$, using the unit step function $s(t)$, through $\widehat{g}_1(\boldsymbol{\omega}, t) = \frac{\sin(ct|\boldsymbol{\omega}|)}{c|\boldsymbol{\omega}|} s(t)$. Inverse Fourier transform yields

$$g_1(\mathbf{x}, t) = \begin{cases} \frac{\delta(|\mathbf{x}| - ct)}{4\pi c^2 t} & \text{for } t > 0, \\ 0 & \text{for } t \le 0, \end{cases} \quad \text{where } \delta(|\mathbf{x}| - ct) \text{ denotes the Lebesgue}$$

surface measure on the sphere around the origin with radius $ct$. Considering the D'Alembertian operator instead, the obtained fundamental solution $g$ for $t > 0$ is (cf. Eq. (12.9)) $g(\mathbf{x}, t) = c^2 g_1(\mathbf{x}, t)$. This fundamental solution $g$ is called the *retarded fundamental solution*, i.e., delayed due to the finite propagation speed $c$. The convolution integral exists for locally integrable $F(\mathbf{x}, t)$, and the solution in the *Kirchhoff formula*, the so-called *retarded potential* at time $t$ is given by

$$u(\mathbf{x}, t) = (F * g)(\mathbf{x}, t) = \int_{\mathbb{R}^3} \int_0^\infty F(\mathbf{x} - \mathbf{y}, t - s) g(\mathbf{y}, s) \, ds \, d\lambda^3(\mathbf{y}).$$

This expression relates $u$ to the values of $F$ at earlier times $t - s$, $s > 0$, i.e., it is a causal solution.

Adding the solution from formula (12.8), p. 427, yields the solution of the wave equation in $\mathbb{R}^3$ for nonvanishing initial conditions.

The distribution $\widetilde{g}(\mathbf{x}, t) = g(\mathbf{x}, -t)$ is also a fundamental solution. It is the so-called *advanced fundamental solution*. For a discussion of that see, e.g., the Feynman Lectures, available from the Caltech website.

The solution for $\Box u = A \sin(\omega t) s(t) \otimes \delta(\mathbf{x})$ is immediately seen by the fundamental solution: $u(\mathbf{x}, t) = A \sin(\omega(t - |\mathbf{x}|/c))/(4\pi |\mathbf{x}|) \, s(t - |\mathbf{x}|/c)$ (see Fig. 12.24 and Eq. (12.9)). For the transformation of the integrals in (12.9), p. 429, use the substitution $c(t - s)\mathbf{n} = \mathbf{w}$. Then the expression with the surface integral is transformed into $\dfrac{1}{4\pi} \displaystyle\int_{\mathbb{R}^3} \dfrac{f(\mathbf{x} + \mathbf{w}, t - \frac{|\mathbf{w}|}{c})}{|\mathbf{w}|} \, d\lambda^3(\mathbf{w})$, which

gives with $\mathbf{y} = \mathbf{x} + \mathbf{w}$ the convolution integral for $f * \dfrac{\delta(t - \frac{|\mathbf{x}|}{c})}{4\pi |\mathbf{x}|}$.

(**A8**) Analogous proceeding as in A7 leads with $k = \hbar/(2m)$ for $t > 0$ to the differential equation $\frac{\partial}{\partial t}\widehat{\psi}(\boldsymbol{\omega}, t) + jk\widehat{\psi}(\boldsymbol{\omega}, t) = 0$, $\widehat{\psi}(\boldsymbol{\omega}, 0) = \widehat{\psi}_0(\boldsymbol{\omega})$. Inverse Fourier transformation of the solution yields for $t > 0$

$$\psi(\mathbf{x}, t) = (4\pi k t)^{-3/2} e^{-j3\pi/4} \int_{\mathbb{R}^3} \psi_0(\mathbf{x} - \mathbf{y}) e^{j|\mathbf{y}|^2/(4kt)} \, d^3\mathbf{y}.$$

Since this book cannot serve as an introduction to the theory of partial differential equations and their significance in physics but aims to present the Fourier transform as one of many tools in this field, interested readers are referred to the many excellent textbooks mentioned in the text on such equations for further work.

# References


Albrecht, E., & Neumann, M. (1979). Automatische Stetigkeitseigenschaften einiger Klassen linearer Operatoren. *Mathematische Annalen, 240*, 251–280. New York: Springer.

Albrecht, E. (2011). Causal discrete shift-invariant liner systems are convolution systems. Private Communication.

Atkinson, K., & Han, W. (2005). *Theoretical numerical analysis.* New York: Springer.

Banach, S. (1932). *Opérations Linéaires, Monografje Matematyczne, I.* Warszawa.

Blatter, C. (2003). *Wavelets—Eine Einführung.* Wiesbaden: Vieweg.

Bracewell, R. N. (1999). *The fourier transform and its applications.* New York: McGraw-Hill.

Braess, D. (1992). *Finite elemente.* Berlin: Springer.

Brass, H., & Petras, K. (2011). *Quadrature theory.* Providence, RI: AMS.

Briggs, W. L., & Van Emden H. (1995). *The DFT, an owners manual for the discrete fourier transform.* Philadelphia: SIAM.

Butterworth, S. (1930). *On the theory of filter amplifiers* (Vol. 7, pp. 536–554). London: Wireless Engineer.

Butzer, P. L., & Stens, R. L. (1992). *Sampling theory for not necessarily band-limited functions: A historical overview*(Vol. 34, No. 1, 40–53). New York: SIAM REVIEW.

Butzer, P. L., Stens, R. L., & Splettstößer, W. (1988). The sampling theorem and linear prediction in signal analysis. *Jahresbericht der Deutschen Mathematiker-Vereinigung, 90*, 1–70.

Carleson, L. (1966). On the convergence and growth of partial sums of Fourier series. *Acta Mathematica, 116*, 135–157.

Cassel, K. W. (2013). *Variational methods with applications in science and engineering.* Cambridge University Press.

Champeney, D. C. (1989). *A handbook of Fourier theorems.* Cambridge University Press.

Chandrasekharan, K. (1989). *Classical Fourier transforms.* Berlin: Springer.

Chang, R. W. (1966). High-speed multichannel data transmission with bandlimited orthogonal signals. *Bell System Technical Journal, 45*, 1775–1796.

Choi, J., Cvijović, D. (2010). Corrigendum, Values of the polygamma functions at rational arguments. *Journal of Physics A: Mathematical and Theoretical, 43*, 239801.

Chui, C. K. (1992). *An introduction to wavelets.* Boston: Academic Press.

Clenshaw, C. W., & Curtis, A. R. (1960). A method for numerical integration on an automatic computer. *Numerische Mathematik, 2*, 197–205.

Cohen, A., Daubechies, I., & Faveau, J. C. (1992). Biorthogonal bases of compactly supported wavelets. *Communications on Pure and Applied Mathematics, 45*, 485–500.

Cooley, J., & Tukey, J. (1965). An algorithm for the machine calculation of complex Fourier series. *Mathematics of Computation, 19*, 297–301.

Couch, L. W. (2012) *Digital and analog communication systems*. Upper Saddle River: Pearson Prentice Hall.

Courant, R., & Hilbert, D. (1993). *Methods of mathematical physics*. New York: Wiley-VCH.

Cox, I., Miller, M., Bloom, J., Fridrich, J., & Kalker, T. (2008). *Digital watermarking and steganography*. Amsterdam/Boston: Morgan Kaufmann Publishers.

Christensen, O. (2000). *An introduction to frames and Riesz Bases*. Berlin: Birkhäuser.

Cumming, I. G., & Wong, F. H. (2005). Digital signal processing of synthetic aperture radar data: Algorithms and implementation. Boston: Artech House.

Dahlke, S., Novak, E., & Sickel, W. (2010). Optimal approximation of elliptic problems by linear and nonlinear mappings IV: Errors in $L_2$ and other norms. *Journal of Complexity, 26*, 102–124.

Daubechies, I. (1992). *Ten lectures on wavelets*. Philadelphia: SIAM.

Dautray, R., & Lions, J. L. (1992). *Mathematical analysis and numerical methods for science and technology*, 6 Bd. Berlin: Springer.

Day, M. M. (1961). Fixed-Point theorems for compact convex sets. *Illinois Journal of Mathematics, 5*, 585–590.

Dehmer, M. (2006). On the location of zeros of complex polynomials. *Journal of Inequalities in Pure and Applied Mathematics, 7*(1), Article 26.

Dirac, P. A. M. (1958). *The principles of Quantum mechanics* (4th ed.). Oxford University Press.

Dieudonné, J. (2006). *Foundations of modern analysis*. Hong Kong: Hesperides Press.

Dirichlet, P. L. (1829). Sur la convergence des séries trigonométriques qui servent á représenter une fonction arbitraire entre des limites données. *Journal fur die Reine und Angewandte Mathematik, 4*, 157–169.

Duhamel, P., & Vetterli, M. (1990). Fast Fourier transforms: A tutorial review and the state of the art. *Signal Processing, 19*, 259–299.

Dym, H., & McKean, H. P. (1985). *Fourier series and integrals*. New York: Academic Press.

Edwards, R. E. (1982). *Fourier Series, a modern introduction* (Vol. 2). Berlin: Springer.

Ehrenpreis, L. (1954). Solution of some problems of division. Part I. Division by a polynomial derivation. *American Journal of Mathematics, 76*, 883–903.

Engl, H., & Groetsch, C. W. (1987). *Inverse and Ill-posed problems*. New York: Academic Press.

Erdös, P., & Turàn, P. (1937). *On interpolation, I. Quadrature and mean convergence in the Lagrange interpolation*. Annals of Mathematics, 38, 142–155.

Faber, G. (1912). Über die interpolatorische Darstellung stetiger Funktionen. *Deutsche Mathematik Jahr, 23*, 192–210.

Feichtinger, H. G., & Strohmer, T. (2003). *Advances in gabor analysis*. Boston: Birkhäuser.

Fejér, L. (1904). Untersuchungen über Fouriersche Reihen. *Mathematische Annalen, 58*, 501–569.

Fine, B., & Rosenberger, G (1997). *The fundamental theorem of algebra*. New York: Springer.

Fleming, I. (2009). *Molecular orbitals and organic chemical reactions*. New York: Wiley.

Folland, G. B. (1992). *Fourier analysis and its applications*. Pacific Grove: Wadsworth.

Folland, G. B. (1995). *Introduction to Partial Differential equations*. Princeton, N.J.: Princeton University Press.

Fourier, J. B. (2009). *The analytical theory of heat*. Cambridge University Press.

Gabor, D. (1946). *Theory of communication. Journal of the Institution of Electrical Engineers, London, 93*, 429–457.

Gardner, W. A., Napoletano, A., & Paura, L. (2008). Cyclostationarity: Half a century of research. *Signal Processing, 86*(4), 639–697.

Gel'fand, I. M., Shilov, G. E., & Vilenkin, N. Y. (1964). *Generalized functions* (Vol. 1–4). New York: Academic Press.

Gibbs, J. W. (1898). Fourier's series. *Nature, 59*, 200.

Goodman, A., & Newman, D. J. (1984). A Wiener type theorem for Dirichlet series. *Proceedings of AMS, 92*, 521–527.

Gottlieb, D., & Shu, C.-W. (1997). On the Gibbs phenomenon and its resolution. *SIAM Review, 39*(4), 644–668.

Grafakos, L. (2008). *Classical Fourier analysis*. New York: Springer.

Grafakos, L. (2010). *Modern Fourier analysis*. New York: Springer.

Gröchenig, K. (2001). Foundations of time-frequency analysis. Boston: Birkhäuser.

Groetsch, C. W. (1993). *Inverse problems in the mathematics sciences*. Wiesbaden: Vieweg.

Grosjean, C. C. (1984). Formulae concerning the computation of the Clausen integral Cl. *Journal of Computational and Applied Mathematics, 11*, 331–342.

Hadamard, J. (1932). *Le problème de Cauchy et les équations aux dérivées partielles linéaires hyperbolic*. Paris: Hermann.

Harris, F. J. (1978). On the use of windows for harmonic analysis with Discrete Fourier transform. *Proceedings of the IEEE, 66* (1), 51–83.

Helmberg, G., & Wagner, P. (1997). Manipulating Gibbs' Phenomenon for Fourier interpolation. *Journal of Functional Analysis, 89*, 308–320.

Henrici, P. (1979). Fast Fourier methods in computational complex analysis. *SIAM Review, 21*(4), 481–527.

Hirata, Y., & Ogata, H. (1958). *Journal of Science of the Hiroshima University*, Ser. A, Vol. 22, No. 3.

Hörmander, L. (2003). *The analysis of linear partial differential operators*. New York: Springer.

Holschneider, M. (1999). *Wavelets: An analysis tool*. Oxford: Oxford University Press.

Horváth, J. (1966). *Topological vector spaces and distributions*. New York: Dover Publications.

Huheey, J. E., Keiter, E. A., & Keiter, R. L. (1993). *Inorganic chemistry: Principles of structure and reactivity*. New York: HarperCollins College.

Hurd, W. J. (1997). Optimum and practical noncausal smoothing filters for estimating carrier phase with phase process noise. In *The Telecommunications and Data Acquisition Progress Report* (Vols. 42–128, pp. 1–7). California: Jet Propulsion Laboratory.

Jackson, D. (1912). On approximation by trigonometric sums and polynomials. *Transactions of the AMS, 14*, 491–515.

Jerri, A. (1977). The Shannon sampling theorem—its various extensions and applications: A tutorial review. *Proceedings of IEEE, 65*(11), 1565–1596.

John, F. (1981) *Partial Differential equations*. Berlin: Springer.

Jury, E. I. (1973) Theory and Application of the $z$-Transform, Krieger Pub. Co., Malabar

Kammeyer, K.-D., & Kroschel, K. (2012). *Digitale Signalverarbeitung*. Wiesbaden: Springer Vieweg.

Kawata, T. (1972). *Fourier analysis in probability theory*. New York: Academic Press.

Kincaid, D., & Cheney, W. (2002). *Numerical analysis*. Brooks/Cole, Pacific Grove.

Koepf, W. (1998). *Hypergeometric summation*. Wiesbaden: Vieweg.

König, H. (1959). Zur Theorie der linearen dissipativen Transformationen. *Archiv Mathematics, 10*, 447–451.

König, H. (1994). An explicit formula for fundamental solutions of linear partial differential operators with constant coefficients. *Proceedings of AMS, 120*, 1315–1318.

Kreyszig, E. (2011). *Advanced engineering mathematics*. Hoboken, NJ: Wiley.

Landau, E. (1933). Über eine trigonometrische Ungleichung. *Mathematics Zeitschrift, 37*, 36.

Leinfelder, H. (2012). *On a new proof of the Malgrange-Ehrenpreis Theorem*. Private Communication.

Leobacher, G., & Pillichshammer, F. (2014). *Introduction to Quasi-Monte Carlo integration and applications*. New York: Birkhäuser.

Ma, Y., Yang, Q., Tang, Y., Chen, S., & Shieh, W. (2010). 1-Tb/s single-channel coherent optical OFDM transmission with orthogonal-band multiplexing and subwavelength bandwidth access. *Journal of Lightwave Technology, 28*(4), 308–315.

Malgrange, B. (1956). Éxistence et approximation des solutions des équations aux dérivées partielles et des équations de convolution. *Annales de l'institut Fourier, 6*, 271–355.

Mallat, S. (1989). Multiresolution approximations and wavelet orthonormal bases of $L^2(\mathbb{R})$. *Transactions of the AMS, 315* (1), 69–87.

Mallat, S. (2009) *A wavelet tour of signal processing*. London: Academic Press.

Mason, J. C., & Handscomb, D. (2002) *Chebyshev polynomials*. Boca Raton: CRC Press.

Messiah, A. (2003). *Quantum mechanics*. New York: Dover.

Meyer, Y. (1993). *Wavelets, algorithms and applications*. Philadelphia: SIAM.

Meyer, Y. (1995). *Wavelets and operators*. Cambridge University Press.

Moler, C., & Van Loan, C. (2003). *Nineteen Dubious ways to compute the exponential of a matrix, twenty five years later. SIAM Review, 45*(1), 3–49.

Montreuil, L., Prodan, R., & Kolze, T. (2013). *Broadcom Recommendations Tx Symbol Shaping*. www.ieee802.org/3/bn/public/jan13/montreuil_01a_0113.pdf

Neumann, M. (1980). Automatic continuity of linear operators. In K.-D. Bierstedt, & B. Fuchssteiner (Eds.), *Functional Analysis, Surveys and Recent Results II*. Amsterdam: North Holland.

Newman, D. J. (1975). A Simple Proof of Wiener's $1/f$ Theorem. *Proceedings of the AMS, 48*, 264–265.

Novak, E., Ritter, K., Schmitt, R., & Steinbauer A. (1999). On a recent interpolatory method for high dimensional integration. *Journal of Computational and Applied Mathematics, 112*, 215–228.

Nussbaumer, H. J. (1982). *Fast Fourier transform and Convolution Algorithms*. Berlin: Springer.

Oberguggenberger, M. (1992). Multiplication of distributions and applications to Partial Differential Equations. In *Pitman Research Notes in Mathematics, Longman, Harlow, U.K.* (Vol. 259)

Oppenheim, A. V. (1978). *Applications of digital signal processing*. Englewood Cliffs: Prentice Hall.

Oppenheim, A V., & Schafer, R. W. (2013). Discret-time signal processing. Harlow: Pearson.

Ortner, N., & Wagner, P. (1994). A short proof of the Malgrange-Ehrenpreis theorem, Functional Analysis—Trier. In S. Dierolf et al. (Eds.), *Proceedings of International Workshop* (pp. 343–352). Berlin: De Gruyter.

Ortner, N., & Wagner, P. (1997). A survey on explicit representation formulae for fundamental solutions of linear partial differential operators. *Acta Applicandae Mathematicae, 47*, 101–124.

Ortner, N., & Wagner, P. (2015). *Fundamental Solutions of Linear Partial Differential operators*. Cham: Springer.

Paley, R., & Wiener, N. (1934). Fourier transforms in the complex domain. Providence: AMS.

Papoulis, A. (1968). *Systems and transforms with applications in optics*. New York: McGraw Hill.

Papoulis, A. (1977). Signal analysis. Tokyo: McGraw-Hill.

Papoulis, A. (1987). *The Fourier integral and its applications*. New York: McGraw-Hill.

Partington, J. R. (2004). Linear operators and linear systems (2004). In *London Mathematical Society Student Texts* (Vol. 60). Cambridge University Press.

Pohl, V., & Boche, H. (2010). *Advanced topics in system and signal theory*. Berlin: Springer.

Proakis, J. G., & Salehi, M. (2013). *Fundamentals of communication systems*. London: Pearson.

Råde, L., & Westergren, B. (2004). *Mathematics handbook for science and engineering*, Springer, Berlin.

Rao, K. R., & Hwang, J. J. (1996). *Techniques & standards for image, video & audio coding*. Upper Saddle River: Prentice Hall.

Rees, C. S., Shah, S. M., & Stanojević, C. V. (1981). *Theory and applications of Fourier analysis*. New York: Dekker.

Rivlin, T. J. (1974). *The chebyshev polynomials*. New York: Wiley.

Rivlin, T. J. (2010). *An introduction to the approximation of functions*. New York: Dover.

Rudin, W. (1972) Invariant means on $L^\infty$. *Studia Mathematics, 44*, 219–227.

Rudin, W. (1991) *Functional analysis*. New York: McGraw-Hill.

Rullière, C. (Ed.) (1998). *Femtosecond Laser Pulses*. New York: Springer.

Runge, C. (1901). Über empirische Funktionen und die Interpolation zwischen äquidistanten Ordinaten. *Zeitschrift für Mathematik und Physik, 46*, 224–243.

Salditt, T., Aspelmeier, T., & Aeffner, S. (2017). *Biomedical imaging*. Berlin: de Gruyter.

Sandberg, I. W. (2001). A note on the convolution scandal. *IEEE Signal Processing Letters, 8*(7), 210–211.

Schwartz, L. (1957). *Théorie des distributions*. Paris: Hermann.

Schwartz, L. (1966). Mathematics for the physical sciences. Paris: Hermann.

Shaik, S. S., & Hiberty, P. C. (2010). *A Chemist's guide to valence bond theory*. New York: Wiley Interscience.

Shannon, C. E. (1949). Communication in the presence of noise. *Proceedings of the IRE, 37*, 10–21.

Shapiro, V. (2019). *Fourier series in several variables with applications to Partial Differential Equations*. Boca Raton: Chapman and Hall/CRC.

Shilov, G. E. (1968). *Generalized functions and Partial Differential Equations*. New York: Gordon and Breach.

Slepian, D. (1983). *Some comments on Fourier Analysis. Uncertainty and Modelling. SIAM Review, 25*(3), 379–393.

Sobolev, S. L. (1964). *Einige Anwendungen der Funktionalanalysis auf Gleichungen der mathematischen Physik*. Berlin: Akademie-Verlag.

Stein, E. M., & Weiss, G. (1971). *Introduction to Fourier analysis on Euclidean spaces*. Princeton University Press.

Stoer, J., & Bulirsch, R. (1992). *Introduction to numerical analysis*. New York: Springer.

Strang, G. (2017). *Calculus*. Wellesley, MA: Wellesley-Cambridge.

Strichartz, R. (1993). How to make wavelets. *The American Mathematical Monthly, 100*, 539–556.

Strichartz, R. (1994). A guide to distribution theory and Fourier Transforms. Boca Raton: CRC Press.

Strohmer, T. (2000). Numerical analysis of the non-uniform sampling problem. *Computational and Applied Mathematics, 122*(1–2), 297–316.

Taubman, D. S., & Marcellin, M. (2001). *JPEG2000: Image compression fundamentals, standards and practice*. Amsterdam: Kluwer.

Tian, J., & Wells Jr., R. O. (1996). A Lossy Image Codec Based on Index Coding. In J. Storer & M. Cohn (Eds.), *Proceedings of the IEEE data compression conference, Snowbird, Utah*. IEEE Computer Society Press.

Tietze, U., & Schenk, C. (2008). *Electronic circuits*. New York: Springer.

Tolstov, G. P. (1976). *Fourier series*. New York: Dover.

Trefethen, L. N. (2008). Is Gauss quadrature better than Clenshaw-Curtis?. *SIAM Review, 50*(1), 67–87.

Triebel, H. (1986). *Analysis and mathematical physics*. Berlin: Springer.

Triebel, H. (1992). *Higher analysis*. Leipzig: J. A. Barth.

Unser, M. (2000). Sampling—50 years after Shannon. *Proceedings of the IEEE, 88*(4), 569–587.

Vladimirov, V. S. (2002). *Methods of the theory of generalized functions*. London: Taylor and Francis.

Wagner, P. (2009). A new constructive proof of the Malgrange-Ehrenpreis theorem. *The American Mathematical Monthly, 116*, 457–462.

Waldvogel, J. (2003). Fast construction of the Fejér and Clenshaw-Curtis quadrature rules. *BIT Numerical Mathematics, 43*(1), 1–18.

Walker, J. S. (1988) *Fourier analysis*. Oxford: Oxford University Press.

Weidmann, J. (1980). *Linear operators in Hilbert spaces*. Berlin: Springer.

Weinstein, S. B., & Ebert, P. M. (1971). Data transmission by frequency-division multiplexing using the Discrete Fourier Transform. *IEEE Transactions on Communication Technology, COM-19*(5), 628–634.

Weinstein, S. B. (2009). The history of orthogonal frequency-division multiplexing. *IEEE Communications Magazine, 2009*, 26–35.

Weiss, G. (1991). Representation of shift-invariant operators on $L_2$ by $H_\infty$ transfer functions. *Mathematics of Control, Signals, and Systems, 4*, 193–203.

Wheeden, R. L., & Zygmund, A. (1977). *Measure and integral, an introduction to real analysis*. New York: Dekker.

Wiener, N. (1933). *The Fourier integral and certain of its applications*. New York: Cambridge University Press.

Wilbraham, H. (1848). On a certain periodic function. *Cambridge and Dublin Mathematical Journal, 3*, 198–201.

Young, R. M. (1980). *An introduction to nonharmonic Fourier series*. New York: Academic Press.

Zemanian, A. H. (2010). *Distribution theory and transform analysis*. New York: Dover.

Zemanian, A. H. (1995). *Realizability theory for continuous linear systems*. New York: Dover.

Zygmund, A. (2003). *Trigonometric series*. Cambridge: Cambridge University.

# Index

**Symbols**

$L^p(\mathbb{R})$, $L^\infty(\mathbb{R})$, 323
$\Re(z)$, $\Im(z)$, 9
$\dot{T}$, $T^{(k)}$, 172
$\mathcal{D}(\Omega)$, $\mathcal{D}'(\Omega)$, 187
$\mathcal{E}'$, $\mathcal{D}'_R$, 202
$\mathrm{vp}(f)$, 167
$\partial_i$, $\partial^k$, 187
$l_d^p$, $l_d^\infty$, 324
$z$-Transform, 358, 360
$\langle T(t), \varphi(t) \rangle$, 164
$L^2([0, T])$, 62
$L^2(\mathbb{R})$, 307
$L_w^2([-1, 1])$, 108
$T * G$, 194
$\Delta u$, 233
$\mathcal{D}$, 155
$\mathcal{D}'$, 164
$\mathcal{D}'_R = \bigcup_{r \in \mathbb{R}} \mathcal{D}'_r$, 322
$\mathcal{D}'_+$, 205
$\mathcal{E}'$, 322
$\mathcal{F}$, 270
$\mathcal{O}'_C$, 301, 322
$\mathcal{O}_M$, 290
$\mathcal{S}'_R = \bigcup_{r \in \mathbb{R}} \mathcal{S}'_r$, 322
$\mathrm{pf}(t^{-m})$, 168
$\mathrm{pv}(f)$, 167
$\widehat{c}_k$, 86
$\mathcal{S}$, 286
$\mathcal{S}'$, 289

**A**

AC circuit calculation, 66
Almost everywhere, 493

**Amplitude modulation, 45**
Amplitude response, 329, 371

**B**

Baseband signal, 88
Bernoulli, 1, 261
Bessel inequality, 49
Boundary value problem, 68
Butterworth lowpass filter, 232, 334
    impulse response, 232

**C**

Cauchy-Schwarz inequality, 308, 449
Causal, 217, 322, 357
Chebyshev abscissae, 110
Chebyshev lowpass filter, 113, 127, 128
Chebyshev polynomials, 107
Circuit
    differentiator, 157
Clenshaw-Curtis quadrature, 104
Compact, 188
Complex AC circuit calculation, 66
Conservation of energy, 75
Convergence
    pointwise, 20
    uniformly, 20
Convolution, 193
    impulse sequence, 199
    impulse trains, 303
    periodic, 63, 96, 213
    periodic convolution, 68
    properties, 195